Advances in Modal Logic

volume 6

Advances in Modal Logic

volume 6

Edited by

Guido Governatori, Ian Hodkinson and Yde Venema

©College Publications, London 2006. All rights reserved.

ISBN 1-904987-20-6

College Publications
Department of Computer Science
Strand, London WC2R 2LS, UK.

Printed by Lightning Source, UK
Cover design by Richard Fraser, Avalon Arts, UK

All rights reserved. No part of this publication may be reproduced, stored in a retrieval system or transmitted, in any form, or by any means, electronic, mechanical, photocopying, recording or otherwise, without prior permission, in writing, from the publisher.

CONTENTS

Preface — vii

RENATE A. SCHMIDT
Developing Modal Tableaux and Resolution Methods
via First-Order Resolution — 1

VALENTIN SHEHTMAN
Completeness and incompleteness in first-order modal logic: an overview — 27

NATASHA ALECHINA AND DMITRY SHKATOV
Logics with an existential modality — 31

PHILIPPE BALBIANI
An expressive two-sorted spatial logic for plane projective geometry — 49

P. BALBIANI, I. SHAPIROVSKY, V. SHEHTMAN
Every world can see a Sahlqvist world — 69

JOHAN VAN BENTHEM AND ERIC PACUIT
The Tree of Knowledge in Action: Towards a Common Perspective — 87

KAI BRÜNNLER
Deep Sequent Systems for Modal Logic — 107

ALEXANDER CHAGROV AND LILIA CHAGROVA
The Truth About Algorithmic Problems in Correspondence Theory — 121

GAËLLE FONTAINE
ML is not finitely axiomatizable over **Cheq** — 139

TIM FRENCH
Bisimulation Quantified Modal Logics: Decidability — 147

OLIVIER GASQUET, ANDREAS HERZIG, MOHAMAD SAHADE
Terminating modal tableaux with simple completeness proof — 167

S. GHILARDI, C. LUTZ, F. WOLTER AND M. ZAKHARYASCHEV
Conservative extensions in modal logic — 187

ROBERT GOLDBLATT
A Kripke-Joyal Semantics for Noncommutative Logic in Quantales — 209

ROBERT GOLDBLATT AND EDWIN D. MARES
A General Semantics for Quantified Modal Logic   227

IGOR GORBUNOV
A decidable modal logic that is finitely undecidable   247

BERNHARD HEINEMANN
Regarding Overlaps in 'Topologic'   259

LLOYD HUMBERSTONE
Weaker-to-Stronger Translational Embeddings in Modal Logic   279

B. KONEV, R. KONTCHAKOV, F. WOLTER AND
M. ZAKHARYASCHEV
Dynamic topological logics over spaces with continuous functions   299

ANDREY KUDINOV
Topological Modal Logics with Difference Modality   319

TADEUSZ LITAK
Isomorphism via translation   333

ERIC MARTIN
Quantification over names and modalities   353

LINH ANH NGUYEN
On the Deterministic Horn Fragment of Test-free PDL   373

MIKHAIL RYBAKOV
Complexity of intuitionistic and Visser's basic and formal logics
in finitely many variables   393

ILYA SHAPIROVSKY
Downward-directed transitive frames with universal relations   413

M. SHEREMET, D. TISHKOVSKY, F. WOLTER AND
M. ZAKHARYASCHEV
From topology to metric:
modal logic and quantification in metric spaces   429

RICARDO SOUSA SILVESTRE
Modality, Paraconsistency and Paracompleteness   449

HIROKI TAKAMURA
The variety of modal **FL**$_{ew}$-algebras is generated by
its finite simple members   469

TERO TULENHEIMO AND MERLIJN SEVENSTER
On Modal Logic, IF Logic, and IF Modal Logic   481

HEINRICH WANSING
Tableaux for multi-agent deliberative-stit logic   503

PREFACE

Advances in Modal Logic (AiML) is an initiative founded in 1995 and aimed at presenting an up-to-date picture of the state of the art in modal logic and its many applications. It consists of a conference series together with volumes based on the conferences. The conference is the main international forum at which research on all aspects of modal logic is presented. The first one was held in 1996 in Berlin, Germany, and since then, it has been organised biennially, with meetings in 1998 in Uppsala, Sweden, in 2000 in Leipzig, Germany (jointly with ICTL-2000), in 2002 in Toulouse, France, and in 2004 in Manchester, UK.

The sixth conference in the series was the first outside Europe. It was hosted by the School of Information Technology and Electrical Engineering, University of Queensland, and held at Australis Noosa Lakes Conference Centre, at Noosaville, Noosa, Queensland, Australia, on 25–28 September 2006. The invited speakers were

- Renate Schmidt,

- Valentin Shehtman,

- Igor Walukiewicz,

- Alberto Zanardo.

We received 56 submitted papers, of which 27 were accepted and appear in these Proceedings. In addition, we asked the invited speakers whether they wanted to contribute to the proceedings, the results of which, the paper by Renate Schmidt and the abstract by Valentin Shehtman, begin this volume. The volume includes papers on the theory of modal logic itself, on process theory, multi-agent systems and spatial reasoning, and work on quantified modal logic, modal reasoning methods, and philosophical issues. On this occasion, for the first time in AiML history, the proceedings were prepared in book form in advance of the meeting, to be distributed to participants on their arrival. It is intended that all the papers will be made available on the AiML website www.aiml.net.

The steering committee of AiML for 2004–2006 consisted of Guido Governatori (who was the local organiser of this meeting), Ian Hodkinson, Mark Reynolds, Renate Schmidt, Nobu-Yuki Suzuki, Yde Venema, Heinrich Wansing, Frank Wolter, and Michael Zakharyaschev. The programme committee for the meeting was co-chaired by Ian Hodkinson and Yde Venema. It comprised the steering committee together with Alessandro Artale, Alexandru Baltag, Guram Bezhanishvili, Julian Bradfield, Melvin Fitting,

Silvio Ghilardi, Robert Goldblatt, Valentin Goranko, Rajeev Goré, Ramon Jansana, Alexander Kurz, Carsten Lutz, Maarten Marx, Martin Otto, Ildikó Sain, and Jerry Seligman. These people were an essential part of the process. They chose the invited speakers, and, together with 66 additional reviewers, they took great care over the (completely anonymous) reviewing of submissions, submitting reports of such astuteness and detail as might be received from a Journal. We have chosen not to list the additional reviewers here, in order to preserve their anonymity. We hope they will understand this and appreciate our reasons. We are extremely grateful to them and to the programme committee members for their hard work, and we believe all the conference participants will join us in thanking them for being instrumental in creating a high quality meeting and proceedings. And of course, we are most grateful to the authors, for contributing such interesting work and preparing it so carefully. Without them, there would be no meeting at all.

We would like to offer our grateful thanks to ARIA-QLD (Association for Research between Italy and Australasia) for partially supporting Alberto Zanardo's attendance at the meeting, and to the Institute for Integrated and Intelligent Systems (IIIS), Griffith University, Queensland, for financial assistance. Special thanks go to Michael Zakharyaschev, for helping several of the Russian authors with the presentation of their papers, and for managing AiML's finances over the years. We are grateful to have had the use of the EasyChair conference management system, which made our lives much easier. Finally, we thank Richard Fraser of Avalon Arts for the cover design, and Jane Spurr for bringing the book to press in such a professional way in time for the meeting.

This has been a draining but worthwhile year-and-a-half for us. Those busy times reviewing, editing, plotting, and the late-night emails and phone calls are now nearly all in the past. We believe that there is a strong meeting to look forward to in Noosa, and we are sure from what we have seen that AiML is a prestigious conference series with a bright future.

Guido Governatori, Ian Hodkinson, and Yde Venema
July 2006

Developing Modal Tableaux and Resolution Methods via First-Order Resolution

Renate A. Schmidt

ABSTRACT. This paper explores the development of calculi using different proof methods for propositional dynamic modal logics. Dynamic modal logics are *PDL*-like extended modal logics which are closely related to description logics. We show how tableau systems, modal resolution systems and Rasiowa-Sikorski systems, which are dual tableau systems, can be developed and studied by using standard principles and methods of first-order theorem proving. We translate modal logic reasoning problems to first-order clausal form and then use a suitable refinement of resolution to construct and mimic derivations using the desired proof method. The inference rules of the calculus can then be read off from the clausal form used.

Keywords: Modal logic, deduction, resolution, tableaux, dual resolution, dual tableaux, modal resolution, decidability.

1 Introduction

In this paper we discuss and extend an approach of developing tableaux calculi for modal logics that has been suggested and followed in our previous work [8, 13, 15, 16, 18, 32]. Although resolution calculi apparently operate considerably differently from tableau calculi, we have shown that it is possible to linearly simulate many forms of modal logic or description logic tableau calculi with standard techniques in resolution-based theorem proving. In [16] we have shown how derivations and search in standard tableau algorithms of the description logic $\mathcal{ALC}$ can be linearly simulated by resolution. This corresponds to local satisfiability testing in the basic multi-modal logic $K_{(m)}$. Using redundancy elimination techniques and a blocking rule we have shown in [15] how to mimic standard tableau algorithms for $\mathcal{ALC}$ with respect to non-empty TBoxes. This corresponds to local satisfiability in multi-modal logic $K_{(m)}$ with respect to a background theory of modal logic formulae (i.e. a set of non-logical axioms). In [18] details can be found of how to simulate derivations in the prefixed single-step tableau calculi of Massacci [20]. These simulation results show that it is possible to use first-order resolution in a way that it simulates modal logic and description logic tableau procedures. The close connection exhibited in these papers between tableau and a certain instance of resolution is exploited in [8] in order to develop a tableau calculus for a logic that has not been considered before. The logic considered was the dynamic modal logic $K_{(m)}(\wedge,\vee,\smile)$. Dynamic modal logics are *PDL*-like modal logics in which the parameters of the modal operators can be relational formulae, which are interpreted as actions or programs in *PDL*, and are closely related to description logics [33]. $K_{(m)}(\wedge,\vee,\smile)$ is the multi-modal logic defined over frames in which the relations are closed under intersection, union and converse. The logic corresponds to the description logic $\mathcal{ALC}$ in which conjunction, disjunction and converse of roles are allowed. In [8] we have shown how a tableau calculus can essentially be 'read off' from the clausal form of the translation of formulae in $K_{(m)}(\wedge,\vee,\smile)$.

It is this *'develop via first-order resolution'* approach which we explore and extend in this paper. We consider in more detail how tableau calculi can be developed for modal logics via a suitable translation to first-order logic and resolution. However we also show that the approach can be used to develop other kinds of deduction methods. In particular, we show how the approach can be extended to develop Rasiowa-Sikorski systems. These are tableau-style calculi for testing the validity of formulae [25]. In addition, we consider the development of modal resolution systems which operate directly on modal logic formulae.

We show that all three types of calculi (tableau, Rasiowa-Sikorski, modal resolution) can be obtained naturally via translation to first-order logic and standard techniques of resolution theorem proving. Key to the 'develop via first-order resolution' approach are two aspects: (i) An effective, sound and complete reduction to first-order logic that retains enough information about the input formula in the source logic so that the inference rules can simply be read off from the clausal form. (ii) A refinement of first-order resolution which performs inferences exactly like the kind of system we want to develop.

The form and property of the calculus one obtains depends very much on the reduction and the refinement used. Small modifications result in different styles and variations of calculi. In this paper we focus on the development of *ground semantic calculi*. By this we mean calculi which operate on labelled modal formulae. For each operator in the logic there is a decomposition rule which basically 'breaks down' formulae into less complex formulae on the basis of the semantics of the top-level operator in one of the premises. The labels are given by constants (or ground Skolem terms) which represent states in the underlying Kripke model. Currently, ground semantic tableau calculi appear to be the preferred style of tableau calculus used in the area, and many modal and description logic theorem provers are based on ground semantic calculi.

The cited previous work shows that ground semantic tableau calculi can be simulated by a structural reduction into range-restricted clauses in combination with hyperresolution. A clause is range-restricted if all variables of the clause occur in the negative literals of that clause. Hyperresolution on range-restricted clauses has the property that all positive premises are ground clauses and all conclusions are ground clauses. This is precisely the property which, when using a structural reduction, allows us to interpret the negative, non-ground premise as an inference rule $\mathcal{I}$ of the ground calculus. The positive, ground premises of a hyperresolution inference step represent then the premises of the rule $\mathcal{I}$, and the conclusions represent the conclusions of the rule. Combined with splitting, hyperresolution allows us to simulate and develop ground semantic tableau calculi. This is shown in Section 4. In Section 5 we discuss how dual hyperresolution with splitting allows us to simulate and develop ground semantic Rasiowa-Sikorski calculi. Furthermore, hyperresolution without a splitting rule produces modal resolution calculi, as is discussed in Section 6.

To illustrate the approach we focus on the development of calculi for the dynamic modal logic $K_{(m)}(\wedge, \vee, \smile, \upharpoonright)$, i.e. multi-modal logic defined over relations closed under intersection, union, converse and domain restriction. As far as we know this logic has not been considered before and certainly no tableau calculus has been described for it in the literature. $K_{(m)}(\wedge, \vee, \smile, \upharpoonright)$ is subsumed by Peirce logic, for which tableau calculi are defined in [22, 34]. However, the relational disjunction operator and the domain restriction operator of $K_{(m)}(\wedge, \vee, \smile, \upharpoonright)$ are not explicit operators in Peirce logic, so there are no tableau rules for relational disjunction and domain restriction in the existing tableau calculi for Peirce logic. Sound rules can be easily defined for these operators, but, on the one hand, the logic lacks the symmetry that Peirce logic has because of the absence of relational negation in $K_{(m)}(\wedge, \vee, \smile, \upharpoonright)$. On the other hand, since $K_{(m)}(\wedge, \vee, \smile, \upharpoonright)$ is decidable and Peirce logic is not decidable, $K_{(m)}(\wedge, \vee, \smile, \upharpoonright)$ is a good candidate for exploring the possibilities of developing special-purpose decision procedures for it via the approach we are purporting in this paper.

In the next section we recall standard definitions of resolution-based first-order theorem proving. Section 3 introduces dynamic modal logics and $K_{(m)}(\wedge, \vee, \smile, \upharpoonright)$. It also defines the structural translation and reduction to clausal form that we use. The main parts are Sections 4–6, in which we describe, in turn, how ground semantic tableau calculi, ground semantic Rasiowa-Sikorski calculi, and ground modal resolution calculi can be developed in a systematic way via first-order resolution. The final section discusses the significance and consequences of the method, and mentions future work.

Throughout the paper we use the notation and terminology of our previous papers, cf. e.g. the surveys [31, 33].

2 First-Order Resolution

The resolution calculus operates on sets of clauses. Clauses are quantifier free disjunctions of literals which may contain function symbols. The variables are implicitly assumed to be universally quantified.

THEOREM 1. *There is a polynomial or linear reduction* Cls *of any first-order formula to clause logic such that φ is valid in first-order logic, i.e.* $\models \varphi$, *iff* Cls$(\neg \varphi)$ *is unsatisfiable.*

This says that any first-order formula can be transformed efficiently into a satisfiability equivalent set of clauses. The clausal form is obtained by transformation to conjunctive normal form, Skolemisation and crucially involves structural transformation which introduces new predicate symbols and definitions. Since resolution is a refutation calculus, instead of proving theoremhood, resolution attempts to refute the negation of a given formula.

The basic (unrefined) resolution calculus consists of two inference rules, the resolution rule and the factoring rule, and no axioms. For propositional logic the resolution rule is just the operation that infers a clause $C \lor D$ from two clauses $C \lor A$ and $D \lor \neg A$. The factoring rule is a contraction rule, i.e. it is a form of simplification which eliminates multiple copies of the same literal from one clause, that is, it infers $C \lor A$ from $C \lor A \lor A$. These two rules provide a sound and refutationally complete calculus for propositional logic and sets of ground clauses. We obtain a sound and refutationally complete inference system for full first-order logic and clause logic, if we augment the rules with unification. This calculus, the *basic resolution calculus* [29], is sound and complete for full first-order logic and clause logic. It is however very prolific in generating new clauses. This was noticed already in the very early days of the development of first-order resolution methods. The first papers, by Robinson and others, on refinements of resolution appeared in the same year that Robinson published his famous paper which introduced resolution. Since the mid-sixties the advances have been impressive. The current generation of theorem provers, which include SPASS [36], E [35] and VAMPIRE [27] (in order of creation), are based on the modern framework of saturation-based resolution and superposition. In the following, when we refer to *resolution* we mean this framework (cf. e.g. [3, 23]).

The main ingredients of the framework are refinements of the inference rules which restrict their applicability and a general notion of redundancy. Refinements of inference rules are defined in terms of two parameters: an ordering $\succ$ and a selection function S. The idea is that inferences do not need to be performed (but can), unless they are on literals maximal under the given ordering or on (negative) literals selected by the selection function S. The selection function can override the ordering. That is, if a literal is selected then it is the preferred candidate for an inference step even though there may be 'larger' literals in the clause. The ordering and selection function are used to limit the number of possible inferences. It is clear that, in general, if we can reduce the number of possible inferences without losing completeness then a refutation proof can be found more quickly as the search space for the proof is reduced. There is a general completeness proof which requires only weak conditions for the admissibility of orderings and selection functions [3].

Simplification and deletion rules are important regardless of the style of deduction one uses. In the resolution framework these are based on a general notion of redundancy, which is based on considerations of the model construction which is at the centre of the completeness proof. Standard simplification rules like elimination of duplicate literals within a clause, tau-

Deduce: $\dfrac{N}{N \cup \{C\}}$ if C is a factor or resolvent of premises in N.

Delete: $\dfrac{N \uplus \{C\}}{N}$ if C is redundant with respect to N.

Simplify: $\dfrac{N}{(N\backslash M) \cup M'}$ if $(N\backslash M) \cup M'$ is satisfiable when N is satisfiable and every clause in M is redundant with respect to $(N\backslash M) \cup M'$.

Split: $\dfrac{N \uplus \{C \vee D\}}{N \cup \{C\} \mid N \cup \{D\}}$ if C and D are variable-disjoint.

Resolvents and factors are computed with:

Ordered resolution: $\dfrac{C \vee A \quad \neg B \vee D}{(C \vee D)\sigma}$

provided (i) σ is the most general unifier of A and B, (ii) no literal is selected in C, and $A\sigma$ is strictly $\succ$-maximal with respect to $C\sigma$, and (iii) $\neg B$ is either selected, or $\neg B\sigma$ is maximal with respect to $D\sigma$ and no literal is selected in D. The left (right) premise is called the *positive (negative) premise*.

Ordered factoring: $\dfrac{C \vee A \vee B}{(C \vee A)\sigma}$

provided (i) σ is the most general unifier of A and B, and (ii) no literal is selected in C and $A\sigma$ is $\succ$-maximal with respect to $C\sigma$.

Figure 1. The resolution calculus $R_{\text{sp}}^{\text{red}}$.

tology deletion, subsumption deletion (forward and backward subsumption deletion), condensing, etc, are instances of this notion [3].

Let $R_{\text{sp}}^{\text{red}}$ be the resolution calculus defined by the rules of Figure 1. (The meaning of 'red' in the notation is 'with redundancy' and the meaning of 'sp' is 'with splitting'. $\uplus$ denotes disjoint union.) In our presentation we distinguish four kinds of rules. The Deduce rules are the ordered resolution and positive factoring rules. The ordering $\succ$ is a parameter which can be any admissible ordering and S is any selection function of negative literals. The Delete and Simplify rules are deletion and replacement rules compatible with the general notion of redundancy [3]. Essentially, a ground clause is redundant with respect to a set N and the ordering $\succ$, if it follows from smaller instances of clauses in N. A non-ground clause is redundant in N if all its ground instances are redundant in N. Testing for redundancy in its general form is an expensive operation; in first-order logic general redundancy elimination is undecidable. For this reason one does not find theorem provers that implement redundancy elimination in full generality, instead only effectively computable instances of the Delete and Simplify rules are implemented.

The Split rule is a rule familiar from DPLL algorithms and tableau calculi. Instead of refuting $N \cup \{C \vee D\}$ one refutes both $N \cup \{C\}$ and $N \cup \{D\}$ (alternatively, it is possible to use the complement splitting rule, which

Ordered hyperresolution:

$$\frac{C_1 \vee A_1 \quad \ldots \quad C_n \vee A_n \quad \neg B_1 \vee \ldots \vee \neg B_n \vee D}{(C_1 \vee \ldots \vee C_n \vee D)\sigma}$$

provided (i) σ is the most general unifier such that $A_i\sigma = B_i\sigma$ for every i, $1 \leq i \leq n$, (ii) $A_i\sigma$ is strictly $\succ$-maximal with respect to $C_i\sigma$, and the C_i are positive clauses, for every i, $1 \leq i \leq n$, and (iii) for every i, $1 \leq i \leq n$, $\neg B_i$ is selected and D is a positive clause. The rightmost premise in the rule is referred to as the *negative premise* and all other premises are referred to as *positive premises*.

Ordered factoring: $\dfrac{C \vee A \vee B}{(C \vee A)\sigma}$

provided (i) σ is the most general unifier of A and B, and (ii) C is positive and $A\sigma$ is $\succ$-maximal with respect to $C\sigma$.

Figure 2. The Deduce rules of ordered hyperresolution.

means that instead of refuting $N \cup \{C \vee D\}$ one refutes both $N \cup \{C\}$ and $N \cup \{\neg C, D\}$). The splitting rule is don't know non-deterministic and usually requires backtracking. However, in the resolution framework an alternative to explicit splitting is splitting through new propositional variables [7, 26] implemented in the theorem prover VAMPIRE [27] or the generalisation called separation in [30].

The calculus without the splitting rule is denoted by R^{red} and R is the calculus with just Deduce rules.

THEOREM 2 (Bachmair et al [3, 4]). $R_{\text{sp}}^{\text{red}}$, R^{red} and R are sound and complete refutation systems for clause sets.

The *(ordered) hyperresolution calculus* is based on maximal selection of negative literals. This means the selection function selects exactly the set of all negative literals in any non-positive clause. Let $OH_{\text{sp}}^{\text{red}}$ be the calculus based on maximal selection and an ordering $\succ$, where the Deduce rules are given by the rules in Figure 2. This means the rules are the hyperresolution rule, positive factoring, redundancy elimination and splitting. Similar as above, OH^{red} denotes the calculus $OH_{\text{sp}}^{\text{red}}$ but without the splitting rule, OH denotes the calculus just consisting of Deduce rules, and OH_{sp} is OH with splitting. For completeness an ordering refinement is optional. We use the notation $H_{\text{sp}}^{\text{red}}$, H^{red}, H and H_{sp} for the unordered versions.

COROLLARY 3. $OH_{\text{sp}}^{\text{red}}$, OH^{red}, OH_{sp}, OH, $H_{\text{sp}}^{\text{red}}$, H^{red}, H_{sp} and H are sound and complete refutation systems for clause sets.

3 Dynamic Modal Logics

A dynamic modal logic is an extension of the multi-modal logic $K_{(m)}$ in which the modal operators are parameterised by relational formulae [33].

Given countably many propositional variables denoted by p_j, and countably many relational variables, denoted by r_i, *dynamic modal formulae* and *relational formulae* are defined inductively as follows. Every propositional variable is a dynamic modal formula and every relational variable is a relational formula. If ϕ, ψ are dynamic modal formulae and α, β are relational formulae, then $\neg\phi$, $\phi \wedge \psi$, $[\alpha]\phi$ are dynamic modal formulae and $\alpha \wedge \beta$, $\alpha \vee \beta$, $\alpha^\smile$ and $\alpha{\upharpoonright}\phi$ are relational formulae. In $K_{(m)}$ the only relational formulae are relational variables.

Thus the language of dynamic modal logics consists of two syntactic types: dynamic modal formulae and relational formulae. The logical connectives are (i) the connectives of the basic multi-modal logic $K_{(m)}$, with the difference that the modal operators are indexed with relational formulae, and (ii) a finite set of relational operators. A dynamic modal logic with relational operators $\star_1, \ldots, \star_k$ is denoted by $K_{(m)}(\star_1, \ldots, \star_k)$.

The semantics of a dynamic modal logic is defined in terms of frames. A frame is a tuple (W, R) of a non-empty set W (of worlds) and a mapping R from relational formulae to binary relations over W. A model is given by a triple $\mathcal{M} = (W, R, v)$, where (W, R) is a frame and v is a mapping from propositional variables to subsets of W satisfying the conditions (R_α is the preferred notation for $R(\alpha)$):

$\mathcal{M}, x \models p$ iff $x \in v(p)$, $\qquad\qquad \mathcal{M}, x \models \neg\phi$ iff $\mathcal{M}, x \not\models \phi$,
$\mathcal{M}, x \models \phi \wedge \psi$ iff both $\mathcal{M}, x \models \phi$ and $\mathcal{M}, x \models \psi$,
$\mathcal{M}, x \models [\alpha]\phi$ iff $(x, y) \in R_\alpha$ implies $\mathcal{M}, y \models \phi$, for any $y \in W$.

In addition, the following conditions are satisfied:

$R_{\alpha \wedge \beta} = R_\alpha \cap R_\beta, \qquad R_{\alpha \vee \beta} = R_\alpha \cup R_\beta,$
$R_{\alpha^\smile} = R_\alpha^\smile, \qquad R_{\alpha{\upharpoonright}\phi} = \{(x,y) \mid (x,y) \in R_\alpha \wedge \mathcal{M}, x \models \phi\}.$

$R^\smile$ denotes the converse (or inverse) of a relation R. We can define the range restriction operator by $\alpha{\downharpoonright}\phi =^{\text{def}} (\alpha^\smile{\upharpoonright}\phi)^\smile$.

If $\mathcal{M}, x \models \varphi$ holds then φ is *(locally) true* at x in $\mathcal{M}$ and $\mathcal{M}$ *(locally) satisfies* φ. A modal formula φ is *(locally) satisfiable* iff there exists a model $\mathcal{M}$ and a world x in $\mathcal{M}$ such that $\mathcal{M}, x \models \varphi$. A modal formula is *(locally) valid* iff it is (locally) satisfiable in every world of all models.

The following result is a consequence of decidability results in [16] and also of the decidability of the two-variable fragment of first-order logic.

THEOREM 4. *Let L be a dynamic modal logic with any subset of $\{\wedge, \vee ,^\smile, \upharpoonright, \downharpoonright\}$ as relational operators. The local (and global) satisfiability problem in L is decidable.*

We know that the result remains true when we allow negation as a relational operator [16], or relational composition and identity [21]. However, any dynamic modal logic with relational conjunction, relational negation and (negative occurrences of) composition is undecidable [31, 33].

Structural Transformation into Clausal Form

For efficiency reasons and in order to be able to have better control over the resolution inferences performed on the clausal form, we use a *structural version of the relational translation* of dynamic modal logics into first-order logic. The translation is similar to the one used in [32]; other structural translations have been used in e.g. [8, 15, 16, 18].

Throughout the paper, we assume that all occurrences of double negation are eliminated from modal formulae and conjunction is an idempotent operator. For any formula F, let $\sim F$ denote G if $F = \neg G$, and $\neg F$ otherwise. Let Def be a transformation of dynamic modal formulae and relational formulae defined as follows.

$$\mathrm{Def}(\psi) =^{\mathrm{def}} \forall x\, (Q_\psi(x) \to \pi(\psi, x)) \land \forall x\, (Q_{\sim\psi}(x) \to \pi(\sim\psi, x))$$
$$\land\, \forall x\, (Q_\psi(x) \to \neg Q_{\sim\psi}(x)),$$
$$\mathrm{Def}(\alpha) =^{\mathrm{def}} \forall x\, y\, (R_\alpha(x, y) \to \pi(\alpha, x, y)) \land \forall x\, y\, (\pi'(\alpha, x, y) \to R_\alpha(x, y)).$$

Def(ψ) is the *definition* of Q_ψ, which is a new predicate symbol uniquely associated with the modal formula ψ. Similarly, Def(α) is the definition of the new symbol R_α uniquely associated with the relational formula α. π is defined as follows (z is any variable distinct from x):

$$\pi(p, x) = \top,$$
$$\pi(\neg p, x) = \neg Q_p(x),$$
$$\pi(\psi \land \phi, x) = Q_\psi(x) \land Q_\phi(x),$$
$$\pi(\neg(\psi \land \phi), x) = Q_{\sim\psi}(x) \lor Q_{\sim\phi}(x),$$
$$\pi([\alpha]\psi, x) = \forall z\, (R_\alpha(x, z) \to Q_\psi(z)),$$
$$\pi(\neg[\alpha]\psi, x) = \exists z\, (R_\alpha(x, z) \land Q_{\sim\psi}(z)),$$
$$\pi(r, x, y) = \pi'(r, x, y) = R_r(x, y),$$
$$\pi(\alpha \land \beta, x, y) = \pi'(\alpha \land \beta, x, y) = R_\alpha(x, y) \land R_\beta(x, y),$$
$$\pi(\alpha \lor \beta, x, y) = \pi'(\alpha \lor \beta, x, y) = R_\alpha(x, y) \lor R_\beta(x, y),$$
$$\pi(\alpha^\smile, x, y) = \pi'(\alpha^\smile, x, y) = R_\alpha(y, x),$$
$$\pi(\alpha \mid \phi, x, y) = R_\alpha(x, y) \land Q_\phi(x),$$
$$\pi'(\alpha \mid \phi, x, y) = R_\alpha(x, y) \land \neg Q_{\sim\phi}(x).$$

Let Def(X) $=^{\mathrm{def}}$ {Def(F) | $F \in X$}, if X denotes a set of modal and relational formulae. By Sf(F) we denote the set of all modal and relational subformulae of F.

THEOREM 5. *Let L be a dynamic modal logic defined over the operators $\{\land, \lor, \smile, \mid\}$, and let φ be any modal formula. Suppose N is the set of clauses obtained from $\varphi' =^{\mathrm{def}} \exists x\, Q_\varphi(x) \land \bigwedge \mathrm{Def}(\mathrm{Sf}(\varphi))$ by transformation into conjunctive normal form, inner Skolemisation, and clausifying the Skolemised formula. Then: (i) Each clause in N is either a unit clause $Q(a)$, for some Skolem constant a, or it is an instance of a definitional clause given in Figure 3. (ii) φ is locally satisfiable in L iff φ' is first-order satisfiable iff N is first-order satisfiable.*

Note that the set N is computable in linear time, but may contain some obvious redundancies. These can be avoided by more careful introduction

Subformula θ	Definitional clauses for $\text{Def}(\theta)$
(shortcut)	$\neg Q_{\neg \psi}(x)^+ \vee \neg Q_\psi(x)^+$
$\psi \wedge \phi$	$\neg Q_{\psi \wedge \phi}(x)^+ \vee Q_\psi(x)$
	$\neg Q_{\psi \wedge \phi}(x)^+ \vee Q_\phi(x)$
$\neg(\psi \wedge \phi)$	$\neg Q_{\neg(\psi \wedge \phi)}(x)^+ \vee Q_{\sim\psi}(x) \vee Q_{\sim\phi}(x)$
$[\alpha]\psi$	$\neg Q_{[\alpha]\psi}(x)^+ \vee \neg R_\alpha(x,y)^+ \vee Q_\psi(y)$
$\neg[\alpha]\psi$	$\neg Q_{\neg[\alpha]\psi}(x)^+ \vee R_\alpha(x, f_{\neg[\alpha]\psi}(x))$
	$\neg Q_{\neg[\alpha]\psi}(x)^+ \vee Q_{\sim\psi}(f_{\neg[\alpha]\psi}(x))$
$\alpha \wedge \beta$	$\neg R_{\alpha \wedge \beta}(x,y)^+ \vee R_\alpha(x,y)$
	$\neg R_{\alpha \wedge \beta}(x,y)^+ \vee R_\beta(x,y)$
	$R_{\alpha \wedge \beta}(x,y) \vee \neg R_\alpha(x,y)^+ \vee \neg R_\beta(x,y)^+$
$\alpha \vee \beta$	$\neg R_{\alpha \vee \beta}(x,y)^+ \vee R_\alpha(x,y) \vee R_\beta(x,y)$
	$R_{\alpha \vee \beta}(x,y) \vee \neg R_\alpha(x,y)^+$
	$R_{\alpha \vee \beta}(x,y) \vee \neg R_\beta(x,y)^+$
$\alpha^\smile$	$\neg R_{\alpha^\smile}(x,y)^+ \vee R_\alpha(y,x)$
	$R_{\alpha^\smile}(x,y) \vee \neg R_\alpha(y,x)^+$
$\alpha\mathord{\upharpoonright}\phi$	$\neg R_{\alpha\mathord{\upharpoonright}\phi}(x,y)^+ \vee R_\alpha(x,y)$
	$\neg R_{\alpha\mathord{\upharpoonright}\phi}(x,y)^+ \vee Q_\phi(x)$
	$R_{\alpha\mathord{\upharpoonright}\phi}(x,y) \vee \neg R_\alpha(x,y)^+ \vee Q_{\sim\phi}(x)$

Figure 3. Definitional clausal forms for $K_{(m)}(\wedge, \vee, \smile, \mathord{\upharpoonright})$.

of new symbols using a polynomial time algorithm. For efficiency reasons it is also sensible to take the polarities of all the occurrences of a subformula in the input problem into account in the specification of Def.

4 Tableau Systems

Numerous tableau methods have been developed, studied and also implemented for traditional-style modal logics, c.f. e.g. [6, 14, 20]. Tableau methods for some dynamic modal logics or logics equivalent to dynamic modal logics can be found in the description logic literature, e.g. [2, 21] and also here [8, 22, 34].

Simulating tableau systems for $K_{(m)}$

In this section, let us first look at how resolution can simulate ground semantic tableau for local satisfiability in the basic multi-modal logic $K_{(m)}$.

For this we define the notions of step-wise simulation, p-simulation and bisimulation. Let L_1 and L_2 be two logics and suppose Π is a sound and complete translation of (sets of) formulae in L_1 to (sets of) formulae in L_2, i.e. $\models N$ iff $\models \Pi(N)$ for any set N of formulae in L_1. In addition we assume that Π is computable in linear or polynomial time. Let C_1 be a calculus, or proof procedure, for L_1, and let C_2 be a calculus, or proof procedure, for L_2. By definition, C_2 *step-wise simulates* C_1 *(with respect to Π)* iff there

$$(\bot) \frac{s : \psi, s : \neg\psi}{\bot}$$

$$(\wedge)_1 \frac{s : \psi \wedge \phi}{s : \psi} \qquad (\wedge)_2 \frac{s : \psi \wedge \phi}{s : \phi} \qquad (\neg\wedge) \frac{s : \neg(\psi \wedge \phi)}{s : \sim\psi \mid s : \sim\phi}$$

$$(\neg\Box_i)_1 \frac{s : \neg\Box_i\psi}{(s,t) : R_i} \qquad (\neg\Box_i)_2 \frac{s : \neg\Box_i\psi}{t : \sim\psi} \qquad (\Box_i) \frac{s : \Box_i\psi, (s,t) : R_i}{t : \psi}$$

The side conditions of the $(\neg\Box_i)_j$ rules ($j \in \{1,2\}$) are that t is a constant uniquely associated with the premise $s : \neg\Box_i\psi$.

Figure 4. Tableau calculus for $K_{(m)}$.

is an n such that for every inference step in C_1 there are at most n inference steps in C_2 which derive the corresponding conclusion. More precisely, if N_i and N_{i+1} are consecutive sets in any C_1-derivation then $\Pi(N_{i+1})$ can be obtained by at most n inference steps in C_2 from $\Pi(N_i)$. If rather than '$\models N$ iff $\models \Pi(N)$' we have that '$\models N$ iff $\neg\Pi(N)$ is unsatisfiable', then we say C_2 *step-wise simulates* C_1 (*with respect to* Π) *in a dual manner*. We say that C_2 *p-simulates* (proofs in) C_1 iff the following condition holds: There is a function g computable in polynomial time which maps any proof (or refutation) in C_1 to a proof (or refutation) in C_2. If two calculi p-simulate each other, we say there is a *bisimulation* between them.

A ground semantic tableau calculus for $K_{(m)}$ is given by the rules in Figure 4 (c.f. e.g. [9, 14]). Definitions in the literature often include only one rule for conjunctions and negated box formulae, but for reasons which become obvious once we look at the simulation by resolution we choose the definitions given.

A *tableau derivation* is a finitely branching tree whose nodes are sets of labelled formulae. Given that φ is a formula to be tested for local satisfiability the root node is the set $\{a : \varphi\}$, where a denotes a constant. Successor nodes are constructed in accordance with a set of expansion rules. Expansion rules have the general form $X / X_1 | \ldots | X_n$, where X is the set of premises and the X_i are sets of conclusions. An expansion rule is applicable to a selected labelled formula F in a node of the tableau if F, together with other formulae in the node, are simultaneous instantiations of all the premises of the rule. n successor nodes are created which contain the formulae of the current node and the appropriate instances of X_i. As usual it is assumed that in a derivation no rule is applied twice to the same set of instances of premises of a rule.

Now consider the clausal form of the structural translation of φ as defined in Theorem 5 (Figure 3). In particular, suppose N is the clausal form of $\exists x\, Q_\varphi(x) \wedge \bigwedge \text{Def}(\text{Sf}(\varphi))$. The definitional clauses relevant for $K_{(m)}$ are in the top-half of the table. Notice that for each of these definitional clauses there is one corresponding tableau expansion rule in Figure 4, and vice versa. The connection is the following: Every application of a tableau rule can be simulated by a hyperresolution inference step, or by a hyperresolution inference step followed immediately by splitting. If φ is the input formula, the derivation in hyperresolution starts with an inference with the (only positive) clause $Q_\varphi(a)$. This corresponds to the root of the tableau derivation given by $\{a : \varphi\}$. Subsequent inference steps are hyperresolution inference steps with definitional clauses as negative premises, possibly combined with splitting steps.

Let H_{sp} denote the unordered hyperresolution calculus with splitting. Further, assume that deletion rules, splitting, and the deduction rules are applied in this order, except that splitting is not applied to clauses which contain a selected literal.

LEMMA 6 ([8]). *H_{sp} on the structural transformation defined in Theorem 5 step-wise simulates ground labelled tableau for local satisfiability in $K_{(m)}$.*

THEOREM 7 ([8]). *There is a linear bisimulation between the tableau calculus in Figure 4 and H_{sp} on the structural transformation for local satisfiability in $K_{(m)}$.*

In fact, not only can corresponding proofs be constructed in a systematic manner, also, corresponding derivations can be constructed in the same way (i.e. even derivations that are not proofs). In addition the overhead is negligible. Hence the search complexity of the tableau system and H_{sp} are identical. Similar simulation results have been shown in [18] for prefixed single-step tableau calculi of K, and K extended with the axioms D, T and B. (The results are also true for multi-modal $K_{(m)}$ with D, T and B modalities.) Using the axiomatic translation method [32] the results can be strengthened to prefixed single-step tableau calculi of other traditional modal logics.

Developing tableau systems for $K_{(m)}(\wedge, \vee, \smile, 1)$

Let us now illustrate how the principles of the 'develop via first-order resolution' approach can be applied to the dynamic modal logic $K_{(m)}(\wedge, \vee, \smile, 1)$.

We can prove the following:

THEOREM 8. *Let φ be any $K_{(m)}(\wedge, \vee, \smile, 1)$-formula and let N be the clausal form of the structural transformation defined in Theorem 5. Then: (i) Any H_{sp}-derivation from N terminates. (ii) φ is locally unsatisfiable in $K_{(m)}(\wedge, \vee, \smile, 1)$ iff the H_{sp}-saturation of N contains the empty clause.*

Part (i) can be shown using an argument similar to the proof of Theorem 6.6 in [18] and Theorem 7.7 in [8]. Part (ii) is a consequence of Theorem 5 and Corollary 3.

The definitional clauses of the input set have the form as specified in Figure 3. The literals selected by the selection function of the calculus H_{sp} are marked with $^+$. The only other clause in any input set is a ground unit clause of the form $Q_\varphi(a)$, where φ is the dynamic modal formula we want to test for satisfiability (see Theorem 5). Since all the definitional clauses are range-restricted clauses, i.e. all variables of a clause occur in the negative part of the clause, all conclusions of hyperresolution and factoring inferences are ground clauses. Since in H_{sp} splitting is applied before the Deduce rules, all non-unit ground clauses are split before they are used as premises in inference steps. This means that the positive premises of any hyperresolution inference step are always ground clauses. These ground clauses have the form $Q_\psi(s)$ or $R_\alpha(t,u)$, where ψ is some dynamic modal formula, α is some relational formula and s,t,u are ground Skolem terms. $Q_\psi(s)$ and $R_\alpha(t,u)$ translate directly to the labelled formulae $s:\psi$ and $(t,u):\alpha$, where the s,t,u are now viewed as constants. We refer to $Q_\psi(s)$ and $R_\alpha(t,u)$ as the *formulae associated* with $Q_\psi(s)$ and $R_\alpha(t,u)$. Every hyperresolution inference step in H_{sp} involves one (or two) positive premises $C_1(,C_2)$ and a negative premise D from Figure 3. The positive premises C_1 and C_2 are ground, unit clauses of the form $Q_\psi(s)$ or $R_\alpha(t,u)$. Following from what we have just said, the conclusion is a positive clause again, and it is either a ground unit clause of the same form, or it is a positive, splittable clause of ground literals of that form. Now it does not take much to see how to write down the tableau rule which performs exactly the hyperresolution inference just described, possibly followed by splitting. In particular, take a definitional clause $C = \neg A_1[\,\vee\,\neg A_2] \vee D$, where A_1, A_2 denote atoms and D is the largest positive subclause of C. If C is a negative clause we let $D = \bot$. C contains at most two variables. Substitute these with s and t, i.e. apply the substitution $\sigma = \{x/s, y/t\}$ to C. Now write $C\sigma$ as the rule $F_1(,F_2)/G$, where F_1 and F_2 are the labelled formulae associated with $A_1\sigma$ and $A_2\sigma$. Similarly D becomes G, but if G is not a unit clause disjunction is replaced by $|$. For example, the definitional clause for $[\alpha]\psi$, $\neg Q_{[\alpha]\psi}(x)^+ \vee \neg R_\alpha(x,y)^+ \vee Q_\psi(y)$, is turned into the rule $s:[\alpha]\psi, (s,t):\alpha\,/\,t:\psi$. All rules in Figure 5 can be obtained in this way from Figure 3. Let *Tab* be the tableau calculus for $K_{(m)}(\wedge,\vee,\smile,1)$ given by these rules.

LEMMA 9. *H_{sp} step-wise simulates Tab with respect to the structural transformation defined in Theorem 5.*

THEOREM 10. *There is a linear bisimulation between Tab and H_{sp} on the structural transformation for local satisfiability in $K_{(m)}(\wedge,\vee,\smile,1)$.*

That H_{sp} p-simulates *Tab* is a direct consequence of Lemma 9. For simulation in the other direction we can show that there are no H_{sp}-refutations that contain any steps that have no counter-part in a tableau derivation. The only rule application that has no counter-part in *Tab* is the factoring rule, but since we assume that $\wedge$ is an idempotent operation, factoring is never applied. Assuming idempotency of $\wedge$ is closely related to assuming that clauses are sets of literals rather than multisets of literals. When viewing clauses as sets we know that when using hyperresolution on range-restricted clauses, factoring is optional for completeness. In the case of $K_{(m)}(\wedge,\vee,\smile,1)$ the clauses we are using are range-restricted, and the issue of whether to make the idempotency assumption or add contraction rules to *Tab* is a minor technical matter. Implemented tableau theorem

$$(\bot) \frac{s : \psi, s : \neg\psi}{\bot}$$

$$(\wedge)_1 \frac{s : \psi \wedge \phi}{s : \psi} \qquad (\wedge)_2 \frac{s : \psi \wedge \phi}{s : \phi} \qquad (\neg\wedge) \frac{s : \neg(\psi \wedge \phi)}{s : \sim\psi \mid s : \sim\phi}$$

$$(\neg[\cdot])_1 \frac{s : \neg[\alpha]\psi}{(s,t) : \alpha} \qquad (\neg[\cdot])_2 \frac{s : \neg[\alpha]\psi}{t : \sim\psi}$$

$$([\cdot]) \frac{s : [\alpha]\psi, (s,t) : \alpha}{t : \psi}$$

$$(\smile) \frac{(s,t) : \alpha^{\smile}}{(t,s) : \alpha} \qquad (\smile)_I \frac{(t,s) : \alpha}{(s,t) : \alpha^{\smile}}$$

$$(\wedge)_1^r \frac{(s,t) : \alpha \wedge \beta}{(s,t) : \alpha} \qquad (\wedge)_2^r \frac{(s,t) : \alpha \wedge \beta}{(s,t) : \beta} \qquad (\wedge)_I^r \frac{(s,t) : \alpha, (s,t) : \beta}{(s,t) : \alpha \wedge \beta}$$

$$(\vee)^r \frac{(s,t) : \alpha \vee \beta}{(s,t) : \alpha \mid (s,t) : \beta}$$

$$(\vee)_{I,1}^r \frac{(s,t) : \alpha}{(s,t) : \alpha \vee \beta} \qquad (\vee)_{I,2}^r \frac{(s,t) : \beta}{(s,t) : \alpha \vee \beta}$$

$$(1)_1 \frac{(s,t) : \alpha \mid \phi}{(s,t) : \alpha} \qquad (1)_2 \frac{(s,t) : \alpha \mid \phi}{s : \phi} \qquad (1)_I \frac{(s, \check{\imath}) : \alpha}{(s,t) : \alpha^{\smile} \phi \mid s : \sim\phi}$$

The side conditions of the $(\neg[\cdot])_j$ rules ($j \in \{1,2\}$) are that t is a constant uniquely associated with the premise $s : \neg[\alpha]\psi$. For the rules $(\smile)_I$, $(\wedge)_{I,j}^r$, $(\vee)_I^r$ and $(1)_I$ the side conditions are that the relational formulae in the conclusions, occur as subformulae of the relational formula γ of a box formula $s : [\gamma]\psi$ on the current branch.

Figure 5. Tableau calculus for $K_{(m)}(\wedge, \vee, \smile, 1)$.

provers use rewrite rules to replace obvious redundancies like duplication of the kind $\psi \wedge \psi$ even if no contraction rule is part of the calculus. Even when clauses are interpreted as multisets where duplication matters, factoring is optional for completeness for range-restricted clauses when using hyperresolution with splitting. However then identical branches might be created and potentially considerable work is repeated during the deduction process. Thus it does make sense for efficiency reasons to extend *Tab* with contraction rules. The following are the contraction rules corresponding to the factoring steps performed in H_{sp}.

$$(contr) \frac{s : \neg(\psi \wedge \psi)}{s : \sim\psi} \qquad (contr)^r \frac{(s,t) : \alpha \vee \alpha}{(s,t) : \alpha}$$

It is on conclusions with the definitional clauses for negated conjunctions of modal formulae and positive occurrences of disjunctions of relational formulae that factoring will be performed.

Soundness and refutational completeness of *Tab* (with or without contraction) is now a consequence of the soundness and refutational completeness of the H_{sp} and Theorem 5 (and Theorem 8(ii)).

The close link to resolution provides justification to enhance the tableau calculus with corresponding notions of redundancy [8]. We say a labelled formula F is *redundant* in a node, if the node contains labelled formulae $F_1, \ldots, F_n$ which are smaller than F and $\models_L (F_1 \wedge \ldots \wedge F_n) \to F$ (for $n \geq 0$). We can base the definition of an ordering on the subformula or subterm ordering, but a more general definition similar to admissible orderings in the resolution framework (cf. [3]) may be chosen. For example, all tautologies in L are redundant according to this definition. The application of a rule is *redundant* if its conclusions are redundant in the current node. For example, if a node includes $s : \neg p$ and $s : \neg(p \wedge q)$, then the $(\neg\wedge)$ rule need not be applied, and creating a new branch can be avoided. (The inference step is redundant because in the corresponding hyperresolution derivation the clauses $Q_{\neg p}(s)$ and $Q_{\neg(p\wedge q)}(s)$ are present and the conclusion $Q_{\neg p}(s) \vee Q_{\neg q}(s)$ with the definitional clause for $Q_{\neg(p\wedge q)}$ is subsumed by $Q_{\neg p}(s)$ and therefore redundant.) We can also use the link to resolution to introduce notions of *redundant tableau inference rules*. An inference rule is redundant with respect to a set X of labelled formulae and a calculus C if the definitional clause associated with the rule is redundant with respect to the union of all definitional clauses associated with the calculus (and X), and the clauses associated with the formulae in X. We can use this notion to define redundant inference rules in a calculus by letting $X = \emptyset$. Observe that this notion allows an inference rule which is not redundant in a calculus to be redundant with respect the calculus and some (derived) formulae.

THEOREM 11. *A formula φ is locally satisfiable in $K_{(m)}(\wedge, \vee, \smile, 1)$ iff a tableau derivation containing a branch $\mathcal{B}$ can be constructed in Tab (modulo redundancy, and with or without contraction) such that $\mathcal{B}$ does not contain $\bot$ and each rule application is redundant.*

This result states soundness and completeness of the calculus Tab. Notice the formulation is stronger than those normally found in the literature. The result extends Theorem 8.1 in [8].

The calculus Tab is unusual for a tableau calculus in that it requires the rules for the relational operators to be applied in two directions. Since the calculus comprises both elimination rules and introduction rules, the calculus can also be viewed as a restricted form of natural deduction calculus. In general, uncontrolled use of introduction rules can jeopardise decidability, but the side conditions restrict the applications of the rules in such a way that no formulae are introduced that do not occur in the input problem. Notice that the side conditions follow from the clausal form of the translated problem. The side conditions imply that the calculus has the subformula property. Since the clausal form does not include a positive non-ground clause the calculus is cut-free. Consequently, any procedure based on this calculus is a decision procedure, and no loop detection mechanism is necessary to ensure termination.

COROLLARY 12. *Any (fair) procedure based on Tab is a decision procedure* for local satisfiability in $K_{(m)}(\wedge, \vee, \smile, \uparrow)$.

This result is also a direct consequence of Theorem 8(i). The following results are extensions of results in [8, 18].

Let N_∞ denote the *limit* of a path $(N =) N_0, N_1, \ldots$ in a resolution derivation starting with N. By definition, N_∞ is the set $\bigcup_{j \geq 0} \bigcap_{k \geq j} N_k$ of persistent clauses in the path.

LEMMA 13. *Let φ be any $K_{(m)}(\wedge, \vee, \smile, \uparrow)$-formula. Let N be the clausal form of the structural transformation of φ. Let I be the set of positive ground unit clauses in the limit N_∞ of a complete open branch in a H_{sp}-derivation starting with N. Then: (i) I is a (Herbrand) model of N_∞ and N, if N_∞ does not contain the empty clause. (ii) A $K_{(m)}(\wedge, \vee, \smile, \uparrow)$-model of φ can be read off from I.*

THEOREM 14. *(i) A finite modal model for any modal formula locally satisfiable in $K_{(m)}(\wedge, \vee, \smile, \uparrow)$ can be effectively constructed with any (fair) procedure based on H_{sp}. (ii) $K_{(m)}(\wedge, \vee, \smile, \uparrow)$ has the finite model property.*

Due to the exact correspondence between clauses and formulae in H_{sp}- and *Tab*-derivations we can conclude:

COROLLARY 15. *A finite modal model for any modal formula locally satisfiable in $K_{(m)}(\wedge, \vee, \smile, \uparrow)$ can be effectively constructed with any (fair) procedure based on Tab.*

The results for $K_{(m)}(\wedge, \vee, \smile, \uparrow)$ in this section hold for all dynamic modal logics defined over the operators $\{\wedge, \vee, \smile, \uparrow, \downarrow\}$.

5 Rasiowa-Sikorski Systems

The 'develop via first-order resolution' approach is not limited to the development of tableau procedures. In this section we use the approach to develop a calculus for validity testing. More specifically, we develop a Rasiowa-Sikorski proof calculus for the dynamic modal logic $K_{(m)}(\wedge, \vee, \smile, \uparrow)$ and show that it is a decision procedure and can be used as the basis for a generator of counter-models.

Rasiowa-Sikorski proof systems [25] are dual tableau systems [34]. Given a formula F, they aim to prove its validity, or, if F is not valid, they aim to construct a counter-model, i.e. a model for the complement of the formula. Starting with the given formula F, this is done by systematic case analysis until fundamental validities are found. Fundamental validities are normally obvious validities such as the law of excluded middle (i.e. $\neg F \vee F$). Rasiowa-Sikorski expansion rules have the same form, $X / X_1 | \ldots | X_n$, as tableau rules and are also applied top-down. The definition of a Rasiowa-Sikorski derivation, and its construction by application of rules, is the same as for a tableau derivation. There is a slight variation in notation and crucially the interpretation of the rules is different. More precisely, X, X_i denote sets of labelled formulae, as in the previous section, but sets of formulae are now interpreted as disjunctions of formulae, whereas branching is interpreted conjunctively.

$$(\neg\bot)\ \frac{s:\psi, s:\neg\psi}{\neg\bot}$$

$$(\neg\wedge)_1\ \frac{s:\neg(\psi\wedge\phi)}{s:\sim\psi} \qquad (\neg\wedge)_2\ \frac{s:\neg(\psi\wedge\phi)}{s:\sim\phi} \qquad (\wedge)\ \frac{s:\psi\wedge\phi}{s:\psi\mid s:\phi}$$

$$([\cdot])_1\ \frac{s:[\alpha]\psi}{(s,t):\sim\alpha} \qquad ([\cdot])_2\ \frac{s:[\alpha]\psi}{t:\psi}$$

$$(\neg[\cdot])\ \frac{s:\neg[\sim\alpha]\psi, (s,t):\alpha}{t:\sim\psi}$$

$$(\neg\breve{})\ \frac{(s,t):\neg(\alpha^\breve{})}{(t,s):\sim\alpha} \qquad (\neg\breve{})_I\ \frac{(t,s):\alpha}{(s,t):\neg((\sim\alpha)^\breve{})}$$

$$(\neg\wedge)_1^r\ \frac{(s,t):\neg(\alpha\wedge\beta)}{(s,t):\sim\alpha} \qquad (\neg\wedge)_2^r\ \frac{(s,t):\neg(\alpha\wedge\beta)}{(s,t):\sim\beta}$$

$$(\neg\wedge)_I^r\ \frac{(s,t):\alpha, (s,t):\beta}{(s,t):\neg(\sim\alpha\wedge\sim\beta)}$$

$$(\neg\vee)^r\ \frac{(s,t):\neg(\alpha\vee\beta)}{(s,t):\sim\alpha\mid (s,t):\sim\beta}$$

$$(\neg\vee)_{I,1}^r\ \frac{(s,t):\alpha}{(s,t):\neg(\sim\alpha\vee\sim\beta)} \qquad (\neg\vee)_{I,2}^r\ \frac{(s,t):\beta}{(s,t):\neg(\sim\alpha\vee\sim\beta)}$$

$$(\neg1)_1\ \frac{(s,t):\neg(\alpha1\phi)}{(s,t):\sim\alpha} \qquad (\neg1)_2\ \frac{(s,t):\neg(\alpha1\phi)}{s:\sim\phi}$$

$$(\neg1)_I\ \frac{(s,t):\alpha}{(s,t):\neg((\sim\alpha)1\phi)\mid s:\phi}$$

The side conditions of the $([\cdot])_j$ rules ($j\in\{1,2\}$) are that t is a constant uniquely associated with the premise $s:[\alpha]\psi$. For the rules $(\breve{})_I$, $(\wedge)_{I,j}^r$, $(\vee)_I^r$ and $(1)_I$ the side conditions are that the relational formulae in the conclusions, occur as subformulae of the relational formula γ of a box formula $s:[\gamma]\psi$ on the current branch.

Figure 6. Rasiowa-Sikorski calculus for $K_{(m)}(\wedge,\vee,\breve{},1)$.

A Rasiowa-Sikorski calculus for local validity in $K_{(m)}(\wedge,\vee,\breve{},1)$ is presented in Figure 6. Let the calculus be denoted by RS. Notice that the rules are dual to the rules of the tableau calculus Tab in Figure 5. To see this, inductively define a function g such that any labelled formula $s:\psi$ in a Tab-derivation, starting with $a:\varphi$, is mapped to $g(s):\sim\psi$ in the corresponding RS-derivation, starting with $a:\sim\varphi$ (where $g(a)=a$ and any term s which was introduced for $\neg[\alpha]\phi$ is mapped by g to the term introduced for $[\alpha]\phi$). In addition, any labelled relational formula $(s,t):\alpha$ is mapped to $(g(s),g(t)):\sim\alpha$. Then extend the definition to a mapping of Tab-inferences to RS-inferences. We can show:

LEMMA 16. *The calculi Tab and RS step-wise simulate each other in a dual manner.*

THEOREM 17. *There is a linear bisimulation between* **Tab** *and* **RS** *for local satisfiability/validity in* $K_{(m)}(\wedge, \vee, \smile, 1)$.

A detailed analysis of the duality between tableau and Rasiowa-Sikorski calculi for Peirce logic can be found in [34].

Lemma 16 implies that all the properties of the tableau calculus transfer to the Rasiowa-Sikorski calculus. With the notion of redundancy dualised in the expected way, we can state:

COROLLARY 18. *A formula φ is locally valid in $K_{(m)}(\wedge, \vee, \smile, 1)$ iff a RS-derivation containing a branch $\mathcal{B}$ can be constructed (modulo redundancy, and with or without the dual contraction rules) such that $\mathcal{B}$ does not contain $\neg\bot$ and each rule application is redundant.*

COROLLARY 19. *Any (fair) procedure based on* **RS** *is a decision procedure for local validity in* $K_{(m)}(\wedge, \vee, \smile, 1)$.

COROLLARY 20. *A finite modal counter-model for any modal formula which is locally $K_{(m)}(\wedge, \vee, \smile, 1)$-invalid can be effectively constructed with any (fair) procedure based on* **RS**.

It is also possible to obtain the rules of the **RS**-calculus via resolution; this time we use resolution in dual form. The dual form of resolution is not very well-known but a little reflection will convince the reader that it is an obvious alternative interpretation of resolution.

Dual resolution calculi operate exactly like resolution calculi with the difference that clauses are obtained by transformation into disjunction normal form and dual Skolemisation, i.e. Skolem terms are used to eliminate universal quantifiers. Also, the dual form of structural transformation is used. The dual clause form is a set of conjunctions of literals and the set is interpreted as a disjunction. The empty clause is interpreted as $\top$. The definition of dual ordered resolution with selection is exactly the same as $R_{\text{sp}}^{\text{red}}$, except that the disjunction in clauses is viewed as a conjunction and branching in the splitting rule is interpreted conjunctively. For example the dual resolution rule for propositional logic derives $C \wedge D$ from $C \wedge A$ and $\neg A \wedge D$. In fact, all techniques and results of classical resolution carry over to dual resolution. As a consequence, by simply interpreting clauses dually, and also all transformation and derivation steps as well as all deletion steps, we can use any resolution prover as a dual resolution prover for validity testing of formulae.

Dual Structural Transformation into Clausal Form

Since we are now interested in showing the validity of a problem we need to base our reduction to first-order logic on the following *dual structural*

Subformula θ	Definitional dual clauses for $\mathrm{Def}^d(\theta)$
(shortcut)	$\neg Q_{\neg\psi}(x) \wedge \neg Q_\psi(x)$
$\psi \wedge \phi$	$\neg Q_{\psi\wedge\phi}(x) \wedge Q_\psi(x) \wedge Q_\phi(x)$
$\neg(\psi \wedge \phi)$	$\neg Q_{\neg(\psi\wedge\phi)}(x) \wedge Q_{\sim\psi}(x)$
	$\neg Q_{\neg(\psi\wedge\phi)}(x) \wedge Q_{\sim\phi}(x)$
$[\alpha]\psi$	$\neg Q_{[\alpha]\psi}(x) \wedge \neg R_\alpha(x, f_{[\alpha]\psi}(x))$
	$\neg Q_{[\alpha]\psi}(x) \wedge Q_\psi(f_{[\alpha]\psi}(x))$
$\neg[\alpha]\psi$	$\neg Q_{\neg[\alpha]\psi}(x) \wedge R_{\sim\alpha}(x,y) \wedge Q_{\sim\psi}(y)$
(shortcut)	$\neg R_{\neg\alpha}(x,y) \wedge \neg R_\alpha(x,y)$
	$R_{\neg\alpha}(x,y) \wedge R_\alpha(x,y)$
$\alpha \wedge \beta$	$\neg R_{\alpha\wedge\beta}(x,y) \wedge R_\alpha(x,y) \wedge R_\beta(x,y)$
	$R_{\alpha\wedge\beta}(x,y) \wedge \neg R_\alpha(x,y)$
	$R_{\alpha\wedge\beta}(x,y) \wedge \neg R_\beta(x,y)$
$\alpha \vee \beta$	$\neg R_{\alpha\vee\beta}(x,y) \wedge R_\alpha(x,y)$
	$\neg R_{\alpha\vee\beta}(x,y) \wedge R_\beta(x,y)$
	$R_{\alpha\vee\beta}(x,y) \wedge \neg R_\alpha(x,y) \wedge \neg R_\beta(x,y)$
$\alpha^{\smile}$	$\neg R_{\alpha^{\smile}}(x,y) \wedge R_\alpha(y,x)$
	$R_{\alpha^{\smile}}(x,y) \wedge \neg R_\alpha(y,x)$
$\alpha{\restriction}\phi$	$\neg R_{\alpha{\restriction}\phi}(x,y) \wedge R_\alpha(x,y) \wedge Q_\phi(x)$
	$R_{\alpha{\restriction}\phi}(x,y) \wedge \neg R_\alpha(x,y)$
	$R_{\alpha{\restriction}\phi}(x,y) \wedge Q_{\sim\phi}(x)$

Figure 7. Definitional dual clausal forms for $K_{(m)}(\wedge, \vee, {}^{\smile}, {\restriction})$

translation of formulae in dynamic modal logic:

$$\mathrm{Def}^d(\psi) =^{\mathrm{def}} \forall x\, (\pi(\psi, x) \to Q_\psi(x)) \wedge \forall x\, (\pi(\sim\psi, x) \to Q_{\sim\psi}(x))$$
$$\wedge\, \forall x\, (\neg Q_{\sim\psi}(x) \to Q_\psi(x)),$$
$$\mathrm{Def}^d(\alpha) =^{\mathrm{def}} \forall x\, y\, (R_\alpha(x,y) \to \pi(\alpha, x, y)) \wedge \forall x\, y\, (\pi'(\alpha, x, y) \to R_\alpha(x,y))$$
$$\wedge\, \forall x\, y\, (R_{\sim\alpha}(x,y) \leftrightarrow \neg R_\alpha(x,y)).$$

The mapping π is defined as in Section 3. We can prove:

THEOREM 21. *Let L be a dynamic modal logic defined over the operators $\{\wedge, \vee, {}^{\smile}, {\restriction}\}$, and let φ be any modal formula. Suppose N is the set of dual clauses obtained from $\varphi' =^{\mathrm{def}} \bigwedge \mathrm{Def}^d(\mathrm{Sf}(\varphi)) \to \forall x\, Q_\varphi(x)$ by transformation into disjunctive normal form, inner dual Skolemisation, and clausifying the Skolemised formula. Then: (i) Each clause in N is either a unit clause $Q(a)$, for some Skolem constant a, or it is an instance of a dual definitional clause given in Figure 7. (ii) φ is locally valid in L iff $\models \varphi'$ iff $\models N$.*

Let DH_{sp} denote the unordered dual hyperresolution calculus $DH_{\mathrm{sp}}^{\mathrm{red}}$ with splitting.

Subformula θ	Definitional dual clauses for $\mathrm{Def}^d(\theta)$
(shortcut)	$\neg Q_{\neg\psi}(x)^+ \wedge \neg Q_\psi(x)^+$
$\psi \wedge \phi$	$\neg Q_{\psi\wedge\phi}(x)^+ \wedge Q_\psi(x) \wedge Q_\phi(x)$
$\neg(\psi \wedge \phi)$	$\neg Q_{\neg(\psi\wedge\phi)}(x)^+ \wedge Q_{\sim\psi}(x)$
	$\neg Q_{\neg(\psi\wedge\phi)}(x)^+ \wedge Q_{\sim\phi}(x)$
$[\alpha]\psi$	$\neg Q_{[\alpha]\psi}(x)^+ \wedge R_{\sim\alpha}(x, f_{[\alpha]\psi}(x))$
	$\neg Q_{[\alpha]\psi}(x)^+ \wedge Q_\psi(f_{[\alpha]\psi}(x))$
$\neg[\alpha]\psi$	$\neg Q_{\neg[\alpha]\psi}(x)^+ \wedge \neg R_\alpha(x,y)^+ \wedge Q_{\sim\psi}(y)$
$\alpha \wedge \beta$	$R_{\neg(\alpha\wedge\beta)}(x,y) \wedge \neg R_{\sim\alpha}(x,y)^+ \wedge \neg R_{\sim\beta}(x,y)^+$
	$\neg R_{\neg(\alpha\wedge\beta)}(x,y)^+ \wedge R_{\sim\alpha}(x,y)$
	$\neg R_{\neg(\alpha\wedge\beta)}(x,y)^+ \wedge R_{\sim\beta}(x,y)$
$\alpha \vee \beta$	$R_{\neg(\alpha\vee\beta)}(x,y) \wedge \neg R_{\sim\alpha}(x,y)^+$
	$R_{\neg(\alpha\vee\beta)}(x,y) \wedge \neg R_{\sim\beta}(x,y)^+$
	$\neg R_{\neg(\alpha\vee\beta)}(x,y)^+ \wedge R_{\sim\alpha}(x,y) \wedge R_{\sim\beta}(x,y)$
$\alpha^{\smile}$	$\neg R_{\neg(\alpha^{\smile})}(x,y)^+ \wedge R_{\sim\alpha}(y,x)$
	$R_{\neg(\alpha^{\smile})}(x,y) \wedge \neg R_{\sim\alpha}(y,x)^+$
$\alpha{\upharpoonright}\phi$	$R_{\neg(\alpha\upharpoonright\phi)}(x,y) \wedge \neg R_{\sim\alpha}(x,y)^+ \wedge Q_\phi(x)$
	$\neg R_{\neg(\alpha\upharpoonright\phi)}(x,y)^+ \wedge R_{\sim\alpha}(x,y)$
	$\neg R_{\neg(\alpha\upharpoonright\phi)}(x,y)^+ \wedge Q_{\sim\phi}(x)$

Figure 8. Definitional dual clausal forms for $K_{(m)}(\wedge, \vee, {}^{\smile}, \upharpoonright)$ in range-restricted form.

By duality, soundness and completeness for validity are a consequence of Corollary 3 (and also Theorem 2).

COROLLARY 22. $\mathsf{DH}_{\mathrm{sp}}^{\mathrm{red}}$ and $\mathsf{DH}_{\mathrm{sp}}$ are sound and complete proof systems for sets of dual clauses.

The same result is true for dual ordered hyperresolution and dual ordered resolution with selection calculi.

Looking at Figure 7 we note that the clauses are not all range-restricted. It is in particular the clause $\neg Q_{\neg[\alpha]\psi}(x) \wedge R_{\sim\alpha}(x,y) \wedge Q_{\sim\psi}(y)$ associated with subformulae of the form $\neg[\alpha]\psi$ that is not range-restricted. This is a minor issue, because using a standard shifting transformation it is possible to transform the clauses into range-restricted clauses. Shifting essentially switches the signs of all binary literals; more precisely, we can use shifting to replace $R_\alpha(s,t)$- and $\neg R_\alpha(s,t)$-literals by $\neg R_{\neg\alpha}(s,t)$- and $R_{\neg\alpha}(s,t)$-literals, respectively. This transforms the clauses into range-restricted clauses. The clauses obtained are given in Figure 8.

Alternatively, we take the input set N computed in accordance with Theorem 21. With ordered resolution restricted to inferences which involve at least one of the relational shortcut clauses

$$\neg R_{\neg\alpha}(x,y) \wedge \neg R_\alpha(x,y)^* \quad \text{and} \quad R_{\neg\alpha}(x,y) \wedge R_\alpha(x,y)^*$$

as a premise we obtain the clauses in Figure 8 as conclusions. We need to use an ordering under which the binary literals are larger than unary literals, and in the shortcut clauses the R_α-literals (indicated with a *) are maximal. As now no more inferences are possible on these, the relational shortcut clauses and clauses containing R_α-literals become redundant and can be deleted. Let N' denote the 'pre-saturated and purified' set of clauses obtained in this way from N. Since the obtained clause set is dual to the clause set obtained for satisfiability of the negated problem (compare Figures 7 and 8), we can now state:

THEOREM 23. *Let φ be any $K_{(m)}(\wedge, \vee, \smile, 1)$-formula and let N' be the set of clauses obtained by a pre-saturation and purification from the dual clausal form of the dual structural transformation. Then: (i) Any DH_{sp}-derivation from N' terminates. (ii) φ is locally valid in $K_{(m)}(\wedge, \vee, \smile, 1)$ iff the DH_{sp}-saturation of N' contains the empty clause $\top$.*

LEMMA 24. *DH_{sp} step-wise simulates RS with respect to set N' obtained as defined in Theorem 23.*

THEOREM 25. *There is a linear bisimulation between RS and DH_{sp} on the structural transformation for local validity in $K_{(m)}(\wedge, \vee, \smile, 1)$.*

Using duality and Theorems 8 and 14 we can now give alternative proofs of Corollaries 18–20, i.e. soundness and completeness of RS, decidability and finite counter-model generation for $K_{(m)}(\wedge, \vee, \smile, 1)$.

6 Modal Resolution Systems

Refutation calculi without a splitting or branching rule need an explicit representation of disjunction. This section shows that if we omit splitting from tableau simulating hyperresolution procedures we get (labelled) modal resolution calculi of the kind described in Areces et al [1].

We focus again on the dynamic modal logic $K_{(m)}(\wedge, \vee, \smile, 1)$. Figure 9 presents a calculus we can read off from the structural encoding in Figure 3 and hyperresolution without splitting. Observe how the closure rule ($\bot$) has become an instance of the standard, ground resolution rule. Because we view clauses as multisets it is necessary to add factoring rules. Note however positive factoring suffices for completeness. As clauses are always ground, factoring has the effect of eliminating duplicate literals from clauses. It is thus easy to see that it is not necessary to add factoring rules if clauses are viewed as sets. Denote the calculus defined in Figure 9 as Res.

LEMMA 26. *H (without splitting) step-wise simulates Res with respect to the structural transformation defined in Theorem 5.*

THEOREM 27. *There is a linear bisimulation between Res and H on the structural transformation for local satisfiability in $K_{(m)}(\wedge, \vee, \smile, 1)$.*

As before, the proofs exploit the correspondence between inference steps in the Res-derivation and the H-derivation on the structural transformation of the given modal formula.

$$(res) \frac{C \vee s : \psi, D \vee s : \neg\psi}{C \vee D}$$

$$(\wedge)_1 \frac{C \vee s : \psi \wedge \phi}{C \vee s : \psi} \qquad (\wedge)_2 \frac{C \vee s : \psi \wedge \phi}{C \vee s : \phi}$$

$$(\neg\wedge) \frac{C \vee s : \neg(\psi \wedge \phi)}{C \vee s : {\sim}\psi \vee s : {\sim}\phi}$$

$$(\neg[\cdot])_1 \frac{C \vee s : \neg[\alpha]\psi}{C \vee (s,t) : \alpha} \qquad (\neg[\cdot])_2 \frac{C \vee s : \neg[\alpha]\psi}{C \vee t : {\sim}\psi}$$

$$([\cdot]) \frac{C \vee s : [\alpha]\psi, D \vee (s,t) : \alpha}{C \vee D \vee t : \psi}$$

$$(\smile) \frac{C \vee (s,t) : \alpha^{\smile}}{C \vee (t,s) : \alpha} \qquad (\smile)_I \frac{C \vee (t,s) : \alpha}{C \vee (s,t) : \alpha^{\smile}}$$

$$(\wedge)_1^r \frac{C \vee (s,t) : \alpha \wedge \beta}{C \vee (s,t) : \alpha} \qquad (\wedge)_2^r \frac{C \vee (s,t) : \alpha \wedge \beta}{C \vee (s,t) : \beta}$$

$$(\wedge)_I^r \frac{C \vee (s,t) : \alpha, D \vee (s,t) : \beta}{C \vee D \vee (s,t) : \alpha \wedge \beta}$$

$$(\vee)^r \frac{C \vee (s,t) : \alpha \vee \beta}{C \vee (s,t) : \alpha \vee (s,t) : \beta}$$

$$(\vee)_{I,1}^r \frac{C \vee (s,t) : \alpha}{C \vee (s,t) : \alpha \vee \beta} \qquad (\vee)_{I,2}^r \frac{C \vee (s,t) : \beta}{C \vee (s,t) : \alpha \vee \beta}$$

$$(1)_1 \frac{C \vee (s,t) : \alpha | \phi}{C \vee (s,t) : \alpha} \qquad (1)_2 \frac{C \vee (s,t) : \alpha | \phi}{C \vee s : \phi}$$

$$(1)_I \frac{C \vee (s,t) : \alpha}{C \vee (s,t) : \alpha | \phi \vee s : {\sim}\phi}$$

$$(fact) \frac{C \vee s : \psi \vee s : \psi}{C \vee s : \psi} \qquad (fact)^r \frac{C \vee (s,t) : \alpha \vee (s,t) : \alpha}{C \vee (s,t) : \alpha}$$

The side conditions of the $(\neg\Box_i)_j$ rules ($j \in \{1,2\}$) are that t is a constant uniquely associated with the premise $s : \neg\Box_i\psi$. For the rules $(\smile)_I$, $(\wedge)_{I,i}^r$, $(\vee)_I^r$ and $(1)_I$ the side conditions are that the relational formulae in the conclusions, occur as subformulae of the relational formula γ of a box formula $s : [\gamma]\psi$ in the current clause set.

Figure 9. Modal resolution calculus for $K_{(m)}(\wedge, \vee, {\smile}, 1)$.

COROLLARY 28. *Res is sound and refutationally complete for testing the local satisfiability of formulae in $K_{(m)}(\wedge, \vee, \smile, 1)$.*

This illustrates that the step from tableau-style systems to systems with an explicit resolution rule is not big. If we view the expansion rules as lazy translation to first-order logic then semantic tableau without splitting (i.e. modal resolution) can be viewed as hyperresolution with lazy translation to first-order logic.

What about decidability? We can prove Theorem 8(i) also holds for hyperresolution without splitting, and any refinement of hyperresolution without splitting.

THEOREM 29. *Let φ be any $K_{(m)}(\wedge, \vee, \smile, 1)$-formula and let N be the clausal form of the structural transformation. Then any H-, H^{red}-derivation from N terminates.*

This means that the calculus *Res* provides a decision procedure as well.

THEOREM 30. *Any (fair) procedure based on Res is a decision procedure for local satisfiability in $K_{(m)}(\wedge, \vee, \smile, 1)$.*

An advantage of the 'develop via first-order resolution' approach is that it is possible to transfer any refinements compatible with the simulating resolution procedure to the newly developed calculus. This may appear to be quite a strong claim, but let us now apply the 'develop via first-order resolution' approach by using ordered hyperresolution *OH*, instead of unordered hyperresolution *H*, and see which rules 'fall out'. Suppose the ordering used is an arbitrary ordering $\succ$ admissible in the sense of [3]. Recall the rules of ordered hyperresolution from Figure 2. Since the selection function overrides the ordering, the ordering does not change which literals are resolved upon in a negative premise. However since no literals can be selected in the positive premises, the ordering restricts inferences to literals strictly maximal with respect to the ordering. These restrictions transfer as follows to the modal resolution calculus. Suppose $\succ'$ is the ordering on labelled modal formulae that corresponds exactly to the ordering $\succ$ used in *OH*. Let *ORes* be the *ordered modal resolution calculus* based on $\succ'$ which is given by the rules of *Res* (cf. Figure 9) but which have side conditions which say that, for each rule except for the factoring rules, the explicitly given literals in every premise are strictly maximal with respect to $\succ'$. The side condition of the factoring rules is that the explicitly given literals are maximal with respect to $\succ'$. This means that any inferences that do not satisfy these side-conditions need not be performed (but can be, as is established in Corollary 28).

As before, it is possible to transfer any instances of redundancy elimination to the newly developed calculus. This includes tautology deletion, subsumption deletion and condensing. Similar as for tableau and Rasiowa-Sikorski systems we can define different forms of redundancy specifically for the modal resolution calculus. We refer to the extension of *ORes* with notions of redundancy as *ORes*^{red}.

LEMMA 31. *OH (OH*^{red}*) based on an admissible ordering $\succ$ step-wise simulates ORes (ORes*^{red}*) based on the corresponding ordering $\succ'$ on labelled formulae with respect to the structural transformation defined in Theorem 5.*

THEOREM 32. *There is a linear bisimulation between ORes*^{red} *and OH*^{red} *on the structural transformation for local satisfiability in $K_{(m)}(\wedge, \vee, \smile, 1)$.*

THEOREM 33. *ORes and ORes*^{red} *are sound and refutationally complete for testing the local satisfiability of formulae in $K_{(m)}(\wedge, \vee, \smile, 1)$.*

THEOREM 34. *Any (fair) procedure based on ORes*$^{(red)}$ *is a decision procedure for local satisfiability in $K_{(m)}(\wedge, \vee, \smile, 1)$.*

Analogously, we can write down (unordered and ordered) modal dual resolution calculi for validity in $K_{(m)}(\wedge, \vee, \smile, 1)$ (with or without redundancy) and prove soundness, completeness and decidability by dual arguments. We leave it to the reader to work out the details.

7 Concluding Remarks

Of course, it is possible to come up with the calculi developed in this paper in an independent way and prove soundness, completeness and decidability based on traditional methods. By taking a slightly unusual approach we can however obtain new insights and results for (familiar and less familiar) logics and the different proof methods that could be used for theorem proving. The 'develop via first-order resolution' approach allows us to develop different calculi and directly compare them in a common framework. We have seen that small variations in the translation mapping and the resolution refinement used, result in notably different styles of calculi. For example, there is only a very small difference between the tableau calculus and the modal resolution calculus for $K_{(m)}(\wedge, \vee, \smile, 1)$ in the simulations (the omission of the splitting rule). This shows that Tab and Res are closely related. We have seen that by interpreting resolution dually we can even devise calculi for proving the validity of formulae. The approach allows us to transfer refinements and techniques such as redundancy elimination to the new calculi, and it enables the formulation of strong soundness and completeness results.

The approach is applicable to more expressive logics and straightforwardly yields sound and (refutationally) complete tableau, Rasiowa-Sikorski or resolution calculi for many, if not all, first-order definable (dynamic) modal logics. The approach applies also to other first-order definable logics and fragments of first-order logic. For example, linear bisimulation results between tableau systems and resolution have been obtained in [13] for decidable fragments of first-order logic which are closely related to the guarded fragment. It is clear that the 'develop via resolution' approach allows us to immediately pull out sound, refutationally complete and terminating tableau procedures for the generalisations obtained in the paper [13]. Similarly, tableau decision procedure can be defined for the solvable class BU introduced and studied in [12].

Soundness and completeness of the extracted calculi is in general automatic as long as the reduction to first-order logic and the refinement used have the properties identified in the Introduction so that bisimulation results can be proved. Apart from finding a suitable combination of a reduction and refinement, the main challenge is to develop ways to guarantee termination and prove decidability. In the case of $K_{(m)}(\wedge, \vee, \smile, 1)$ there are existing decidability results for hyperresolution which we have extended.

It is possible to derive in a systematic way sound and complete calculi via non-standard translation methods, for instance, translations based on functional translation approaches (cf. [24, 33]) or the axiomatic translation approach [32]. In fact the results in [32] suggest that it is possible to obtain tableau calculi in a systematic and semi-automatic way from the Hilbert axiomatisations of traditional modal logics. The approach followed there also belongs to the 'develop via first-order resolution' approach. There is actually a lot of flexibility and potential in the approach which we have not explored due to lack of space.

There are good reasons why the approach discussed in this paper is combined with first-order resolution. Currently no other proof method exhibits the level of sophistication that the resolution framework has. It combines and integrates numerous principles and techniques that have been developed over many years in the area of automated reasoning. Previous work has shown that the available concepts of refinement and redundancy, in particular, mean that it is well suited for solving decidability problems of non-classical logics and fragments of first-order logic, cf. the surveys [10, 19, 33]. Furthermore, the framework is well suited for developing and studying model builders, cf. e.g. [5, 10, 11]. As we have seen in this paper refinements and redundancy are crucial ingredients of the 'develop via first-order resolution' approach.

One of the attractions of situating the approach within the framework of resolution is that it is easy to implement procedures based on the extracted calculi with existing first-order resolution theorem provers. Hyperresolution, or essentially equivalent refinements, are standardly implemented in all of the well-known first-order resolution theorem provers. Moreover, ordering restrictions can be flexibly defined in these and splitting is currently available in at least (M)SPASS [17, 36] and VAMPIRE [27]. With modest implementation effort it is therefore possible to use these provers as essentially modal tableau provers, modal Rasiowa-Sikorski provers, or modal resolution provers. All that is necessary, is to implement the structural transformations used and then choose the correct combination of flag settings so that the prover uses the simulating refinement. This provides a simple approach of implementing special-purpose procedures. In addition, it allows us to compare them also experimentally in a uniform framework. Examples of some empirical research performed in this framework can be found in [5, 18].

Further work consists of applying the approach to other logics, other styles of deduction methods as well as other forms of inference problems. We are presently investigating the simulation and development of natural deduction calculi for modal logics and first-order logic [28]. It would also be of interest if a resolution prover can be directly used as a special-purpose prover without any extra implementation effort. A student project to use MSPASS as a tableau prover for modal logics is currently under way.

Acknowledgements

The work has benefitted from discussions with Ullrich Hustadt and David Robinson. I also thank Yde Venema for useful comments.

BIBLIOGRAPHY

[1] C. Areces, M. de Rijke, and H. de Nivelle. Resolution in modal, description and hybrid logic. *J. Logic Computat.*, 11(5):717–736, 2001.

[2] F. Baader and U. Sattler. An overview of tableau algorithms for description logics. *Studia Logica*, 69:5–40, 2001.

[3] L. Bachmair and H. Ganzinger. Resolution theorem proving. In A. Robinson and A. Voronkov, editors, *Handbook of Automated Reasoning*, vol. I, pp. 19–99. Elsevier, 2001.

[4] L. Bachmair, H. Ganzinger, and U. Waldmann. Superposition with simplification as a decision procedure for the monadic class with equality. In *Proc. KGC'93*, vol. 713 of *LNCS*, pp. 83–96. Springer, 1993.

[5] P. Baumgartner and R. A. Schmidt. Blocking and other enhancements for bottom-up model generation methods. In *Proc. IJCAR 2006*, LNAI. Springer, 2006. To appear.

[6] M. A. Castilho, L. Fariñas del Cerro, O. Gasquet, and A. Herzig. Modal tableaux with propagation rules and structural rules. *Fundamenta Informaticae*, 3–4(32):281–297, 1997.

[7] H. de Nivelle. Splitting through new proposition symbols. In *Proc. LPAR 2001*, vol. 2250 of *LNAI*, pp. 172–185. Springer, 2001.

[8] H. De Nivelle, R. A. Schmidt, and U. Hustadt. Resolution-based methods for modal logics. *Logic J. IGPL*, 8(3):265–292, 2000.

[9] L. Fariñas del Cerro and O. Gasquet. A general framework for pattern-driven modal tableaux. *Logic J. IGPL*, 10(1):51–83, 2002.

[10] C. Fermüller, A. Leitsch, U. Hustadt, and T. Tammet. Resolution decision procedures. In A. Robinson and A. Voronkov, editors, *Handbook of Automated Reasoning*, pp. 1791–1849. Elsevier, 2001.

[11] L. Georgieva, U. Hustadt, and R. A. Schmidt. Computational space efficiency and minimal model generation for guarded formulae. In *Proc. LPAR 2001*, vol. 2250 of *LNAI*, pp. 85–99. Springer, 2001.

[12] L. Georgieva, U. Hustadt, and R. A. Schmidt. A new clausal class decidable by hyperresolution. In *Proc. CADE-18*, vol. 2392 of *LNAI*, pp. 260–274. Springer, 2002.

[13] L. Georgieva, U. Hustadt, and R. A. Schmidt. Hyperresolution for guarded formulae. *J. Symbolic Computat.*, 36(1–2):163–192, 2003.

[14] R. Goré. Tableau methods for modal and temporal logics. In M. D'Agostino, D. Gabbay, R. Hähnle, and J. Posegga, editors, *Handbook of Tableau Methods*, pp. 297–396. Kluwer, 1999.

[15] U. Hustadt and R. A. Schmidt. On the relation of resolution and tableaux proof systems for description logics. In *Proc. IJCAI'99*, pp. 110–115. Morgan Kaufmann, 1999.

[16] U. Hustadt and R. A. Schmidt. Issues of decidability for description logics in the framework of resolution. In *Automated Deduction in Classical and Non-Classical Logics*, vol. 1761 of *LNAI*, pp. 191–205. Springer, 2000.

[17] U. Hustadt and R. A. Schmidt. MSPASS: Modal reasoning by translation and first-order resolution. In *Proc. TABLEAUX 2000*, vol. 1847 of *LNAI*, pp. 67–71. Springer, 2000.

[18] U. Hustadt and R. A. Schmidt. Using resolution for testing modal satisfiability and building models. *J. Automated Reasoning*, 28(2):205–232, 2002.

[19] U. Hustadt, R. A. Schmidt, and L. Georgieva. A survey of decidable first-order fragments and description logics. *J. Relational Meth. in Computer Sci.*, 1:251–276, 2004.

[20] F. Massacci. Single step tableaux for modal logics: Computational properties, complexity and methodology. *J. Automated Reasoning*, 24(3):319–364, 2000.

[21] F. Massacci. Decision procedures for expressive description logics with intersection, composition, converse of roles and role identity. In *Proc. IJCAI 2001*, pp. 193–198. Morgan Kaufmann, 2001.

[22] K. Nellas. Reasoning about sets and relations: A tableaux-based automated theorem prover for Peirce logic. Master's thesis, Univ. Manchester, UK, 2001.

[23] R. Nieuwenhuis and A. Rubio. Paramodulation-based theorem proving. In A. Robinson and A. Voronkov, editors, *Handbook of Automated Reasoning*, pp. 371–443. Elsevier, 2001.

[24] H. J. Ohlbach, A. Nonnengart, M. de Rijke, and D. Gabbay. Encoding two-valued non-classical logics in classical logic. In A. Robinson and A. Voronkov, editors, *Handbook of Automated Reasoning*, pp. 1403–1486. Elsevier, 2001.

[25] H. Rasiowa and R. Sikorski. *The Mathematics of Metamathematics*, vol. 41 of *Monografie matematyczne*. Polish Scientific Publ., Warsaw, 1963.

[26] A. Riazanov and A. Voronkov. Splitting without backtracking. In *Proc. IJCAI 2001*, pp. 611–617. Morgan Kaufmann, 2001.

[27] A. Riazanov and A. Voronkov. The design and implementation of VAMPIRE. *Artificial Intelligence Commun.*, 15(2-3):91–110, 2002.

[28] D. Robinson and R. A. Schmidt. Resolution and natural deduction. In preparation, 2006.

[29] J. A. Robinson. A machine-oriented logic based on the resolution principle. *J. ACM*, 12(1):23–41, 1965.

[30] R. A. Schmidt and U. Hustadt. A resolution decision procedure for fluted logic. In *Proc. CADE-17*, vol. 1831 of *LNAI*, pp. 433–448. Springer, 2000.

[31] R. A. Schmidt and U. Hustadt. Mechanised reasoning and model generation for extended modal logics. In *Theory and Applications of Relational Structures as Knowledge Instruments*, vol. 2929 of *LNCS*, pp. 38–67. Springer, 2003.

[32] R. A. Schmidt and U. Hustadt. The axiomatic translation principle for modal logic. *ACM Trans. Computational Logic*, 2005. To appear.

[33] R. A. Schmidt and U. Hustadt. First-order resolution methods for modal logics. In *Volume in Memoriam of Harald Ganzinger*, LNCS. Springer, 2006. To appear.

[34] R. A. Schmidt, E. Orlowska, and U. Hustadt. Two proof systems for Peirce algebras. In *Proc. RelMiCS 7*, vol. 3051 of *LNCS*, pp. 238–251. Springer, 2004.

[35] S. Schulz. E: A brainiac theorem prover. *J. AI Comm.*, 15(2–3):111–126, 2002.

[36] C. Weidenbach, U. Brahm, T. Hillenbrand, E. Keen, C. Theobald, and D. Topic. SPASS version 2.0. In *Proc. CADE-18*, vol. 2392 of *LNAI*, pp. 275–279. Springer, 2002.

Renate A. Schmidt

School of Computer Science, University of Manchester, Oxford Road, Manchester M13 9PL, UK.

Renate.Schmidt@manchester.ac.uk

Completeness and incompleteness in first-order modal logic: an overview

VALENTIN SHEHTMAN

In this talk we describe the present nontrivial situation in semantics for first-order modal and superintuitionistic logics. The same topic was considered in our talk of 1990, cf. [11]; nowadays the field has become more developed, but perhaps less clear.

We consider normal monomodal logics extending **QK**, the pure quantified version of **K** and also the logics with equality. A logic is a set of formulas in the full first-order language with a single modal connective $\Box$ and without function symbols and constants; this set should be closed under Substitution, Modus Ponens, Generalization, and Necessitation and it should contain the standard axioms of classical logic (with equality — if necessary) and the axioms of **K**.

We also consider superintuitionistic logics; such a logic is a set of formulas in the same first-order language without $\Box$, containing the axioms of first-order intuitionistic calculus (in an appropriate form) and closed under Substitution, Modus Ponens, and Generalization.

QL denotes the pure quantified extension of a propositional logic **L**.

The main problem is to find a reasonable balance between syntax and semantics. Kripke semantics gives a plausible solution in the field of propositional modal logic (with some exceptions). But the whole situation in the first-order case can be briefly characterized as a big gap: simply axiomatizable logics may have a complicated semantical description, and the other way round — "relatively simple" classes of frames may correspond to non-axiomatizable logics.

Here are some issues to be discussed.

1. In standard Kripke semantics there are incompleteness results for large families of logics. For example, every logic **QL**, for any $L \supseteq S4$ such that $S5 \not\subseteq L$ and $L \not\subseteq S4.3$ is incomplete [5]. Another family consists of the logics **QL** for any intermediate propositional logic **L** of finite depth ≥ 2, cf. [8].

2. Unlike the propositional case, elementarity and Kripke-completeness are not closely related; moreover, within a certain class of first-order modal formulas every elementary formula axiomatizes an incomplete logic [2].

3. Sahlqvist's completeness theorem does not transfer to Kripke semantics of first-order modal logics; a simple example of an incomplete logic with Sahlqvist axioms is $\mathbf{QK} + \Diamond\top + \Box\Box p \to \Box p$, cf. [4]. Moreover, Kripke-complete logics are usually not canonical (in a natural sense); so completeness proofs in Kripke semantics are rather nontrivial, cf. [3], [13].

4. Unlike the propositional case, one should not expect axiomatizability for logics of non-elementary classes of frames, cf. general results in [15], [14]. Typical examples — the superintuitionistic logic of all frames based on an arbitrary Nötherian poset of unrestricted height is Π_1^1-hard; the superintuitionistic logic of a class of Kripke frames based on a class of finite posets is recursively axiomatizable iff this logic is characterized by a single finite frame.

5. The hyperdoctrinal semantics, where all first-order modal logics are complete [12], is rather complicated, so its real applicability is questionable. Recent results on completeness with respect to general Kripke-type frames presented at this conference [7] are of different flavour — this semantics is "first-order" in Thomason's sense, i.e. valuations do not range over all subsets.

6. There exists a Kripke-type ("second order") semantics, for which an analogue of Sahlqvist Theorem can be proved [10]. This is "simplicial" (or "metaframe") semantics introduced in [16]. However the convenience of simplicial frames is dubious. Functor semantics (introduced by S. Ghilardi) seems more efficient, but here a general completeness theorem is known only for superintuitionistic logics of the form $\mathbf{QL}$, with $\mathbf{L}$ canonical [6]. Generally speaking, simplicial frames are not representable of by set-valued functors, and we do not know any sufficient condition for representability. For modal logics functor semantics is weaker than simplicial [16], but we do not know if this is the case for superintuitionistic logics.

7. Algebraic semantics with equality also seems a good candidate to deal with – note that it is adequate for all propositional logics. Nevertheless we do not know if every first-order modal logic is algebraically complete.

8. Modal and superintuitionistic logics are not so closely related as in the propositional case. It seems unclear, whether every superintuitionistic logic is a fragment of some modal logic extending $\mathbf{QS4}$. For example, this problem arises for simplicially complete logics that are not Kripke-complete, because intuitionistically sound simplicial frames may be not modally sound [16].

9. Very little is known about topological semantics, which is between (standard) Kripke and algebraic. For some recent completeness results

see [1]. The logics of familiar spaces (like the real line) with the constant domain are still unknown.

10. In many cases for simple, but Kripke-incomplete logics, nothing is known about completeness even in simplicial semantics – e.g. for **QS4.1** or **QH** $+ \neg\neg\exists x P(x) \to \exists x \neg\neg P(x)$. In other cases, when simplicial semantics is adequate, it is unclear, what happens in more convenient semantics (like functor or topological) – a typical example is **QS4.2** $+ \forall x \Box P(x) \to \Box \forall x P(x)$.

BIBLIOGRAPHY

[1] Arló-Costa, H., Pacuit, E.: *First-order classical modal logic: applications in logics of knowledge and probability.* In: Van der Meyden, R., ed. Proceedings of TARK X (2005), July 10-12, 2005, Singapore. www.tark.org/proceedings/tark_jul10_05/proceedings.html

[2] Astretsov, S.: *Elementarity and incompleteness for predicate modal logics with constant domains.* International Conference "Computer Science applications of modal logic", September 5-9, 2005, Abstracts. Poncelet Math. Laboratory, 2005, Moscow. www.mccme.ru/poncelet/

[3] Corsi, G., Ghilardi, S.: *Directed frames.* Archive for math. logic, **29** (1989), 53–67.

[4] Gasquet, O.: *A new incompleteness result in Kripke semantics.* Fundamenta Informatica, **24** (1995), 407–415.

[5] Ghilardi, S.: *Incompleteness results in Kripke semantics.* J. Symbolic Logic, **56** (1991), 517–538.

[6] Ghilardi, S.: *Quantified extensions of canonical propositional intermediate logics.* Studia Logica, **51** (1992), 195–214.

[7] Goldblatt, R., Mares, E.: *A general semantics for quantified modal logic* (2006), see this Volume.

[8] Ono, H.: *Incompleteness of semantics for intermediate predicate logics. I. Kripke's semantics.* Proc. Jap. Acad., **49:** (1973), 711–713.

[9] Ono, H.: *Model extension theorem and Craig's interpolation theorem for intermediate predicate logics.* Reports in Math. Logic, No. 15 (1983), 41–58.

[10] Shehtman, V.: *Sahlqvist-type completeness theorem in modal predicate logic.* In: 10th International Congress in Logic, Methodology and Philosophy of Science, Florence, 1995, Abstracts, 220.

[11] Shehtman, V., Skvortsov, D.: *Semantics of non-classical first order predicate logics.* In Petkov, P., ed., Proceedings of Summer School and Conference on Mathematical Logic, Sept. 13–23 in Chaika, Bulgaria, 105–116. Plenum Press, 1990.

[12] Shirasu, H.: *Hyperdoctrinal semantics for non-classical predicate logics.* PhD thesis, Japan Advanced Institute of Science and Technology, 1997.

[13] Skvortsov, D.: *On the predicate logic of linear Kripke frames and some of its extensions.* Studia Logica, **81** (2005), 261–282.

[14] Skvortsov, D.: *The superintuitionistic predicate logic of finite Kripke frames is not recursively axiomatizable.* J. Symbolic Logic, **70** (2005), 451–459.

[15] Skvortsov, D.: *On non-axiomatizability of superintuitionistic predicate logics of some classes of well-founded and dually well-founded Kripke frames.* Journal of Logic and Computation (2006), to appear.

[16] Skvortsov, D., Shehtman, V.: *Maximal Kripke-type semantics for modal and superintuitionistic predicate logics.* Annals of Pure and Applied Logic, **63** (1993), 69–101.

Valentin Shehtman

Institute for Information Transmission Problems
Russian Academy of Sciences

B.Karetny 19, Moscow, Russia, 127994
shehtman@netscape.net

Logics with an existential modality

NATASHA ALECHINA AND DMITRY SHKATOV

ABSTRACT. We consider multi-modal logics interpreted over edge-labelled graphs with a modality $\langle \# \rangle$, where $\langle \# \rangle \varphi$ means 'φ is accessible by an edge with *some* label'. In a logic with finitely many edge labels, $\langle \# \rangle$ is definable, but if the set of labels is infinite, it is an independent modality. We axiomatise multi-modal K, deterministic multi-modal K, and PDL with converse and a single nominal, extended with #. The latter gives an axiomatisation of the logic PDL^{path} introduced in [3].

Keywords: Existential modality, multi-modal K, PDL, determinism

1 Introduction

In this paper, we consider multi-modal logics interpreted over edge-labelled graphs with a modality $\langle \# \rangle$, where $\langle \# \rangle \varphi$ means 'φ is accessible by an edge with *some* label'. We start by explaining what motivated our interest in the existential modality: namely, logics for modelling semi-structured data, such as data on the Web [1]. A collection of web pages can be represented as a graph with labelled edges. Edge labels come from some set I, which is either finite but very large, or even countably infinite. For example, I could be the set of all URLs, or all possible phrases in English (link names). Suppose we want to reason about constraints on possible paths in a graph, expressed as inclusions of regular expressions (inclusion constraints were introduced by Abiteboul and Vianu in [2]):

$$a;(b+c);\#;d^* \subseteq e;f$$

(if a data item is reachable by a path defined by $a;(b+c);\#;d^*$, that is: an a link followed by either a b or a c link, followed by an arbitrary link, followed by finitely many d links, then it is also reachable by a path $e;f$). We can study the implication problem for inclusion constraints (whether a set of constraints implies a constraint) by expressing it in a logical language; in [3], a logic called PDL^{path} was introduced for this purpose. The only unusual feature of PDL^{path} compared to other flavours of PDL, see e.g. [6, 5], is the wild card, or existential modality $\langle \# \rangle$, standing for 'any label'. In this paper, we consider axiomatisation and decidability problems for $\langle \# \rangle$, since, as far as we know, it has not been studied extensively before. The only reference we could find is in [4], where $\langle \# \rangle$ is used to make DPDL with intersection of programs badly undecidable. Clearly, in a language with finitely many edge labels $\langle \# \rangle$ is definable. It is easy to show (we do it below) that if the set of labels is infinite, then $\langle \# \rangle$ is not definable. But even if the set of labels I is finite, we may not want to write the very large formulas required to define $\langle \# \rangle$ (disjunction over all possible labels in I). For example, if I is the set of all URLs (which is finite but contains billions of elements), consider the difference between saying 'this web page does not have any outgoing links' as $\neg \langle \# \rangle \top$ and the alternative expression which involves a disjunction over all possible URLs.

The paper is organised as follows. First we study the logic $\mathsf{K}_\#$, obtained by adding $\langle \# \rangle$ to multi-modal K. We give a complete axiomatisation of $\mathsf{K}_\#$ and show that its complexity is the same as that of multi-modal K. However, an essential technical device we use in those proofs would not work with deterministic graphs. The device is as follows: given a formula φ and a model M for that formula, replace all edge labels in M which do not occur in φ by a single fresh edge label, and show that the resulting model still satisfies φ. This construction would not work if M were a deterministic model; we therefore believe that the completeness and decidability proof for deterministic multi-modal K, $\mathsf{DK}_\#$ is of an independent interest. This proof is given in section 4. In section 5 we briefly describe PDL^{path} introduced in [3] and in section 6 give a sound and weakly complete axiomatisation for it; decidability was proved in [3].

2 Logics $\mathsf{K}_\#$ and $\mathsf{DK}_\#$

Consider the propositional modal language $\mathcal{L}_\#^I$ containing (1) a countable set of propositional parameters **Par**; (2) propositional connectives $\neg$ ("not") and $\vee$ ("or"); (3) for every element i of the countable set I of modal indices, a modal operator $\langle i \rangle$; and (4) a modal operator $\langle \# \rangle$. All the other connectives, including the dual modalities $[i]$ and $[\#]$, can be defined in the usual way. The formulas of $\mathcal{L}_\#^I$ are defined by

$$\varphi := p \mid \neg \varphi \mid \varphi_1 \vee \varphi_2 \mid \langle i \rangle \varphi \mid \langle \# \rangle \varphi,$$

where $p \in \mathbf{Par}$ and $i \in I$. These formulas are evaluated on $\mathcal{L}_\#^I$-models.

DEFINITION 1. An $\mathcal{L}_\#^I$-model is a tuple $\mathcal{M} = (W, \{\mathcal{R}_i\}_{i \in I}, \mathcal{R}_\#, V)$, where $W \neq \emptyset$, $\mathcal{R}_i \subseteq W \times W$, $\mathcal{R}_\# = \bigcup_{i \in I} \mathcal{R}_i$, and V is a function from **Par** into 2^W.

$\mathcal{M}$ is deterministic if, for every $w \in W$ and every $i \in I$, there is no more than one v such that $w\mathcal{R}_i v$.

The truth definitions for formulas of $\mathcal{L}_\#^I$ are standard; in particular,

- $\mathcal{M}, w \Vdash \langle i \rangle \varphi$ iff $\exists v \in W (w\mathcal{R}_i v$ and $\mathcal{M}, v \Vdash \varphi)$

- $\mathcal{M}, w \Vdash \langle \# \rangle \varphi$ iff $\exists v \in W (w\mathcal{R}_\# v$ and $\mathcal{M}, v \Vdash \varphi)$.

It is easy to see that $\langle \# \rangle$ increases the expressive power of only those languages that contain at least a countable set I of modal indices; otherwise, $\langle \# \rangle \varphi$ can be defined as a finite disjunction of formulas of the form $\langle i \rangle \varphi$.

LEMMA 2. *Let φ be a formula not containing $\langle \# \rangle$, and let all labels occurring in φ be in the set L. Then φ is preserved with respect to the operation of removing R_i edges (where $i \notin L$).*

Proof. Let $\mathcal{M}, w \Vdash \varphi$, and let $\mathcal{M}'$ be obtained from $\mathcal{M}$ by removing all edges with labels not in L. Then $\mathcal{M}$ and $\mathcal{M}'$ are bisimilar with respect to $\{R_i : i \in L\}$, hence $\mathcal{M}', w \Vdash \varphi$. ∎

LEMMA 3. *Let φ be a formula which does contain $\langle \# \rangle$, and let all labels occurring in φ be in the set L. Then,*

1. *φ is not guaranteed to be preserved with respect to removing non-L edges;*

2. *φ is preserved with respect to renaming non-L edges, provided that the new names are also not in L.*

COROLLARY 4. *$\langle \# \rangle$ is not definable in a language with an infinite set of labels I.*

Let us denote the logic of all $\mathcal{L}_\#^I$-models as $\mathsf{K}_\#$ and the logic of all deterministic $\mathcal{L}_\#^I$-models as $\mathsf{DK}_\#$. We are now going to formulate Hilbert-style axiomatisations of $\mathsf{K}_\#$ and $\mathsf{DK}_\#$. Since $\langle \# \rangle$ resembles the existential quantifier of first-order logic, it's not hard to see that the axiomatisation of $\mathsf{K}_\#$ should look as follows (π stands for either an arbitrary $i \in I$ or $\#$):

Axiom schemata:
(A0) All classical tautologies;
(K) $[\pi](\varphi \to \psi) \to ([\pi]\varphi \to [\pi]\psi)$;
(ER) $\langle i \rangle \varphi \to \langle \# \rangle \varphi$.

Inference rules:

(MP) $\dfrac{\vdash \varphi \to \psi, \vdash \varphi}{\vdash \psi}$;   **(N)** $\dfrac{\vdash \varphi}{\vdash [\pi]\varphi}$;

(EL) $\dfrac{\vdash \langle i \rangle \varphi \to \psi}{\vdash \langle \# \rangle \varphi \to \psi}$, provided i does not occur in ψ.

Also, it is not difficult to guess that the axiomatisation of $\mathsf{DK}_\#$ can be obtained by adding to the axiom schemata above the 'axiom of determinism':

D $\langle i \rangle \varphi \to [i]\varphi$.

It is easy to show the following.

THEOREM 5. *$\mathsf{K}_\#$ is sound with respect to the class of all $\mathcal{L}_\#$-models. $\mathsf{DK}_\#$ is sound with respect to the class of all deterministic $\mathcal{L}_\#$-models.*

It is also easy to see that both $\mathsf{K}_\#$ and $\mathsf{DK}_\#$ are non-compact (consider the set $\Gamma = \{ \langle \# \rangle p, \neg \langle i \rangle p : i \in I \}$), and hence don't have strongly-complete axiomatisations. In the next two sections, we prove weak completeness of

$\mathsf{K}_\#$ and $\mathsf{DK}_\#$, showing that every $\mathsf{K}_\#$-consistent and every $\mathsf{DK}_\#$-consistent formula has a model.

3 Completeness for $\mathsf{K}_\#$

We use the completeness-via-finite-models technique described in detail in [4].

Let us define $\sim \varphi$, the pseudo-negation of φ, as follows: if φ is $\neg\psi$, $\sim\varphi$ is ψ; otherwise, it is $\neg\varphi$. By the closure of the set of formulas Σ, we mean the smallest set $\mathsf{CL}(\Sigma)$ containing all the subformulas of formulas of Σ and their pseudo-negations. It is easy to see that $\mathsf{CL}(\Sigma)$ is finite whenever Σ is. A finite canonical model for a $\mathsf{K}_\#$-consistent formula φ is built out of $\{\varphi\}$-atoms, maximally consistent subsets of $\mathsf{CL}(\{\varphi\})$. (In general, for an arbitrary set of formulas Σ, a Σ-atom is a maximally consistent subset of $\mathsf{CL}(\Sigma)$.) The following is straightforward:

LEMMA 6. *If $\varphi \in \mathsf{CL}(\Sigma)$ is $\mathsf{K}_\#$-consistent, then there exists an atom A over Σ such that $\varphi \in A$.*

It is easy to see that every Σ-atom A has the following properties:

1. For every $\varphi \in \mathsf{CL}(\Sigma)$, exactly one of φ and $\sim\varphi$ belongs to A.

2. For every $\varphi \vee \psi \in \mathsf{CL}(\Sigma)$, $\varphi \in \mathsf{CL}(\Sigma)$ or $\psi \in \mathsf{CL}(\Sigma)$.

Now we define finite canonical models for $\mathsf{K}_\#$.

DEFINITION 7. Let Σ be a finite set of $\mathcal{L}_\#^I$-formulas and let $a \in I$ be such that a does not occur in Σ. The finite canonical model over Σ, $\mathcal{M}^\Sigma$, is the tuple $(At(\Sigma), \{\mathcal{R}_i^\Sigma\}_{i \in I}, \mathcal{R}_\#^\Sigma, V^\Sigma)$, where

1. $At(\Sigma)$ is the set of all atoms over Σ;

2. $A\mathcal{R}_i^\Sigma A'$ iff $i \in \Sigma$ or $i = a$ and $\widehat{A} \wedge \langle i \rangle \widehat{A'}$ is consistent ($\widehat{X}$ stands for $\bigwedge_{\varphi \in X} \varphi$);

3. $A\mathcal{R}_\#^\Sigma A'$ iff $\widehat{A} \wedge \langle \# \rangle \widehat{A'}$ is consistent;

4. For every $p \in \mathbf{Par}$, $V^\Sigma(p) = \{\, A \in At(\Sigma) : p \in A \,\}$.

In a standard way (for details, see [4]), we can prove the following lemma.

LEMMA 8. *Let Σ be a set of $\mathcal{L}_\#^I$-formulas, A be an atom over Σ, and π be either an index occurring in Σ or $\pi = \#$. Then, for all $\langle\pi\rangle\varphi \in \mathsf{CL}(\Sigma)$, $\langle\pi\rangle\varphi \in A$ iff there is an atom A' such that $A\mathcal{R}_\pi A'$ and $\varphi \in A'$.*

From lemma 8 and the properties of atoms, we immediately obtain the following lemma.

LEMMA 9. *Let Σ be a set of $\mathcal{L}_\#$ formulas, $\mathcal{M}^\Sigma$ be the finite canonical model over Σ, and $\psi \in \mathsf{CL}(\Sigma)$. Then, for every $A \in At(\Sigma)$, $\mathcal{M}^\Sigma, A \Vdash \psi$ iff $\psi \in A$.*

All that remains to be done is to show that finite canonical models are $\mathcal{L}_\#^I$-models.

LEMMA 10. *Every finite canonical model* $\mathcal{M}^\Sigma = (At(\Sigma), \{\mathcal{R}_i^\Sigma\}_{i \in I}, \mathcal{R}_\#^\Sigma, V^\Sigma)$ *is an* $\mathcal{L}_\#$-*model.*

Proof. All we have to prove is that $\mathcal{R}_\#^\Sigma = \bigcup_{i \in I} \mathcal{R}_i^\Sigma$. First, right-to-left. Suppose that, for some $i \in I$, $A\mathcal{R}_i^\Sigma A'$ but $A\mathcal{R}_\#^\Sigma A'$ does not hold. Then, by definition 7, $\widehat{A} \wedge \langle i \rangle \widehat{A'}$ is consistent, but $\widehat{A} \wedge \langle \# \rangle \widehat{A'}$ is inconsistent. Then, $\vdash \langle \# \rangle \widehat{A'} \to \neg \widehat{A}$ and hence, by (ER), $\vdash \langle i \rangle \widehat{A'} \to \neg \widehat{A}$, which is impossible since $\widehat{A} \wedge \langle i \rangle \widehat{A'}$ is consistent.

Now, left-to-right. Suppose that $A\mathcal{R}_\#^\Sigma A'$. If, for some $i \in I$, $A\mathcal{R}_i^\Sigma A'$, then we are done. So, let us assume that for no $i \in I$ does $A\mathcal{R}_i^\Sigma A'$ hold. Then, $A\mathcal{R}_a^\Sigma A'$ holds. Indeed, if we suppose otherwise, then $\widehat{A} \wedge \langle \# \rangle \widehat{A'}$ is consistent and $\widehat{A} \wedge \langle a \rangle \widehat{A'}$ inconsistent. But then $\vdash \langle a \rangle \widehat{A'} \to \neg \widehat{A}$, and hence, by (EL) (note that $a \notin \Sigma$), $\vdash \langle \# \rangle \widehat{A'} \to \neg \widehat{A}$, which is impossible since $\widehat{A} \wedge \langle \# \rangle \widehat{A'}$ is consistent. ∎

THEOREM 11. $\mathsf{K}_\#$ *is complete with respect to the class of all* $\mathcal{L}_\#$ *models.*

Proof. Immediately follows from lemmas 6, 9, and 10. ∎

REMARK 12. If we had not added to the indices occurring in Σ a "new" index a, we would not have been able to prove that finite canonical models are $\mathcal{L}_\#$-models. As a counterexample, consider the set $\Sigma = \{\langle \# \rangle p \wedge \neg \langle b \rangle p\}$. Since $\langle \# \rangle p \wedge \neg \langle b \rangle p$ is consistent, in $\mathcal{M}^\Sigma$ there is an atom A such that $\langle \# \rangle p \wedge \neg \langle b \rangle p \in A$. Then, for some $B \in \mathcal{M}^\Sigma$ such that $p \in B$, we have $A\mathcal{R}_\#^\Sigma B$, but for no index $c \in \Sigma$ do we have $A\mathcal{R}_c^\Sigma B$.

In the proof of Theorem 11, we have constructed a model for a consistent formula φ which is of size $2^{|\varphi|}$. This gives us decidability of $\mathsf{K}_\#$. However, the complexity of checking whether φ is satisfiable by examining all models of size $2^{|\varphi|}$ is NEXPTIME. We can do better than that and show that the complexity of satisfiability problem for $\mathsf{K}_\#$ is no worse than that of multi-modal **K**.

THEOREM 13. *The satisfiability problem for* $\mathsf{K}_\#$ *is PSPACE-complete.*

Proof. There is a polynomial reduction from $\mathsf{K}_\#$ satisfiability to multi-modal **K** satisfiability (which is PSPACE-complete). A $\mathsf{K}_\#$ formula φ which contains labels $i_1, \ldots, i_n$ is satisfiable if and only if a formula φ' is satisfiable, where φ' is obtained from φ by replacing each subformula $\langle \# \rangle \psi$ with $\langle i_1 \rangle \psi \vee \ldots \vee \langle i_{n+1} \rangle \psi$. From Lemma 3: rename all labels which do not occur in φ to be i_{n+1}. The resulting model satisfies φ'. The other direction: let $\mathcal{M}, w \Vdash \varphi'$. φ' does not contain $\langle \# \rangle$, so it is satisfied in model $\mathcal{M}'$ where all labels not in the set $\{i_1, \ldots, i_{n+1}\}$ are removed (by Lemma 2). But on $\mathcal{M}'$, φ and φ' are equivalent, so $\mathcal{M}', w \Vdash \varphi$.

PSPACE-hardness follows from PSPACE-completeness of multi-modal **K** and the fact that $\mathsf{K}_\#$ includes multi-modal **K**. ∎

4 Completeness for $\mathsf{DK}_\#$

The above proof cannot be turned into a completeness proof for $\mathsf{DK}_\#$ in a straightforward way. If we simply replace in the definition of finite canonical models $\mathsf{K}_\#$-consistency by $\mathsf{DK}_\#$-consistency, we will not be able to prove that models so defined are deterministic, as the following example shows.

EXAMPLE 14. Consider the formula $\varphi = p \wedge \langle i \rangle q$ and the finite canonical model $\mathcal{M}^\varphi$ over φ. Then, among the points of $\mathcal{M}^\varphi$ (that is, among $\mathsf{DK}_\#$-atoms over $p \wedge \langle i \rangle q$) are $A = \{p, q, \langle i \rangle q, p \wedge \langle i \rangle q\}$ and $A' = \{\neg p, q, \langle i \rangle q, \neg(p \wedge \langle i \rangle q)\}$. Then, both $\widehat{A} \wedge \langle i \rangle \widehat{A}$ and $\widehat{A} \wedge \langle i \rangle \widehat{A'}$ are $\mathsf{DK}_\#$-consistent, which means that $A\mathcal{R}_i^\varphi A$ and $A\mathcal{R}_i^\varphi A'$.

Nevertheless, we will be able to reshape finite canonical models for $\mathsf{DK}_\#$ into deterministic models. First, we need to slightly adjust the definition of closure from the previous section. Given a subformula ψ of φ, we call the number of modal operators whose scope contains ψ the modal depth of ψ within φ, symbolically $\mathsf{md}_\varphi(\psi)$. Now, by the deterministic closure of the set of formulas Σ, we mean the smallest set $\mathsf{DCL}(\Sigma)$ containing (1) all the subformulas of formulas of Σ, (2) their pseudo-negations, and (3) for every $\varphi \in \Sigma$ and ψ such that $\mathsf{md}_\varphi(\psi) > 0$, if index i occurs in Σ, then $\langle i \rangle \psi \in \mathsf{DCL}(\Sigma)$ and $\langle i \rangle \sim \psi \in \mathsf{DCL}(\Sigma)$. It is easy to see that $\mathsf{DCL}(\Sigma)$ is finite whenever Σ is. In this section, Σ-atoms are taken to be maximally $\mathsf{DK}_\#$-consistent subsets of $\mathsf{DCL}(\Sigma)$.

DEFINITION 15. Let Σ be a finite set of $\mathcal{L}_\#^I$-formulas. The finite canonical model over Σ, $\mathcal{M}^\Sigma$, is the tuple $(At(\Sigma), \{\mathcal{R}_i^\Sigma\}_{i \in I}, \mathcal{R}_\#^\Sigma, V^\Sigma)$, where

1. $At(\Sigma)$ is the set of all atoms over Σ;

2. $A\mathcal{R}_i^\Sigma A'$ iff $\widehat{A} \wedge \langle i \rangle \widehat{A'}$ is consistent;

3. $A\mathcal{R}_\#^\Sigma A'$ iff $\widehat{A} \wedge \langle \# \rangle \widehat{A'}$ is consistent;

4. For every $p \in \mathbf{Par}$, $V^\Sigma(p) = \{A \in At(\Sigma) : p \in A\}$.

Proceeding exactly as in the completeness proof for $\mathsf{K}_\#$, we get the following two lemmas.

LEMMA 16. Let Σ be a set of $\mathcal{L}_\#^I$ formulas, $\mathcal{M}^\Sigma$ be the finite canonical model for $\mathsf{DK}_\#$ over Σ, and $\psi \in \mathsf{DCL}(\Sigma)$. Then, for every $A \in At(\Sigma)$, $\mathcal{M}^\Sigma, A \Vdash \psi$ iff $\psi \in A$.

LEMMA 17. Every finite canonical model $\mathcal{M}^\Sigma = (At(\Sigma), \{\mathcal{R}_i^\Sigma\}_{i \in I}, \mathcal{R}_\#^\Sigma, V^\Sigma)$ is an $\mathcal{L}_\#$-model.

Now, we have to reshape $\mathcal{M}^\Sigma$ into a deterministic model. We do so in two stages: first, we get rid of non-determinism with respect to indices not occurring in Σ, and then with respect to indices occurring in Σ. The first stage is easy.

LEMMA 18. *Let* $\mathcal{M}^\Sigma = (At(\Sigma), \{\mathcal{R}_i^\Sigma\}_{i \in I}, \mathcal{R}_\#^\Sigma, V^\Sigma)$ *be a finite canonical model for* DK$_\#$. *Then, there exists a model* $\mathcal{M}'^\Sigma = (At(\Sigma), \{\mathcal{R}'^\Sigma_i\}_{i \in I}, \mathcal{R}_\#^\Sigma, V^\Sigma)$ *such that (1) for every* $i \notin \Sigma$ *and every* $A, B, B' \in At(\Sigma)$, *if* $A\mathcal{R}'^\Sigma_i B$ *and* $A\mathcal{R}'^\Sigma_i B'$ *then* $B = B'$, *(2) the number of* $\mathcal{R}_i$ *edges with* $i \notin \Sigma$, *is at most* $|At(\Sigma)|^2$, *and (3) for every* $\psi \in \mathsf{DCL}(\Sigma)$ *and every* $X \in At(\Sigma)$, $\mathcal{M}'^\Sigma, X \Vdash \psi$ *iff* $\mathcal{M}^\Sigma, X \Vdash \psi$.

Proof. First, note that, by definition 15, if $A\mathcal{R}_\#^\Sigma B$ then $A\mathcal{R}_i^\Sigma B$ holds for every i not occurring in Σ. For every pair of atoms A and B connected by $\mathcal{R}_\#^\Sigma$, choose a single fresh link i_{AB} between them not occurring in Σ, and rename all other non-Σ links between A and B to be i_{AB}. The renaming of the links cannot affect the truth values of formulas in $\mathsf{DCL}(\Sigma)$, as the renamed links are indexed by indices not occurring in Σ and hence $\mathsf{DCL}(\Sigma)$. ∎

At the second stage, we proceed as follows. We will take a submodel of $\mathcal{M}'^\varphi$ generated by the atom A_φ containing φ, unravel this submodel into a tree-like model with root A_φ and then prune the resultant tree, leaving only one $\mathcal{R}_i$ branch for every $i \in I$. We show that such a model still satisfies φ, since φ cannot tell apart the points on the branch we leave in the tree from the pruned ones. We need the versions of tree-likeness and unravelling that are slightly different from the standard ones.

DEFINITION 19. *Let* $\mathcal{M} = (W, \{\mathcal{R}_i\}_{i \in I}, \mathcal{R}_\#, V)$ *be a* $\mathcal{L}_\#^I$-*model*. $\mathcal{M}$ *is tree-like if the structure* $(W, \mathcal{R}_\#)$ *is an irreflexive tree*. $\mathcal{M}$ *is strongly tree-like if* $\mathcal{M}$ *is tree-like and, for every* $(w, v) \in \mathcal{R}_\#$, *there exists exactly one* $i \in I$ *such that* $(w, v) \in \mathcal{R}_i$.

THEOREM 20. *Let* $\mathcal{M} = (W, \{\mathcal{R}_i\}_{i \in I}, \mathcal{R}_\#, V)$ *be a rooted* $\mathcal{L}_\#^I$-*model with root* w. *Then, there exists a strongly tree-like* $\mathcal{L}_\#^I$-*model* $\mathcal{M}^T = (W^T, \{\mathcal{R}_i^T\}_{i \in I}, \mathcal{R}_\#^T, V^T)$ *with root* w *such that, for every* $\mathcal{L}_\#^I$-*formula* φ, $\mathcal{M}, w \Vdash \varphi$ *iff* $\mathcal{M}^T, w \Vdash \varphi$.

Proof. First, consider model $\mathcal{M}' = (W', \{\mathcal{R}'_i\}_{i \in I}, \mathcal{R}'_\#, V')$, where

1. W is the set of all possible sequences of the form $(w, w_1^{i_1}, \ldots, w_n^{i_n})$, where $w_1, \ldots, w_{n \geq 0} \in W$ and $i_1, \ldots, i_n \in I$;

2. $(w, w_1^{i_1}, \ldots, w_n^{i_n})\mathcal{R}'_j(w, w_1^{i_1}, \ldots, w_n^{i_n}, w_{n+1}^{i_{n+1}})$ if $w_n \mathcal{R}_j w_{n+1}$ and $j = i_{n+1}$;

3. $\mathcal{R}'_\# = \bigcup_{i \in I} \mathcal{R}'_i$;

4. $V'(p) = \{(w, w_1^{i_1}, \ldots, w_n^{i_n}) : w_n \in V(p)\}$, for every $p \in \mathbf{Par}$.

Next, take the submodel of $\mathcal{M}'$ generated by w. Let us call it $\mathcal{M}^T$. It is clear that $\mathcal{M}^T$ is strongly tree-like (the last member of the sequence serving as the second argument of each $\mathcal{R}'_i$ bears exactly one superscript). The truth-preservation is guaranteed by the existence of a bisimulation $Z \subseteq W \times W^T$ defined by $vZ(w, w_1^{i_1}, \ldots, w_n^{i_n})$ iff $w_n = v$, which connects the roots of the two models. ∎

Now we show that, in tree-like models, for every formula φ, the value of φ at the root does not change if we replace a point v accessible from the root in k steps with another point v' such that v and v' agree on all the subformulas of φ of modal depth k. (In the statement of the following lemma, we use $w\mathcal{R}_\#^k v$ to mean that there are $u_1, \ldots, u_{k-1}$ such that $w\mathcal{R}_\# u_1 \mathcal{R}_\# \ldots \mathcal{R}_\# u_{k-1} \mathcal{R}_\# v$; in particular, $w\mathcal{R}_\#^0 v$ means that $w = v$. $\mathsf{Sub}(\varphi)$ stands for the set of all subformulas of φ.)

LEMMA 21. *Let φ be a $\mathcal{L}_\#^I$-formula, $\mathcal{M} = (W, \{\mathcal{R}_i\}_{i \in I}, \mathcal{R}_\#, V)$ a tree-like $\mathcal{L}_\#^I$-model, $w \in W$, and $v \in W$ such that $w\mathcal{R}_\#^k v$. Let $\mathcal{M}'$ be obtained from $\mathcal{M}$ by replacing the subtree generated by v by another subtree, with root v', such that, for every $\psi \in \mathsf{Sub}(\varphi)$ with $\mathsf{md}_\varphi(\psi) = k$, $\mathcal{M}, v \Vdash \psi$ iff $\mathcal{M}', v' \Vdash \psi$. Then, $\mathcal{M}, w \Vdash \varphi$ iff $\mathcal{M}', w' \Vdash \varphi$.*

Proof. By induction on k

Let $k = 0$. Then, $w = v$. Moreover, v and v' agree on all $\psi \in \mathsf{Sub}(\varphi)$ with $\mathsf{md}_\varphi(\psi) = 0$. As $\mathsf{md}_\varphi(\varphi) = 0$, w and v' agree on φ.

Assume that the statement of the lemma is true for $k = n$. We show that it is true for $k = n+1$. Suppose that it is not. Then, v and v' agree on all $\psi \in \mathsf{Sub}(\varphi)$ with $\mathsf{md}_\varphi(\psi) = n + 1$ and $\mathcal{M}, w \Vdash \varphi$, but $\mathcal{M}', w' \not\Vdash \varphi$ (the other case is symmetrical). Since no changes have been made to w itself, φ should have a subformula $\langle i \rangle \chi$ with $\mathsf{md}_\varphi(\langle i \rangle \chi) = 0$ such that, for some u such that $w\mathcal{R}_i u$ and $u \in path(w, v)$, $\mathcal{M}, u \Vdash \chi$ but $\mathcal{M}', u \not\Vdash \chi$ (the other case is symmetrical). Now, $\mathsf{md}_\varphi(\chi) = \mathsf{md}_\varphi(\langle i \rangle \chi) + 1$ and $\mathsf{Sub}(\chi) \subseteq \mathsf{Sub}(\varphi)$; therefore, v and v' agree on all $\psi \in \mathsf{Sub}(\chi)$ with $\mathsf{md}_\chi(\psi) = n$. As $u\mathcal{R}_\#^n v$, applying the inductive hypothesis to the tree generated by u, we get $\mathcal{M}, u \Vdash \chi$ iff $\mathcal{M}', u \Vdash \chi$, a contradiction. ∎

LEMMA 22. *Let $\mathcal{M}^{\varphi T}$ be a strongly tree-like model obtained from the canonical model over φ, $\mathcal{M}^\varphi$, by unravelling the submodel of $\mathcal{M}^\varphi$ generated by an atom A_φ containing φ. Then, for every $B, B' \in \mathcal{M}^{\varphi T}$ such that, for some C, $C\mathcal{R}_i B$ and $C\mathcal{R}_i B'$, and every ψ such that $\mathsf{md}_\varphi(\psi) > 0$, we have $\mathcal{M}^{\varphi T}, B \Vdash \psi$ iff $\mathcal{M}^{\varphi T}, B' \Vdash \psi$.*

Proof. Assume that there exist B and B' such that $C\mathcal{R}_i B$ and $C\mathcal{R}_i B'$, $\mathcal{M}^{\varphi T}, B \Vdash \psi$, and $\mathcal{M}^{\varphi T}, B' \not\Vdash \psi$. Then, $\mathcal{M}^{\varphi T}, C \Vdash \langle i \rangle \psi$ and $\mathcal{M}^{\varphi T}, C \Vdash \langle i \rangle \sim \psi$. Therefore, since $\langle i \rangle \psi, \langle i \rangle \sim \psi \in \mathsf{DCL}(\varphi)$, by lemma 16, $\langle i \rangle \psi, \langle i \rangle \sim \psi \in C$. This, however, is impossible, since by axiom (F), $\vdash_{\mathsf{DK}_\#} \langle i \rangle \psi, \langle i \rangle \sim \psi \to \bot$. ∎

Now we can prove the completeness theorem.

THEOREM 23. *$\mathsf{DK}_\#$ is weakly complete with respect to the class of deterministic $\mathcal{L}_\#$-models.*

Proof. Let φ be a $\mathsf{DK}_\#$-consistent formula. Build the finite canonical model $\mathcal{M}^\varphi$ over φ. There is in $\mathcal{M}^\varphi$ an atom A_φ such that $\varphi \in A_\varphi$. By lemma 16, $\mathcal{M}^\varphi, A_\varphi \Vdash \varphi$. Remove, using the construction of lemma 18, all the "redundant" atomic links in $\mathcal{M}^\varphi$ indexed by i not occurring in Σ. Next, unravel $\mathcal{M}'^\varphi$ into a strongly tree-like model $\mathcal{M}'^{\varphi T}$ using the construction of theorem 20. Now, level by level, for every point C and label i at level n such that C can reach several points $B_1, \ldots, B_m$ by an edge labelled i, replace all B_js by B_1. Denote the resultant model by $\mathcal{M}'^{\varphi T'}$. By the lemmas above, $\mathcal{M}'^{\varphi T'}, A_\varphi \Vdash \varphi$. Lastly, construct $\mathcal{M}'^{\varphi T''}$ by replacing all identical copies of B_1 produced in construction of $\mathcal{M}'^{\varphi T'}$ by a single point B_1. $\mathcal{M}'^{\varphi T'}$ and $\mathcal{M}'^{\varphi T''}$ are obviously bisimilar, so $\mathcal{M}'^{\varphi T''}, A_\varphi \Vdash \varphi$. It is clear that $\mathcal{M}'^{\varphi T''}$ is deterministic. ∎

The model for φ we have constructed in the proof of Theorem 23 is possibly infinite (it is an unravelling of a possibly cyclic model $\mathcal{M}^\varphi$ of size $2^{|\varphi|}$). We can however make it finite by pruning the tree at depth k, where k is the maximal depth of nesting of modal operators in φ (note that k is bound by $|\varphi|$). The branching factor of the tree is $|\varphi| + 2^{|\varphi|}$ (from every node A, there are at most $|\varphi|$ edges using labels from φ, and at most $2^{|\varphi|}$ fresh links - at most one for every other node B in the original canonical model). Given the branching factor and the depth of the finite tree model, the maximal number of nodes there is $(|\varphi| + 2^{|\varphi|})^{|\varphi|}$. This effective bounded model property for DK$_\#$ gives us the following theorem:

THEOREM 24. *DK$_\#$ is decidable.*

The best upper bound we have is NEXPTIME (guess a model for φ of size at most $(|\varphi| + 2^{|\varphi|})^{|\varphi|}$, which is $O(2^{|\varphi|^2})$, and verify that it satisfies φ).

5 Logic PDL^{path}

The language of PDL^{path} is an extension of the language of PDL, propositional dynamic logic. The language of PDL has two kinds of primitive symbols: propositional parameters and atomic transitions (or, modality indices). Indices are used to label edges in the transition system. Compound path expressions are built out of indices using binary operators ○ (composition), ∪ (union) and a unary operator ∗ (finite iteration). In addition to these, the language of PDL^{path}, introduced in [3], has the modal identity constant *id*, the unary converse operator $\cdot^-$ and the wild card modality #. Moreover, the language of PDL^{path} has a single nominal (a propositional letter that is true at exactly one point of a model) r, which is meant to mark the root of the graph. In the literature, PDL with the converse operator is referred to as **converse** PDL or CPDL. Thus, PDL^{path} is a fragment (since we have only one nominal) of CPDL with nominals augmented with the existential modality #.

In this paper, we give a complete Hilbert-style axiomatisation for PDL^{path}. To that end, we need to extend the language of PDL^{path} as introduced in [3] with the "at" modality @ of hybrid logics, which we will need to axiomatically describe the behaviour of the nominal r.

DEFINITION 25. Given a countable set of indices $I = \{i_1, i_2, \ldots, i_n, \ldots\}$, path expressions over I are defined by the following BNF expression:

$$\Lambda_I := I \mid id \mid \# \mid \Lambda_I \circ \Lambda_I \mid \Lambda_I \cup \Lambda_I \mid \Lambda_I^* \mid \Lambda_I^-.$$

PDL^{path}-formulas over the set of path expressions Λ_I are defined as follows:

$$\varphi := \top \mid \bot \mid r \mid \neg\varphi \mid \varphi \vee \varphi \mid \langle \Lambda_I \rangle \varphi \mid @_r\varphi.$$

PDL^{path}-formulas are evaluated on path models.

DEFINITION 26. A path model $\mathcal{M}$ over the set of labels Λ_I is a tuple $(W, \{\mathcal{R}_\pi\}_{\pi \in \Lambda_I}, V)$, where

1. $W \neq \emptyset$;

2. V is a function assigning some $\{w\} \subseteq W$ to r.

3. $\{\mathcal{R}_\pi\}_{\pi \in \Lambda_I}$ is a collection of binary relations over W satisfying the following conditions:

    (a) $\mathcal{R}_\# = \bigcup_{i \in I} \mathcal{R}_i$;
    (b) $\mathcal{R}_{id} = \{(w,w) : w \in W\}$ (identity relation);
    (c) $\mathcal{R}_{\pi^-} = \mathcal{R}_\pi^-$ (converse);
    (d) $\mathcal{R}_{\pi_1 \circ \pi_2} = \mathcal{R}_{\pi_1} \circ \mathcal{R}_{\pi_2}$ (composition);
    (e) $\mathcal{R}_{\pi_1 \cup \pi_2} = \mathcal{R}_{\pi_1} \cup \mathcal{R}_{\pi_2}$ (union);
    (f) $\mathcal{R}_{\pi^*} = \mathcal{R}_\pi^*$ (reflexive-transitive closure);
    (g) For every $w, v \in W$, there is a sequence of points $u_1, \ldots, u_n$ such that (1) $w = u_1$, (2) $v = u_n$, and (3) for every $1 \leq i \leq n-1$, either, for some $i \in I$, $u_i \mathcal{R}_i u_{i+1}$, or, for some $i \in I$, $u_{i+1} \mathcal{R}_i u_i$ (connectedness).

The truth of PDL^{path}-formulas at a point in a path model is defined as follows.

DEFINITION 27. Let $\mathcal{M} = (W, \{\mathcal{R}_\pi\}_{\pi \in \Lambda_I}, V)$ be a path model, $w, v \in W$. Then,

$\mathcal{M}, w \Vdash \top$		always;
$\mathcal{M}, w \Vdash \bot$		never;
$\mathcal{M}, w \Vdash r$	iff	$V(r) = \{w\}$;
$\mathcal{M}, w \Vdash \neg\varphi$	iff	$\mathcal{M}, w \nVdash \varphi$;
$\mathcal{M}, w \Vdash \varphi \vee \psi$	iff	$\mathcal{M}, w \Vdash \varphi$ or $\mathcal{M}, w \Vdash \psi$;
$\mathcal{M}, w \Vdash \langle \pi \rangle \varphi$	iff	for some $v \in W$, $w\mathcal{R}_\pi v$ and $\mathcal{M}, v \Vdash \varphi$;
$\mathcal{M}, w \Vdash @_r\varphi$	iff	$\mathcal{M}, v \Vdash \varphi$ and $V(r) = \{v\}$.

Here are some examples of properties definable in PDL^{path}: r defines the root; $\neg\langle\#\rangle\top$ defines leaf nodes; $\langle(\# \cup \#^-)^*\rangle r$ defines nodes connected to the root. To express a path constraint $\pi_1 \subseteq \pi_2$ (everything reachable from the root by a path π_1 is reachable by a path π_2), we can say $@_r[\pi_1]\langle \pi_2^-\rangle r$. Note that on connected graphs, $@_r\varphi$ is definable as $\langle(\# \cup \#^-)^*\rangle(r \wedge \varphi)$.

In [3], it was proved that PDL^{path} is decidable (the proof is similar to the proof of Theorem 13, reducing the satisfiability problem for PDL^{path} to the satisfiability problem for CPDL with nominals. CPDL with nominals is decidable in EXPTIME [5].

Now, we describe a Hilbert-style axiomatisation of PDL^{path}. Axiom schemata of PDL^{path} can be logically divided into four parts.

The first part describes the behaviour of propositional connectives and conventional modal operators $\langle \pi \rangle$ and $[\pi]$:

(A0) all classical tautologies;

(K) $[\pi](\varphi \rightarrow \psi) \rightarrow ([\pi]\varphi \rightarrow [\pi]\psi)$;

(A1) $\langle \pi \rangle \varphi \leftrightarrow \neg[\pi]\neg\varphi$.

The second part describes the properties of path expression operators:

(A2) $\langle \pi_1 \circ \pi_2 \rangle \varphi \leftrightarrow \langle \pi_1 \rangle \langle \pi_2 \rangle \varphi$;

(A3) $\langle \pi_1 \cup \pi_2 \rangle \varphi \leftrightarrow \langle \pi_1 \rangle \varphi \vee \langle \pi_2 \rangle \varphi$;

(A4) $\langle \pi^* \rangle \varphi \leftrightarrow \varphi \vee \langle \pi \rangle \langle \pi^* \rangle \varphi$;

(A5) $[\pi^*](\varphi \rightarrow [\pi]\varphi) \rightarrow (\varphi \rightarrow [\pi^*]\varphi)$;

(A6) $\varphi \rightarrow [\pi^-]\langle \pi \rangle \varphi$;

(A7) $\varphi \rightarrow [\pi]\langle \pi^- \rangle \varphi$;

(A8) $\varphi \leftrightarrow \langle id \rangle \varphi$;

(ER) $\langle i \rangle \varphi \rightarrow \langle \# \rangle \varphi$.

The third part describes properties of $@_r$ operator:

(A9) $@_r(\varphi \rightarrow \psi) \rightarrow (@_r\varphi \rightarrow @_r\psi)$;

(A10) $@_r\varphi \leftrightarrow \neg@_r\neg\varphi$;

(A11) $r \wedge \varphi \rightarrow @_r\varphi$;

(A12) $@_r r$;

(A13) $\langle \pi \rangle @_r\varphi \rightarrow @_r\varphi$.

Finally, the following axiom pertains to connectedness:

(A14) $\langle (\# \cup \#^-)^* \rangle r$.

The inference rules are:

$$(\text{MP}) \; \frac{\vdash \varphi \rightarrow \psi, \vdash \varphi}{\vdash \psi}; \quad (\text{N}) \; \frac{\vdash \varphi}{\vdash [\pi]\varphi}; \quad (\text{NN}) \; \frac{\vdash \varphi}{\vdash @_r\varphi};$$

(EL) $\dfrac{\vdash \langle i \rangle \varphi \to \psi}{\vdash \langle \# \rangle \varphi \to \psi}$, provided i does not occur in ψ.

In addition to the above axiom schemata and rules of inference, in the course of the following completeness proof, we will appeal to two additional rules of inference pertaining to the converse operator, whose derivability we establish in the following lemma.

LEMMA 28. *The following rules of inference are derivable in* PDL^{path} :

$$\dfrac{\vdash \varphi \to [\pi]\neg\psi}{\vdash \psi \to [\pi^-]\neg\varphi}; \quad \dfrac{\vdash \varphi \to [\pi^-]\neg\psi}{\vdash \psi \to [\pi]\neg\varphi}.$$

Proof. The first rule can be derived as follows.

1. $\varphi \to [\pi]\neg\psi$ – premise
2. ψ – assumption
3. $[\pi^-](\varphi \to [\pi]\neg\psi)$ – by (N) from 1
4. $\psi \to [\pi^-]\langle\pi\rangle\psi$ – (A6)
5. $[\pi^-]\langle\pi\rangle\psi$ – by (MP) from 2, 4
6. $[\pi^-](\langle\pi\rangle\psi \land (\varphi \to [\pi]\neg\psi))$ – from 3, 5 by (K)
7. $[\pi^-](\neg\varphi \lor (\langle\pi\rangle\psi \land [\pi]\neg\psi))$ – by (K) and propositional reasoning from 6
8. $[\pi^-]\neg\varphi$ – by (A1) and propositional reasoning from 7
9. $\psi \to [\pi^-]\neg\varphi$ – from 2, 8.

The second rule can be derived analogously, relying on axiom (A7). ∎

6 Completeness for PDL^{path}

In this section, we prove completeness of the above axiomatisation of PDL^{path} (its soundness is straightforward). As the language of PDL^{path} contains $\langle \# \rangle$ and $\langle \pi^* \rangle$, both of which give rise to non-compact logics, we can only prove weak completeness for PDL^{path}. As in the completeness proofs for $\mathsf{K}_\#$ and $\mathsf{DK}_\#$, we use the completeness-via-finite-models technique.

DEFINITION 29. Let Σ be a set of PDL^{path}-formulas over Λ_I. The closure of Σ, $\mathsf{CL}(\Sigma)$, is the smallest set such that

- if $\varphi \in \Sigma$ then $\mathsf{Sub}(\varphi) \subseteq \mathsf{CL}(\Sigma)$;
- if $\langle \pi^- \rangle \varphi \in \Sigma$ then $[\pi]\langle \pi^- \rangle \varphi \in \mathsf{CL}(\Sigma)$ (here and below, π ranges over all path labels);

- if $\langle \pi_1 \circ \pi_2 \rangle \varphi \in \mathsf{CL}(\Sigma)$ then $\langle \pi_1 \rangle \langle \pi_2 \rangle \varphi \in \mathsf{CL}(\Sigma)$;
- if $\langle \pi_1 \cup \pi_2 \rangle \varphi \in \mathsf{CL}(\Sigma)$ then $\langle \pi_1 \rangle \varphi \vee \langle \pi_2 \rangle \varphi \in \mathsf{CL}(\Sigma)$;
- if $\langle \pi^* \rangle \varphi \in \mathsf{CL}(\Sigma)$ then $\langle \pi \rangle \langle \pi^* \rangle \varphi \in \mathsf{CL}(\Sigma)$;
- if $\psi \in \mathsf{CL}(\Sigma)$ and $\psi \neq @_r\chi$ and $\psi \neq \neg @_r\chi$, then $@_r\psi \in \mathsf{CL}(\Sigma)$;
- $@_r r \in \mathsf{CL}(\Sigma)$;
- $\langle (\# \cup \#^-)^* \rangle r \in \mathsf{CL}(\Sigma)$;
- if $\varphi \in \mathsf{CL}(\Sigma)$, then $\sim \varphi \in \mathsf{CL}(\Sigma)$.

LEMMA 30. *Let Σ be a set of PDL^{path}-formulas. If Σ is finite, then $\mathsf{CL}(\Sigma)$ is finite, too.*

PDL^{path}-atoms are defined exactly as $\mathsf{K}_\#$-atoms. It is easy to show that, in addition to the properties satisfied by $\mathsf{K}_\#$-atoms, PDL^{path}-atoms have the following ones:

- for all $\langle \pi^- \rangle \varphi \in \mathsf{CL}(\Sigma)$, if $\varphi \in A$ then $[\pi]\langle \pi^- \rangle \varphi \in A$;
- for all $\langle \pi_1 \circ \pi_2 \rangle \varphi \in \mathsf{CL}(\Sigma)$, $\langle \pi_1 \circ \pi_2 \rangle \varphi \in A$ iff $\langle \pi_1 \rangle \langle \pi_2 \rangle \varphi \in A$;
- for all $\langle \pi_1 \cup \pi_2 \rangle \varphi \in \mathsf{CL}(\Sigma)$, $\langle \pi_1 \cup \pi_2 \rangle \varphi \in A$ iff $\langle \pi_1 \rangle \varphi \vee \langle \pi_2 \rangle \varphi \in A$;
- for all $\langle \pi^* \rangle \varphi \in \mathsf{CL}(\Sigma)$, $\langle \pi^* \rangle \varphi \in A$ iff $\langle \pi \rangle \langle \pi^* \rangle \varphi \in A$;
- for all $\langle id \rangle \varphi \in \mathsf{CL}(\Sigma)$, $\langle id \rangle \varphi \in A$ iff $\varphi \in A$.

LEMMA 31. *If $\varphi \in \mathsf{CL}(\Sigma)$ is PDL^{path}-consistent, then there exists an atom A over Σ such that $\varphi \in A$.*

Now we define the finite canonical PDL^{path}-model over Σ.

DEFINITION 32. Let Σ be a finite set of PDL^{path}-formulas over the set of path expressions Λ_I and let a be an index such that $a \in I$ but a does not occur in $\mathsf{CL}(\Sigma)$. First, define a family of binary relations $\{S_\pi\}$ on the set $At(\Sigma)$ of atoms over Σ, as follows:

- For all atoms $A, A' \in At(\Sigma)$, $AS_\pi A'$ iff $\pi \in \mathsf{CL}(\Sigma)$ or $\pi = a$, and $\widehat{A} \wedge \langle \pi \rangle \widehat{A'}$ is consistent.

Now, the finite canonical model $\mathcal{M}^\Sigma$ over Λ_I is a tuple $(W^\Sigma, \{\mathcal{R}_\pi^\Sigma\}_{\pi \in \Lambda_I}, V^\Sigma)$ such that

1. $W = At(\Sigma)$;
2. $V(r) = \{ A \in At(\Sigma) : r \in A \}$;
3.
    - for every atomic c such that $c \in \mathsf{CL}(\Sigma)$ or $c = a$, $\mathcal{R}_c^\Sigma = S_c$;
    - $\mathcal{R}_\#^\Sigma = S_\#$;

- $\mathcal{R}_{id}^{\Sigma} = \{ (A, A) : A \in At(\Sigma) \}$;
- $\mathcal{R}_{\rho^-}^{\Sigma} = (\mathcal{R}_{\rho}^{\Sigma})^-$;
- $\mathcal{R}_{\pi_1 \circ \pi_2}^{\Sigma} = \mathcal{R}_{\pi_1}^{\Sigma} \circ \mathcal{R}_{\pi_2}^{\Sigma}$;
- $\mathcal{R}_{\pi_1 \cup \pi_2}^{\Sigma} = \mathcal{R}_{\pi_1}^{\Sigma} \cup \mathcal{R}_{\pi_2}^{\Sigma}$;
- $\mathcal{R}_{\pi^*}^{\Sigma} = (\mathcal{R}^{\Sigma})_{\pi}^*$.

It is easy to see that finite canonical models for PDL^{path} satisfy conditions (3a)–(3f) required by definition 26 of path models (indeed, conditions (3b)–(3f) are satisfied because of definition 32, and condition (3a) can be shown to be satisfied in the same way as in the proof of lemma 10); thus, finite canonical models are regular. This is enough to prove the existence lemma and the truth lemma for finite canonical models.

To prove the existence lemma for finite canonical models, we first need to show that, for every $\pi \in \Lambda_I$, $S_\pi \subseteq \mathcal{R}_\pi^\Sigma$.

LEMMA 33. *For every $\pi \in \Lambda_I$, $S_\pi \subseteq \mathcal{R}_\pi^\Sigma$.*

Proof. By induction on the complexity of π.

(0) The cases $\pi \in I$ and $\pi = \#$ are obvious, since for $\pi \in I \cup \{\#\}$, $\mathcal{R}_\pi^\Sigma = S_\pi$.

(1) Let π be id. Suppose that $AS_{id}B$, that is $\widehat{A} \wedge \langle id \rangle \widehat{B}$ is consistent. By (A8), $\widehat{A} \wedge \widehat{B}$ is consistent. Since both A and B are atoms, this is only possible if $A = B$. Therefore, $A\mathcal{R}_{id}^\Sigma B$.

(2) Let π be ρ^-. Suppose that $AS_{\rho^-}B$, that is $\widehat{A} \wedge \langle \rho^- \rangle \widehat{B}$ is consistent. This implies consistency of $\widehat{B} \wedge \langle \rho \rangle \widehat{A}$. Indeed, if we suppose otherwise, then $\vdash \widehat{B} \to \neg \langle \rho \rangle \widehat{A}$ and hence $\vdash \widehat{B} \to [\rho] \neg \widehat{A}$. Then, by lemma 28, $\vdash \widehat{A} \to [\rho^-] \neg \widehat{B}$, which means that, contrary to the assumption, $\widehat{A} \wedge \langle \rho^- \rangle \widehat{B}$ is inconsistent. Thus, $\widehat{B} \wedge \langle \rho \rangle \widehat{A}$ is consistent and hence $BS_\rho A$. By the inductive hypothesis, $B\mathcal{R}_\rho^\Sigma A$ and therefore $A\mathcal{R}_{\rho^-}^\Sigma B$, as required.

(3)–(5) The other cases are proved exactly as for PDL. ∎

LEMMA 34 (Existence lemma). *Let Σ be a set of PDL^{path} -formulas over Λ_I, A be an atom over Σ, and $\pi \in \Lambda_I$. Then, for all $\langle \pi \rangle \psi \in CL(\Sigma)$, $\langle \pi \rangle \psi \in A$ iff there is an atom A' such that $A\mathcal{R}_\pi^\Sigma A'$ and $\psi \in A'$.*

Proof. The left-to-right direction can be proved using the standard "forcing choices" technique: picking, for every $\psi \in CL(\Sigma)$, either ψ itself of its pseudo-negation (in a consistency-preserving way), build an atom A' such that $\widehat{A} \wedge \langle \pi \rangle \widehat{A'}$ is consistent. Then, by lemma 33, $A\mathcal{R}_\pi^\Sigma A'$. The right-to-left direction is proved by induction on the complexity of π.

(0) $\pi \in I$. Suppose that there is an atom A' such that $\varphi \in A'$ and $A\mathcal{R}_\pi A'$. Then, by definition 32, $AS_\pi A'$, which means that $\widehat{A} \wedge \langle \pi \rangle \widehat{A'}$ is consistent. Then, as $\varphi \in A'$ and, thus, φ is one of the conjuncts of $\widehat{A'}$, $\widehat{A} \wedge \langle \pi \rangle \varphi$ is consistent, too. Then, as $\langle \pi \rangle \varphi \in CL(\Sigma)$ and A is an atom, $\langle \pi \rangle \varphi \in A$.

(1) $\pi = \#$. Analogously to (0).

(2) $\pi = \rho^-$. Suppose that $A\mathcal{R}_{\rho^-}^\Sigma A'$ and $\psi \in A'$. Then, $A'\mathcal{R}_\rho^\Sigma A$. As we know, $[\rho]\langle\rho^-\rangle\psi \in A'$. But this implies $\langle\rho^-\rangle\psi \in A$; indeed, if we suppose otherwise then $\neg\langle\rho^-\rangle\psi \in A$ and so, by inductive hypothesis, $\langle\rho\rangle\neg\langle\rho^-\rangle\psi \in A'$, which is impossible since then A' would be inconsistent.

(3)-(5) As for PDL. ∎

It is now easy to prove the truth lemma.

LEMMA 35. *Let Σ be a set of PDL^{path}-formulas, $\mathcal{M}^\Sigma$ be the finite canonical model over Σ, and $\psi \in \mathsf{CL}(\Sigma)$. Then, for every $A \in At(\Sigma)$, $\mathcal{M}^\Sigma, A \Vdash \psi$ iff $\psi \in A$.*

What remains to be done is ensure that we can reshape $\mathcal{M}^\Sigma$ into a model with exactly one root in a truth-preserving way. To that end, we will show that, given an atom $A \in \mathcal{M}^\Sigma$, if we take a submodel $\mathcal{M}_A^\Sigma$ of $\mathcal{M}^\Sigma$ generated by A, then $\mathcal{M}_A^\Sigma$ contains at most one root. This will be enough to prove weak completeness for PDL^{path}, since axiom (A14) ensures that $\mathcal{M}_A^\Sigma$ contains *at least one* root.

First, note the following simple fact.

LEMMA 36. *Let $\mathcal{M}$ be a regular model and $w \in \mathcal{M}$. Then, the submodel of $\mathcal{M}$ generated by w is also regular.*

Next, we prove that all the atoms of the submodel of $\mathcal{M}^\Sigma$ generated by A agree on formulas beginning with $@_r$.

LEMMA 37. *Let A be an atom, $\mathcal{M}_A^\Sigma$ be a submodel of $\mathcal{M}^\Sigma$ generated by A, and B and B' be atoms such that $B, B' \in \mathcal{M}_A^\Sigma$. Then, for every $@_r\psi \in \mathsf{CL}(\Sigma)$, $@_r\psi \in B$ iff $@_r\psi \in B'$.*

Proof. Assume that $@_r\psi \in B$ and $@_r\psi \notin B'$ (the other case is symmetrical) and, hence $\neg @_r\psi \in B'$.

Notice that, for any two atoms $X, X' \in \mathcal{M}_A^\Sigma$, if $X\mathcal{R}_{i \in I}^\Sigma X'$ and $@_r\psi \in X'$, then $@_r\psi \in X$. Indeed, otherwise, $\neg @_r\psi \in X$, which is impossible since, on the one hand, by (A13), $\neg @_r\psi \wedge \langle i\rangle @_r\psi$ is inconsistent and hence $\widehat{X} \wedge \langle i\rangle \widehat{X'}$ is inconsistent, and on the other, by definition 32, $X\mathcal{R}_i^\Sigma X'$ holds only if $\widehat{X} \wedge \langle i\rangle \widehat{X'}$ is consistent. Analogously, for any two atoms $X, X' \in \mathcal{M}_A^\Sigma$, if $X\mathcal{R}_i^\Sigma X'$ and $\neg @_r\psi \in X'$, then $\neg @_r\psi \in X$. For otherwise, $@_r\psi \in X$, which is impossible since, on the one hand, by (A10) and (A13), $@_r\psi \wedge \langle i\rangle \neg @_r\psi$ is inconsistent and hence $\widehat{X} \wedge \langle i\rangle \widehat{X'}$ is inconsistent, and on the other, by definition 32, $X\mathcal{R}_a^\Sigma X'$ holds only if $\widehat{X} \wedge \langle i\rangle \widehat{X'}$ is consistent. From the foregoing, it also follows that, for any $X, X' \in \mathcal{M}_A^\Sigma$ such that $X'\mathcal{R}_i^\Sigma X$, if $@_r\psi \in X'$ then $@_r\psi \in X$ and $\neg @_r\psi \in X'$ then $\neg @_r\psi \in X$.

Since $\mathcal{M}^\Sigma$, and hence, by lemma 36, also $\mathcal{M}_A^\Sigma$ are regular, $B \in \mathcal{M}_A^\Sigma$ implies that there is a chain of atomic transitions $\mathcal{R}_i^\Sigma$ connecting A and B (so that, to reach B from A, we can move forward as well as backward along $\mathcal{R}_i^\Sigma$'s in the chain). It follows that from $@_r\psi \in B$ we can infer $@_r\psi \in A$ (using the argument of the preceding paragraphs, "pull back" $@_r\psi$ along the chain connecting A and B). Analogously, from $\neg @_r\psi \in B'$ we can infer $\neg @_r\psi \in A$. This is impossible since A is an atom. ∎

Next, we can show that $\mathcal{M}_A^\Sigma$ has at most one root.

LEMMA 38. *Let A be an atom, $\mathcal{M}_A^\Sigma$ be a submodel of $\mathcal{M}^\Sigma$ generated by A, and B and B' be atoms such that (1) $B, B' \in \mathcal{M}_A^\Sigma$ and (2) $B \neq B'$. Then, at most one of B and B' contains r.*

Proof. Assume that $r \in B$ and $r \in B'$. Since $B \neq B'$, there is $\psi \in \mathsf{CL}(\Sigma)$ such that $\psi \in B$ and $\sim \psi \in B'$. There are two cases to consider: (1) $@_r\psi \in \mathsf{CL}(\Sigma)$ and (2) $@_r\psi \notin \mathsf{CL}(\Sigma)$ and, hence, either $\psi = @_r\chi$ or $\psi = \neg @_r\chi$.

(1) Suppose that $@_r\psi \in \mathsf{CL}(\Sigma)$, and hence, $\neg @_r\psi \in \mathsf{CL}(\Sigma)$. As $\psi \in B$ and $r \in B$, we also have $@_r\psi \in B$ (due to (A11), otherwise B would be inconsistent). Analogously, $\sim \psi \in B$ and $r \in B$ imply $\neg @_r\psi \in B'$. However, since $@_r\psi \in B$ (by lemma 37), we also have $@_r\psi \in B'$, which is impossible.

(2) Suppose that $@_r\psi \notin \mathsf{CL}(\Sigma)$ and, hence, either (2a) $\psi = @_r\chi$ or (2b) $\psi = \neg @_r\chi$. The case (2a) is analogous to the case (1), and the case (2b) is symmetrical. ∎

Now we can show that $\mathcal{M}_A^\Sigma$ is a path model.

LEMMA 39. *Let A be an atom and $\mathcal{M}_A^\Sigma$ be a submodel of $\mathcal{M}^\Sigma$ generated by A. Then, $\mathcal{M}_A^\Sigma$ is a path model.*

Proof. By lemma 36, $\mathcal{M}_A^\Sigma$ is regular and, by lemma 38, it has no more than one root. Moreover, (A14) guarantees that it has at least one root. ∎

The foregoing gives us the following theorem.

THEOREM 40. *PDL^{path} is complete with respect to the class of all path models.*

7 PDL^{path} without connectedness

Now, we consider what happens if we want to drop from the semantic definition of PDL^{path} the requirement that path models should be connected. It is easy to see that all we have to do to axiomatise PDL^{path} without connectedness is to drop from the above axiomatisation of PDL^{path} axiom (A14). Then, we can still show that every consistent formula has a model with *exactly one* root.

The only difference between the completeness proof for PDL^{path} and the completeness proof for PDL^{path} without connectedness is that, in the latter case, we cannot prove the analogue of lemma 39, as the following example shows.

EXAMPLE 41. Consider the formula $\varphi = \neg\langle(\# \cup \#^-)^*\rangle r$. Since now path models are allowed to be unconnected, it is consistent, and hence, there is, in the finite canonical model $\mathcal{M}^{\{\varphi\}}$ over $\{\varphi\}$, an atom A_φ such that $\varphi \in A_\varphi$. It is easy to see that the submodel $\mathcal{M}^{\{\varphi\}}_{A_\varphi}$ of $\mathcal{M}^{\{\varphi\}}$ generated by A_φ, does not contain an atom B such that $r \in B$.

Nevertheless, as the following lemma shows, given a finite canonical model for PDL^{path} without connectedness $\mathcal{M}^\Sigma$ and an atom A, we can always reshape $\mathcal{M}_A^\Sigma$ into a path model.

LEMMA 42. *Let A be an atom and $\mathcal{M}_A^\Sigma$ be a submodel of $\mathcal{M}^\Sigma$ generated by A such that no $X \in \mathcal{M}_A^\Sigma$ contains r. Then, there exists $\mathcal{M'}_A^\Sigma$ such that (1) $\mathcal{M'}_A^\Sigma$ is a path model, and (2) for every $X \in \mathcal{M}_A^\Sigma$ and every $\psi \in \mathsf{CL}(\Sigma)$, $\mathcal{M'}_A^\Sigma, X \Vdash \psi$ iff $\mathcal{M}_A^\Sigma, X \Vdash \psi$.*

Proof. Let us take an arbitrary atom $B \in \mathcal{M}_A^\Sigma$ and form the set $B_r = \{\chi : @_r\chi \in B\}$ (because of lemma 37, it does not matter which B we take).

First, note that B_r is consistent. Indeed, suppose that $\chi_1 \wedge \ldots \wedge \chi_n$ is inconsistent, where $\{\chi_1, \ldots, \chi_n\} = B_r$. Then, $\vdash \neg(\chi_1 \wedge \ldots \wedge \chi_n)$, and hence, by (NN), $\vdash @_r\neg(\chi_1 \wedge \ldots \wedge \chi_n)$. Therefore, due to (A10), $\vdash \neg@_r(\chi_1 \wedge \ldots \wedge \chi_n)$ and, due to (K) and **PL**, $\vdash \neg(@_r\chi_1 \wedge \ldots \wedge @_r\chi_n)$, which is impossible since then B would be inconsistent. Secondly, note that, as every $X \in \mathcal{M}_A^\Sigma$ contains $@_r r$ (due to (A12)), $r \in B_r$. Since B_r is consistent, by lemma 31, there exists an atom C such that $B_r \subseteq C$.

Next, obtain $\mathcal{M'}_A^\Sigma$ by adding to $\mathcal{M}_A^\Sigma$ the submodel $\mathcal{M}_C^\Sigma$ of $\mathcal{M}^\Sigma$ generated by C. It is easy to see that $\mathcal{M'}_A^\Sigma$ is a disjoint union of $\mathcal{M}_A^\Sigma$ and $\mathcal{M}_C^\Sigma$. Indeed, if for some $X \in \mathcal{M}_A^\Sigma$, some $X' \in \mathcal{M}_C^\Sigma$, and some $i \in I$ we would have either $X\mathcal{R}_i^\Sigma X'$ or $X'\mathcal{R}_i^\Sigma X$, then C would be in $\mathcal{M}_A^\Sigma$, which contradicts our assumption that no atom in $\mathcal{M}_A^\Sigma$ contains r. Now, first, by lemma 38, $\mathcal{M'}_A^\Sigma$ contains exactly one atom containing r (namely, C). Moreover, as both $\mathcal{M}_A^\Sigma$ and $\mathcal{M}_C^\Sigma$ are, by lemma 36, regular (since they are generated submodels of a regular model $\mathcal{M}^\Sigma$), $\mathcal{M'}_A^\Sigma$, being their disjoint union, is also regular. Therefore, $\mathcal{M'}_A^\Sigma$ is a path model. Secondly, as $\mathcal{M'}_A^\Sigma$ is a disjoint union of $\mathcal{M}_A^\Sigma$ and $\mathcal{M}_C^\Sigma$, for every $X \in \mathcal{M}_A^\Sigma$ and every $\psi \in \mathsf{CL}(\Sigma)$, $\mathcal{M'}_A^\Sigma, X \Vdash \psi$ iff $\mathcal{M}_A^\Sigma, X \Vdash \psi$. ∎

Using the preceding lemma, we can prove the following theorem.

THEOREM 43. *PDL^{path} without axiom (A14) is complete with respect to the class of all (not necessarily connected) path models.*

8 Conclusions and future work

We have proved completeness and decidability of extensions of multi modal K and DK with the existential modality, and axiomatised PDL^{path}, which also contains this modality. In future work, we plan to investigate the decidability of deterministic PDL^{path}. This involves investigating an open problem of decidability of deterministic CPDL with nominals.

Acknowledgements We thank Brian Logan and the anonymous AiML referees for useful comments and suggestions. This research was supported by the EPSRC grant GR/M98050/01. Natasha Alechina would like to thank the Isaac Newton Institute for Mathematical Sciences for hospitality during work on this paper.

BIBLIOGRAPHY

[1] Serge Abiteboul, Peter Buneman, and Dan Suciu. *Data on the Web: From Relations to Semistructured Data and XML*. Morgan Kaufmann, 1999.

[2] Serge Abiteboul and Victor Vianu. Regular path queries with constraints. In *Proceedings of the Sixteenth ACM SIGACT-SIGMOD-SIGART Symposium on Principles of Database Systems (PODS'97)*, pages 122–133, 1997.

[3] Natasha Alechina, Maarten de Rijke, and Stéphane Demri. A modal perspective on path constraints. *Journal of Logic and Computation*, 13(6):939–956, 2003.

[4] Patrick Blackburn, Maarten de Rijke, and Yde Venema. *Modal Logic*. Cambridge University Press, 2001.

[5] Giuseppe De Giacomo. *Decidability of Class-Based Knowledge Representation Formalisms*. PhD thesis, Universitá degli Studi di Roma "La Sapienza", 1995.

[6] David Harel, Dexter Kozen, and Jerzy Tiuryn. *Dynamic Logic*. MIT Press, 2000.

Natasha Alechina
School of Computer Science
University of Nottingham
NG8 1BB Nottingham, UK.
nza@cs.nott.ac.uk

Dmitry Shkatov
Institute of Philosophy
Volkhonka 14
Moscow 119992, Russia
d_shkatov@hotmail.com

An expressive two-sorted spatial logic for plane projective geometry

PHILIPPE BALBIANI

ABSTRACT. The aim of this paper is to introduce an expressive two-sorted spatial logic for plane projective geometry. We firstly attack the problem of the definability of multifarious classes of incidence frames and the problem of the reductions between several sublanguages of our language. We secondly attack the problem of the axiomatization of the logics of some natural classes of incidence frames.

1 Introduction

Projective planes are one of the most classical relational structures in geometry. They consist of either a nonempty set of points and a ternary relation of collinearity or a nonempty set of points, a nonempty set of lines and a binary relation of incidence. In the latter case any two distinct points are incident with precisely one common line and any two distinct lines are incident with exactly one common point. In general it is also assumed that any point is incident with at least two lines and any line is incident with at least two points. See Beutelspacher [6] for a survey. The aim of this paper is to introduce an expressive two-sorted spatial logic for plane projective geometry.
A modal treatment of plane projective geometry has been considered for the first time by Balbiani [1] and Venema [18] who interpreted the spatial modalities [I] and [O] by the incidence relation between points and lines, [I]a being read "in all lines x on the current point, a is realized at x" and [O]A being read "in all points X in the current line, A is realized at X". In this paper we consider the new spatial modalities [01] and [10], [01](a, b) being read "in all lines x, y on the current point, if x and y are different then a is realized at x or b is realized at y" and [10](A, B) being read "in all points X, Y in the current line, if X and Y are different then A is realized at X or B is realized at Y".
The paper has three major sections. Section 2 introduces our language based on [01] and [10] and interprets it in projective planes. Section 3 attacks the problem of the definability of multifarious classes of incidence frames and the problem of the reductions between several sublanguages of our language. Section 4 attacks the problem of the axiomatization of the logics of some natural classes of incidence frames. In the annex we prove some of our results. We assume the reader is at home with tools and techniques in modal logic such as rules for the undefinable and the canonical

model construction. For more on these see Blackburn et al [7], Chagrov and Zakharyaschev [8] or Kracht [13].

2 Syntax and semantics

2.1 Syntax

Now it is time to present the language we will be working with. Let PV be a countable set of *point variables*, with typical members denoted P, Q, etc, and LV be a countable set of *line variables*, with typical members denoted p, q, etc. We define the set of all *point formulas*, with typical members denoted A, B, etc, and the set of all *line formulas*, with typical members denoted a, b, etc, by the rules

- $A := P \mid \bot \mid \neg A \mid (A \vee B) \mid [01](a,b)$ and
- $a := p \mid \bot \mid \neg a \mid (a \vee b) \mid [10](A,B)$.

We adopt the standard rules for omission of the parentheses. Point formulas and line formulas constitute the formulas of our language, with typical representatives denoted ϕ, ψ, etc. The point formula $[01](a,b)$ means "in all lines x, y on the current point, if x and y are different then a is realized at x or b is realized at y" and the line formula $[10](A,B)$ means "in all points X, Y in the current line, if X and Y are different then A is realized at X or B is realized at Y". We make use of the abbreviations

- $[\in]a := [01](a,a)$, i.e. "at the very most, one line on the current point does not realize a",
- $[I]a := [01](a,\bot)$, i.e. "in all lines on the current point, a",
- $[\cdot]A := [01]([10](A,\bot),\bot)$, i.e. "in all points, A",
- $\odot A := \neg A \wedge [01]([10](A,A),\bot)$, i.e. "in all points but the current one, A",
- $[\ni]A := [10](A,A)$, i.e. "at the very most, one point in the current line does not realize A",
- $[O]A := [10](A,\bot)$, i.e. "in all points in the current line, A",
- $[-]a := [10]([01](a,\bot),\bot)$ i.e. "in all lines, a", and
- $\ominus a := \neg a \wedge [10]([01](a,a),\bot)$ i.e. "in all lines but the current one, a".

Remark that $[\cdot]A$ is $[I][O]A$, $\odot A$ is $\neg A \wedge [I][\ni]A$, $[-]a$ is $[O][I]a$ and $\ominus a$ is $\neg a \wedge [O][\in]a$. The other constructs are defined as usual. In particular we have dual operators for our boxes which are defined by $\langle 01 \rangle(a,b) := \neg[01](\neg a, \neg b)$, $\langle \in \rangle a := \neg[\in]\neg a$, $\langle I \rangle a := \neg[I]\neg a$, $\langle \cdot \rangle A := \neg[\cdot]\neg A$, $\langle 10 \rangle(A,B) := \neg[10](\neg A, \neg B)$, $\langle \ni \rangle A := \neg[\ni]\neg A$, $\langle O \rangle A := \neg[O]\neg A$ and $\langle - \rangle a := \neg[-]\neg a$.

2.2 Semantics

Now it is time to interpret our language in relational structures. An *incidence frame* is a relational structure

- $\mathcal{F} = (Po, Li, I, O)$

where Po is a nonempty set of "points", with typical members denoted X, Y, etc, Li is a nonempty set of "lines", with typical members denoted x, y, etc, I is a binary relation between points and lines and O is a binary relation between lines and points. If $I(X, x)$, i.e. the relation I holds for point X and line x, we say that "X is in x" and if $O(x, X)$, i.e. the relation O holds for line x and point X, we say that "x is on X". We introduce special notations for the ternary relations

- $In(X, y, z) := (I(X, y) \wedge I(X, z) \wedge y \neq z)$ and
- $On(x, Y, Z) := (O(x, Y) \wedge O(x, Z) \wedge Y \neq Z)$

which will be used further. $\mathcal{F}$ is said to be *rooted* if

- $\exists X((I \circ O)^\star(X) = Po \wedge (O \circ I)^\star(I(X)) = Li)$ or
- $\exists x((I \circ O)^\star(O(x)) = Po \wedge (O \circ I)^\star(x) = Li)$

where $I(X) = \{x \colon I(X, x)\}$, $O(x) = \{X \colon O(x, X)\}$, $\circ$ denotes composition and * denotes iteration. For all natural integers n such that $n \geq 2$, we shall say that $\mathcal{F}$ is *n-rich* if

- $\forall X(Card(I(X)) \geq n)$ and
- $\forall x(Card(O(x)) \geq n)$.

$\mathcal{F}$ is said to be ω-*rich* if for all natural integers n such that $n \geq 2$, $\mathcal{F}$ is n-rich. We shall say that $\mathcal{F}$ is *correct* if

- $\forall X \forall x(I(X, x) \rightarrow O(x, X))$ and
- $\forall x \forall X(O(x, X) \rightarrow I(X, x))$.

$\mathcal{F}$ is said to be *projective* if

- $\forall X \forall Y(I(X) \cap I(Y) \neq \emptyset)$ and
- $\forall x \forall y(O(x) \cap O(y) \neq \emptyset)$.

We shall say that $\mathcal{F}$ is *normal* if

- $\forall X \forall Y \forall z \forall t(I(X, z) \wedge I(Y, z) \wedge I(X, t) \wedge I(Y, t) \rightarrow X = Y \vee z = t)$ and
- $\forall x \forall y \forall Z \forall T(O(x, Z) \wedge O(y, Z) \wedge O(x, T) \wedge O(y, T) \rightarrow x = y \vee Z = T)$.

$\mathcal{F}$ is said to be *Arguesian* if, as a projective plane, it satisfies the theorem of Desargues. See [6] at page 109 for details. We shall say that $\mathcal{F}$ is *Pappian* if, as a projective plane, it satisfies the theorem of Pappus. See [6] at page 117 for details. An *incidence model* based on $\mathcal{F}$ is an ordered triple

- $\mathcal{M} = (\mathcal{F}, \pi, \lambda)$

where π is a function assigning to each point variable $P \in PV$ a subset $\pi(P)$ of Po and λ is a function assigning to each line variable $p \in LV$ a subset $\lambda(p)$ of Li. We define the notion of a point formula A being *true* at a point $X \in Po$, in symbols $\mathcal{M}, X \models A$, and the notion of a line formula a being *true* at a line $x \in Li$, in symbols $\mathcal{M}, x \models a$, by

- $\mathcal{M}, X \models P$ iff $X \in \pi(P)$,
- $\mathcal{M}, X \not\models \bot$,
- $\mathcal{M}, X \models \neg A$ iff $\mathcal{M}, X \not\models A$,
- $\mathcal{M}, X \models A \vee B$ iff $\mathcal{M}, X \models A$ or $\mathcal{M}, X \models B$,
- $\mathcal{M}, X \models [01](a,b)$ iff for all lines y, z, if $In(X,y,z)$ then $\mathcal{M}, y \models a$ or $\mathcal{M}, z \models b$,
- $\mathcal{M}, x \models p$ iff $x \in \lambda(p)$,
- $\mathcal{M}, x \not\models \bot$,
- $\mathcal{M}, x \models \neg a$ iff $\mathcal{M}, x \not\models a$,
- $\mathcal{M}, x \models a \vee b$ iff $\mathcal{M}, x \models a$ or $\mathcal{M}, x \models b$ and
- $\mathcal{M}, x \models [10](A,B)$ iff for all points Y, Z, if $On(x,Y,Z)$ then $\mathcal{M}, Y \models A$ or $\mathcal{M}, Z \models B$.

Remark that the interpretation of the point formula $[01](a,b)$ involves the relation of inequality between lines and the interpretation of the line formula $[10](A,B)$ involves the relation of inequality between points.

LEMMA 1. *Let*

- $SEP_0 := [01](p,q) \to [I](p \vee r) \vee [I](q \vee \neg r)$ *and*
- $SEP_1 := [10](P,Q) \to [O](P \vee R) \vee [O](Q \vee \neg R)$.

Then

- *for all points $X \in Po$, $\mathcal{M}, X \models SEP_0$ and*
- *for all lines $x \in Li$, $\mathcal{M}, x \models SEP_1$.*

Par abus de langage let $\pi(A) = \{X : \mathcal{M}, x \models A\}$ and $\lambda(a) = \{x : \mathcal{M}, x \models a\}$. Operators like $[\in]$ and $[\ni]$ remind one of operators that are sometimes called graded modalities [10], seeing that

LEMMA 2.

- *For all points $X \in Po$, $\mathcal{M}, X \models [\in]a$ iff there exists a line x such that $I(X) \setminus \{x\} \subseteq \lambda(a)$ and*

- for all lines $s \in Li$, $\mathcal{M}, x \models [\ni]A$ iff iff there exists a point X such that $O(x) \setminus \{X\} \subseteq \pi(A)$.

The operators $[I]$ and $[O]$ were introduced by Balbiani [1] and Venema [18] in the course of their work on spatial logics, seeing that

LEMMA 3. *If $\mathcal{M}$ is 2-rich then*

- *for all points $X \in Po$, $\mathcal{M}, X \models [I]a$ iff $I(X) \subseteq \lambda(a)$ and*
- *for all lines $x \in Li$, $\mathcal{M}, x \models [O]A$ iff $O(x) \subseteq \pi(A)$.*

Operators like $[\cdot]$ and $[-]$ remind one of operators that are sometimes called universal modalities [12], seeing that

LEMMA 4. *If $\mathcal{M}$ is 2-rich, correct and projective then*

- *for all points $X \in Po$, $\mathcal{M}, X \models [\cdot]A$ iff $\pi(A) = Po$ and*
- *for all lines $x \in Li$, $\mathcal{M}, x \models [-]a$ iff $\lambda(a) = Li$.*

Operators like $\odot$ and $\ominus$ remind one of operators that are sometimes called difference modalities [14], seeing that

LEMMA 5. *If $\mathcal{M}$ is 2-rich, correct and projective then*

- *for all points $X \in Po$, $\mathcal{M}, X \models \odot A$ iff $\pi(A) = Po \setminus \{X\}$ and*
- *for all lines $x \in Li$, $\mathcal{M}, x \models \ominus a$ iff $\lambda(a) = Li \setminus \{x\}$.*

We shall say that point formula A is *valid* in $\mathcal{M}$, in symbols $\mathcal{M} \models A$, if $\pi(A) = Po$ and line formula a is *valid* in $\mathcal{M}$, in symbols $\mathcal{M} \models a$, if $\lambda(a) = Li$. Now consider an incidence frame $\mathcal{F} = (Po, Li, I, O)$. A formula ϕ is said to be *valid* in $\mathcal{F}$, in symbols $\mathcal{F} \models \phi$, if ϕ is valid in all incidence models based on $\mathcal{F}$. We shall say that a set Φ of formulas is *valid* in $\mathcal{F}$, in symbols $\mathcal{F} \models \Phi$, if all formulas of Φ are valid in $\mathcal{F}$. Now consider a class Σ of incidence frames. A formula ϕ is said to be *valid* in Σ, in symbols $\Sigma \models \phi$, if ϕ is valid in all incidence frames in Σ. We shall say that a set Φ of formulas is *valid* in Σ, in symbols $\Sigma \models \Phi$, if all formulas of Φ are valid in all incidence frames in Σ.

3 Definability and reductions

3.1 Definability

In this section we attack the problem of the definability of multifarious classes of incidence frames. A class Σ of incidence frames is said to be (*finitely*) *definable* within a class Δ of incidence frames if there exists a (finite) set Φ of formulas such that for all incidence frames $\mathcal{F}$ in Δ, $\mathcal{F}$ is in Σ iff $\mathcal{F} \models \Phi$.

PROPOSITION 6. *Let n be a natural integer such that $n \geq 2$. The class Σ^n of all n-rich incidence frames is finitely definable within the class Σ_{all} of all incidence frames. More precisely let*

- $RICH_0^n := [01](p_1, q_1) \wedge \ldots \wedge [01](p_{n-1}, q_{n-1}) \to [I]p_1 \vee \ldots \vee [I]p_{n-1} \vee \langle I \rangle (q_1 \wedge \ldots \wedge q_{n-1})$ and

- $RICH_1^n := [10](P_1, Q_1) \wedge \ldots \wedge [10](P_{n-1}, Q_{n-1}) \to [O]P_1 \vee \ldots \vee [O]P_{n-1} \vee \langle O \rangle (Q_1 \wedge \ldots \wedge Q_{n-1})$.

Then for all incidence frames $\mathcal{F}$, $\mathcal{F}$ is in Σ^n iff $\mathcal{F} \models \{\langle 01 \rangle(\top, \top), \langle 10 \rangle(\top, \top), RICH_0^n, RICH_1^n\}$.

The question lies open whether for all natural integers n such that $n \geq 2$, Σ^n is finitely definable within Σ_{all} in the language based on $[I]$ and $[O]$.

PROPOSITION 7. *The class Σ^ω of all ω-rich incidence frames is definable within Σ_{all}. More precisely for all incidence frames $\mathcal{F}$, $\mathcal{F}$ is in Σ^ω iff $\mathcal{F} \models \{\langle 01 \rangle(\top, \top), \langle 10 \rangle(\top, \top)\} \cup \{RICH_0^n : n \geq 2\} \cup \{RICH_1^n : n \geq 2\}$.*

The question lies open whether Σ^ω is definable within Σ_{all} in the language based on $[I]$ and $[O]$. Is Σ^ω finitely definable within Σ_{all}?

PROPOSITION 8. *The class Σ_c of all correct incidence frames is finitely definable within Σ^2. More precisely let*

- $C_0 := P \to [I]\langle O \rangle P$ and

- $C_1 := p \to [O]\langle I \rangle p$.

Then for all incidence frames $\mathcal{F}$ in Σ^2, $\mathcal{F}$ is in Σ_c iff $\mathcal{F} \models \{C_0, C_1\}$.

PROPOSITION 9. *The class Σ_p of all projective incidence frames is finitely definable within the class Σ_{rc}^2 of all rooted, 2-rich and correct incidence frames. More precisely let*

- $P_0 := [\cdot]P \to [\cdot][\cdot]P$ and

- $P_1 := [-]p \to [-][-]p$.

Then for all incidence frames $\mathcal{F}$ in Σ_{rc}^2, $\mathcal{F}$ is in Σ_p iff $\mathcal{F} \models \{P_0, P_1\}$.

PROPOSITION 10. *The class Σ_n of all normal incidence frames is finitely definable within the class Σ_{cp} of all correct and projective incidence frames. More precisely let*

- $N_0 := \odot P \to [\cdot](P \to [01]([O]P, [O]P))$ and

- $N_1 := \ominus p \to [-](p \to [10]([I]p, [I]p))$.

Then for all incidence frames $\mathcal{F}$ in Σ_{cp}, $\mathcal{F}$ is in Σ_n iff $\mathcal{F} \models \{N_0, N_1\}$.

The question lies open whether Σ_n is finitely definable within Σ_{cp} in the language based on $[I]$ and $[O]$.

PROPOSITION 11. *The class Σ_{Ar} of all Arguesian incidence frames is finitely definable within the class Σ_{cp}^2 of all 2-rich, correct and projective incidence frames. More precisely there exists a point formula Ar_0 and a line formula Ar_1 such that for all incidence frames $\mathcal{F}$ in Σ_{cp}^2, $\mathcal{F}$ is in Σ_{Ar} iff $\mathcal{F} \models \{Ar_0, Ar_1\}$.*

PROPOSITION 12. *The class Σ_{Pa} of all Pappian incidence frames is finitely definable within Σ_{cp}^2. More precisely there exists a point formula Pa_0 and a line formula Pa_1 such that for all incidence frames $\mathcal{F}$ in Σ_{cp}^2, $\mathcal{F}$ is in Σ_{Pa} iff $\mathcal{F} \models \{Pa_0, Pa_1\}$.*

3.2 Reductions

In this section we attack the problem of the reductions between several sublanguages of the language based on [01], [∈], [I], [·], ⊙, [10], [∋], [O], [−] and ⊖. It may be of interest to note that for all incidence frames $\mathcal{F}$,

- $\mathcal{F} \models [01](p,q) \leftrightarrow [I]p \vee [I]q \vee [\in](p \wedge q)$ and
- $\mathcal{F} \models [10](P,Q) \leftrightarrow [O]P \vee [O]Q \vee [\ni](P \wedge Q)$.

Hence [01] and [10] can be defined as abbreviations in the language based on [∈], [I], [∋] and [O] within Σ_{all}. Nevertheless

PROPOSITION 13. *[01] and [10] cannot be defined as abbreviations in the language based on [I], [·], ⊙, [O], [−] and ⊖ on the real projective plane $I\!R(2)$.*

PROPOSITION 14. *[01] and [10] cannot be defined as abbreviations in the language based on [∈], [·], ⊙, [∋], [−] and ⊖ on $I\!R(2)$.*

It may be of interest to note that for all incidence frames $\mathcal{F}$,

- $\mathcal{F} \models [\cdot]P \leftrightarrow [I][O]P$ and
- $\mathcal{F} \models [-]p \leftrightarrow [O][I]p$.

Hence [·] and [−] can be defined as abbreviations in the language based on [I] and [O] within Σ_{all}. Nevertheless

PROPOSITION 15.

1. *[I] and [O] cannot be defined as abbreviations in the language based on [∈] and [∋] on $I\!R(2)$.*

2. *[·] and [−] cannot be defined as abbreviations in the language based on [∈] and [∋] on $I\!R(2)$.*

3. *⊙ and ⊖ cannot be defined as abbreviations in the language based on [∈] and [∋] on $I\!R(2)$.*

4. *[∈] and [∋] cannot be defined as abbreviations in the language based on [I] and [O] on $I\!R(2)$.*

5. *⊙ and ⊖ cannot be defined as abbreviations in the language based on [I] and [O] within the class Σ_{cpn}^ω of all ω-rich, correct, projective and normal incidence frames.*

6. *[∈] and [∋] cannot be defined as abbreviations in the language based on [·] and [−] on $I\!R(2)$.*

7. $[I]$ and $[O]$ cannot be defined as abbreviations in the language based on $[\cdot]$ and $[-]$ on $I\!R(2)$.

8. $\odot$ and $\ominus$ cannot be defined as abbreviations in the language based on $[\cdot]$ and $[-]$ on $I\!R(2)$.

9. $[\in]$ and $[\ni]$ cannot be defined as abbreviations in the language based on $\odot$ and $\ominus$ on $I\!R(2)$.

10. $[I]$ and $[O]$ cannot be defined as abbreviations in the language based on $\odot$ and $\ominus$ on $I\!R(2)$.

11. $[\cdot]$ and $[-]$ cannot be defined as abbreviations in the language based on $\odot$ and $\ominus$ on $I\!R(2)$.

The question lies open whether $\odot$ and $\ominus$ can be defined as abbreviations in the language based on $[I]$ and $[O]$ on $I\!R(2)$.

4 Axiomatization and completeness

4.1 Axiomatization

The set of all formulas that are valid in a class Σ of incidence frames, in symbols $L(\Sigma)$, is called the *logic* of Σ. In this section we attack the problem of the axiomatization of $L(\Sigma_{cpn}^{\omega})$. An *inference rule* (Φ, ϕ) consists of a set Φ of formulas and a formula ϕ. (Φ, ϕ) is said to be *finitary* if Φ is finite. An *inference system* (Φ, Ψ) consists of a recursive set Φ of formulas and a finite set Ψ of inference rules. We shall say that (Φ, Ψ) is *finite* if Φ is finite. (Φ, Ψ) is said to be *finitary* if all inference rules in Ψ are finitary. Let

- T_0 be the set of all point tautologies,
- T_1 be the set of all line tautologies,
- $K_0 := [I](p \to q) \to ([I]p \to [I]q)$ and
- $K_1 := [O](P \to Q) \to ([O]P \to [O]Q)$.

Let

- $MP_0 := (\{A, A \to B\}, B)$,
- $MP_1 := (\{a, a \to b\}, b)$,
- $US_0 := (\{A\}, A')$ where A' is obtained from A by uniformly replacing point variables in A by arbitrary point formulas and line variables in A by arbitrary line formulas,
- $US_1 := (\{a\}, a')$ where a' is obtained from a by uniformly replacing point variables in a by arbitrary point formulas and line variables in a by arbitrary line formulas,
- $G_0 := (\{a\}, [I]a)$ and

- $G_1 := (\{A\}, [O]A)$.

We shall say that a formula ϕ is *derivable* in (Φ, Ψ), in symbols $(\Phi, \Psi) \vdash \phi$, if ϕ belongs to the least set Φ' of formulas such that $\Phi \cup T_0 \cup T_1 \cup \{K_0, K_1\} \subseteq \Phi'$ and Φ' is closed under the inference rules in $\Psi \cup \{MP_0, MP_1, US_0, US_1, G_0, G_1\}$. A class Σ of incidence frames is said to be (*finitely*) *axiomatizable* if there exists a (finite) inference system (Φ, Ψ) such that for all formulas ϕ, ϕ is in $L(\Sigma)$ iff $(\Phi, \Psi) \vdash \phi$. Our aim is to find the axiomatization of the logics of some natural classes of incidence frames. Let $\Phi_{cpn}^\omega := \{SEP_0, SEP_1, \langle 01 \rangle(\top, \top), \langle 10 \rangle(\top, \top)\} \cup \{RICH_0^n: n \geq 2\} \cup \{RICH_1^n: n \geq 2\} \cup \{C_0, C_1, P_0, P_1, N_0, N_1\}$ and $\Psi_{cpn}^\omega := \emptyset$. Of course $(\Phi_{cpn}^\omega, \Psi_{cpn}^\omega)$ is *sound* with respect to Σ_{cpn}^ω, i.e. $\{\phi: (\Phi_{cpn}^\omega, \Psi_{cpn}^\omega) \vdash \phi\} \subseteq L(\Sigma_{cpn}^\omega)$. Nevertheless the question lies open whether $(\Phi_{cpn}^\omega, \Psi_{cpn}^\omega)$ is *complete* with respect to Σ_{cpn}^ω, i.e. $L(\Sigma_{cpn}^\omega) \subseteq \{\phi: (\Phi_{cpn}^\omega, \Psi_{cpn}^\omega) \vdash \phi\}$. In particular the proof of the truth lemma for $(\Phi_{cpn}^\omega, \Psi_{cpn}^\omega)$ does not seem to be feasible. How shall we overcome this difficulty? Our solution is based on the following

PROPOSITION 16. *Let p be a line variable that occurs neither in a nor in b and P be a point variable that occurs neither in A nor in B. Then*

- $\Sigma_{all} \models \forall p([I](a \vee p) \vee [I](b \vee \neg p)) \to [01](a, b)$ *and*
- $\Sigma_{all} \models \forall P([O](A \vee P) \vee [O](B \vee \neg P)) \to [10](A, B)$.

In the language without propositional quantifiers, we can imitate the above implications by additional inference rules:

- $R_0 := (\{A \to [I](a \vee p) \vee [I](b \vee \neg p): p \in LV\}, A \to [01](c, b))$ and
- $R_1 := (\{a \to [O](A \vee P) \vee [O](B \vee \neg P): P \in PV\}, a \to [10](A, B))$.

Remark that R_0 and R_1 are similar to the inference rule of intersection for the propositional dynamic logic with intersection of programs studied by Balbiani and Vakarelov [5]. Let $\Psi_{cpn}^{\omega+} := \Psi_{cpn}^\omega \cup \{R_0, R_1\}$. Of course $(\Phi_{cpn}^\omega, \Psi_{cpn}^{\omega+})$ is sound with respect to Σ_{cpn}^ω. The main effect of R_0 and R_1 is to make possible the proof of the truth lemma for $(\Phi_{cpn}^\omega, \Psi_{cpn}^{\omega+})$. Nevertheless the question lies open whether $(\Phi_{cpn}^\omega, \Psi_{cpn}^{\omega+})$ is complete with respect to Σ_{cpn}^ω. In particular the proof that the rooted subframes of the canonical frame for $(\Phi_{cpn}^\omega, \Psi_{cpn}^{\omega+})$ are in Σ_{cpn}^ω does not seem to be feasible. How shall we overcome this difficulty? Our solution is based on the following

PROPOSITION 17. *Let P be a point variable that does not occur in A and p be a line variable that does not occur in a. Then*

- $\Sigma_{all} \models \forall P(\odot P \to A) \to A$ *and*
- $\Sigma_{all} \models \forall p(\ominus p \to a) \to a$.

In the language without propositional quantifiers, we can imitate the above implications by additional inference rules:

- $D_0 := (\{\odot P \to A \colon P \in PV\}, A)$ and
- $D_1 := (\{\ominus p \to a \colon p \in LV\}, a)$.

Remark that D_0 and D_1 are similar to the inference rule of irreflexivity for the modal logic of inequality studied by de Rijke [14]. Let $\Psi_{cpn}^{\omega}{}^{++}$ be $\Psi_{cpn}^{\omega}{}^{+} \cup \{D_0, D_1\}$. Of course $(\Phi_{cpn}^{\omega}, \Psi_{cpn}^{\omega}{}^{++})$ is sound with respect to Σ_{cpn}^{ω}. The main effect of D_0 and D_1 is to make possible the proof that the rooted subframes of the canonical frame for $(\Phi_{cpn}^{\omega}, \Psi_{cpn}^{\omega}{}^{++})$ are in Σ_{cpn}^{ω}. Let us see how.

4.2 Completeness

The SD-theorem proved by Venema [17] implies the completeness of a wide variety of normal logics containing the inference rule of irreflexivity for the difference modality and satisfying some special conditions. Its proof uses techniques which are not necessary when we want to prove the completeness of a concrete normal logic. Moreover Venema's theorem has been obtained within the context of a one-sorted language. Hence it does not apply directly to our two-sorted language. In order to make this paper independent from Venema's paper we shall follow the line of reasoning suggested by Balbiani et al [2] and present our own proof of the completeness of $(\Phi_{cpn}^{\omega}, \Psi_{cpn}^{\omega}{}^{++})$ using ideas similar to Venema's ideas. In particular the proofs of lemmas 18–23 and theorem 24 are adaptations of the proofs of lemmas 3.5–3.11 in Balbiani et al [2]. Let $\flat$ and $\sharp$ be new symbols. Following the line of reasoning suggested by Goldblatt [11] we inductively define the *necessity forms* as follows:

- $\flat$ is a $(0,0)$-necessity form,
- $\sharp$ is a $(1,1)$-necessity form,
- if φ is a $(0,0)$-necessity form then $(A \to \varphi)$ is a $(0,0)$-necessity form and $[O]\varphi$ is a $(1,0)$-necessity form,
- if φ is a $(0,1)$-necessity form then $(A \to \varphi)$ is a $(0,1)$-necessity form and $[O]\varphi$ is a $(1,1)$-necessity form,
- if φ is a $(1,0)$-necessity form then $(a \to \varphi)$ is a $(1,0)$-necessity form and $[I]\varphi$ is a $(0,0)$-necessity form and
- if φ is a $(1,1)$-necessity form then $(a \to \varphi)$ is a $(1,1)$-necessity form and $[I]\varphi$ is a $(0,1)$-necessity form.

Note that each $(0,0)$-necessity form has a unique occurrence of $\flat$, each $(0,1)$-necessity form has a unique occurrence of $\sharp$, each $(1,0)$-necessity form has a unique occurrence of $\flat$ and each $(1,1)$-necessity form has a unique occurrence of $\sharp$. Now we introduce the following versions of R_0, R_1, D_0 and D_1:

- $R_{00} := (\{\varphi(A \to [I](a \vee p) \vee [I](\flat \vee \neg p)) \colon p \in LV\}, \varphi(A \to [01](a,b)))$ where φ is a $(0,0)$-necessity form,

- $R_{01} := (\{\varphi(a \to [O](A \lor P) \lor [O](B \lor \neg P)): P \in PV\}, \varphi(a \to [10](A,B)))$ where φ is a $(0,1)$-necessity form,

- $R_{10} := (\{\varphi(A \to [I](a \lor p) \lor [I](b \lor \neg p)): p \in LV\}, \varphi(A \to [01](a,b)))$ where φ is a $(1,0)$-necessity form,

- $R_{11} := (\{\varphi(a \to [O](A \lor P) \lor [O](B \lor \neg P)): P \in PV\}, \varphi(a \to [10](A,B)))$ where φ is a $(1,1)$-necessity form,

- $D_{00} := (\{\varphi(\odot P \to A): P \in PV\}, \varphi(A))$ where φ is a $(0,0)$-necessity form,

- $D_{01} := (\{\varphi(\ominus p \to a): p \in LV\}, \varphi(a))$ where φ is a $(0,1)$-necessity form,

- $D_{10} := (\{\varphi(\odot P \to A): P \in PV\}, \varphi(A))$ where φ is a $(1,0)$-necessity form and

- $D_{11} := (\{\varphi(\ominus p \to a): p \in LV\}, \varphi(a))$ where φ is a $(1,1)$-necessity form.

LEMMA 18. R_{00}, R_{01}, R_{10}, R_{11}, D_{00}, D_{01}, D_{10} and D_{11} are admissible in $(\Phi^\omega_{cpn}, \Psi^\omega_{cpn}{}^{++})$, i.e. $\{\phi: (\Phi^\omega_{cpn}, \Psi^\omega_{cpn}{}^{++}) \vdash \phi\}$ is closed under R_{00}, R_{01}, R_{10}, R_{11}, D_{00}, D_{01}, D_{10} and D_{11}.

A set X of point formulas will be called a *point theory* if $\{A: (\Phi^\omega_{cpn}, \Psi^\omega_{cpn}{}^{++}) \vdash A\} \subseteq X$ and X is closed under MP_0, R_{00}, R_{01}, D_{00} and D_{01} and a set x of line formulas will be called a *line theory* if $\{a: (\Phi^\omega_{cpn}, \Psi^\omega_{cpn}{}^{++}) \vdash a\} \subseteq x$ and x is closed under MP_1, R_{10}, R_{11}, D_{10} and D_{11}. Now consider a point theory X and a line theory x. X is said to be *consistent* if $\bot \notin X$ and x is said to be *consistent* if $\bot \notin x$. We shall say that X is *maximal* if it is consistent and for all point formulas A, $A \in X$ or $\neg A \in X$ and x is *maximal* if it is consistent and for all line formulas a, $a \in x$ or $\neg a \in x$. It can be proved that each maximal theory is a maximal consistent set of formulas in the classical sense and each maximal consistent set of formulas which is closed under the inference rules considered in lemma 18 is a maximal theory. Given a set X of point formulas and a set x of line formulas, let $X + A := \{B: A \to B \in X\}$, $x + a := \{b: a \to b \in x\}$, $[I]X := \{a: [I]a \in X\}$ and $[O]x := \{A: [O]A \in x\}$.

LEMMA 19. *Let X be a point theory and x be a line theory. Then*

- $X + A$ *is a point theory,*

- $x + a$ *is a line theory,*

- $[I]X$ *is a line theory and*

- $[O]x$ *is a point theory.*

Moreover $X + A$ is consistent iff $\neg A \notin X$ and $x + a$ is consistent iff $\neg a \notin x$.

LEMMA 20. *Let X be a point theory and x be a line theory. Then*

- If X is consistent then it can be extended to a maximal point theory and
- if x is consistent then it can be extended to a maximal line theory.

LEMMA 21. *Let X be a point theory and x be a line theory. Then*

- *If $[01](a,b) \notin X$ then there exists maximal line theories y, z such that $[I]X \subseteq y$, $[I]X \subseteq z$, $y \neq z$, $a \notin y$ and $b \notin z$ and*
- *if $[10](A,B) \notin x$ then there exists maximal point theories Y, Z such that $[O]x \subseteq Y$, $[O]x \subseteq Z$, $Y \neq Z$, $A \notin Y$ and $B \notin Z$.*

The *canonical frame* for $(\Phi^\omega_{cpn}, \Psi^\omega_{cpn}{}^{++})$ is the relational structure $\mathcal{F}_c = (Po_c, Li_c, I_c, O_c)$ where Po_c is the set of all maximal point theories, Li_c is the set of all maximal line theories, I_c is the binary relation between maximal point theories and maximal line theories defined by $I_c(X,x)$ iff $[I]X \subseteq x$ and O_c is the binary relation between maximal line theories and maximal point theories defined by $O_c(x,X)$ iff $[O]x \subseteq X$. Let π_c be the function assigning to each point variable $P \in PV$ the subset $\{X: P \in X\}$ of Po_c and λ_c be the function assigning to each line variable $p \in LV$ the subset $\{x: p \in x\}$ of Li_c.

LEMMA 22. *Let X be a maximal point theory and x be a maximal line theory. Then*

- $(\mathcal{F}_c, \pi_c, \lambda_c), X \models A$ *iff* $A \in X$ *and*
- $(\mathcal{F}_c, \pi_c, \lambda_c), x \models a$ *iff* $a \in x$.

LEMMA 23. *Let $\mathcal{F} = (Po, Li, I, O)$ be a rooted subframe of $\mathcal{F}_c$. Then $\mathcal{F}$ is in Σ^ω_{cpn}.*

THEOREM 24. $(\Phi^\omega_{cpn}, \Psi^\omega_{cpn}{}^{++})$ *is sound and complete with respect to Σ^ω_{cpn}.*

Nevertheless

THEOREM 25. $L(\Sigma^\omega_{cpn})$ *cannot be finitely axiomatized with the inference rules in $\Psi^\omega_{cpn}{}^{++}$ in the language based on [01] and [10].*

To conclude this section we extend the results obtained so far to the problem of the axiomatization of the logic of multifarious classes of incidence frames. Let $\Phi^\omega_{cpnAr} := \Phi^\omega_{cpn} \cup \{Ar_0, Ar_1\}$ and $\Phi^\omega_{cpnPa} := \Phi^\omega_{cpn} \cup \{Pa_0, Pa_1\}$.

THEOREM 26.

- $(\Phi^\omega_{cpnAr}, \Psi^\omega_{cpn}{}^{++})$ *is sound and complete with respect to the class Σ^ω_{cpnAr} of all ω-rich, correct, projective, normal and Arguesian incidence frames and*
- $(\Phi^\omega_{cpnPa}, \Psi^\omega_{cpn}{}^{++})$ *is sound and complete with respect to the class Σ^ω_{cpnPa} of all ω-rich, correct, projective, normal and Pappian incidence frames.*

Nevertheless

THEOREM 27. $L(\Sigma^\omega_{cpnAr})$ and $L(\Sigma^\omega_{cpnPa})$ cannot be finitely axiomatized with the inference rules in $\Psi^\omega_{cpn}{}^{++}$ in the language based on [01] and [10].

Obviously $\mathbb{R}(2)$ is ω-rich, correct, projective, normal, Arguesian and Pappian. The question lies open whether $\mathbb{R}(2)$ is axiomatizable. For all natural integers n such that $n \geq 2$, let $\Phi^n_{cpn} = \{SEP_0, SEP_1, \langle 01\rangle(\top,\top), \langle 10\rangle(\top,\top), RICH^n_0, RICH^n_1, C_0, C_1, P_0, P_1, N_0, N_1\}$, $\Phi^n_{cpnAr} := \Phi^n_{cpn} \cup \{Ar_0, Ar_1\}$ and $\Phi^n_{cpnPa} := \Phi^n_{cpn} \cup \{Pa_0, Pa_1\}$.

THEOREM 28.

- $(\Phi^n_{cpn}, \Psi^\omega_{cpn}{}^{++})$ is sound and complete with respect to the class Σ^n_{cpn} of all n-rich, correct, projective and normal incidence frames,

- $(\Phi^n_{cpnAr}, \Psi^\omega_{cpn}{}^{++})$ is sound and complete with respect to the class Σ^n_{cpnAr} of all n-rich, correct, projective, normal and Arguesian incidence frames and

- $(\Phi^n_{cpnPa}, \Psi^\omega_{cpn}{}^{++})$ is sound and complete with respect to the class Σ^n_{cpnPa} of all n-rich, correct, projective, normal and Pappian incidence frames.

5 Conclusion

Our language can express multifarious modalities. The question lies open whether the modalities $\Box(A, B)$ "in all points X and Y collinear with and different from the current point, if X and Y are different then A is realized at X or B is realized at Y" and $\Box(a, b)$ "in all lines x and y coincident with and different from the current line, if x and y are different then a is realized at x or b is realized at y" can be expressed in our language.
Using the inference rules R_0, R_1, D_0 and D_1 we have given sound and complete axiomatizations of $L(\Sigma^\omega_{cpn})$, $L(\Sigma^\omega_{cpnAr})$ and $L(\Sigma^\omega_{cpnPa})$ and for all natural integers n such that $n \geq 2$, we have given sound and complete axiomatizations of $L(\Sigma^n_{cpn})$, $L(\Sigma^n_{cpnAr})$ and $L(\Sigma^n_{cpnPa})$. The question lies open whether the inference rules R_0, R_1, D_0 and D_1 can be eliminated in these axiomatizations and replaced with finitely many formulas.
Let us remark that according to a result of Venema [18] $NEXPTIME$ is a lower bound for the complexity of the satisfiability problems in Σ^ω_{cpn}, Σ^ω_{cpnAr} and Σ^ω_{cpnPa} and for all natural integers n such that $n \geq 2$, $NEXPTIME$ is a lower bound for the complexity of the satisfiability problems in Σ^n_{cpn}, Σ^n_{cpnAr} and Σ^n_{cpnPa}. The question lies open whether these satisfiability problems are decidable.

Acknowledgement

Special acknowledgement is heartly granted to the referees who made several helpful comments for improving the readability of the paper.

BIBLIOGRAPHY

[1] P. Balbiani, *The modal multilogic of geometry*. Journal of Applied Non-Classical Logics **8** (1998) 259–281.

[2] P. Balbiani, L. Fariñas del Cerro, T. Tinchev, and D. Vakarelov, *Modal logics for incidence geometries*. Journal of Logic and Computation **7** (1997) 59–78.

[3] P. Balbiani and V. Goranko, *Modal logics for parallelism, orthogonality, and affine geometries*. Journal of Applied Non-Classical Logics **12** (2002) 365–397.

[4] P. Balbiani and T. Tinchev, *Line-based affine reasoning in Euclidean plane*. J. Alferes and J. Leite (editors): *Logics in Artificial Intelligence*. Springer-Verlag (2004) 474–486.

[5] P. Balbiani and D. Vakarelov, *Iteration-free PDL with intersection: a complete axiomatization*. Fundamenta Informaticæ **45** (2001) 1–22.

[6] A. Beutelspacher, *Projective planes*. F. Buekenhout (editor): *Handbook of Incidence Geometry: Buildings and Foundations*. Elsevier Science (1995) 107–136.

[7] P. Blackburn, M. de Rijke, and Y. Venema, *Modal Logic*. Cambridge University Press (2001).

[8] A. Chagrov and M. Zakharyaschev, *Modal Logic*. Oxford University Press (1997).

[9] C. Chang and H. Keisler, *Model Theory*. Elsevier Science (1990).

[10] M. Fattorosi-Barnaba and F. de Caro, *Graded modalities. I*. Studia Logica **44** (1985) 197–221.

[11] R. Goldblatt, *Axiomatising the Logic of Computer Programming*. Springer-Verlag (1982).

[12] V. Goranko and S. Passy, *Using the universal modality: gains and questions*. Journal of Logic and Computation **2** (1992) 5–30.

[13] M. Kracht, *Tools and Techniques in Modal Logic*. Elsevier Science (1999).

[14] M. de Rijke, *The modal logic of inequality*. Journal of Symbolic Logic **57** (1992) 566–584.

[15] V. Rybakov, *Admissibility of Logical Inference Rules*. Elsevier Science (1997).

[16] V. Stebletsova and Y. Venema, *Undecidable theories of Lyndon algebras*. Journal of Symbolic Logic **66** (2001) 207–224.

[17] Y. Venema, *Derivation rules as anti-axioms in modal logic*. Journal of Symbolic Logic **58** (1993) 1003–1034.

[18] Y. Venema, *Points, lines and diamonds: a two-sorted modal logic for projective planes*. Journal of Logic and Computation **9** (1999) 601–621.

Annex

Proof of lemmas 1–5 and propositions 6–12. Left to the reader. ∎

Proof of proposition 13. Suppose there exists a point formula $A(p,q)$ in the language based on $[I]$, $[\cdot]$, $\odot$, $[O]$, $[-]$ and $\ominus$ such that $\mathbb{R}(2) \models [01](p,q) \leftrightarrow A(p,q)$. Let X be a point in $\mathbb{R}(2)$ and $\mathcal{M} = (\mathbb{R}(2), \pi, \lambda)$ be an incidence model based on $\mathbb{R}(2)$ such that $\lambda(p) = Li \setminus I(X)$ and $\lambda(q) = Li \setminus I(X)$. Let y be a line in $\mathbb{R}(2)$ and $\mathcal{M}' = (\mathbb{R}(2), \pi', \lambda')$ be an incidence model based on $\mathbb{R}(2)$ such that y is not on X and $\lambda'(p) = Li \setminus (I(X) \cup \{y\})$ and $\lambda'(q) = Li \setminus (I(X) \cup \{y\})$. Let Z be a point in $\mathbb{R}(2)$ such that Z is in y. Remark that $\mathcal{M}, Z \models [01](p,q)$ and $\mathcal{M}', Z \not\models [01](p,q)$. We omit the easy proof that for all line formulas a in $\{p, \neg p, q, \neg q\}$, $\mathcal{M}$ and $\mathcal{M}'$ agree on $[I]a$, $[-]a$ and $\ominus a$. Hence $\mathcal{M}, Z \models A(p,q)$ iff $\mathcal{M}', Z \models A(p,q)$. Consequently $\mathcal{M}, Z \models [01](p,q)$ iff $\mathcal{M}', Z \models [01](p,q)$: a contradiction. ∎

Proof of proposition 14. Suppose there exists a point formula $A(p,q)$ in the language based on $[\in]$, $[\cdot]$, $\odot$, $[\ni]$, $[-]$ and $\ominus$ such that $\mathbb{R}(2) \models [01](p,q) \leftrightarrow A(p,q)$. Let X be a point in $\mathbb{R}(2)$ and $\mathcal{M} = (\mathbb{R}(2), \pi, \lambda)$ be an incidence model based on $\mathbb{R}(2)$ such that $\lambda(p) = \emptyset$ and $\lambda(q) = I(X)$. Let y be a line in $\mathbb{R}(2)$ and $\mathcal{M}' = (\mathbb{R}(2), \pi', \lambda')$ be an incidence model based on $\mathbb{R}(2)$ such that y is on X and $\lambda'(p) = \emptyset$ and $\lambda'(q) = I(X) \setminus \{y\}$. Remark that $\mathcal{M}, X \models [01](p,q)$ and $\mathcal{M}', X \not\models [01](p,q)$. It is readily checked that for all line formulas a in $\{p, \neg p, q, \neg q\}$, $\mathcal{M}$ and $\mathcal{M}'$ agree on $[\in]a$, $[-]a$ and $\ominus a$. Hence $\mathcal{M}, X \models A(p,q)$ iff $\mathcal{M}', X \models A(p,q)$. Consequently $\mathcal{M}, X \models [01](p,q)$ iff $\mathcal{M}', X \models [01](p,q)$: a contradiction. ■

Proof of proposition 15. 1. By proposition 14.
2. Suppose there exists a point formula $A(P)$ in the language based on $[\in]$ and $[\ni]$ such that $\mathbb{R}(2) \models [\cdot]P \leftrightarrow A(P)$. Let $\mathcal{M} = (\mathbb{R}(2), \pi, \lambda)$ be an incidence model based on $\mathbb{R}(2)$ such that $\pi(P) = Po$. Let X be a point in $\mathbb{R}(2)$ and $\mathcal{M}' = (\mathbb{R}(2), \pi', \lambda')$ be an incidence model based on $\mathbb{R}(2)$ such that $\pi'(P) = Po \setminus \{X\}$. Remark that $\mathcal{M}, X \models [\cdot]P$ and $\mathcal{M}', X \not\models [\cdot]P$. We omit the easy proof that for all point formulas A in $\{P, \neg P\}$, $\mathcal{M}$ and $\mathcal{M}'$ agree on $[\ni]A$. Hence $\mathcal{M}, X \models A(P)$ iff $\mathcal{M}', X \models A(P)$. Consequently $\mathcal{M}, X \models [\cdot]P$ iff $\mathcal{M}', X \models [\cdot]P$: a contradiction.
3. Suppose there exists a point formula $A(P)$ in the language based on $[\in]$ and $[\ni]$ such that $\mathbb{R}(2) \models \odot P \leftrightarrow A(P)$. Let X be a point in $\mathbb{R}(2)$ and $\mathcal{M} = (\mathbb{R}(2), \pi, \lambda)$ be an incidence model based on $\mathbb{R}(2)$ such that $\pi(P) = Po \setminus \{X\}$. Let $\mathcal{M}' = (\mathbb{R}(2), \pi', \lambda')$ be an incidence model based on $\mathbb{R}(2)$ such that $\pi'(P) = Po$. Remark that $\mathcal{M}, X \models \odot P$ and $\mathcal{M}', X \not\models \odot P$. It is readily checked that for all point formulas A in $\{P, \neg P\}$, $\mathcal{M}$ and $\mathcal{M}'$ agree on $[\ni]A$. Hence $\mathcal{M}, X \models A(P)$ iff $\mathcal{M}', X \models A(P)$. Consequently $\mathcal{M}, X \models \odot P$ iff $\mathcal{M}', X \models \odot P$: a contradiction.
4. By proposition 13.
5. Suppose there exists a point formula $A(P)$ in the language based on $[I]$ and $[O]$ such that $\Sigma^\omega_{cpn} \models \odot P \leftrightarrow A(P)$. Let X be a point in $\mathbb{R}(2)$ and $\mathcal{M} = (\mathbb{R}(2), \pi, \lambda)$ be an incidence model based on $\mathbb{R}(2)$ such that $\pi(P) = Po \setminus \{X\}$. Let $\mathcal{M}_1 = (Po_1, Li_1, I_1, O_1, \pi_1, \lambda_1)$ and $\mathcal{M}_2 = (Po_2, Li_2, I_2, O_2, \pi_2, \lambda_2)$ be disjoint copies of $\mathcal{M}$ and $\mathcal{M}' = (Po', Li', I', O', \pi', \lambda')$ be their disjoint unions. It is easy to see that $\mathcal{M}$ is a bounded morphic image of $\mathcal{M}'$ in the language based on $[I]$ and $[O]$. Obviously $\mathcal{M}'$ is ω-rich, correct and normal. Hence following the line of reasoning suggested by Venema [18] the reader may easily verify that $\mathcal{M}$ is a bounded morphic image of an ω-rich, correct, projective and normal incidence model $\mathcal{M}''$ containing $\mathcal{M}'$. Let X'' be a point in $\mathcal{M}''$ corresponding to X. Remark that $\mathcal{M}, X \models \odot P$ and $\mathcal{M}'', X'' \not\models \odot P$. Moreover for all point formulas A in the language based on $[I]$ and $[O]$, $\mathcal{M}, X \models A$ iff $\mathcal{M}'', X'' \models A$. Hence $\mathcal{M}, X \models A(P)$ iff $\mathcal{M}'', X'' \models A(P)$. Consequently $\mathcal{M}, X \models \odot P$ iff $\mathcal{M}'', X'' \models \odot P$: a contradiction.
6. Suppose there exists a point formula $A(p)$ in the language based on $[\cdot]$ and $[-]$ such that $\mathbb{R}(2) \models [\in]p \leftrightarrow A(p)$. Let x be a line in $\mathbb{R}(2)$

and $\mathcal{M} = (\mathbb{R}(2), \pi, \lambda)$ be an incidence model based on $\mathbb{R}(2)$ such that $\lambda(p) = Li \setminus \{x\}$. Let y be a line in $\mathbb{R}(2)$ and $\mathcal{M}' = (\mathbb{R}(2), \pi', \lambda')$ be an incidence model based on $\mathbb{R}(2)$ such that $x \neq y$ and $\lambda'(p) = Li \setminus \{x, y\}$. Let Z be the unique point in $\mathbb{R}(2)$ such that Z is in x and Z is in y. Remark that $\mathcal{M}, Z \models [\in]p$ and $\mathcal{M}', Z \not\models [\in]p$. We omit the easy proof that for all line formulas a in $\{p, \neg p\}$, $\mathcal{M}$ and $\mathcal{M}'$ agree on $[-]a$. Hence $\mathcal{M}, Z \models A(p)$ iff $\mathcal{M}', Z \models A(p)$. Consequently $\mathcal{M}, Z \models [\in]p$ iff $\mathcal{M}', Z \models [\in]p$: a contradiction.

7. Suppose there exists a point formula $A(p)$ in the language based on $[\cdot]$ and $[-]$ such that $\mathbb{R}(2) \models [I]p \leftrightarrow A(p)$. Let X be a point in $\mathbb{R}(2)$ and $\mathcal{M} = (\mathbb{R}(2), \pi, \lambda)$ be an incidence model based on $\mathbb{R}(2)$ such that $\lambda(p) = I(X)$. Let x be a line in $\mathbb{R}(2)$ and $\mathcal{M}' = (\mathbb{R}(2), \pi', \lambda')$ be an incidence model based on $\mathbb{R}(2)$ such that x is on X and $\lambda'(p) = I(X) \setminus \{x\}$. Remark that $\mathcal{M}, X \models [I]p$ and $\mathcal{M}', X \not\models [I]p$. It is readily checked that for all line formulas a in $\{p, \neg p\}$, $\mathcal{M}$ and $\mathcal{M}'$ agree on $[-]a$. Hence $\mathcal{M}, X \models A(p)$ iff $\mathcal{M}', X \models A(p)$. Consequently $\mathcal{M}, X \models [I]p$ iff $\mathcal{M}', X \models [I]p$: a contradiction.

8. Suppose there exists a point formula $A(P)$ in the language based on $[\cdot]$ and $[-]$ such that $\mathbb{R}(2) \models \odot P \leftrightarrow A(P)$. Let X be a point in $\mathbb{R}(2)$ and $\mathcal{M} = (\mathbb{R}(2), \pi, \lambda)$ be an incidence model based on $\mathbb{R}(2)$ such that $\pi(P) = Po \setminus \{X\}$. Let Y be a point in $\mathbb{R}(2)$ and $\mathcal{M}' = (\mathbb{R}(2), \pi', \lambda')$ be an incidence model based on $\mathbb{R}(2)$ such that $X \neq Y$ and $\pi'(P) = Po \setminus \{X, Y\}$. Remark that $\mathcal{M}, X \models \odot P$ and $\mathcal{M}', X \not\models \odot P$. It is easy to see that for all point formulas A in $\{P, \neg P\}$, $\mathcal{M}$ and $\mathcal{M}'$ agree on $[\cdot]A$. Hence $\mathcal{M}, X \models A(P)$ iff $\mathcal{M}', X \models A(P)$. Consequently $\mathcal{M}, X \models \odot P$ iff $\mathcal{M}', X \models \odot P$: a contradiction.

9. Suppose there exists a point formula $A(p)$ in the language based on $\odot$ and $\ominus$ such that $\mathbb{R}(2) \models [\in]p \leftrightarrow A(p)$. Let X be a point in $\mathbb{R}(2)$, y be a line in $\mathbb{R}(2)$ and $\mathcal{M} = (\mathbb{R}(2), \pi, \lambda)$ be an incidence model based on $\mathbb{R}(2)$ such that X is in y and $\lambda(p) = I(X) \setminus \{y\}$. Let z be a line in $\mathbb{R}(2)$ and $\mathcal{M}' = (\mathbb{R}(2), \pi', \lambda')$ be an incidence model based on $\mathbb{R}(2)$ such that z is on X, $y \neq z$ and $\lambda'(p) = I(X) \setminus \{y, z\}$. Remark that $\mathcal{M}, X \models [\in]p$ and $\mathcal{M}', X \not\models [\in]p$. It is easy to see that for all line formulas a in $\{p, \neg p\}$, $\mathcal{M}$ and $\mathcal{M}'$ agree on $\ominus a$. Hence $\mathcal{M}, X \models A(p)$ iff $\mathcal{M}', X \models A(p)$. Consequently $\mathcal{M}, X \models [\in]p$ iff $\mathcal{M}', X \models [\in]p$: a contradiction.

10. Suppose there exists a point formula $A(p)$ in the language based on $\odot$ and $\ominus$ such that $\mathbb{R}(2) \models [I]p \leftrightarrow A(p)$. Let X be a point in $\mathbb{R}(2)$ and $\mathcal{M} = (\mathbb{R}(2), \pi, \lambda)$ be an incidence model based on $\mathbb{R}(2)$ such that $\lambda(p) = I(X)$. Let y be a line in $\mathbb{R}(2)$ and $\mathcal{M}' = (\mathbb{R}(2), \pi', \lambda')$ be an incidence model based on $\mathbb{R}(2)$ such that y is on X and $\lambda'(p) = I(X) \setminus \{y\}$. Remark that $\mathcal{M}, X \models [I]p$ and $\mathcal{M}', X \not\models [I]p$. It is easy to see that for all line formulas a in $\{p, \neg p\}$, $\mathcal{M}$ and $\mathcal{M}'$ agree on $\ominus a$. Hence $\mathcal{M}, X \models A(p)$ iff $\mathcal{M}', X \models A(p)$. Consequently $\mathcal{M}, X \models [I]p$ iff $\mathcal{M}', X \models [I]p$: a contradiction.

11. Suppose there exists a point formula $A(P)$ in the language based on $\odot$ and $\ominus$ such that $\mathbb{R}(2) \models [\cdot]P \leftrightarrow A(P)$. Let $\mathcal{M} = (\mathbb{R}(2), \pi, \lambda)$ be an

incidence model based on $\mathbb{R}(2)$ such that $\pi(P) = Po$. Let X, Y be points in $\mathbb{R}(2)$ and $\mathcal{M}' = (\mathbb{R}(2), \pi', \lambda')$ be an incidence model based on $\mathbb{R}(2)$ such that $X \neq Y$ and $\pi'(P) = Po \setminus \{X, Y\}$. Remark that $\mathcal{M}, X \models [\cdot]P$ and $\mathcal{M}', X \not\models [\cdot]P$. It is easy to see that for all point formulas A in $\{P, \neg P\}$, $\mathcal{M}$ and $\mathcal{M}'$ agree on $\odot A$. Hence $\mathcal{M}, X \models A(P)$ iff $\mathcal{M}', X \models A(P)$. Consequently $\mathcal{M}, X \models [\cdot]P$ iff $\mathcal{M}', X \models [\cdot]P$: a contradiction. ∎

Proof of proposition 16. Suppose $\Sigma_{all} \not\models \forall p([I](a \vee p) \vee [I](b \vee \neg p)) \to [01](a, b)$. Hence there exists an incidence frame $\mathcal{F} = (Po, Li, I, O)$, an incidence model $\mathcal{M} = (\mathcal{F}, \pi, \lambda)$ based on $\mathcal{F}$ and a point X such that $\mathcal{M}, X \models \forall p([I](a \vee p) \vee [I](b \vee \neg p))$ and $\mathcal{M}, X \not\models [01](a, b)$. Consequently there exists lines y, z such that $In(X, y, z)$, $\mathcal{M}, y \not\models a$ and $\mathcal{M}, z \not\models b$. Let λ' be the function assigning to each line variable $q \in LV$ different from p the subset $\lambda(q)$ of Li and assigning to p the subset $\{z\}$ of Li. Thus $(\mathcal{F}, \pi, \lambda'), X \models [I](a \vee p) \vee [I](b \vee \neg p)$. If $(\mathcal{F}, \pi, \lambda'), X \models [I](a \vee p)$ then $\mathcal{M}, y \models a$: a contradiction. If $(\mathcal{F}, \pi, \lambda'), X \models [I](b \vee \neg p)$ then $\mathcal{M}, z \models b$: a contradiction. ∎

Proof of proposition 17. Suppose $\Sigma_{all} \not\models \forall P(\odot P \to A) \to A$. Hence there exists an incidence frame $\mathcal{F} = (Po, Li, I, O)$, an incidence model $\mathcal{M} = (\mathcal{F}, \pi, \lambda)$ based on $\mathcal{F}$ and a point X such that $\mathcal{M}, X \models \forall P(\odot P \to A)$ and $\mathcal{M}, X \not\models A$. Let π' be the function assigning to each point variable $Q \in PV$ different from P the subset $\pi(Q)$ of Po and assigning to P the subset $Po \setminus \{X\}$ of Po. Consequently $(\mathcal{F}, \pi', \lambda), X \models \odot P \to A$. Thus $\mathcal{M}, X \models A$: a contradiction. ∎

Proof of lemma 18. The proof is based on the fact that Φ^ω_{cpn} contains C_0 and C_1. It can be done by induction on the length of necessity forms. ∎

Proof of lemma 19. The proof is based on the fact that point theories are closed under MP_0, R_{00}, R_{01}, D_{00} and D_{01} and line theories are closed under MP_1, R_{10}, R_{11}, D_{10} and D_{11}. ∎

Proof of lemma 20. This is the Lindenbaum's lemma for $(\Phi^\omega_{cpn}, \Psi^\omega_{cpn}{}^{++})$. Let X be a consistent point theory. Assuming that the set of all point formulas is arranged in some determinate order $A_1, A_2, \ldots$, we define the sequence $X_0, X_1, \ldots$ of consistent point theories as follows. Let $X_0 = X$. Consider a positive integer n. If $X_{n-1} + A_n$ is consistent then we consider two cases. In the first case A_n cannot be represented as the negation of a conclusion of R_{00}, R_{01}, D_{00} and D_{01}. Hence let $X_n = X_{n-1} + A_n$. In the second case A_n can be represented as the negation of finitely many conclusions $\varphi_1(\phi_1), \ldots, \varphi_k(\phi_k)$ of R_{00}, R_{01}, D_{00} and D_{01}. We define the sequence $X_n^0, \ldots, X_n^k$ of consistent point theories as follows. Let $X_n^0 = X_{n-1}$. Consider a positive integer i such that $i \leq k$. If $\varphi_i(\phi_i)$ is a conclusion of R_{00} then ϕ_i is a point formula of the form $A \to [01](a, b)$. Let $p \in LV$ be a line variable such that $X_n^{i-1} + \neg \varphi_i(A \to [I](a \vee p) \vee [I](b \vee \neg p))$ is consistent. Consequently let $X_n^i = X_n^{i-1} + \neg \varphi_i(A \to [I](a \vee p) \vee [I](b \vee \neg p))$. If $\varphi_i(\phi_i)$ is a conclusion of R_{01} then ϕ_i is a line formula of the form $a \to [10](A, B)$. Let $P \in PV$ be a point variable such that $X_n^{i-1} + \neg\varphi_i(a \to [O](A \vee P) \vee [O](B \vee \neg P))$ is consistent. Thus let $X_n^i = X_n^{i-1} + \neg\varphi_i(a \to [O](A \vee P) \vee [O](B \vee \neg P))$. If $\varphi_i(\phi_i)$ is a conclusion of D_{00} then ϕ_i is a point formula of the form A. Let $P \in PV$ be a point variable such that $X_n^{i-1} + \neg\varphi_i(\odot P \to A)$ is consistent. Therefore let $X_n^i = X_n^{i-1} + \neg\varphi_i(\odot P \to A)$. If $\varphi_i(\phi_i)$ is a conclusion of D_{01} then ϕ_i is a line formula of the form a. Let $p \in LV$ be a line variable such that $X_n^{i-1} + \neg\varphi_i(\ominus p \to a)$ is consistent. Hence let $X_n^i = X_n^{i-1} + \neg\varphi_i(\ominus p \to a)$. Proceed further in this way until k and let $X_n = X_n^k$. If $X_{n-1} + A_n$ is not consistent then let $X_n = X_{n-1}$. Then let $X' = X_0 \cup X_1 \cup \ldots$. The reader may easily verify that X' is a maximal point theory extending X. ∎

Proof of lemma 21. This is the diamond lemma for $(\Phi_{cpn}^\omega, \Psi_{cpn}^{\omega\ ++})$. Let X be a point theory. If $[01](a, b) \notin X$ then there exists a line variable $p \in LV$ such that $[I](a \vee p) \vee [I](b \vee \neg p) \notin X$. Hence $[I](a \vee p) \notin X$ and $[I](b \vee \neg p) \notin X$. Let $y = [I]X + \neg(a \vee p)$ and $z = [I]X + \neg(b \vee \neg p)$. Consequently y is a consistent line theory such that $[I]X \subseteq y$ and $\neg(a \vee p) \in y$ and z is a consistent line theory such that $[I]X \subseteq z$ and $\neg(b \vee \neg p) \in z$. By lemma 20 there exists maximal line theories y', z' such that $[I]X \subseteq y'$, $[I]X \subseteq z'$, $y' \neq z'$, $a \notin y'$ and $b \notin z'$. ∎

Proof of lemma 22. This is the truth lemma for $(\Phi_{cpn}^\omega, \Psi_{cpn}^{\omega\ ++})$. The proof can be done by induction on the length of formulas. The base case follows from the definition of π_c and λ_c. The Boolean cases follow from the classical properties of maximal consistent sets of formulas. It remains to deal with the modalities. The right to left direction is more or less immediate from the interpretation of $[01]$ and the interpretation of $[10]$. If $(\mathcal{F}_c, \pi_c, \lambda_c), X \not\models [01](a, b)$ then there exists maximal line theories y, z such that $[I]X \subseteq y$, $[I]X \subseteq z$, $y \neq z$, $(\mathcal{F}_c, \pi_c, \lambda_c), y \not\models a$ and $(\mathcal{F}_c, \pi_c, \lambda_c), z \not\models b$. Hence there exists a line formula c such that $a \vee c \notin y$ and $b \vee \neg c \notin z$. Consequently $[I](a \vee c) \vee [I](b \vee \neg c) \notin X$. Since $[01](a, b) \to [I](a \vee c) \vee [I](b \vee \neg c) \in X$, then $[01](a, b) \notin X$. For the left to right direction suppose $[01](a, b) \notin X$. By lemma 21 there exists maximal line theories y, z such that $[I]X \subseteq y$, $[I]X \subseteq z$, $y \neq z$, $a \notin y$ and $b \notin z$. Thus $(\mathcal{F}_c, \pi_c, \lambda_c), y \not\models a$ and $(\mathcal{F}_c, \pi_c, \lambda_c), z \not\models b$. Therefore $(\mathcal{F}_c, \pi_c, \lambda_c), X \not\models [01](a, b)$. ∎

Proof of lemma 23. Firstly let us prove that $\mathcal{F}$ is 2-rich. If $\mathcal{F}$ is not 2-rich then there exists a point X such that $Card(I(X)) < 2$ or there exists a line x such that $Card(O(x)) < 2$. Without loss of generality let us assume that there exists a point X such that $Card(I(X)) < 2$. Since $\langle 01 \rangle(\top, \top) \in X$, then by lemma 22 $(\mathcal{F}_c, \pi_c, \lambda_c), X \models \langle 01 \rangle(\top, \top)$. Hence there exists lines y, z such that $In_c(X, y, z)$. Consequently $Card(I_c(X)) \geq 2$: a contradiction. Secondly let us prove that $\mathcal{F}$ is correct. If $\mathcal{F}$ is not correct then there exists a point X and a line x such that $I(X, x)$ and not $O(x, X)$ or there exists a line x and a point X such that $O(x, X)$ and not $I(X, x)$. Without loss of generality let us assume that there exists a point X and a line x such that

$I(X,x)$ and not $O(x,X)$. Hence $[I]X \subseteq x$ and $[O]x \not\subseteq X$. Consequently there exists a point formula A such that $[O]A \in x$ and $\neg A \in X$. Since $\neg A \to [I]\neg[O]A \in X$, then $[I]\neg[O]A \in X$. Thus $[O]A \notin x$: a contradiction. Thirdly let us prove that $\mathcal{F}$ is projective. If $\mathcal{F}$ is not projective then there exists points X, Y such that $I(X) \cap I(Y) = \emptyset$ or there exists lines x, y such that $O(x) \cap O(y) = \emptyset$. Without loss of generality let us assume that there exists points X, Y such that $I(X) \cap I(Y) = \emptyset$. Since $\mathcal{F}$ is 2-rich, then $X \neq Y$. Let $Q \in PV$ be a point variable such that $\odot Q \in Y$. Since $\mathcal{F}$ is rooted and correct, then there exists a natural integer n such that $Y \in (I \circ O)^n(X)$. Hence $\langle \cdot \rangle^n \odot Q \in X$. Since $\langle \cdot \rangle^n \odot Q \to \neg[01](\neg \langle O \rangle \odot Q, \bot) \in X$, then $[01](\neg \langle O \rangle \odot Q, \bot) \notin X$. By lemma 21 there exists a maximal line theory z such that $[I]X \subseteq z$ and $\langle O \rangle \odot Q \in z$. If $[I]Y \not\subseteq z$ then there exists a line formula a such that $[I]a \in Y$ and $a \notin z$. Consequently $\langle \cdot \rangle^n(\odot Q \wedge [I]a) \in X$. Since $\langle \cdot \rangle^n(\odot Q \wedge [I]a) \to [I][O](\odot Q \to [I]a) \in X$, then $[I][O](\odot Q \to [I]a) \in X$. Thus $[O](\odot Q \to [I]a) \in z$. Therefore $\langle O \rangle [I]a \in z$. Since $\langle O \rangle [I]a \to a \in z$, then $a \in z$: a contradiction. Hence $I(X) \cap I(Y) \neq \emptyset$: a contradiction.

Fourthly let us prove that $\mathcal{F}$ is normal. If $\mathcal{F}$ is not normal then there exists points X, Y and lines z, t such that $I(X,z), I(Y,z), I(X,t), I(Y,t)$, $X \neq Y$ and $z \neq t$ or there exists lines x, y and points Z, T such that $O(x,Z), O(y,Z), O(x,T), O(y,T)$, $x \neq y$ and $Z \neq T$. Without loss of generality let us assume that there exists points X, Y and lines z, t such that $I(X,z), I(Y,z), I(X,t), I(Y,t), X \neq Y$ and $z \neq t$. Let $P \in PV$ be a point variable such that $\odot P \in X$. Hence $P \notin X$ and $[01]([10](P,P),\bot) \in X$. By lemma 22 $(\mathcal{F}_c, \pi_c, \lambda_c), X \not\models P$ and $(\mathcal{F}_c, \pi_c, \lambda_c), X \models [01]([10](P,P),\bot)$. Consequently $(\mathcal{F}_c, \pi_c, \lambda_c), z \models [10](P,P)$. Thus $(\mathcal{F}_c, \pi_c, \lambda_c), Y \models P$. By lemma 22 $P \in Y$. Since $\odot P \to [I][O](P \to [01]([O]P,[O]P)) \in X$, then $[I][O](P \to [01]([O]P,[O]P)) \in X$. Therefore $[O](P \to [01]([O]P,[O]P)) \in z$. Hence $P \to [01]([O]P,[O]P) \in Y$. Consequently $[01]([O]P,[O]P) \in Y$. By lemma 22 $(\mathcal{F}_c, \pi_c, \lambda_c), Y \models [01]([O]P,[O]P)$. Thus $(\mathcal{F}_c, \pi_c, \lambda_c), z \models [O]P$ or $(\mathcal{F}_c, \pi_c, \lambda_c), t \models [O]P$. If $(\mathcal{F}_c, \pi_c, \lambda_c), z \models [O]P$ then $(\mathcal{F}_c, \tau_c, \lambda_c), X \models P$: a contradiction. If $(\mathcal{F}_c, \pi_c, \lambda_c), t \models [O]P$ then $(\mathcal{F}_c, \pi_c, \lambda_c).X \models P$: a contradiction.

Fifthly let us prove that $\mathcal{F}$ is ω-rich. If $\mathcal{F}$ is not ω-rich then there exists a natural integer n such that $n \geq 2$ and $\mathcal{F}$ is not n-rich. Hence there exists a point X such that $Card(I(X)) < n$ or there exists a line x such that $Card(O(x)) < n$. Without loss of generality let us assume that there exists a point X such that $Card(I(X)) < n$. Consequently there exists lines $y_1, \ldots, y_{n-1}$ such that $I(X) = \{y_1, \ldots, y_{n-1}\}$. Since $\mathcal{F}$ is 2-rich, then $n-1 \geq 2$. Let $p_1 \in LV, \ldots, p_{n-1} \in LV$ be line variables such that $\ominus p_1 \in y_1$, $\ldots, \ominus p_{n-1} \in y_{n-1}$. The reader may easily verify that $(\mathcal{F}_c, \pi_c, \lambda_c), X \models [01](p_1,p_1) \wedge \ldots \wedge [01](p_{n-1},p_{n-1})$ and $(\mathcal{F}_c, \pi_c, \lambda_c), X \not\models [I]p_1 \vee \ldots \vee [I]p_{n-1} \vee \langle I \rangle (p_1 \wedge \ldots \wedge p_{n-1})$. By lemma 22 $[01](p_1,p_1) \wedge \ldots \wedge [01](p_{n-1},p_{n-1}) \in X$. Since $[01](p_1,p_1) \wedge \ldots \wedge [01](p_{n-1},p_{n-1}) \to [I]p_1 \vee \ldots \vee [I]p_{n-1} \vee \langle I \rangle (p_1 \wedge \ldots \wedge p_{n-1}) \in X$, then $[I]p_1 \vee \ldots \vee [I]p_{n-1} \vee \langle I \rangle (p_1 \wedge \ldots \wedge p_{n-1}) \in X$. By lemma 22 $(\mathcal{F}_c, \pi_c, \lambda_c), X \models [I]p_1 \vee \ldots \vee [I]p_{n-1} \vee \langle I \rangle (p_1 \wedge \ldots \wedge p_{n-1})$: a contradiction. ∎

Proof of theorem 24. Let A be a point formula such that $(\Phi^\omega_{cpn}, \Psi^\omega_{cpn}{}^{++}) \nvdash A$. By lemmas 19 and 20 there exists a maximal point theory X such that $A \notin X$. By lemma 22 $(\mathcal{F}_c, \pi_c, \lambda_c), X \nvDash A$. Let $\mathcal{F}$ be the subframe of $\mathcal{F}_c$ rooted at X. Hence $\mathcal{F} \nvDash A$. By lemma 23 $\mathcal{F}$ is in Σ^ω_{cpn}. Consequently $\Sigma^\omega_{cpn} \nvDash A$. ∎

Proof of theorem 25. If $L(\Sigma^\omega_{cpn})$ can be finitely axiomatized with the inference rules in $\Psi^\omega_{cpn}{}^{++}$ in the language based on [01] and [10] then there exists a point formula A and a line formula a such that for all formulas ϕ, ϕ is in $L(\Sigma^\omega_{cpn})$ iff $(\{A, a\}, \Psi^\omega_{cpn}{}^{++}) \vdash \phi$. Hence A is in $L(\Sigma^\omega_{cpn})$ and a is in $L(\Sigma^\omega_{cpn})$. Consequently $(\Phi^\omega_{cpn}, \Psi^\omega_{cpn}{}^{++}) \vdash A$ and $(\Phi^\omega_{cpn}, \Psi^\omega_{cpn}{}^{++}) \vdash a$. Let n be the greatest natural integer such that $n \geq 2$ and $RICH_0^n$ or $RICH_1^n$ are necessarily used in the derivations of A or a in $(\Phi^\omega_{cpn}, \Psi^\omega_{cpn}{}^{++})$ and $\Phi^n_{cpn} = \{SEP_0, SEP_1, \langle 01 \rangle(\top, \top), \langle 10 \rangle(\top, \top), RICH_0^n, RICH_1^n, C_0, C_1, P_0, P_1, N_0, N_1\}$. Thus for all incidence frames $\mathcal{F}$, if $\mathcal{F} \models \Phi^n_{cpn}$ then $\mathcal{F} \models \Phi^\omega_{cpn}$. Let p be the least prime number such that $p \geq n$. It is a well-known fact that there exists a p-rich, correct, projective and normal incidence frame $\mathcal{F}$ such that $\mathcal{F}$ is not ω-rich. By lemma 1 and propositions 6, 8, 9 and 10 $\mathcal{F} \models \Phi^n_{cpn}$. Therefore $\mathcal{F} \models \Phi^\omega_{cpn}$. By proposition 7 $\mathcal{F} \nvDash \Phi^\omega_{cpn}$: a contradiction. ∎

Proof of theorem 26. Let $\mathcal{F}_{cAr}$ be the canonical frame for $(\Phi^\omega_{cpnAr}, \Psi^\omega_{cpn}{}^{++})$, $\mathcal{F}_{cPa}$ be the canonical frame for $(\Phi^\omega_{cpnPa}, \Psi^\omega_{cpn}{}^{++})$, $\mathcal{F}_{Ar}$ be a rooted subframe of $\mathcal{F}_{cAr}$ and $\mathcal{F}_{Pa}$ be a rooted subframe of $\mathcal{F}_{cPa}$. Obviously lemmas 18–22 still hold for $(\Phi^\omega_{cpnAr}, \Psi^\omega_{cpn}{}^{++})$ and $(\Phi^\omega_{cpnPa}, \Psi^\omega_{cpn}{}^{++})$. Hence following the line of reasoning suggested in the proof of lemma 23 the reader may easily verify that $\mathcal{F}_{Ar}$ is Arguesian and $\mathcal{F}_{Pa}$ is Pappian. Consequently following the line of reasoning suggested in the proof of theorem 24 the reader may easily verify that $(\Phi^\omega_{cpnAr}, \Psi^\omega_{cpn}{}^{++})$ is sound and complete with respect to Σ^ω_{cpnAr} and $(\Phi^\omega_{cpnPa}, \Psi^\omega_{cpn}{}^{++})$ is sound and complete with respect to Σ^ω_{cpnPa}. ∎

Proof of theorem 27. The proof can be easily adapted from the proof of theorem 25. ∎

Proof of theorem 28. The proof can be easily adapted from the proof of theorems 24 and 26. ∎

Philippe Balbiani
Université Paul Sabatier
Institut de recherche en informatique de Toulouse
31062 Toulouse Cedex 9
France
balbiani@irit.fr

Every world can see a Sahlqvist world

P. BALBIANI, I. SHAPIROVSKY, V. SHEHTMAN

ABSTRACT. We introduce a class of propositional modal logics axiomatized by infinite sequences of formulas in special form. Two logics of this type are known from [10] and [13]. Although the axioms are beyond Sahlqvist class and its generalization defined in [8], all the resulting logics are still complete with respect to elementary classes of frames. For two particular examples of these logics related to the McKinsey axiom, we also study elementarity and finite model property.

1 Introduction

The starting point for our research is the well-known Sahlqvist Theorem. For about thirty years this result was considered as the strongest one giving a syntactic sufficient condition for completeness and first-order definability (elementarity) in modal logic. More recent studies show that Sahlqvist class can be extended to a larger class of "inductive" modal formulas inheriting both completeness and elementarity [8].

Now there is natural question: what happens beyond this new class? It is well-known that completeness or definability may be lost. Perhaps the simplest counterexample is given by the McKinsey axiom $\Box\Diamond p \to \Diamond\Box p$, which is non-elementary, but still complete and even has the finite model property. So after a slight variation of Sahlqvist formulas we may still hope to preserve at least some of nice properties.

On the other hand, recently Ian Hodkinson has found a precise description of quasi-elementary logics (i.e. those complete with respect to elementary classes of Kripke frames) [9]: he defines a translation from a first-order theory T into a set of hybrid modal formulas, and next – into a set of pure modal formulas axiomatizing exactly the class of models of T. Moreover, the corresponding hybrid formulas are identified as "quasipositive". This general result is quite impressive, but the suggested axiomatization method may be not optimal in particular cases.

The class of modal logics studied in this paper is certainly covered by Hodkinson's theorem, but we propose a simpler description, which does not directly follow from [9]. Namely, consider different versions of a modal formula $\varphi(p_1, \ldots, p_k)$ with disjoint sets of proposition letters:

$$\varphi' = \varphi(p_{k+1}, \ldots, p_{2k}), \ \varphi'' = \varphi(p_{2k+1}, \ldots, p_{3k}), \ldots$$

and put

$$\varphi^n_\Diamond = \Diamond(\varphi \wedge \varphi' \wedge \cdots \wedge \varphi^{(n-1)}),$$

$$L_\Diamond(\varphi) = \mathbf{K} + \{\varphi_\Diamond^n \mid n \geq 0\}.$$

As we show, every logic $L_\Diamond(\varphi)$ is quasi-elementary, whenever φ is a Sahlqvist formula (or even an inductive formula). By Fine – Van Benthem Theorem (cf. [4, Theorem 10.19]) this implies canonicity. An appropriate first-order condition can be obtained in a standard way:

$$\forall x \exists t (xRt \land \Phi(t)),$$

where Φ is the first-order correspondent of φ, i.e. "every world sees a φ-world". However this condition does not always characterize the frames for $L_\Diamond(\varphi)$ — in general we cannot state that this logic is elementary, and it may be non-axiomatizable by inductive formulas. On the other hand, we show that it is finitely elementary, i.e. all its finite frames satisfy the above mentioned condition.

Also note that $L_\Diamond(\varphi)$ is the union of the increasing sequence of logics $L_\Diamond^n(\varphi) = \mathbf{K} + \varphi_\Diamond^n$, so it may be not finitely axiomatizable.

Two examples of such logics have been known from the literature, both are non-elementary and non-finitely axiomatizable.

The first logic $L_\Diamond(\Box p \to p)$ was introduced and studied by Hughes [10], who also proved the finite model property and decidability.

The second example is Lemmon's logic $KM^\infty = L_\Diamond(\Diamond p \to \Box p)$ [13]. This logic is even worse in some respect — recent results show that although KM^∞ is canonical itself, it cannot be axiomatized by canonical formulas and all the logics $\mathbf{K} + \varphi_\Diamond^n$ are not canonical (for $\varphi = \Diamond p \to \Box p$) [7]. Note that $\varphi_\Diamond^1 = \Diamond(\Diamond p_1 \to \Box p_1)$ is an equivalent form of the McKinsey axiom, for which the non-canonicity was obtained earlier [6]. To add to the list of negative results, we show that any logic between $\mathbf{K} + \varphi_\Diamond^1$ and KM^∞ is non-elementary. On the other hand, KM^∞ has the fmp, by Fine's Theorem on uniform logics, cf. [4]. This result can be improved and in fact KM^∞ is PSPACE-complete, as we are going to show in the second part of this paper.

In this paper we also consider a third example, viz. $L_\omega^{fw} = L_\Diamond(p \to \Box p)$. This logic happens to be "more regular": it is canonical, elementary and modally defines "McKinsey property" (which is a first-order equivalent of the McKinsey axiom in the transitive case, cf. [2]; moreover, L_ω^{fw} is finitely axiomatizable, has the fmp and therefore is decidable (and as we shall prove in the sequel, PSPACE-complete)

Note that all our logics are axiomatized by modal formulas of the form $\Diamond(\varphi_1 \land \cdots \land \varphi_n)$, where φ_i are "good" (say, Sahlqvist). The use of only $\Diamond$ and $\land$ seems "relatively safe" in this context, and so there is a hope to extend our results to a larger class of formulas. In the final section we briefly discuss some further results and related open problems.

2 Basic concepts

We assume the reader is familiar with tools and techniques in modal logic such as the canonical model construction and the filtration method. We also assume the reader is at home with well-known results such as Sahlqvist's theorem and Ladner's theorem. For more on these see Blackburn, de Rijke and Venema [3], Chagrov and Zakharyaschev [4] and Kracht [11]. Still in this Section we recall some basic notions, for the sake of terminology and notation.

2.1 Syntax

Let $PV = \{p_1, p_2, \dots\}$ be a countable set of proposition letters, with typical members denoted by p, q, etc. Modal formulas over PV are built using the constant $\bot$, the unary connective $\Box$ and the binary connective $\to$. Other constructs are defined as usual, in particular $\Diamond \varphi$ is the abbreviation for $\neg \Box \neg \varphi$.

For a formula φ, $cl(\varphi)$ denotes the set of all subformulas of φ. Put $PV(\varphi) = cl(\varphi) \cap PV$. The notation $\varphi(p_1, \dots, p_n)$ means $\{p_1, \dots, p_n\} \supseteq PV(\varphi)$. For formulas $\psi_1, \dots, \psi_n$, $\varphi(\psi_1, \dots, \psi_n)$ denotes the result of simultaneous substitution of $\psi_1, \dots, \psi_n$ for $p_1, \dots, p_n$ in φ. For a set x of formulas, let $x^\Box = \{\varphi \mid \Box\varphi \in x\}$.

A *(normal) modal logic* is a set L of formulas that contains all tautologies, the formula $\Box(p \to q) \to (\Box p \to \Box q)$ and that is closed under the standard rules: Modus ponens, Uniform substitution, and Generalization (given φ infer $\Box\varphi$).

For a set Γ of formulas, $L+\Gamma$ denotes the smallest modal logic containing $(L \cup \Gamma)$.

A formula φ is said to be *L-deducible* from a set Γ of formulas, in symbols $\Gamma \vdash_L \varphi$ if there exists formulas $\varphi_1, \dots, \varphi_n \in \Gamma$ such that $(\varphi_1 \wedge \dots \wedge \varphi_n \to \varphi) \in L$. A set Γ of formulas is called *L-consistent* if $\Gamma \nvdash_L \bot$.

Recall the inductive definition of *uniform modal formulas*. Any formula without modal operators is a uniform formula of degree 0; uniform formulas of degree $n+1$ are built from the set

$$\{\Box\psi \mid \psi \text{ is a uniform formula of degree } n\}$$

using boolean connectives.

A logic is called *uniform* if it can be axiomatized by uniform formulas.

2.2 Semantics

As usual, a *(Kripke) frame* is a pair $\mathcal{F} = (W, R)$, where W is a non-empty set of worlds and R is a binary relation on W. A world $x \in W$ is called *final* in $\mathcal{F}$ if $R(x) \subseteq \{x\}$ and *deterministic* in $\mathcal{F}$ if $card(R(x)) \leq 1$. $\mathcal{F}$ is called *serial* if for all worlds x in W, $R(x) \neq \varnothing$.

For a world x in W and sets $W_1, \dots, W_n \subseteq W$, we say that $\{W_1, \dots, W_n\}$ is a *cover* of $R(x)$ if $R(x) \subseteq W_1 \cup \dots \cup W_n$.

A *(Kripke) model* based on $\mathcal{F}$ is a pair $\mathcal{M} = (\mathcal{F}, V)$, where V is a function assigning to each proposition letter p a subset $V(p)$ of W. The inductive definition of the truth value of a formula φ at a world x in a model $\mathcal{M}$ is standard. $\mathcal{M}, x \vDash \varphi$ denotes that φ is true at x in $\mathcal{M}$.

A formula φ is called *true* (respectively *satisfiable*) in a model $\mathcal{M} = (W, R, V)$, in symbols $\mathcal{M} \vDash \varphi$, if φ is true at any (resp., some) world in W; φ is *valid* in a frame $\mathcal{F}$, in symbols $\mathcal{F} \vDash \varphi$, if φ is true in all models based on $\mathcal{F}$. φ is *valid* at $\mathcal{F}, x$ (notation: $\mathcal{F}, x \vDash \varphi$) if it is true at x in all models based on $\mathcal{F}$. A formula φ is said to be valid in a class $\mathcal{C}$ of frames, in symbols $\mathcal{C} \vDash \varphi$, if φ is valid in all frames in $\mathcal{C}$. We say that a set L of formulas is valid in a frame $\mathcal{F}$, in symbols $\mathcal{F} \vDash L$, if all formulas from L are valid in $\mathcal{F}$.

Every frame can also be regarded as a first-order structure, and we use the same sign $\vDash$ to denote the truth of a first-order formula in this structure.

The *modal logic of a class of frames* $\mathcal{C}$ is defined as $\mathbf{L}(\mathcal{C}) = \{\varphi \mid \mathcal{C} \vDash \varphi\}$. A logic L is called *complete* with respect to $\mathcal{C}$ if $L = \mathbf{L}(\mathcal{C})$. L is said to have the *finite model property* (fmp) if it is complete with respect to a class of finite frames.

We say that a set of modal formulas Γ *modally defines* the class of frames $\mathbf{Fr}(\Gamma) := \{\mathcal{F} \mid \mathcal{F} \vDash \Gamma\}$; Γ *modally defines* $\mathcal{C}$ *within a class* $\mathcal{C}'$ if $\mathcal{C} = \mathbf{Fr}(\Gamma) \cap \mathcal{C}'$. We say that a formula φ modally defines $\mathcal{C}$ [within $\mathcal{C}'$] if the set $\{\varphi\}$ does.

A modal logic L is called *strongly complete* with respect to a class $\mathcal{C}$ of frames if for any L-consistent set Γ of formulas, there exists a model $\mathcal{M}$ based on a frame from $\mathcal{C}$ such that all formulas from Γ are simultaneously true at some world in $\mathcal{M}$.

Recall that a formula φ is *locally elementary* if for some first-order formula Φ with only one free variable t we have: for any frame $\mathcal{F}$ and any world a in $\mathcal{F}$, $\mathcal{F}, a \vDash \varphi$ iff $\mathcal{F} \vDash \Phi(a)$. In this case Φ is called a *local first-order correspondent* of φ.

A set Γ of modal formulas is called *elementary* (respectively, Δ-*elementary*) if the class $\mathbf{Fr}(\Gamma)$ is elementary (respectively, Δ-elementary), i.e. if $\mathbf{Fr}(\Gamma)$ is the class of models of some first-order formula (resp., theory). Every locally elementary formula is obviously elementary.

A modal logic of the form $\mathbf{L}(\mathcal{C})$, for an elementary $\mathcal{C}$, is called *quasi-elementary* (or *elementarily generated*).

3 Every world can see a φ-world

In this section we describe a family of quasi-elementary and finitely elementary logics.

For a tuple of proposition letters $\mathbf{p} = (p_1, \ldots, p_k)$ let

$$\mathbf{p}^n := (p_{kn+1}, \ldots, p_{kn+k})$$

for $n \geq 0$. So $\mathbf{p}^0 = \mathbf{p}$ and all the tuples $\mathbf{p}^n$ are disjoint.

For a modal formula $\varphi(\mathbf{p})$ put

$$\varphi_\Diamond^n := \Diamond(\varphi(\mathbf{p}^0) \wedge \cdots \wedge \varphi(\mathbf{p}^{n-1}))$$

(in particular, $\varphi_\Diamond^0 := \Diamond\top$), and also

$$L_\Diamond(\varphi) := \mathbf{K} + \{\varphi_\Diamond^n \mid n \geq 0\},$$

$$L_\diamond^n(\varphi) := \mathbf{K} + \varphi_\diamond^n.$$

The following is rather trivial.

PROPOSITION 1. $L_\diamond^0(\varphi) \subseteq L_\diamond^1(\varphi) \subseteq L_\diamond^2(\varphi) \ldots \subseteq L_\diamond(\varphi).$

Now let φ be a locally elementary formula, and let $\Phi(t)$ be its local first-order correspondent. Consider the class of frames

$$\mathcal{C}_\diamond(\varphi) := \{\mathcal{F} \mid \mathcal{F} \vDash \forall x \exists t (xRt \wedge \Phi(t))\}.$$

LEMMA 2. *For a locally elementary formula* φ, $\mathcal{C}_\diamond(\varphi) \subseteq \mathbf{Fr}(L_\diamond(\varphi))$.

Proof. Suppose $\mathcal{F} \in \mathcal{C}_\diamond(\varphi(\mathbf{p}))$. This means that for any $a \in \mathcal{F}$ there exists $b \in R(a)$ such that $\mathcal{F} \vDash \Phi(b)$, which is equivalent to $\mathcal{F}, b \vDash \varphi$, by the definition of Φ. But the latter implies $\mathcal{F}, b \vDash \varphi(\mathbf{p}^n)$ for any n, hence $\mathcal{F}, a \vDash \varphi_\diamond^n$, and thus $\mathcal{F}, a \vDash L_\diamond(\varphi)$. ∎

A formula φ is *locally d-persistent* if for any descriptive (general) frame $(\mathcal{F}, \mathcal{D})$ and any world a, $(\mathcal{F}, \mathcal{D}), a \vDash \varphi$ implies $\mathcal{F}, a \vDash \varphi$ (the notion of a *descriptive frame* is defined in a standard way, see e.g. [4]).

THEOREM 3. *Let φ be a locally elementary and locally d-persistent modal formula. Then*

1. *the canonical frame for $L_\diamond(\varphi)$ is in $\mathcal{C}_\diamond(\varphi)$;*

2. $L_\diamond(\varphi)$ *is canonical and therefore strongly complete with respect to* $\mathcal{C}_\diamond(\varphi)$.

Proof. Let us prove (1); then (2) readily follows from Lemma 2 and the properties of the canonical model.

Let $\mathcal{F} = (W, R)$ be the canonical frame for $L_\diamond(\varphi(\mathbf{p}))$, $\mathbf{p} = (p_1, \ldots, p_k)$. For any $a \in W$, put

$$a^+ := a^\square \cup \{\varphi(\psi_1, \ldots, \psi_k) \mid \psi_1, \ldots, \psi_k \text{ are arbitrary modal formulas}\}.$$

CLAIM 1 a^+ is $L^\diamond(\varphi)$-consistent.

Suppose the contrary. Then for some formulas $\gamma_1, \ldots, \gamma_m \in a^\square$ and for some k-tuples of formulas $\overline{\psi}_1, \ldots, \overline{\psi}_n$ we have $L_\diamond(\varphi) \vdash \gamma_1 \wedge \cdots \wedge \gamma_m \wedge \varphi(\overline{\psi}_1) \wedge \cdots \wedge \varphi(\overline{\psi}_n) \to \bot$.

Since $\diamond(\varphi(\overline{\psi}_1) \wedge \cdots \wedge \varphi(\overline{\psi}_n))$ is a substitution instance of $\varphi_\diamond^n$, we have $\diamond(\varphi(\overline{\psi}_1) \wedge \cdots \wedge \varphi(\overline{\psi}_n)) \in a$, so for some $b \in R(a)$ we have $\varphi(\overline{\psi}_1), \ldots, \varphi(\overline{\psi}_n) \in b$. Since every γ_i is in $a^\square$, we also have $\gamma_1, \ldots, \gamma_m \in b$, so $\bot \in b$, which is a contradiction. Q.e.d.

By the Lindenbaum Lemma, there exists $b \in W$ such that $a^+ \subseteq b$. So $a^\square \subseteq b$, thus aRb.

Now consider the general canonical frame $(\mathcal{F}, \mathcal{D})$ for $L_\diamond(\varphi)$; recall that

$$\mathcal{D} = \{V_0(\psi) \mid \psi \text{ is a modal formula}\},$$

where V_0 is the canonical valuation.

CLAIM 2 $(\mathcal{F}, \mathcal{D}), b \vDash \varphi$.

In fact, for a valuation V in $(\mathcal{F}, \mathcal{D})$ let us show that $b \in V(\varphi)$. By definition of $\mathcal{D}$, for every i there exists a modal formula ψ_i such that $V(p_i) = V_0(\psi_i)$; let $\overline{\psi} = (\psi_1, \ldots, \psi_k)$. Then by induction it follows that

$$V(\varphi(\mathbf{p})) = V_0(\varphi(\overline{\psi})).$$

But $b \supseteq a^+$, so $\varphi(\overline{\psi}) \in b$, and thus $b \in V_0(\varphi(\overline{\psi})) = V(\varphi)$. Q.e.d.

Since $(\mathcal{F}, \mathcal{D})$ is descriptive and φ is locally d-persistent, from Claim 2 we obtain $\mathcal{F}, b \vDash \varphi$, i.e. $\mathcal{F} \vDash \Phi(b)$; thus $\mathcal{F} \vDash \forall x \exists t(xRt \wedge \Phi(t))$. ∎

For a set Γ of modal formulas we can also define

$$L_\Diamond(\Gamma) := \mathbf{K} + \{\varphi_\Diamond^n \mid \varphi \in \Gamma,\ n \geq 0\},$$

$$\mathcal{C}_\Diamond(\Gamma) := \bigcap_{\varphi \in \Gamma} \mathcal{C}_\Diamond(\varphi).$$

So we have

COROLLARY 4. *If Γ is a set of locally elementary and locally d-persistent modal formulas, then $L_\Diamond(\Gamma) = \mathbf{L}(\mathcal{C}_\Diamond(\Gamma))$.*

COROLLARY 5. *If Γ is a set of inductive modal formulas, then $L_\Diamond(\Gamma) = \mathbf{L}(\mathcal{C}_\Diamond(\Gamma))$, and thus this logic is canonical and strongly complete with respect to $\mathcal{C}_\Diamond(\Gamma)$.*

Proof. In fact, every inductive formula is locally elementary and locally d-persistent [8]. ∎

DEFINITION 6. A modal logic L is called *finitely elementary* if there exists a first-order formula Φ such that for any finite frame $\mathcal{F}$, $\mathcal{F} \vDash L$ iff $\mathcal{F} \vDash \Phi$.

THEOREM 7. *If φ is a locally elementary modal formula, then $L_\Diamond(\varphi)$ is finitely elementary.*

Proof. For a formula $\varphi(\mathbf{p})$, $\mathbf{p} = (p_1, \ldots, p_k)$ with a local first-order correspondent Φ, let us prove that $\mathcal{F} \vDash L_\Diamond(\varphi)$ iff $\mathcal{F} \in \mathcal{C}_\Diamond(\varphi)$ for any finite frame $\mathcal{F}$. The direction "if" is already proved in Lemma 2.

So consider a finite frame $\mathcal{F} = (W, R)$ with $card(W) = n$, such that $\mathcal{F} \vDash L_\Diamond(\varphi)$. We have to show that $\mathcal{F} \vDash \forall x \exists t(xRt \wedge \Phi(t))$.

Consider the following Kripke model $\mathcal{M} = (\mathcal{F}, V)$. Take the N-element set $\mathcal{P}(W)^k$ of all k-tuples of subsets of W (where $N = 2^{nk}$) and put it in some order: $\mathbf{W}^0, \mathbf{W}^1, \ldots, \mathbf{W}^{N-1}$. If $\mathbf{W}^j = (W_1^j, \ldots, W_k^j)$, we define

$$V(p_{jk+i}) := W_i^j.$$

So we have defined $V(p_m)$ for $m = 1, \ldots, Nk$, and we assume that $V(p_m)$ is arbitrary for all other m.

Now take any $a \in W$. Since $\mathcal{M}, a \vDash \varphi_\diamond^N$, there exists $b \in R(a)$ such that $\mathcal{M}, b \vDash \varphi(\mathbf{p}^0) \wedge \cdots \wedge \varphi(\mathbf{p}^{N-1})$. Then we claim that $\mathcal{F}, b \vDash \varphi$. In fact, consider an arbitrary valuation V' in $\mathcal{F}$. By our construction, there exists j such that $\mathbf{W}^j = (V'(p_1), \ldots, V'(p_k))$, i.e.

$$V'(p_i) = V(p_{jk+i})$$

whenever $1 \leq i \leq k$. Hence by induction it easily follows that

$$V'(\varphi(\mathbf{p})) = V(\varphi(\mathbf{p}^j)).$$

Now since $\mathcal{M}, b \vDash \varphi(\mathbf{p}^j)$, we obtain $(\mathcal{F}, V'), b \vDash \varphi$. Thus $\mathcal{F}, b \vDash \varphi$, which is equivalent to $\mathcal{F} \vDash \Phi(b)$. Eventually $\mathcal{F} \vDash \forall x \exists y (xRy \wedge \Phi(y))$. ∎

The following simple fact is motivated by Proposition 6.2 from [9], though it is not formulated explicitly in that paper.

PROPOSITION 8. *If a recursively axiomatizable and finitely elementary modal logic has the fmp, then it is decidable.*

Proof. In fact, if finite frames for L are defined by a certain first-order sentence, then the set of finite L-frames (whose worlds are identified with integers) is decidable. Together with the fmp, this implies the co-enumerability of L. So since L is recursively enumerable, it is decidable. ∎

Note that in general a recursively axiomatizable logic with the fmp can be undecidable (the three-dimensional logic $\mathbf{K}^3$ is a typical example), but this does not affect the logics considered in the present paper.

4 Every world can see a final world
4.1 Definitions
In this section we consider a particular case of the above construction, when $\varphi = (p_1 \to \Box p_1)$. The corresponding first-order formula is $\Phi(t) = \forall x (tRx \to t = x)$.

Let

$$\alpha_n := \varphi_\diamond^n = \diamond((p_1 \to \Box p_1) \wedge \ldots \wedge (p_n \to \Box p_n)).$$

Consider the modal logics (for $0 \leq n < \omega$):

$$L_n^{fw} := \mathbf{K} + \alpha_n, \ L_\omega^{fw} := L_\diamond(\varphi) = \mathbf{K} + \{\alpha_0, \alpha_1, \ldots\}.$$

Note that $L_0^{fw} = \mathbf{D} = \mathbf{K} + \diamond\top$ and $L_1^{fw} = \mathbf{K} + \diamond(p \to \Box p) = \mathbf{K} + \Box p \to \diamond \Box p$. Then $\mathcal{C}_\diamond(\varphi)$ is the class $\mathcal{C}_\omega^{fw}$ of all frames $\mathcal{F} = (W, R)$ satisfying *possible finality condition*[1]

- for all worlds x in W, there exists a world y in $R(x)$ such that $R(y) \subseteq \{y\}$,

---
[1] In the transitive case it is also called "McKinsey property" [14]

i.e. every world can see a final world.

Let us also consider for every $n \geq 0$, the class $\mathcal{C}_n^{fw}$ of all frames $\mathcal{F} = (W, R)$ such that

- for all worlds x in W and for all covers $\{W_1, \ldots, W_n\}$ of $R(x)$, there exists a world y in $R(x)$ such that for all i in $\{1, \ldots, n\}$, if $y \in W_i$ then $R(y) \subseteq W_i$.

PROPOSITION 9. $\mathcal{C}_0^{fw} \supseteq \mathcal{C}_1^{fw} \ldots \supseteq \mathcal{C}_\omega^{fw}$.

Proof. Trivial. ∎

4.2 Weakly condensed frames

Remark that $\mathcal{C}_0^{fw}$ is nothing but the class of all serial frames and $\mathcal{C}_1^{fw}$ is the class of all *weakly condensed* frames $\mathcal{F} = (W, R)$, i.e. such that:

- for all worlds $x \in W$, there exists $y \in R(x)$ such that $R(y) \subseteq R(x)$.

For a frame $\mathcal{F} = (W, R)$, consider the relation $R^\to \subseteq R$:

$$xR^\to y := \{y\} \cup R(y) \subseteq R(x)$$

Thus $\mathcal{F}$ is weakly condensed iff $(W, R^\to)$ is serial. One can easily see that $R^\to$ is transitive: if $xR^\to y R^\to z$, then $R(z) \cup \{z\} \subseteq R(y) \subseteq R(x)$.

By a straightforward argument, $\mathcal{F}$ is weakly condensed iff $\mathcal{F} \vDash \alpha_1$. Since the Sahlqvist formula $\Box p \to \Diamond \Box p$ is an equivalent form of α_1, we obtain

LEMMA 10. *If $\mathcal{F}$ is the canonical frame for a logic $L \supseteq L_1^{fw}$, then $\mathcal{F}$ is weakly condensed.*

4.3 Completeness

THEOREM 11. *The canonical frame for L_2^{fw} is in $\mathcal{C}_\omega^{fw}$.*

Proof. Let $\mathcal{F} = (W, R)$ be the canonical frame for $L = L_2^{fw}$.
For any $x \in W$, put $x^+ := x^\Box \cup \{\Box\varphi \mid \Box\varphi \in x\}$.
CLAIM 1 For any worlds x, y in the canonical model

$$R(x) \supseteq R(y) \text{ iff } y \supseteq \{\Box\varphi \mid \Box\varphi \in x\}.$$

In fact, if $y \supseteq \{\Box\varphi \mid \Box\varphi \in x\}$ and yRz, then by the definition of R, $\Box\varphi \in x$ implies $\varphi \in z$, i.e. xRz. The other way round, if $y \not\supseteq \{\Box\varphi \mid \Box\varphi \in x\}$, then for some $\Box\varphi \in x$ we have $\Box\varphi \notin y$, and so there exists $z \in R(y)$ such that $\varphi \notin z$. But $\Box\varphi \in x$, hence $z \notin R(x)$. Therefore $R(y) \not\subseteq R(x)$. Q.e.d.
CLAIM 2 $xR^\to y$ iff $y \supseteq x^+$.
In fact, by definition, xRy iff $y \supseteq x^\Box$, and $R(x) \supseteq R(y)$ iff $y \supseteq \{\Box\varphi \mid \Box\varphi \in x\}$, by Claim 1. Q.e.d.

Due to Lemma 10, $\mathcal{F}$ is weakly condensed, and thus by Claim 2, x^+ is an L-consistent set for any $x \in W$.

For the proof of our theorem, consider an arbitrary world x in W. Assuming that all formulas are arranged in some fixed order φ_1, φ_2, ..., we define the sequence y_0, y_1, ... of L-consistent sets of formulas by induction:

- $y_0 = x^+$

- $$y_{n+1} = \begin{cases} y_n \cup \{\Box\varphi_n\} & \text{if this set is } L-\text{consistent,} \\ y_n & \text{otherwise} \end{cases}$$

Let $y' = y_0 \cup y_1 \cup \ldots$. Note that y' is L-consistent, and for all formulas φ, if $y' \cup \{\Box\varphi\}$ is L-consistent, then $\Box\varphi \in y'$. By the Lindenbaum Lemma, there exists a maximal L-consistent set y'' such that $y' \subseteq y''$. Since $y'' \supseteq x^+$, then by Claim 2, $xR^\rightarrow y''$.

CLAIM 3 If $y''R^\rightarrow v$ then $R(v) = R(y'')$.

Given $v \in R(y'')$ and $R(v) \subseteq R(y'')$, let us show that $R(y'') \subseteq R(v)$. So we assume $t \in R(y'')$ and show that vRt. For this we further assume $\Box\varphi \in v$ and show that $\varphi \in t$. Since v is a world in $R(x)$ such that $R(v) \subseteq R(x)$, we have $y' \subseteq v$. In fact, note that $x^\Box \subseteq v$ since xRv; all other formulas in y' are of the form $\Box\psi$, and $\Box\psi \in y' \subseteq y''$ implies $y'' \models \Box\psi$, and thus $v \models \Box\psi$ (since $R(v) \subseteq R(y'')$), i.e. $\Box\psi \in v$.

Hence $y' \cup \{\Box\varphi\}$ is an L-consistent set of formulas, and thus $\Box\varphi \in y'$ by our construction. Therefore $\Box\varphi \in y''$. Since $y''Rt$, we eventually obtain $\varphi \in t$. Q.e.d.

CLAIM 4 $card(R(y'')) \leq 1$.

In fact, let z, t be worlds in $R(y'')$ such that $z \neq t$. Then there exists a formula $\varphi \in z$ such that $\varphi \notin t$. Thus

$$u = (y'')^+ \cup \{\Diamond\varphi \to \Box\varphi\}$$

is an L-consistent set of formulas. For otherwise there exists formulas $\varphi_1, \ldots, \varphi_{m+n}$ such that $\Box\varphi_1, \ldots, \Box\varphi_{m+n} \in y''$ and

(1) $\neg(\varphi_1 \wedge \ldots \wedge \varphi_m \wedge \Box\varphi_{m+1} \wedge \ldots \wedge \Box\varphi_{m+n} \wedge (\Diamond\varphi \to \Box\varphi)) \in L$.

Let

$$\varphi' = \varphi_1 \wedge \ldots \wedge \varphi_m, \quad \varphi'' = \varphi_{m+1} \wedge \ldots \wedge \varphi_{m+n}.$$

So $\Box\varphi' \wedge \Box\varphi''$ is in y''.

On the other hand, by (1) and classical logic

$$\neg(\varphi' \wedge \Box\varphi'' \wedge (\Box\neg\varphi \vee \Box\varphi)) \in L,$$

which is equivalent to

$$\varphi' \to \neg(\Box(\varphi'' \wedge \varphi) \vee \Box(\varphi'' \wedge \neg\varphi)) \in L.$$

Hence

$$\Box\varphi' \to \Box\neg(\Box(\varphi'' \wedge \varphi) \vee \Box(\varphi'' \wedge \neg\varphi)) \in L,$$

which is equivalent to

(2) $\Box\varphi' \to \Box(\neg\Box(\varphi'' \wedge \varphi) \wedge \neg\Box(\varphi'' \wedge \neg\varphi)) \in L$.

But

(3) $\Diamond((\varphi'' \wedge \varphi \to \Box(\varphi'' \wedge \varphi)) \wedge (\varphi'' \wedge \neg\varphi \to \Box(\varphi'' \wedge \neg\varphi))) \in L$

as a substitution instance of α_2. Since $\Box\varphi' \in y''$, from (2) and (3) we obtain for some $w \in R(y'')$:

$$\neg(\varphi'' \wedge \varphi) \wedge \neg(\varphi'' \wedge \neg\varphi) \in w.$$

Then $\neg\varphi'' \in w$, which contradicts $\Box\varphi'' \in y''$.

Therefore u is L-consistent. By the Lindenbaum Lemma, there exists a maximal L-consistent set $u' \supseteq u$. By the above construction and Claim 2, $y''R^{\to}u'$, and also $\Diamond\varphi \to \Box\varphi \in u'$. So by Claim 3, $R(u') = R(y'')$. Therefore z, t are worlds in $R(u')$. Hence $\Diamond\varphi \in u'$, and consequently $\Box\varphi$ is in u'. Thus φ is in t: a contradiction. Q.e.d.

Since $\mathcal{F}$ is weakly condensed, for some z we have $y''R^{\to}z$, i.e., $y''Rz$ and $R(z) \subseteq R(y'')$. By Claim 4, $R(y'') = \{z\}$, thus $R(z) \subseteq \{z\}$. $xR^{\to}y''$ and $y''R^{\to}z$ implies xRz, since $R^{\to}$ is transitive and $R^{\to} \subseteq R$. Thus $\mathcal{F} \in \mathcal{C}_\omega^{fw}$. ■

COROLLARY 12. L_2^{fw} *is strongly complete with respect to* $\mathcal{C}_\omega^{fw}$.

COROLLARY 13. *The* McKinsey *formula* $\Box\Diamond p \to \Diamond\Box p$ *is in* L_2^{fw}.

Proof.[2] In fact, $\mathcal{C}_\omega^{fw} \models \Box\Diamond p \to \Diamond\Box p$. ■

PROPOSITION 14. $L_2^{fw} = L_3^{fw} \ldots = L_\omega^{fw}$.

Proof. Follows from Theorems 3 and 11.

However let us give a syntactic proof of this fact proposed by Max Cresswell in a private communication.

It is sufficient to show that $L_n^{fw} \vdash \alpha_{n+1}$ for $n \geq 2$. Let

$$\pi_1 := (p_1 \to \Box p_1), \quad \pi_2 := (p_2 \to \Box p_2) \wedge \cdots \wedge (p_n \to \Box p_n),$$
$$\pi_3 := (p_{n+1} \to \Box p_{n+1}).$$

Now we argue in L_n^{fw}:

(1)	$\Diamond(\pi_1 \wedge (\neg q \to \Box\neg q))$	$\alpha_2, Subst$
(2)	$\Diamond(\pi_1 \wedge (\Diamond q \to q))$	(1), equivalent replacement
(3)	$\Diamond(\pi_1 \wedge (\Diamond(\pi_2 \wedge \pi_3) \to (\pi_2 \wedge \pi_3)))$	(2), $Subst$
(4)	$\Diamond(\pi_2 \wedge \pi_3)$	$\alpha_n, Subst$
(5)	$\Box\Diamond(\pi_2 \wedge \pi_3)$	(4), Gen
(6)	$\Diamond(p \wedge (q \to r)) \to (\Box q \to \Diamond(p \wedge r))$	derivable in **K**
(7)	$\Diamond(\pi_1 \wedge \pi_2 \wedge \pi_3)$	(6), $Subst$, (3), MP, (5), MP

Note that together with Theorem 3, this argument provides an alternative proof of Theorem 11. ■

---

[2] Note that there also exists a simple syntactic proof of this corollary.

4.4 Elementarity

PROPOSITION 15. Let $n \geq 2$. Then for any frame $\mathcal{F}$, $\mathcal{F} \in \mathcal{C}_n^{fw}$ iff $\mathcal{F} \models L_n^{fw}$.

Proof. Let $\mathcal{F} = (W, R)$ be a frame such that $\mathcal{F} \models L_n^{fw}$ and let us show that $\mathcal{F} \in \mathcal{C}_n^{fw}$. Take an arbitrary world x in W and a cover $\{W_1, \ldots, W_n\}$ of $R(x)$. Let $\mathcal{M} = (\mathcal{F}, V)$ be a model based on $\mathcal{F}$ such that $V(p_1) = W_1$, ..., $V(p_n) = W_n$. Since $x \models \alpha_n$, there exists a world y in $R(x)$ such that for all i in $\{1, \ldots, n\}$, $\mathcal{M}, y \models p_i \to \Box p_i$. Then for any i in $\{1, \ldots, n\}$, if $y \in W_i$, then $R(y) \subseteq W_i$.

Now suppose $\mathcal{F} = (W, R)$ is a frame such that $\mathcal{F} \not\models L_n^{fw}$. Then there exists a model $\mathcal{M} = (\mathcal{F}, V)$ and a world $x \in W$ such that for any $y \in R(x)$, there exists i in $\{1, \ldots, n\}$ such that $\mathcal{M}, y \not\models p_i \to \Box p_i$. Let $W_i = V(p_i)$. Then $\{W_1, \ldots, W_n\}$ is a cover of $R(x)$ and for any $y \in R(x)$ there exists i in $\{1, \ldots, n\}$ such that $y \in W_i$, but $R(y) \not\subseteq W_i$. Hence $\mathcal{F} \notin \mathcal{C}_n^{fw}$. ∎

COROLLARY 16. $\mathcal{C}_n^{fw} = \mathcal{C}_2^{fw}$ for any $n \geq 2$.

Proof. By Propositions 14 and 15. ∎

By Theorems 3, 7, and Proposition 14, L_2^{fw} is quasi-elementary and finitely elementary. The following theorem states the elementarity of $\bar{L}_2^{fw}$.

THEOREM 17. For any frame $\mathcal{F}$, $\mathcal{F} \in \mathcal{C}_\omega^{fw}$ iff $\mathcal{F} \models L_2^{fw}$; thus α_2 modally defines $\mathcal{C}_\omega^{fw}$.

Proof. By Lemma 2, it is sufficient to show that $\mathbf{Fr}(L_2^{fw}) \subseteq \mathcal{C}_\omega^{fw}$.

So let $\mathcal{F} = (W, R) \models L_2^{fw}$. Then by Proposition 15 and Corollary 16, $\mathcal{F} \in \mathcal{C}_2^{fw} = \mathcal{C}_3^{fw}$. Hence we obtain

CLAIM 1 $(W, R^\to)$ is in $\mathcal{C}_2^{fw}$.

In fact, let $x \in W$ and let $\{W_1, W_2\}$ be a cover of $R^\to(x)$. Also let

$$V_1 := W_1 \cap R^\to(x), \ V_2 := W_2 \cap R^\to(x), \ V_3 = R(x) \setminus (V_1 \cup V_2).$$

So $\{V_1, V_2, V_3\}$ is a cover of $R(x)$ and there exists a world y in $R(x)$ such that for all i in $\{1, 2, 3\}$, if $y \in V_i$, then $R(y) \subseteq V_i \subseteq R(x)$. Therefore $y \in R^\to(x)$. Next, for $i = 1, 2$ we have: if $y \in W_i$, then $y \in V_i$, so $R^\to(y) \subseteq R(y) \subseteq V_i \subseteq W_i$. Hence $(W, R^\to)$ is in $\mathcal{C}_2^{fw}$. Q.e.d.

CLAIM 2 $(W, R^\to)$ is in $\mathcal{C}_\omega^{fw}$.

In fact, $\Box \Diamond p \to \Diamond \Box p$ is in L_2^{fw} (Corollary 13); then by Proposition 15 and Claim 1, $(W, R^\to) \models \Box \Diamond p \to \Diamond \Box p$. Since $R^\to$ is transitive, $(W, R^\to)$ is in $\mathcal{C}_\omega^{fw}$. Q.e.d.

By Claim 2, for any world x there exists a world y in $R^\to(x)$ such that $R^\to(y) \subseteq \{y\}$. Let $W_1 = \{y\}$ and $W_2 = R(y) \setminus \{y\}$. Then $\{W_1, W_2\}$ is a cover of $R(y)$, and so there exists a world $z \in R(y)$ such that for all i in $\{1, 2\}$, if $z \in W_i$, then $R(z) \subseteq W_i$.

Suppose $z \in W_2$. Then $R(z) \cup \{z\} \subseteq R(y)$. So $yR^{\to}z$ and $z \neq y$. Consequently $R^{\to}(y) \not\subseteq \{y\}$: a contradiction.

Hence $z \in W_1$, i.e. $z = y$, and so $R(z) \subseteq \{y\}$. Since $y \in R^{\to}(x) \subseteq R(x)$, y is a final world in $R(x)$. ∎

4.5 Fmp and decidability

THEOREM 18. L_2^{fw} has the fmp and therefore is decidable.

Proof. Given an L_2^{fw}-consistent formula φ, let us show that φ is satisfiable in some finite frame from $\mathcal{C}_\omega^{fw}$.

Let $\mathcal{M} = (W, R, V)$ be the canonical model for L_2^{fw}. Then $\mathcal{M}, x_0 \models \varphi$ for some $x_0 \in W$ and $(W, R) \in \mathcal{C}_\omega^{fw}$ by Theorem 11. Let W^f be the set of all final worlds in $\mathcal{M}$. Take a new proposition letter $q \notin cl(\varphi)$, and let U be a valuation such that for all $p \in cl(\varphi)$, $U(p) = V(p)$, and $U(q) = W^f$. For $\mathcal{N} := (W, R, U)$ we obviously have $\mathcal{N}, x_0 \models \phi$.

Let $\mathcal{N}' = (W', R', U')$ be the minimal filtration of $\mathcal{N}$ through $\Psi = cl(\varphi) \cup \{q\}$. Recall that W' is the quotient set under the equivalence relation

$x \sim y$ iff the same formulas from Ψ are true at $\mathcal{N}, x$ and $\mathcal{N}, y$,

and for $\overline{y}, \overline{z} \in W'$

$$\overline{y} R' \overline{z} \text{ iff } (y_1 R z_1 \text{ for some } y_1 \in \overline{y}, z_1 \in \overline{z}),$$

where $\overline{x}$ denotes the class of x modulo $\sim$. So $\mathcal{N}'$ is finite, and $\mathcal{N}', \overline{x_0} \models \varphi$, by the Filtration Lemma.

Let us show that $(W', R') \in \mathcal{C}_\omega^{fw}$. Consider $\overline{x} \in W'$. Since $(W, R) \in \mathcal{C}_\omega^{fw}$, there exists $y \in W^f \cap R(x)$. Then $\overline{x} R' \overline{y}$. Suppose $\overline{y} R' \overline{z}$, so $y_1 R z_1$ for some $y_1 \in \overline{y}$, $z_1 \in \overline{z}$. Since $y \in W^f$ we have $\mathcal{N}, y \models q$, so $\mathcal{N}, y_1 \models q$ and $y_1 \in W^f$. Hence $y_1 R z_1$ implies $z_1 = y_1$, thus $\overline{z} = \overline{y}$. It follows that $R'(\overline{y}) = \{\overline{y}\}$, thus $(W', R') \in \mathcal{C}_\omega^{fw}$. ∎

5 Every world can see a deterministic world

In our second example, we put $\varphi := \Diamond p_1 \to \Box p_1$. Let

$$\begin{aligned}\beta_n &:= \varphi_\Diamond^n = \Diamond((\Diamond p_1 \to \Box p_1) \wedge \ldots \wedge (\Diamond p_n \to \Box p_n)); \\ L_\omega^{dw} &:= L_\Diamond(\varphi) = \mathbf{K} + \{\beta_0, \beta_1, \ldots\} \ (= KM^\infty); \\ L_n^{dw} &:= L_\Diamond^n(\varphi) = \mathbf{K} + \beta_n;\end{aligned}$$

for $n \geq 0$.

Note that L_0^{dw} is nothing but $\mathbf{K} + \Diamond\top$, and $L_1^{dw} = \mathbf{K} + \Box\Diamond p \to \Diamond\Box p$.

The class $\mathcal{C}_\omega^{dw} := \mathcal{C}_\Diamond(\varphi)$ is the class of all frames $\mathcal{F} = (W, R)$ with the following property:

- for all worlds x in W, there exists a world y in $R(x)$ such that $card(R(y)) = 1$, i.e., every world in W can see a deterministic world in $\mathcal{F}$.

Let us also consider for $n \geq 1$, the class $\mathcal{C}_n^{dw}$ of all frames $\mathcal{F} = (W, R)$ such that

- for all worlds x in W and for all sets $W_1, \ldots, W_n \subseteq W$, there exists a world y in $R(x)$ such that for all i in $\{1, \ldots, n\}$, if $R(y) \cap W_i \neq \emptyset$ then $R(y) \subseteq W_i$.

Let $\mathcal{C}_0^{dw}$ be the class of all serial frames. One can easily see that

PROPOSITION 19. $\mathcal{C}_0^{dw} \supseteq \mathcal{C}_1^{dw} \supseteq \mathcal{C}_2^{dw} \ldots \supseteq \mathcal{C}_\omega^{dw}$.

PROPOSITION 20. Let $n \geq 1$. Then $\mathcal{C}_n^{dw} = \mathbf{Fr}(L_n^{dw})$.

Proof. Given a frame $\mathcal{F} = (W, R) \in \mathcal{C}_n^{dw}$, let us show that for any model $\mathcal{M} = (\mathcal{F}, V)$, for any $x \in W$ we have $\mathcal{M}, x \vDash \beta_n$. Put $W_i := V(p_i)$. Then for some $y \in R(x)$ we have: for all i in $\{1, \ldots, n\}$, $R(y) \cap W_i = \emptyset$ or $R(y) \subseteq W_i$. If $\mathcal{M}, y \vDash \Diamond p_i$ then $R(y) \cap W_i \neq \emptyset$, so $R(y) \subseteq W_i$, thus $\mathcal{M}, y \vDash \Box p_i$. Hence $\mathcal{M}, x \vDash \beta_n$.

For the converse, suppose $\mathcal{F} \vDash L_n^{dw}$ and show that $\mathcal{F} \in \mathcal{C}_n^{dw}$. Let $x \in W$, $W_1, \ldots, W_n \subseteq W$. Consider a model $\mathcal{M} = (\mathcal{F}, V)$ such that $V(p_i) = W_i$, $1 \leq i \leq n$. Since $\mathcal{M}, x \vDash \beta_n$, for some $y \in R(x)$ for all $i \in \{1, \ldots n\}$ we have: $\mathcal{M}, y \vDash \Diamond p_i \to \Box p_i$, so $R(y) \cap W_i \neq \emptyset$ implies $R(y) \subseteq W_i$. Thus $\mathcal{F} \in \mathcal{C}_n^{dw}$. ∎

From [7] it follows that L_ω^{dw} is not finitely-axiomatizable (this proof is based on constructions from [5], [6]).

It is well known that L_1^{dw} is not Δ-elementary; this was proved independently in [1] and [5]; a proof can also be found in [3]. Next, [7] proves that L_ω^{dw} is not Δ-elementary (and thus certainly $\mathcal{C}_\omega^{dw} \neq \mathbf{Fr}(L_\omega^{dw})$). Let us now prove the following generalization of this fact:

THEOREM 21. If L is a modal logic, $L_1^{dw} \subseteq L \subseteq L_\omega^{dw}$, then L is not Δ-elementary.

Proof. The main idea is the same as in [1]. First, for any Kripke model $\mathcal{M}$ and $n \geq 1$ we define the relation

$$z_1 \equiv_n^{\mathcal{M}} z_2 := \mathcal{M}, z_1 \vDash p_i \Leftrightarrow \mathcal{M}, z_2 \vDash p_i \text{ for all } i \in \{1 \ldots n\}.$$

Clearly, $\equiv_n^{\mathcal{M}}$ is an equivalence relation on $\mathcal{M}$, and the quotient set $W/\equiv_n^{\mathcal{M}}$ is finite. By a straightforward argument,

$$\mathcal{M}, y \vDash (\Diamond p_1 \to \Box p_1) \wedge \ldots \wedge (\Diamond p_n \to \Box p_n) \text{ iff } z_1 \equiv_n^{\mathcal{M}} z_2 \text{ for all } z_1, z_2 \in R(y)$$

Now let us define a certain frame $\mathcal{F} = (W, R)$. For a countable set $Z = \{z_i \mid i \in \mathbb{N}\}$, take the uncountable set

$$Y := \{y_U \mid U \subseteq Z, U \text{ is infinite}\},$$

and put $W := Z \cup Y \cup \{x\}$, where $x \notin Z \cup Y$. R is defined as follows (Fig. 1):

$$R(x) := Y, \ R(y_U) := U, \ R(z_i) := \{z_i\}.$$

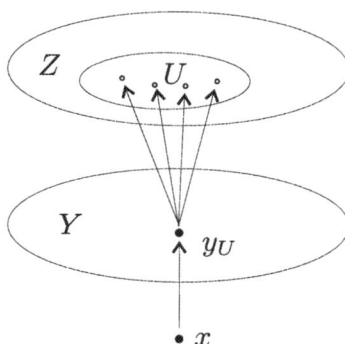

Figure 1.

CLAIM 1 $\mathcal{F} \vDash L_\omega^{dw}$.

It is sufficient to show that for all $n \geq 0$ and for all models $\mathcal{M}$ based on $\mathcal{F}$, $\mathcal{M}, x \vDash \beta_n$. Since $W/\equiv_n^\mathcal{M}$ is finite, there exists an $\equiv_n^\mathcal{M}$-equivalence class U_0 such that $U := U_0 \cap Z$ is infinite. Then

$$\mathcal{M}, y_U \vDash (\Diamond p_1 \to \Box p_1) \wedge \ldots \wedge (\Diamond p_n \to \Box p_n)$$

and therefore $\mathcal{M}, x \vDash \beta_n$. Q.e.d.

Now suppose the class $\mathbf{Fr}(L)$ is Δ-elementary, i.e., definable by a first-order theory T in the language $\{R, =\}$. Since $\mathcal{F} \vDash L_\omega^{dw}$, we have $\mathcal{F} \vDash L$, thus $\mathcal{F} \vDash T$. By the Löwenheim-Skolem theorem, there exists a countable subframe $\mathcal{F}' = (W', R')$ of $\mathcal{F}$ such that $W' \supseteq Z \cup \{x\}$ and $\mathcal{F}' \vDash T$. Then $\mathcal{F}' \vDash L$, and thus $\mathcal{F}' \vDash \beta_1$

However by [6, Theorem 1], $\neg \beta_1$ is satisfiable at a point x in a frame (W, R) if for all y in $R(x)$, $R(y)$ is infinite and $card(R(y)) \geq card(R(x))$. Thus $\mathcal{F}', x \nvDash \beta_1$.

This contradiction proves the theorem. ∎

COROLLARY 22. $\mathcal{C}_\omega^{dw}$ *is not modally definable.*

Proof. In fact, if an elementary class of frames $\mathcal{C}$ is modally definable, then its modal logic $\mathbf{L}(\mathcal{C})$ is elementary — just because $\mathbf{Fr}(\mathbf{L}(\mathcal{C}))$ is the smallest modally definable class containing $\mathcal{C}$. ∎

For $\mathbf{v} = \{v_1, \ldots, v_n\} \in \{0,1\}^n$, put $\mathbf{p^v} := \bigwedge_{1 \leq i \leq n} p_i^{v_i}$, where $p^1 := p$, $p^0 := \neg p$. It is not difficult to check that for all $n > 0$, β_n is equivalent to $\bigvee_{\mathbf{v} \in \{0,1\}^n} \Diamond \Box \mathbf{p^v}$. Thus the logics $L_1^{dw}, L_2^{dw}, \ldots, L_\omega^{dw}$ are uniform. Since all serial uniform logics have the fmp (see e.g. [4]), we have the following

THEOREM 23. *The logics* $L_1^{dw}, L_2^{dw}, \ldots, L_\omega^{dw}$ *have the fmp.*

COROLLARY 24. *The logics* $L_1^{dw}, L_2^{dw}, \ldots, L_\omega^{dw}$ *are decidable.*

Proof. For finite n L_n^{dw} is finitely axiomatizable. The decidability of L_ω^{dw} follows by Theorem 23 and Proposition 8. ∎

6 Complexity

It is well-known that all logics between **K** and **S4** are PSPACE-hard [12]. This result can be easily extended to all logics between **K** and $\mathbf{S4} + \Box\Diamond p \to \Diamond\Box p$ (see e.g. [15]). Thus L_2^{fw}, L_ω^{dw} are PSPACE-hard. PSPACE-decidability of the logics L_2^{fw}, L_ω^{dw} was recently obtained by the first author, the proof will be published in the sequel.

7 Conclusion

Stepping aside from a familiar field leads us to various nontrivial questions. We hope to address some of them in the second part of this paper. Let us only mention several topics for further study.

1. We see that $L_\Diamond(\varphi)$ (for a Sahlqvist formula φ) is sometimes finitely axiomatizable. Does there exist a reasonable criterion (or at least a sufficient condition) for that?

2. Basing on his investigation in first-order modal logic, Sergei Astretsov (Moscow State University) proposed the following conjecture: $L_\Diamond(\varphi)$ is finitely axiomatizable iff it is elementary. This conjecture is consistent with the three examples from the present paper.

3. Does the fmp always transfer from $\mathbf{K} + \varphi$ to $L_\Diamond(\varphi)$?

4. What are the properties of $L_\Diamond(\varphi)$ for well-known formulas φ, such as transitivity, non-branching, symmetry? As for transitivity, there is some progress: recently Stanislav Kikot (Moscow State University) has proved that $\mathcal{C}_\Diamond(\Box p \to \Box^2 p)$ is not modally definable. This leads to another question: does the elementarity of $L_\Diamond(\varphi)$ imply the modal definability of $\mathcal{C}_\Diamond(\varphi)$? Note that the converse is rather trivial, see the proof of Corollary 22.

5. The logics $L_\Diamond(\varphi)$ are not Δ-elementary for $\varphi = \Box p \to p$, $p \to \Box p$. What happens to other non-elementary logics $L_\Diamond(\varphi)$ from Theorem 3? Note that according to a result by Van Benthem [2], every *finitely axiomatizable* Δ-elementary modal logic is elementary.

6. Note that Theorem 21 on non-elementarity of "approximants" $L_\Diamond^n(\varphi)$ holds for KM^∞, but its analogue fails for Hughes' logic $L_\Diamond(\Box p \to p)$, since $L_\Diamond^1(\Box p \to p) = \mathbf{K} + \Box\Box p \to \Diamond p$ is axiomatized by a Sahlqvist formula. What happens in the general case – is it true that if $L_\Diamond(\varphi)$ (for a Sahlqvist φ) is non-elementary, then $L_\Diamond^n(\varphi)$ are also non-elementary for sufficiently large n?

7. Is it true that all the logics $L_\Diamond^n(\varphi)$ are complete (again for a Sahlqvist φ)? We do not know this even for simple cases, like $n = 2$, $\varphi = p \to \Box p$.

8. Is it possible to extend the result on non-canonicity of approximants for $L_\Diamond(\varphi)$ by Goldblatt–Hodkinson [7] to a larger class of logics $L_\Diamond(\varphi)$? What about noncompactness (for the McKinsey axiom this was established in [16])?

9. What happens in the transitive case? More exactly, consider logics of the form $\mathbf{K4} + L_\Diamond(\varphi)$ for locally elementary and locally d-persistent φ. Our

results show that they are quasi-elementary. Are they elementary? finitely axiomatizable? Note that e.g. $\mathbf{K4} + L_\Diamond(p \to \Box p) = \mathbf{K4} + \Box\Diamond p \to \Diamond\Box p$ is elementary.

10. Theorem 3 can probably be extended to a larger class of logics. Of course, it survives in the polymodal case. Moreover, instead of the prefix $\Diamond$ one can take a prefix $\Diamond_{k_1} \ldots \Diamond_{k_n}$, with the corresponding condition "every world sees a φ-world via the composed relation $R_{k_1} \circ \cdots \circ R_{k_n}$". And furthermore, we can consider more complicated conditions, like "every world sees a world seeing a φ-world and a ψ-world". It would be interesting to find a natural general result of this kind.

Acknowledgements

Many thanks to Mikhail Rybakov, Tinko Tinchev and Dimiter Vakarelov who made several helpful comments for improving the readability of the paper. Special thanks to three anonymous referees and also to Ian Hodkinson and Yde Venema who made essential remarks on the earlier version of the paper. We also thank Max Cresswell who kindly sent us his syntactic proof of Proposition 14.

The work of the second and the third authors was supported by Poncelet Laboratory (UMI 2615 of CNRS and Independent University of Moscow), by RFBR-NWO grant 047.011.2004.040, and by Russian presidential program of support of leading scientific schools 5351.2006.1.

BIBLIOGRAPHY

[1] Van Benthem, J.: *A note on modal formulae and relational properties*. J. Symb. Log. 40(1): 55-58 (1975)
[2] Van Benthem, J.: *Modal logic and classical logic*. Bibliopolis (1983)
[3] Blackburn, P., de Rijke, M., Venema, Y.: *Modal Logic*. Cambridge University Press (2001).
[4] Chagrov, A., Zakharyaschev, M.: *Modal Logic*. Oxford University Press (1997).
[5] Goldblatt, R.: *First-order definability in modal logic*. J. Symb. Log. 40(1): 35-40 (1975)
[6] Goldblatt, R.: *The McKinsey axiom is not canonical*. Journal of Symbolic Logic **56** (1991) 554–562.
[7] Goldblatt, R., Hodkinson, I.: *The McKinsey–Lemmon logic is barely canonical*. Submitted, (2006).
[8] Goranko, V., Vakarelov, D. *Elementary canonical formulae: extending Sahlqvist's theorem*. Annals of Pure and Applied Logic, **141** (2006), 180–217.
[9] Hodkinson, I.: *Hybrid formulas and elementarily generated modal logics*. (2006) http://www.doc.ic.ac.uk/~imh/ Notre Dame J. Formal Logic, to appear.
[10] Hughes, G.: *Every world can see a reflexive world*. Studia Logica **49** (1989) 175–181.
[11] Kracht, M.: *Tools and Techniques in Modal Logic*. Elsevier Science (1999).
[12] Ladner, R.: *The computational complexity of provability in systems of modal logic*. SIAM Journal on Computing **6** (1977) 467–480.
[13] Lemmon, E.: *The "Lemmon Notes": An Introduction to Modal Logic*. Basil Blackwell, (1977).
[14] Segerberg, K. *Decidability of S4.1*. Theoria, **34** (1968), 7–20.
[15] Shapirovsky, I.: *On PSPACE-decidability in transitive modal logics*. In: Schmidt, R., Pratt-Hartmann, I., Reynolds, M., Wansing, H., eds. *Advances in Modal Logic*. Volume 5. King's College Publications (2005) 269–287.

[16] Wang, X.: *The McKinsey axiom is not compact.* Journal of Symbolic Logic **57** (1992), 1230–1238.

Philippe Balbiani
Université Paul Sabatier
Institut de recherche en informatique de Toulouse
31062 Toulouse Cedex 9
France
Balbiani@irit.fr

Ilya Shapirovsky
Institute for Information Transmission Problems
Russian Academy of Sciences
B.Karetny 19, Moscow, Russia, 127994
shapirov@mccme.ru

Valentin Shehtman
Institute for Information Transmission Problems
Russian Academy of Sciences
B.Karetny 19, Moscow, Russia, 127994
shehtman@netscape.net

The Tree of Knowledge in Action: Towards a Common Perspective

JOHAN VAN BENTHEM AND ERIC PACUIT

ABSTRACT. We survey a number of decidability and undecidability results concerning epistemic temporal logic. The goal is to provide a general picture which will facilitate the 'sharing of ideas' from a number of different areas concerned with modeling agents in interactive social situations.

1 Introduction

When thinking about rational agents facing choices, one appealing mathematical model recurs in the literature. From Borges' story 'The Garden of Forking Paths' to a host of technical paradigms, sometimes at war, sometimes at peace, all invoke the picture of a branching tree of finite sequences of events with epistemic indistinguishability relations for agents between these sequences, reflecting their limited powers of observation. Indeed, tree models for computation, with branches standing for process evolutions over time, have long been studied in computer science, cf. [32, 33, 7, 2, 14]. Philosophers have studied similar models, now enriched with epistemic relations, for the behavior of intelligent human agents facing choices: see Thomason & Gupta [38], Belnap et al. [5] and Horty [20]. Epistemic models of events over time have also been used in computer science by various authors, witness Fagin et al. [8] and Parikh & Ramanujam [29, 30]. Such trees model not only processes, but also games (see Abramsky [1], Halpern [15] and van Benthem [40]). And finally, 'dynamic logics' of communication and information flow in the tradition of Baltag, Moss & Solecki [3] have tree models of events as their natural broader habitat.

Bringing together knowledge and temporal change is a natural move in modeling, but it is also a potentially dangerous one from a complexity perspective, as has been shown forcefully in Halpern & Vardi [16]. The context is clear from the literature cited just now. Rabin's Theorem tells us that the full monadic second-order logic of the tree of events ordered by the relation of 'initial segment', and provided with some finite set of successor functions is decidable [33]. This explains the decidability of purely temporal logics of events such as **CTL**, and others. Likewise, the tree-like nature of models explains the decidability of many modal logics (see [24]). In a slogan, 'Trees are Safe'. But, we also know that the monadic second-order logic, indeed, even the monadic Π_1^1-theory of the grid $\mathbb{N} \times \mathbb{N}$ is undecidable (see [17]). A grid is like a tree, but successors meet, and the resulting confluent structure is known to cause high complexity in many areas of modal logic ([22]), witness in particular the work on 'product models' [11, 34, 12, 19]. In one more slogan: 'Grids are Dangerous'.

Now, epistemic temporal logics live at a dangerous edge here. On top of Rabin-style tree models, they introduce epistemic indistinguishability relations which generate a 'second dimension', and if the language gets too powerful, enough grid structure can be encoded to cause undecidability. Illustrations for this again come from a wide range of papers. E.g., Thomas [37] points out, following Läuchli, how introducing a relation of 'simultaneity' into the Rabin tree makes the monadic second-order logic undecidable. Likewise, Halpern & Vardi show how epistemic-temporal logics of agents with Perfect Recall and No Learning can become undecidable [16] (cf. also [19]). But the situation is delicate, as small changes in an epistemic temporal language or its intended class of models can affect the complexity of the resulting logic in drastic ways.

This is the view 'from above', viewing epistemic temporal models as a Grand Stage where events unfold. There is also the view 'from below', found in 'dynamic epistemic logics' which construct successive new event models in definable stages (cf. Baltag, Moss and Solecki [4] and van Benthem, van Eijck and Kooi [43]). The logics tend to be decidable (though cf. [43] and [25]) and this, too, calls for explanation.

In this paper, we position ourselves close to the edge of undecidability in a straightforward system of epistemic temporal logic. We will discuss a number of complexity results, on both sides of the edge, while pointing out how results from all different traditions mentioned here help illuminate the landscape. As a result, we are also able to 'place' dynamic epistemic logics as species of epistemic temporal ones — and find room for comparing ideas from both traditions, e.g., in process algebra and game analysis.

In doing all this, we also have a broader aim. The area that we are describing consists of a number of different frameworks, whose practitioners either do not know about relevant work by others, or are not even on speaking terms. We feel that this is an unfortunate situation, since much is to be gained by seeing the commonality of one area of research here. As we shall see, issues are often the same, and notions and techniques can be borrowed freely. Our paper is one such contribution toward a merge[1].

2 Epistemic Temporal Logic

This section describes the basic models for our study, whose typical interpretations are conversations or games. We are interested in how the agents' knowledge about the situation may change over time. Let Σ be a set of **events**. An event might be a move in some game, or a message sent from one agent to others. Not all agents need be aware of all events. Also, there is a global discrete clock, labelled by natural numbers, which agents may or may not be aware of. Agents do have a finite capacity to remember events, perhaps unbounded.

---

[1] We emphasize only main lines: cf. [42] for details, here and throughout this paper.

2.1 Epistemic temporal models and structural conditions

We first settle on some notation for the 'playgrounds'. Let Σ be any set of **events**. Given any set X, X^* is the set of finite strings over X and X^ω is the set of infinite strings over X. Elements of $\Sigma^* \cup \Sigma^\omega$ will be called **histories**. Given $H \in \Sigma^* \cup \Sigma^\omega$, $\mathsf{len}(H)$ is the **length** of H, i.e. the number of characters (possibly infinite) in H. Given $H, H' \in \Sigma^* \cup \Sigma^\omega$, we write $H \preceq H'$ if H is a *finite* prefix of H'. If $H \preceq H'$ we call H an **initial segment** of H' and H' an **extension** of H. Given an event $e \in \Sigma$, we write $H \prec_e H'$ if $H' = He$. Finally, let ϵ be the empty string and $\mathsf{FinPre}(\mathcal{H}) = \{H \mid \exists H' \in \mathcal{H} \text{ such that } H \preceq H'\}$ be the set of finite prefixes of the elements of $\mathcal{H}$ and $\mathsf{FinPre}_{-\epsilon}(\mathcal{H}) = \mathsf{FinPre}(\mathcal{H}) - \{\epsilon\}$.

DEFINITION 1. Let Σ be any set of events. A set $\mathcal{H} \subseteq \Sigma^* \cup \Sigma^\omega$ is called a **protocol** provided $\mathsf{FinPre}_{-\epsilon}(\mathcal{H}) \subseteq \mathcal{H}$. A **rooted protocol** is any set $\mathcal{H} \subseteq \Sigma^* \cup \Sigma^\omega$ where $\mathsf{FinPre}(\mathcal{H}) \subseteq \mathcal{H}$.

Intuitively, a protocol is the set of all possible ways an interactive situation may evolve. Given a protocol $\mathcal{H}$ and a finite history $H \in \mathcal{H}$, $\mathsf{Ext}_\mathcal{H}(H) = \{H' \mid H' \in \mathcal{H}, H \preceq H'\}$ is the set of extensions of H from $\mathcal{H}$. If no confusion arises, we write $\mathsf{Ext}(H)$ instead of $\mathsf{Ext}_\mathcal{H}(H)$. Also, $\mathsf{Ext}^{<\omega}(H)$ is the set of **finite extensions** of H and $\mathsf{Ext}^\omega(H)$ the **infinite extensions** of H. Given $t \in \mathbb{N}$ and a history H, H_t is the unique initial segment of H of length t.

Once the underlying temporal structure is in place, we can add the uncertainty of the agents. The most general models we have in mind are 'forests' with epistemic relations between finite branches.

DEFINITION 2. An **ETL frame** is a tuple $\langle \Sigma, \mathcal{H}, \{\sim_i\}_{i \in \mathcal{A}} \rangle$ where Σ is a (finite or infinite) set of events, $\mathcal{H}$ is a protocol, and for each $i \in \mathcal{A}$, $\sim_i$ is an equivalence relation on the set of finite strings in $\mathcal{H}$.

Making assumptions about the underlying event structure corresponds to "fixing the playground" where the agents will interact. The assumptions of interest are as follows: Let $\mathcal{F} = \langle \Sigma, \mathcal{H}, \{\sim_i\}_{i \in \mathcal{A}} \rangle$ be an *ETL* frame. If Σ is assumed to be finite, then we say that $\mathcal{F}$ is **finitely branching**. If $\mathcal{H}$ is a rooted protocol, $\mathcal{F}$ is a **tree frame**. We will be interested in **protocol frames** which satisfy both of these conditions. These are finitely branching trees with epistemic relations between the finite branches.

REMARK 3. *Three Equivalent Approaches*: There are at least two further approaches to uncertainty in the literature. The first, discussed in [29], represents agents' "observational" power. That is, each agent i has a set E_i of events it *can* observe[2]. For simplicity, we can assume $E_i \subseteq \Sigma$ but this is not necessary. A **local view** function is a map $\lambda_i : \mathsf{FinPre}(\mathcal{H}) \to E_i^*$. Given a finite history $H \in \mathcal{H}$, the intended interpretation of $\lambda_i(H)$ is "the sequence of events observed by agent i at H". The second approach comes from Fagin et al. [8]. Each agent has a set L_i of **local states** (if necessary, one can also assume a set L_e of environment states). Events e are tuples of local states (one for each agent) $\langle l_1, \ldots, l_n \rangle$ where for each $i = 1, \ldots, n$,

---
[2]This may be different from what the agent *does* observe in a given situation.

$l_i \in L_i$. Then two finite histories H and H' are i-equivalent provided the local state of the last of event on H and H' is the same for agent i. From a technical point of view, the three approaches to modeling uncertainty are equivalent ([27] provides the relevant intertranslations). However, they may still be different for modeling purposes.

2.2 Agent oriented conditions

Now we turn from the "playground" to the "players". Various types of agents place constraints on the interplay between the epistemic and temporal relations. We survey some conditions from the literature.

DEFINITION 4. Fix an epistemic temporal frame $\langle \Sigma, \mathcal{H}, \{\sim_i\}_{i\in\mathcal{A}}\rangle$. An agent $i \in \mathcal{A}$ satisfies the property **No Miracles** (sometimes called, somewhat misleadingly, **No Learning**) if for all finite histories $H, H' \in \mathcal{H}$ and events $e \in \Sigma$ with $He \in \mathcal{H}$ and $H'e \in \mathcal{H}$, if $H \sim_i H'$ then $He \sim_i H'e$.

Thus, unless a 'miracle' happens, uncertainty of agents cannot be erased by the same event. The next condition is the dual property.

DEFINITION 5. An agent $i \in \mathcal{A}$ satisfies the property **Perfect Recall** provided for all finite histories $H, H' \in \mathcal{H}$ and events $e \in \Sigma$ with $He \in \mathcal{H}$ and $H'e \in \mathcal{H}$, if $He \sim_i H'e$ then $H \sim_i H'$.

Perfect Recall means that the histories an agent considers possible can only decrease or remain the same, unless new indistinguishable events occur.

DEFINITION 6. An agent $i \in \mathcal{A}$ is **synchronized** provided for all finite histories $H, H' \in \mathcal{H}$, if $H \sim_i H'$ then $\mathsf{len}(H) = \mathsf{len}(H')$.

Intuitively, if an agent is synchronized, then that agent knows the value of the global clock (this may or may not be expressible in the formal language). For other assumptions that can be made about the interaction between the epistemic relation and time, the reader is referred to [8, 41]. Finally, note that in general we do not assume that all agents have the same reasoning capabilities. When they do, we say, for example, that a frame $\mathcal{F}$ is synchronous if all agents are synchronized.

2.3 Formal languages and truth in a model

Different modal languages can reason about the above structures (see the Handbook chapter [18]), with 'branching' or 'linear' variants. Here we give just the bare necessities.

Let At be a countable set of atomic propositions. We are interested in languages with various combinations of the following modalities: $P\phi$ (ϕ is true *sometime* in the past), $F\phi$ (ϕ is true *sometime* in the future), $Y\phi$ (ϕ is true at *the* previous moment), $N\phi$ (ϕ is true at *the* next moment), $K_i\phi$ (agent i knows ϕ) and $C_B\phi$ (the group $B \subseteq \mathcal{A}$ commonly knows ϕ). Dual operators are written as usual (eg., $\langle i \rangle \phi = \neg K_i \neg \phi$). If X is a sequence of modalities from $\{P, F, Y, N\}$ let $\mathcal{L}_n^X$ be the language with n knowledge modalities $K_1, \ldots, K_n$ together with the modalities from X. For a sequence of modalities X, $\mathcal{L}_C^X$ is the language $\mathcal{L}_n^X$ closed under the common knowledge modality C. Let $\mathcal{L}_{ETL}$ be the full epistemic temporal language, i.e., it contains all of the above temporal and knowledge operators.

Regardless of whether the language has *branching time* or *linear time* temporal operators, formulas express properties about finite histories. The difference lies in the format of the satisfaction relation. In a linear temporal setting, formulas are interpreted at pairs H, t where H is a 'maximal' (possibly infinite) history and t an element of $\mathbb{N}$. The intended interpretation of $H, t \models \phi$ is that *on the branch H at time t, ϕ is true.* In the branching time setting, we only need the moment, and formulas can be interpreted at finite histories H. In the interest of a unified approach we will interpret formulas at branch-time pairs. However, it will sometimes be useful to take the branching time interpretation. This helps draw parallels with results in temporal modal logic and products of modal logics [11].

DEFINITION 7. An **ETL model** based on an ETL frame $\langle \Sigma, \mathcal{H}, \{\sim_i\}_{i \in \mathcal{A}} \rangle$ is a tuple $\langle \Sigma, \mathcal{H}, \{\sim_i\}_{i \in \mathcal{A}}, V \rangle$ where V is a valuation $V : \mathsf{At} \to 2^{\mathsf{FinPre}(\mathcal{H})}$.

Formulas are interpreted at pairs H, t where $t \in \mathbb{N}$ and $H \in \mathcal{H}$ has length longer than t (finite or infinite). Truth for the languages $\mathcal{L}_n^X$ is defined as usual: see [8] and [18] for details. We only remind the reader of the definition of the knowledge and some temporal operators:

- $H, t \models P\phi$ iff there exists $t' \leq t$ such that $H, t' \models \phi$
- $H, t \models F\phi$ iff there exists $t' \geq t$ such that $H, t' \models \phi$
- $H, t \models K_i\phi$ iff for each $H' \in \mathcal{H}$ and $m \geq 0$ if $H_t \sim_i H'_m$ then $H', m \models \phi$

Of course, in addition to our epistemic temporal formulas, there are also the standard logical languages appropriate to these models, such as first-order logic, second-order logic, and other well-known systems.

3 Living at the Edge

Having set up our basic framework, we now want to demonstrate some key facts about the borderline between decidable and undecidable epistemic temporal logics. The previous section did highlight a number of dimensions which may lead undecidability, and even much higher complexity:

1. Expressivity of the formal language. Does the language include a common knowledge operator? A future operator? Both?

2. Structural conditions on the underlying event structure. Do we restrict to protocol frames (finitely branching trees)? Or forests?

3. Conditions on the reasoning abilities of the agents. Do the agents satisfy Perfect Recall? No Miracles? Synchronization?

Instead of setting up a huge grid of possible model classes and languages, we highlight a few major stages, including (in Section 4) one new highly undecidable epistemic tree logic. The main line of our observations is not all that new by itself, but our presentation and variety of sources is.

3.1 Purely temporal reasoning on protocol models

In this section we fix the underlying event structure and vary other dimensions. The Rabin Tree ([33]) consists of all finite sequences of events from a given finite set, with the binary relation of 'initial subsequence' plus successor functions taking a sequence H to He, for each $e \in \Sigma$.

THEOREM 8 (Rabin [33]). *The monadic second-order logic of the Rabin Tree is decidable.*

This landmark result explains the decidability of many modal and temporal logics, as first pointed out by Gabbay[3] [10]. It applies particularly well to our setting here, since the Rabin Tree has both points and branches, represented as special sets of points. Here is a well-known consequence:

THEOREM 9. *The satisfiability problem for $\mathcal{L}_{TL}$ with respect to TL tree models without epistemic structure is decidable.*

Proof. A formula ϕ involving finitely many events e is true in all protocol models if $\forall A(\text{'}subtree(A)\text{'} \Rightarrow (\phi)_A)$ is true on the corresponding Rabin Tree. Here $(\phi)_A$ is the syntactic relativization of ϕ to the unary predicate A, and '$subtree(A)$' says that A is closed under taking initial segments. ∎

A number of authors have noted that seemingly simple extensions to the Rabin tree language leads to undecidability. For example, Läuchli proved that the first-order theory of the Rabin tree expanded with a binary 'equilevel' predicate[4] for nodes is undecidable. Upon first inspection, this appears to be bad news for for the innocent assumption of synchronous communication. However, Thomas [37] provides a more fine-grained perspective: he shows that the monadic second-order theory of the Rabin Tree with an 'equilevel' predicate remains decidable provided that we let the second-order quantification run over *linear chains*, rather than arbitrary subsets. More succinctly: 'Path Logic' over the Rabin Tree with an equilevel predicate is decidable. Path Logic extends our temporal languages, since these talk about initial segments and extensions of the current finite history.

3.2 $\mathcal{L}_{ETL}$ over arbitrary models

First, consider arbitrary ETL tree models ('forests') and the full epistemic-temporal language $\mathcal{L}_{ETL}$. The logic remains simple. Indeed the 'fusion' of epistemic logic (**S5**) with common knowledge plus a complete temporal logic with past time operators (cf. [11]) will be such an axiomatization. This result is standard, so we only give some relevant details.

THEOREM 10. *The validity problem for arbitrary ETL frames is RE.*

Proof. (Sketch) Any non-theorem of the fusion of an epistemic logic and a temporal logic has a bimodal (Kripke) counter-model $\mathbb{M}$ with one accessibility relation for the temporal modalities and one for the epistemic modalities.

---

[3] Also relevant here is the emphasis in [45] on the *bounded tree property* as the source of decidability for temporal logics.

[4] That is, the nodes have the same distance from the root.

In order to generate a standard ETL model, we unravel the Kripke model *at each point*. This creates a forest where each tree is rooted by a state from the Kripke structure where we set $H \sim_i H'$ iff $\mathsf{last}(H) \sim_i \mathsf{last}(H')$ in $\mathbb{M}$. The relation from points s in $\mathbb{M}$ to histories with s their last element is a bisimulation. Thus the unraveled $\mathbb{M}$ is an ETL counter model. ∎

We do not know if this general logic is decidable, though we suspect that it is, by the general results on transfer of decidability for fusions of modal logics in Gabbay et al. [11], Kurucz [21].

ETL *tree* models will validate some principles not valid in the fusion of epistemic and temporal logic. The first is structural — tree models have a root. It is not hard to find axioms for this (see French, van der Meyden and Reynolds [9] for completeness theorems under this assumption). The second principle enforces that each agent knows the underlying protocol. The formula $\langle i \rangle \phi \to PF\phi$ says that any epistemic alternative is reachable in the tree by going down and moving up again.

THEOREM 11. *The satisfiability problem for the language $\mathcal{L}_{ETL}$ over ETL tree models is RE.*

Proof. (Sketch) The logic of ETL *tree* frames is the fusion of epistemic logic with common knowledge and temporal logic together with the principles discussed above. Starting with a Kripke counter model, we can unravel at the root only, making the above principle true in the model. ∎

Of course, behaviour of specific agents will take place in models satisfying additional epistemic-temporal constraints. As we will see in the next sections this can lead to high undecidability results.

3.3 Ideal epistemic agents have a highly undecidable tree logic

Let us now consider the usual idealizations of epistemic logic. For example, Agents have perfect memory, and seeing new events will not confuse them: that is, we have the above Perfect Recall, and No Miracles properties. The resulting interaction of temporal and epistemic structure makes trees look more like grids, and indeed, undecidability strikes. We highlight this result, because it is indicative of the 'danger zone' that we are in. The following result is one of many from a landmark publication:

THEOREM 12 (Halpern & Vardi [16]). *The validity problem for $\mathcal{L}_{ETL}$ on arbitrary ETL frames with No Miracles or Perfect Recall is Π_1^1 complete.*

In fact, these results hold whether or not one assumes that the frames are synchronous (see [16] for details). Essentially, these results show that if we fix the underlying event structure to be an ETL frame (i.e., a forest with arbitrary branching), then any practically any idealization lead to high undecidability as long as we are working in a language with common knowledge and arbitrary future modalities.

One may suspect that the Π_1^1-completeness is due to the underlying event structure and that things are better on event trees instead of forests. However, high complexity still strikes.

THEOREM 13 (Halpern & Vardi [16]). *The validity problem for $\mathcal{L}_{ETL}$ on ETL tree frames with Perfect Recall and No Miracles is Π_1^1-complete.*

On certain playgrounds, these idealizations turn out to be less dangerous.

THEOREM 14 (Halpern & Vardi [16]). *The validity problem for $\mathcal{L}_{ETL}$ with respect to ETL trees that satisfy the no miracles property is co-RE.*

Indeed, under synchronous communication the validity problem even becomes decidable (see [16] for details). These results indicate that when working in a language with both a common knowledge operator and arbitrary future modality there is an interesting interplay between structural assumptions about the underlying event structure and structural about the epistemic capabilities of the agents. Of course, for a full analysis of the situation we need to get our hands dirty and analyze the Π_1^1-completeness results. This will be the topic of Section 4.

This concludes our survey of typical results on decidability and undecidability over epistemic temporal tree models. Not surprisingly, the boundary has to do with the transition from mere trees to grid encoding using the additional epistemic structure. The epistemic setting adds some special flavor, however, in that the small differences which affect complexity represent very concrete assumptions about agents' capabilities, and what we can say about these. Moreover, we have shown how one can learn about relevant results from traditions that look prima facie quite different: epistemic temporal logic, tree languages in the foundations of computation, and (as we shall see in Section 5) current work on products of modal logics.

3.4 Bounded agents have a simple logic

Special agents may also have easier epistemic temporal logics. At the opposite extreme of Perfect Recall, agents with *bounded memory* have some finite bound to the number of preceding events which they can remember. Now, epistemic relations can be defined in terms of temporal ones.

THEOREM 15. *The epistemic temporal logic of memory bounded agents over arbitrary ETL frames is decidable.*

The key observation is that with a finite number of events, the modality $K_i \phi$ is definable. For convenience, we do the case of memory bound one:

$$K_i \leftrightarrow \bigvee_e (P_e \top \wedge U(P_e \top \rightarrow \phi))$$

where U is the *universal modality* and $P_e \top$ says the last event was e. The result follows the decidability for the purely temporal language.

4 High Undecidability on Trees

In the previous section we saw that, for a language with a common knowledge and a future operator, varying the underlying event structure and epistemic assumptions about the agents has drastic effects on the decidability of the logic. Now we investigate the tension between the underlying event structure, idealization of the agents and the formal language.

4.1 Tiling arguments

Imagine a finite set of tiles where each side has a different color. Let $\mathcal{T}$ be such a finite set of tile types and for $T \in \mathcal{T}$, let $right(T)$, $left(T)$, $up(T)$ and $down(T)$ be the colors of T. The *tiling problem* (for the first quadrant) asks if there a function $t : \mathbb{N} \times \mathbb{N} \to \mathcal{T}$ such that for each $n, m \in \mathbb{N}$

$$right(t(n,m)) = left(t(n+1,m))$$
$$up(t(n,m)) = down(t(n,m+1))$$

That is, can we place the tiles on the $\mathbb{N} \times \mathbb{N}$ plane so that the colors of the edges match. The function t is called a **tiling** of $\mathbb{N} \times \mathbb{N}$. Prima facie this problem looks highly complex (monadic Σ_1^1) as it asserts the existence of a function. However, by appealing to König's Lemma it can be seen to be Π_1^0: it is enough to show the existence of tilings of arbitrarily large *finite* planes. More formally, call any function $t^{(n)} : \{(i,j) \mid 0 \leq i \leq n, 0 \leq j \leq n\} \to \mathcal{T}$ that satisfies the above conditions (i.e., tiles match vertically and horizontally) a $(n \times n)$-**tiling** of the plane. Two tilings $t^{(n)}$ and $t^{(m)}$ are **consistent** if one extends the other. Thus each $(n \times n)$-tiling can be thought of as a sequence of partial *consistent* tilings.

LEMMA 16. *Suppose that for each $n > 0$, there is at least one (but only finitely many) partial tilings $t^{(n)}$. Then there is a tiling of the entire plane.*

However, David Harel showed [17] that small changes to the problem greatly increases the complexity. For example, the **recurrent tiling** problem asks, given a set of tiles $\mathcal{T}$ with a distinguished tile $T_1 \in \mathcal{T}$, if there is a tiling t such that T_1 occurs infinitely often in the first row.

THEOREM 17 ([17]). *The recurrent tiling problem is Σ_1^1-complete.*

Thus if there is a formula in the desired language that is satisfiable iff there is a recurrent tiling of the plane, then the satisfiability problem with respect to that language (on the relevant frames) is Σ_1^1-complete. For concreteness, assume that $\mathcal{T} = \{T_1, \ldots, T_k\}$ is a finite set of tiles and $t_1, \ldots, t_k$ is a set of propositional variables.

4.2 A PDL-style tree language

In this section we will use a **PDL**-style language which capture features of both linear and branching time languages, and which refers explicitly to events. Let $\mathcal{A}$ be a (finite) set of agents and recall that Σ is a (finite) set of events. Define $\mathcal{L}_\Sigma(\mathcal{A})$ inductively as follows:

$$\phi := p \mid \neg\phi \mid \phi \wedge \psi \mid \langle \alpha \rangle \phi$$

$$\alpha := a \mid ?p \mid \alpha;\beta \mid \alpha \cup \beta \mid \alpha^*$$

where $p \in At$, $a \in \Sigma \cup \mathcal{A}$ and $\sigma \in \Sigma$. Let $\mathcal{L}_\Sigma(\mathcal{A})^-$ be the language $\mathcal{L}_\Sigma(\mathcal{A})$ which allows expressions of the forms $\langle \sigma^- \rangle \phi$.

This language is (strictly) stronger than those described above, as we allow mixing of temporal and epistemic steps under the scope of the *-

operator. For example, $\langle(i;e)^*\rangle\phi$ is a well-formed expression of the above grammar, whereas it is not an element of $\mathcal{L}_{ETL}$.

Before defining truth in a model we introduce a relation R_α on the set $\mathsf{FinPre}(\mathcal{H})$, where α is defined by the above grammar. Let H, H' be finite sequences of events and V a valuation (assigning sets of atomic propositions to finite sequences). Suppose $\sigma \in \Sigma$ and $i \in \mathcal{A}$.

- $HR_\sigma H'$ iff $H' = H\sigma$ if $\sigma \in \Sigma$
- $HR_i H'$ iff $H \sim_i H'$
- $HR_{\sigma^-} H'$ iff $\mathsf{len}(H) \geq 1$ and $H = H'\sigma$
- $HR_{?p} H'$ iff $H = H'$ and $p \in V(H)$.

Clauses for the PDL operators are as usual. Truth is also defined as usual, we only give the definition of the modal operator:

- $H, t \models \langle\alpha\rangle\phi$ iff there exists $H' \in \mathcal{H}$ and $m \in \mathbb{N}$ such that $H_t R_\alpha H'_m$ and $H', m \models \phi$

Under the assumption that there are only finitely many events and using a well-known translation of epistemic logic into PDL (with a converse operator), we see that $\mathcal{L}_{ETL}$ is a fragment of $\mathcal{L}_\Sigma(\mathcal{A})$. We will write $G\phi$ for $[(\cup_{e \in \Sigma} e)^*]\phi$ and $C\phi$ for[5] $[(\cup_{i \in \mathcal{A}} i)^*]\phi$.

4.3 High complexity over arbitrary ETL frames

We first reprove one of Halpern and Vardi's results from [16] using a tiling argument. [16] use a reduction of the *recurrent Turning machine problem*. They comment that a tiling argument "cannot be straightforwardly applied" in their setting (p. 208). Our argument works thanks to our formulation of the No Miracles and Perfect Recall properties.

THEOREM 18 (Halpern & Vardi [16]). *The validity problem for the $\mathcal{L}_{ETL}$ fragment of $\mathcal{L}_\Sigma(\mathcal{A})$ on finitely branching ETL frames with No Miracles (with at least two agents) is Π_1^1 complete.*

The first step in any tiling argument is to identify a *universal* modality. The combination of the universal temporal and the common knowledge operator (GC) will serve this purpose. The second step is to encode a grid.

The 'x-axis' will be encoded by occurrences of a distinguished event $e \in \Sigma$. The formula $\phi_1 := GC\langle e\rangle\top$ says that each accessible finite history has an extension consisting of an infinite sequence of e's. As in [16], the epistemic relations encode the 'y-axis'. Let p be a new propositional variable. Consider the following two formulas: $\phi_2 := GC((p \rightarrow Gp) \wedge (\neg p \rightarrow G\neg p))$ and $\phi_3 := GC(p \rightarrow \langle 1\rangle p) \wedge (\neg p \rightarrow \langle 2\rangle p)$. If $H, t \models \phi_2 \wedge \phi_3$, then we can think

---

[5] Of course this only works if there are finitely many events and finitely many agents. If there are not finitely many events, we assume that F is a primitive operator defined as in Section 2.3. Furthermore, note that we do not need the converse operator here, since we are assuming that the agent's accessibility relations are equivalence relations.

of the histories reachable from H_t as being labeled by p and $\neg p$. Furthermore, there are 1-accessibility relations between from p to $\neg p$ histories and 2-accessibility relations from $\neg p$ to p histories. Thus an 'up-step' is represented by the program $\alpha_u := (?p; 1; ?\neg p; 2)$. Now, the No Miracle property imposes a grid condition on the relations R_e and R_{α_u}.

LEMMA 19. *Suppose that $\mathcal{M}$ is an arbitrary ETL model with no miracles and $H, t \models \phi_1 \wedge \phi_2 \wedge \phi_3$. If H_1, H_2 and H_3 are finite histories reachable from H_t, $H_1 R_{\alpha_u} H_2$ and $H_1 R_e H_3$, then there is an H_4 such that H_4 is reachable from H_t, $H_3 R_{\alpha_u} H_4$ and $H_2 R_e H_4$.*

To complete the proof of Theorem 18, we find a formula that is satisfiable iff there is a recurrent tiling of the plane. The next section sketches how to do this in an analogous case.

4.4 High complexity over ETL protocol frames

Halpern & Vardi mainly consider models where the initial model may be infinite, or there may be infinite branching. In this case, even the 'unmixed' language of Section 2.3 above led to undecidability with No Miracles or Perfect Recall. In this section, we consider finitely branching trees.

Our goal in this section is to sketch a proof of the following theorem.

THEOREM 20. *The satisfiability problem of $\mathcal{L}_\Sigma(\mathcal{A})$ with respect to ETL protocol frames that satisfy No Miracles is Σ_1^1-complete.*

For concreteness, assume $\Sigma = \{l, r\}$ and $\mathcal{A} = \{1, 2\}$. We must find a formula $\phi_\mathcal{T}$ that is satisfiable iff there is a recurrent tiling of $\mathbb{N} \times \mathbb{N}$ using the tiles from $\mathcal{T}$. We begin by describing the formula $\phi_\mathcal{T}$. The formula $\phi_\mathcal{T}$ consists of three parts: 1. a formula which forces the extensions of a finite history to have a particular structure, 2. a formula which forces a grid structure and 3. a formula which places tiles on the grid.

To that end, let ϕ_S be the conjunction of the following formulas: Only $r^* - l^*$ paths: $[r^*; l; l^*] \neg \langle r \rangle \top$; infinite l-paths: $[r^*; l^*] \langle l \rangle \top$; Infinite r-path: $[r^*] \langle r \rangle \top$; Even p paths: $[(r; r)^*][l^*]p$; and Odd $\neg p$ paths: $[r; (r; r)^*][l^*] \neg p$. Then if $H, t \models \phi_S$, the extensions of H_t can be pictured as follows:

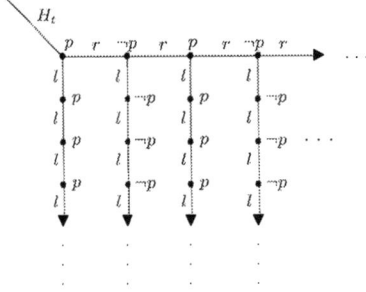

This model represents half of the $\mathbb{N} \times \mathbb{N}$ grid. The idea is to think of the infinite r-path as the y-axis and the first infinite l-path as the x-axis (the fact that the truth value of p alternates between the paths will be used below). We now show how to force the second half of the grid. That is, we need a formula that will be satisfied if there are infinitely many infinite "up" paths. The trick will be to consider the following program:

$$\alpha_u := ?p; l; 1; ?\neg p; l; 2; ?p.$$

Making a step of the above program corresponds to making a 'zig-zag' move through the tree between points which will cross between different branches.

Let ϕ_u be the conjunction of the following two formulas

1. "Up moves" are always possible: $[r^*; l^*]((p \to \langle 1 \rangle \neg p) \wedge (\neg p \to \langle 2 \rangle p))$

2. There are infinitely many "Up moves": $[l^*][\alpha_u^*]\langle \alpha_u \rangle \top$

Finally, we place the tiles on our tree.

1. One tile at each node: $\phi_1 := [(r;r)^*; l^*](\bigvee_{i=1}^k t_i \wedge \bigwedge_{1 \leq i < j \leq k} \neg(t_i \wedge t_j))$

2. Place tiles going across: $\phi_2 := [(r;r)^*; l^*](\bigvee_{right(T_i)=left(T_j)} (t_i \wedge \langle l \rangle t_j))$

3. Place tiles going up: $\phi_3 := [(r;r)^*; l^*](\bigvee_{up(T_i)=down(T_j)} (t_i \wedge \langle \alpha_u \rangle t_j))$

4. Enough tiles: $\phi_4 := [(r;r)^*; l^*](\bigwedge_{up(T_i) \neq down(T_j)} (t_i \to \neg \langle \alpha_u \rangle t_j))$

Let $\phi_\mathcal{T} := \phi_S \wedge \phi_u \wedge \phi_1 \wedge \phi_2 \wedge \phi_3 \wedge \phi_4$.

We first note that any tiling of $\mathbb{N} \times \mathbb{N}$ induces a model for $\phi_\mathcal{T}$. The key idea is to remove all vertical lines from the grid and treat the remaining structure as a tree rooted at $(0,0)$. Next, as in the previous section, the No Miracles property imposes a grid structure on the relations H_l and H_{α_u}.

LEMMA 21. *On epistemic temporal frames $\langle \mathcal{H}, \{\sim_i\}_{i \in \mathcal{A}} \rangle$ with No Miracles, if HR_lH' and $HR_{\alpha_u}H''$, there is a H''' with $H''R_lH'''$ and $H'R_{\alpha_u}H'''$.*

Proof of Theorem 20 We need only show that under the assumption of No Miracles, there is a tiling of the plane from a model for $\phi_\mathcal{T}$. Indeed, we find a function f from $\mathbb{N} \times \mathbb{N}$ into $\mathsf{Ext}^{<\omega}(H_t)$ such that

- $t(n,m) = T_i$ iff$_{\text{def}}$ $f(n,m) \models t_i$.

is a tiling of $\mathbb{N} \times \mathbb{N}$. We will show that such a function can be extracted from a satisfying model. First of all, since $H, t \models \phi_1$, f is well-defined. Start by defining $f(0,0) = T$ where $H, t \models t$. Once $f(n,m)$ has been defined, we can define $f(n+1,m)$ and $f(n,m+1)$:

LEMMA 22. *Let $H, t \models \phi_\mathcal{T}$, $(n,m) \in \mathbb{N} \times \mathbb{N}$, $f(n,m) = H_tH'$ and $H_tH' \models t_i$. Then there are finite extensions of H_t, H^r and H^u, with (1) $H_tH^r \models t_j$, and $right(T_i) = left(T_j)$, and (b) $H_tH^u \models t_k$, and $up(T_i) = down(T_k)$.*

All that remains is to "complete the square". Let $f(n,m)$, $f(n,m+1)$ and $f(n+1,m)$ have finite extensions of H_t as in the Lemma 22: $H_{(n,m)}$, $H_{(n,m+1)}$ and $H_{(n+1,m)}$ respectively, with matching properties for a tiling. Then there is a finite extension $H_{(n+1,m+1)}$ of H_t with the right properties.

LEMMA 23. *Suppose that $H_{(n,m)}, H_{(n,m+1)}$ and $H_{(n+1,m)}$ have been defined as above. Then there is a finite history $H_{(n+1,m+1)} \in \text{Ext}^{<\omega}(H_t)$ such that there is a unique tile proposition with $H_{(n+1,m+1)} \models t$ with $right(T_j) = left(T)$ and $up(T_k) = down(T)$.*

Finally, we must show that there is a formula such that, if satisfiable, implies that a particular tile occurs infinitely often along the x-axis. Let $T_0 \in \mathcal{T}$ be a tile and t_0 the corresponding propositional variable. Consider the formula $\phi_{t_0} := [l^*]\langle l; l^* \rangle t_0$ Note that $H, t \models \phi_{t_0}$ implies that t_0 is true infinitely often on the first branch extending H_t. This proves Theorem 20.

Note that we have focused only on the assumption of No Miracles. For systems satisfying Perfect Recall alone epistemic uncertainty is propagated backwards towards the root. Thus, the above arguments will not work. However, by appealing to Lemma 16, we need only show that for each $n \in \mathbb{N}$ there is a tiling of the finite $n \times n$ plane (cf. [42]).

THEOREM 24. *The satisfiability problem with respect to synchronous ETL protocol frames that satisfy perfect recall is Σ_1^1-complete.*

4.5 Synchronized agents on ETL trees

In more special settings we can still get positive results from other areas:

THEOREM 25. *The logic of $\mathcal{L}_C^{P,F}$ over synchronous trees is decidable.*

Proof. Thomas [37] embeds Path Logic into the monadic second-order theory of the Rabin tree, by sending chains to pairs of subsets (A, B) where A encodes the left-most branch on which the chain lies, while B encodes which nodes are on the chain. More precisely, B 'goes left' at levels not represented on the chain, and it 'goes right' at levels where the chain has a node. The equilevel predicate for two nodes is then expressed by saying that they are one-element chains, whose B-sequences go right at the same place. Now, our epistemic relations are subrelations of the 'equilevel' predicate, which latter corresponds roughly to the transitive closure of their union. We can encode this into Path Logic by modifying the chain representation. ■

5 Dynamic Epistemic Logic

Our take on epistemic temporal logic is within the tradition of Fagin et al. [8] and Parikh & Ramanujam [29, 30]. One current paradigm which diverges from these, though addressing similar phenomena, is 'dynamic epistemic logic' ('DEL', [13, 31, 4]). Here, epistemic actions such as announcing a true proposition, or more complex forms of communication and observation, are encoded explicitly in 'action models' $\mathbb{A}$ consisting of the relevant events and the preconditions for their occurrence, plus agents' epistemic relations over these, representing their partial powers of observation. In ur-DEL, preconditions for events are defined by purely epistemic formulas. What agents learn in such a setting, given some current epistemic model $\mathbb{M}$, is encoded by a new 'product model' $\mathbb{M} \times \mathbb{A}$, where agents are uncertain between worlds (s, e) and (t, f) iff they were uncertain between both the old worlds s, t and the observed events e, f. The language for these models has the usual epistemic operators, plus dynamic modalities $\langle \mathbb{A}, e \rangle \phi$:

$$\mathbb{M}, s \models \langle \mathbb{A}, e \rangle \phi \text{ iff } \mathbb{M} \times \mathbb{A}, (s, e) \models \phi$$

The resulting logic is decidable, and it revolves around 'reduction axioms' for compositional analysis of effects of epistemic events. E.g., a typical reduction axiom analyzes agents' knowledge after public announcement:

$$[!P] K_i \phi \leftrightarrow (P \rightarrow K_i [!P] \phi)$$

For more precise definitions and complete dynamic-epistemic logics, cf. van Benthem [40], Baltag & Moss [3], van Benthem [39], van Benthem, van Eijck & Kooi [43], and van Ditmarsch, van der Hoek & Kooi [44].

Prima facie, DEL does not look like ETL: there is no explicit mention of time. And unlike DEL, ETL does not explicitly describe the events that make up its models. But appearances are misleading, and a 'convergence' is easy to find, making mutual borrowing easy and natural.

5.1 Representing DEL models inside ETL models

Product update involves three major ingredients with a logical 'reflection', as was first observed in van Benthem [40]:

(a) Product update implies Perfect Recall: $(x, a) \sim_i y$ implies there exists a b such that $y = (y', b)$ and $x \sim_i y'$, and

(b) 'No Miracles' holds uniformly: if $(x, a) \sim_i (y, b)$, then, whenever $u \sim_i v$, also $(u, a) \sim_i (v, b)$ if the latter events can occur.

Now, any initial model $\mathbb{M}$ and action model $\mathbb{A}$ induce a natural epistemic model $Tree(\mathbb{M}, \mathbb{A})$. Nodes are finite sequences of events, and the iterated epistemic products with $\mathbb{A}$ are the horizontal levels of the tree. Events only take place when their precondition is satisfied. $Tree(\mathbb{M}, \mathbb{A})$ is a 'forest model', whose initial epistemic model at the root can be arbitrary. (Van Benthem [39], Sadzik [36] explore the special case of trees with finitely many events, looking for epistemic bisimulations between different finite levels.)

The epistemic decoration of the tree models $Tree(\mathbb{M}, \mathbb{A})$ is rather special, since it obeys the above three constraints. Indeed, van Benthem & Liu [41] prove the following representation result[6]:

THEOREM 26. *An epistemic tree model $\mathcal{M}$ is bisimilar to a model of the form $Tree(\mathbb{M}, \mathbb{A})$ if and only if it satisfies (a) Perfect Recall, (b) Uniform No Miracles, and (c) for any event e, the set of nodes where e can take place is closed under epistemic bisimulations inside $\mathbb{M}$.*

Thus, product update may be viewed as a special epistemic temporal logic. Indeed, we can unpack the above conditions to the usual axioms

---

[6]Theorem 26 suggests a broader representation theory in a epistemic-temporal universe by varying its assumptions on perfect agents and uniform event models.

for Perfect Recall ($K_i[a]\phi \to [a]K_i\phi$) and No Miracles, where the uniform version (c) requires the use of universal modalities $E\phi$, $U\phi$ stating that ϕ holds at some world, at all worlds, resp. (cf. [6]):

$$E(\langle a^-\rangle\top \wedge \neg K_i\neg\langle b^-\rangle\top) \to U(\langle a^-\rangle\neg K_i\neg\phi \to K_i[b^-]\phi)$$

Thus, dynamic-epistemic logic describes special idealized agents on epistemic temporal trees, much as discussed in preceding sections. But DEL has further special features. E.g., No Miracles is a much more plausible way of propagating ignorance than the usual No Learning, which seems to say that the passage of time never helps increase knowledge. Here, DEL gives a deeper analysis of the processes that drive information change, instead of merely describing the Grand Stage where all agents live in time. Even so, in the light of Section 3, the very decidability of DEL calls for explanation!

5.2 Why is *basic* DEL decidable? The ETL answer.

In the setting of Theorem 27, DEL-style languages are fragments of our earlier epistemic temporal ones:

THEOREM 27. *Over models $Tree(\mathbb{M}, \mathbb{A})$, DEL is the ETL language of epistemic logic plus one-step future operators $\langle e \rangle$.*

Thus, we do not reach the expressive power needed for undecidability in our earlier arguments. Indeed, for standard DEL, the reduction axioms translate every formula into an equivalent one without action modalities, and one then uses the decidability of the purely epistemic language. But something stronger holds. Let events have preconditions in the epistemic temporal language, as in communication scenarios which refer to the past of the current conversation. In its most blunt form, the precondition for e to occur is then just this: $\langle e \rangle \top$.

THEOREM 28. *The logic of the epistemic temporal language with only operators $\langle e \rangle$ on models with Perfect Recall and No Miracles is decidable.*

Proof. This can be derived from the decidability result for the modal 'product logic' **PDL** × $\mathbf{K}_m$ in Gabbay, Wolter et al. [11], by embedding our language into it. The models for this logic may be viewed as grids with an 'epistemic' **PDL** direction and a 'temporal' one-step **K**-direction. Still, no embedding of tiling problems is possible, because the language does not contain a true universal modality or transitive closure modality accessing all points of the grid (cf. also Marx & Mikulas [23]). ∎

Thus DEL can be stronger than it is now, and still stay decidable. To get this positive result, we have appealed to one more tradition in the area of epistemic temporal structure, viz. combining logics through *product constructions*. This model-constructing approach fits quite well with DEL.

5.3 Program structure, true future, and undecidability

DEL style logics do become undecidable when the complete future is added. This would happen, e.g., if one adds sequential structure to action models, modeling, say, conversational processes involving composition and iteration. The landmark paper Miller & Moss [25] proves many relevant results, including this surprising effect of combining two decidable logics:

THEOREM 29 (Miller & Moss). *The dynamic-epistemic logic of public announcement with program iterations is undecidable.*

This shows that the undecidability phenomena already noted in Halpern & Vardi for the language with common knowledge and true future even occur in very restricted settings, where events are just announcements. Indeed, Miller & Moss show that iterated announcement of one single proposition $\Diamond\top$ suffices. However, in the light of our ETL-based Section 3, their analysis also leaves open questions. One of them has to do, again, with our assumption of finite levels. It is unknown whether their undecidability results hold when initial models are assumed to be finite.

5.4 Decidable fragments; tomorrow and yesterday

Our epistemic temporal analysis also suggests various additions to the standard language of DEL which still remain decidable. One typical illustration is the addition of the temporal past of an epistemic process (van Benthem [39]). Yap [46] and Sack [35] analyze such additions, and propose valid axioms. Here is the general setting.

THEOREM 30. *The logic of the epistemic temporal language with one-step future operators $\langle e \rangle$ and one-step past operators $\langle e^- \rangle$ on models with Perfect Recall and No Miracles is decidable.*

Proof. This follows from a simple modification of the above-mentioned decidability proof for **PDL** × **K** given in Gabbay, Wolter et al. [11]. ∎

We conjecture that decidability still holds when we add many-step Past operators, at least, on our rooted finite-event trees. Such logics can express what agents knew earlier on, but they can also state preconditions for events that reach back in time, such as "say P if you have not already done so".

5.5 Protocols and model constructions

But there are still further features to the comparison of DEL and ETL. First, it is sometimes claimed that DEL lacks an essential resource available in our epistemic temporal models, viz. the choice of a *protocol*, i.e., a set of 'relevant histories'. Now, this is not true, since the above models $Tree(\mathbb{M}, \mathbb{A})$ do have explicit restrictions on their available runs, since events can only occur when their precondition is satisfied. Thus, DEL has an explicit calculus for preconditions, as these are encoded in the action models, and these are again available inside the formal language through the modalities $\langle \mathbb{A}, e \rangle \phi$. On the other hand, given the special epistemic format for these preconditions, one can only define special protocols, via local restrictions, that must be stated in a purely epistemic language. A more general approach here would merge the two ideas. On the one hand, it seems a good idea to make the protocols explicit in the language, as DEL does. On the other hand, one needs a richer repertoire of definitions for realistic protocols, including temporal operators in their formulation. This is achieved in the following 'Logic of Protocols' (cf. the earlier $\mathcal{L}_\Sigma(\mathcal{A})$):

We first introduce a **PDL** style language with 'protocols' defined by:

$$e \mid \alpha;\beta \mid \alpha \cup \beta \mid \alpha^* \mid \text{skip} \mid \phi?$$

with $e \in \Sigma$ an event and ϕ a formula of $\mathcal{L}_{ETL}$. For example, the protocol $(e \cup \text{skip})^*$ represents the set of histories that contain the event e. A test-free protocol α is a regular expression; and so it represents the set of histories that match α. A more interesting example is a 'Liar Protocol'. Let $\text{send}(i, B, p)$ be the event "agent i sends the message p to the group of agents B", i.e., "i announces p to the set of agents B". Then $((K_i \neg p?; \text{send}(i, \mathcal{A}, p)) \cup (K_i p?; \text{send}(i, \mathcal{A}, \neg p)) \cup \text{skip})^*$ represents a liar protocol. That is, if i knows p then i publically announces $\neg p$, if i knows $\neg p$ then i publically announces p, or i does not say anything.

For each protocol α introduce a modal operator N_α to the language. The intended interpretation of $N_\alpha \phi$ is that ϕ is true at the next moment in *all* extensions of the current history compatible with the protocol α. Thus truth is defined as $H, t \models N_\alpha \phi$ if $H' \in \mathcal{H}, H \preceq_\alpha H'$ and $H', t \models \phi$, where $\preceq_\alpha$ is an extension relation much like the previously defined R_α relations. This addition, though very useful in practice, is arguably a matter of convenience:

THEOREM 31. *ETL with explicit protocols is no more expressive than ETL by an effective translation.*

Introducing explicit protocols is also akin to the use of 'knowledge programs' in Fagin et al. [8]. We forgo the precise connection here.

Our conclusion is that older and newer approaches to dynamic actions and epistemic logic all meet in the same arena of epistemic temporal logic, and that insights can be transferred in illuminating ways.

6 Logics for Model Change

DEL is a calculus of piece-meal model construction, while ETL assumes the 'playground' or Grand Stage has been given as a temporal universe of all possible histories. An update $!P$ is then a minimal move to some available future state where one knows that P- and likewise for belief revision. In this setting, no definable explicit construction takes place for 'the next model' as in DEL - and it is the externally supplied temporal model which decides where things can go. This view is also that of *extensive games*.

As a logic of model construction, DEL, fits well with other modal logics, such as the London-style product logics mentioned before. But note that product update is more like 'direct product', and this shows that there is a large variety of possible constructions. One area where this has been studied in great generality, close to modal bisimulation-based paradigms, is *Process Algebra*, which deals with 'composition', 'choice', 'parallel composition', and 'iteration'. PA gives mainly algebraic calculi for process equivalence, but adding a temporal language would be a very natural step, allowing one to discuss what running a process actually achieves over time. Moreover, adding an epistemic component to current process theories would also make sense, witness the central role of *communication* between processes, as in Milner's recent work ([26]).

Model construction is also found in *game semantics* for logics and programming languages, cf. [28] and [1]. Here operations forming new games reflect the interactions between players. Interestingly, key structures in this case are again branching tree models. Neither Parikh nor Abramsky introduce ETL languages to talk about games, but it makes just as much sense as for the process theory. We can then describe players' activities over time, and define many of their strategies explicitly. As for complexity: the main concern in this paper, basic game logics tend to be decidable - but certain repertoires lead to undecidability: witness the ! of linear logic. We cannot go into details in the context of this paper, but again, we think that the parallels are striking, and worth systematic investigation.

7 Conclusions

This paper shows that epistemic temporal models are a natural meeting place for logicians. In Section 2, we defined basic structures that recur in most major studies of agents' interaction and information. In Section 3, we discussed the decidable/undecidable boundary where many interesting issues live concerning diversity of agent behaviour. This led to a natural merge of insights from different traditions: epistemic temporal logics in computer science, logics of computation, modal logics of products, dynamic-epistemic logics - and eventually also, process algebra and game semantics.

Concerning the relation between all these meeting frameworks, our view is this. Epistemic logics in the style of Fagin et al, and Parikh et al. are largely the same, even up to mutual mathematical representation (cf. [27]). The link between these systems and dynamic-epistemic logics is a bit more complex, but Section 4 has shown some natural merges. Thus, playing up differences between these approaches as different 'paradigms' seems both pointless from a mathematical viewpoint, and harmful from a conceptual or a practical point of view, as it impedes mutual flow of ideas.

Indeed, many further examples of mutual traffic can be found, which we had to leave out here. E.g., we have new results on more exciting scenarios where bounded and ideal agents meet. And, the explicit treatment of model constructions in dynamic-epistemic logic suggests new forms of *process algebra* for agents that can display rational intelligent behavior.

Thus, our approach suggests a turn from competition between frameworks to cooperation. Compare the situation in the 1930s, when many different models were proposed for computation. Instead of creating different churches, logicians started looking for similarities and equivalences (at some appropriate level), and the result was Church's Thesis, usually taken to mean that the field had a stable and mathematically respectable topic. Likewise, convergence, if not downright equivalence, between 'epistemic temporal logics' of agents might signal to a broader world that there is a core notion of genuine interest here concerning 'intelligent interaction', rather than a set of warring religions. Seeing differences may make for short-term gains, seeing analogies leads to a long-term common cause.

BIBLIOGRAPHY

[1] S. Abramsky and R. Jagadeesan. Games and full completeness for multiplicative linear logic. In R. Shyamsundar, editor, *Foundations of Software Technology and Theoretical Computer Science*, pages 291–301. Springer-Verlag, 1992.

[2] R. Alur, T. A. Henzinger, and O. Kupferman. Alternating-time temporal logic. *Journal of the ACM*, 2002.

[3] A. Baltag and L. Moss. Logics for epistemic programs. *Synthese: Knowledge, Rationality, and Action*, 2:165 – 224, 2004.

[4] A. Baltag, L. Moss, and S. Solecki. The logic of public announcements, common knowledge and private suspicions. In *Proceedings of TARK 1998*, 1998.

[5] N. Belnap, M. Perloff, and M. Xu. *Facing the Future*. Oxford University Press, 2001.

[6] P. Blackburn, M. de Rijke, and Y. Venema. *Modal Logic*. Cambridge University Press, Cambridge, 2002.

[7] E. A. Emerson. Temporal and modal logics. In J. van Leeuwen, editor, *Handbook of Theoretical Computer Science*. 1990.

[8] R. Fagin, J. Halpern, Y. Moses, and M. Vardi. *Reasoning about Knowledge*. The MIT Press, Boston, 1995.

[9] T. French, R. van der Meyden, and M. Reynolds. Axioms for logics of knowledge and past time: Synchrony and unique initial states. In *Proceedings of AiML 2005*. King's College Press, to appear. appeared in the pre-proceedings of the Conference on Advances in Modal Logic, Manchester, Sept 2004.

[10] D. Gabbay. *Investigations into modal and tense logics, with applications to problems in linguistics and philosophy*. Reidel Dordrecht, 1976.

[11] D.M. Gabbay, A. Kurucz, F. Wolter, and M. Zakharyaschev. *Many-Dimensional Modal Logics: Theory and Applications*. Elsevier, 2003.

[12] D. Gabelaia, A. Kurucz, F. Wolter, and M. Zakharyaschev. Products of 'transitive' modal logics. *Journal of Symbolic Logic*, 70(3), 2005.

[13] J. Gerbrandy. *Bisimulations on Planet Kripke*. PhD thesis, ILLC, 1998.

[14] V. Goranko and G. van Drimmelen. Complete axiomatization and decidability of alternating-time temporal logic. *Theoretical Computer Science*, to appear 2006.

[15] Joseph Halpern. A computer scientist looks at game theory. *Games and Economic Behavior*, 45(1):114 – 131, 2003.

[16] Joseph Halpern and Moshe Vardi. The complexity of reasoning about knowledge and time. *J. Computer and System Sciences*, 38:195 – 237, 1989.

[17] David Harel. Recurring dominoes: Making the highly undecidable highly understandable. *Annals of Discrete Mathematics*, 24:51 — 72, 1985.

[18] I. Hodkinson and M. Reynolds. Temporal logic. In *Handbook of Modal Logic*. forthcoming.

[19] I. Hodkinson, F. Wolter, and M. Zakharyaschev. Decidable and undecidable fragments of first-order branching temporal logics. In *Proc. 17th Annual IEEE Symposium on Logic in Computer Science (LICS 2002)*, pages 393–402.

[20] J. Horty. *Agency and Deontic Logic*. Oxford University Press, 2001.

[21] A. Kurucz. *Handbook of Modal Logic*, chapter Combining Modal Logics, page forthcoming. Elsevier, 2006.

[22] M. Marx. *Handbook of Modal Logic*, chapter Complexity of Modal Logics, page forthcoming. Elsevier, 2006.

[23] M. Marx and S. Mikulas. Products, or how to create modal logics of high complexity. *Logic Journal of the IGPL*, pages 77 –88, 2001.

[24] M. Marx and Y. Venema. Local variations on a loose theme: modal logic and decidability. In M. Vardi and S. Weinstein, editors, *Finite-Model Theory and Its Applications*, Texts in Theoretical Computer Science. Springer, 2005.
[25] J. Miller and L. Moss. The undecidability of iterated modal relativization. *Studia Logica*, 79(3), 2005.
[26] R. Milner. *Communicating and Mobile Systems: The Pi Calculus*. MIT Press, 1999.
[27] E. Pacuit. Some comments on history based structures. *Journal of Applied Logic*, forthcoming.
[28] R. Parikh. The logic of games and its applications. *Annals of Disc. Math.*, 1985.
[29] Rohit Parikh and R. Ramanujam. Distributed processes and the logic of knowledge. In *Logic of Programs*, volume 193 of *Lecture Notes in Computer Science*, pages 256 – 268. Springer, 1985.
[30] Rohit Parikh and R. Ramanujam. A knowledge based semantics of messages. *Journal of Logic, Language and Information*, 12:453 – 467, 2003.
[31] J. Plaza. Logics of public communications. In *Proceedings 4th International Symposium on Methodologies for Intelligent Systems*, 1989.
[32] A. Pnueli. The temporal logic of programs. In *Proc. 18th Symp. Foundations of Computer Science*, pages 46 — 57, 1977.
[33] M. O. Rabin. Decidability of second-order theories and automata on infinite trees. *Transactions of the American Mathematical Society*, 141:1 – 35, 1969.
[34] M. Reynolds and M. Zakharyaschev. On the products of linear modal logics. *Journal of Logic and Computation*, 11(6):909 – 931, 2001.
[35] J. Sack. Temporal language for epistemic programs. Unpublished manuscript, Indiana at Bloomington, 2006.
[36] T. Sadzik. Exploring the update universe. ILLC Report, Amsterdam, 2006.
[37] W. Thomas. Infinite trees and automaton definable relations over omega-words. *Theoretical Computer Science*, 103(1):143 – 159, 1992.
[38] Richmond Thomason and Anil Gupta. A theory of conditionals in the context of branching time. *The Philosophical Review*, 80:65–90, 1980.
[39] J. van Benthem. One is a lonely number. To appear in Logic Colloquim 2002, Editors S. Buss et al., AMS Publications.
[40] J. van Benthem. Games in dynamic epistemic logic. *Bulletin of Economic Research*, 53:216 – 248, 2001.
[41] J. van Benthem and F. Liu. Diversity of logical agents in games. *Philosophia Scientiae*, 8(2):163 – 178, 2004.
[42] J. van Benthem and E. Pacuit. The tree of knowledge in action: Towards a common perspective. Technical report, ILLC, University of Amsterdam, 2006.
[43] J. van Benthem, J. van Eijck, and B. Kooi. Logics of communication and change. In *Proceedings of TARK 2005*, 2005.
[44] H. van Ditmarsch, W. van der Hoek, and B. Kooi. *Dynamic Epistemic Logic*. Springer, forthcoming.
[45] M. Vardi. Why is modal logic so robustly decidable? DIMACS Workshop, 1996.
[46] A. Yap. Product update and looking backward. Unpublished manuscript, Stanford University, 2005.

Johan van Benthem
ILLC, University of Amsterdam
johan@science.uva.nl

Eric Pacuit
ILLC, University of Amsterdam
epacuit@science.uva.nl

Deep Sequent Systems for Modal Logic

KAI BRÜNNLER

ABSTRACT. We see a systematic set of cut-free axiomatisations for all the basic normal modal logics formed from the axioms t, b, 4, 5. They employ a form of deep inference but otherwise stay very close to Gentzen's sequent calculus, in particular they enjoy a subformula property in the literal sense. No semantic notions are used inside the proof systems, in particular there is no use of labels. All their rules are invertible, contraction is admissible and they allow for straightforward terminating proof search procedures.

Keywords: sequent calculus, modal logic, deep inference

1 Introduction

Numerous extensions of the sequent calculus have been proposed in order to give cut-free axiomatisations of modal logic. They are divided into two classes: *labelled* formalisms, which incorporate Kripke semantics in the proof system, and *unlabelled* formalisms, which do not. Prominent examples of unlabelled formalisms are the hypersequent calculus [1] and the display calculus [2, 13]. These and more can be found in the survey by Wansing [14]. A recent account of labelled sequent systems which also includes more references can be found in Negri [9].

The labelled approach seems to have become more prominent and according to several criteria has been more successful than the unlabelled approaches. It allows to capture a wide class of modal logics and does so systematically. In many important cases it yields systems which are natural and easy to use, which have good structural properties like contraction-admissibility and invertibility of all rules, and which give rise to decision procedures.

However, there are concerns about incorporating the semantics into the syntax of a proof system, see for example [1]. For motivating the present work I would just like to take it as a given that it is an interesting question whether something that has been achieved using labels can also be achieved without them. My goal here is to develop proof systems with the same good properties of Negri's labelled systems but to do so without using labels.

There is a closely related current research effort by Hein, Stewart and Stouppa which has the same aim [6, 10, 11]. To make the property of "not using labels" a bit more precise we call a proof system *pure* if each sequent has an equivalent formula. Sequent systems for modal logic are clearly pure: just read the comma on the left as conjunction, the comma on the right as disjunction, and the turnstile as implication. Hypersequents are also pure. A labelled sequent, on the other hand, does not generally have an equivalent modal formula.

Hein, Stewart and Stouppa use the *calculus of structures* [3, 5, 6, 10] to give pure systems for modal logics. This formalism is based on *deep inference*, which is the ability to apply rules deep inside of a formula. So far the calculus of structures has captured essentially those modal logics which can also be captured using the sequent calculus or hypersequents. In particular that does not include B and K5. In the present work I introduce the formalism of *deep sequents* which uses deep inference (like the calculus of structures) but maintains tree-shaped derivations and a distinction between logical and structural connectives (like the sequent calculus). Deep sequent systems capture all the normal logics formed from the axioms t, b, 4, 5, thus in particular B and K5. They can be easily embedded into corresponding systems in the calculus of structures, so this answers questions from [10].

The plan of this paper is as follows: after some preliminaries I present deep sequent systems and prove invertibility of rules and admissibility of contraction. Then I show that they are sound and complete for the respective Kripke semantics. The completeness proof constructs a countermodel from the failure of a terminating proof search procedure. Some discussion of related formalisms and of future work ends this paper.

2 Preliminaries

Formulas and models. Propositions p and their negations $\bar{p}$ are *atoms*, with $\bar{\bar{p}}$ defined to be p. Atoms are denoted by a, b, c and so on. *Formulas*, denoted by A, B, C, D are given by the grammar

$$A ::= p \mid \bar{p} \mid (A \vee A) \mid (A \wedge A) \mid \Diamond A \mid \Box A \quad .$$

Given a formula A, its *negation* $\bar{A}$ is defined as usual using the De Morgan laws, $A \supset B$ is defined as $\bar{A} \vee B$ and $\bot$ is defined as $p \wedge \bar{p}$ for some proposition p. A *frame* is a pair $(S, \rightarrow)$ of a nonempty set S of *states* and a binary relation $\rightarrow$ on it. A *model* $\mathcal{M}$ is a triple $(S, \rightarrow, V)$ where $(S, \rightarrow)$ is a frame and V is a a mapping which assigns a subset of S to each proposition, and which is called a *valuation*. A model $\mathcal{M}$ as given above induces a relation $\models$ between states and formulas which is defined as usual. In particular we have $s \models p$ iff $s \in V(p)$, $s \models \bar{p}$ iff $s \notin V(p)$, $s \models A \vee B$ iff $s \models A$ or $s \models B$, $s \models A \wedge B$ iff $s \models A$ and $s \models B$, $s \models \Diamond A$ iff there is a state t such that $s \rightarrow t$ and $t \models A$, and $s \models \Box A$ iff for all t if $s \rightarrow t$ then $t \models A$. Further, a formula A is valid in a model $\mathcal{M}$, denoted $\mathcal{M} \models A$, iff for all states s of $\mathcal{M}$ we have $s \models A$. A formula A is valid in a frame $(S, \rightarrow)$, denoted $(S, \rightarrow) \models A$, iff for all valuations V we have $(S, \rightarrow, V) \models A$.

Deep sequents. A *(deep) sequent* is a finite multiset of formulas and boxed sequents. A *boxed sequent* is an expression $[\Gamma]$ where Γ is a sequent. Sequents are denoted by $\Gamma, \Delta, \Lambda, \Pi, \Sigma$. A sequent is always of the form

$$A_1, \ldots, A_m, [\Delta_1], \ldots, [\Delta_n] \quad ,$$

where, as usual, the comma denotes multiset union and there is no distinction between a singleton multiset and its element. The *corresponding formula* of the above sequent is $\bot$ if $m = n = 0$ and otherwise

$$A_1 \vee \cdots \vee A_m \vee \Box(D_1) \vee \cdots \vee \Box(D_n) \quad,$$

where $D_1 \ldots D_n$ are the corresponding formulas of the sequents $\Delta_1 \ldots \Delta_n$. Often we do not distinguish between a sequent and its corresponding formula, e.g. a model of a sequent is a model of its corresponding formula. A sequent has a *corresponding tree* whose nodes are marked with multisets of formulas. The corresponding tree of the above sequent is

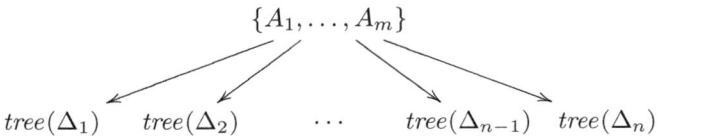

where $tree(\Delta_1) \ldots tree(\Delta_n)$ are the corresponding trees of $\Delta_1 \ldots \Delta_n$. Often we do not distinguish between a sequent and its corresponding tree, e.g. the root of a sequent is the root of its corresponding tree. In particular, a sequent Δ is an *immediate subtree* of a sequent Γ if there is a sequent Λ such that $\Gamma = \Lambda, [\Delta]$. It is a *proper subtree* if it is an immediate subtree either of Γ or of a proper subtree of Γ, and it is a *subtree* if it is either a proper subtree of Γ or $\Delta = \Gamma$. The set of all subtrees of Γ is denoted by $st(\Gamma)$. A formula A is *in* a sequent Γ if $A \in \Gamma$ and it is *inside* Γ if there is a subtree Δ of Γ such that $A \in \Delta$. The *set sequent* of the above sequent is the underlying set of

$$A_1, \ldots, A_m, [\Lambda_1], \ldots, [\Lambda_n] \quad,$$

where $\Lambda_1 \ldots \Lambda_n$ are the set sequents of $\Delta_1 \ldots \Delta_n$. Clearly the set sequent of a given sequent is again a sequent since a set is a multiset.

A *(sequent) context* is a sequent with exactly one occurrence of the symbol { }, the *hole*, which does not occur inside formulas. It is denoted by $\Gamma\{\ \}$. The sequent $\Gamma\{\Delta\}$ is obtained by replacing { } inside $\Gamma\{\ \}$ by Δ. A *double context*, denoted by $\Gamma\{\ \}\{\ \}$ is a triple of contexts $\Gamma_0\{\ \}, \Gamma_1\{\ \}, \Gamma_2\{\ \}$. Then $\Gamma\{\Delta\}\{\ \}$ denotes the context $\Gamma_0\{\Gamma_1\{\Delta\}, \Gamma_2\{\ \}\}$ and $\Gamma\{\ \}\{\Delta\}$ denotes the context $\Gamma_0\{\Gamma_1\{\ \}, \Gamma_2\{\Delta\}\}$. The *depth* of a context $\Gamma\{\ \}$, denoted $depth(\Gamma\{\ \})$ is defined as $depth(\Gamma, \{\ \}) = 0$ and $depth(\Gamma, [\Delta\{\ \}]) = depth(\Delta\{\ \}) + 1$.

3 The Modal Systems

Figure 1 shows the set of rules from which we form our deductive systems. *System* K includes the rules $\{\wedge, \vee, \Box, k\}$. We will look at extensions of System K with sets of rules $X \subseteq \{t, b, 4, 5\}$. The 5-rule carries the proviso that the depth of $\Gamma\{\ \}\{\emptyset\}$ is greater than zero.

$$\Gamma\{a,\bar{a}\} \qquad \wedge \frac{\Gamma\{A\} \quad \Gamma\{B\}}{\Gamma\{A \wedge B\}} \qquad \vee \frac{\Gamma\{A,B\}}{\Gamma\{A \vee B\}} \qquad \Box \frac{\Gamma\{[A]\}}{\Gamma\{\Box A\}}$$

$$\mathsf{k} \frac{\Gamma\{\Diamond A, [\Delta, A]\}}{\Gamma\{\Diamond A, [\Delta]\}} \qquad \mathsf{t} \frac{\Gamma\{\Diamond A, A\}}{\Gamma\{\Diamond A\}} \qquad \mathsf{b} \frac{\Gamma\{[\Delta, \Diamond A], A\}}{\Gamma\{[\Delta, \Diamond A]\}}$$

$$\mathsf{4} \frac{\Gamma\{\Diamond A, [\Delta, \Diamond A]\}}{\Gamma\{\Diamond A, [\Delta]\}} \qquad \mathsf{5} \frac{\Gamma\{\Diamond A\}\{\Diamond A\}}{\Gamma\{\Diamond A\}\{\emptyset\}}$$

Figure 1. System K+{t,b,4,5}

$$\mathsf{nec} \frac{\Gamma}{[\Gamma]} \qquad \mathsf{wk} \frac{\Gamma\{\emptyset\}}{\Gamma\{\Delta\}} \qquad \mathsf{ctr} \frac{\Gamma\{\Delta, \Delta\}}{\Gamma\{\Delta\}} \qquad \mathsf{cut} \frac{\Gamma\{A\} \quad \Gamma\{\bar{A}\}}{\Gamma\{\emptyset\}}$$

Figure 2. Necessitation, weakening, contraction and cut

In an instance of an inference rule $\rho \frac{\Gamma}{\Delta}$ we call Γ its *premise* and Δ its *conclusion*, and similarly for the $\wedge$-rule, which has two premises. A *system*, denoted by $\mathcal{S}$, is a set of rules. A *derivation* in a system $\mathcal{S}$ is a upward-growing finite tree whose nodes are labelled with sequents and which is built according to the inference rules from $\mathcal{S}$. The sequent at the root is the *conclusion* and the sequents at the leaves which are not instances of the axiom $\Gamma\{a,\bar{a}\}$ are the *premises* of the derivation. A *proof* of a sequent Γ in a system is a derivation in this system with conclusion Γ and without premises. We write $\mathcal{S} \vdash \Gamma$ if there is a proof of Γ in system $\mathcal{S}$. An inference rule ρ is *(depth-preserving) admissible* for a system $\mathcal{S}$ if for each proof in $\mathcal{S} \cup \{\rho\}$ there is a proof in $\mathcal{S}$ with the same conclusion (and with at most the same depth). For each rule ρ there is its *inverse*, denoted by $\bar{\rho}$, which is obtained by exchanging premise and conclusion. The $\bar{\wedge}$-rule allows both $\Gamma\{A\}$ and $\Gamma\{B\}$ as conclusions of $\Gamma\{A \wedge B\}$. An inference rule ρ is *(depth-preserving) invertible* for a system $\mathcal{S}$ if $\bar{\rho}$ is (depth-preserving) admissible for $\mathcal{S}$.

LEMMA 1. *For each system* K + X *with* X $\subseteq$ {t, b, 4, 5} *the following hold:*
(i) The necessitation and weakening rules are depth-preserving admissible.
(ii) All its rules are depth-preserving invertible.
(iii) The contraction rule is depth-preserving admissible.

Proof. (i) follows from a routine induction on the depth of the proof. The same works for the $\wedge$, $\vee$ and $\Box$-rules in (ii). The inverses of all other rules are just weakenings. For (iii) we also proceed by induction on the depth of the proof tree, using depth-preserving invertibility of the rules. The cases are easy for the propositional rules and for the $\Box$, t-rules. For the k rule we

t:	reflexive	$\forall s\ s \to s$	$A \supset \Diamond A$
b:	symmetric	$\forall st\ s \to t \supset t \to s$	$A \supset \Box \Diamond A$
4:	transitive	$\forall stu\ s \to t \wedge t \to u \supset s \to u$	$\Box A \supset \Box \Box A$
5:	euclidean	$\forall stu\ s \to t \wedge s \to u \supset t \to u$	$\Diamond A \supset \Box \Diamond A$

Figure 3. Frame conditions and modal axioms

consider the formula $\Diamond A$ from its conclusion $\Gamma\{\Diamond A, [\Delta]\}$ and its position inside the premise of contraction $\Lambda\{\Sigma, \Sigma\}$. We have the cases 1) $\Diamond A$ is inside Σ or 2) $\Diamond A$ is inside $\Lambda\{\ \}$. We have three subcases for case 1: 1.1) $[\Delta]$ inside $\Lambda\{\ \}$, 1.2) $[\Delta]$ inside Σ, 1.3) Σ, Σ inside $[\Delta]$. There are two subcases of case 2: 2.1) $[\Delta]$ inside $\Lambda\{\ \}$ and 2.2) $[\Delta]$ inside Σ. All cases are either simpler than or similar to case 1.2, which is as follows:

$$\mathsf{ctr}\frac{\mathsf{k}\dfrac{\Lambda'\{\Diamond A, \Sigma', [\Delta, A], \Sigma', [\Delta]\}}{\Lambda'\{\Diamond A, \Sigma', [\Delta], \Sigma', [\Delta]\}}}{\Lambda'\{\Diamond A, \Sigma', [\Delta]\}} \quad \rightsquigarrow \quad \mathsf{k}\frac{\mathsf{ctr}\dfrac{\overline{\mathsf{k}}\dfrac{\Lambda'\{\Diamond A, \Sigma', [\Delta, A], \Sigma', [\Delta]\}}{\Lambda'\{\Diamond A, \Sigma', [\Delta, A], \Sigma', [\Delta, A]\}}}{\Lambda'\{\Diamond A, \Sigma', [\Delta.\ A]\}}}{\Lambda'\{\Diamond A, \Sigma', [\Delta]\}},$$

where the instance of $\overline{\mathsf{k}}$ in the proof on the right is removed because it is depth-preserving admissible and the instance of contraction is removed by the induction hypothesis. The case for the 4-rule works the same way.

For the b-rule we make a case analysis based on the position of $[\Delta, \Diamond A]$ from its conclusion $\Gamma\{[\Delta, \Diamond A]\}$ inside the premise of contraction $\Lambda\{\Sigma, \Sigma\}$. We have three cases: 1) $[\Delta, \Diamond A]$ inside $\Lambda\{\ \}$, 2) $[\Delta, \Diamond A]$ in Σ and 3) Σ, Σ inside $[\Delta, \Diamond A]$. Case 3 has two subcases: either $\Diamond A \in \Sigma$ or not. All cases are trivial except for case 2 where invertibility of the b-rule is used.

For the 5 rule we make a case analysis based on the positions of the sequent occurrences $\Diamond A$ and $\emptyset$ from its conclusion $\Gamma\{\Diamond A\}\{\emptyset\}$ inside the premise of contraction $\Lambda\{\Sigma, \Sigma\}$. We have two cases: 1) $\emptyset$ inside $\Lambda\{\ \}$, 2) $\emptyset$ inside Σ. The first case is trivial, in the second we have two subcases: 1) $\Diamond A$ inside $\Lambda\{\ \}$ and 2) $\Diamond A$ inside Σ. Cases 2.1 and 2.2 are similar to the transformation shown above. ∎

4 Soundness

Each name of a rule in $\{\mathsf{t}, \mathsf{b}, 4, 5\}$ corresponds both to a frame condition and to a modal Hilbert-style axiom as shown in Figure 3. The k-rule corresponds to the modal axiom $\Box(A \vee B) \supset (\Box A \vee \Diamond B)$.

For a subset $\mathsf{X} \subseteq \{\mathsf{t}, \mathsf{b}, 4, 5\}$ we call a frame an X-frame if it satisfies all the conditions determined by the names in X. A formula is X-*valid* if it is valid in all X-frames and just *valid* if it is valid in all frames.

LEMMA 2. *The 5-rule is derivable for* {5a, 5b, 5c, ctr}, *where* 5a, 5b, 5c *are the rules*

$$5a \frac{\Gamma\{[\Delta], \Diamond A\}}{\Gamma\{[\Delta, \Diamond A]\}} \quad , \quad 5b \frac{\Gamma\{[\Delta], [\Lambda, \Diamond A]\}}{\Gamma\{[\Delta, \Diamond A], [\Lambda]\}} \quad , \quad 5c \frac{\Gamma\{[\Delta, [\Lambda, \Diamond A]]\}}{\Gamma\{[\Delta, \Diamond A, [\Lambda]]\}} \quad .$$

Proof. Seen bottom-up, the 5-rule allows to put a formula $\Diamond A$ which occurs at a node different from the root into an arbitrary node. We can use contraction to duplicate $\Diamond A$ and move one copy either to the root or to some child of the root by 5a. By 5b we can move it to any child of the root and by 5c into any descendant of a child of the root. ∎

LEMMA 3. *For all contexts* $\Gamma\{\ \}$ *and formulas* A, B *the formula* $(A \supset B) \supset (\Gamma\{A\} \supset \Gamma\{B\})$ *is valid.*

Proof. By induction on $\Gamma\{\ \}$ using propositional reasoning and the implication $(A \supset B) \supset (\Box A \supset \Box B)$. ∎

THEOREM 4 (Soundness). *Let* Γ, Γ', Δ *be sequents and let* $\mathsf{X} \subseteq \{\mathsf{t}, \mathsf{b}, 4, 5\}$. *Then the following hold:*

(i) For any rule $\rho \in \mathsf{K}$ *if* $\rho \dfrac{\Gamma \ (\Gamma')}{\Delta}$ *then* $\Gamma(\wedge \Gamma') \supset \Delta$ *is valid.*

(ii) For any rule $\rho \in \{\mathsf{t}, \mathsf{b}, 4, 5\}$ *if* $\rho \dfrac{\Gamma}{\Delta}$ *then* $\Gamma \supset \Delta$ *is* $\{\rho\}$-*valid.*

(iii) If $\mathsf{K} + \mathsf{X} \vdash \Gamma$ *then* Γ *is* X-*valid.*

Proof. The axiom is valid in all frames which follows from an induction on $\Gamma\{\ \}$ where necessitation is used in the induction step. Thus (i) and (ii) imply (iii). Most cases of (i) are trivial, for the ∧-rule it follows from an induction on the context and uses the implication $\Box A \wedge \Box B \supset \Box(A \wedge B)$. Lemma 3 used together with the k-axiom yields that the premise of the k-rule implies its conclusion. The cases from (ii) for the $\{\mathsf{t}, \mathsf{b}, 4\}$-rules are similar to the k-rule, using the corresponding modal axiom and for the corresponding frames.

For the soundness of the 5-rule we use Lemma 2 and show soundness of the rules 5a, 5b, 5c. For 5c we show that a euclidean countermodel for the conclusion is also a countermodel for the premise, the other cases are similar. A countermodel for $[\Delta, \Diamond A, [\Lambda]]$ has to contain states $s \to t \to u$ such that $t \not\models \Delta$, $u \not\models \Lambda$ and $v \not\models A$ for any v with $t \to v$. We need to show that for any w with $u \to w$ we have $w \not\models A$. By euclideanness we obtain, in this order: $t \to t, u \to t, t \to w$. Thus $w \not\models A$. ∎

5 Completeness

We will not directly prove completeness of the systems $\mathsf{K} + \mathsf{X}$, but of different, equivalent systems $(\mathsf{K} + \mathsf{X})^*$ that we define now. For each rule

$\rho \in \{\wedge, \vee, \square, \mathsf{k}, \mathsf{t}, \mathsf{b}, 4, 5\}$ we define a rule ρ' which keeps the main formula from the conclusion. For most rules $\rho = \rho'$ except for

$$\wedge' \frac{\Gamma\{A \wedge B, A\} \quad \Gamma\{A \wedge B, B\}}{\Gamma\{A \wedge B\}}$$

$$\vee' \frac{\Gamma\{A \vee B, A, B\}}{\Gamma\{A \vee B\}} \qquad \square' \frac{\Gamma\{\square A, [A]\}}{\Gamma\{\square A\}} \quad .$$

Then for each rule $\rho \in \{\wedge, \vee, \square, \mathsf{k}, \mathsf{t}, \mathsf{b}, 4, 5\}$ we define a rule ρ^*, which is ρ' and carries the proviso that for all of its premises the set sequent is different from the set sequent of the conclusion. Given a system $\mathcal{S} \subseteq \{\wedge, \vee, \square, \mathsf{k}, \mathsf{t}, \mathsf{b}, 4, 5\}$ the system $\mathcal{S}'(\mathcal{S}^*)$ is obtained by replacing each rule $\rho \in \mathcal{S}$ by $\rho'(\rho^*)$.

LEMMA 5. *For all systems* $\mathsf{X} \subseteq \{\mathsf{t}, \mathsf{b}, 4, 5\}$ *and for all sequents* Γ

$$\mathsf{K} + \mathsf{X} \vdash \Gamma \qquad \textit{iff} \qquad (\mathsf{K} + \mathsf{X})' \vdash \Gamma \qquad \textit{iff} \qquad (\mathsf{K} + \mathsf{X})^* \vdash \Gamma \quad .$$

Proof. The way right to middle is obvious, all other ways work by induction on the proof tree, using weakening admissibility for $\mathsf{K} + \mathsf{X}$ from left to middle and contraction admissibility for $\mathsf{K} + \mathsf{X}$ from middle to left. Middle to right uses contraction admissibility for $(\mathsf{K} + \mathsf{X})'$, which is established in the same way as for $\mathsf{K} + \mathsf{X}$. ∎

Before proving completeness, we first need to characterise the euclidean and the transitive-euclidean closure of a relation.

DEFINITION 6. *Let* $\to$ *be a binary relation on a set* S. *Then* $\leftarrow$ *denotes its inverse,* $\leftrightarrow$ *its symmetric closure,* $\to^+$ *its transitive closure and* $\to^*$ *its reflexive-transitive closure. For* $\mathsf{X} \subseteq \{\mathsf{t}, \mathsf{b}, 4, 5\}$ $\to^\mathsf{X}$ *denotes the smallest relation that includes* $\to$ *and has the properties in* X.

DEFINITION 7. *Let* $\to$ *be a binary relation on a set* S *and let* $s, t \in S$. *A euclidean connection for* $\to$ *from* s *to* t *is a nonempty sequence* $s_1 \ldots s_n$ *of elements of* S *such that we have*

$$s \leftarrow s_1 \leftrightarrow s_2 \leftrightarrow \cdots \leftrightarrow s_n \to t \quad .$$

A *transitive-euclidean connection* is defined likewise but such that

$$s = s_1 \leftrightarrow s_2 \leftrightarrow \cdots \leftrightarrow s_n \to t \quad .$$

We write $s \to_{(4)5} t$ if there is a (transitive-)euclidean connection for $\to$ from s to t.

LEMMA 8. *Let* $\to$ *be a binary relation on a set* S. *Then the following hold:*
(i) For all $\mathsf{X} \subseteq \{\mathsf{t}, \mathsf{b}, 4, 5\}$ *the relation* $\to^\mathsf{X}$ *is well-defined.*
(ii) The relation $\to \cup \to_5$ *is the least euclidean relation that contains* $\to$.
(iii) The relation $\to_{45}$ *is the least transitive and euclidean relation that contains* $\to$.

Proof. (i) is easy to check except for the cases for $\{5\}$ and $\{4,5\}$, which follow from (ii) and (iii).

(ii) Euclideanness is easy to check. For leastness we show that any euclidean relation $\to'$ that includes $\to$ also includes $\to_5$. If $s \to_5 t$ then $s\to'_5 t$. We show $s\to'_5 t$ for a euclidean connection of length n implies $s\to' t$ by induction on n. Assume there is an s_i in the euclidean connection such that $s_{i-1}\to' s_i \leftarrow' s_{i+1}$. Then we have two smaller euclidean connections to which we apply the induction hypothesis and obtain $s\to' t$ by euclideanness. If there is no such s_i then the euclidean connection looks as follows:

$$s = s_0 \leftarrow' s_1 \leftarrow' \ldots \leftarrow' s_j \to' \ldots \to' s_n \to' s_{n+1} = t \quad ,$$

and by euclideanness we have $s_{j-1}\to' s_{j+1}$ and thus removing s_j yields a smaller euclidean connection from s to t which by induction hypothesis implies $s\to' t$.

(iii) Euclideanness and transitivity are easy to check. For leastness we show that any transitive-euclidean relation $\to'$ that includes $\to$ also includes $\to_{45}$. If $s \to_{45} t$ then $s\to'_{45} t$. If there is no s_i in the transitive-euclidean such that $s_i \leftarrow' s_{i+1}$, then $s\to' t$ follows by transitivity. Otherwise, choose the first such s_i. We have a euclidean connection from s_i to t, thus similarly to (ii) obtain $s_i\to' t$ and by transitivity $s\to' s_i$ and $s\to' t$. ∎

DEFINITION 9. A leaf of a sequent is called *cyclic* if there is an inner node in the sequent that carries the same set of formulas. A leaf is called *extensible* if it contains a formula of the form $\Box A$. We define a procedure $prove(\Gamma, \mathsf{X})$, which takes a sequent Γ and a system $\mathsf{X} \subseteq \{\mathsf{t}, \mathsf{b}, 4, 5\}$ and builds a derivation tree for Γ by applying rules from $(\mathsf{K} + \mathsf{X})^*$ to non-axiomatic leaves in a bottom-up fashion as follows:

1. simultaneously apply the $\Box^*$ rule once wherever possible,

2. keep applying the rules in $((\mathsf{K} + \mathsf{X}) \setminus \Box)^*$ as long as possible.

Repeat the above until each non-axiomatic leaf in the derivation tree carries a sequent such that each extensible leaf in the sequent tree is cyclic. If $prove(\Gamma, \mathsf{X})$ terminates and all leaves in the derivation tree are axiomatic then it *succeeds* and if it terminates and there is a non-axiomatic leaf then it *fails*.

LEMMA 10. *For all sets* $\mathsf{X} \subseteq \{\mathsf{t}, \mathsf{b}, 4, 5\}$ *and for all sequents* Γ *the procedure* $prove(\Gamma, \mathsf{X})$ *terminates.*

Proof. Consider a sequence of sequents along a given branch of the derivation and starting from the root. A rule application in step 2 does not create new nodes in the sequent and causes the set of formulas at some node in the sequent to strictly grow. By the subformula property only finitely many formulas can occur in a node, so step 2 terminates. If there is an extensible leaf in a sequent then the size of the sequent strictly grows in step 1. Since there are only finitely many sets of formulas that can occur, once the sequent is large enough each extensible leaf has to be cyclic. ∎

The current set of modal rules does not allow a modular completeness result of the form "if Γ is X-valid then $\mathsf{K} + \mathsf{X} \vdash \Gamma$". In particular we have $\mathsf{K} + \{\mathsf{t},5\} \not\vdash \Box A \supset \Box\Box A$ and $\mathsf{K} + \{\mathsf{b},4\} \not\vdash \Diamond A \supset \Box\Diamond A$. However, we obtain a weaker form of modularity. We define a set $\mathsf{X} \subseteq \{\mathsf{t},\mathsf{b},4,5\}$ to be *maximal* if it is closed under implication when the elements are read as frame conditions, or, more precisely, for each $\rho \in \{\mathsf{t},\mathsf{b},4,5\} \setminus \mathsf{X}$ there is an X-frame which does not satisfy ρ. The set $\{\mathsf{b},4\}$ is not maximal, for example, while $\{\mathsf{b},4,5\}$ is. Our completeness result will hold for maximal X.

THEOREM 11 (Completeness). *For all maximal sets $\mathsf{X} \subseteq \{\mathsf{t},\mathsf{b},4,5\}$ and for all sequents Γ the following hold:*
(i) If Γ is X-valid then $\mathsf{K} + \mathsf{X} \vdash \Gamma$.
(ii) If $prove(\Gamma, \mathsf{X})$ fails then Γ is not X-valid.

Proof. The contrapositive of (i) follows from (ii): if $\mathsf{K} + \mathsf{X} \not\vdash \Gamma$ then by Lemma 5 also $(\mathsf{K} + \mathsf{X})^* \not\vdash \Gamma$ and thus in particular $prove(\Gamma, \mathsf{X})$ cannot yield a proof and by Lemma 10 has to fail. For (ii) we define a model $\mathcal{M}$ on an X-frame for which we prove that it is a countermodel for Γ. Let Γ^* be the set sequent of the non-axiomatic sequent obtained with all extensible leaves cyclic. Let Y be the set of all extensible leaves in Γ^*. Let $S = st(\Gamma^*) \setminus Y$. Let $f : Y \to S$ be some function which maps an extensible leaf to a sequent in S whose root carries the same set of formulas and extend f to $st(\Gamma^*)$ by the identity on S. Define a binary relation $\to$ on S such that $\Delta \to \Lambda$ iff either 1) Λ is an immediate subtree of Δ or 2) Δ has an immediate subtree $\Sigma \in Y$ and $f(\Sigma) = \Lambda$. Let $V(p) = \{\Delta \in S \mid \bar{p} \in \Delta\}$. Let $\mathcal{M} = (S, \to^{\mathsf{X}}, V)$. We prove three claims about $\mathcal{M}$, each claim depending on the next. Since all rules seen top-down preserve countermodels Claim 1 implies that $\mathcal{M} \not\models \Gamma$.

Claim 1 $\forall \Delta \in st(\Gamma^*) \quad \mathcal{M}, f(\Delta) \not\models \Delta$

By induction on the depth of Δ. For depth zero this follows from Claim 2 and the fact that a formula is in Δ iff it is in $f(\Delta)$. So let $\Delta = A_1, \ldots, A_m, [\Delta_1], \ldots, [\Delta_n]$ and $n > 0$. Then $f(\Delta) = \Delta$. We have $\mathcal{M}, f(\Delta) \not\models A_i$ for all $i \leq m$ by Claim 2 and $\mathcal{M}, \Delta \not\models [\Delta_i]$ because $\Delta \to f(\Delta_i)$ and by induction hypothesis $\mathcal{M}, f(\Delta_i) \not\models \Delta_i$.

Claim 2 $\forall \Delta \in S \, \forall A \in \Delta \quad \mathcal{M}, \Delta \not\models A$

By induction on the depth of A. For atoms it is clear from the definition of $\mathcal{M}$ and since Γ^* is not axiomatic. For the propositional connectives it is clear from the shape of the $\wedge, \vee$-rules. If $A = \Box B$ then by the $\Box$-rule we have some $[\Lambda] \in \Delta$ with $B \in \Lambda$. By induction hypothesis we have $\mathcal{M}, \Lambda \not\models B$ and thus $\mathcal{M}, \Delta \not\models \Box B$. If $A = \Diamond B$ then by Claim 3 we have $B \in \Lambda$ for all Λ with $\Delta \to^{\mathsf{X}} \Lambda$, and thus $\mathcal{M}, \Lambda \not\models B$. Thus $\mathcal{M}, \Delta \not\models \Diamond B$.

Claim 3 For all Δ, Λ with $\Delta \to^{\mathsf{X}} \Lambda \, \forall A \quad \Diamond A \in \Delta \implies A \in \Lambda$

K $\mathsf{X} = \emptyset$: By the definition of $\to$ there is an immediate subtree of Δ whose root node carries the same set of formulas as the root node of Λ. By the k-rule we have A in (the root node of) all immediate subtrees of Δ.

T $\mathsf{X} = \{\mathsf{t}\}$: $\Delta \to^{\{\mathsf{t}\}} \Lambda$ iff $\Delta \to \Lambda$ or $\Delta = \Lambda$. In the second case $A \in \Lambda$ follows from the t-rule.

KB $X = \{b\}$: $\Delta \to^{\{b\}} \Lambda$ iff $\Delta \to \Lambda$ or $\Lambda \to \Delta$. In the second case $A \in \Lambda$ follows by the b-rule.

K4 $X = \{4\}$: $\Delta \to^{\{4\}} \Lambda$ iff there is a sequence

$$\Delta = \Delta_0 \to \Delta_1 \to \Delta_2 \to \cdots \to \Delta_n = \Lambda \,,$$

with $n \geq 1$. An induction on i gives us that $\Diamond A \in \Delta_i$ for $0 \leq i \leq n$ by the 4-rule. That $A \in \Delta_i$ for $1 \leq i \leq n$ follows from that by the k-rule.

K5 $X = \{5\}$: By Lemma 8 we have $\Delta \to^{\{5\}} \Lambda$ iff $\Delta \to \Lambda$ or there is a euclidean connection from Δ to Λ. In the second case there are sequents Π, Σ such that $\Pi \to \Delta$ and $\Sigma \to \Lambda$. Thus there is a subtree Δ' of Π with the same formulas as Δ and a subtree Λ' of Σ with the same formulas as Λ. Since $\Diamond A \in \Delta$ we have $\Diamond A \in \Delta'$ and since $\Delta' \neq \Gamma^*$ by the 5-rule we have $\Diamond A \in \Sigma$. Thus by the k-rule we have A in Λ' and thus in Λ.

K45 $X = \{4, 5\}$: By Lemma 8 we have $\Delta \to^{\{4,5\}} \Lambda$ iff $\Delta \to \Lambda$ or there is a transitive-euclidean connection from Δ to Λ. In the second case there is a sequent Σ such that $\Sigma \to \Lambda$ and thus a subtree Λ' of Σ with the same formulas as Λ. Since $\Diamond A \in \Delta$, by the 5- and 4-rules we have $\Diamond A$ in every subtree of Γ^* and thus also in Σ, and by the k-rule we have A in Λ' and thus in Λ. (The 5a-rule instead of the 5-rule is enough.)

KB5 $X = \{b, 4, 5\}$: $\Delta \to^{\{b,4,5\}} \Lambda$ iff $\Delta \leftrightarrow^+ \Lambda$. Thus there is a sequent Σ such that either $\Sigma \to \Lambda$ or $\Sigma \leftarrow \Lambda$. Rule 4, 5 imply that $\Diamond A$ is in every subtree of Γ^* and thus in particular in Σ. We have $A \in \Lambda$ in the first case by the k-rule and in the second case by the b-rule.

KTB $X = \{b, t\}$: $\Delta \to^{\{b,t\}} \Lambda$ iff $\Delta \to \Lambda$ or $\Delta \leftarrow \Lambda$ or $\Delta = \Lambda$. In these cases $A \in \Lambda$ respectively follows from the k- or b- or t-rule.

S4 $X = \{t, 4\}$: $\Delta \to^{\{t,4\}} \Lambda$ iff $\Delta \to^+ \Lambda$ or $\Delta = \Lambda$. In the first case $A \in \Lambda$ follows from the 4-rule and in the second case from the t-rule.

S5 $X = \{t, b, 4, 5\}$: $\Delta \to^{\{t,b,4,5\}} \Lambda$ iff $\Delta \leftrightarrow^* \Lambda$. We have $\Diamond A$ in all subtrees of Γ^* by the rules 4, 5 and thus also A by the t-rule. (Again, the 5a-rule is enough and the b-rule is not needed.) ∎

6 Discussion

Our goal was to give pure proof systems with the good properties of Negri's labelled sequent systems. To some extent we have succeeded: while we do not (yet) have cut-free systems for all the logics which have cut-free labelled systems, we have captured all logics formed from t,b,4,5 and thus several important cases. Our systems are systematic in the sense that there is a one-to-one correspondence between the modal rules and the frame conditions considered, they enjoy invertibility, contraction admissibility and a terminating proof search procedure. The main conceptual difference between labelled and deep sequents is that the structural level in labelled systems is more general: it can form an arbitrary graph, while deep sequents are always trees. I hope that this restriction will help in using deep sequent systems for interpolation proofs, for which labelled systems do not seem to be well-suited.

Relation to hypersequents. Deep sequents are a natural generalisation of (modal) hypersequents, in allowing arbitrary nestings of boxed disjunctions instead of just a disjunction of boxed disjunctions. I am not aware of hypersequent systems for K5 or B nor of hypersequent systems with invertible rules and contraction admissibility for the modal logics treated here. In fact, deep sequent systems came out of unsuccessfully trying to design an invertible (hyper)sequent system for S4. A notational simplification I enjoy with respect to hypersequent systems is that the two kinds of context in inference rules (sequent context and hypersequent context) are merged into one.

Relation to the display calculus. Display sequents are closely related, in particular the idea of simply allowing □ as a structural connective is common to display sequents and deep sequents. However, the proof systems are rather different. Loosely speaking, in the display calculus one has to make a formula bubble up to the top by using the structural rules in order to apply a logical rule to it, while in deep sequent systems one can apply the rule on the spot. This leads to deductive systems with fewer rules and shortens derivations. On the other hand the display calculus so far has captured more modal logics than deep sequents and also enjoys a general cut elimination result, which for deep sequents is subject of current research. As with hypersequents, I am not aware of display systems with invertible rules and contraction admissibility.

The decision procedures induced by the sequent systems are very simple and thus easily seen to be correct and complete, but of course by design not particularly fast or efficient. Readers looking for efficient decision procedures should look e.g. at [4] in this volume or at [7].

Also, it is not my current goal to give a computational interpretation of cut elimination procedures in these systems, the classical setting and the fact that we have such procedures only for very few systems are obstacles to that. Readers interested in such questions are referred to [8].

We now turn to some future work and open questions.

Seriality. The set of modal axioms treated does not include seriality. I chose to put it aside for the time being, not because it is particularly problematic, but because it does not quite follow the same scheme as the other rules. The candidate for the d-rule is

$$d \frac{\Gamma\{\Diamond A, [A]\}}{\Gamma\{\Diamond A\}}$$

Clearly it increases the size of the sequent going up, so in the proof search procedure it has to be applied together with □. The notion of serial closure also has to be defined differently from the other closures. The conjecture is as follows:

CONJECTURE 12. For each sequent Γ and each maximal $X \subseteq \{d, t, b, 4, 5\}$ we have $K + X \vdash \Gamma$ iff Γ is X-valid.

Modularity. Our systems are not modular in the sense that each combination of modal rules is complete for the corresponding class of frames

$$\text{ser}\,\frac{\Gamma\{[\emptyset]\}}{\Gamma\{\emptyset\}} \qquad \text{refl}\,\frac{\Gamma\{\Sigma,[\Sigma]\}}{\Gamma\{\Sigma\}} \qquad \text{sym}\,\frac{\Gamma\{\Sigma,[\Delta,[\Sigma]]\}}{\Gamma\{\Sigma,[\Delta]\}}$$

$$\text{trans}\,\frac{\Gamma\{[\Delta,[\Sigma,\Lambda]],[\Sigma]\}}{\Gamma\{[\Delta,[\Sigma,\Lambda]]\}} \qquad \text{euc}\,\frac{\Gamma\{[\Delta,[\Sigma]],[\Lambda,\Sigma]\}}{\Gamma\{[\Delta],[\Lambda,\Sigma]\}}$$

Figure 4. Modal axioms as structural rules

(labelled systems are modular in that sense). The modal rules presented are all $\Diamond$-rules, in the sense that the active formula in the conclusion has $\Diamond$ as main connective. I am not sure whether modularity can be achieved with this style of rules. However, there is also the possibility of formulating the modal rules as structural rules (structural in the sense of not affecting connectives of formulas), shown in Figure 4. I conjecture that these systems are modular. Notice the absence of the word "maximal":

CONJECTURE 13. *For each sequent Γ and $\mathsf{X} \subseteq \{\mathsf{ser}, \mathsf{refl}, \mathsf{sym}, \mathsf{trans}, \mathsf{euc}\}$ we have $\mathsf{K} + \mathsf{X} \vdash \Gamma$ iff Γ is X-valid.*

Robert Hein independently came up with essentially the same rules and the same conjecture for the calculus of structures.

Syntactic cut elimination. A cut elimination procedure is easily defined in the cases where the rules 4 and 5 are absent. It works like the standard procedure for system G3 for first-order predicate logic, see for example [12]. Invertibility is used in all the passive cases and in the active case for the k-rule the proof

$$\text{cut}\,\frac{\Box\,\dfrac{\overset{1}{\Gamma\{[\Delta],[A]\}}}{\Gamma\{[\Delta],\Box A\}} \qquad \mathsf{k}\,\dfrac{\overset{2}{\Gamma\{[\Delta,\bar{A}],\Diamond\bar{A}\}}}{\Gamma\{[\Delta],\Diamond\bar{A}\}}}{\Gamma\{[\Delta]\}}$$

is replaced by the proof

$$\text{cut}\,\frac{\text{ctr}\,\dfrac{\text{wk}^2\,\dfrac{\overset{1}{\Gamma\{[\Delta],[A]\}}}{\Gamma\{[\Delta,A],[\Delta,A]\}}}{\Gamma\{[\Delta,A]\}} \qquad \text{cut}\,\dfrac{\text{wk},\Box\,\dfrac{\overset{1}{\Gamma\{[\Delta],[A]\}}}{\Gamma\{[\Delta,\bar{A}],\Box A\}} \qquad \overset{2}{\Gamma\{[\Delta,\bar{A}],\Diamond\bar{A}\}}}{\Gamma\{[\Delta,\bar{A}]\}}}{\Gamma\{[\Delta]\}}$$

where the lower cut has a lower rank and the upper cut decreased in the sum of the depths of its two subproofs. The cases for t and b also work, but for 4 and 5 some more refined measure is needed which is subject of current research.

BIBLIOGRAPHY

[1] Arnon Avron. The method of hypersequents in the proof theory of propositional non-classical logics. In Wilfrid Hodges, Martin Hyland, Charles Steinhorn, and John Truss, editors, *Logic: from foundations to applications. Proc. Logic Colloquium, Keele, UK, 1993*, pages 1–32. Oxford University Press, New York, 1996.

[2] Nuel D. Belnap, Jr. Display logic. *Journal of Philosophical Logic*, 11:375–417, 1982.

[3] Kai Brünnler. *Deep Inference and Symmetry in Classical Proofs*. PhD thesis, Technische Universität Dresden, September 2003.

[4] Olivier Gasquet, Andreas Herzig and Mohamad Sahade. Terminating modal tableaux with simple completeness proof. In *Proceedings of Advances in Modal Logic*, this volume, 2006.

[5] Alessio Guglielmi. A system of interaction and structure. Technical Report WV-02-10, Technische Universität Dresden, 2002. To appear in ACM Transactions on Computational Logic.

[6] Robert Hein and Charles Stewart. Purity through unravelling. In Paola Bruscoli, François Lamarche, and Charles Stewart, editors, *Structures and Deduction*, pages 126–143. Technische Universität Dresden, 2005.

[7] A. Heuerding, M. Seyfried, H. Zimmermann. Efficient loop-check for backward proof search in some non-classical propositional logics. In Proceedings *Tableaux 96*, pages 210–225, 1996.

[8] Simone Martini and Andrea Masini. A computational interpretation of modal proofs. In *Proof theory of modal logic*, volume 2 of *Applied logic series*, pages 213–241. Kluwer, 1996.

[9] Sara Negri. Proof analysis in modal logic. *Journal of Philosophical Logic*, 34(5 – 6):507 – 544, 2005.

[10] Charles Stewart and Phiniki Stouppa. A systematic proof theory for several modal logics. In Renate Schmidt, Ian Pratt-Hartmann, Mark Reynolds, and Heinrich Wansing, editors, *Advances in Modal Logic*, volume 5 of *King's College Publications*, pages 309–333, 2005.

[11] Phiniki Stouppa. A deep inference system for the modal logic S5. To appear in Studia Logica, 2006.

[12] Anne Sjerp Troelstra and Helmut Schwichtenberg. *Basic Proof Theory*. Cambridge University Press, 1996.

[13] Heinrich Wansing. *Displaying Modal Logic*, volume 3 of *Trends in Logic Series*. Kluwer Academic Publishers, Dordrecht, 1998.

[14] Heinrich Wansing. Sequent systems for modal logics. In D. Gabbay and F. Guenther, editors, *Handbook of Philosophical Logic, 2nd edition*, volume 8, pages 61–145. Kluwer, Dordrecht, 2002.

Kai Brünnler
Institut für angewandte Mathematik und Informatik,
Neubrückstr. 10, CH – 3012 Bern, Switzerland
kai@iam.unibe.ch

The Truth About Algorithmic Problems in Correspondence Theory

ALEXANDER CHAGROV AND LILIA CHAGROVA

ABSTRACT. We present proofs that the following three major algorithmic problems in Correspondence Theory are undecidable: first-order definability of propositional modal formulas, modal (propositional) definability of first-order (classical predicate) formulas, and the correspondence problem (equivalence on Kripke frames) for propositional modal formulas and first-order formulas. These results have been known since Chagrova's PhD thesis [4] published in 1989. However, [4] considered (propositional) intuitionistic formulas which are much harder to deal with than the modal ones. Although the proofs from [4] provided some extra subtle results (such as the undecidability of the set of propositional formulas that are first-order definable on the class of countable frames, but not on the class of all frames; see also [9]), the technique employed was rather involved and the intuition behind was not clear (which has been mentioned by many of our colleagues).

The main aim of this paper is to give simple and direct proofs for the modal case and to consider some variations of the three major problems (e.g., the case of countable frames). Our second goal is to show that in the proof of the undecidability of modal definability of first-order formulas, it is sufficient to use one simple modal formula, namely $\bot$.

1 Introduction

The question about the model-theoretic correspondence between nonclassical (modal, in the first place) propositional formulas and classical first-order formulas arose practically simultaneously with the creation of Kripke semantics: it turned out that models of first-order formulas in the signature of one binary predicate (and equality) can serve as 'model structures' (later called frames) for modal formulas. Already early papers in this area, say [13], observed that many natural conditions expressible in such structures by first-order formulas correspond to validity of some propositional modal formulas. We only remind the reader of two banal examples:

first-order formula	modal formula
$\forall x\, xRx$ (reflexivity)	$\Box p \to p$
$\forall x\, \forall y\, \forall z\, (xRy\ \&\ yRz \Rightarrow xRz)$ (transitivity)	$\Box p \to \Box\Box p$.

Those observations led to the creation of Correspondence Theory (see [1, 2]), which has become an integral part of the branch of mathematical logic known as 'Modal Logic.' Moreover, some authors consider modal logic as just one of the languages for describing properties of relational structures.[1]

We are not going to discuss here the role of Correspondence Theory in the general theory of modal logic; this topic has already been analysed by many authors; see, e.g., [1, 2]. The problems we consider here are much more modest: Can we, given a modal formula, recognize whether this formula has an equivalent first-order condition on modal frames? Conversely, is it possible, given a condition expressible by a first-order formula, to recognise whether it has a propositional modal equivalent? One more important problem is to recognise, for a first-order formula and a modal formula, whether they determine the same class of frames. (Here we take a rather wide view of modal formulas: they can be, for example, polymodal formulas or propositional dynamic formulas with the intended semantics.) It should also be noted that in this paper we only consider the so-called global variants of definability. However, our methods can easily be extended to the local cases.

A solution to the problems above for the (uni)modal language has been known since Chagrova's PhD thesis [4][2] where it was proved, in an indirect way, that all of them are undecidable. For example, [4] proves that definability of first-order formulas by intuitionistic propositional formulas (on the class of intuitionistic frames) is undecidable. To show that modal definability is undecidable as well, it suffices to replace the intuitionistic formulas used in the proof with their Gödel T-translations; for definitions see [10]. The resulting reduction of a certain undecidable problem to definability of first-order formulas by means of modal formulas, which contain as conjuncts the reflexivity and transitivity of the relation R, leads to the following dilemma: either the given first-order formula has no modal equivalent at all, or its modal equivalent is the T-translation of some intuitionistic formula with two extra conjuncts $\Box p \to p$ and $\Box p \to \Box\Box p$. We shall not go into further details here because this proof from [4] was a 'side effect' of rather cumbersome constructions required to solve more difficult and subtle problems such as first-order definability of intuitionistic formulas on the class of countable frames. In fact, as we show below, the direct proof of the undecidability of modal definability is not so complex. Note also that, according to the remark in [2] on page 398, the undecidability of modal definability of first-order sentences was proved by Frank Wolter.[3]

We were encouraged to write this paper by our friends and colleagues Ph. Balbiani, T. Tinchev and D. Vakarelov who discussed with us the algorithmic problems in Correspondence Theory during our stay at the Institut de recherche en informatique de Toulouse (Universite Paul Sabastier) in October–November 2003. Thus, Ph. Balbiani raised the algorithmic problem of definability of first-order formulas by propositional formulas on the

---

[1] See, for example, [3], p. 24: "So far we have been viewing modal languages as tools for talking about models."

[2] Parts of the thesis were also published as preprints [5, 6, 7].

[3] In fact, [2] refers to [15] which contains no such result. Moreover, Frank Wolter kindly informed us that he never published his proof, but discussed it in some seminars.

class of frames for propositional dynamic logic, D. Vakarelov was interested in definability of first-order formulas by variable-free modal formulas. The results of this paper are related to both extensions of modal languages and the number of variables used in modal formulas. T. Tinchev helped us to amend the original proofs. We are grateful to our colleagues. We would also like to thank the anonymous referees for their useful comments. Especially we are grateful to our friend Michael Zakharyaschev whose remarks and suggestions substantially helped to improve the presentation

2 Undecidability of first-order definability of modal formulas

In this section we prove the following:

THEOREM 1. *The problem of first-order definability of modal formulas is undecidable.*

Proof. Here and in what follows our 'master' problems are some standard undecidable problems for Minsky machines; for definitions see Section 5.1. We are going to construct a set of formulas denoted by

$$B(\mathcal{P}, (\alpha, m, n), (\beta, k, l))$$

with the following properties:

1. $B(\mathcal{P}, (\alpha, m, n), (\beta, k, l))$ is constructed effectively for a given Minsky machine program $\mathcal{P}$ and configurations (α, m, n), (β, k, l).

2. The following equivalence holds:

$$B(\mathcal{P}, (\alpha, m, n), (\beta, k, l)) \text{ is first-order definable}$$
$$\Updownarrow$$
$$\mathcal{P} : (\alpha, m, n) \to (\beta, k, l).$$

As the problem '$\mathcal{P} : (\alpha, m, n) \to (\beta, k, l)$?' is undecidable, these two properties will prove Theorem 1.

According to the second item, if $\mathcal{P} : (\alpha, m, n) \to (\beta, k, l)$, then the formula $B(\mathcal{P}, (\alpha, m, n), (\beta, k, l))$ must be first-order definable. We ensure this by choosing our formula to be deductively equivalent in this case to the formula $\neg H$, where $H = \Diamond(\Box^2 \bot \wedge \Diamond \neg p \wedge \Diamond p)$. This formula is clearly first-order definable. (Perhaps some readers would prefer the equivalent formula $\Box(\Box^2 \bot \to \Box p \vee \Box \neg p)$.)

Observe that to be able to refute $\neg H$ (or to satisfy H) a frame $\mathfrak{F}$ should satisfy the following condition (note a clear similarity between this condition and the formula $\neg H$ itself):

$$\mathfrak{F} \models \exists x \exists y \, (xRy \wedge \neg \exists z \, yR^2 z \wedge \exists u \exists v(yRu \wedge yRv \wedge u \neq v)).$$

Conversely, if $\mathfrak{F}$ satisfies this condition, then one can define a valuation in $\mathfrak{F}$ satisfying H. Indeed, let x, y, u, v be the points the existence of which is asserted by this condition. Take $u \not\models p$ and $v \models p$. The condition $\neg \exists z\, yR^2z$ guarantees that $y \models \Box^2 \bot$. Together with yRu and yRv, this gives $y \models \Box^2\bot \wedge \Diamond \neg p \wedge \Diamond p$, from which we obtain $x \models H$ because xRy.

Thus we have:

LEMMA 2. $\neg H$ is first-order definable by the formula

$$\neg \exists x \exists y\, (xRy \wedge \neg \exists z\, yR^2z \wedge \exists u \exists v (yRu \wedge yRv \wedge u \neq v)).$$

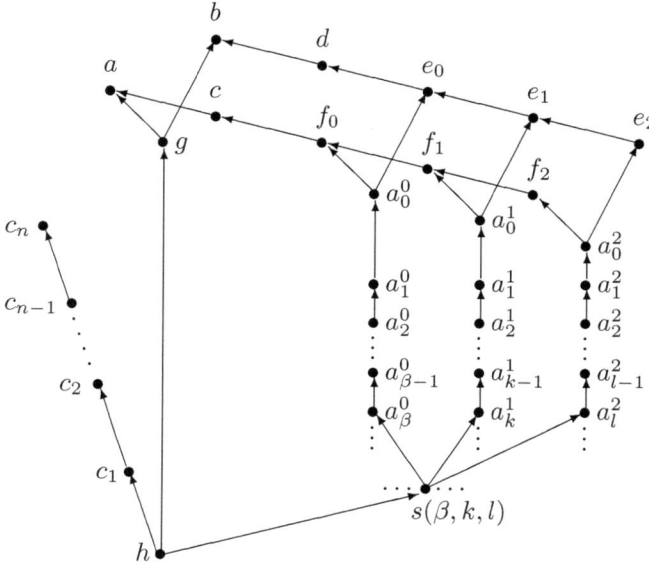

Figure 1. Frame $\mathfrak{F}_n$.

Now fix a Minsky machine $\mathcal{P}$ and a configuration (α, m, n). To understand the meaning of the formulas below the reader should consult the frame $\mathfrak{F}_n$ depicted in Fig. 1. We briefly explain its construction. First, all the points • are supposed to be irreflexive. The accessibility relation xRy in $\mathfrak{F}_n$ is the transitive closure of the relation defined by the arrows $\to$. The points of the form $a_0^i, a_1^i, a_2^i, \ldots$ have natural numbers as subscripts. The meaning of these subscripts is as follows: a_β^0 represents the state of $\mathcal{P}$, a_k^1 represents the information on the first tape, and a_l^2 the information on the second tape. The points $s(\beta, k, l)$ form the set $\{s(\beta, k, l) \mid \mathcal{P} : (\alpha, m, n) \to (\beta, k, l)\}$, that is, we include in $\mathfrak{F}_n$ the point $s(\beta, k, l)$ only if the configuration (β, k, l) is reachable from (α, m, n) by $\mathcal{P}$. We also assume that

$$s(\beta, k, l)Ra_\beta^0, \quad s(\beta, k, l)Ra_k^1, \quad s(\beta, k, l)Ra_l^2,$$

but none of the relations

$$s(\beta, k, l)Ra_{\beta+1}^0, \quad s(\beta, k, l)Ra_{k+1}^1, \quad s(\beta, k, l)Ra_{l+1}^2$$

holds. Note that $\mathfrak{F}_n$ refutes the formula $\neg H$. Moreover, if $\mathfrak{F}_n \not\models \neg H$ then

either $\quad a \models \Box\bot \wedge p,\ b \models \Box\bot \wedge \neg p \quad$ or $\quad a \models \Box\bot \wedge \neg p,\ b \models \Box\bot \wedge p$.

Since these two cases are symmetric, we will only consider the former.

Now define the formulas:

$$A = \Box\bot \wedge p,\ B = \Box\bot \wedge \neg p,$$

$$C = \Diamond A \wedge \neg \Diamond\Diamond A \wedge \neg \Diamond B,\ D = \Diamond B \wedge \neg \Diamond\Diamond B \wedge \neg \Diamond A,$$

$$G = \Box^2 \bot \wedge \Diamond \neg p \wedge \Diamond p,\ H = \Diamond G,$$

$$E_0 = \Diamond D \wedge \neg \Diamond\Diamond D,\ E_1 = \Diamond E_0 \wedge \neg \Diamond\Diamond E_0 \wedge \neg \Diamond C,$$

$$E_2 = \Diamond E_1 \wedge \neg \Diamond\Diamond E_1 \wedge \neg \Diamond C,\ F_0 = \Diamond C \wedge \neg \Diamond\Diamond C,$$

$$F_1 = \Diamond F_0 \wedge \neg \Diamond\Diamond F_0 \wedge \neg \Diamond D,\ F_2 = \Diamond F_1 \wedge \neg \Diamond\Diamond F_1 \wedge \neg \Diamond D,$$

$$A_0^0 = \Diamond E_0 \wedge \Diamond F_0 \wedge \neg \Diamond\Diamond E_0 \wedge \neg \Diamond\Diamond F_0,\ A_0^1 = \Diamond E_1 \wedge \Diamond F_1 \wedge \neg \Diamond\Diamond E_1 \wedge \neg \Diamond\Diamond F_1,$$

$$A_0^2 = \Diamond E_2 \wedge \Diamond F_2 \wedge \neg \Diamond\Diamond E_2 \wedge \neg \Diamond\Diamond F_2,$$

$$A_{j+1}^i = \Diamond A_j^i \wedge \neg \Diamond^2 A_j^i \wedge \bigwedge_{i \neq k = 0}^{2} \neg \Diamond A_0^k,$$

where $i \in \{0, 1, 2\}$, $j \geq 0$, and let

$$S(\beta, \varphi, \psi) = \Diamond A_\beta^0 \wedge \neg \Diamond A_{\beta+1}^0 \wedge \Diamond \varphi \wedge \neg \Diamond\Diamond \varphi \wedge \Diamond \psi \wedge \neg \Diamond\Diamond \psi$$

for any formulas φ, ψ, and $\beta \geq 0$.

LEMMA 3. *Let a valuation in $\mathfrak{F}_n$ be such that $\mathfrak{F}_n \not\models \neg H$. Then the following hold:*

1. $\{x \mid x \models H\} = \{h\}$;
2. $\{x \mid x \models A\} = \{a\}$, $\{x \mid x \models B\} = \{b\}$;
3. $\{x \mid x \models C\} = \{c\}$, $\{x \mid x \models D\} = \{d\}$;
4. $\{x \mid x \models E_i\} = \{e_i\}$, $\{x \mid x \models F_i\} = \{f_i\}$ for $i \in \{0, 1, 2\}$;
5. $\{x \mid x \models A_0^i\} = \{a_0^i\}$ for $i \in \{0, 1, 2\}$;
6. $\{x \mid x \models A_j^i\} = \{a_j^i\}$ for $i \in \{0, 1, 2\}$, $j > 0$;
7. for $\beta, k, l \geq 0$

$$\{x \mid x \models S(\beta, A_k^1, A_l^2)\} = \begin{cases} \{s(\beta, k, l)\}, & \text{if } \mathcal{P} : \langle \alpha, m, n \rangle \to \langle \beta, k, l \rangle; \\ \emptyset, & \text{if } \mathcal{P} : \langle \alpha, m, n \rangle \not\to \langle \beta, k, l \rangle \ . \end{cases}$$

Proof. The straightforward proof is left to the reader (the proof of item 6 is by induction on j). ∎

REMARK 4. If $\neg H$ is refuted under a valuation with $a \models \Box\bot \land \neg p$ and $b \models \Box\bot \land p$, then items 2, 3, 4 of the lemma above should be rewritten as follows:

2. $\{x \mid x \models A\} = \{b\}$, $\{x \mid x \models B\} = \{a\}$;
3. $\{x \mid x \models C\} = \{d\}$, $\{x \mid x \models D\} = \{c\}$;
4. $\{x \mid x \models E_i\} = \{f_i\}$, $\{x \mid x \models F_i\} = \{e_i\}$ for $i \in \{0,1,2\}$.

These changes have no influence on what follows.

Now define the following formulas:

$$Q_1 = (\Diamond A_0^1 \lor A_0^1) \land \neg \Diamond A_0^0 \land \neg \Diamond A_0^2 \land p_1 \land \neg \Diamond p_1,$$
$$Q_2 = \Diamond A_0^1 \land \neg \Diamond A_0^0 \land \neg \Diamond A_0^2 \land \Diamond p_1 \land \neg \Diamond \Diamond p_1,$$
$$R_1 = (\Diamond A_0^2 \lor A_0^2) \land \neg \Diamond A_0^0 \land \neg \Diamond A_0^1 \land p_2 \land \neg \Diamond p_2,$$
$$R_2 = \Diamond A_0^2 \land \neg \Diamond A_0^0 \land \neg \Diamond A_0^1 \land \Diamond p_2 \land \neg \Diamond \Diamond p_2.$$

Some properties of these formulas are formulated in the following lemma.

LEMMA 5. (i) *Let a valuation in $\mathfrak{F}_n$ be such that $x \models Q_1$ for some point x. Then, for some $i \geq 0$,*

$$\{x \mid x \models Q_1\} = \{a_i^1\}, \ \{x \mid x \models Q_2\} = \{a_{i+1}^1\}.$$

(ii) *Let a valuation in $\mathfrak{F}_n$ be such that $x \models Q_2$ for some point x. Then, for some $i \geq 1$,*

$$\{x \mid x \models Q_2\} = \{a_i^1\}, \ \{x \mid x \models Q_1\} = \{a_{i-1}^1\}.$$

(iii) *Let a valuation in $\mathfrak{F}_n$ be such that $x \models R_1$ for some point x. Then, for some $i \geq 0$,*

$$\{x \mid x \models R_1\} = \{a_i^2\}, \ \{x \mid x \models R_2\} = \{a_{i+1}^2\}.$$

(iv) *Let a valuation in $\mathfrak{F}_n$ be such that $x \models R_2$ for some point x. Then, for some $i \geq 1$*

$$\{x \mid x \models R_2\} = \{a_i^2\}, \ \{x \mid x \models R_1\} = \{a_{i-1}^2\}.$$

Proof. Note first that if a valuation satisfies one of (i)–(iv) then $\mathfrak{F}_n \not\models \neg H$, more precisely $h \models H$.

As the proofs of all four claims are similar, we only consider (iv). By Lemma 3, $x \models R_2$ means that

$$x R a_0^2 \text{ and } \neg x R a_0^0, \neg x R a_0^1.$$

Therefore, by the structure of $\mathfrak{F}_n$, we obtain $x \in \{a_u^2 \mid u \geq 1\}$. Let i be such that $x = a_i^2$. Again, $x \models R_2$ implies

$$a_i^2 \models \Diamond p_2 \text{ and } a_i^2 \not\models \Diamond\Diamond p_2,$$

which means that there is a point accessible from a_i^2 where p_2 is true, and p_2 is false at any point that is accessible from a_i^2 in two (or more) steps. By the structure of $\mathfrak{F}_n$, it follows that a_{i-1}^2 is the only point in the frame accessible from a_i^2 where p_2 is true. This means, apart from other things, that a_i^2 is the only point in the set $\{a_u^2 \mid u \geq 1\}$ such that $a_u^2 \models \Diamond p_2 \land \neg \Diamond \Diamond p_2$ and a_{i-1}^2 is the only point in the set $\{a_u^2 \mid u \geq 1\}$ such that $a_u^2 \models p_2 \land \neg \Diamond p_2$. Therefore, by Lemma 3, we are done. ∎

We are now in a position to define a set of formulas AxI which corresponds to the instruction set of the program $\mathcal{P}$:

- if $I = \gamma \to \langle \delta, 1, 0 \rangle$ then

$$AxI = H \land \Diamond S(\gamma, Q_1, R_1) \to H \land \Diamond S(\delta, Q_2, R_1) ;$$

- if $I = \gamma \to \langle \delta, 0, 1 \rangle$ then

$$AxI = H \land \Diamond S(\gamma, Q_1, R_1) \to H \land \Diamond S(\delta, Q_1, R_2) ;$$

- if $I = \gamma \to \langle \delta_1, -1, 0 \rangle (\langle \delta_2, 0, 0 \rangle)$ then

$$\begin{aligned} AxI = \ & (H \land \Diamond S(\gamma, Q_2, R_1) \to H \land \Diamond S(\delta_1, Q_1, R_1)) \land \\ & \land (H \land \Diamond S(\gamma, A_0^1, R_1) \to H \land \Diamond S(\delta_2, A_0^1, R_1)) ; \end{aligned}$$

- if $I = \gamma \to \langle \delta_1, -1, 0 \rangle (\langle \delta_2, 0, 0 \rangle)$ then

$$\begin{aligned} AxI = \ & (H \land \Diamond S(\gamma, Q_1, R_2) \to H \land \Diamond S(\delta_1, Q_1, R_1)) \land \\ & \land (H \land \Diamond S(\gamma, Q_1, A_0^2) \to H \land \Diamond S(\delta_2, Q_1, A_0^2)) . \end{aligned}$$

Finally, define

$$Ax\mathcal{P} = \bigwedge_{I \in \mathcal{P}} AxI .$$

Denote by $\mathbf{K} \oplus Ax\mathcal{P}$ the closure of $\mathbf{K} \cup \{Ax\mathcal{P}\}$ under modus ponens, substitution and necessitation (and use a similar notation in similar cases). $L \vdash \varphi$ and $\varphi \in L$ are regarded to be synonymous.

LEMMA 6. *If*

$$P : (\alpha, m, n) \to (\beta, k, l)$$

then

$$\mathbf{K} \oplus Ax\mathcal{P} \vdash H \land \Diamond S(\alpha, A_m^1, A_n^2) \to H \land \Diamond S(\beta, A_k^1, A_l^2) .$$

Proof. The proof is by induction on the length of the computation from (α, m, n) to (β, k, l).

For example, consider a step of the computation where we use the instruction I of the form 'in state q_γ, add 1 to s_1, and go to state q_δ.' The resulting configuration will be $(\delta, s_1 + 1, s_2)$. Using substitution, from the formula AxI corresponding to I, that is

$$H \wedge \Diamond S(\gamma, Q_1, R_1) \to H \wedge \Diamond S(\delta, Q_2, R_1),$$

we obtain a formula that is equivalent to

$$H \wedge \Diamond S(\gamma, A^1_{s_1}, A^2_{s_2}) \to H \wedge \Diamond S(\delta, A^1_{s_1+1}, A^2_{s_2}).$$

Therefore,

$$\mathbf{K} \oplus Ax\mathcal{P} \vdash H \wedge \Diamond S(\gamma, A^1_{s_1}, A^2_{s_2}) \to H \wedge \Diamond S(\delta, A^1_{s_1+1}, A^2_{s_2}).$$

Other instructions are considered in the same way. ∎

LEMMA 7. *For any $0 < n < \omega$, $\mathfrak{F}_n \models Ax\mathcal{P}$.*

Proof. A straightforward proof using Lemmas 3 and 5 is left to the reader. ∎

Define the following formula

$$\begin{aligned}B(\mathcal{P}, (\alpha, m, n), (\beta, k, l)) &= Ax\mathcal{P} \wedge \\ &((H \wedge \Diamond S(\alpha, A^1_m, A^2_n) \to H \wedge \Diamond S(\beta, A^1_k, A^2_l)) \to \neg H) \wedge \\ &(\neg H \vee (\Box(\Box q \to q) \to \Box q)).\end{aligned}$$

The definition of $B(\mathcal{P}, (\alpha, m, n), (\beta, k, l))$ is clearly constructive, which gives property 1 on page 123. Let us show that property 2 holds as well.

LEMMA 8. *If $\mathcal{P} : (\alpha, m, n) \to (\beta, k, l)$ then $B(\mathcal{P}, (\alpha, m, n), (\beta, k, l))$ is first-order definable.*

Proof. If $\mathcal{P} : (\alpha, m, n) \to (\beta, k, l)$ then, by Lemma 6, we have

$$\mathbf{K} \oplus Ax\mathcal{P} \vdash H \wedge \Diamond S(\alpha, A^1_m, A^2_n) \to H \wedge \Diamond S(\beta, A^1_k, A^2_l),$$

and, by modus ponens,

$$\mathbf{K} \oplus B(\mathcal{P}, (\alpha, m, n), (\beta, k, l)) \vdash \neg H.$$

Besides, we have

$$\mathbf{K} \oplus \neg H \vdash B(\mathcal{P}, (\alpha, m, n), (\beta, k, l)).$$

This means that $B(\mathcal{P}, (\alpha, m, n), (\beta, k, l))$ is deductively equivalent to $\neg H$ and, therefore, is first-order definable by Lemma 2. ∎

LEMMA 9. *If $\mathcal{P} : (\alpha, m, n) \not\to (\beta, k, l)$ then $B(\mathcal{P}, (\alpha, m, n), (\beta, k, l))$ is not first-order definable.*

Proof. Let $\mathcal{P} : (\alpha, m, n) \not\to (\beta, k, l)$. Then $B(\mathcal{P}, (\alpha, m, n), (\beta, k, l))$ is valid in the frame $\mathfrak{F}_n$, for any $0 < n < \omega$. Indeed,

$\mathfrak{F}_n \models Ax\mathcal{P}$ by Lemma 7;

$\mathfrak{F}_n \models (H \wedge \Diamond S(\alpha, A_m^1, A_n^2) \to H \wedge \Diamond S(\beta, A_k^1, A_l^2)) \to \neg H$ by Lemma 3;

$\mathfrak{F}_n \models \neg H \vee (\Box(\Box q \to q) \to \Box q)$ because there is no infinite ascending chain in this frame.

Suppose that $B(\mathcal{P}, (\alpha, m, n), (\beta, k, l))$ is equivalent to some first-order formula φ. Then we have $\mathfrak{F}_n \models \varphi$, for any $0 < n < \omega$.

Consider the formula

$$G(x, y) = xRy \wedge \neg \exists z\, yR^2z \wedge \exists u \exists v (yRu \wedge yRv \wedge u \neq v).$$

In the frame $\mathfrak{F}_n$, this formula can select the points h and g; more precisely, we have

$$\mathfrak{F}_n \models G(x, y) \iff x = h \text{ and } y = g.$$

Using this fact, we obtain that the following formulas are true in the $\mathfrak{F}_n$, for any n:

$\forall x\, \neg xRx,$ \hfill (1)
$\forall x \forall y \forall z\, (xRy \wedge yRz \to xRz),$ \hfill (2)
$\exists x \exists y\, G(x, y),$ \hfill (3)
$\forall x \forall y\, (G(x, y) \to \exists ! z\, G(x, z)),$ \hfill (4)
$\forall x\, (\exists y\, G(x, y) \to \forall z\, (x = z \vee xRz)).$ \hfill (5)

Consider the formulas

$$S(k) = \forall x \forall y \forall u \forall v\, (xRy \wedge \neg \exists z yR^2z \wedge yRu \wedge yRv \wedge u \neq v \to$$
$$\exists w_1 \ldots \exists w_k\, (xRw_1 \ldots w_{k-1}Rw_k \wedge \neg(w_kRu \vee w_kRv \vee w_k = u \vee w_k = v) \wedge$$
$$\forall z\, (\neg(zRu \vee zRv \vee z = u \vee z = v) \to z = w_1 \vee \cdots \vee z = w_k \vee w_kRz))).$$

We have $\mathfrak{F}_n \models S(k)$, for any $n \geq k$.

Now we form the ultraproduct of all $\mathfrak{F}_n$ with respect to some nonprincipal ultrafilter over $\omega - \{0\}$ and denote it by $\mathfrak{F}$. By the Łoś theorem[4], for any $k \in \omega - \{0\}$,

$$\mathfrak{F} \models \varphi \wedge (1) \wedge (2) \wedge (3) \wedge (4) \wedge (5) \wedge S(k).$$

By (3), we have in $\mathfrak{F}$ four points h, g, a, b such that hRg, gRa, gRb (there is no z such that gR^2z), and h, g are unique with these properties by (4) and (5).

By $S(k)$, we obtain a sequence $c_1, c_2, \ldots, c_k, \ldots$ such that $\neg c_kRa, \neg c_kRb$, $c_k \neq a, c_k \neq b$, and

$$c_1Rc_2Rc_3 \ldots c_{k-1}Rc_kRc_{k+1} \ldots.$$

---
[4]For notions and results of classical model theory one can consult [11].

Define V on $\mathfrak{F}$ as follows: $V(p) = \{a\}$, $V(q) = W - \{c_1, c_2, \ldots\}$. Then we have $a \models \Box\bot \land p$, $b \models \Box\bot \land \neg p$, $g \models \Box^2\bot \land \Diamond p \land \Diamond \neg p$, $h \models H$, so

$$h \not\models \neg H.$$

Besides,
$$h \not\models \Box(\Box q \to q) \to \Box q.$$

Therefore,
$$\mathfrak{F} \not\models B(\mathcal{P}, (\alpha, m, n), (\beta, k, l)),$$

contrary to $\mathfrak{F} \models \varphi$. ∎

By Lemmas 8 and 9, we obtain that property 2 on page 123 is satisfied. This completes the proof of Theorem 1. ∎

3 Undecidability of modal definability

In this section we give a solution to the problem that is in some sense inverse to the problem considered in Section 2: is it possible to check algorithmically whether a given first-order formula is modally definable? In other words, is modal definability of first-order formulas decidable? However, this case is considerably simpler, since the language of first-order logic is more natural and, therefore, easier to use than the language of propositional modal logic.

THEOREM 10. *Modal definability of first-order formulas in the signature of one binary predicate and equality is algorithmically undecidable.*

Proof. We will use the transitive irreflexive frame $\mathfrak{F}$ (which is also a model for first-order predicate logic) that is obtained from $\mathfrak{F}_n$ in Fig. 1 by replacing the finite chain $c_1 R c_2 R \ldots c_{n-1} R c_n$ with the infinite chain $c_1 R c_2 R \ldots c_{n-1} R c_n \ldots$; see Fig. 2.

The first-order formulas associated with this frame can be found in Section 5.2. Here we only give their English translations. The first-order formula AxI imitating an instruction I of a Minsky machine means 'given a configuration, apply I and move to the next configuration.' It can be constructed analogously to its modal counterpart AxI from Section 2. The formula $Ax\mathcal{P}$ contains some additional conjuncts expressing some properties of $\mathfrak{F}$:

$$Ax\mathcal{P} = \bigwedge_{I \in \mathcal{P}} AxI \land IRR \land TRANS \land LIN_1 \land LIN_2,$$

where

- IRR says that the accessibility relation is irreflexive,

- $TRANS$ says that the accessibility relation is transitive,

- LIN_1 says that the set $\{a_i^1 \mid i \geq 0\}$ is linearly ordered by the accessibility relation,

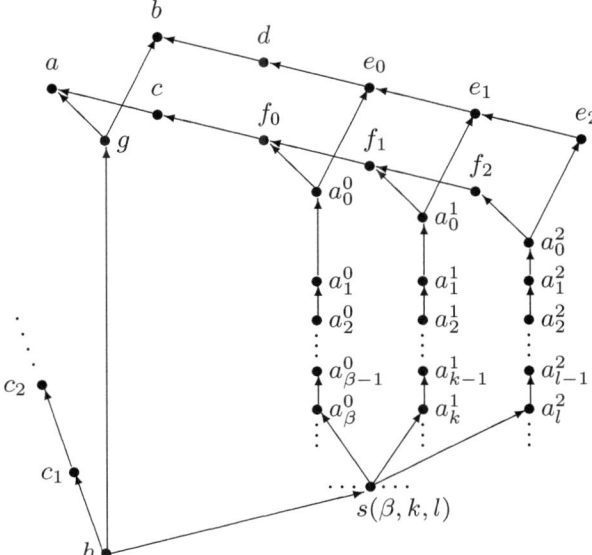

Figure 2. Frame $\mathfrak{F}$.

- LIN_2 says that the set $\{a_i^2 \mid i \geq 0\}$ is linearly ordered by the accessibility relation.

In Lemma 11 below we only need the first conjunct $\bigwedge_{I \in \mathcal{P}} AxI$ of $Ax\mathcal{P}$ and a first-order formula $A(\langle s_1, i_1, j_1 \rangle, \langle s_n, i_n, j_n \rangle)$ which says that if there is a point of the form $s(s_1, i_1, j_1)$ in $\mathfrak{F}$, then there is a point of the form $s(s_n, i_n, j_n)$ as well.

LEMMA 11. *If*
$$\mathcal{P} : \langle s_1, i_1, j_1 \rangle \to \langle s_n, i_n, j_n \rangle,$$
then
$$Ax\mathcal{P} \models A(\langle s_1, i_1, j_1 \rangle, \langle s_n, i_n, j_n \rangle).$$

Proof. The straightforward proof by induction on n (the key case is $n = 2$) is left to the reader. ∎

We are now in a position to prove Theorem 10, where the following formula is required.

Consider a machine $\mathcal{P}$ with configurations $\langle s_1, i_1, j_1 \rangle$ and $\langle s_n, i_n, j_n \rangle$, and let

$$B(\mathcal{P}, \langle s_1, i_1, j_1 \rangle, \langle s_n, i_n, j_n \rangle) = Ax\mathcal{P} \wedge \neg A(\langle s_1, i_1, j_1 \rangle, \langle s_n, i_n, j_n \rangle).$$

Since this formula is constructed effectively, our theorem follows immediately from the undecidability of the configuration problem and the following lemma:

LEMMA 12. *For every Minsky machine $\mathcal{P}$ and all configurations $\langle s_1, i_1, j_1 \rangle$, $\langle s_n, i_n, j_n \rangle$ the following equivalence holds:*

$$B(\mathcal{P}, \langle s_1, i_1, j_1 \rangle, \langle s_n, i_n, j_n \rangle) \text{ is modally definable}$$
$$\Updownarrow$$
$$\mathcal{P} : \langle s_1, i_1, j_1 \rangle \to \langle s_n, i_n, j_n \rangle.$$

Proof. ($\Uparrow$) follows from Lemma 11. In the case $\mathcal{P} : \langle s_1, i_1, j_1 \rangle \to \langle s_n, i_n, j_n \rangle$ the formula $B(\mathcal{P}, \langle s_1, i_1, j_1 \rangle, \langle s_n, i_n, j_n \rangle)$ is trivially modally definable by $\bot$.

To show ($\Downarrow$), suppose the contrary: we assume that φ is a modal equivalent of $B(\mathcal{P}, \langle s_1, i_1, j_1 \rangle, \langle s_n, i_n, j_n \rangle)$, yet $\mathcal{P} : \langle s_1, i_1, j_1 \rangle \not\to \langle s_n, i_n, j_n \rangle$.

Consider the frame $\mathfrak{F}$ in Fig. 2. In this frame, all points $s(\beta, k, l)$ are such that $\mathcal{P} : \langle s_1, i_1, j_1 \rangle \to \langle \beta, k, l \rangle$. One can easily check that

$$\mathfrak{F} \models B(\mathcal{P}, \langle s_1, i_1, j_1 \rangle, \langle s_n, i_n, j_n \rangle),$$

and so

$$\mathfrak{F} \models \varphi.$$

Now take the frame $\mathfrak{F}'$ that is reconstructed from $\mathfrak{F}$ by replacing the infinite chain $c_1 R c_2 R \ldots$ with a single reflexive point c. Clearly, $\mathfrak{F}'$ is a p-morphic image of $\mathfrak{F}$, from which

$$\mathfrak{F}' \models \varphi.$$

On the other hand, by the conjunct IRR, we have

$$\mathfrak{F}' \not\models B(\mathcal{P}, \langle s_1, i_1, j_1 \rangle, \langle s_n, i_n, j_n \rangle),$$

which is a contradiction. ∎

This completes the proof of Theorem 10. ∎

4 Discussion

Since the main problem solved in Section 2 was concerned with the model-theoretic (equivalence) relation between modal formulas and first-order formulas, it would be useful to recall that, for first-order sentences φ and ψ, the following statements are equivalent (by the well-known Gödel completeness theorem and Löwenheim-Skolem-Tarski theorem):

- φ and ψ are true in the same models;

- φ and ψ are true in the same countable models;

- φ and ψ are deductively equivalent;

- the formula $\varphi \leftrightarrow \psi$ is derivable in the predicate calculus.

It is now natural to ask whether the same holds for modal formulas. The answer has been known for some time now: the behaviour of modal formulas is quite different. Here we mean various incompleteness effects (for the exact definitions and references see [10]): Kripke incompleteness of some modal calculi and the fact that countable frames are not sufficient for some Kripke complete modal logics. It would be fair to mention that the last statement is still proved only for logics: there are a normal modal logic L and a formula φ such that L is Kripke complete, $L \nvdash \varphi$, and φ is true in all countable frames of L. It is not known whether a *logic* in this statement can be replaced with a *calculus* (a finitely axiomatisable logic).

In the context of this paper it would be relevant to investigate first-order definability of modal formulas on frames of bounded cardinality. In particular, what can be said about (the algorithmic problem of) first-order definability of modal formulas on the class of countable frames?

The answer to this question is almost obvious; the situation with such definability is the same as with first-order definability in the general case.

THEOREM 13. *The problem of first-order definability of modal formulas on countable frames is undecidable.*

Proof. First, we do not have to change anything in the proof of Theorem 1 in the case when $\mathfrak{F}$ is not just an ultraproduct of the frames $\mathfrak{F}_n$ with respect to some nonprincipal ultrafilter over $\omega - \{0\}$, but some countable elementary submodel of it. Such a submodel exists by the Löwenheim-Skolem-Tarski theorem, since the formula $\varphi \wedge (1) \wedge (2) \wedge (3) \wedge (4) \wedge (5) \wedge S(k)$ is still true.

Now, Theorem 13 can be proved simultaneously with Theorem 1. Indeed, in that proof, uncountable frames can be used (and are used) only in one place, where the ultraproduct is considered in Lemma 9. However, instead of the ultraproduct we can use its countable elementary submodel. Therefore, Lemma 9 can be reformulated in a stronger form: *if* $\mathcal{P} : (\alpha, m, n) \nrightarrow (\beta, k, l)$ *then* $B(\mathcal{P}, (\alpha, m, n), (\beta, k, l))$ *is not first-order definable on the class of all countable frames.*

This change is enough to prove Theorem 13. ∎

The reader may wonder whether Theorem 1 and Theorem 13 are equivalent. Can we obtain either theorem as a simple corollary of the other one (or is at least one of the theorems a corollary of the other one)? The answer to this question is negative, and the following example demonstrates why.

EXAMPLE 14. It is shown in K. Doets' PhD thesis [12] that K. Fine's formula
$$\Diamond \Box (p \vee q) \rightarrow \Diamond (\Box p \vee \Box q)$$
is not first-order definable, although it is first-order definable on countable frames. An analogous intuitionistic formula was constructed in the PhD thesis [4]. It was shown that the set of intuitionistic formulas that have this property is algorithmically undecidable.

Let us now analyse the proof of Theorem 10. The fact that we were dealing with *modal* formulas was actually used only once in this proof, namely, at the very end we used persistence of modal formulas under p-morphisms. Analogous effects can be obtained by using other model-theoretic constructions. For example, we can use persistence with respect to generated submodels: just replace the increasing chain $hRc_1Rc_2R\ldots$ with the decreasing chain $\ldots Rc_2Rc_1Rh$ and the conjunct IRR of $Ax\mathcal{P}$ with the formula $\forall x\, \exists y\, yRx$. Besides, we can replace the condition of irreflexivity with the condition of reflexivity. These modifications give numerous analogs of Theorem 10: for definability of first-order formulas by (poly)modal formulas on the classes of **Grz**-frames, **GL**-frames, **PDL**-frames, etc., for definability of first-order formulas by intuitionistic formulas, and many others.

Note also that in the proof of Theorem 10 we used only one modal formula: the formula $\bot$. Therefore, we can replace 'modal definability' with 'definability by variable-free modal formulas' or even 'definability by modal formula $\bot$' everywhere in the proof. This immediately gives us the following two results.

THEOREM 15. *Definability of first-order formulas in the signature of one binary predicate and equality by variable-free modal formulas is undecidable.*

THEOREM 16. *Definability of first-order formulas in the signature of one binary predicate and equality by the modal formula $\bot$ is undecidable.*

Furthermore, in these theorems we can replace 'decidability' with 'recursive separability.' Thus, the following variant of Theorem 16 holds:

THEOREM 17. *Let A be the set of first-order formulas in the signature of one binary predicate and equality which are definable by modal formula $\bot$, and let B be the set of first-order formulas in the signature of one binary predicate and equality that are not modally definable. Then the sets A and B are not recursively separable.*

Other theorems above can also be reformulated in the same way. (We only did it for Theorem 16 because it is an immediate consequence of Church's theorem on the undecidability of the first-order predicate logic in the given signature.)

Using a slight modification of the frame $\mathfrak{F}$ in Fig. 1 and first-order equivalents of modal formulas of the form (see Lemma 2 for $n=1$)

$$\Box(\Box^2\bot \to \bigvee_{\langle i_1,\ldots,i_n\rangle\in\{0,1\}^n} \Box(\neg^{i_1}p_1 \wedge \cdots \wedge \neg^{i_n}p_n))$$

where $\neg^1 = \neg$ and $\neg^0$ is blank, we can prove some extension of Theorem 15.

THEOREM 18. *For any $0 < n < \omega$, the set of first-order formulas in the signature of one binary predicate and equality which are definable by modal formulas in n variables, but are not definable by modal formulas in $n-1$ variables is undecidable.*

Finally, we observe that the proofs of Theorems 1 and 10 give a negative solution to the correspondence problem, that is the question whether a given modal formula and a given first-order formula are equivalent on Kripke frames:

THEOREM 19. *The correspondence problem of modal formulas and first-order formulas is undecidable.*

5 Appendix

5.1 Minsky machines

A two-tape *Minsky machine* (today a Minsky machine would be described as a register machine with two registers) is a finite set of instructions (program $\mathcal{P}$) for transforming triples $\langle s, m, n \rangle$ of natural numbers, called *configurations*. The intended meaning of the components in the current configuration $\langle s, m, n \rangle$ is as follows: s is the number (label) of the current machine state, i.e. the number of the instruction to be executed at the next step, and m, n represent the current state of information (i.e. records on the first and the second tapes). Each instruction has one of the four possible forms:

$$s \to \langle t, 1, 0 \rangle, \ s \to \langle t, 0, 1 \rangle, \ s \to \langle t, -1, 0 \rangle\,(\langle t', 0, 0 \rangle), \ s \to \langle t, 0, -1 \rangle\,(\langle t', 0, 0 \rangle).$$

The last of them, for instance, means: transform $\langle s, m, n \rangle$ into $\langle t, m, n-1 \rangle$ if $n > 0$ and into $\langle t', m, n \rangle$ if $n = 0$. The meaning of the others is defined analogously. So, the instruction $s \to \langle t, 1, 0 \rangle$ means: transform $\langle s, m, n \rangle$ into $\langle t, m+1, n \rangle$.

If $\mathcal{P}$ is a Minsky machine then the notation $\mathcal{P} : \langle s, m, n \rangle \to \langle t, k, l \rangle$ means that starting with $\langle s, m, n \rangle$ and applying the instructions in $\mathcal{P}$, in finitely many steps we can reach the configuration $\langle t, k, l \rangle$.

As is well-known, the configuration problem for Minsky machines is undecidable: *there is no algorithm which, given a program $\mathcal{P}$ and configurations $\langle s, m, n \rangle$ and $\langle t, k, l \rangle$, can decide whether $\mathcal{P} : \langle s, m, n \rangle \to \langle t, k, l \rangle$ holds.*

This problem may be used for establishing a lot of undecidability results, but not all of them. Sometimes it is much more convenient to use a variant of the configuration problem with fixed suitable $\mathcal{P}$ and $\langle s, m, n \rangle$. It is called the *second configuration problem: there exist a program $\mathcal{P}$ and a configuration $\langle s, m, n \rangle$ such that there is no algorithm which is capable of deciding, given a configuration $\langle t, k, l \rangle$, whether $\mathcal{P} : \langle s, m, n \rangle \to \langle t, k, l \rangle$.*

For more details consult [10] and/or [14].

5.2 First-order formulas used in the proof of theorem 10

The following formulas AxI simulate the instructions I of a Minsky machine in the frame $\mathfrak{F}$ in Fig. 2:

- if $I = \gamma \to \langle \delta, 1, 0 \rangle$, then

$$AxI = \forall h, g, a, b, y_1, y_2 (hR_1g \wedge \neg \exists z\, gR_2z \wedge gR_1a \wedge gR_1b \wedge c \neq b \wedge$$
$$P(y_1) \wedge Q(y_2) \wedge \exists x \exists y_0 (xR_1y_0 \wedge xR_1y_1 \wedge xR_1y_2 \wedge A_\gamma^0(y_0)) \to$$
$$\exists x, y_0, z (xR_1y_0 \wedge xR_1zR_1y_1 \wedge xR_1y_2 \wedge A_\delta^0(y_0) \wedge P(z)));$$

- if $I = \gamma \to \langle \delta, 0, 1 \rangle$, then

$$AxI = \forall h,g,a,b,y_1,y_2(hR_1g \wedge \neg\exists z\, gR_2z \wedge gR_1a \wedge gR_1b \wedge a \neq b \wedge$$
$$P(y_1) \wedge Q(y_2) \wedge \exists x \exists y_0(xR_1y_0 \wedge xR_1y_1 \wedge xR_1y_2 \wedge A_\gamma^0(y_0)) \to$$
$$\exists x,y_0,z(xR_1y_0 \wedge xR_1y_1 \wedge xR_1zR_1y_2 \wedge A_\delta^0(y_0) \wedge Q(z)));$$

- if $I = \gamma \to \langle \delta_1, -1, 0\rangle\, (\langle \delta_2, 0, 0\rangle)$, then
$$AxI = \forall h,g,a,b,y_1,y_2(hR_1g \wedge \neg\exists z\, gR_2z \wedge gR_1a \wedge gR_1b \wedge a \neq b \wedge P(y_1) \wedge$$
$$Q(y_2) \wedge \exists x,y_0,z(xR_1y_0 \wedge xR_1zR_1y_1 \wedge xR_1y_2 \wedge A_\gamma^0(y_0) \wedge P(z)) \to$$
$$\exists x, y_0(xR_1y_0 \wedge xR_1y_1 \wedge xR_1y_2 \wedge A_{\delta_1}^0(y_0))) \wedge$$
$$\forall h,g,a,b,y_1,y_2(hR_1g \wedge \neg\exists z\, gR_2z \wedge gR_1a \wedge gR_1b \wedge a \neq b \wedge A_0^1(y_1) \wedge$$
$$Q(y_2) \wedge \exists x, y_0(xR_1y_0 \wedge xR_1y_1 \wedge xR_1y_2 \wedge A_\gamma^0(y_0)) \to$$
$$\exists x,y_0(xR_1y_0 \wedge xR_1y_1 \wedge xR_1y_2 \wedge A_{\delta_2}^0(y_0)));$$

- if $I = \gamma \to \langle \delta_1, 0, -1\rangle\, (\langle \delta_2, 0, 0\rangle)$, then
$$AxI = \forall h,g,a,b,y_1,y_2(hR_1g \wedge \neg\exists z\, gR_2z \wedge gR_1a \wedge gR_1b \wedge a \neq b \wedge P(y_1) \wedge$$
$$Q(y_2) \wedge \exists x, y_0 \exists z(xR_1y_0 \wedge xR_1y_1 \wedge xR_1zR_1y_2 \wedge A_\gamma^0(y_0) \wedge Q(z)) \to$$
$$\exists x \exists y_0(xR_1y_0 \wedge xR_1y_1 \wedge xR_1y_2 \wedge A_{\delta_1}^0(y_0))) \wedge$$
$$\forall h,g,a,b,y_1,y_2(hR_1g \wedge \neg\exists z\, gR_2z \wedge gR_1a \wedge gR_1b \wedge a \neq b \wedge P(y_1) \wedge$$
$$A_0^2(y_2) \wedge \exists x, y_0(xR_1y_0 \wedge xR_1y_1 \wedge xR_1y_2 \wedge A_\gamma^0(y_0)) \to$$
$$\exists x, y_0(xR_1y_0 \wedge xR_1y_1 \wedge xR_1y_2 \wedge A_{\delta_2}^0(y_0))).$$

Here we use the following abbreviations:

- 'accessible in exactly k steps:'
$$xR_k y = xR^k y \wedge \neg xR^{k+1}y, \quad (k \geq 1),$$
where, as usual,
$$xR^0 y = (x = y), \quad xR^{k+1}y = \exists z\,(xR^k z \wedge zRy);$$

- 'the point is of the type a_0^0:'
$$A_0^0(x) = \exists y(xR_1 y \wedge yR_2 a \wedge \neg yRb) \wedge \exists y(xR_1 y \wedge yR_2 b \wedge \neg yRa);$$

- 'the point is of the type a_0^1:'
$$A_0^1(x) = \exists y(xR_1 y \wedge yR_3 a \wedge \neg yRb) \wedge \exists y(xR_1 y \wedge yR_3 b \wedge \neg yRa);$$

- 'the point is of the type a_0^2:'
$$A_0^2(x) = \exists y(xR_1 y \wedge yR_4 a \wedge \neg yRb) \wedge \exists y(xR_1 y \wedge yR_4 b \wedge \neg yRa);$$

- 'the point is of the type a_s^0:'
$$A_s^0(x) = \exists y(xR_s y \wedge A_0^0(y)) \wedge \neg \exists y(xRy \wedge A_0^1(y)) \wedge \neg \exists y(xRy \wedge A_0^2(y));$$

- 'the point is of the type a_t^1:'

$$A_t^1(x) = \exists y(xR_t y \wedge A_0^1(y)) \wedge \neg\exists y(xRy \wedge A_0^0(y)) \wedge \neg\exists y(xRy \wedge A_0^2(y));$$

- 'the point is of the type a_u^2:'

$$A_u^2(x) = \exists y(xR_u y \wedge A_0^2(y)) \wedge \neg\exists y(xRy \wedge A_0^0(y)) \wedge \neg\exists y(xRy \wedge A_0^1(y));$$

- 'the point is of the type a_t^1 with some arbitrary t:'

$$P(x) = (A_0^1(x) \vee \exists y(xRy \wedge A_0^1(y))) \wedge \neg\exists y(xRy \wedge A_0^0(y)) \wedge \neg\exists y(xRy \wedge A_0^2(y));$$

- 'the point is of the type a_u^2 with some arbitrary u:'

$$Q(x) = (A_0^2(x) \vee \exists y(xRy \wedge A_0^2(y))) \wedge \neg\exists y(xRy \wedge A_0^0(y)) \wedge \neg\exists y(xRy \wedge A_0^1(y)).$$

$Ax\mathcal{P}$ is defined as the conjunction of the following formulas:

- $IRR = \forall x \neg xRx$;
- $TRANS = \forall x, y, z \, (xRy \wedge yRz \rightarrow xRz)$;
- $LIN_1 = \forall h, g, a, b, y_1, y_2 (hR_1 g \wedge \neg \exists z \, gR_2 z \wedge gR_1 a \wedge gR_1 b \wedge a \neq b \rightarrow$
$\forall x \forall y (P_1(x) \wedge P_1(y) \rightarrow x = y \vee xRy \vee yRx));$
- $LIN_2 = \forall h, g, a, b, y_1, y_2 (hR_1 g \wedge \neg \exists z \, gR_2 z \wedge gR_1 a \wedge gR_1 b \wedge a \neq b \rightarrow$
$\forall x, y (Q_1(x) \wedge Q_1(y) \rightarrow x = y \vee xRy \vee yRx));$
- AxI, for every instruction I of a machine $\mathcal{P}$.

Finally, the formula saying that starting with a configuration we can eventually reach $\langle s_1, i_1, j_1 \rangle$ is defined as follows:

$A(\langle s_1, i_1, j_1 \rangle, \langle s_n, i_n, j_n \rangle) =$
$\forall h, g, a, b (hR_1 g \wedge \neg \exists z \, gR_2 z \wedge gR_1 a \wedge gR_1 b \wedge a \neq b \wedge$
$\exists x, y_0, y_1, y_2 (xR_1 y_0 \wedge xR_1 y_1 \wedge xR_1 y_2 \wedge A_{s_1}^0(y_0) \wedge A_{i_1}^1(y_1) \wedge A_{j_1}^2(y_2)) \rightarrow$
$\exists x, y_0, y_1, y_2 (xR_1 y_0 \wedge xR_1 y_1 \wedge xR_1 y_2 \wedge A_{s_n}^0(y_0) \wedge A_{i_n}^1(y_1) \wedge A_{j_n}^2(y_2))).$

BIBLIOGRAPHY

[1] J. F. A. K. van Benthem. *Modal Logic and Classical Logic*. Bibliopolis, 1983.
[2] J. F. A. K. van Benthem. Correspondence Theory. In: D. M. Gabbay and F. Guenthner (eds.). *Handbook of Philosophical Logic*. 2nd edition. Kluwer Academic Publishers, 325–408, 2001.
[3] P. Blackburn, M. de Rijke and Y. Venema *Modal Logic* Cambridge University Press, 2001.
[4] L. A. Chagrova. *On the problem of definability of propositional formulas of intuitionistic logic by formulas of classical first order logic*. PhD, Kalinin State University, 1989. (In Russian)

[5] L. A. Chagrova. *A superintuitionistic calculus simulating Minsky's machine*. Technical report, Kalinin State University, 1989. (In Russian)
[6] L. A. Chagrova. *First order definability of some superintuitionistic calculi simulating Minsky's machines*. Technical report, Kalinin State University, 1989. (In Russian)
[7] L. A. Chagrova. *Undecidable problems related to the first order definability of intuitionistic formulas*. Technical report, Kalinin State University, 1989. (In Russian)
[8] L. A. Chagrova. An undecidable problem in correspondence theory. *Journal of Symbolic Logic*, 56:1261–1272, 1991.
[9] A. V. Chagrov and L. A. Chagrova. Algorithmic problems concerning first order definability of modal formulas on the class of all finite frames. *Studia Logica*, 55:421–448, 1995.
[10] A. Chagrov and M. Zakharyaschev. *Modal Logic*. Oxford University Press, 1997.
[11] C. C. Chang and H. Jerome Keisler. *Model Theory*. Elsevier Science Publishers, 1990.
[12] K. Doets. *Completeness and Definability*. PhD, Universiteit van Amsterdam, 1987.
[13] S. Kripke. Semantical Analysis of Modal Logic I: Normal Propositional Calculi. *Zeitschrift für Mathematische Logik und Grundlagen der Mathematik*, 9:67–96, 1963.
[14] A. I. Mal'cev. *Algorithms and Recursive Functions*. Wolters-Noordhoff, Groningen, 1970.
[15] F. Wolter. *Lattices of Modal Logics*. PhD, Freie Universität Berlin, 1993. (Parts of this thesis are in [16])
[16] F. Wolter. The Structure of Lattices of Subframe Logics. *Annals of Pure and Applied Logic*, 86:47–100, 1997.

Alexander Chagrov, Lilia Chagrova
Department of Mathematics
Tver State University
Zhelyabova Street, 33
Tver
170000
Russia
chagrov@mail.ru

ML is not finitely axiomatizable over Cheq

Gaëlle Fontaine

ABSTRACT. We show that the Medvedev logic **ML** is not finitely axiomatizable over the logic **Cheq** of chequered subsets of $\mathbb{R}^\infty$. This gives a negative solution to one of the questions raised by Litak [2].

1 Introduction

In 1962 Medvedev [4] introduced the intermediate logic of "finite problems". It became known as the Medvedev logic **ML**. It is known that **ML** has the finite model property, the disjunction property, contains the Kreisel-Putnam and Scott logics, and is contained in the logic of weak excluded middle (see, e.g., [1]). In the late 1970's Maksimova et al. [3] showed that **ML** is not finitely axiomatizable by any set of formulas in finitely many variables. The question whether **ML** is decidable is one of the most long-standing open problems in the field of intermediate logics; see, e.g., [1, §16] for further discussion.

Recently, van Benthem et al. [5] introduced the modal logic of chequered subsets of $\mathbb{R}^\infty$ and showed that it has the finite model property. Litak [2] denoted the intermediate companion of this modal logic by **Cheq** and showed that it has the disjunction property, contains the Scott logic and is contained in the Medvedev logic **ML**. In fact, every finite rooted **Cheq**-frame is a p-morphic image of a finite rooted **ML**-frame. However, it is still unknown whether **Cheq** is finitely axiomatizable and/or decidable.

Litak [2] raised a question whether **ML** is finitely axiomatizable over **Cheq**. If this were the case, it would imply that **Cheq** is not finitely axiomatizable. In this note we give a negative solution to Litak's question by proving that **ML** is not finitely axiomatizable over **Cheq**. Thus, the connection between the Medvedev logic and **Cheq** is not as strong as it first appeared. It still remains an open problem whether **Cheq** is finitely axiomatizable.

2 ML is not finitely axiomatizable

We assume the reader's familiarity with basics of Kripke semantics for intermediate logics and refer to Chagrov and Zakharyaschev [1] for the details.

DEFINITION 1 (Maksimova et al. [3]). For a finite non-empty set D, let $\mathcal{P}^0(D)$ denote the Kripke frame

$$\mathcal{P}^0(D) = \langle \{X \subseteq D | X \neq \emptyset\}, \supseteq \rangle.$$

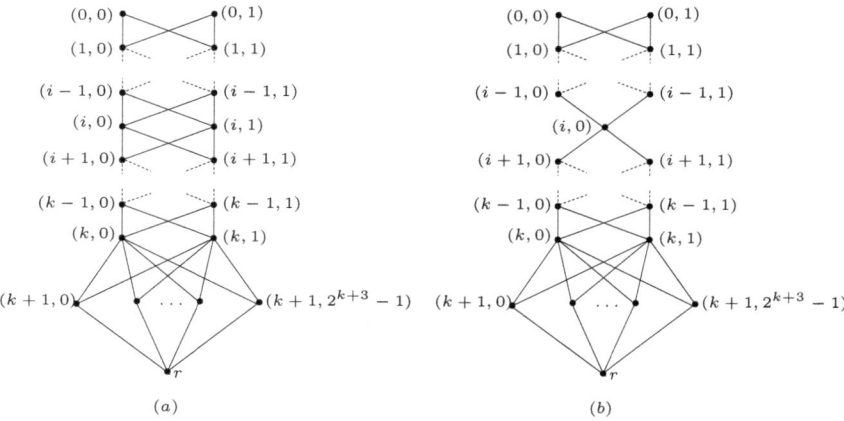

Figure 1. The frames $\mathcal{G}_k$ and $\mathcal{G}_k^i$.

We call $\mathcal{P}^0(D)$ a *Medvedev frame*. The intermediate logic **ML** is the logic of all Medvedev frames. As usual, a frame $\mathcal{F}$ is called an **ML**-*frame* if all the theorems of **ML** are valid in $\mathcal{F}$.

For each natural number $k \neq 0$ and each $i \leq k$, let $\mathcal{G}_k$ and $\mathcal{G}_k^i$ be the frames shown in Figure 1 (a) and (b), respectively. The following lemma is proved in [3].

LEMMA 2.

(a) *For each natural number $k > 0$, the frame $\mathcal{G}_k$ is not an **ML**-frame.*

(b) *For each natural number $k > 0$ and each $i \leq k$, the frame $\mathcal{G}_k^i$ is an **ML**-frame.*

(c) *Let φ be a formula with k variables. There exists a natural number $i \leq k$ such that*
$$\mathcal{G}_k \Vdash \varphi \text{ iff } \mathcal{G}_k^i \Vdash \varphi.$$

It is an easy corollary of Lemma 2 that **ML** is not finitely axiomatizable. Indeed, suppose there is a finite set of formulas axiomatizing **ML**. Without loss of generality we may assume that **ML** is axiomatized by a single formula φ with k variables (for some natural number k). By Lemma 2(c), there exists a natural number $i \leq k$ such that φ is valid in $\mathcal{G}_k$ iff φ is valid in $\mathcal{G}_k^i$. By Lemma 2(b), $\mathcal{G}_k^i$ is an **ML**-frame. Thus, φ is valid in $\mathcal{G}_k^i$. Therefore, φ is valid in $\mathcal{G}_k$. But $\mathcal{G}_k$ is not an **ML**-frame by Lemma 2(a). This contradiction proves that such a φ does not exist. Thus, we arrive at the following theorem.

THEOREM 3 (Maksimova et al. [3]). *The logic **ML** is not finitely axiomatizable. Moreover, it is not axiomatizable by any set of formulas in finitely many variables.*

Figure 2. The frame $\mathcal{F}_1$.

3 ML is not finitely axiomatizable over Cheq

DEFINITION 4 (van Benthem et al. [5]). Let $\mathcal{F}$ denote the two-fork Kripke frame shown in Figure 2. Let $\mathcal{F}_n = \underbrace{\mathcal{F} \times \cdots \times \mathcal{F}}_{n \text{ times}}$. The intermediate logic **Cheq** is the logic of $\{\mathcal{F}_n : n \in \mathbb{N}\}$.

THEOREM 5 (Litak [2]). **ML** *is a proper extension of* **Cheq**.

Our main goal is to show that **ML** is not finitely axiomatizable over **Cheq**. For an n-tuple x, let $N_i(x)$ denote the number of w_i that occur in x ($i = 0, 1, 2$). We denote the j-th component of x by $x(j-1)$. For a Kripke frame $\langle W, \leq \rangle$ and $w, v \in W$, we say that v is an *immediate successor* of w if $w \neq v$, $w \leq v$ and there is no $u \notin \{w, v\}$ such that $w \leq u$ and $u \leq v$. Note that if $x \in \mathcal{F}_n$ is an immediate successor of $(w_0, \ldots, w_0)$, it has only one component $x(i_0)$ that differs from w_0 and we denote it by $\delta(x)$.

For every $k > 1$ and $l > 0$, let $\mathcal{G}_{k,l}$ denote the frame shown in Figure 3 (note that $\mathcal{G}_k = \mathcal{G}_{k, 2^{k+3}-1}$).

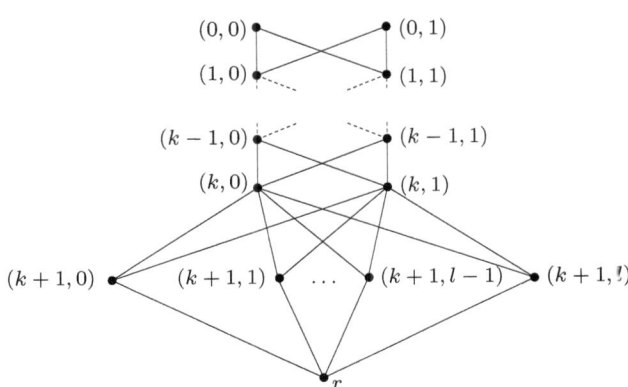

Figure 3. The frame $\mathcal{G}_{k,l}$.

PROPOSITION 6. *For each $l > 0$, there exists n such that $\mathcal{G}_{2,l}$ is a p-morphic image of $\mathcal{F}_n$. Moreover, there is a p-morphism f from $\mathcal{F}_n$ onto $\mathcal{G}_{2,l}$ such that $f^{-1}\{(3,i) | i \leq l\}$ is the set of all immediate successors of $(w_0, \ldots, w_0)$.*

Proof. Fix $l > 0$ and an arbitrary n so that $2n \geq l+1$ and $n > 3$. We show that there is a p-morphism f from $\mathcal{F}_n$ onto $\mathcal{G}_{2,l}$ such that $f^{-1}\{(3,i)|i \leq l\}$ is the set of all immediate successors of $(w_0, \ldots, w_0)$. Since $2n \geq l+1$, there is a map g from the set of immediate successors of $(w_0, \ldots, w_0)$ onto $\{(3,i)|i \leq l\}$.

Define f by

$$f(x) = \begin{cases} r & \text{if } x = (w_0, \ldots, w_0) \\ g(x) & \text{if } x \text{ is an immediate successor of } (w_0, \ldots, w_0) \\ (2,0) & \text{if } N_0(x) = n-2, x(i) = w_1, x(j) = w_2 \text{ and } i+j \text{ is even} \\ (2,1) & \text{if } N_0(x) = n-2, x(i) = w_1, x(j) = w_2 \text{ and } i+j \text{ is odd} \\ (1,0) & \text{if } x \text{ is not maximal}, N_1(x) > 1 \text{ and } N_2(x) \leq 1 \\ (1,1) & \text{if } x \text{ is not maximal}, N_2(x) > 1 \text{ and } N_1(x) \leq 1 \\ (0,0) & \text{if } x \text{ is maximal and either } N_1(x) = 1 \text{ or } N_2(x) = 1 \\ (0,1) & \text{otherwise.} \end{cases}$$

Obviously, f is a well-defined onto map such that $f^{-1}\{(3,i)|i \leq l\}$ is the set of all immediate successors of $(w_0, \ldots, w_0)$. We show that f is a p-morphism; that is, if $f(x) \leq u$, then there is y such that $x \leq y$ and $f(y) = u$ and if $x \leq y$, then $f(x) \leq f(y)$. First, we verify the former condition.

For $x \in \mathcal{F}_n$ and $u \in \mathcal{G}_{2,l}$, let $f(x) \leq u$. Then we need to find a $y \in \mathcal{F}_n$ such that $x \leq y$ and $f(y) = u$. If $f(x) = u$, then define y as x. From now on we assume that $f(x) \neq u$.

Since $\mathcal{G}_{2,l}$ is finite, there are $k \in \mathbb{N}$ and $u_0, \ldots, u_k \in \mathcal{G}_{2,l}$ such that $f(x) \leq u_0 \leq \cdots \leq u_k = u$, u_0 is an immediate successor of $f(x)$ and each u_{i+1} is an immediate successor of u_i. We show the existence of y by induction on k. If $k = 0$, u is an immediate successor of $f(x)$ and there are nine cases possible.

1. $x = (w_0, \ldots, w_0)$. Take any y such that $f(y) = u$.

2. x is an immediate successor of $(w_0, \ldots, w_0)$ and $u = (2,0)$. Without loss of generality we may assume that $x(i_0) = w_1$. Since $n > 3$, there is an index $i_1 \neq i_0$ such that $i_0 + i_1$ is even. Then take y such that $y(i_1) = w_2$ and $y(i) = x(i)$ for all $i \neq i_1$.

3. x is an immediate successor of $(w_0, \ldots, w_0)$ and $u = (2,1)$. Then the argument is similar to case (2).

4. $N_0(x) = n - 2$, $x(i) = w_1$, $x(j) = w_2$ and $u = (1,0)$. Since $n > 3$, there is an index i_0 such that $x(i_0) = w_0$. Then take y such that $y(i_0) = w_1$ and $y(i) = x(i)$ for all $i \neq i_0$.

5. $N_0(x) = n - 2$, $x(i) = w_1$, $x(j) = w_2$ and $u = (1,1)$. Then the argument is similar to case (4).

6. $N_1(x) > 1$, $N_2(x) \leq 1$ and $u = (0,0)$. If $N_2(x) = 1$, then there exists i_0 such that $x(i_0) = w_2$. Then take y such that $y(i_0) = w_2$

and $y(i) = w_1$ for all $i \neq i_0$. If $N_2(x) = 0$, fix an index i_0 such that $x(i_0) = w_0$ and take y such that $y(i_0) = w_2$ and $y(i) = w_1$ for all $i \neq i_0$.

7. $N_2(x) > 1$, $N_1(x) \leq 1$ and $u = (0,0)$. Then the argument is similar to case (6).

8. $N_1(x) > 1$, $N_2(x) \leq 1$ and $u = (0,1)$. If $N_2(x) = 0$, then define y as $(w_1, \ldots, w_1)$. If $N_2(x) = 1$, then there exists i_0 such that $x(i_0) = w_0$ and take y such that $y(i_0) = w_2$ and $y(i) = x(i)$ for all $i \neq i_0$.

9. $N_2(x) > 1$, $N_1(x) \leq 1$ and $u = (0,1)$. Then the argument is similar to case (8).

Next suppose that $k = k' + 1$. By the induction hypothesis, there is $y' \in \mathcal{F}_n$ such that $x \leq y'$ and $f(y') = u_{k'}$. Recall that $u = u_{k'+1}$ is an immediate successor of $u_{k'} = f(y')$. In the same way as we showed above, we can prove that there is $y \in \mathcal{F}_n$ such that $y' \leq y$ and $f(y) = u$. Therefore, we obtain that $x \leq y' \leq y$ and $f(y) = u$.

Finally we verify that if $x \leq y$, then $f(x) \leq f(y)$. Suppose $x, y \in \mathcal{F}_n$ are two distinct points such that $x \leq y$. We show that $f(x) \leq f(y)$. There are six cases possible.

1. $x = (w_0, \ldots, w_0)$. Then $f(x) = r$ and $r \leq f(y)$.

2. x is an immediate successor of $(w_0, \ldots, w_0)$. By the definition of f, we have $f(x)$ is equal to some $(3, i)$. Since y is not an immediate successor of $(w_0, \ldots, w_0)$, $f(y)$ is also not an immediate successor of r. Hence, $f(x) \leq f(y)$.

3. $N_0(x) = n - 2$, $x(i) = w_1$ and $x(j) = w_2$. By the definition of f $f(x)$ is either $(2,0)$ or $(2,1)$. Since $x \leq y$, we can deduce that either $N_1(y) > 1$ or $N_2(y) > 1$. In both cases this implies that $f(y)$ belongs to $\{(1,0), (1,1), (0,0), (0,1)\}$. So $f(x) \leq f(y)$.

4. x is not maximal, $N_1(x) > 1$ and $N_2(x) \leq 1$. From the definition of f it follows that $f(x) = (1,0)$. Moreover, since $x \leq y$, we also have that $N_1(y) > 1$. So $f(y)$ belongs to $\{(1,0), (0,0), (0,1)\}$. In any case, $f(x) \leq f(y)$.

5. x is not maximal, $N_2(x) > 1$ and $N_1(x) \leq 1$. Then the argument is similar to case (4).

6. $N_1(x) > 1$ and $N_2(x) > 1$. By the definition of f, we have that $f(x) = (0,1)$. Moreover, $x \leq y$ implies $N_1(y) > 1$ and $N_2(y) > 1$. So $f(y)$ is also equal to $(0,1)$.

PROPOSITION 7. *For each $k > 1$ and $l > 0$, there exists $n > 2$ such that $\mathcal{G}_{k,l}$ is a p-morphic image of $\mathcal{F}_n$. Moreover, there is a p-morphism f from $\mathcal{F}_n$ onto $\mathcal{G}_{k,l}$ such that $f^{-1}\{(k+1,i)|i \leq l\}$ is the set of all immediate successors of $(w_0, \ldots, w_0)$.*

Proof. The proof is by induction on k. If $k = 2$, apply Proposition 6. Suppose $k = k'+1$ and there is a p-morphism f from $\mathcal{F}_n$ onto $\mathcal{G}_{k',l}$ such that $f^{-1}\{(k'+1,i)|i \leq l\}$ is the set of all immediate successors of $(w_0, \ldots, w_0)$.
If $x \in \mathcal{F}_{n+1}$, $x' = (x(0), \ldots, x(n-1))$ and $j \in \{1, 2\}$, define $g(x)$ by

$$g(x) = \begin{cases} r & \text{if } x = (w_0, \ldots, w_0) \\ (k'+2, 0) & \text{if } x = (w_0, \ldots, w_0, w_j) \\ (k'+2, i) & \text{if } x \neq (w_0, \ldots, w_0, w_j),\ N_0(x) = n \text{ and} \\ & f(x') = (k'+1, i) \\ (k'+1, 0) & \text{if } N_0(x) = n-1,\ N_0(x') = n-1 \text{ and } \delta(x') = x(n) \\ (k'+1, 1) & \text{if } N_0(x) = n-1,\ N_0(x') = n-1 \text{ and } \delta(x') \neq x(n) \\ f(x') & \text{if } N_0(x') < n-1. \end{cases}$$

Intuitively, the frame $\mathcal{G}_{k'+1,l}$ is obtained from the frame $\mathcal{G}_{k',l}$ by adding two points between the points of depth $k'+1$ and the points of depth $k'+2$. In general, if $x \in \mathcal{F}_{n+1}$ and if $x = (x', w)$, we just map x to the same point that x' was mapped to before. The only exceptions are when $w \neq w_0$ and x' is either $(w_0, \ldots, w_0)$ or an immediate successor of $(w_0, \ldots, w_0)$. In the case x' is equal to $(w_0, \ldots, w_0)$ and w is either w_1 or w_2, we map x to an arbitrary immediate successor of r. In the case x' is an immediate successor of $(w_0, \ldots, w_0)$ and w is either w_1 or w_2, we map x to one of the two added points.

Obviously $g : \mathcal{F}_{n+1} \to \mathcal{G}_{k'+1,l}$ is a well-defined onto map such that $g^{-1}\{(k+1,i)|i \leq l\}$ is the set of all immediate successors of $(w_0, \ldots, w_0)$. We check that g is a p-morphism. For $x \in \mathcal{F}_{n+1}$ and $u \in \mathcal{G}_{k,l}$, let $g(x) \leq u$. Then we need to find a $y \in \mathcal{F}_{n+1}$ such that $x \leq y$ and $g(y) = u$. As in the previous proof we may assume that u is an immediate successor of $g(x)$. There are eight cases possible.

1. $x = (w_0, \ldots, w_0)$ and $u = (k'+2, i)$. By the induction hypothesis, there is a t such that $N_0(t) = 1$ and $f(t) = (k'+1, i)$. Then put $y = (t, w_0)$.

2. $x = (x', w_0)$, $N_0(x) = n$ and $u = (k'+1, 0)$. Then put $y = (x', \delta(x'))$.

3. $x = (x', w_0)$, $N_0(x) = n$ and $u = (k'+1, 1)$. Then the argument is similar to case (2).

4. $x = (w_0, \ldots, w_0, w_i)$, i is either 1 or 2 and $u = (k'+1, 0)$. Then put $y = (w_i, w_0, \ldots, w_0, w_i)$.

5. $x = (w_0, \ldots, w_0, w_i)$, i is either 1 or 2 and $u = (k'+1, 1)$. Then the argument is similar to case (4).

6. $x = (x', w_i)$, i belongs to $\{1,2\}$, $N_0(x') = n - 1$, $u = (i_1, i_2)$ and $i_1 \leq k'$. Recall that $f(x')$ has to be equal to some $(k'+1, i)$. Since f is a p-morphism, there is $s \in \mathcal{F}_n$ such that $x' \leq s$ and $f(s) = u$. We put $y = (s, w_i)$.

7. $x = (x', w_i)$, $N_0(x') < n-1$, $f(x') = (i_1, i_2)$ and $u = (i_1-1, 0)$. By the definition of g, we have that $g(x) = (i_1, i_2)$. Since f is a p-morphism, there is $s \in \mathcal{F}_n$ such that $x' \leq s$ and $f(s) = u$. We put $y = (s, w_i)$.

8. $x = (x', w_i)$, $N_0(x') < n-1$, $f(x') = (i_1, i_2)$ and $u = (i_1 - 1, 1)$. Then the argument is similar to case (7).

Next suppose that $x, y \in \mathcal{F}_{n+1}$ are two distinct points such that $x \leq y$. We show that $g(x) \leq g(y)$. Let x', y', i and i' be such that $x = (x', w_i)$ and $y = (y', w_{i'})$. There are four cases possible.

1. $x = (w_0, \ldots, w_0)$. Then $g(x) = r$ and $r \leq g(y)$.

2. x is an immediate successor of $(w_0, \ldots, w_0)$. It is easy to see that $g^{-1}\{(k'+2, i) | i \leq l\}$ is the set of all immediate successors of $(w_0, \ldots, w_0)$. Thus, $g(x) \leq g(y)$.

3. $N_0(x) = n-2$ and $i \in \{1,2\}$. By the definition of g, $g(x)$ is either $(k'+1, 0)$ or $(k'+1, 1)$. Now since $x' \leq y'$ and $x' \neq y'$, we also have that $N_0(y') < n-1$. So $g(y)$ is equal to $f(y')$ and from our assumption on f, we can deduce that $f(y')$ is equal to some (i_1, i_2), where $i_1 \leq k'$. It follows that $g(x) \leq f(y')$.

4. $N_0(y) < n - 1$. By the definition of g, $g(x)$ is equal to $f(x')$. Also since $x' \leq y'$, we have that $N_0(y') < n - 1$ and so $g(y) = f(y')$. Using the fact that f is a p-morphism, we obtain that $f(x') \leq f(y')$. ∎

COROLLARY 8.

(a) For each $k > 1$, the frame $\mathcal{G}_k$ is a p-morphic image of some $\mathcal{F}_n$.

(b) For each $k > 1$, $\mathcal{G}_k$ is a **Cheq**-frame.

Proof. The result follows from Proposition 7. ∎

THEOREM 9. *The logic* **ML** *is not finitely axiomatizable over* **Cheq**. *In fact,* **ML** *is not axiomatizable over* **Cheq** *by any set of formulas in finitely many variables.*

Proof. Suppose there is a finite set of formulas that axiomatizes **ML** over **Cheq**. Without loss of generality we may assume that there is a single formula φ with k variables such that **ML** = **Cheq** + φ. By Lemma 2(c), there exists a natural number $i \leq k$ such that φ is valid in $\mathcal{G}_k$ iff φ is valid in $\mathcal{G}_k^i$. By Lemma 2(b), $\mathcal{G}_k^i$ is an **ML**-frame. Thus, φ is valid in $\mathcal{G}_k^i$. Therefore, φ is valid in $\mathcal{G}_k$. By Corollary 8, $\mathcal{G}_k$ is a **Cheq**-frame. Thus, $\mathcal{G}_k$ is an **ML**-frame. But this contradicts Lemma 2(a). ∎

4 Conclusion

We proved that **ML** is not finitely axiomatizable over **Cheq**. Thus, the two logics **ML** and **Cheq** are not as closely related as previously thought. It still remains an open problem whether **Cheq** is finitely axiomatizable and/or decidable. At present we can only show that **Cheq** is not axiomatizable in four variables. Of course, the decidability of **ML** still remains an interesting (but difficult) open problem.

Acknowledgments. I would like to thank Nick Bezhanishvili for introducing me to the subject and supervizing me, Guram Bezhanishvili for his help and support in writing this paper and Tadeusz Litak for many valuable suggestions. I would also like to thank the anonymous referees for their comments and the *Fonds national de la recherche scientifique en Belgique* for supporting this research.

BIBLIOGRAPHY

[1] A. Chagrov and M. Zakharyaschev. *Modal Logic*. Clarendon Press, 1997.
[2] T. Litak. Some notes on the superintuitionistic logic of chequered subsets of $\mathbb{R}^\infty$. *Bulletin of the Section of Logic*, 33:81–86, 2004.
[3] L. Maksimova, V. Shehtman, and D. Skvorcov. The impossibility of a finite axiomatization of Medvedev's logic of finitary problems. *Soviet Math. Dokl.*, 20:394–398, 1979.
[4] Ju. T. Medvedev. Finite problems. *Soviet Math. Dokl.*, 3:227–230, 1962.
[5] J. van Benthem, G. Bezhanishvili, and M. Gehrke. Euclidian hierarchy in modal logic. *Studia Logica*, 75:327–344, 2003.

Gaëlle Fontaine
ILLC (Institute for Logic, Language and Computation)
Plantage Muidergracht 24
1018 TV Amsterdam
The Netherlands.
gfontain@science.uva.nl

Bisimulation Quantified Modal Logics: Decidability

TIM FRENCH

ABSTRACT. Bisimulation quantifiers are a natural extension of modal logics. They preserve the bisimulation invariance of modal logic, while allowing monadic second-order expressivity. Unfortunately, it is not the case that extension by bisimulation quantifiers always preserves the decidability of a modal logic. Here we examine a general class of modal logics for which decidability is preserved under extension by bisimulation quantifiers.

1 Introduction

Bisimulation quantifiers extend pure modal logics with a form of weak propositional quantification. They were introduced by Visser [21] and Ghilardi and Zawadowski [11] to semantically characterize uniform interpolants for modal logics, and they are related to Pitt's quantifiers [18] in intuitionistic propositional logic. Bisimulation quantifiers can often increase expressivity allowing us to represent monadic second-order properties not expressible in pure modal logic. They also preserve many of the intuitions of pure modal logic, such as bisimulation invariance [4].

Standard propositional quantification for modal logics was considered by Fine [6], and was shown to make several simple modal logics, such as **K**, undecidable. Bisimulation quantifiers weaken standard propositional quantification by allowing us to quantify "modulo bisimulation". That is, instead of quantifying over all interpretations of a propositional atom in a given structure, we quantify over all interpretations of a propositional atom in all bisimilar structures.

Modal logics are bisimulation invariant. That is, two bisimilar models will validate exactly the same set of modal formulas. One of the advantages of modal logic has been the natural abstraction inherent in this bisimulation invariance. A model is not required to be a precise representation of a real-world system. A model simply has to faithfully represent the salient properties of the system up to bisimulation. When we seek to extend the expressive power of modal logic, it is therefore important to preserve this bisimulation invariance. If we recognize a model to be an *abstraction* of a system, then when we quantify over the interpretation of an atom, we should quantify over that atom, not only in the given abstraction, but in all equally good abstractions. That is, we apply bisimulation quantification.

Another advantage of bisimulation quantification is that it gives us some of the expressive capabilities of the modal μ-calculus, but retains a correspondence to natural language, not evident in fixed-point operators. A bisimulation quantifier, $\exists x$, can be read as "hypothetically, x could be such that...". The bisimulation quantifier preserves all modal properties that are independent of x, so they quantify over all interpretations of x consistent with all other properties of the model.

Bisimulation quantifiers have been investigated in a number of logics. In the case of **GL**, **Grz** and **K** they have been used to characterize uniform interpolants [21, 11]. In the case of **S4** and **K4** they have been used to show that the logics do not have uniform interpolation. In [1] bisimulation quantifiers have been used to characterize uniform interpolants for the modal μ-calculus. More recently, a sound and complete axiomatization has been provided for an expressive bisimulation quantified modal logic [2] and in [3] bisimulation quantifiers are considered for transitive modal logics. Also, in [10] bisimulation quantifiers are used to characterize conservative extensions in description logics.

In [7] it was shown that bisimulation quantifiers for modal logics defined over an arbitrary class of frames may not always be well-behaved. Particularly examples were given where

1. bisimulation quantifiers may not preserve the intuitions of propositional quantification (so the standard axioms for propositional quantification were not valid),

2. bisimulation quantifiers did not preserve the decidability of the modal logic.

A sufficient condition, amalgamation, was given for classes of frames where bisimulation quantifiers validate the axioms of propositional quantification. Unfortunately, no simple condition was found that ensures extension by bisimulation quantifiers preserve the decidability of the pure logic.

In this paper we describe an interpretation for bisimulation quantifiers in various modal logics and describe a relatively general class of modal logics which is well-behaved under extension by bisimulation quantifiers.

2 Preliminaries

In this section we give the basic notation, definitions and lemmas required throughout this paper.

2.1 Pure modal logic

DEFINITION 1. Let Λ be a non-empty set. A Λ-*frame* is a tuple (S, R) where S is some set and $R : \Lambda \longrightarrow \wp(S \times S)$.

A Λ-frame (also referred to as a *Kripke frame*) is the basic relational structure of our abstraction. The set S corresponds to set of possible worlds, and for each $a \in \Lambda$, $R(a)$ refers to some relation between the possible worlds. To define a model we are required to assign atomic propositions to the set of possible worlds. A model adds a valuation to a Λ-frame which defines which atomic propositions are true at each element of S, and identifies *current world.* The current world is required since modal logic is always evaluated with respect to a single world.

DEFINITION 2. A Λ-$\mathcal{V}$-*model* is a tuple (S, R, ρ, s) where (S, R) is a Λ-frame, $\rho : S \longrightarrow \wp(\mathcal{V})$, and $s \in S$.

To make use of such a model we must define a syntax. We will assume that $\mathcal{V}$ is fixed as some countably infinite set, and make Λ explicit in the logic by referring to the following syntax as $\mathcal{ML}_\Lambda$:

(1) $\quad \alpha ::= x \mid \neg \alpha \mid \alpha_1 \vee \alpha_2 \mid \Diamond_a \alpha$

where $x \in \mathcal{V}$ and $a \in \Lambda$. Correspondingly, we will refer to Λ-$\mathcal{V}$-models as Λ-models when there is no ambiguity about $\mathcal{V}$. We let the abbreviations $\wedge, \rightarrow, \leftrightarrow, \top$ and $\bot$ be defined as usual and let $\Box_a \alpha$ abbreviate $\neg \Diamond_a \neg \alpha$. We let $var(\alpha)$ be the set of propositional atoms appearing in α.

The modal logic L_Λ (referred to as the *pure modal logic of Λ*) augments the syntax $\mathcal{ML}_\Lambda$ with the following semantics. Given some Λ-model $M = (S, R, \rho, s)$ for each $t \in S$ let $M_t = (S, R, \rho, t)$. We then define:

$$\begin{aligned} M \models x &\iff x \in \rho(s) \\ M \models \neg \alpha &\iff M \not\models \alpha \\ M \models \alpha \vee \alpha' &\iff M \models \alpha \text{ or } M \models \alpha' \\ M \models \Diamond_a \alpha &\iff \text{for some } (s,t) \in R(a),\ M_t \models \alpha. \end{aligned}$$

From a purely semantic view of logic, we can define a variety of modal logics by restricting the class of models over which formulas are evaluated.

DEFINITION 3. Given some set Λ, a Λ-*class* $\mathcal{C}$ is a class of Λ-frames. The logic $L_\mathcal{C}$ is the set of all $\mathcal{ML}_\Lambda$ formulas which are true on all Λ-$\mathcal{V}$-models, (S, R, ρ, s), where $(S, R) \in \mathcal{C}$.

In an abuse of notation we may refer to $M \in \mathcal{C}$, where $M = (S, R, \rho, s)$ is a Λ-frame, $\mathcal{C}$ is a Λ-class and $(S, R) \in \mathcal{C}$.

2.2 The modal μ-calculus

The language, μL_Λ is recursively defined as follows:

(2) $\quad \alpha ::= x \mid \alpha \vee \alpha \mid \Diamond_a \alpha \mid \mu x \alpha \mid \neg \alpha$

where $x \in \mathcal{V}$, $a \in \Lambda$ and in the recursion for $\mu x \alpha$ we require that x only occurs in the scope of an even number of negations in α. In addition to the usual abbreviations for propositional and temporal logic, we also let $\nu x \alpha$ be an abbreviation for $\neg \mu x \neg \alpha[x \backslash \neg x]$, (where $\alpha[x \backslash \neg x]$ is α with every free occurrence of x replaced by $\neg x$). Note in the case that Λ is a singleton we will refer to μL_Λ as μL, and we let the phrase *modal μ-calculus* refer to μL_Λ for arbitrary sets Λ.

The models of the modal μ-calculus are identical to the models of modal logic. The operators μ and ν are the fixed point operators. In $\mu x \alpha$ the operator μ refers to the least of all assignments of x such that x is true exactly where α is true, and the operator ν refers to the greatest such assignment. Given a model $M = (S, R, \rho, s)$, $x \in \mathcal{V}$ and some set $T \subseteq S$, we let the model $M^{[x \to T]} = (S, R, \tau, s)$ where for all $u \in S$

(3) $\tau(u) = \begin{cases} \rho(u) \cup \{x\} & \text{if } u \in T \\ \rho(u) \setminus \{x\} & \text{otherwise.} \end{cases}$

Let $\|\alpha\| = \{u \mid M_u \models \alpha\}$, and for $T \subseteq S$ let $\|\alpha\|_{[x \to T]} = \{u \mid M_u^{[x \to T]} \models \alpha\}$.
Formally, for $M = (S, R, \rho, s)$, the semantics for L_μ is given by:

$$M \models x \iff x \in \rho(s)$$
$$M \models \Diamond_a \alpha \iff M_t \models \alpha \text{ for some } (s, t) \in R(a)$$
$$M \models \mu x \alpha \iff s \in \bigcap \{T \subseteq S \mid \|\alpha\|_{[x \to T]} \subseteq T\}$$

where the propositional operators have their usual meaning.

The modal μ-calculus was introduced by Kozen in [14]. In the same work an axiomatization was proposed, and the satisfiability problem was shown to be decidable in non-elementary time. Emerson and Jutla presented an exponential decision procedure for the modal μ-calculus in [5].

2.3 Bisimulation Quantifiers

We will first present the syntax $\mathcal{QL}_\Lambda$ for a modal logic $\mathrm{QL}_\mathcal{C}$ (where $\mathcal{C}$ is a Λ-frame) and we will then give the semantics of $\mathrm{QL}_\mathcal{C}$. Let $\mathcal{V}$ be a countable infinite set of atomic propositions and suppose that $\mathcal{C}$ is a Λ-class for some countable, non-empty set Λ. We recursively define the formulas of $\mathcal{QL}_\Lambda$ as follows:

(4) $\alpha ::= x \mid \neg \alpha \mid \alpha_1 \vee \alpha_2 \mid \Diamond_a \alpha \mid \exists x \alpha$

where $x \in \mathcal{V}$ and $a \in \Lambda$. We let the abbreviations $\wedge$, $\to$, $\leftrightarrow$, $\top$, $\bot$ and $\forall$ be defined as usual and let $\Box_a \alpha$ abbreviate $\neg \Diamond_a \neg \alpha$.

DEFINITION 4. Let $M = (S, R, \rho, s_0)$ and $N = (T, P, \lambda, t_0)$ be Λ-$\mathcal{V}$-models and suppose $\Theta \subseteq \mathcal{V}$. We say the models M and N are Θ-*bisimilar* (written $M \cong_\Theta N$) if there is some relation $B \subseteq S \times T$ such that:

1. $(s_0, t_0) \in B$;

2. for all $(s, t) \in B$, $\rho(s) \setminus \Theta = \lambda(t) \setminus \Theta$;

3. for all $(s, t) \in B$, for all $a \in \Lambda$, for all $u \in S$ such that $(s, u) \in R(a)$ there is some $v \in T$ such that $(u, v) \in B$ and $(t, v) \in P(a)$;

4. for all $(s, t) \in B$, for all $a \in \Lambda$, for all $v \in T$ such that $(t, v) \in P(a)$ there is some $u \in S$ such that $(u, v) \in B$ and $(s, u) \in R(a)$.

We call such a relation B a Θ-*bisimulation* from M to N and if $M \cong_{\{x\}} N$ we say M and N are x-bisimilar (written $M \cong_x N$). In the case $\Theta = \emptyset$ we refer to B as a bisimulation.

The semantics of $\mathrm{QL}_\mathcal{C}$ interpret formulas of $\mathcal{QL}_\Lambda$ as follows:

$$M_s \models_\mathcal{C} x \iff x \in \rho(s)$$
$$M_s \models_\mathcal{C} \Diamond_a \alpha \iff \text{for some } (s,t) \in R(a),\ M_t \models_\mathcal{C} \alpha$$
$$M \models_\mathcal{C} \exists x \alpha \iff \text{for some } \mathcal{C}\text{-model } N,\ \text{s.t. } M \cong_x N,\ N \models_\mathcal{C} \alpha.$$

We note that the class $\mathcal{C}$ appears explicitly in the semantics, and thus has a direct effect on the meaning of formulas. For example, we might have some Λ-model M and some formula α such that for two classes, $\mathcal{C}$ and $\mathcal{D}$ $M \models_\mathcal{C} \alpha$ and $M \models_\mathcal{D} \neg\alpha$. Consider, for example, the formula $\alpha = \forall x(x \to \Diamond_a x)$, and suppose M is a $\{a\}$-model consisting of a single world related to itself by $R(a)$.

- In the class $\mathcal{C}$ of all reflexive $\{a\}$-frames we would have $M \models_\mathcal{C} \alpha$.

- However, in the class of all $\{a\}$-frames, $\mathcal{D}$, we have $M \models_\mathcal{D} \neg\alpha$, since M is $\{x\}$-bisimilar to some model (T, P, τ, t) where $x \in \tau(t)$ and $(t,t) \notin P(a)$.

This is in contrast to the case of pure modal logic, where there is one semantics for all logics, so that $L_\mathcal{C}$ and $L_\mathcal{D}$ will always agree on the meaning of a formula (i.e. whether a given formula is satisfied by a given model), but they may differ on whether a formula is valid.

The definition of bisimulation quantifiers for pure modal logic can also be extended to propositional dynamic logic (PDL). We will find the added expressivity of dynamic modalities convenient for defining decidable bisimulation quantified modal logics. We let the syntax, $\mathcal{BQDL}_\Lambda$, for the *bisimulation quantified dynamic logic over $\mathcal{C}$* (BQDL$_\mathcal{C}$, where $\mathcal{C}$ is a Λ-class) be defined by the mutual recursions

(5) $\quad \alpha ::= x \mid \alpha \vee \alpha \mid \neg\alpha \mid \langle\pi\rangle\alpha \mid \exists x \alpha$

(6) $\quad \pi ::= a \mid \pi;\pi \mid \pi \cup \pi \mid \pi^* \mid \alpha?$

where $x \in \mathcal{V}$ and $a \in \Lambda$. In the mutual recursions we refer to α as the *proposition* and π as a *program*. We let $Prog_\Lambda$ refer to the set of programs defined with respect to Λ. Note, this logic (restricted to the class of all Λ-frames) was referred to as **BQL** in [1]. In [8] the language of BQDL$_\mathcal{C}$ is extended to include the converse operator in the definition of programs.

The logic BQDL$_\mathcal{C}$ extends basic propositional logic with modalities corresponding to programs and bisimulation quantifiers. The semantic interpretation of the formulas of BQDL$_\mathcal{C}$ requires a concurrent induction over the complexity of programs and propositions. To define modalities corresponding to programs we extend the function R to map all programs, π, to a relation over S inductively where the base case is given by $R(c)$ for $a \in \Lambda$, and:

$$R(\pi;\pi') = \{(u,w) | \exists v \text{ s.t. } (u,v) \in R(\pi), (v,w) \in R(\pi')\}$$

$$R(\pi \cup \pi') = R(\pi) \cup R(\pi')$$

$$R(\pi^*) = \{(u_0, u_i) | \exists u_1, \ldots, u_{i-1} \in S \text{ s.t. } \forall j < i, \ (u_j, u_{j+1}) \in R(\pi)\}$$

$$R(\alpha?) = \{(u, u) | M_u \models_{\mathcal{C}} \alpha\}$$

Since $R(\alpha?)$ is defined with respect to the interpretation of α the following inductive definition is required

$$\begin{align}
M \models_{\mathcal{C}} x &\iff x \in \rho(s) \\
M \models_{\mathcal{C}} \alpha \vee \beta &\iff M \models_{\mathcal{C}} \alpha \text{ or } M \models_{\mathcal{C}} \beta \\
M \models_{\mathcal{C}} \neg \alpha &\iff M \not\models_{\mathcal{C}} \alpha \\
M \models_{\mathcal{C}} \langle \pi \rangle \alpha &\iff \exists (s, t) \in R(\pi) \text{ s.t. } M_t \models_{\mathcal{C}} \alpha \\
M \models_{\mathcal{C}} \exists x \alpha &\iff \text{for some } \mathcal{C}\text{-model, } N, \text{ s.t. } M \cong_x N, \ N \models_{\mathcal{C}} \alpha.
\end{align}$$

In the remainder of this paper we will let QL_Λ (respectively, BQDL_Λ) refer to the logic $\text{QL}_\mathcal{C}$ (respectively, $\text{BQDL}_\mathcal{C}$) where C is the class of all Λ-frames.

3 Bisimulations in modal logic

Bisimulations were introduced in the context of process algebras by Milner [16] and Park [17]. Prior to this, an equivalent notion of the *p-relation* was given in the context of modal logic by van Benthem [4]. The concept of bisimulation is very natural in modal logic. Bisimilar models are, for the purposes of modal logic, equivalent. That is, bisimilar models will validate exactly the same set of modal formulas.

DEFINITION 5. Let P be some *property* of Λ-models, so that for every Λ-model, P is either true or false. We say P is *bisimulation invariant* if the set of models for which P is true is closed under bisimulation. Given a logic L defined over Λ-models we say L is *bisimulation invariant* if for every L-formula, α, for all L-models, M and N where $M \cong_{\mathcal{V}-var(\alpha)} N$, $M \models \alpha$ if and only if $N \models \alpha$.

It is well-known that modal logic is bisimulation invariant. Furthermore, in [4] it is shown that the properties definable in modal logic are exactly the bisimulation invariant properties that are definable in first order logic. This result has been extended by Janin and Walukiewicz [13] to show that properties definable in the modal μ-calculus, μL_Λ, are exactly the bisimulation invariant properties definable in monadic second-order logic.

LEMMA 6. *A property is expressible in the modal μ-calculus if and only if it is expressible in monadic second-order logic and it is bisimulation invariant.*

Ironically, it is not necessary that bisimulation quantified modal logics will be bisimulation invariant. In [7] a class $\mathcal{C}$ is defined such that two $\mathcal{C}$-models may be Θ-bisimilar, but still disagree on the interpretation of some formula that does not contain atoms from Θ. However, the restriction to amalgamative classes below (Section 3.2) is sufficient to ensure bisimulation invariance.

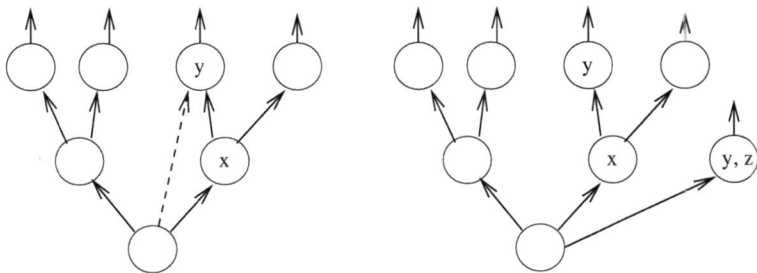

Figure 1. Two z-bisimilar $S4$-models, illustrating the validity of formula (7). The transitive arrows have been ignored, except for the salient one which is dashed.

3.1 Examples

Having established the formal definitions for bisimulation quantified modal logics, we can now examine them in more detail. An example of the interpretation of bisimulation quantifiers is given in Figure 1, where the class of frames is taken to be the class of all reflexive, transitive frames (those frames corresponding to **S4**). The figure illustrates the validity of the formula

(7) $\forall x \forall y ((\neg x \land \Diamond(x \land \neg y \land \Diamond(y \land \neg x))) \rightarrow \exists z (\Diamond(z \land y) \land \Box(x \rightarrow \Box \neg z)))$

Another interesting example of the application of bisimulation quantifiers is in the logic **KD45**, which is Kripke complete for the class of serial, transitive, Euclidean frames (a frame is Euclidean if any two successors of a world are related to each other). In QL_{KD45},

(8) $\Diamond(\exists x (x \land \Box \neg x))$

is a validity, but $\exists x \Diamond(x \land \Box \neg x)$ is a contradiction. The unsatisfiability of the second follows from the fact that $\Box(x \rightarrow \Diamond x)$ is a validity of **KD45**. Figure 2 demonstrates how formula (8) is satisfied. Here, M is a **KD45** model, and $M_t \cong_x N_{t'}$, but there is no **KD45**-model $K = (U, Q, \gamma, u)$ where $K \cong_x M$ and $K_v \cong N_{t'}$ for some v where $(u, v) \in Q$. That is, the bisimulation quantifier does not necessary commute with modalities. This may appear to be a flaw in the interpretation of bisimulation quantifiers. However, **KD45** is the logic of belief and in this context it makes sense. An agent *believes* that some set of worlds are possible, and refuses to believe any other world is possible. Suppose that we are dealing with an informed agent that knows the difference between belief and knowledge. The agent would *believe* that in principle it is *possible* that he believes some (hypothetical) property to be true, even though that property is not satisfied at the current world. That is, $\Diamond \exists x (\neg x \land \Box x)$ is valid. However the agent would still not concede that any of the real, non-hypothetical properties (i.e. unquantified propositions) that he believes are false. For example, I believe that whenever Australia loses a football match, the referee is at fault. I concede that it is

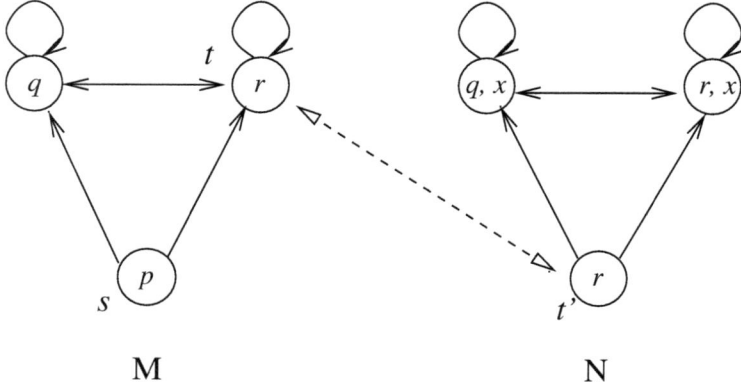

Figure 2. A depiction of the validity (8) with respect to the class of **KD45** frames.

hypothetically possible that Australia might lose a match because it is not the better team, and I still blame the refereeing. However, I'm not aware of any such case.

3.2 Amalgamation

It was shown in [7] that bisimulation quantifiers are not well behaved for all classes of frames. Particularly we can define a class of frames $\mathcal{C}$ such that $\mathrm{QL}_\mathcal{C}$ does not validate the standard rules of propositional quantification (see for example [18]). These rules are:

1. *Existential elimination*: If $\phi \to \psi$ is a validity and ψ does not contain free occurrences of the variable x, then $\exists x \phi \to \psi$ should also be a validity.

2. *Existential introduction*: Suppose α is a formula such that β is free for x in α. Then $\alpha[x\backslash\beta] \to \exists x \alpha$ is a validity.

Here $\alpha[x\backslash\beta]$ is the formula α with every free occurrence of the variable x replaced by the formula β, and β is *free for x in α* if and only if for every free variable, y, of β the variable x is not in the scope of a quantifier, $\exists y$, in α.

Several classes of frames that do not validate these rules are presented in [7]. These rules are sound for all logics, $\mathrm{QL}_\mathcal{C}$, where $\mathcal{C}$ has the *amalgamation property*.

DEFINITION 7. We say the class of frames $\mathcal{C}$ has the *amalgamation property* (or $\mathcal{C}$ is *amalgamative*) if and only if for any $\Theta_1, \Theta_2 \subset \mathcal{V}$, for any $\mathcal{C}$-models M and N such that $M \cong_{\Theta_1 \cup \Theta_2} N$, there is some $\mathcal{C}$-model, K such that $M \cong_{\Theta_1} K$ and $N \cong_{\Theta_2} K$.

A related concept of amalgamation has been studied in the context of modal algebras (see, for example, [15]). The formulation of amalgamation above appeared in Lemma 3.3 of [1], and is also presented in [3]. The same property is referred to as *safety* in [7].

LEMMA 8. *For any Λ, the class of all Λ-frames is amalgamative.*

Proof. Suppose that $M = (S, R, \rho, s)$ and $N = (T, P, \tau, t)$ are Λ-models such that $M \cong_{\Theta \cup \Gamma} N$ via the $\Theta \cup \Gamma$-bisimulation, $B \subseteq S \times T$. We construct a Λ-model $K = (U, Q, \upsilon, u)$ where:

- $U = B$;

- for all $a \in \Lambda$, $(v, w) \in Q(a)$ if and only if $v, w \in B$, $(v_1, w_1) \in R(a)$ and $(v_2, w_2) \in P(a)$ (where $v = (v_1, v_2)$ and $w = (w_1, w_2)$);

- for all $v \in B$, $\upsilon(v) = (\rho(v_1) \backslash \Theta) \cup (\tau(v_2) \backslash \Gamma)$ (where $v = (v_1, v_2)$);

- $u = (s, t)$.

It is then straightforward to check that $K \cong_\Theta M$ via the Θ-bisimulation $\{(v, w) \mid v_1 = w\}$ and $K \cong_\Gamma N$ via the Γ-bisimulation $\{(v, w) \mid v_2 = w\}$. ∎

The following was shown in [7].

LEMMA 9. *Given $\mathcal{C}$ enjoys amalgamation, the rules existential elimination and existential introduction are sound for* $\text{QL}_\mathcal{C}$.

3.3 Decidability of BQDL_Λ

In this section we give a number of translations between the modal μ-calculus and bisimulation quantified modal logics.

From [1] we have:

LEMMA 10. *The logics BQDL_Λ and μL_Λ are expressively equivalent. Furthermore there is a computable translation in each direction that preserves the meaning of formulas. That is, for every BQDL_Λ formula α there is a μL_Λ formula α^μ that is satisfied by exactly the models that satisfy α.*

The translation from μL_Λ formulas to BQDL_Λ formulas is relatively straightforward and a generalization of this construction is presented below in Lemma 11. The equivalent BQDL_Λ formula is linear in the length of the original μL_Λ formula. The translation from BQDL_Λ formulas to μL_Λ formulas is not as simple. The translation is given via μ-automata [13], and the resulting μL_λ formula may be non-elementary in the length of the original BQDL_Λ formula. In [8] it is shown that there is no elementary translation from μL_Λ to BQDL_Λ.

As a corollary to Lemmas 6 and 10 we have that BQDL_Λ can express exactly the properties definable in monadic-second order logic that are bisimulation invariant, and that BQDL_Λ is decidable. An alternative proof of the decidability of BQDL_Λ (including converse operators) is presented in [8].

We can generalize one direction of Lemma 10 from the class of all Λ-frames to any class of Λ-frames that enjoys amalgamation. Since μL_Λ and $\text{BQDL}_\mathcal{C}$ agree on the semantics for all common operators, it suffices to show that the fixed-point formula $\mu x \alpha$ is equivalent to some $\text{BQDL}_\mathcal{C}$ formula, α^*. The translation can then be applied recursively. More precisely, we let $(\cdot)^*$ be a transformation such that $x^* = x$, $(\Diamond_a \alpha)^* = \langle a \rangle (\alpha^*)$ and likewise for the propositional operators. We define

(9) $(\mu x.\alpha)^* = \forall x ([*](\alpha^* \to x) \to x)$,

where $\langle * \rangle$ is an abbreviation for $\langle (\bigcup_{a \in \Lambda} a)^* \rangle$. We note the translation given from $\mu L_\mathcal{C}$ to $\text{BQDL}_\mathcal{C}$ is linear.

LEMMA 11. *Let $\mathcal{C}$ be any Λ-class. Then for every μL_Λ formula, α, for all $\mathcal{C}$-models M, $M \models_\mathcal{C} \alpha^*$ if and only if $M \models \alpha$.*

Proof. This lemma is proven by induction over the complexity of formulas where our induction hypothesis is that for all $\mathcal{C}$-models N we have $N \models \alpha$ if and only if $N \models_\mathcal{C} \alpha^*$. The only non-trivial case of this induction is to show that $\mu x \alpha$ and $(\mu x \alpha)^*$ have the same interpretation. First, suppose that $M \not\models_\mathcal{C} (\mu x \alpha)^*$. Then there is some $\mathcal{C}$-model $N = (T, P, \tau, t)$ such that $N \cong_x M$ and $N \models_\mathcal{C} [*](\alpha^* \to x) \wedge \neg x$. Let $U \subseteq T$ be the set $\{u \in T \mid x \in \tau(u)\}$, and let $V = \{v \in T \mid (t, v) \in P(*)\}$. Applying the induction hypothesis we have, $N \models_\mathcal{C} [*](\alpha \to x)$ so for all $v \in V$, $N_v \models \alpha \to x$. Let $N' = (V, P, \tau, t)$ (where P and τ are appropriately restricted). Clearly $N' \cong N$ so by the bisimulation invariance of μL_Λ (Lemma 6) we have for all $v \in V$, $N'_v \models \alpha \to x$. By the semantic definition (Section 2.2) of $\mu x \alpha$ it follows that for all $v \in V$, $N'_v \models \mu x \alpha$ only if $v \in U$. Since $x \notin \tau(t)$, we have $N'_v \models \neg \mu x \alpha$. Finally, since $M \cong_x N'$ and μL_Λ is bisimulation invariant it follows that $M \models \neg \mu x \alpha$.

Conversely suppose $M \models_\mathcal{C} (\mu x \alpha)^*$. Therefore, $M \models_\mathcal{C} \forall x ([*](\alpha^* \to x) \to x)$ and thus for every $T \subset S$, if $M_u^{[x \to T]} \models_\mathcal{C} \alpha^*$ only for $u \in T$, then $s \in T$. By the induction hypothesis, $M_u^{[x \to T]} \models_\mathcal{C} \alpha^*$ if and only if $M_u^{[x \to T]} \models_\mathcal{C} \alpha$. Thus, for every $T \subseteq S$ where $M_u^{[x \to T]} \models \alpha$ implies $u \in T$, we have $s \in T$. By the semantic definition of $\mu x \alpha$ it follows that $M \models \mu x \alpha$. ∎

Note that the other direction of Lemma 10 cannot be generalized to arbitrary, or even amalgamative classes of frames. In [3] a class of transitive well-founded frames, $GL3$, is given and it is shown that QL_{GL3} is strictly more expressive than μL_{GL3}.

4 Idempotent transductions

We now investigate classes of frames, $\mathcal{C}$ such that $\text{QL}_\mathcal{C}$ is decidable and amalgamative. The classes we define, the *idempotent transduction classes*, are built by using dynamic modalities to define constraints on frames. This technique is often used for pure modal logics (see [9]), where some logic may be translated into PDL. Bisimulation quantifiers add an extra layer of complexity. It is not enough to be able to express a transformation as a set of dynamic modalities, since we also have to ensure that the transformation commutes with bisimulations.

DEFINITION 12. Given some finite set Λ, a Λ-*transduction* is a function, $\Pi : \Lambda \longrightarrow Prog_\Lambda$, where $Prog_\Lambda$ is the set of programs defined over the set of atomic programs Λ (see Section 2.3). We say a Λ-transduction, Π, is *closed* if for every $a \in \Lambda$, $\Pi(a)$ does not contain any free propositions.

A Λ-transduction induces a function that acts on Λ-frames. For every $a \in \Lambda$ the transduction $\Pi(a)$ replaces the relation $R(a)$ in a frame with the relation $R(\Pi(a))$. In this way we are able to "construct" frames satisfying certain constraints.

DEFINITION 13. Given a closed Λ-transduction, Π, and a Λ-frame $F = (S,R)$ we define $F^\Pi = (S, R^\Pi)$ where for all $u,v \in S$, $(u,v) \in R^\Pi(a)$ if and only if for some Λ-model $M = (S, R, \rho, s)$, $(u,v) \in R(\Pi(a))$ (see Section 2.3). We define the *transduction class of* Π (written $\mathcal{C}_\Pi$) to be the class of Λ-frames F^Π where F is any Λ-frame.

Note that if a program $\Pi(a)$ contains a bisimulation quantifier, the bisimulation quantifier is interpreted with respect to the class of all Λ-frames. Given a closed Λ-transduction, Π, and a Λ-model $M = (S, R, \rho, s)$ we define M^Π to be the $\mathcal{C}_\Pi$-model $(S, R(\Pi), \rho, s)$.

DEFINITION 14. A closed Λ-transduction Π is an *idempotent transduction* if for all Λ-models M, we have $(M^\Pi)^\Pi = M^\Pi$. If Π is an idempotent transduction, then $\mathcal{C}_\Pi$ is an *idempotent transduction class* of frames and the logic $(Q)L_{\mathcal{C}_\Pi}$ is a *(bisimulation quantified) idempotent transduction logic*.

In [8] a weaker definition of idempotent transductions is given that only required that $(M^\Pi)^\Pi \cong M^\Pi$. This allowed the *converse operator* to appear in transductions, which increases the complexity of the required proofs of decidability, but it does allow us to define idempotent transductions corresponding to logics such as **S5** and **KD45**. It was shown that for programs not containing the converse operator, both notions of idempotence are equivalent.

Idempotence is a strong condition that ensures that bisimulations and transductions can commute. Particularly we would like to have, if $M \cong_\Theta N$, then $M^\Pi \cong_\Theta N^\Pi$. This is shown in Lemma 15. This leads to the main results of this paper: that idempotent transduction classes are amalgamative; and that idempotent transduction logics are decidable.

4.1 Examples

Let us now consider some examples of idempotent transduction logics. The pure modal logic **S4** is evaluated over frames with a single reflexive, transitive relation. We will suppose that this single modality is labelled as a. Then the class of **S4** frames can be seen to be the idempotent transduction class corresponding to the $\{a\}$-transduction $\Pi(a) = a^*$. We note that for all $\{a\}$-models, M, M^Π is transitive and reflexive. It is also easy to see that $(M^\Pi)^\Pi = M^\Pi$.

As a less trivial example of an idempotent transduction logic we consider the *Gödel Löb* logic, **GL**. The logic **GL** was shown by Segerberg [19] to be Kripke complete for all frames (S,R) where R is a Noetherian strict partial orders (i.e. all irreflexive, transitive frames containing no infinite path). Let us refer to this class of frames as GL. We can define an idempotent $\{a\}$-transduction Γ such that $\mathcal{C}_\Gamma$ is exactly the class of frames GL. This transduction is defined by

(10) $\Gamma(a) = (\forall x(x \to \langle a^*\rangle(x \land [a]\neg x)))?; a; a^*$

Given a model $M = (S, R, \rho, s)$, for all $u, v \in S$, we have $(u,v) \in R^\Gamma(a)$ if and only if

1. $M_u \models_\Lambda \forall x(x \to \langle a^*\rangle(x \land [a]\neg x))$ (it is easy to check that this is so if and only if every path beginning at u is well-founded), and

2. v is reachable from u via a path of length at least one.

Therefore, the transduction Γ will remove every pair $(u, v) \in R(a)$ where u appears in some infinite path, and take the transitive closure of the remaining pairs. An example of this transduction is given in Figure 3. We can see that the transduction Γ removes the pairs (a, b), (a, c) and (b, b) from the model M since both a and b belong to an infinite path. The transduction Γ takes the transitive closure of the remaining pairs so the pairs (c, f) and (c, g) are added to create the model M^Γ. The logic **GL** has uniform interpolation [20] and has been shown to be equivalent to QL_{GL} in [21, 11].

As a final example, we note that given any idempotent Λ-transduction, Π, we can always add the reflexive, transitive closure of all the modalities and still have an idempotent transduction. That is, we can define the $\Lambda \cup \{o\}$ transduction, Π', where $\Pi'(o) = (\bigcup_{a \in \Lambda} \Pi(a))^*$ and $\Pi'(a) = \Pi(a)$ for all $a \in \Lambda$.

4.2 Amalgamation for idempotent transductions

In this section we show that all bisimulation quantified idempotent transduction logics preserve our intuitions of propositional quantification. That is, they enjoy amalgamation (Definition 7) and thus satisfies the rules of existential elimination and existential introduction (Section 3.2). The following lemma is well-known.

LEMMA 15. *Suppose that Π is a Λ-transduction, and M and N are Λ-models such that $N \cong_\Theta M$ for some Θ where for all $a \in \Lambda$, no atom of Θ appears free in $\Pi(a)$. Then $N^\Pi \cong_\Theta M^\Pi$.*

Proof. Suppose that $M = (S, R, \rho, s)$ and $N = (T, P, \tau, t)$ are Θ-bisimilar Λ-models and $B \subseteq S \times T$ is a Θ-bisimulation from M to N. We will show that B is also a bisimulation from M^Π to N^Π by induction over the complexity of programs. Suppose that for all sub-programs π' of π we have, for all $(u, u') \in B$,

1. for all $v \in S$ where $(u, v) \in R(\pi')$, there is some $v' \in T$ where $(u', v') \in P(\pi')$ and $(v, v') \in B$, and

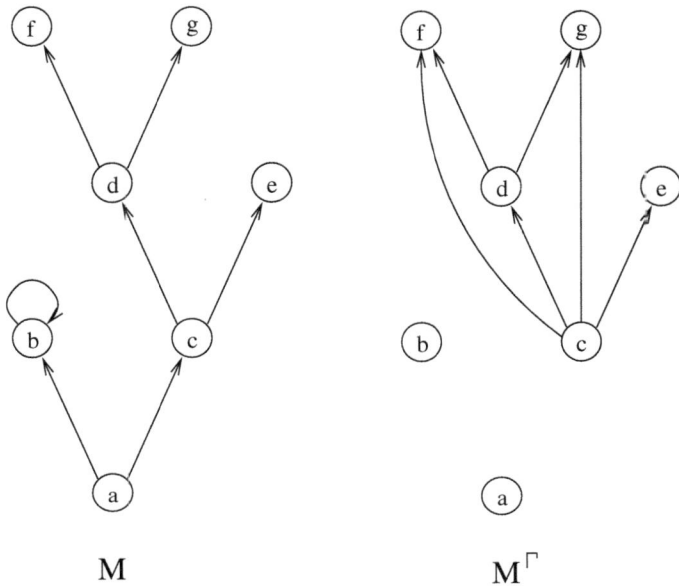

Figure 3. An example of how the transduction Γ defines the class of frames GL.

2. for all $v' \in T$ where $(u', v') \in P(\pi')$, there is some $v \in S$ where $(u, v) \in R(\pi')$ and $(v, v') \in B$.

Then we show by induction that for all $(u, u') \in B$,

Zig: for all $v \in S$ where $(u, v) \in R(\pi)$, there is some $v' \in T$ where $(u', v') \in P(\pi)$ and $(v, v') \in B$, and

Zag: for all $v' \in T$ where $(u', v') \in P(\pi)$, there is some $v \in S$ where $(u, v) \in R(\pi)$ and $(v, v') \in B$.

The inductive cases are for the *Zig* clauses below, and the proofs for *Zag* are symmetric.

1. Let $\pi = \pi_1; \pi_2$. Then if $(u, u') \in B$ and $(u, v) \in R(\pi)$, there is some $w \in S$ such that $(u, w) \in R(\pi_1)$ and $(w, v) \in R(\pi_2)$. By the inductive hypothesis there exists $w' \in T$ where $(u', w') \in P(\pi_1)$ and $(w, w') \in B$. Applying the induction hypothesis a second time there is some $v' \in T$ such that $(w', v') \in P(\pi_2)$ and $(v, v') \in B$. Therefore, there is some $v' \in T$ such that $(u', v') \in P(\pi)$ and $(v, v') \in B$, as required.

2. Let $\pi = \pi_1 \cup \pi_2$. Then if $(u, u') \in B$ and $(u, v) \in R(\pi)$, there either $(u, v) \in R(\pi_1)$ or $(u, v) \in R(\pi_2)$. Suppose without loss of generalization, that $(u, v) \in R(\pi_1)$. Then by the induction hypothesis, there

must be some $v' \in T$ such that $(u', v') \in P(\pi_1)$ and $(v, v') \in B$. Therefore, there is some $v' \in T$ such that $(u', v') \in P(\pi)$ and $(v, v') \in B$, as required.

3. Let $\pi = \pi_1^*$. Then if $(u_0, u_0') \in B$ and $(u_0, v) \in R(\pi)$, for some $k \geq 0$ there exists $u_1, \ldots, u_k \in S$ such that for all $i < k$, $(u_i, u_{i+1}) \in R(\pi_1)$ and $u_k = v$. By the induction hypothesis, there exists $u_1' \in T$ such that $(u_0', u_1') \in P(\pi_1)$ and $(u_1, u_1') \in B$. Likewise, we can show there exists $u_2' \in T$ such that $(u_1', u_2') \in P(\pi_1)$ and $(u_2, u_2') \in B$, and by induction it follows that there exists $u_1', \ldots, u_k' \in T$ where for all $i < k$, $(u_i', u_{i+1}') \in P(\pi_1)$ and for all $i \leq k$, $(u_i, u_i') \in B$. By the definition of $P(\pi)$ we have $(u_0', u_k') \in P(\pi)$ and $(v, u_k') \in B$.

4. Suppose $\pi = \alpha?$. Then if $(u, u') \in B$ and $(u, v) \in R(\pi)$, we have $v = u$ and $M_u \models \alpha$. Since $M_u \cong_\Theta N_{u'}$ and no atom from Θ appears free in α, we have $N_{u'} \models \alpha$ (as BQDL_Λ is bisimulation invariant). Therefore $(u', u') \in P(\pi)$, and $(u, u') \in B$ as required.

We can apply this inductive argument for all programs $\Pi(a)$ where $a \in \Lambda$. It follows from the definitions that for any $(u, u') \in B$, for all $v \in S$, if $(u, v) \in R^\Pi(a)$, we must have $(u, v) \in R(\Pi(a))$. By the reasoning above, there is some $v' \in T$ such that $(u', v') \in P(\Pi(a))$ (so also $(u', v') \in P^\Pi(a)$) and $(v, v') \in B$. As the Zag case is symmetric we have shown

- $(s, t) \in B$, and for all $(u, v) \in B$, $\rho(u)\backslash\Theta = \tau(v)\backslash\Theta$ (since B is a bisimulation from M to N),

- (Zig) for all $(u, u') \in B$, for all $v \in S$ where $(u, v) \in R^\Pi(a)$, there is some $v' \in T$ such that $(u', v') \in P^\Pi(a)$ and $(v, v') \in B$, and

- (Zag) for all $(u, u') \in B$, for all $v' \in T$ where $(u', v') \in P^\Pi(a)$, there is some $v \in S$ where $(u, v) \in R^\Pi(a)$ and $(v, v') \in B$.

Therefore B is a Θ-bisimulation from M^Π to N^Π. ∎

An equivalent result was first shown by Hollenberg [12]. In addition, Hollenberg found that the set of all binary relations that are definable in monadic second-order logic and preserve bisimulation (the *bisimulation-safe fragment of MSO*) are exactly the binary relations that can be described by the programs of BQDL_Λ (Section 2.3). Lemma 15 does not use the full bisimulation-safe fragment of MSO as the programs $\Pi(a)$ cannot contain any free variables, and the notion of bisimulation is generalized to Θ-bisimulation, for some set of atoms Θ. Lemma 15 is applied below to show that idempotent transduction classes are amalgamative, and idempotent transduction logics are decidable.

LEMMA 16. *Every idempotent transduction class is amalgamative.*

Proof. Suppose that Π is an idempotent Λ-transduction, $M = (S, R, \rho, s)$ and $N = (T, P, \tau, t)$ are $\mathcal{C}_\Pi$-models and $M \cong_{\Theta \cup \Gamma} N$. Since M and N are

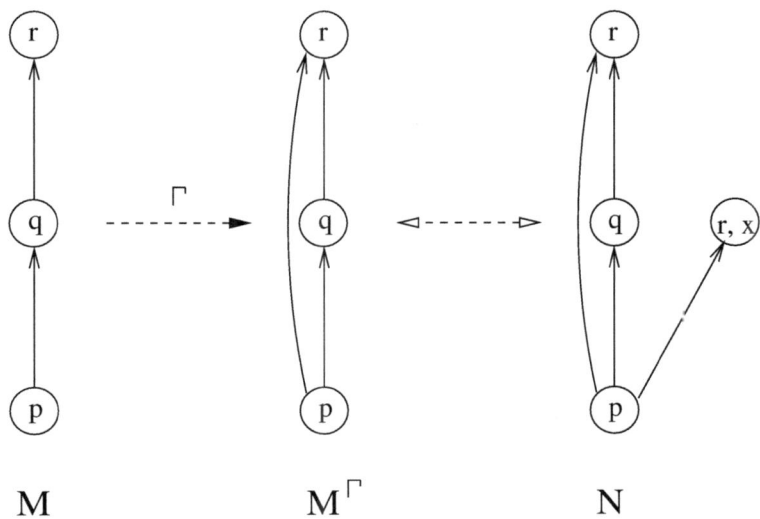

Figure 4. An example of the transduction for $K4$, Γ. We note that there is no Λ-model, K, x-bisimilar to M such that K^Γ is bisimilar to N.

Λ-models and the class of all Λ-models is amalgamative (Lemma 8) there is some Λ-model K such that $M \cong_\Theta K$ and $K \cong_\Gamma N$. By Lemma 15 and the fact that Π is idempotent,

$$M = M^\Pi \cong_\Theta K^\Pi \cong_\Gamma N^\Pi = N,$$

so $\mathcal{C}_\Pi$ is amalgamative. ∎

4.3 Decidability for idempotent transduction logics

In this section we apply the results of Sections 3.3 and 4.2 to show the decidability of all bisimulation quantified idempotent transduction logics. To do this we define a translation taking a QL_Λ formula α to a BQDL_Λ formula, α^Π, such that for all Λ-models, M, $M \models_\Lambda \alpha^\Pi$ if and only if $M^\Pi \models_{\mathcal{C}_\Pi} \alpha$. Therefore the satisfiability problem for $\mathrm{QL}_{\mathcal{C}_\Pi}$ reduces to the satisfiability problem for BQDL_Λ which is known to be decidable (see [8] or [1]).

Before we proceed to this quite complicated translation, we will first see why the more obvious translation (that simply replaces modalities $\Diamond_a$ with programs $\langle \Pi(a) \rangle$) will not work. Consider the set of all transitive Λ-frames $K4$, where $\Lambda = \{a\}$. It is not hard to see that this set of frames is equal to $\mathcal{C}_\Gamma$ where Γ is the idempotent Λ-transduction defined by, $\Gamma(a) = a; a^*$. In Figure 4 we see how the Λ model M is modified by Γ. We can also see that

(11) $\quad M^\Gamma \models_{\mathcal{C}_\Gamma} \exists x(\Diamond_a(x \wedge r) \wedge \Box_a(q \to \Box_a \neg x))$

as the $K4$ model N satisfies the unquantified fragment of the formula, and N and M^Γ are clearly x-bisimilar. However, it is not hard to show that

(12) $M \not\models_\Lambda \exists x(\langle a;a^*\rangle(x \wedge r) \wedge [a;a^*](q \to [a;a^*]\neg x))$.

We can see that there is no Λ-model K that is x-bisimilar to M such that K^Γ is bisimilar to N.

Naively replacing every modality $\Diamond_a$ with the program $\langle\Pi(a)\rangle$ is not sufficient to encode the satisfiability problem for $\mathrm{QL}_{\mathcal{C}_\Pi}$. However, the translation can encode the satisfiability problem for a related class.

DEFINITION 17. Given an idempotent Λ-transduction, Π, let $\overline{\Lambda}$ be the set $\{a, a' \mid a \in \Lambda\}$ where for each $a \in \Lambda$, a' is a new distinct element. We define the *rigid class of* Π to be the $\overline{\Lambda}$-class of frames (S, R) where for all $a \in \Lambda$, $R(a) = R(\Pi(a'))$. We will denote this class as $\mathcal{D}_\Pi$.

The rigid class of an idempotent transduction still allows us to define new modalities corresponding to Λ-programs, but it retains the original modalities that are used to define the Λ-programs. This removes a lot of the flexibility associated with the bisimulation quantifiers, and makes the satisfiability problem a lot easier. Given the QL_Λ formula α, let $\alpha[\Lambda\backslash\Pi(\Lambda)]$ be the BQDL_Λ formula that results when every occurrence of any subformula $\Diamond_a \gamma$ is replaced by $\langle\Pi(a)\rangle\gamma$. Likewise for every BQDL_Λ formula α we can define $\alpha[\Lambda\backslash\Pi(\Lambda)]$ to be the BQDL_Λ formula α where every occurrence of a in every program in α is replaced by $\Pi(a)$.

LEMMA 18. *Let Π be an idempotent Λ-transduction and let $M = (S, R, \rho, s)$ be a Λ-model. We define the $\mathcal{D}_\Pi$ model $M^{\overline{\Pi}} = (S, \overline{R}, \rho, s)$ where for all $a \in \Lambda$, $\overline{R}(a) = R(\Pi(a))$ and $\overline{R}(a') = R(a)$. Then for every QL_Λ (BQDL_Λ) formula α,*

$$M^{\overline{\Pi}} \models_{\mathcal{D}_\Pi} \alpha \iff M \models_\Lambda \alpha[\Lambda\backslash\Pi(\Lambda)]$$

Proof. The proof is given by induction over the complexity of formulas. The cases for propositional atoms and propositional operators are trivial, so suppose that for all models M, $M^{\overline{\Pi}} \models_{\mathcal{D}_\Pi} \beta$ if and only if $M \models_\Lambda \beta[\Lambda\backslash\Pi(\Lambda)]$. If $M^{\overline{\Pi}} \models_{\mathcal{D}_\Pi} \Diamond_a \beta$, then there is some $t \in S$ such that $(s, t) \in R(\Pi(a))$ and $M_t^{\overline{\Pi}} \models_{\mathcal{D}_\Pi} \beta$. By the induction hypothesis, $M_t \models_\Lambda \beta[\Lambda\backslash\Pi(\Lambda)]$, and as $(s, t) \in R(\Pi(a))$ we have $M \models_\Lambda (\Diamond_a \beta)[\Lambda\backslash\Pi(\Lambda)]$. Conversely, if $M \models_\Lambda (\Diamond_a \beta)[\Lambda\backslash\Pi(\Lambda)]$, then there is some $(s, t) \in R(\Pi(a))$ such that $M_t \models_\Lambda \beta[\Lambda\backslash\Pi(\Lambda)]$. From the induction hypothesis and the definition of $M^{\overline{\Pi}}$ it follows that $M^{\overline{\Pi}} \models_{\mathcal{D}_\Pi} \Diamond_a \beta$. In the case of BQDL_Λ formulas, (where we have programs rather than modalities) a similar proof can be applied. The details are left to the reader.

If $M^{\overline{\Pi}} \models_{\mathcal{D}_\Pi} \exists x \beta$, then there is some $\mathcal{D}_\Pi$-model $N = (T, P, \tau, t)$ such that $N \cong_x M^{\overline{\Pi}}$ and $N \models_{\mathcal{D}_\Pi} \beta$. Let $K = (T, P', \tau, t)$ be the Λ-model where for all $a \in \Lambda$, $P'(a) = P(a')$. It follows from Definition 4 that if B is an x-bisimulation from N to $M^{\overline{\Pi}}$ then B is also an x-bisimulation from K to M, so $K \cong_x M$. Also, since $K^{\overline{\Pi}} = N$ we can apply the induction hypothesis to show $K \models_\Lambda \beta[\Lambda\backslash\Pi(\Lambda)]$. Therefore $M \models_\Lambda (\exists x \beta)[\Lambda\backslash\Pi(\Lambda)]$ as required. Conversely, if $M \models_\Lambda (\exists x \beta)[\Lambda\backslash\Pi(\Lambda)]$, then there is some Λ-model N such that

$M \cong_x N$ and $N \models \beta[\Lambda \backslash \Pi(\Lambda)]$. By the same argument used in Lemma 15, if $M \cong_x N$, then $M^{\overline{\Pi}} \cong_x N^{\overline{\Pi}}$. Applying the induction hypothesis we have $N^{\overline{\Pi}} \models_{\mathcal{D}_\Pi} \beta$, so $M^{\overline{\Pi}} \models_{\mathcal{D}_\Pi} \exists x \beta$, completing the proof. ∎

Lemma 18 can be applied to show any logic $\text{QL}_{\mathcal{D}_\Pi}$ is decidable where Π is idempotent. This is not a surprising result as translating formulas of $\text{QL}_{\mathcal{D}_\Pi}$ to BQDL_Λ is a trivial exercise. By combining Lemma 18 with Lemmas 10 and 11 we get a much stronger result. Specifically, we already know of the translations: $(\cdot)^\mu$ (from Lemma 10) that translates a BQDL_Λ formula to an equivalent μL_Λ formula; and $(\cdot)^*$ (from Lemma 11) that translates a μL_Λ formula to an equivalent BQDL_Λ formula. These two translations are the inverse of each other, so it follows that $(\alpha^\mu)^*$ is simply equivalent to α in BQDL_Λ. However, when we consider $\text{BQDL}_\mathcal{C}$ for different classes of $\mathcal{C}$ the equivalence does not necessarily hold. Rather, we have the following lemma.

LEMMA 19. *Let $\mathcal{C}$ be a Λ-class, suppose that α is some formula of BQDL_Λ, and let the translations $(\cdot)^\mu$ and $(\cdot)^*$ be the translations described in Lemma 10 and Lemma 11 respectively. Then for any $\mathcal{C}$-model M,*

$$M \models_\mathcal{C} (\alpha^\mu)^* \iff M \models_\Lambda \alpha.$$

Proof. From Lemma 11 we have $M \models_\mathcal{C} (\alpha^\mu)^*$ if and only if $M \models \alpha^\mu$. We note here that α^μ is a μL_Λ formula and consequently it is independent of the Λ-class $\mathcal{C}$. From Lemma 10 we have $M \models \alpha^\mu$ if and only if $M \models_\Lambda \alpha$. Therefore $M \models_\mathcal{C} (\alpha^\mu)^*$ if and only if $M \models_\Lambda \alpha$. ∎

The complexity of Lemma 19 is hidden in the proof of Lemma 10. The result is quite powerful, allowing us to syntactically "unwind" any $\mathcal{C}$-model into a Λ-model. That is, we can consider the interpretation of a formula independent of the restraints of the Λ-class $\mathcal{C}$. Furthermore, the translation $(\alpha^\mu)^*$ is independent of the class $\mathcal{C}$. We let $(\cdot)^{\mu*}$ refer to the composition of the translations $(\cdot)^\mu$ and $(\cdot)^*$ (so that $\alpha^{\mu*} = (\alpha^\mu)^*$).

We define the translation $(\cdot)^\Pi$ from QL_Λ to BQDL_Λ inductively as follows

1. $x^\Pi = x$, where x is an atomic proposition;
2. $(\alpha \vee \beta)^\Pi = \alpha^\Pi \vee \beta^\Pi$;
3. $(\neg \alpha)^\Pi = \neg(\alpha^\Pi)$;
4. $(\Diamond_a \alpha)^\Pi = \langle \Pi(a) \rangle \alpha^\Pi$;
5. $(\exists x \alpha)^\Pi = ((\exists x(\alpha^\Pi))^{\mu*}[\Lambda \backslash \Pi(\Lambda)]$.

We will first examine how the translation $(\cdot)^\Pi$ works with respect to the example of Figure 4. We let Γ be the transduction for $K4$ defined as before, and suppose $\alpha = \Diamond_a(x \wedge r) \wedge \Box_a(q \to \Box_a \neg x)$. Given the model M, to decide if $M^\Gamma \models_{\mathcal{C}_\Gamma} \exists x \alpha$ we ask whether there is any $\overline{\Lambda}$-model N that is both

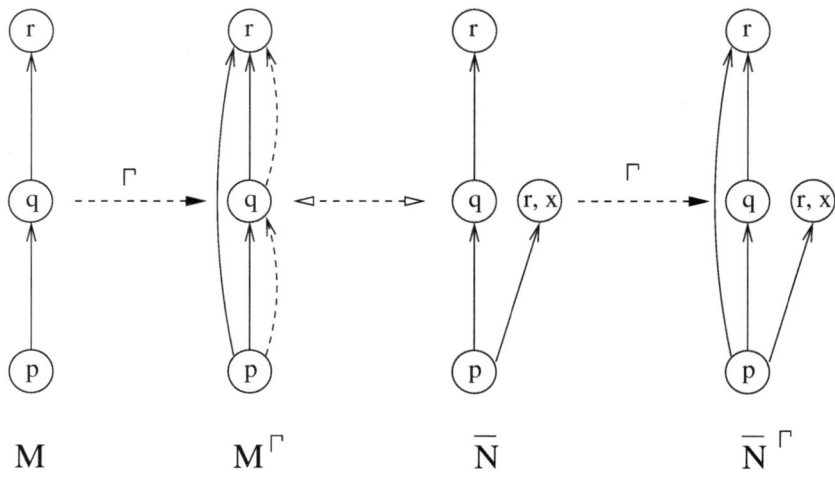

Figure 5. An example of models used to evaluate the translation $(\cdot)^\Gamma$ applied to the idempotent transduction logic $\mathrm{QL}_{\mathcal{C}_\Gamma}$. The dashed arrows correspond to the modality a'.

x-bisimilar to $M^{\overline{\Gamma}}$ and that satisfies α^Γ. By Lemma 20 (below) this implies $\overline{N}^\Gamma \models_{\mathcal{C}_\Gamma} \alpha$ (where $\overline{N}$ is the model N restricted to the modality a.) Finally, it can be shown that by the idempotence of Γ, $M^\Gamma \cong_x \overline{N}^\Gamma$ as required. This is represented in Figure 5.

LEMMA 20. *Let Π be an idempotent Λ-transduction and let M be a Λ-model. Then for every formula α of QL_Λ,*

$$M^\Pi \models_{\mathcal{C}_\Pi} \alpha \iff M \models_\Lambda \alpha^\Pi.$$

Proof.
We are required to show $M \models_\Lambda \alpha^\Pi$ if and only if $M^\Pi \models_{\mathcal{C}_\Pi} \alpha$. Proceeding by induction over the complexity of α, it is trivial to see that the induction is correct for the first three cases cases (atoms, disjunctions and negations).

For the case of modalities, suppose for all models $M = (S, R, \rho, s)$, $M^\Pi \models_{\mathcal{C}_\Pi} \beta$ if and only if $M \models_\Lambda \beta^\Pi$. Then if $M^\Pi \models_{\mathcal{C}_\Pi} \Diamond_a \beta$, for some $t \in S$ where $(s,t) \in R^\Pi(a)$ we have $M_t^\Pi \models_{\mathcal{C}_\Pi} \beta$. By the induction hypothesis it follows $M_t \models_\Lambda \beta^\Pi$ and by the definition of M^Π, we have $(s,t) \in R(\Pi(a))$. Therefore, $M \models_\Lambda \langle\Pi(a)\rangle\beta^\Pi$. Conversely, if $M \models_\Lambda \langle\Pi(a)\rangle\beta^\Pi$, then there is some $t \in S$ where $(s,t) \in R(\Pi(a))$ and $M_t \models \beta^\Pi$. By the induction hypothesis, $(M_t)^\Pi \models_{\mathcal{C}_\Pi} \beta$. Clearly $(M_t)^\Pi = (M^\Pi)_t$, so since $(s,t) \in R(\Pi(a)) = R^\Pi(a)$ we have $M^\Pi \models_{\mathcal{C}_\Pi} \Diamond_a \beta$.

We are now left to deal with bisimulation quantifiers. Assume inductively that $M \models_\Lambda \beta^\Pi$ if and only if $M^\Pi \models_{\mathcal{C}_\Pi} \beta$. We will prove that $M \models_\Lambda (\exists x \beta)^\Pi$

if and only if $M^\Pi \models_{\mathcal{C}_\Pi} \exists x \beta$ through a chain of equivalences. The first equivalence is

(13) $M^\Pi \models_{\mathcal{C}_\Pi} \exists x \beta \iff M^\Pi \models_\Lambda \exists x(\beta^\Pi)$.

To show this, first assume that $M^\Pi \models_{\mathcal{C}_\Pi} \exists x \beta$. Then there is some $\mathcal{C}_\Pi$-model, N such that $N \models_{\mathcal{C}_\Pi} \beta$ and $N \cong_x M^\Pi$. Since N is a $\mathcal{C}_\Pi$-model, we have $N^\Pi = N$, so $N^\Pi \models_{\mathcal{C}_\Pi} \beta$ and by the induction hypothesis $N \models_\Lambda \beta^\Pi$. From this it is immediate that $M^\Pi \models_\Lambda \exists x(\beta^\Pi)$.

Conversely, suppose that $M^\Pi \models_\Lambda \exists x(\beta^\Pi)$. Then for some Λ-model N with $N \cong_x M^\Pi$ we have $N \models_\Lambda \beta^\Pi$. By the induction hypothesis we obtain $N^\Pi \models_{\mathcal{C}_\Pi} \beta$. Also by Lemma 15 we have $N^\Pi \cong_x (M^\Pi)^\Pi = M^\Pi$, so $M^\Pi \models_{\mathcal{C}_\Pi} \exists x \beta$. This proves (13).

Since only the modalities from Λ appear in the formula β^Π it follows from the semantics of BQDL$_\Lambda$ that

(14) $M^\Pi \models_\Lambda \exists x(\beta^\Pi) \iff M^{\overline{\Pi}} \models_{\overline{\Lambda}} \exists x(\beta^\Pi)$.

Applying Lemma 19 to the $\overline{\Lambda}$-class, $\mathcal{D}_\Pi$ we have

(15) $M^{\overline{\Pi}} \models_{\overline{\Lambda}} \exists x(\beta^\Pi) \iff M^{\overline{\Pi}} \models_{\mathcal{D}_\Pi} (\exists x(\beta^\Pi))^{\mu*}$.

From [1] we know that $(\exists x(\beta^\Pi))^{\mu*}$ will be a formula of BQDL$_A$ so by applying Lemma 18 we have

(16) $M^{\overline{\Pi}} \models_{\mathcal{D}_\Pi} (\exists x(\beta^\Pi))^{\mu*} \iff M \models_\Lambda (\exists x(\beta^\Pi))^{\mu*})[\Lambda \backslash \Pi(\Lambda)]$

From the equivalences (13), (14), (15) and (16) we have $M^\Pi \models_{\mathcal{C}_\Pi} \exists x \beta$ if and only if $M \models_\Lambda (\exists x \beta)^\Pi$, as required. This completes the proof. ∎

As a corollary to Lemma 20 we have main theorem of this paper.

THEOREM 21. *Given any idempotent Λ-transduction, Π, QL$_{\mathcal{C}_\Pi}$ is decidable.*

To decide whether a formula α of QL$_{\mathcal{C}_\Pi}$ is satisfiable (where Π is idempotent), we simply compute α^Π and apply a decision procedure for BQDL$_\Lambda$. (See for example [1, 8]).

5 Conclusion

Here we have just begun to investigate the range of logics for which bisimulation quantifiers can meaningfully applied. In [8] the class of idempotent transductions is extended to allow for programs containing the converse operator. This allows us to define idempotent transduction classes corresponding to logics such as **KD45** and **S5**. However, it is shown that even with the converse operator some simple logics, such as **Grz** are not definable using idempotent transductions. In future work we will examine further generalizations of the idempotent transduction classes.

Acknowledgements: The author would like to thank Mark Reynolds, Yde Venema, Ian Hodkinson and the anonymous reviewers for their many helpful comments and suggestions.

BIBLIOGRAPHY

[1] G. D'Agostino and M. Hollenberg. Logical questions concerning the μ-calculus: interpolation, Lyndon and Los-Tarski. *J. Symb. Log.*, 65(1):310–332, 2000.

[2] G. D'Agostino and G. Lenzi. An axiomatization of bisimulation quantifiers via the μ-calculus. *Theor. Comput. Sci.*, 338:64–95, 2005.

[3] G. D'Agostino and G. Lenzi. Bisimulation quantifiers and fixed points over K4, 2005. unpublished.

[4] J. van Benthem. Correspondence theory. *Handbook of Philosophical Logic*, 2:167–247, 1984.

[5] E. A. Emerson and C. S. Jutla. Tree automata, μ-calculus and determinacy. In *Proceedings of the 32nd annual symposium on Foundations of computer science*, pages 368–377, Los Alamitos, CA, USA, 1991. IEEE Computer Society Press.

[6] K. Fine. Propositional quantifiers in modal logic. *Theoria*, 36:336–346, 1970.

[7] T. French. Bisimulation quantified logics: undecidability. In *Proceedings of FSTTCS '05*, volume 3821 of *LNCS*, pages 396–407, 2005.

[8] T. French. *Bisimulation quantifiers for modal logic*. PhD thesis, The University of Western Australia, 2006. submitted, available from http://people.csse.uwa.edu.au/tim/.

[9] D. Gabbay, A. Kurucz, F. Wolter, and M. Zakharyaschev. *Many Dimensional Modal Logics: Theory and Applications*. Elsevier, 2003.

[10] S. Ghilardi, C. Lutz, and F. Wolter. Did I damage my ontology? A case for conservative extensions in description logic. In *Proceedings of the Tenth International Conference on Principles of Knowledge Representation and Reasoning (KR2006)*, Lake District, UK, 2006. To appear.

[11] S. Ghilardi and M. Zawadowski. Undefinability of propositional quantifiers in the modal system S4. *Studia Logica*, 55:259–271, 1995.

[12] M. Hollenberg. *Logic and Bisimulation*. PhD thesis, University of Utrecht, 1998.

[13] D. Janin and I. Walukiewicz. Automata for the modal μ-calculus and related results. In *MFCS '95: Proceedings of the 20th International Symposium on Mathematical Foundations of Computer Science*, pages 552–562, London, UK, 1995. Springer-Verlag.

[14] D. Kozen. Results on the propositional mu-calculus. *Theoretical Computer Science*, 27, 1983.

[15] L. Maksimova. Craig's theorem in superintuitionistic logics and amalgamable varieties of pseudoboolean algebras. *Algebra Logika*, 16(6):643–681, 1977.

[16] R. Milner. A calculus of communicating systems. *Lecture Notes in Computer Science*, 92, 1980.

[17] D. Park. Concurrency and automata on infinite sequences. *Lecture Notes in Computer Science*, 104:167–183, 1981.

[18] A. M. Pitts. On an interpretation of second order quantification in first order intuitionistic propositional logic. *Jour. Symbolic Logic*, 57:33–52, 1992.

[19] K. Segerberg. An essay in classical modal logic. *Philosophical Studies*, 13, 1971.

[20] V. Yu. Shavrukov. *Subalgebras of diagonalizable algebras of theories containing arithmetic*. Dissertationes Mathematicae CCCXXIII. Polska Akademia Nauk, Mathematical Institute, Warszawa, 1993.

[21] A. Visser. Uniform interpolation and layered bisimulation. In *Gödel '96*, volume 6 of *Lecture Notes in Logic*, pages 139–164, 1996.

Tim French
School of Computer Science and Software Engineering,
The University of Western Australia
tim@csse.uwa.edu.au

Terminating modal tableaux with simple completeness proof

OLIVIER GASQUET, ANDREAS HERZIG, MOHAMAD SAHADE

ABSTRACT. In this paper we study formal tools for the definition of tableaux systems for modal logics. We begin by some preliminaries about graphs. Then we present graph rewriting rules together with control constructs for defining strategies of application of these rules. We establish fair strategies that are sound, complete and terminating. These results cover almost all basic modal logics such as K, T, S4, KB4, K+Confluence etc., and can be generalized for logics such LTL and PDL. Our framework provides the theoretical basis for our generic tableau theorem prover LoTREC.

Keywords: modal logics, tableau methods, automatic proof.

1 Introduction

Basic modal logics are normal modal logics defined by any combination of axioms D, T, B, 4, 5 [4]. For all these logics the complexity of satisfiability is in PSPACE, and research (in particular in the field of description logics) has focussed on tableau decision procedures for them that stay within that boundary. For all these logics one can find sound, complete and terminating tableau algorithms in the standard textbooks [5, 7], and implemented tableau theorem provers such as LogicWorkBench (LWB) [8] are readily available.

This contrasts with the situation for more complex modal logics that are used to reason about mental states of artificial agents, about knowledge and communication, or norms and regulations. In these domains we have to deal with BDI logics (that is, logics capturing the notions of belief, desires and intentions), logics of knowledge, common knowledge and time, logics of action and obligation, etc. Such logics are not as 'stable' as the traditional modal logics we find in textbooks: for example, it often happens that one would like to slightly modify the interplay between the modalities (preservation of beliefs through actions, influence of actions on obligations, interplay between belief and knowledge, etc.), or to add a new modality to an existing logic (add actions or time to a logic of belief, add goals to a logic of belief and action, add obligations to BDI logics, etc.) The variety and complexity of these logics make that in practice we do not always have at our disposal tableau algorithms that enjoy soundness, completeness and termination.

The contributions of our paper are the following:

1. We provide a general formal framework for tableau systems, within which virtually all tableau algorithms for modal logics can be implemented. This framework is based on rooted directed acyclic graphs (RDAGs) in order to stay close to possible worlds models, and its basic mechanism is graph rewriting. It provides a simple language to define tableau rules and a simple language to define strategies. We formally define the semantics of the usually rather implicit or purely operational notions of rule and strategy.

2. Then we give simple and uniform completeness and soundness proof for a large family of modal logics.

3. We state general termination theorems for two classes of strategies. These classes contain strategies for the modal logics K, T, S4, S5, KB4, and even LTL and PDL.

This paper presents the theoretical elements which are the basis of the generic tableau prover LoTREC that we develop in the LILaC group of IRIT. LoTREC is available at: *http://www.irit.fr/LILAC/Lotrec*

In particular there is a version of LoTREC that is executable via the web. A system description can be found in [6].

2 Modal logics and relational properties

The modal logics we consider are all obtained by extending the basic modal logic K by one or several of the well-known axioms T, B, 4, 5, D, and C (axiom of confluence: $\Diamond\Box p \to \Box\Diamond p$). Thus KDC4 denotes the modal logic obtained by adding the axioms D, C and 4 to the basic system K.

With each of these axioms there is a associated relational property of the accessibility relation of the Kripke models. The axioms: T:$\Box p \to p$, 4: $\Box p \to \Box\Box p$, B:$\Diamond\Box p \to p$, 5:$\Diamond\Box p \to \Box p$, D:$\Box p \to \Diamond p$ and C:$\Diamond\Box p \to \Box\Diamond p$ correspond respectively to: reflexivity (*Ref*), transitivity (*Tr*), symmetry (*Sym*), euclideanity (*Eucl*), seriality (*Ser*), confluence (*Conf*).

As a consequence of Sahlqvist's theorem [11], a modal logic based on K plus any combination of these axioms is characterized by the class of Kripke models whose accessibility relation satisfies the corresponding properties. Thus, KD4 is characterized by Kripke models where the accessibility relation is both serial and transitive.

Let group 1 be {*Ref, Tr, Sym, Eucl*} and group 2 be {*Ser, Conf* }.

DEFINITION 1. Given a set ρ of relational properties among group 1 and 2, a ρ-*model* is a Kripke model whose accessibility relation satisfies ρ. A formula is ρ-satisfiable iff it is satisfiable in a ρ-model. It is ρ-valid iff it is valid in the class of all ρ-models, which will be noted $\models_\rho A$. Thus A is a theorem of a system denoted by a set ρ of properties iff it is ρ-valid.

3 Rooted Directed Graph

In the paper [3] we have advocated that tableaux systems should work on rooted directed acyclic graphs (RDAGs). Such a choice not only has the advantage of generality: as RDAGs are closer to Kripke models than sequent and trees, it also makes the way from an open tableau to a countermodel shorter. Here we pursue that approach.

DEFINITION 2. A labelled ρ-directed graph is a triple $\langle \mathcal{N}, \Sigma, \text{FOR} \rangle$ where:

- $\langle \mathcal{N}, \Sigma \rangle$ is a directed graph: $\mathcal{N}$ is some set and $\Sigma \subseteq \mathcal{N} \times \mathcal{N}$
- $\langle \mathcal{N}, \Sigma \rangle$ satisfies all properties of ρ;
- FOR is a function that associates a set of formulae with each of the nodes.

A labelled *rooted directed graph* ρ-(RGRAPH) is a couple $\langle G, root \rangle$ where $G = \langle \mathcal{N}, \Sigma, \text{FOR} \rangle$ is a labelled directed graph, and $root \in \mathcal{N}$ is a root of G, that can access every other node in the transitive closure of Σ.

The elements $\langle x, x' \rangle$ of Σ are also written $x\Sigma x'$. $x\Sigma x'$ means that x' is accessible from x via the relation Σ. For $\langle x, A \rangle \in$ FOR we sometimes also write $A \in \text{FOR}(x)$. $A \in \text{FOR}(x)$ means that the formula A is associated with node x in the graph. Graphs are denoted by $G, g, g', \ldots$. Small letters will be used for subgraphs.

DEFINITION 3. (RDAG) A labelled *rooted directed acyclic graph*(ρ-RDAG) is a labelled RGRAPH where Σ contains no cycles.

4 Closure of RGRAPH

We define the following closure operation on RGRAPH:

DEFINITION 4. Let $\langle G, root \rangle$ be an RGRAPH over a set and ρ a set of relational properties from group 1; the ρ-closure of Σ (denoted by Σ^ρ) is the least RGRAPH that contains Σ and which satisfies every property of ρ.

This ρ-closure always exists for properties among $\{Ref, Tr, Sym, Eucl\}$. A very important point is that for properties of group 1, the closure can be expressed in terms of the initial RGRAPH. For example, the transitive closure of an RGRAPH $\langle G, root \rangle$ is defined by: $(x, y) \in \Sigma^{Tr}$ iff $\exists n \geq 1$ such that $(x, y) \in \Sigma^n$. Note that we do not consider here properties of group 2: is not so easy to define closure under a property of group 2. This is the reason why they will be handled in a different way: as we will see later no propagation rule can simulate them.

LEMMA 5 (Relational Closure Lemma).
Let Σ be an RGRAPH over a set $\mathcal{N}$ of nodes:

- $(x, y) \in \Sigma^{Ref}$ iff $(x, y) \in \Sigma$ or $x = y$.
- $(x, y) \in \Sigma^{Sym}$ iff $(x, y) \in \Sigma$ or $(y, x) \in \Sigma$.
- $(x, y) \in \Sigma^{Tr}$ iff $\exists n \geq 1$ such that $(x, y) \in \Sigma^n$.
- $(x, y) \in \Sigma^{Eucl}$ iff $(x, y) \in \Sigma$ or $\exists u \in \mathcal{N}$ $\exists n \geq 1$ $\exists m \geq 1$ such that $(u, x) \in \Sigma^n$ and $(u, y) \in \Sigma^m$.

- $(x,y) \in \Sigma^{Ref,Sym}$ iff $(x,y) \in \Sigma$ or $x=y$ or $(y,x) \in \Sigma$.
- $(x,y) \in \Sigma^{Ref,Tr}$ iff $\exists n \geq 0$ such that $(x,y) \in \Sigma^n$.
- $(x,y) \in \Sigma^{Ref,Eucl}$ iff $\exists n \geq 0 \ \exists x_0 = x, \ldots x_i, x_{i+1}, \ldots, x_n = y$: $(x_i, x_{i+1}) \in \Sigma$ or $(x_{i+1}, x_i) \in \Sigma$.
- $(x,y) \in \Sigma^{Sym,Tr}$ iff $\exists n \geq 1 \ \exists x_0 = x, \ldots x_i, x_{i+1}, \ldots, x_n$: $(x_i, x_{i+1}) \in \Sigma$ or $(x_{i+1}, x_i) \in \Sigma$.
- $(x,y) \in \Sigma^{Tr,Eucl}$ iff $\exists u \in \mathcal{N} \ \exists n \geq 0 \ \exists m \geq 1$ such that $(u,x) \in \Sigma^n$ and $(u,y) \in \Sigma^m$.

LEMMA 6. *The remaining cases are reducible to those of the previous lemma:*

- $\Sigma^{Sym,Eucl} = \Sigma^{Sym,Tr,Eucl} = \Sigma^{Sym,Tr}$;
- $\Sigma^{Ref,Sym,Tr} = \Sigma^{Ref,Tr,Eucl} = \Sigma^{Ref,Sym,Eucl} = \Sigma^{Ref,Sym,Tr,Eucl} = \Sigma^{Ref,Eucl}$.

The above lemmas will be a powerful tool for proving completeness: it will allow to define a model for a formula from an open tableau. But this is not the whole story. As we previously said, some properties are handled structurally; roughly speaking seriality and confluence are treated by generating the RDAG underlying of the tableau.

LEMMA 7 (Structural Lemma). *Let ρ_2 be a subset of group 2, ρ_1 a subset of group 1 and let Σ be a ρ_2-RGRAPH over a set $\mathcal{N}$ of nodes. Then Σ^{ρ_1} is also a ρ_2-RGRAPH, and hence is a $(\rho_1 \cup \rho_2)$-RGRAPH.*

For the proof of the above three lemmas see ([3]).

5 A simple graph-rewriting language $\mathcal{RL}$

As we would like to handle complex modal logics, the tableau rules that we need must be able to perform at least the following operations on graphs: add a formula A to a node x, add a link with some label R between two nodes x, x', create a new node x, duplicate the graph (to take into account disjunctions), mark and unmark a node x by an expression E, mark and unmark a formula A in a node x by an expression E.

Note that links have to be labelled in order to be able to handle multiple modalities. Note also that marking is not needed for the simplest modal logics (the basic modal logics defined by any combination of axioms D, T, B, as well as $K45, KD45$ and $S5$).

Such rewriting steps are executed under some conditions. We need at least the following ones: node x contains a formula A (or not), node x is linked to node x' by a link having label R (or not), node x is an ancestor of node x', node x is marked by an expression E (or not).

Note that only the first two conditions are necessary for the simplest modal logics.

We here choose a minimal rule language that is restricted to exactly the above conditions and actions. It is also the rule language of LoTREC, with self-explaining conditions such as: if a node x contains a formula A, or does not contain a formula A, ... and actions such as: add an expression E to a node x, link two nodes x, x'...

DEFINITION 8. (pattern) A pattern is a connected graph.

DEFINITION 9. A *tableau rule* r is a couple $(p, \{g_1, \ldots, g_n\})$ where p is a pattern (the pattern on which the rule can be applied) and $g_i, 1 \leq i \leq n$ are the graphs resulting from the application of the rule on patterns which match p.

The rule r is *analytic* if every formula appearing in $g_i, 1 \leq i \leq n$ is a subformula of some formula appearing in p. r is strictly analytic if every formula appearing in $g_i, 1 \leq i \leq n$ is a strict subformula of some formula appearing in p.

Note that in the sequel I use the terms: *conditions part* of a rule to represent the pattern p and *action part* of a rule to represent the modifications done which lead to the new graphs $\{g_1, \ldots, g_n\}$.

Application of a rule r to a pattern p_0 is defined by:

$$(p, \{g_1, \ldots, g_n\})(p_0) = \begin{cases} \{\nu(g_1), \ldots, \nu(g_n)\} & \text{if } \exists \nu \text{ such that } \nu(p) = p_0 \\ \bot & \text{otherwise} \end{cases}$$

Note that ν is an isomorphism on the structure of the graph, but it is a homomorphism on the set of formulas.

We extend the partial function r to a total function (application of r on a graph) by stipulating one of the following definitions:

1− $r(G) = \{G \cup (\bigcup_{p \subseteq G} \delta) : \delta \in r(p)\}$ where $\forall p, r(p)$ is applicable

2− $r(G) = \{(G \cup (\bigcup_{\substack{p \subseteq G \\ r(p) \neq \bot}} \delta)) : \delta \in r(p)\}$

Note that a graph is a set of patterns (and may be only one pattern). Application of a rule r to a set of graphs Γ is defined by:

$$r(\Gamma) = \{\Gamma \cup (\bigcup_{G \subseteq \Gamma} r(G))$$

As every rule $(p, \{g_1, \ldots, g_n\})$ is monotonic (in the sense that that $p \subseteq g_i, 1 \leq i \leq n$), we only give the new expressions of $g_i, 1 \leq i \leq n$.

REMARK 10. p and $g_i, 1 \leq i \leq n$ may contain node variables and formula variables (that are instantiated via ν when the rule is applied). We can think of such rules as the (infinite) set of their instances.

REMARK 11. Of course, $r(p)$ may contain new nodes and/or edges that do not belong to p and that have to be named. We suppose that such new nodes will be denoted by ground first-order terms: a Skolem function denoted by the name of the rule (and possibly with an integer subscript if the rule introduces more that one new node) and with elements of g as arguments. As an example, if some node x contains a formula $\Diamond A$ then applying the rule for $\Diamond$ will introduce the new node $\Diamond(x, A)$.

6 Tableau rules

In order to define a tableau calculus for a modal logic denoted by $KA_1, \ldots, A_n$ corresponding to a set of properties ρ_1 form group 1 and a set of properties ρ_2 of group 2, we must associate a set of rules with it. All the tableaux calculi we are going to define contain: the classical rules, createSuccessor and propagateFormulas rules which will be written in $\mathcal{RL}$ as $(\langle \{n_0\}, \emptyset, \{(n_0, \{\neg \Box A\})\}\rangle, \{\langle \emptyset, \{(n_0, n)\}, \{(n, \{\neg A\})\}\rangle\})$, $(\langle \{n_0, n_1\}, \{(n_0, n_1)\}, \{(n_0, \{\Box A\})\}\rangle, \{\langle \emptyset, \emptyset, \{(n_1, \{A\})\}\rangle\})$ respectively (see also appendix B). As these rules are common to all tableaux calculi, we will henceforth omit them. More over we have some structural and propagation rules.

A tableau calculus for a logic defined by $(\rho_1 \cup \rho_2)$ is obtained by taking (in addition to classical rules, createSuccessor and propagateFormulas rules) the rules corresponding to properties of $(\rho_1 \cup \rho_2)$; this correspondence is given in the figure below.

	Properties	Rules
Group 1	Ref	propagateNec2ActualWorld
	Sym	addFormula2Parent
	Tr	propagateNec2Child
	Eucl	copyNecIfLinked, copyNec2Parent, copyNec2Brothers
Group 2	Ser	addPosTrue
	Conf	confluence

In the case of logics which contain the relational properties: Eucl or Tr, to stop the possibly infinite computation we perform an inclusion test called loop test. It tests if a node x is included in some ancestor. In this case we don't create any successor for x and we stop the computation on the tableau which contain x.

REMARKS 12.

It is worth noticing that our three rules handling euclideanity are reminiscent of the three sequent rules used in [2]. This connection between a tableau approach and a sequent approach probably deserves to be investigated further.

DEFINITION 13. A $(\rho_1 \cup \rho_2)$-tableau for a formula A is the limit of a sequence of RDAGs $\Upsilon_0, \ldots, \Upsilon_i, \Upsilon_{i+1}, \ldots$ such that[1]:

- Υ_0 is an RDAG consisting of only one node whose associated set of formulas is $\{A\}$: $\Upsilon_0 = \langle root, \emptyset, \{(root, \{A\})\}\rangle$;

---
[1] this limit being defined as $\bigcup_{i \geq 0} \Upsilon_i$

- Υ_{i+1} is obtained from Υ_i by applying either a classical rule, or the `createSuccessor` rule, or the rule `propagateFormulas`, or a rule for $(\rho_1 \cup \rho_2)$;

- every applicable rule has been applied.

Note that what we call a tableau here is usually called a tableau branch in the literature. While this is appropriate for classical logic, it might lead to confusions to have branches that are in fact graphs.

DEFINITION 14. A tableau is closed if some node in it contains `False`; it is open otherwise. A formula is $(\rho_1 \cup \rho_2)$-closed iff all its $(\rho_1 \cup \rho_2)$-tableaux are closed [2].

7 A simple strategy language

Suppose we have defined a set of rewriting rules $RULES$ for our logic **L**, and we now want to design a strategy and prove it terminates.

Strategies define how the rules from $RULES$ are applied, and in which order. Given a set of graphs, the simplest way of applying rule $r \in RULES$ is to apply it to *every* formula in *every* node in *every* graph. It is also the most radical definition. It follows that in order to avoid loops (in the case of modal logics beyond the simplest ones), rule application has to be blocked by means of appropriate marking conditions. For example in S4, `createSuccessor`-rules application in node x is blocked if x is included in some ancestor node.

7.1 Combining rule applications

One of the simplest ways of combining rule applications is by means of regular expressions. We here provide a language that is made up of such expressions, and that is inspired by the concepts that are used in the Isabelle and HOL provers for higher-order logics [10].

We choose the following constructs: `allRules`, `firstRule` and `repeat` to combine rules for defining strategies. They work as follows, where each r_i is itself a strategy, possibly consisting of only one rule:

- (`allRules` $r_1 \ldots r_n$) sequentially applies each of the applicable rules to the current graph, starting with r_1.

- (`firstRule` $r_1 \ldots r_n$) applies only the first applicable rule among the listed ones to the current tableau.

- (`repeat` r_1) repeatedly applies r_1 until a fixpoint is reached.

The semantic of our constructs `allRules`, `firstRule` and `repeat` for strategies is:

DEFINITION 15. (semantics of a strategy)

---

[2] Due to the rule $\vee$, which is equivalent to the rule `NotAnd` in the language of LoTREC, a formula may have several distinct tableaux.

$$\text{allRules } (r_1 \ldots r_k)(G) = (r_k(\ldots(r_2(r_1(G)))) \ldots)$$

$$\text{firstRule } (nil)(G) = G, \text{ where } nil \text{ is the empty sequence of rules}$$

$$\text{firstRule } (r_1 \ldots r_k)(G) = r_1(G) \qquad \text{if } k \geq 1 \text{ and } r_1(G) \neq G$$

$$\text{firstRule } (r_1 \ldots r_k)(G) = \text{firstRule } (r_2 \ldots r_k)(G) \text{ otherwise}$$

$$\text{repeat } (r)(G) = \text{lfp}(r, G)$$

Thus repeat-loops build a least fix-point.

To apply a strategy to a formula A means to apply the strategy to the initial graph $G = (\text{root}, \emptyset, \{(\text{root}, \{A\})\}$.

Of course, a repeat-loop may make the resulting strategy loop forever if r is not designed carefully.

8 Completeness

In this section we prove the completeness of our tableau calculi[3]. We show how, from a given open **L**-tableau for A we can construct a ρ-model for A, where ρ is the set of accessibility relations which characterize the logic **L**.

DEFINITION 16 (End Node). x is an end node iff neither createSuccessor nor addPosTrue nor confluence rule are applicable.

DEFINITION 17 (Loop Node). A node x is a loop node in a tableau Υ iff:

1. the rule createsuccessor or addPosTrue or confluence is applicable to x and

2. $\exists y \colon y\Sigma^+ x$ and $x \subseteq y$ (such y will be denoted by *loop(x)*).

DEFINITION 18 (Looping Tableau). A tableau $\Upsilon = (\mathcal{N}, \Sigma, \text{FOR})$ is a *looping tableau* iff each of its leafs is either an end node or a looping node.
The set of loop nodes of Υ is $loop(\Upsilon)$. Note that looping tableaux are finite, since the computation is stopped either because no create new node rules are applicable or due to the inclusion test.

DEFINITION 19 (Folded Tableau).

Let Υ be a looping open **L**-tableau for A such that Tr or $Eucl$ are in **L**, where we stop the execution by making the inclusion test.

The Folded Tableau $\mathcal{F}\Upsilon = (\mathcal{F}\mathcal{N}, \mathcal{F}\Sigma, \mathcal{F}\text{FOR})$ corresponding to Υ is defined as follows:

- $\mathcal{F}\mathcal{N} = \mathcal{N} - loop(\Upsilon)$;

- $\mathcal{F}\Sigma = \Sigma - \{(x,y)/y \in loop(\Upsilon)\} \cup \{(x, loop(y))/y \in loop(\Upsilon)\}$ where $x \in \mathcal{N}$;

- $\mathcal{F}\text{FOR}$ is the restriction of FOR to $\mathcal{F}\mathcal{N}$.

---
[3] We make the usual assumption of *fairness*.

Let Υ be an open Folded **L**-tableau for A. Now let $\mathcal{M} = (W, R, \tau)$ be the Kripke model defined as follows:

DEFINITION 20 (Folded Model).
- $W = \mathcal{FN}$
- R is the ρ_1-closure of $\mathcal{F}\Sigma$, i.e. $R = \mathcal{F}\Sigma^{\rho_1}$
- for all $w \in W$, $w \in \tau(p)$ iff $p \in \mathcal{F}\text{FOR}(w)$.

To prove the completeness of a **L**-tableaux we will consider for a formula A a **L**-folded open tableau $\mathcal{F}\Upsilon = (\mathcal{FN}, \mathcal{F}\Sigma, \mathcal{F}\text{FOR})$ and then give a **L**-model for this formula from Υ (as described in Definition 20). Of course, folded models are finite. We first establish the following important lemma:

LEMMA 21 (Box Lemma). *Let* $\mathcal{F}\Upsilon = (\mathcal{FN}, \mathcal{F}\Sigma, \mathcal{F}\text{FOR})$ *be a folded* $(\rho_1 \cup \rho_2)$-*tableau. Let* x, y *be such that* $(x, y) \in \mathcal{F}\Sigma^{\rho_1}$ *and* $\Box A \in x$; *then* $A \in y$.

Proof. There are several cases, depending on $(\rho_1 \cup \rho_2)$; we only prove the lemma for some complex cases involving euclideanity:

- $\rho_1 = \{Eucl\}$: if $(x, y) \in \mathcal{F}\Sigma^{\rho_1}$ then by the Relational Closure Lemma (Lemma 4) in the case of a folded tableau, we have:
  - either $(x, y) \in \Sigma$ and then $A \in y$ by rule propagateFormulas
  - or $\exists u\ \exists n \geq 1\ \exists m \geq 1$ such that $(u, x) \in \mathcal{F}\Sigma^n$ and $(u.y) \in \mathcal{F}\Sigma^m$. Hence
    * $\exists x_0 = x, \ldots, x_i, x_{i+1}, \ldots, x_n = u$:
      for $0 \leq i \leq n$ we have either:
      · $(x_{i+1}, x_i) \in \Sigma$; then $A, \Box A \in x_i$ (by rules propagateFormulas and CopyNec2Parent)
      · $(x_{i+1}, x_i) \in \mathcal{F}\Sigma$ and $(x_{i+1}, x_i) \notin \Sigma$; then $\exists z \in \mathcal{N}$: $(x_{i+1}, z) \in \Sigma$ and $z \subseteq x_i$ then $A, \Box A \in z$ (by rules propagateFormulas and CopyNecIfLinked) and $\Box A \in x_i$ since $z \subseteq x_i$.
      In particular: $A, \Box A \in x_n = u$.

    * $\exists y_0 = u, \ldots, y_i, y_{i+1}, \ldots, y_m = y$:
      · if $m = 1$ then $y = y_1$ and $A \in y$ by rule propagateFormulas.
      We have $\Box A \in y_1$ (by rule copyNec2Brothers since $\Box A \in x_{n-1}$). For $1 \leq i \leq m$ we have :
      · either $(y_i, y_{i+1}) \in \Sigma$; then $A, \Box A \in y_i$ (by rules propagateFormulas and CopyNecIfLinked)
      · or $(y_i, y_{i+1}) \in \mathcal{F}\Sigma$; and $(y_i, y_{i+1}) \notin \Sigma$: then $\exists z \in \mathcal{N}$: $(y_i, z) \in \Sigma$ and $z \subseteq y_{i+1}$ then $A, \Box A \in z$ (by rules propagateFormulas and CopyNecIfLinked) and $A, \Box A \in y_{i+1}$ since $z \subseteq y_{i+1}$.

Since $\Box A \in x_n = u = y_0$ we have: $\Box A \in y_i$ for $0 \leq i \leq m$. Hence $A \in y_i$ for $1 \leq i \leq m$ (by rule propagateFormulas), in particular $A \in y$.

- $\rho_1 = \{Tr, Eucl\}$: if $(x, y) \in \mathcal{F}\Sigma^{\rho_1}$ then by the Relational Closure Lemma, we have $\exists u \in \mathcal{N}$ $\exists n \geq 0$ $\exists m \geq 1$ such that $(u, x) \in \mathcal{F}\Sigma^n$ and $(u, y) \in \mathcal{F}\Sigma^m$. This implies that:

  - $\exists n \geq 0$ $\exists x_0 = x, \ldots, x_i, x_{i+1}, \ldots, x_n = u$:
    * $(x_{i+1}, x_i) \in \Sigma$; then $\Box A \in x_i$ implies $\Box A \in x_{i+1}$ (by rule CopyNec2Parent)
    * $(x_{i+1}, x_i) \in \mathcal{F}\Sigma$ and $(x_{i+1}, x_i) \notin \Sigma$; then $\exists z \in \mathcal{N}$: $(x_{i+1}, z) \in \Sigma$ and $z \subseteq x_i$ then $\Box A \in x_i$ implies $\Box A \in z$ (by rule CopyNecIfLinked) and $\Box A \in x_{i+1}$ since $z \subseteq x_{i+1}$.

    In particular $\Box A \in x_0 = u$.

  - $\exists m \geq 1$ $\exists y_0 = u, \ldots, y_i, y_{i+1}, \ldots, y_{m+1} = y$: :
    * if $m = 1$ then $y = y_1$ and $A \in y$ by rule propagateFormulas.
    * $(y_i, y_{i+1}) \in \Sigma$; hence $\Box A \in y_i$ implies $\Box A \in y_{i+1}$ (by rule propagateNec2Child).
    * $(y_i, y_{i+1}) \in \mathcal{F}\Sigma$ and $(y_i, y_{i+1}) \notin \Sigma$: then $\Box A \in y_{i+1}$ (same proof as above)

    In particular, if $\Box A \in y_{m-1}$, then $A \in y$ (by rule propagateFormulas).

- In the case of logics with the axiom D, we build a serial model by adding to every leaf of the folded tableau a reflexive edge. We call this operation reflexive leaf closure (RLC). The folded tableau will be the tableau obtained from the original tableau by RLC.

  Now if $(x, y) \in \mathcal{F}\Sigma$, we have:

  - either $(x, y) \in \Sigma$, so $A \in y$ by rule propagateFormulas;
  - or (x, y) is a RLC edge; in this case $x = y$ is a leaf and it cannot contain a formula of the form $\Box A$. Because if it is the case, the rule addPosTrue adds a $\Diamond \top$ to x and createSuccessor creates a successor to it.

■

To achieve the completeness proof, it remains to prove that:

for all $B \in sf(A)$ if $B \in x$ then $M, x \models B$ for all **L**-models obtained from the tableaux of A.

Where $sf(A)$ is the set of subformulas of A. The following fundamental lemma brings us to the desired conclusion:

LEMMA 22 (Fundamental Lemma). *Let $\mathcal{F}\Upsilon$ be an open ρ_1-tableau for A, let $\mathcal{M}$ be the ρ_1-model defined as in (Definition 20) from $\mathcal{F}\Upsilon$, let $B \in sf(A)$ and let $x \in \mathcal{N} - loop(\mathcal{F}\Upsilon)$. Then (i) $B \in x$ iff $\mathcal{M}, x \models B$.*

Proof. By induction on the structure of B. Without loss of generality suppose that B is written only with $\neg, \bot, \wedge, \square$.

- Induction initialization: let B be an atom, (i) holds by the definition of τ;

- Induction step:

  1. B cannot be False, otherwise Υ would be closed;
  2. If B is a propositional classical formula the proof is straightforward.
  3. Let $B = \neg \square C$.
     $\neg \square C \in x$
     $\Rightarrow \exists y$ such that $(x,y) \in \mathcal{F}\Sigma$ and $\neg C \in y$ (rule createSuccessor or because of the inclusion if x is a loop node)
     $\Rightarrow \exists y$ such that $(x,y) \in R$ and $C \notin y$ ($\mathcal{F}\Sigma \subseteq R$)
     $\Rightarrow \exists y$ such that $(x,y) \in R$ and M$, y \not\models C$
     $\Rightarrow$ M$, x \models \neg \square C$
  4. Let B be $\square C$ and suppose $(x,y) \in R$; then by the Box Lemma 21, $C \in y$. Then by induction hypothesis, $\mathcal{M}, y \models C$ for every y such that $(x,y) \in R$. Hence, $\mathcal{M}, x \models \square C$.

■

As a direct consequence of the previous lemma, we have:

COROLLARY 23. *If A has an open looping $(\rho_1 \cup \rho_2)$-tableau then it has a $(\rho_1 \cup \rho_2)$-model.*

We will show later that for all our logics, and for any formula, either all its tableaux are closed or it has at least one open looping tableau. In other words, a formula cannot have an infinite tableau. In the first case we will show that the formula is unsatisfiable (section 9). If is not the case, we will show that it has an open fair looping ρ-tableau (section 10).

9 Soundness

In this section, we prove the soundness of our tableaux calculi: if a formula A is $(\rho_1 \cup \rho_2)$-satisfiable then there exists a open $(\rho_1 \cup \rho_2)$-tableau for A.

We use pattern(r) to note the pattern described by the left hand side of a rule r. To prove the soundness we start by the following proposition:

PROPOSITION 24. *Let Υ be a tableau. Suppose that RULES is subset of the rules given in appendix B. $\forall \rho \in RULES \ \forall g \in pattern(r)$ we have: $g \subseteq \mathcal{F}\Upsilon$ (folded tableau) iff there is $g' \subseteq \Upsilon$ such that g' is a pattern isomorphic to g.*

Proof. The only new patterns are caused by the redirected edges, but these edges existed before the folding operation. ■

PROPOSITION 25. *Let Φ be a modal formula, we have the following implications:*

1. *If Φ is $(\rho_1 \cup \rho_2)$-satisfiable then there is an infinite open tableau for Φ (without loop test);*

2. *If there is an infinite open tableau for Φ then there is a finite open tableau for Φ (by using the looping test);*

3. *If there is an finite open tableau for Φ then there is an open folded tableau for Φ.*

Proof. For the first implication see [3]. The second implication is trivial since we stop the computation by the loop test, the resulting tableau is included in the infinite one and it can not be closed. The last implication results from Proposition 24. ∎

10 Termination

The setting we are concerned with is a well-delimited one (labelled graphs and monotonic rules, regular expression-like strategies), and it would be desirable to have general termination results. Unfortunately there seems to be no results in the literature that apply to such a case. For example techniques from rewriting theory such as Knuth-Bendix completion cannot be applied here (in particular because rewriting focusses on rule application in any order, Church-Rosser property etc, while we insist on the contrary).

Here we give a formal termination proof of our strategies: In the case where our strategies are formed only by the constructors: `allRules` and `firstRule` the termination is trivial since in the first one there are a finite number of rules and every applicable one will be applied once, and in the second case only the first applicable rule will be applied.

It remains to prove the termination in the case of `repeat`. We here give two termination criteria, which are enough to handle many logics.

10.1 Adding strict sub-formulas only

Our first kind of strategies only adds strict subformulas in the action part. It might also be called strictly analytic.

THEOREM 26. *Let $d(x)$ be the length of the shortest path from the root to the node x, and let RULES be a set of rules such that for every $r \in$ RULES,*

- *r is strictly analytic (only strict subformulas of the original formula are added to nodes in the action part of r);*

- *if the action part of r contains the `createNewNode`-action creating a node x then*
    - *the condition part of r checks for existence of at least one formula in any node among $x_0, \ldots, x_k$, and*

- $d(x) > \max(d(x_0), \ldots, d(x_k))$, i.e. new nodes are strictly farther from the root (ensuring strict decrease of their contents),

where $x_0, \ldots, x_k$ are the nodes referred to in the condition part of r.

Let S be any strategy on RULES. Let G be a finite graph. Then the application of S to input G terminates, and $S(G)$ is a finite graph.

Proof. We prove that

- application of S can never lead to RDAGs with infinite branching factor;
- application of S can never lead to RDAGs of infinite depth.

The proof is part of the proof of Theorem 27. ∎

Theorem 26 covers the case of logics like K, KD, KT, K+confluence. More generally, such strategies can account for many modal logics without transitive accessibility relations, in particular those with symmetric accessibility relations such as KB, KDB, and KTB.

10.2 Testing loops

Our second criterion applies in particular to logics with transitive accessibility relations. In these logics we use a dynamic loop test (DLT): the createSuccessor-rule does not apply in a node if this node is included in some ancestor [4]. We say that createSuccessor-rule is enriched by DLT. We discuss how DLT is implemented in LoTREC in the conclusion.

THEOREM 27. *Let RULES be a set of rules such that for every $r \in$ RULES,*

- *r is analytic (only subformulas of the original formula are added to nodes in the action part of r);*
- *The createSuccessor-rule does not apply in a node which is included in some ancestor.*

Let S be a strategy that is built from RULES. Then the tableaux construction terminates, in other words apply S to any formula leads to a looping tableau.

Proof. First, observe that due to the subformula condition:

- every node only contains a finite number of formulas, and
- there can only be a finite number of nodes with differing associated set of formulas.

---

[4] ...i.e. we have $x_0(\bigcup \Sigma)^+ x$, and every formula A appearing in x also appears in x_0.

We prove that (1) application of S can never lead to RDAGs of infinite depth, and (2) application of S can never lead to RDAGs with infinite branching factor.

For Theorem 26, the argument is that creation of new nodes is subject to non-emptiness, but with our constraints, a branch cannot be of length more than the modal degree of the input formula: because of the strict sub-formula condition, nodes situated farther would be empty, and then the criterion for applying the `createNewNode`-action would no more be fulfilled.

For Theorem 27, the argument is that creation of new nodes is subject to non-inclusion in an ancestor node, which due to the condition on S is always tested before node creation. Therefore a branch of infinite length would contain an infinite number of nodes having different associated formula sets. This cannot be the case because our rules are monotonic: no formulas are erased. This even gives us the classical upper bound: length of branches is bounded by an exponential in the length of the input formula[5].

Infinite branching in node x could only be produced by introducing infinitely many new nodes of the form $\Diamond(x, A)$. But given a node n, the number of $\Diamond$ formulas being bounded by the (linear) number of sub-formulae of the input formula, each node may only have a linear number of successors. Thus the branching factor is bounded. This ends the proof. ∎

The important point of our works is that contrarily to what can be found in the literature, any fair strategy using the set of rules corresponding to a logic ($\rho_1 \cup \rho_2$) is sound, complete and terminating. This contrasts with the rather procedural strategies of the literature with either natural language argument for termination [7] or complex and procedural proof [9].

This theorem covers the case of logics build over the axioms K, T, B, 4, 5, k4+confluence, KDB, in addition to any logic results from the combination of the axioms of group 1 and group 2.

To sum it up, the above theorems give us some general termination criteria for our strategies. These strategies cover all standard modal logics, including Linear Temporal Logic LTL and Propositional Dynamic Logic PDL (see the LoTREC web page for the rules).

For logics that do not fit into the above framework we have to prove termination on a case by case basis.

11 Conclusion

We have defined a graph-based formal language in terms of rules and strategies that enables programming tableau systems for modal logics. We have given a semantics to it, and proved termination results for two important classes of strategies, as well as terminating strategies for the modal logics of confluence. We have proved also the completeness and soundness of all modal logics build from the basic axioms.

---

[5] In fact the standard argument applies that limits the length of paths to a polynomial, since non strict analytic rules are □-propagation ones. Hence along a path the number of □-ed formulae is always increasing

All these strategies can be tested in our generic tableaux theorem prover LoTREC that is accessible via Internet. LoTREC is similar in its aims to the Tableaux Workbench (TWB) [1], which also aims at a generic tableaux theorem prover. They differ in perspective: while LoTREC adopts the more semantic model building view, TWB opts for the more syntactic sequent calculus view.

We implement dynamic loop test in LoTREC by using the marking mechanism. We suppose that the set of rules contains among others the following.

- The rule markIncludedNodes, which in LoTREC takes the form

```
rule markIncludedNodes
    if isAncestor    node0  node
    if contains      node0  node
    do mark          node CONTAINED
end
```

- One or more node-creating rules createSuccessor(pos,R) (with a modal operator pos and an accessibility relation R as parameters), which in LoTREC take the form

```
rule createSuccessor
    if hasElement    node    pos (variable A)
    if isNotMarked   node    CONTAINED
    do createNewNode node1
    do ...
    ...
end
```

Where no other CreateNewNode action appears in the action part.

The first rule blocks every node x whose associated set of formulas is included in another node x_0 that is an ancestor of x by marking x appropriately. The second kind of rules creates a Σ-successor x_1 for every node x and modal formula[6] $\Diamond A$ in x, and adds some formulas and/or marks to x and x_1, *under the condition that x is not marked as being included in some ancestor node.*

But then one should only mark a node as CONTAINED if this node is not subject to further modifications. This restricts the set of terminating strategies.

If one is only interested in the decision problem of satisfiability then instead of testing if a node is included, it suffices to test its depth in the tableau. But if one wishes to build models (as we do), testing node inclusion is necessary.

---

[6] In practice the formula will be of the 'possible'-type (we therefore used a generic $\Diamond$, alias "pos" in LoTREC's notation), and R is the accessibility relation associated to $\Diamond$.

Appendix
A Some rules in the graph language

In this appendix, we give some rules written in $\mathcal{RL}$ language. Recall that a rule in $\mathcal{RL}$ has the following form: $\langle g, \{g_1, \ldots, g_n\}\rangle$.

For sake of clarity, in the right side of the rule, we give only what is new to the graph. For the same reason, the node's names in LoTREC language (node0,node1,...) will be written $n_0, n_1, \ldots$. S is supposed to be a set of formulae, and N a set of nodes, n, n_0, n_1, n_2 are universally quantified.

Rules In LoTREC	Rules in $\mathcal{RL}$
Stop	$(\langle\{n\}, \emptyset, \{(n, \{A, \neg A\})\}\rangle, \{\langle\emptyset, \emptyset, \{(n, \{\bot\})\}\rangle\})$
And	$(\langle\{n\}, \emptyset, \{(n, \{A \wedge B\})\}\rangle, \{\langle\emptyset, \emptyset, \{(n, \{A, B\})\}\rangle\})$
NotNot	$(\langle\{n\}, \emptyset, \{(n, \{\neg\neg A\})\}\rangle, \{\langle\emptyset, \emptyset, \{(n, \{A\})\}\rangle\})$
NotAnd	$(\langle\{n\}, \emptyset, \{(n, \{\neg(A \wedge B)\})\}\rangle, \{\langle\emptyset, \emptyset, \{(n, \{\neg A\})\}\rangle, \langle\emptyset, \emptyset, \{(n, \{\neg B\})\}\rangle\})$

For logics built from the combinations of the axioms $K, T, B, D, 5$ here are their transformations:

Rules In LoTREC	Rules in $\mathcal{RL}$
createSuccessor	$(\langle\{n_0\}, \emptyset, \{(n_0, \{\neg\Box A\})\}\rangle, \{\langle\emptyset, \{(n_0, n)\}, \{(n, \{\neg A\})\}\rangle\})$
propagateFormulas	$(\langle\{n_0, n_1\}, \{(n_0, n_1)\}, \{(n_0, \{\Box A\})\}\rangle, \{\langle\emptyset, \emptyset, \{(n_1, \{A\})\}\rangle\})$
addFormula2Parent	$(\langle\{n_0, n_1\}, \{(n_0, n_1)\}, \{(n_1, \{\Box A\})\}\rangle, \{\langle\emptyset, \emptyset, \{(n_0, \{A\})\}\rangle\})$
addPosTrue	$(\langle\{n_0\}, \emptyset, \{(n_0, \{\Box A\})\}\rangle, \{\langle\emptyset, \emptyset, \{(n_0, \{\neg\Box\neg\top\})\}\rangle\})$
PropagatNec2ActualWorld	$(\langle\{n_0\}, \emptyset, \{(n_0, \{\Box A\})\}\rangle, \{\langle\emptyset, \emptyset, \{(n_0, \{A\})\}\rangle\})$
copyNecIfLinked	$(\langle\{n_0, n_1, n_2\}, \{(n_0, n_1), (n_1, n_2)\}, \{(n_1, \{\Box A\})\}\rangle, \{\langle\emptyset, \emptyset, \{(n_2, \{\Box A\})\}\rangle\})$
copyNec2Parent	$(\langle\{n_0, n_1\}, \{(n_0, n_1)\}, \{(n_1, \{\Box A\})\}\rangle, \{\langle\emptyset, \emptyset, \{(n_0, \{\Box A\})\}\rangle\})$
copyNec2Brothers	$(\langle\{n_0, n_1, n_2\}, \{(n_0, n_1), (n_0, n_2)\}, \{(n_1, \{\Box A\})\}\rangle, \{\langle\emptyset, \emptyset, \{(n_2, \{\Box A\})\}\rangle\})$

B Some rules in the LoTREC Language

Here are the rules we need for dealing with the logics of group 1 and 2 written in LoTREC Language. In the second appendix, we give the equivalence of some of these rules in the formal form given in definition 9(texts after // are comments):

Rules for classical logics

```
rule Stop
   if hasElement node0    (variable A)
   if hasElement node0    not (variable A)
   do add node0 FALSE
```

```
    do stop node0
end

rule NotNot
   if hasElement node0    not not (variable A)
   do add          node0    (variable A)
end

rule And
   if hasElement node0    and (variable A) (variable B)
   do add          node0    (variable A)
   do add          node0    (variable B)
end

rule NotAnd
   if hasElement node0    not and (variable A) (variable B)
   do duplicate  node0    begin node0 node1 end
   do add          node0    not (variable A)
   do add          node1    not (variable B)
end
```

Rules for modal logics

```
rule createSuccessor
   if hasElement      node0          not nec (variable A)
   do createNewNode node0  node1
   do link            node0  node1 R
   do add             node1          not (variable A)
end

rule propagateFormulas
  if hasElement node0 nec variable a
  if isLinked node0 node1   R
  do add node1 variable a
end

rule addFormula2Parent // For axiom B
  if hasElement node0 nec variable A
  if isLinked node1 node0 R
  do add node1 variable A
end

rule addPosTrue // For axiom D
    if hasElement node0    nec variable A
    do add node0    not nec not True
end
```

```
rule propagateNec2ActualWorld // For axiom T
  if hasElement node0    nec (variable A)
  do add         node0   (variable A)
end

// The following three rules are for the axiom 5
rule copyNecIfLinked
  if hasElement  node1  (nec (variable A))
  if isLinked    node0  node1  (R)
  if isLinked    node1  node2  (R)
  do add         node2  (nec (variable A))
end

rule copyNec2Parent
  if isLinked    node0  node1 (R)
  if hasElement  node1  nec (variable A)
  do add         node0  nec (variable A)
end

rule copyNec2Brothers
  if isLinked    node0  node1 (R)
  if hasElement  node1  nec (variable A)
  if isLinked    node0  node2 (R)
  do add         node2  nec (variable A)
end
```

Rules for modal logics with Transitivity

```
rule createSuccessor
    if   hasElement    node0 pos variable A
    if   isNotMarked   node0 CONTAINED
    do   createNewNode node0 node1
    do   link          node0 node1 R
    do   add           node1 variable A
end

//This rule is for the inclusion test
rule loopTest
    if isNewNode node1
    if isAncestor node0 node1
    if contains node0 node1
    do mark node1 CONTAINED
end

//This rule is for the 2 axioms K and 4
rule propagateNec2Child
```

```
    if hasElement node0    (nec (variable A))
    if isLinked node0 node1    (R)
    do add node1    (nec (variable A))
end


//In addition to the rule propagateFormulas given above
```

Rules for modal logics with confluence
```
rule confluence true // true is to say that the rule is
commutative
    if isLinked N0 N1    (R)
    if isLinked N0 N2    (R)
    if areNotIdentical N1 N2
    if hasElement N1 variable A // tests if  N1 is not empty.
    if hasElement N2 variable B // tests if  N2 is not empty.
    do createNewNode    N0 N3
    do link N1 N3(R)
    do link N2 N3 (R)
end
```

BIBLIOGRAPHY

[1] Pietro Abate and Rajeev Goré. The tableaux work bench. In *TABLEAUX*, pages 230–236, 2003.
[2] Kai Brünnler. Deep sequent systems for modal logic. This volume.
[3] Marcos A. Castilho, Luis Farinas del Cerro, Olivier Gasquet, and Andreas Herzig. Modal tableaux with propagation rules and structural rules. *Fundamenta Informaticae*, 32(3-4):281–297, 1997.
[4] Brian Chellas. *Modal logic: An introduction*. Cambridge University Press, 1980.
[5] Melvin Fitting. *Proof methods for modal and intuitionistic logics*. D. Reidel, Dordrecht, 1983.
[6] Olivier Gasquet, Andreas Herzig, Dominique Longin, and Mohamad Sahade. Lotrec: Logical tableaux research engineering companion. In *TABLEAUX*, pages 318–322, 2005.
[7] R.P. Goré. Tableau methods for modal and temporal logics. In M. D'Agostino, D. Gabbay, R. Hähnle, and J. Posegga, editors, *Handbook of Tableau Methods*. Kluwer Academic Publishers, 1999.
[8] A. Heuerding, M. Seyfried, and H. Zimmermann. Efficient loop-check for backward proof search in some non-classical propositional logics. In P. Miglioli, U. Moscato, D. Mundici, and M. Ornaghi, editors, *Proceedings of the 5th International Workshop TABLEAUX'96: Theorem Proving with Analytic Tableaux and Related Methods*, number 1071 in LNAI, pages 210–225. Springer-Verlag, 1996.
[9] Fabio Massacci. Strongly analytic tableaux for normal modal logics. In Alan Bundy, editor, *Proc. 5th Int. Conf. on Automated Deduction (CADE)*, volume 814 of *LNAI*, pages 723–737. Springer-Verlag, 1994.
[10] L. C. Paulson. Isabelle: A generic theorem prover. *LNCS*, pages 748–752, 1994.
[11] H. Sahlqvist. Completeness and correspondence in the first and second order semantics for modal logics. In Stig Kanger, editor, *Proc. 3rd Scandinavian Logic Symposium 1973*, number 82 in Studies in Logic. North Holland, 1975.

Olivier Gasquet, Andreas Herzig and Mohamad Sahade
IRIT- Université Paul Sabatier,
118 route de Narbonne,
31062 Toulouse cedex 04, France.
{gasquet, herzig, sahade} @irit.fr

Conservative extensions in modal logic

S. GHILARDI, C. LUTZ, F. WOLTER AND

M. ZAKHARYASCHEV

ABSTRACT. Every modal logic L gives rise to the consequence relation $\varphi \models_L \psi$ which holds if, and only if, ψ is true in a world of an L-model whenever φ is true in that world. We consider the following algorithmic problem for L. Given two modal formulas φ_1 and φ_2, decide whether $\varphi_1 \wedge \varphi_2$ is a conservative extension of φ_1 in the sense that whenever $\varphi_1 \wedge \varphi_2 \models_L \psi$ and ψ does not contain propositional variables not occurring in φ_1, then $\varphi_1 \models_L \psi$. We first prove that the conservativeness problem is coNExpTime-hard for all modal logics of unbounded alternatives (which have rooted frames with more than N successors of the root, for any $N < \omega$). Then we show that this problem is (i) coNExpTime-complete for **S5** and **K**. (ii) in ExpSpace for **S4** and (iii) ExpSpace-complete for **GL.3** (the logic of finite strict linear orders). The proofs for **S5** and **K** use the fact that these logics have uniform interpolants of exponential size.

1 Introduction

A theory T_2 is said to be a *conservative extension* of a theory T_1 if any consequence of T_2, which only uses symbols from T_1, is a consequence of T_1 as well. This notion plays an important role in mathematical logic and the foundations of mathematics. For example, the result that Bernays–Gödel set theory BG (or BGC) is a conservative extension of Zermelo–Fraenkel set theory ZF (or ZFC) implies the relative consistency of BG(C): if ZF(C) is consistent then BG(C) is also consistent.

Rather surprisingly, in modal logic the notion of conservative extension has hardly been investigated. Indeed, modal theories—similarly to first-order theories—have become fundamental tools for representing various domains. For example, in epistemic logic, modal theories represent the (possibly introspective) knowledge of an agent; in temporal logic, theories serve as specifications of concurrent systems; in description logic, theories (called TBoxes) are ontologies used to fix the terminology of an application domain, etc. In all these examples, the notion of conservative extension can be used to compare different theories and derive important information about their relation to each other: for instance, a temporal specification T_2 can be regarded as a 'safe' refinement of another temporal specification T_1 if, and only if, T_2 is a conservative extension of T_1 (see, e.g., [15]). A description logic ontology T_2 is a 'safe' extension of another description logic ontology T_1 if, and only if, T_2 is a conservative extension of T_1 (see [1, 6]).

One of the main reasons for using modal logic instead of full first-order logic in the applications above is that reasoning in modal logic is quite often decidable. To employ the notion of conservative extension for modal logics, it is therefore crucial to analyse the algorithmic problem of deciding whether one modal theory is a conservative extension of another modal theory.

In this paper, we investigate the notion of conservative extension for a number of basic modal logics and, in particular, determine the computational complexity of the conservativeness problem for these logics.

In modal logic, the notion of conservative extension depends on the consequence relation we are interested in. Of particular importance are the 'local consequence relation' according to which a formula φ follows from a formula ψ if φ is true *in every world* where ψ is true, and the 'global consequence relation' according to which φ follows from ψ if φ is true *everywhere in a model* whenever ψ is true everywhere in this model (see, e.g., [10]). In this paper, we concentrate on the local consequence relation. Some information about the global consequence relation is provided in the final section.

We begin by showing that deciding non-conservativeness is NExpTime-hard for all modal logics of unbounded alternatives (which have rooted frames with more than N successors of the root, for any $N < \omega$). This result covers almost all standard modal logics, for example, **K**, **S4**, **S5**, and **S4.3**. Thus, deciding conservativeness turns out to be much harder than deciding satisfiability. We also observe that for tabular modal logics (see, e.g., [2, 18]) non-conservativeness is NP^{NP}-complete, which coincides with the complexity of non-conservativeness in classical propositional logic [11]. The proof of this result and many other proofs in this paper are based on some elementary facts connecting conservativeness with bisimulations.

Next, to warm up, we consider the modal logic **S5** and show that in this case non-conservativeness is NExpTime-complete by proving that one can construct a uniform interpolant of exponential size in exponential time (in the size of a given formula) and using the fact that **S5**-satisfiability is decidable in NP. This proof is based on a general result connecting conservativeness with uniform interpolation (see [12, 17] for a discussion of this variant of interpolation).

A slightly different technique is used to prove that for **K** non-conservativeness is NExpTime-complete. Here we employ a result from [14] according to which there exist uniform interpolants for **K** of (only) exponential size, and then provide a direct algorithm deciding non-conservativeness without computing the uniform interpolant.

After that we consider the non-conservativeness problem for **S4** and establish an ExpSpace upper bound. As this upper bound (probably) does not match the NExpTime lower bound, we leave the exact complexity as an open problem. The logic **S4** does not have uniform interpolation [8], and therefore a 'direct' algorithm had to be found. Similar arguments show that non-conservativeness is decidable in ExpSpace for **K4**, **Grz** and **GL**.

Finally, we prove that conservativeness is ExpSpace-complete for **GL.3**, the logic of finite strict linear orders. Here we again give direct proofs for both lower and upper bounds. Similar proofs show ExpSpace-completeness of conservativeness for **K4.3** and **S4.3**.

In many cases, full proofs are omitted. The reader can find them in [7].

2 Preliminaries

We consider the language $\mathcal{ML}$ of propositional unimodal logic with countably many propositional variables $p_1, p_2, \ldots$, the Booleans $\wedge$ and $\neg$, and the modal operator $\Box$. Other Boolean operators and the modal diamond $\Diamond$ are defined as usual. Given an $\mathcal{ML}$-formula φ, we denote by $var(\varphi)$ the set of propositional variables occurring in φ.

A *(Kripke) frame* $\mathfrak{F} = (W, R)$ is a nonempty set W of points (or worlds) with a binary relation R on it. A *(Kripke) model* $\mathfrak{M} = (\mathfrak{F}, \mathfrak{V})$ consists of a frame and a *valuation* $\mathfrak{V}$ giving truth-values to propositional variables in the worlds of W. The *satisfaction relation* '$(\mathfrak{M}, w) \models \varphi$' between *pointed models* $(\mathfrak{M}, w)$ (where $w \in W$) and $\mathcal{ML}$-formulas φ is defined as usual. (If $\mathfrak{M}$ is clear from the context, instead of $(\mathfrak{M}, w) \models \varphi$ we often write $w \models \varphi$.) A formula φ is said to be *valid* in a frame $\mathfrak{F}$ if $(\mathfrak{M}, w) \models \varphi$ holds for every model $\mathfrak{M}$ based on $\mathfrak{F}$ and every point w in it.

Consider now a Kripke complete normal modal logic L (i.e., a subset L of $\mathcal{ML}$ for which there exists a class $\mathcal{F}$ of frames such that L is the set of all formulas that are valid in every $\mathfrak{F} \in \mathcal{F}$). The *local consequence relation* '$\varphi_1 \models_L \varphi_2$' for L is defined as follows: $\varphi_1 \models_L \varphi_2$ holds if, and only if, for every pointed model $(\mathfrak{M}, w)$ based on a frame for L, we have $(\mathfrak{M}, w) \models \varphi_2$ whenever $(\mathfrak{M}, w) \models \varphi_1$.

Given a Kripke complete normal modal logic L and two $\mathcal{ML}$-formulas φ_1 and φ_2, we say that $\varphi_1 \wedge \varphi_2$ is a *conservative extension of* φ_1 in L if, for every $\psi \in \mathcal{ML}$ with $var(\psi) \subseteq var(\varphi_1)$, $\varphi_1 \wedge \varphi_2 \models_L \psi$ implies $\varphi_1 \models_L \psi$.

If $\varphi_1 \wedge \varphi_2$ is *not* a conservative extension of φ_1 in L, then there is a formula ψ with $var(\psi) \subseteq var(\varphi_1)$ such that $\varphi_1 \wedge \psi$ is satisfiable in a model based on a frame for L, while $\varphi_1 \wedge \varphi_2 \wedge \psi$ is not satisfiable in any such model. In this case we call ψ a (non-conservativeness) *witness formula* (or simply a *witness*) for the pair (φ_1, φ_2) in L.

The notion of conservative extension turns out to be closely connected with the notion of uniform interpolation. We remind the reader that a modal logic L is said to have *uniform interpolation* if, for every formula φ and every finite set $\mathbf{p}$ of variables, there exists a formula $\exists_L \mathbf{p}.\varphi$ such that

- $var(\exists_L \mathbf{p}.\varphi) \subseteq var(\varphi) \setminus \mathbf{p}$,

- $\varphi \models_L \exists_L \mathbf{p}.\varphi$, and

- $\varphi \models_L \psi$ implies $\exists_L \mathbf{p}.\varphi \models_L \psi$, for every formula ψ with $var(\psi) \cap \mathbf{p} = \emptyset$.

LEMMA 1. *If L has uniform interpolation, then $\varphi_1 \wedge \varphi_2$ is a conservative extension of φ_1 in L iff $\varphi_1 \models_L \exists_L \mathbf{p}.(\varphi_1 \wedge \varphi_2)$, where $\mathbf{p} = var(\varphi_2) \setminus var(\varphi_1)$.*

Proof. Suppose that $\varphi_1 \wedge \varphi_2$ is not a conservative extension of φ_1. Take a formula ψ with $var(\psi) \subseteq var(\varphi_1)$ such that $\varphi_1 \wedge \varphi_2 \models_L \psi$ and $\varphi_1 \not\models_L \psi$. Then we must have $\exists_L \mathbf{p}.(\varphi_1 \wedge \varphi_2) \models_L \psi$, from which $\varphi_1 \not\models_L \exists_L \mathbf{p}.(\varphi_1 \wedge \varphi_2)$.

Conversely, suppose $\varphi_1 \not\models_L \exists_L \mathbf{p}.(\varphi_1 \wedge \varphi_2)$. But then $\varphi_1 \wedge \varphi_2$ cannot be a conservative extension of φ_1 because $\varphi_1 \wedge \varphi_2 \models_L \exists \mathbf{p}.(\varphi_1 \wedge \varphi_2)$. ∎

This lemma suggests the following procedure for deciding the conservativeness problem for a modal logic L with uniform interpolation: given φ_1 and φ_2, construct $\exists_L \mathbf{p}.(\varphi_1 \wedge \varphi_2)$ with $\mathbf{p} = var(\varphi_2) \setminus var(\varphi_1)$ and then check whether $\varphi_1 \models_L \exists_L \mathbf{p}.(\varphi_1 \wedge \varphi_2)$. Below, we will follow this approach for the modal logic **S5**. However, many standard modal logics, such as **S4**, **K4** or **S4.3**, do not have uniform interpolation [8]. In these cases we will provide direct proofs.

An important notion that can be used for analysing conservative extensions (as well as uniform interpolation) is the standard bisimulation between Kripke models [9]: for a finite set $\mathbf{p}$ of propositional variables and two models $(\mathfrak{M}_1, w_1)$ and $(\mathfrak{M}_2, w_2)$, a $\mathbf{p}$-*bisimulation* between $(\mathfrak{M}_1, w_1)$ and $(\mathfrak{M}_2, w_2)$ is a relation between the two models satisfying the standard bisimulation conditions for the variables in $\mathbf{p}$. If $(\mathfrak{M}_1, w_1)$ and $(\mathfrak{M}_2, w_2)$ are $\mathbf{p}$-bisimilar, then we write $(\mathfrak{M}_1, w_1) \leftrightarroweq_\mathbf{p} (\mathfrak{M}_2, w_2)$. The first important property of bisimilar models is that if $(\mathfrak{M}_1, w_1) \leftrightarroweq_\mathbf{p} (\mathfrak{M}_2, w_2)$ then $(\mathfrak{M}_1, w_1) \models \varphi$ iff $(\mathfrak{M}_2, w_2) \models \varphi$, for all formulas φ with $var(\varphi) \subseteq \mathbf{p}$. To discuss the second important property, we remind the reader that a frame $\mathfrak{F} = (W, R)$ (and a model based on $\mathfrak{F}$) is said to be *m-transitive*, for $m \geq 1$, if whenever $uRx_1R\ldots Rx_mRv$ then there exist $k < m$ and points $y_1, \ldots, y_k \in W$ such that $uRy_1R\ldots Ry_kRv$ (in this sense, standard transitivity is 1-transitivity). A Kripke complete normal modal logic L is called *m-transitive* if the frames validating it are m-transitive.

We will be using the following property of m-transitive models: for every finite set $\mathbf{p}$ of variables and every finite pointed m-transitive model $(\mathfrak{M}, w)$, one can construct a formula $\chi_\mathbf{p}(\mathfrak{M}, w)$ containing only variables from $\mathbf{p}$ such that, for every pointed m-transitive model $(\mathfrak{M}', w')$,

$$(\mathfrak{M}', w') \models \chi_\mathbf{p}(\mathfrak{M}, w) \quad \text{iff} \quad (\mathfrak{M}, w) \leftrightarroweq_\mathbf{p} (\mathfrak{M}', w').$$

$\chi_\mathbf{p}(\mathfrak{M}, w)$ is called the *characteristic formula* for $(\mathfrak{M}, w)$ and $\mathbf{p}$. Notice that $\chi_\mathbf{p}(\mathfrak{M}, w)$ is uniquely determined modulo equivalence in the minimal m-transitive modal logic.

LEMMA 2. *For every m-transitive modal logic L with the finite model property, the following conditions are equivalent:*

- *$\varphi_1 \wedge \varphi_2$ is a conservative extension of φ_1 in L,*

- *for every finite model $(\mathfrak{M}, w)$ based on a frame for L, if $(\mathfrak{M}, w) \models \varphi_1$ then there exists a finite $var(\varphi_1)$-bisimilar model $(\mathfrak{M}', w')$ based on a frame for L and such that $(\mathfrak{M}', w') \models \varphi_2$.*

Moreover, if $\varphi_1 \wedge \varphi_2$ is not a conservative extension of φ_1 in L, then there exists a finite pointed model $(\mathfrak{M}, w)$ based on a frame for L such that $\chi_{var(\varphi_1)}(\mathfrak{M}, w)$ is a witness for (φ_1, φ_2) in L.

As a first application of Lemma 2, one can prove the following result for tabular modal logics. (Recall that a modal logic L is called *tabular* if L is the logic of a finite set of finite frames [2].)

THEOREM 3. *The non-conservativeness problem for each tabular logic L is NP^{NP}-complete.*

Intuitively, given formulas φ_1 and φ_2 the NP^{NP}-algorithm guesses a finite model based on a frame for L and satisfying φ_1, and then calls an NP-oracle to verify that no $var(\varphi_1)$-bisimilar model based on a frame for L satisfies φ_2. A full proof can be found in [7].

3 The NEXPTIME lower bound

Say that a Kripke complete modal logic L is of *unbounded alternatives* if, for every $N < \omega$, there exist a Kripke frame $\mathfrak{F} = (W, R)$ for L and a point $w \in W$ such that the number of R-successors of w is at least N, or $|\{v \in W \mid wRv\}| \geq N$, to be more precise. Many standard modal logics such as **K**, **K4**, **S4**, **GL**, **S4.3**, **S5** are clearly of unbounded alternatives—unlike the logics $\mathbf{Alt}_n$.

Given a model $\mathfrak{M}$ and a set $\mathbf{q}$ of variables, we call a model $\mathfrak{M}'$ a $\mathbf{q}$-*variant* of $\mathfrak{M}$ if $\mathfrak{M}'$ can be obtained from $\mathfrak{M}$ by changing the valuation of some variables in $\mathbf{q}$ (but nothing else).

THEOREM 4. *Let L be a Kripke complete normal modal logic of unbounded alternatives. Then the conservativeness problem for L is coNEXPTIME-hard.*

Proof. The proof is by reduction of the complement of the well-known NEXPTIME-complete $2^n \times 2^n$-*bounded tiling problem* (see, e.g., [16]): given $n < \omega$, a finite set $\mathcal{T}$ of tile types and a $t_0 \in \mathcal{T}$, decide whether $\mathcal{T}$ can tile the $2^n \times 2^n$ grid in such a way that t_0 is placed onto $(0, 0)$. More precisely, let H and V be the binary relations on $\mathcal{T} \times \mathcal{T}$ such that $(t, t') \in H$ ($(t, t') \in V$) iff the colours of the right (upper) edge of t and the left (bottom) edge of t' coincide. Then $\mathcal{T}$ is said to *tile* the $2^n \times 2^n$ grid if there is a function $\tau : 2^n \times 2^n \to \mathcal{T}$ such that

- if $\tau(i, j) = t$ and $\tau(i+1, j) = t'$ then $(t, t') \in H$, for all $i < 2^n - 1$, $j < 2^n$,

- if $\tau(i, j) = t$ and $\tau(i, j+1) = t'$ then $(t, t') \in V$, for all $i < 2^n$, $j < 2^n - 1$,

- $\tau(0, 0) = t_0$.

Given a set $\mathcal{T} = \{t_1, \ldots, t_m\}$ of tile types, we are going to construct two formulas φ_1 and φ_2 such that (i) $|\varphi_1|$ and $|\varphi_2|$ are polynomial in m and n, and (ii) $\varphi_1 \wedge \varphi_2$ is a conservative extension of φ_1 in L iff $\mathcal{T}$ cannot tile the $2^n \times 2^n$ grid in such a way that t_0 is placed onto $(0, 0)$.

To construct φ_1 and φ_2, we will use the following propositional variables:

- $\mathbf{p} = \{p_1, \ldots, p_n\}$ and $\mathbf{q} = \{q_1, \ldots, q_n\}$ to represent the points (i, j) of the $2^n \times 2^n$ grid in models by means of the standard binary encoding; for example, $(1, 2)$ is represented by a point w of some model iff

$$w \models p_1 \wedge \neg p_2 \wedge \cdots \wedge \neg p_n \quad \text{and} \quad w \models \neg q_1 \wedge q_2 \wedge \neg q_3 \wedge \cdots \wedge \neg q_n$$

(we will call a **p**-*literal* any conjunction $\neg_1 p_1 \wedge \cdots \wedge \neg_n p_n$ where each $\neg_i$ is either $\neg$ or blank),

- $\mathbf{t} = \{t_1, \ldots, t_m\}$: $w \models t_i$ will mean that the grid point represented by w is covered by a tile of type t_i,

- the set **A** of auxiliary variables $P_1, \ldots, P_n$, $Q_1, \ldots, Q_n$, $T_1, \ldots, T_m$; these variables will occur in φ_2, but not in φ_1.

The formula

$$\varphi_1 = (\neg p_1 \wedge \cdots \wedge \neg p_n) \wedge (\neg q_1 \wedge \cdots \wedge \neg q_n) \wedge t_0 \wedge \Box \bigvee_{i=1}^{m} t_i.$$

is supposed to say that if $(\mathfrak{M}, w) \models \varphi_1$ then w represents $(0,0)$, which is covered by t_0, and each point of the grid represented by some R-successor of w is covered by at least one tile of a type from $\mathcal{T}$.

We say that a pointed model $(\mathfrak{M}, w)$ *represents a proper tiling of the* $2^n \times 2^n$ *grid by* $\mathcal{T}$ if $(\mathfrak{M}, w) \models \rho_{\mathcal{T},n}$, where $\rho_{\mathcal{T},n}$ is the conjunction of the following formulas:

$$\Diamond^+(l_1 \wedge l_2),$$

$$\Diamond^+(l_1 \wedge l_2 \wedge t) \rightarrow \Box^+\left(l_1 \wedge l_2 \rightarrow (t \wedge \neg \bigvee_{t' \neq t} t')\right),$$

$$\Diamond^+(l_1 \wedge l_2 \wedge t) \rightarrow \Diamond^+ \bigvee_{(t,t') \in H} ((l_1 + 1) \wedge l_2 \wedge t')), \text{ for } l_1 < 2^n - 1,$$

$$\Diamond^+(l_1 \wedge l_2 \wedge t) \rightarrow \Diamond^+ \bigvee_{(t,t') \in V} (l_1 \wedge (l_2 + 1) \wedge t'), \text{ for } l_2 < 2^n - 1,$$

for all possible **p**-literals l_1, **q**-literals l_2, and $t, t' \in \mathbf{t}$. Here we use the abbreviations $\Box^+ \psi = \psi \wedge \Box \psi$, $\Diamond^+ \psi = \psi \vee \Diamond \psi$, and if l_i represents a number $k < 2^n - 1$ then $l_i + 1$ represents $k + 1$.

The formula φ_2 to be constructed below will have the property that, for every model $\mathfrak{M}$ with $(\mathfrak{M}, w) \models \varphi_1$, the following conditions are equivalent:

1. there is an **A**-variant $\mathfrak{M}'$ of $\mathfrak{M}$ such that $(\mathfrak{M}', w) \models \varphi_2$,

2. $(\mathfrak{M}, w)$ *does not* represent a proper tiling of the $2^n \times 2^n$ grid by $\mathcal{T}$.

Suppose for the moment that we have managed to construct such a formula φ_2 of length polynomial in m and n. We claim then that $\varphi_1 \wedge \varphi_2$ is a conservative extension of φ_1 in L iff $\mathcal{T}$ cannot tile the $2^n \times 2^n$ grid in such a way that t_0 is placed onto $(0,0)$. Indeed, assume first that $\mathcal{T}$ cannot tile the grid in this way, and consider any model $\mathfrak{M}$ based on a frame for L and satisfying φ_1 at its root w. Then $(\mathfrak{M}, w)$ cannot represent a proper tiling of the grid by means of $\mathcal{T}$, and so we can find an **A**-variant $\mathfrak{M}'$ of $\mathfrak{M}$ such that $(\mathfrak{M}', w) \models \varphi_2$. Clearly, this means that $\varphi_1 \wedge \varphi_2$ is a conservative extension of φ_1 in L.

Now suppose that $\mathcal{T}$ can tile the grid. Clearly, we can satisfy $\varphi_1 \wedge \rho_{\mathcal{T},n}$ in a model based on a frame for L. But then $\rho_{\mathcal{T},n}$ is a witness for (φ_1, φ_2) in L because a model $(\mathfrak{M}, w)$ with $(\mathfrak{M}, w) \models \varphi_1 \wedge \varphi_2 \wedge \rho_{\mathcal{T},n}$ would trivially satisfy the former condition above but not the latter.

Now we construct the required formula φ_2. How to ensure that a point and its successors do not represent a tiling properly? There can be three different types of *defects*:

1. two different tiles cover the same point or two different tiles cover two points representing the same pair (i,j) on the grid,

2. there is a point representing (i,j) but there is no point representing $(i+1,j)$, for $i < 2^n - 1$, or no point representing $(i, j+1)$, for $j < 2^n - 1$,

3. colour mismatch.

To encode the *existence* of at least one of these defects we require our auxiliary variables P_i, Q_i, and T_i which will be used to carry, everywhere in the relevant part of the model, the information that there exists a point representing some grid point v covered by some tile t. The formula

$$\bigwedge_{i=1}^{n} (\Diamond^+ P_i \leftrightarrow \Box^+ P_i) \wedge \bigwedge_{i=1}^{n} (\Diamond^+ Q_i \leftrightarrow \Box^+ Q_i) \wedge \bigwedge_{i=1}^{m} (\Diamond^+ T_i \leftrightarrow \Box^+ T_i) \quad (1)$$

says that each of the P_i, Q_i, and T_i has the same truth-value everywhere in the relevant part of the model. Suppose that (1) is true at the root w of our hypothetical model. Then, by making the formula

$$\exists p, q, t \;=\; \Diamond^+ \Big(\bigwedge_{i=1}^{n} ((p_i \leftrightarrow P_i) \wedge (q_i \leftrightarrow Q_i)) \wedge \bigwedge_{i=1}^{m} (t_i \leftrightarrow T_i) \Big)$$

true at w, we send—via the P_i, Q_i and T_i—to all worlds accessible from w (and w itself) the information that there is a point v in the model representing some grid point and covered by some tiles. In particular, the formula

$$at P, Q \;=\; \bigwedge_{i=1}^{n} ((p_i \leftrightarrow P_i) \wedge (q_i \leftrightarrow Q_i))$$

is true at a point u accessible from the root or the root itself iff u and v represent the same grid point, and

$$at P, Q, T \;=\; \bigwedge_{i=1}^{n} ((p_i \leftrightarrow P_i) \wedge (q_i \leftrightarrow Q_i)) \wedge \bigwedge_{i=1}^{m} (t_i \leftrightarrow T_i)$$

is true at u iff u and v represent the same grid point and are covered by the same tiles. Using these formulas we can now express that the model under consideration contains a defect of type 1:

$$\exists p, q, t \wedge \bigvee_{i \neq j} (\Diamond^+ (at P, Q \wedge t_i) \wedge \Diamond^+ (at P, Q \wedge t_j)). \quad (2)$$

To describe defects of type 2, we require the formulas

$$\exists p^+, q, t \;=\; \Diamond^+\Big(\neg\bigwedge_{k=1}^n p_k \wedge \bigwedge_{k\le n}\Big((\bigwedge_{i<k} p_i \wedge \neg p_k) \to \bigwedge_{i<k}\neg P_i \wedge P_k \wedge \\ \bigwedge_{j=k+1}^n (p_j \leftrightarrow P_j)\Big) \wedge \bigwedge_{i=1}^n (q_i \leftrightarrow Q_i) \wedge \bigwedge_{i=1}^m (t_i \leftrightarrow T_i)\Big)$$

and

$$\exists p, q^+, t \;=\; \Diamond^+\Big(\neg\bigwedge_{k=1}^n q_k \wedge \bigwedge_{k\le n}\Big((\bigwedge_{i<k} q_i \wedge \neg q_k) \to \bigwedge_{i<k}\neg Q_i \wedge Q_k \wedge \\ \bigwedge_{j=k+1}^n (q_j \leftrightarrow Q_j)\Big) \wedge \bigwedge_{i=1}^n (p_i \leftrightarrow P_i) \wedge \bigwedge_{i=1}^m (t_i \leftrightarrow T_i)\Big).$$

The latter, for instance, says that, for some point in the relevant part of the model representing some grid point (k, l) and covered by some tiles t, the variables P_i represent k, the Q_i represent $l+1$, and the T_i represent the same tiles t. (For describing defects of type 2 we do not need the last conjuncts for t_i in these formulas. They are required for defects of type 3.)

The existence of a defect of type 2 can be guaranteed then by the formula

$$(\exists p^+, q, t \wedge \neg\Diamond^+ atP,Q) \vee (\exists p, q^+, t \wedge \neg\Diamond^+ atP,Q), \tag{3}$$

and the existence of a defect of type 3 can be ensured by the formula

$$\Big(\exists p^+, q, t \wedge \Diamond^+\big(atP,Q \wedge \neg\bigvee_{(t_i,t_j)\in H} (T_i \wedge t_j)\big)\Big) \vee \\ \Big(\exists p, q^+, t \wedge \Diamond^+\big(atP,Q \wedge \neg\bigvee_{(t_i,t_j)\in V} (T_i \wedge t_j)\big)\Big). \tag{4}$$

Finally, we define φ_2 by taking

$$\varphi_2 \;=\; (1) \wedge \big((2) \vee (3) \vee (4)\big).$$

It is easy to see that φ_2 is as required. We leave details to the reader. ∎

4 The upper bound for S5

It is well known that the modal logic **S5** has uniform interpolation. One can easily construct a uniform interpolant $\exists_{\mathbf{S5}}q.\varphi$ for a formula φ of double exponential size using the local tabularity of **S5**, according to which there are only $2^{2^{2^{O(n)}}}$ pairwise non-equivalent formulas in **S5** built from n variables. It follows from Lemma 1 and the decidability of **S5** in coNP that the non-conservativeness problem for **S5** is decidable in non-deterministic 2ExpTime.

In this section, we improve this bound by showing that **S5** has uniform interpolants of exponential size (which can be constructed in exponential time). Thus, we obtain, by Lemma 1 and Theorem 4, that the conservativeness problem for **S5** is coNExpTime-complete.

Suppose $\varphi(\mathbf{p},\mathbf{q})$ with disjoint $\mathbf{p}$ and $\mathbf{q}$ is given. The $\mathbf{p}$-literal given by a model $\mathfrak{M}$ and a point x in it will be denoted by $l_{\mathfrak{M}}(x)$ or simply $l(x)$ if $\mathfrak{M}$ is understood. Thus,

$$l(x) \;=\; \bigwedge \{p_i \mid (\mathfrak{M},x) \models p_i\} \cup \{\neg p_i \mid (\mathfrak{M},x) \not\models p_i\}.$$

Denote by $M(\varphi)$ the number of occurrences of the modal operator $\Box$ in φ plus one.

A uniform interpolant $\exists_{\mathbf{S5}}\mathbf{q}.\varphi$ of size at most exponential in $|\varphi|$ can be constructed in the following way. First we take the set of all pairwise non-isomorphic rooted **S5**-models over $\mathbf{p}$ and $\mathbf{q}^1$ with at most $M(\varphi)$ worlds and such that no two distinct worlds in the model validate precisely the same variables from $\mathbf{p}$ and $\mathbf{q}$. The total number of such models is not exceeding $2^{2n \cdot M(\varphi)}$. Then we partition this set of models into (disjoint) subsets, say $\mathcal{K}_1, \ldots, \mathcal{K}_m$, such that all models from the same $\mathcal{K}_i$ validate the same subformulas of φ starting with $\Box$ (recall that we use $\Diamond$ as an abbreviation).

For each $\mathfrak{M} \in \mathcal{K}_i$ based on a frame (W, R) and each point x in $\mathfrak{M}$ with $x \models \varphi$, let

$$\chi_{\mathfrak{M}}(x) \;=\; l(x) \wedge \bigwedge_{w \in W \setminus \{x\}} \Diamond l(w),$$

and let ψ_i be the disjunction of all such $\chi_{\mathfrak{M}}(x)$, for all $\mathfrak{M} \in \mathcal{K}_i$ and x in $\mathfrak{M}$ with $x \models \varphi$. Denote by T_i the set of all $\mathbf{p}$-literals that are not satisfied in any model from $\mathcal{K}_i$, and set

$$\chi_i \;=\; \psi_i \wedge \bigwedge_{\text{line} \in T_i} \neg \Diamond l, \qquad \exists_{\mathbf{S5}}\mathbf{q}.\varphi \;=\; \bigvee_{i=1}^{m} \chi_i.$$

Clearly, the size of $\exists_{\mathbf{S5}}\mathbf{q}.\varphi$ is at most exponential in the size of φ, and it can be constructed in exponential time. So it remains to prove the following:

THEOREM 5. $\exists_{\mathbf{S5}}\mathbf{q}.\varphi$ *is a uniform interpolant for* φ *in* **S5**.

Proof. To show that $\varphi \models_{\mathbf{S5}} \exists_{\mathbf{S5}}\mathbf{q}.\varphi$, consider an arbitrary rooted **S5**-model $\mathfrak{M}$ based on a frame $(W, W \times W)$ and such that $x \models \varphi$ for some $x \in W$. Let $i \in \{1, \ldots, m\}$ be such that $\mathfrak{M}$ validates precisely the same subformulas of φ starting with $\Box$ or $\Diamond$ as the models from $\mathcal{K}_i$. Now observe that, for every point $y \in W$, we can always find a subset $Y \subseteq W$ containing y such that the restriction of $\mathfrak{M}$ to Y is (isomorphic to) some model from $\mathcal{K}_i$ (just pick up one 'witness' satisfying $\neg \psi$ for every subformula $\Box \psi$ of φ such that $y \models \neg \Box \psi$). It follows that $(\mathfrak{M}, x) \models \chi_i$, and so $(\mathfrak{M}, x) \models \exists_{\mathbf{S5}}\mathbf{q}.\varphi$.

---
[1]This means that we restrict valuations in models to the variables in $\mathbf{p}$ and $\mathbf{q}$.

Suppose now that we have a formula ψ with $var(\psi) \cap \mathbf{q} = \emptyset$. We need to prove that if $\exists_{\mathbf{S5}}\mathbf{q}.\varphi \not\models_{\mathbf{S5}} \psi$ then $\varphi \not\models_{\mathbf{S5}} \psi$. Let $\exists_{\mathbf{S5}}\mathbf{q}.\varphi \not\models_{\mathbf{S5}} \psi$. Take a model $\mathfrak{M} = (\mathfrak{F}, \mathfrak{V})$ with $\mathfrak{F} = (W, W \times W)$ such that $(\mathfrak{M}, w) \models \exists_{\mathbf{S5}}\mathbf{q}.\varphi$ and $(\mathfrak{M}, w) \not\models \psi$ for some $w \in W$. By the definition of $\exists_{\mathbf{S5}}\mathbf{q}.\varphi$, we can find $i \in \{1, \ldots, m\}$, $\mathfrak{N} \in \mathcal{K}_i$, and x in $\mathfrak{N}$ such that $(\mathfrak{M}, w) \models \chi_i$ and $(\mathfrak{M}, w) \models \chi_{\mathfrak{N}}(x)$. We know that $(\mathfrak{N}, x) \models \varphi$. Now, with the help of $\mathfrak{M}$, we extend $\mathfrak{N}$ to a model $\mathfrak{K}$ such that $(\mathfrak{K}, x) \models \varphi$ and $(\mathfrak{K}, x) \not\models \psi$.

Let $\mathfrak{N} = (\mathfrak{G}, \mathfrak{U})$ and $\mathfrak{G} = (U, U \times U)$. By the definition of $\chi_{\mathfrak{N}}(x)$, for every $u \in U$ there is $w_u \in W$ such that $l_{\mathfrak{N}}(u) = l_{\mathfrak{M}}(w_u)$. Clearly, we can take $w_x = w$. We can also assume that W is disjoint from U. Now define a new model $\mathfrak{K} = (\mathfrak{H}, \mathfrak{W})$ based on the frame $\mathfrak{H} = (V, V \times V)$, where

$$V = U \cup (W \setminus \{w_u \mid u \in U\})$$

and, for each $u \in U$,

$$\mathfrak{W}(p, u) = \begin{cases} \mathfrak{U}(p, u), & \text{for } p \in \mathbf{p} \cup \mathbf{q} \\ \mathfrak{V}(p, w_u), & \text{otherwise.} \end{cases}$$

For the remaining points of V the valuation $\mathfrak{W}$ is defined as follows. For each $v \in V \setminus U$ we can find, by the second conjunct of χ_i, a model $\mathfrak{N}_v \in \mathcal{K}_i$ (with a valuation $\mathfrak{U}_v$) and a point z_v in it such that $\text{line}_{\mathfrak{N}_v}(z_v) = \text{line}_{\mathfrak{M}}(v)$. Then we set, for all such v,

$$\mathfrak{W}(p, v) = \begin{cases} \mathfrak{U}_v(p, z_v), & \text{for } p \in \mathbf{p} \cup \mathbf{q} \\ \mathfrak{V}(p, v), & \text{otherwise.} \end{cases}$$

We have $(\mathfrak{K}, x) \models \varphi$ because, restricted to the variables from $\mathbf{p} \cup \mathbf{q}$, the model $\mathfrak{K}$ consists of $\mathfrak{N} \in \mathcal{K}_i$ together with some points from other models in $\mathcal{K}_i$ validating, by definition, precisely the same subformulas of φ starting with $\square$. Finally, we have $(\mathfrak{K}, x) \not\models \psi$ because $var(\psi) \cap \mathbf{q} = \emptyset$, and, restricted to the variables in $var(\psi)$, the model $\mathfrak{K}$ is isomorphic to $\mathfrak{M}$ with x corresponding to w. ∎

5 The upper bound for K

As in the case of **S5**, it is known [17, 5] that the modal logic **K** has uniform interpolation, with a uniform interpolant for a given formula and set of variables being constructed in an effective way. Moreover, it has been recently shown in [14] that one can construct a uniform interpolant $\exists_{\mathbf{K}}\mathbf{q}.\varphi$ for φ in exponential time in such a way that the size of $\exists_{\mathbf{K}}\mathbf{q}.\varphi$ is at most exponential in the size $|\varphi|$ of φ — $2^{p(|\varphi|)}$ for a certain polynomial p which does not depend on φ, to be more exact, — and its modal depth does not exceed the modal depth of φ.

Using this result, the fact that the decision problem for **K** is PSPACE-complete, and the algorithm from Section 2, one can obtain an algorithm deciding the conservativeness problem for **K** using exponential *space* in the size of the input formulas. In this section we improve this bound by providing a coNExpTime algorithm. Thus, we obtain

THEOREM 6. *The conservativeness problem for* **K** *is co*NExpTime*-complete.*

The coNExpTime lower bound follows from Theorem 4. Here we present a nondeterministic exponential time algorithm for deciding the complement of the conservativeness problem for **K**. Suppose that we are given formulas $\varphi_1(\mathbf{p})$ and $\varphi_2(\mathbf{p}, \mathbf{q})$ with disjoint $\mathbf{p} = \{p_1, \ldots, p_n\}$ and $\mathbf{q} = \{q_1, \ldots, q_n\}$. Denote by $\mathsf{sub}(\varphi_i)$, $i = 1, 2$, the closure under single negation of the set of all subformulas of φ_i. As usual, by a φ_i-*type* t we mean a Boolean-closed subset of $\mathsf{sub}(\varphi_i)$, i.e.,

- $\psi \in t$ iff $\neg \psi \notin t$, for every $\neg \psi \in \mathsf{sub}(\varphi_i)$,
- $\psi \wedge \chi \in t$ iff $\psi \in t$ and $\chi \in t$, for every $\psi \wedge \chi \in \mathsf{sub}(\varphi_i)$.

Denote by tp_i the set of all types for φ_i; clearly, $|\mathsf{tp}_i| \leq 2^{|\varphi_i|}$. Let d be the maximum of the modal depths of φ_1 and φ_2. The algorithm is presented in Fig. 1.

LEMMA 7. *The algorithm in Fig. 1 accepts input* φ_1, φ_2 *iff* $\varphi_1 \wedge \varphi_2$ *is not a conservative extension of* φ_1.

Proof. ($\Leftarrow$) Suppose that $\varphi_1 \wedge \varphi_2$ is not a conservative extension of φ_1. By Lemma 1, we have $\varphi_1 \not\models_\mathbf{K} \exists_\mathbf{K} \mathbf{q}.(\varphi_1 \wedge \varphi_2)$. According to [14], there exists a uniform interpolant $\exists_\mathbf{K} \mathbf{q}.(\varphi_1 \wedge \varphi_2)$ whose size does not exceed $N' = 2^{p(|\varphi_1 \wedge \varphi_2|)}$ and whose modal depth is $\leq d$. So there is a model $\mathfrak{M}$ based on an irreflexive and intransitive tree $\mathfrak{F} = (W, R)$ of depth $\leq d$ and branching factor $\leq N'$ such that both φ_1 and $\neg \exists_\mathbf{K} \mathbf{q}.(\varphi_1 \wedge \varphi_2)$ are true at the root r of $\mathfrak{F}$. Without loss of generality we may assume that, for every $x \in W$, if $x \models \neg \Box \psi_1 \wedge \cdots \wedge \neg \Box \psi_k$ for pairwise distinct $\neg \Box \psi_i \in \mathsf{sub}(\varphi_2)$ then there are k distinct R-successors $y_1, \ldots, y_k$ of x such that $y_i \models \neg \psi_i$; if this is not the case, we can duplicate some relevant subtrees, thereby increasing the branching factor to $\leq N$. Thus, we may assume that, restricted to the variables in $\mathbf{p}$, the algorithm above guesses the model $\mathfrak{M}$.

Clearly, in Step 2 of the algorithm, we are in the 'else-case.' Suppose now that there is a $t_0 \in \ell(r)$ with $\varphi_2 \in t_0$, that is, the algorithm rejects the input φ_1, φ_2. Define a function f that maps each $x \in W$ to an element of $\ell(x)$ by induction starting from the root:

- Set $f(r) = t_0$.

- If $f(x)$ is already defined, but $f(\cdot)$ is not defined for the R-successors of x, then do the following. Let the negated box formulas in $f(x)$ be $\neg \Box \psi_1, \ldots, \neg \Box \psi_k$ and the box formulas $\Box \vartheta_1, \ldots, \Box \vartheta_h$. As $f(x) \in \ell(x)$, by the definition of ℓ there are distinct R-successors $y_1, \ldots, y_k$ of x and types $t_1 \in \ell(y_1), \ldots, t_k \in \ell(y_k)$ such that $\{\neg \psi_i, \vartheta_1, \ldots, \vartheta_h\} \subseteq t_i$ for $1 \leq i \leq k$. Set $f(y_i) = t_i$ for $1 \leq i \leq k$. For each R-successor y of

---

Input: formulas $\varphi_1(\mathbf{p})$ and $\varphi_2(\mathbf{p}, \mathbf{q})$.

1. Guess a Kripke model $\mathfrak{M}$ over the variables in $\mathbf{p}$ such that
   - $\mathfrak{M}$ is based on an irreflexive intransitive tree $\mathfrak{F} = (W, R)$,
   - the depth of $\mathfrak{F}$ is $\leq d$,
   - the branching factor of $\mathfrak{F}$ is bounded by $N = |\varphi_2| \cdot 2^{p(|\varphi_1 \wedge \varphi_2|)}$.

2. Check whether $(\mathfrak{M}, r) \models \varphi_1$, where r is the root of $\mathfrak{M}$. If this is not the case, reject the input. Else

3. label the points x of $\mathfrak{F}$ with subsets $\ell(x)$ of tp_2 by induction starting from the leafs as follows:
   - ff $x \in W$ is a leaf of $\mathfrak{F}$, then $\ell(x)$ consists of all types $t \in \mathsf{tp}_2$ such that $p_i \in t$ iff $x \models p_i$, for all $p_i \in \mathbf{p}$, and t contains all formulas of the form $\Box \psi \in \mathsf{sub}(\varphi_2)$.
   - Suppose now that $x \in W$ is not a leaf and all R-successors of x have already been labelled. Consider a type $t \in \mathsf{tp}_2$ and let $\Box \vartheta_1, \ldots, \Box \vartheta_m$ be all the box formulas in t and $\neg \Box \psi_1, \ldots, \neg \Box \psi_k$ be all the negated box formulas in t. Then $t \in \ell(x)$ iff the following conditions hold:
     (a) $p_i \in t$ iff $x \models p_i$, for all $p_i \in \mathbf{p}$,
     (b) for each R-successor y of x, there exists a $t' \in \ell(y)$ with $\{\vartheta_1, \ldots, \vartheta_m\} \subseteq t'$,
     (c) there exist pairwise distinct R-successors $y_1, \ldots, y_k$ of x and types $t_1 \in \ell(y_1), \ldots, t_k \in \ell(y_k)$ such that $\{\neg \psi_i, \vartheta_1, \ldots, \vartheta_m\} \subseteq t_i$, $1 \leq i \leq k$.

4. Check whether there is a $t \in \ell(r)$ such that $\varphi_2 \in t$. If this is the case, then reject the input; otherwise accept it.

Figure 1. Deciding non-conservativeness for **K**.

x with $y \notin \{y_1, \ldots, y_k\}$, there is a $t \in \ell(y)$ such that $\{\vartheta_1, \ldots, \vartheta_h\} \subseteq t$. Set $f(y) = t$.

Now define a $\mathbf{q}$-variant $\mathfrak{M}'$ of $\mathfrak{M}$ by taking, for all $x \in W$ and all $p \in \mathbf{q}$,

$$x \models p \quad \text{iff} \quad p \in f(x).$$

We still have $(\mathfrak{M}', r) \models \varphi_1$. Moreover, it can be easily shown by induction that $(\mathfrak{M}', x) \models \bigwedge_{\psi \in f(x)} \psi$ for all $x \in W$. Since $\varphi_2 \in f(r)$, we must have $(\mathfrak{M}', r) \models \varphi_2$. But then $(\mathfrak{M}, r) \models \exists_{\mathbf{K}} \mathbf{q}.(\varphi_1 \wedge \varphi_2)$, which is a contradiction.

($\Rightarrow$) Suppose now that our algorithm accepts φ_1, φ_2. Let $\mathfrak{M}$ be the model guessed by the algorithm and based on a tree $\mathfrak{F} = (W, R)$ with root r. Without loss of generality we may assume that, whenever xRy in $\mathfrak{F}$ then there are $|\varphi_2|$-many distinct R-successors z of x with $t_{\mathfrak{M}}^1(z) = t_{\mathfrak{M}}^1(y)$, where $t_{\mathfrak{M}}^1(x)$ the φ_1-type of x in $\mathfrak{M}$, that is,

$$t_{\mathfrak{M}}^1(x) = \{\psi \in \mathsf{sub}(\varphi_1) \mid (\mathfrak{M}, x) \models \psi\}.$$

With each $x \in W$ we associate inductively a formula $\psi(x)$ over **p** starting from the leaves of $\mathfrak{F}$:

- if $x \in W$ is a leaf, then

$$\psi(x) = \Box\bot \wedge \bigwedge_{x\models p_i} p_i \wedge \bigwedge_{x\not\models p_i} \neg p_i,$$

- if $x \in W$ is a non-leaf and $y_1, \ldots, y_k$ are its R-successors, then

$$\psi(x) = \bigwedge_{x\models p_i} p_i \wedge \bigwedge_{x\not\models p_i} \neg p_i \wedge \bigwedge_{1 \le i \le k} \Diamond \psi(y_i) \wedge \Box \bigvee_{1 \le i \le k} \psi(y_i).$$

Now one can show that $\varphi_1 \wedge \varphi_2 \models_\mathbf{K} \neg \psi$, but $\varphi_1 \not\models_\mathbf{K} \neg \psi$, which means that $\varphi_1 \wedge \varphi_2$ is not a conservative extension of φ_1. ∎

To complete the proof of Theorem 6, we note that the algorithm above runs in exponential time: the guessed model $\mathfrak{M}$ is at most exponentially large, and checking whether φ_1 is true in r can be done in exponential time. It remains to consider the $\ell(\cdot)$ labelling procedure. We have to label $|W|$ points—i.e., at most exponentially many. For each point x, we have to check for $|\mathsf{tp}_2|$ (exponentially) many types whether or not they should be included in $\ell(x)$. Condition (a) can be checked in polynomial time. Condition (b) can be checked in exponential time, since there are at most exponentially many successors in $\mathfrak{M}$. For condition (c) we have to consider all k-tuples of pairs (y, t) with y a successor of x and $t \in \ell(y)$. It is clear that there are at most exponentially many such tuples.

6 The upper bound for S4

In this section we present an algorithm deciding the conservativeness problem for **S4** in EXPSPACE in the size of the input formulas.

Before proceeding to the technical details, we remind the reader that Kripke models for **S4** are based on quasi-orders $\mathfrak{F} = (W, R)$, that is, R is transitive and reflexive. A set $C \subseteq W$ is called a *cluster* in $\mathfrak{F}$ if $C = \{y \in W \mid xRy \ \& \ yRx\}$ for some $x \in W$; in this case we also say that C is the cluster *generated by* x and denote it by $C(x)$. Recall that every rooted model for **S4** is a p-morphic image of a model based on a *tree of clusters*, that is, a rooted quasi-order (W, R) such that, for all $x, y, z \in W$, if xRz and yRz, then either xRy or yRx or $C(x) = C(y)$. Without loss of generality we assume in this section that all our models are based on *finite trees of clusters*. Given a quasi-order $\mathfrak{F} = (W, R)$, we say that a cluster $C(x)$, for $x \in W$, is an *immediate strict predecessor* of a cluster $C(y)$ if xRy, $C(x) \ne C(y)$ and whenever $xRzRy$ then either $C(z) = C(x)$ or $C(z) = C(y)$. By the *depth* of $\mathfrak{F}$ we understand the length n of the longest sequence $C(x_1), \ldots, C(x_n)$ of clusters in $\mathfrak{F}$ such that $C(x_i)$ is an immediate strict predecessor of $C(x_{i+1})$. A point y is a *strict successor* of a point x iff xRy and $C(x) \ne C(y)$. The *branching factor* of $\mathfrak{F}$ is the maximal number of immediate strict successor clusters of a cluster in $\mathfrak{F}$.

Suppose that we are given two formulas φ_1 and φ_2. The central role in our algorithm will be played by the following notion of a realisable triple for φ_1, φ_2. Consider a triple $\mathsf{t} = (t, \Gamma, \Delta)$ where t is a φ_1-type and Γ, Δ are sets of $\varphi_1 \wedge \varphi_2$-types. We call t *realisable* if there exists a pointed model $(\mathfrak{M}, x)$ based on a tree of clusters with root x such that

- $t = t^1_\mathfrak{M}(x)$ (where, as before, $t^1_\mathfrak{M}(x) = \{\psi \in \mathsf{sub}(\varphi_1) \mid (\mathfrak{M}, x) \models \psi\}$),

- Γ is the set of all $\varphi_1 \wedge \varphi_2$-types s such that $\bigwedge_{\sigma \in s} \sigma \wedge \chi_{var(\varphi_1)}(\mathfrak{M}, x)$ is satisfiable;

- Δ is the set of all $\varphi_1 \wedge \varphi_2$-types s for which there exists a point y in $\mathfrak{M}$ such that $\bigwedge_{\sigma \in s} \sigma \wedge \chi_{var(\varphi_1)}(\mathfrak{M}, y)$ is satisfiable.

In this case we say that $\mathsf{t} = (t, \Gamma, \Delta)$ is *realised* by $(\mathfrak{M}, x)$. Observe that if (t, Γ, Δ) is realisable, then $\Gamma \subseteq \Delta$ and, as follows from the main property of characteristic formulas and bisimulations, $t \subseteq s$ for every $s \in \Gamma$. The meaning of realisable triples will become clear from the following easily proved lemma.

LEMMA 8. *The following two conditions are equivalent for any formulas φ_1 and φ_2:*

(1) *$\varphi_1 \wedge \varphi_2$ is not a conservative extension of φ_1,*

(2) *there exists a realisable triple $\mathsf{t} = (t, \Gamma, \Delta)$ (for φ_1, φ_1) such that $\varphi_1 \in t$ but $\varphi_1 \wedge \varphi_2 \notin s$ for any $s \in \Gamma$.*

Moreover, for every finite pointed model $(\mathfrak{M}, x)$ based on a tree of clusters, $\chi_{var(\varphi_1)}(\mathfrak{M}, x)$ is a witness for (φ_1, φ_2) iff the triple t realised by $(\mathfrak{M}, x)$ satisfies condition (2).

Observe that Δ does not play any role in the lemma above. Its role in realisable triples is indicated in the proof of Lemma 9 below.

We use the notion of a realisable triple to show that whenever $\varphi_1 \wedge \varphi_2$ is not a conservative extension of φ_1, then there exists a witness $\chi_{var(\varphi_1)}(\mathfrak{M}, x)$ of a certain bounded size. First, associate with every realisable triple $\mathsf{t} = (t, \Gamma, \Delta)$ the set

$$\Phi_\mathsf{t} = \{\{\Box\psi_1, \ldots, \Box\psi_k\} \subseteq \mathsf{sub}(\varphi_1 \wedge \varphi_2) \mid \forall s \in \Gamma \ \{\Box\psi_1, \ldots, \Box\psi_k\} \not\subseteq s\}.$$

LEMMA 9. *For every realisable triple $\mathsf{t} = (t, \Gamma, \Delta)$ for φ_1, φ_2, there is a realisable triple $\mathsf{t}' = (t, \Gamma', \Delta')$ such that $\Gamma \supseteq \Gamma'$, $\Delta \supseteq \Delta'$, and t' can be realised in a model $\mathfrak{M}'$ based on a tree of clusters $\mathfrak{F}'$ such that*

- *each cluster in $\mathfrak{F}'$ contains at most $2^{|\varphi_1|}$ points,*

- *the branching factor of $\mathfrak{F}'$ is bounded by $2^{|\varphi_1 \wedge \varphi_2|}$,*

- the depth of $\mathfrak{F}'$ is bounded by $1 + 2^{|\varphi_1 \wedge \varphi_2|}$.

Moreover, for any two points x, y such that y is a strict successor of x in $\mathfrak{F}'$ and x is not in the root cluster of $\mathfrak{F}'$, $\Phi_{\mathbf{t}_y} \subsetneq \Phi_{\mathbf{t}_x}$, for the triples $\mathbf{t}_x$ and $\mathbf{t}_y$ realised by $(\mathfrak{M}', x)$ and $(\mathfrak{M}', y)$, respectively.

Proof. (This is a sketch, a full proof can be found in [7].) Suppose that a triple $\mathbf{t} = (t, \Gamma, \Delta)$ for φ_1, φ_2 is realised in a model $(\mathfrak{M}, x)$ based on a finite tree of clusters $\mathfrak{F} = (W, R)$ with root x. The upper bound on the cardinality of clusters follows from the simple fact that if two points y and y' validate the same variables from φ_1 then we can omit one of these points and the resulting models will be $var(\varphi_1)$-bisimilar to the original one.

Consider now some $y \in W$ and denote by $\mathbf{t}_y = (t_y, \Gamma_y, \Delta_y)$ the triple for φ_1, φ_2 realised by $(\mathfrak{M}, y)$. For every $\varphi_1 \wedge \varphi_2$-type $s \notin \Gamma_y$ we take the set $\{\Box \psi_1, \ldots, \Box \psi_k\}$ of all box formulas in s and choose a maximal (with respect to R) strict successor z of y such that there does not exist a type $s' \in \Gamma_z$ with $\{\Box \psi_1, \ldots, \Box \psi_k\} \subseteq s'$, if such a strict R-successor exists. Observe that (since $t_y \subseteq s$ for every $s \in \Gamma_y$) for each $\neg \Box \psi \in t_y$ we have also chosen a maximal strict R-successor z of y with $(\mathfrak{M}, z) \models \neg \Box \psi$, if such a successor exists.

Now remove from the submodel $\mathfrak{M}_y$ of $\mathfrak{M}$ generated by y all clusters $C(u)$ such that $C(u) \neq C(y)$, yRu and for no chosen point z do we have zRu. Denote the resulting model by $\mathfrak{N}$. Denote by $\mathbf{t}' = (t', \Gamma', \Delta')$ the triple realised by $(\mathfrak{N}, y)$. Our aim is to show

- $t_y = t'$;
- $\Gamma' \subseteq \Gamma_y$;
- $\Delta' \subseteq \Delta_y$.

Having shown this, we can replace $\mathfrak{M}_y$ in $\mathfrak{M}$ with $\mathfrak{N}$ where the number of immediate strict successors of $C(y)$ does not exceed $2^{|\varphi_1 \wedge \varphi_2|}$. It can be proved that $t = t''$, $\Gamma'' \subseteq \Gamma$, and $\Delta'' \subseteq \Delta$ for the triple $(t'', \Gamma'', \Delta'')$ realised in the resulting model $(\mathfrak{M}', x)$. For a proof we refer the reader to [7]. Note that the proof requires the third component of realizable triples because the second component alone only encodes the behaviour of the root of a model but not of non-root points.

Now, since every $\neg \Box \psi \in t_y$ has a witness in $\mathfrak{N}$, we clearly have $t_y = t'$. To prove that $\Gamma' \subseteq \Gamma_y$, suppose otherwise. Then we have some $s \in \Gamma'$ such that $s \notin \Gamma_y$. Two cases are possible now. *Case 1*: there is a strict successor z of y in $\mathfrak{M}_y$ such that there does not exist a type $s' \in \Gamma_z$ with $\{\Box \psi_1, \ldots, \Box \psi_k\} \subseteq s'$ (where $\{\Box \psi_1, \ldots, \Box \psi_k\}$ are all box formulas in s). Then a (possibly different) point with this property was chosen for $\mathfrak{N}$. From this one can easily derive a contradiction with s being in Γ'. *Case 2*: there is no such z in $\mathfrak{M}_y$. Let $C(y_1), \ldots, C(y_m)$ be all immediate strict successors of $C(y)$ which do not belong to $\mathfrak{N}$. Denote by $\mathfrak{M}_i$ the submodel of $\mathfrak{M}_y$ generated by y_i. For each y_i there exists a pointed model $(\mathfrak{M}'_i, y'_i)$ which is $var(\varphi_1)$-bisimilar to $(\mathfrak{M}_i, y_i)$ and $(\mathfrak{M}'_i, y'_i) \models \Box \psi_i$ for every $\Box \psi \in s$. Let $\mathfrak{K}$

> Input: formulas φ_1, φ_2.
>
> 1. Choose, non-deterministically, a triple (t, Γ, Δ), where t is a φ_1-type containing φ_1 and Γ, Δ are sets of $\varphi_1 \wedge \varphi_2$-types such that $\varphi_1 \wedge \varphi_2 \notin s$ for any $s \in \Gamma$.
> 2. Accept input φ_1, φ_2 iff $\mathsf{realise}_0(t, \Gamma, \Delta)$ returns 'true.'

Figure 2. Deciding non-conservativeness for **S4**.

be a model based on a tree of clusters with root v which is $var(\varphi_1)$-bisimilar to $(\mathfrak{N}, y)$ and satisfies s at v. Add the models $\mathfrak{M}'_i$ to $\mathfrak{K}$ as immediate strict successors of the points which are $var(\varphi_1)$-bisimilar to y and denote the resulting model by $\mathfrak{K}'$. Clearly, $(\mathfrak{K}', v)$ is $var(\varphi_1)$-bisimilar to $(\mathfrak{M}_y, y)$. On the other hand, s is still satisfied by $(\mathfrak{K}', v)$, contrary to $s \notin \Gamma_y$.

Finally, the inclusion $\Delta' \subseteq \Delta_y$ follows from $\Gamma' \subseteq \Gamma_y$ and the fact that if a point from a non-root cluster is chosen for $\mathfrak{N}$ then all of its successors belong to $\mathfrak{N}$ as well.

By recursively performing this operation whenever possible for some y, we obtain a model with the properties required. ∎

We are now in a position to present a NExpSpace algorithm deciding the non-conservativeness problem for **S4**. The ExpSpace upper bound is then obtained from the fact that NExpSpace = ExpSpace. To formulate the algorithm, we require the following definition. Let $\mathcal{R}$ be some set of realisable triples. We say that a triple t is *obtained in one step from* $\mathcal{R}$ if there exists a pointed model $(\mathfrak{M}, w)$ based on a tree of clusters with root w such that

(a) $(\mathfrak{M}, w)$ realises t,

(b) for every strict immediate successor $C(y)$ of $C(w)$ in $\mathfrak{M}$, there is some $x \in C(y)$ such that the triple realised by $(\mathfrak{M}, x)$ belongs to $\mathcal{R}$,

(c) every triple from $\mathcal{R}$ is realised by $(\mathfrak{M}, y)$, for some strict immediate successor $C(y)$ of $C(w)$.

The algorithm is shown in Fig. 2; it uses the procedure $\mathsf{realise}$ described in Fig. 3.

The following lemma shows that this 'algorithm' can be refined in such a way that it indeed runs in ExpSpace (a proof can be found in [7]).

LEMMA 10. (i) *It can be decided in* ExpSpace *(in the size of φ_1, φ_2) whether a triple (t, Γ, Δ) is realisable in a finite pointed model $(\mathfrak{M}, w)$ based on a cluster.* (ii) *It is decidable in* ExpSpace *(in the size of φ_1, φ_2) whether a triple is obtained in one step from a set $\mathcal{R}$ of triples of cardinality not exceeding $2^{|\varphi_1 \wedge \varphi_2|}$.*

We can now analyse the algorithm in Fig. 2. By Lemma 10 and condition (1) of the procedure $\mathsf{realise}$, the procedures $\mathsf{realise}$ and $\mathsf{realise}_0$ always terminate and require exponential space only.

> Input: a triple $\mathsf{t} = (t, \Gamma, \Delta)$, where t is a φ_1-type and Γ, Δ are sets of $\varphi_1 \wedge \varphi_2$-types.
>
> realise(t) returns 'true' iff t is realisable in a pointed model based on a cluster or there exists a set $\mathcal{R}$ of triples with $|\mathcal{R}| \leq 2^{|\varphi_1 \wedge \varphi_2|}$ such that
>
> (1) for all $\mathsf{t}' \in \mathcal{R}$, $\Phi_{\mathsf{t}'} \subsetneq \Phi_{\mathsf{t}}$,
>
> (2) for all $\mathsf{t}' \in \mathcal{R}$, realise($\mathsf{t}'$) returns 'true,'
>
> (3) t is obtained in one step from $\mathcal{R}$.
>
> The procedure realise$_0$ is defined in the same way as realise with the exception that condition (1) is omitted. (In particular, it still calls realise in (2)).

Figure 3. The procedures realise(t, Γ, Δ) and realise$_0(t, \Gamma, \Delta)$.

Now suppose that $\varphi_1 \wedge \varphi_2$ is not a conservative extension of φ_1. Then there is a realisable triple (t, Γ, Δ) such that $\varphi_1 \in t$ but $\varphi_1 \wedge \varphi_2 \notin s$, for any $s \in \Gamma$. Take a pointed model $(\mathfrak{M}, w)$ with the properties of Lemma 9 which realises a $\mathsf{t}' = (t, \Gamma', \Delta')$ with $\Gamma' \subseteq \Gamma$ and $\Delta' \subseteq \Delta$. Now, let the algorithm in Fig. 2 guess the triple t'. Then it obviously accepts input φ_1, φ_2. Observe that we start with the procedure realise$_0$ instead of realise because we have not proved that $\Phi_{\mathsf{t}'} \supsetneq \Phi_{\mathsf{t}''}$ for every triple t'' realised in a strict successor of w.

In conclusion, we obtain:

THEOREM 11. *The conservativeness problem for* **S4** *is decidable in* EXPSPACE.

7 Conservative extensions in GL.3

Recall that the set of finite rooted frames for **GL.3** coincides with the set of finite strict linear orders (W, R). We are going to show the following

THEOREM 12. *The conservativeness problem for* **GL.3** *is* EXPSPACE-*complete*.

In what follows we will use the observation that for models based on strict linear orders, every **p**-bisimulation between $(\mathfrak{M}, w)$ and $(\mathfrak{M}', w')$ is an isomorphism between the submodels of these models generated by w and w', respectively, and restricted to the variables from **p**. To formulate the decision procedure we require the notion of a realisable pair: a pair (t, Γ), where t is a φ_1-type t and Γ is a set of $\varphi_1 \wedge \varphi_2$-types, is said to be *realised in a pointed model* $(\mathfrak{M}, w)$ if

- $(\mathfrak{M}, w) \models t$ and

- Γ is the set of $\varphi_1 \wedge \varphi_2$-types s such that $s \wedge \chi_{var(\varphi_1)}(\mathfrak{M}, w)$ is satisfiable (in a finite model for **GL.3**).

LEMMA 13. $\varphi_1 \wedge \varphi_2$ *is not a conservative extension of* φ_1 *in* **GL.3** *iff there exists a pointed model* $(\mathfrak{M}, w)$ *based on a finite strict linear order such that*

Input: formulas φ_1, φ_2.

1. Choose, non-deterministically, a pair (t, Γ), where t is a φ_1-type containing φ_1 and Γ is a set of $\varphi_1 \wedge \varphi_2$-types such that $t \subseteq s$ for all $s \in \Gamma$ and $\varphi_1 \wedge \varphi_2 \notin s$ for any $s \in \Gamma$.
2. Call $\mathsf{real}(t, \Gamma, 2^{2^{|\varphi_1 \wedge \varphi_2|}})$.

Definition of procedure $\mathsf{real}(t, \Gamma, n)$ (where n is a natural number coded in binary):

1. If no s in Γ contains a formula of the form $\neg \Box \psi$, then accept input φ_1, φ_2 and stop. Else,
2. if $n = 0$, then reject input φ_1, φ_2 and stop. Else,
3. choose, non-deterministically, a pair (t', Γ'), where t' is a φ_1-type and Γ' is a set of $\varphi_1 \wedge \varphi_2$-types. Check whether (t', Γ') has the following properties:
   - $t' \subseteq s'$, for all $s' \in \Gamma$,
   - (t, t') is a suitable pair,
   - for every $s \in \Gamma$, there exists $s' \in \Gamma'$ such that (s, s') is a suitable pair,
   - for each $\varphi_1 \wedge \varphi_2$-type $s \notin \Gamma$, $t \not\subseteq s$ or there is no $s' \in \Gamma'$ such that (s, s') is a suitable pair.
4. If (t', Γ') does not have these properties, reject the input φ_1, φ_2 and stop. Else,
5. set $n := n - 1$.
6. Call $\mathsf{real}(t', \Gamma', n)$.

Figure 4. Deciding non-conservativeness for **GL.3**.

- *for the pair (t, Γ) realised by $(\mathfrak{M}, w)$, $\varphi_1 \in t$ and $\varphi_1 \wedge \varphi_2 \notin s$, for any $s \in \Gamma$,*
- *for any two points $x \neq y$ in $\mathfrak{M}$, the pair realised by $(\mathfrak{M}, x)$ is different from the pair realised by $(\mathfrak{M}, y)$.*

In particular, the length of the finite strict linear order underlying $\mathfrak{M}$ does not exceed $2^{2^{|\varphi_1 \wedge \varphi_2|}}$.

We say that a pair of types (t, t') is *suitable* if

- $\Box \psi \in t$ implies $\psi, \Box \psi \in t'$, and
- $\neg \Box \psi \in t$ implies $\neg \psi \in t'$ or $\neg \Box \psi \in t'$.

The non-deterministic algorithm deciding non-conservativeness in **GL.3** is shown in Fig. 4. Clearly, this algorithm requires exponential space only. Using the fact that $\mathrm{NEXPSPACE} = \mathrm{EXPSPACE}$, we obtain an EXPSPACE algorithm. The correctness of this algorithm follows from Lemma 13.

EXPSPACE-hardness of the conservativeness problem for **GL.3** is proved by simulating computations of Turing machines $\mathcal{M} = (Q, \Sigma, \Gamma, q_0, \Delta)$ that solve an EXPSPACE-hard problem and consume at most 2^n tape cells if started on an input of length n.

Let $\mathsf{w} = a_0 \cdots a_{n-1} \in \Sigma^*$ be an input to $\mathcal{M}$. In [7], we construct formulas φ_1 and φ_2 (depending on $\mathcal{M}$ and w) such that $\varphi_1 \wedge \varphi_2$ is *not* a conservative extension of φ_1 if, and only if, $\mathcal{M}$ does accept w. More precisely, we construct φ_1 and φ_2 in such a way that, if ψ is a witness for (φ_1, φ_2), then rooted models of $\varphi_1 \wedge \psi$ describe an accepting computation of $\mathcal{M}$ on w. In these models, each point represents a tape cell of a configuration of $\mathcal{M}$, and moving to the immediate successor of a point means moving to the next tape cell in the same configuration, or, if we are already at the end of the configuration, moving to the first tape cell of a successor configuration. Such models will have depth $2^n \cdot 2^{2^n}$ since the length of computations is bounded by 2^{2^n}.

8 Discussion

We have investigated the complexity of the conservativeness problem for the local consequence relation of a number of basic modal logics. One interesting conclusion is that the complexity of deciding conservativeness is not monotonically related to the complexity of the logic in question: for example, the satisfiability problem is NP-complete for **GL.3** and PSPACE-complete for **K**, while the conservativeness problem is (probably) more complex for **GL.3** than for **K**. This resembles the situation with products of modal logics where **GL.3** × **GL.3** is Π_1^1-complete [13], while **K** × **K** is decidable [4].

In this paper, we have considered modal languages with one modal operator only. It is not difficult, however, to modify the proofs above to show that conservativeness (for the local consequence relation) is still NEXPTIME-complete for multimodal **S5** and multimodal **K**. Similarly, for multimodal **S4** and **K4** it is still decidable in EXPSPACE.

For the global consequence relation the results are different: recall that φ *follows globally* from ψ in a modal logic L if φ is true everywhere in a model based on a frame for L whenever ψ is true everywhere in this model. Conservativeness with respect to the global consequence relation is now defined in the obvious way. Of course, for m-transitive modal logics the complexity upper bound for deciding conservativeness with respect to the local consequence is an upper bound for deciding conservativeness relative to the global consequence relation as well. This applies to **S5**, **S4** and **GL.3**. For **K**, however, deciding conservativeness with respect to the global consequence becomes 2EXPTIME-complete, as follows from the investigation of conservative extensions in description logics in [6]. We expect deciding conservativeness with respect to the global consequence in multimodal **S5** and **S4** to be 2EXPTIME-complete as well.

Finally, as indicated by Lemma 2, there is a close connection between studying modal logics extended by bisimulation quantifiers (cf. [3]) and conservative extensions. For many modal logics (it remains to explore which ones exactly) checking conservativeness is equivalent to deciding validity of the Σ_1-fragment of its extension by bisimulation quantifiers. This raises, for example, the question whether the analysis presented here can be extended to higher prenex fragments.

Acknowledgements. The work on this paper was partially supported by U.K. EPSRC grants no. GR/S63175/02, GR/S61973/02, GR/S63182/01, GR/S61966/01.

BIBLIOGRAPHY

[1] G. Antoniou and K. Kehagias. A note on the refinement of ontologies. *International Journal of Intelligent Systems*, 15:623–632, 2000.
[2] A. Chagrov and M. Zakharyaschev. *Modal Logic*. Oxford University Press, Oxford, 1997.
[3] T. French. *Bisimulation Quantifiers for Modal Logics*. PhD thesis, University of Western Australia, 2006.
[4] D. Gabbay and V. Shehtman. Products of modal logics. Part I. *Logic Journal of the IGPL*, 6:73–146, 1998.
[5] S. Ghilardi. An algebraic theory of normal forms. *Annals of Pure and Applied Logic*, 71(3):189–245, 1995.
[6] S. Ghilardi, C. Lutz, and F. Wolter. Did I damage my ontology? A case for conservative extensions in description logics. In *Proceedings of the Tenth International Conference of Principles of Knowledge Representation and Reasoning*, 187–197, 2006.
[7] S. Ghilardi, C. Lutz, F. Wolter, and M. Zakharyaschev. Conservative extensions in modal logic. http://www.csc.liv.ac.uk/~frank/publ/publ.html, 2006.
[8] S. Ghilardi and M. Zawadowski. Undefinability of propositional quantifiers in the modal system S4. *Studia Logica*, 55(2):259–271, 1995.
[9] V. Goranko and M. Otto. Modal model theory. In J. van Benthem, P. Blackburn, and F. Wolter, editors, *Handbook of Modal Logic*. Elsevier, 2006.
[10] M. Kracht. Modal consequence relations. In J. van Benthem, P. Blackburn, and F. Wolter, editors, *Handbook of Modal Logic*. Elsevier, 2006.
[11] C. Papadimitriou. *Computational Complexity*. Addison Wesley, 1995.
[12] A. Pitts. On an interpretation of second-order quantification in first-order intuitionistic propositional logic. *Journal of Symbolic Logic*, 57(1):33–52, 1992.
[13] M. Reynolds and M. Zakharyaschev. On the products of linear modal logics. *Journal of Logic and Computation*, 11:909–931, 2001.
[14] B. ten Cate, W. Conradie, M. Marx, and Y. Venema. Definitorially complete description logics. In *Proceedings of the Tenth International Conference of Principles of Knowledge Representation and Reasoning*, 79–89, 2006.
[15] W.M. Turski and T. Maibaum. *The Specification of Computer Programs*. Addison-Wesley, 1987.
[16] P. van Emde Boas. The convenience of tilings. In A. Sorbi, editor, *Complexity, Logic and Recursion Theory*, volume 187 of *Lecture Notes in Pure and Applied Mathematics*, pages 331–363. Marcel Dekker Inc., 1997.
[17] A. Visser. Uniform interpolation and layered bisimulation. In *Gödel'96 (Brno, 1996)*, volume 6 of *Lecture Notes Logic*, pages 139–164. Springer, Berlin, 1996.
[18] F. Wolter and M. Zakharyaschev. Modal decision problems. In J. van Benthem, P. Blackburn, and F. Wolter, editors, *Handbook of Modal Logic*. Elsevier, 2006.

Silvio Ghilardi
Department of Computer Science
University of Milan, Italy
ghilardi@dsi.unimi.it

Carsten Lutz
Institut für Theoretische Informatik
TU Dresden, Germany
lutz@tcs.inf.tu-dresden.de

Frank Wolter
Department of Computer Science
The University of Liverpool
Liverpool L69 3BX, U.K.
frank@csc.liv.ac.uk

Michael Zakharyaschev
School of Computer Science and Information Systems
Birkbeck College
Malet Street, London WC1E 7HX, U.K.
michael@dcs.bbk.ac.uk

A Kripke-Joyal Semantics for Noncommutative Logic in Quantales

ROBERT GOLDBLATT[1]

ABSTRACT. A structural semantics is developed for a first-order logic, with infinite disjunctions and conjunctions, that is characterised algebraically by quantales. The model structures involved combine the "covering systems" approach of Kripke-Joyal intuitionistic semantics from topos theory with the ordered groupoid structures used to model various connectives in substructural logics. The latter are used to interpret the noncommutative quantal conjunction & ("and then") and its residual implication connectives.

The completeness proof uses the MacNeille completion and the theory of quantic nuclei to first embed a residuated semigroup into a quantale, and then represent the quantale as an algebra of subsets of a model structure.

The final part of the paper makes some observations about quantal modal logic, giving in particular a structural modelling of the logic of closure operators on quantales.

Keywords: quantale, noncommutative conjunction, quantic nucleus, Kripke-Joyal semantics, infinitary proof theory, modality.

1 Introduction

A *locale* is a complete lattice in which finite meets distribute over arbitrary joins, the motivating example being the lattice of open subsets of a topological space. Any locale is a Heyting algebra – with the relative pseudocomplement $a \Rightarrow b$ being the join of $\{x : x \sqcap a \leq b\}$ – so it provides algebraic semantics for infinitary first-order intuitionistic logic, with $\Rightarrow$ interpreting implication, lattice joins and meets interpreting disjunctions $\bigvee$ and conjunctions $\bigwedge$ (possibly infinite), and the quantifiers $\exists$ and $\forall$ being treated as special disjunctions and conjunctions, respectively.

The notion of a *quantale* was introduced by C. J. Mulvey [13] to give a noncommutative extension of the locale concept that could be applied to spaces related to the foundations of quantum theory, such as the spectra of C*-algebras. A quantale is a complete lattice with an associative (but possibly not commutative) operation $a \bullet b$ that distributes over joins in each argument. A locale is then just a quantale in which $a \bullet b$ is the lattice meet of a and b. Mulvey suggested that $\bullet$ should interpret a logical connective & that is a kind of "sequential conjunction" with "a vestige of temporality in its interpretation". A propositional formula $\varphi \& \psi$ is to be read "φ and

---
[1] Supported by grant 03-VUW-048 from the Marsden Fund of the Royal Society of New Zealand.

then ψ". See [14] for a representation of the spectrum of a C*-algebra as the Lindenbaum algebra of a propositional logic of formulas built using this & and $\bigvee$. The paper [16] gives a novel application of propositional logic in quantales to the classification of Penrose tilings of the plane. Further information about the way that quantales generalise locales is given in [15].

Since a quantale is complete, it can still interpret all of $\bigvee, \bigwedge, \exists$ and $\forall$. It no longer has a Heyting implication $\Rightarrow$ (unless • is commutative), but instead • has left and right *residual* operations, $\Rightarrow_l$ and $\Rightarrow_r$, which can be used to interpret two connectives, $\to_l$ and $\to_r$, which we think of as left and right implication. The aim of this paper is to develop a semantics for the logic of all these connectives as interpreted in quantales, by combining the idea of the Kripke-Joyal intuitionistic semantics for $\bigvee, \bigwedge, \exists, \forall$ arising from topos theory [11, 3] with the models for $\&, \to_l, \to_r$ in substructural logics that are based on ordered groupoids $\langle S, \leq, \cdot \rangle$ [19, 18, 6, 7].

It has long been recognised[1] that a binary connective like & can be modelled by a ternary relation R on Kripke-type models, with the semantics

$$x \models \varphi \& \psi \quad \text{iff} \quad \exists y \exists z : Rxyz \text{ and } y \models \varphi \text{ and } z \models \psi.$$

This relates naturally to the "and then" reading of & if we view $Rxyz$ as a relation of relativistic temporal ordering, meaning that y precedes z in time from the perspective of x. In the groupoid formalism, $Rxyz$ becomes the condition that $y \cdot z \leq x$. The Kripke-Joyal semantics uses collections of sets called "covers" in a way that it is formally similar to the neighbourhood semantics of modal logics. We define a notion of *model structure* as a preordered groupoid, a semigroup in fact, with a covering system obeying axioms that interact the covers with the ordering and the semigroup structure.

This infinitary first-order logic of quantales is axiomatised by constructing Lindenbaum algebras that are residuated semigroups and then embedding these in quantales by means of the MacNeille completion and the theory of quantic nuclei. We then show that any quantale can be represented as an algebra of subsets of a model structure, and read off the Kripke-Joyal semantics from this. The final section of the paper makes a foray into the world of quantal modal logic, giving in particular a structural modelling of the logic of closure operators on quantales.

2 Posemigroups and Quantales

Given a *poset* $\langle S, \leq \rangle$, comprising a partial ordering $\leq$ on a set S, we write $\sum X$ for the *join* (=least upper bound), and $\prod X$ for the *meet* (=greatest lower bound), of a set $X \subseteq S$, when these bounds exist. A poset is *complete* if every subset has a join, or equivalently if every subset has a meet.

A *posemigroup* $\mathcal{S} = \langle S, \leq, \bullet \rangle$ has an associative binary operation • that is monotone (i.e. order preserving) in each argument, meaning that $x \leq z$

---

[1] The idea goes back to Jónsson and Tarski. The ternary relation semantics is most associated with the Routley-Meyer semantics of the (commutative) fusion connective in relevant logic. See also [7].

implies $x \bullet y \leq z \bullet y$ and $y \bullet x \leq y \bullet z$. A *quantale* is a complete poset with an associative $\bullet$ in which the equations

$$(\sum X) \bullet a = \sum_{x \in X}(x \bullet a) \qquad (2.1)$$
$$a \bullet (\sum X) = \sum_{x \in X}(a \bullet x) \qquad (2.2)$$

hold for every set $X \subseteq S$. These equations imply that $\bullet$ is monotone.

A posemigroup S is *residuated* if there are binary operations $\Rightarrow_l$ and $\Rightarrow_r$ on S, called the *left and right residuals* of $\bullet$, satisfying

$$x \bullet a \leq b \quad \text{iff} \quad x \leq a \Rightarrow_l b \qquad (2.3)$$
$$a \bullet x \leq b \quad \text{iff} \quad x \leq a \Rightarrow_r b. \qquad (2.4)$$

These two residual operations are identical precisely when $\bullet$ is commutative. In a residuated semigroup, the equations (2.1) and (2.2) hold whenever the joins they refer to exist. This is a well-known fact, and is really an instance of the general categorical result that left adjoint functors preserve colimits: here the map $x \mapsto x \bullet a$ is a functor on the poset category $\langle S, \leq \rangle$ that is left adjoint to $b \mapsto a \Rightarrow_l b$ by (2.3), while $x \mapsto a \bullet x$ is left adjoint to $b \mapsto a \Rightarrow_r b$ by (2.4).

Thus a complete residuated posemigroup is a quantale. The converse is also true: every quantale is residuated with $a \Rightarrow_l b = \sum\{x : x \bullet a \leq b\}$ and $a \Rightarrow_r b = \sum\{x : a \bullet x \leq b\}$.

Any residuated posemigroup can be embedded into a quantale by the famous completion construction of MacNeille [12]. This is shown in Section 4 of [18], where it is inferred from a more abstract result, and in Chapter 8 of [21] in the commutative case. Here we give the details of the proof in a way that emphasises its dependence on both residual operations when $\bullet$ is not commutative. The construction uses the theory of closure operators: a *closure operator* on a poset S is a function $j : S \to S$ that is *monotone*: $x \leq y$ implies $jx \leq jy$; *inflationary*: $x \leq jx$; and *idempotent*: $jjx = jx$. An element x is *j-closed* if $jx = x$. If S is complete, then the set S^j of j-closed elements is closed under meets $\prod X$, and so is complete under the same partial ordering. The join operation $\sum^j$ in S^j is given by $\sum^j X = j(\sum X)$.

Now if X is a subset of a poset S, let lX be the set of all lower bounds, and uX the set of all upper bounds, of X in S. Put $mX = luX$. Then m is a closure operator on the complete poset $\langle \mathcal{P}S, \subseteq \rangle$, where $\mathcal{P}S$ is the powerset of S. Any set of the form lX is m-closed, including the set $x\downarrow = l\{x\} = \{y \in S : y \leq x\}$ for each $x \in S$. So the set $(\mathcal{P}S)^m$ of all m-closed subsets of S is complete under the ordering $\subseteq$, as explained above, with $\prod \mathcal{X} = \bigcap \mathcal{X}$ and $\sum \mathcal{X} = m(\bigcup \mathcal{X})$ in $(\mathcal{P}S)^m$, where $\mathcal{X}$ is any collection of m-closed sets. The function $f_m(x) = x\downarrow$ is an order-invariant injection $f_m : \langle S, \leq \rangle \to \langle (\mathcal{P}S)^m, \subseteq \rangle$ having the crucial property that it *preserves any joins and meets that exist in S* (see [5, pp. 40–44] for a comprehensive discussion of this MacNeille completion construction).

A *quantic nucleus* is a closure operator on a posemigroup that satisfies

$$jx \bullet jy \leq j(x \bullet y). \qquad (2.5)$$

If j is a quantic nucleus on a quantale $\langle Q, \leq, \bullet \rangle$, then the complete poset $\langle Q^j, \leq \rangle$ of j-closed elements is a quantale under the operation $a \bullet_j b = j(a \bullet b)$ [17, Theorem 2.1]. Moreover Q^j is closed under the residuals of $\bullet$, and indeed both $a \Rightarrow_l b$ and $a \Rightarrow_r b$ belong to Q^j whenever $b \in Q^j$ [20, Prop. 3.1.2]. From this it can be shown that the residuals of $\bullet_j$ on Q^j are just the restrictions of the residuals of $\bullet$ on Q to Q^j.

Now from any semigroup $\langle S, \bullet \rangle$ we can construct a quantale $\langle \mathcal{P}S, \subseteq, \bullet \rangle$ on the powerset of S by putting

$$X \bullet Y = \{x \bullet y : x \in X \text{ and } y \in Y\}$$

for all $X, Y \subseteq S$. In this quantale the join $\sum \mathcal{X}$ is the set-theoretic union $\bigcup \mathcal{X}$, and the residuals are given by

$$X \Rightarrow_l Y = \{z \in S : \{z\} \bullet X \subseteq Y\}, \ X \Rightarrow_r Y = \{z \in S : X \bullet \{z\} \subseteq Y\}. \quad (2.6)$$

LEMMA 1. *For any residuated semigroup $\langle S, \leq, \bullet \rangle$, the MacNeille closure operator $mX = luX$ is a quantic nucleus on the quantale $\langle \mathcal{P}S, \subseteq, \bullet \rangle$, and so $\langle (\mathcal{P}S)^m, \subseteq, \bullet_m \rangle$ is a quantale. Moreover the injection $f_m : S \to (\mathcal{P}S)^m$ preserves $\bullet$ and its residuals.*

Proof. We have to show that $(luX) \bullet (luY) \subseteq lu(X \bullet Y)$, so fix any $x \in luX$ and $y \in luY$. Let $z \in u(X \bullet Y)$. We have to show $x \bullet y \leq z$.

Now if $y' \in Y$, then for all $x' \in X$, $x' \bullet y' \leq z$ and hence $x' \leq y' \Rightarrow_l z$. This shows that $y' \Rightarrow_l z \in uX$, so $x \leq y' \Rightarrow_l z$ as $x \in luX$, hence $x \bullet y' \leq z$ and so $y' \leq x \Rightarrow_r z$. Since that holds for all $y' \in Y$, $x \Rightarrow_r z \in uY$, so $y \leq x \Rightarrow_r z$, implying $x \bullet y \leq z$ as required. Thus (2.5) holds when $j = m$.

For preservation of $\bullet$ by f_m, note that since $x \bullet y \in (x\downarrow) \bullet (y\downarrow)$ we get $(x \bullet y)\downarrow \subseteq m((x\downarrow) \bullet (y\downarrow))$. But $(x\downarrow) \bullet (y\downarrow) \subseteq (x \bullet y)\downarrow$, so $m((x\downarrow) \bullet (y\downarrow)) \subseteq m((x \bullet y)\downarrow) = (x \bullet y)\downarrow$. Hence $(x \bullet y)\downarrow = m((x\downarrow) \bullet (y\downarrow)) = (x\downarrow) \bullet_m (y\downarrow)$ as required.

Now the residuals of $\bullet_m$ are just the restrictions of the residuals of $\bullet$, so these are given on $(\mathcal{P}S)^m$ by (2.6). Preservation of $\Rightarrow_l$ thus amounts to the condition that $z \leq x \Rightarrow_l y$ iff $z \bullet (x\downarrow) \subseteq (y\downarrow)$, which follows readily by (2.3) and monotonicity of $\bullet$. Preservation of $\Rightarrow_r$ follows similarly from (2.4). ∎

COROLLARY 2. *Every residuated posemigroup has an isomorphic embedding into the residuated posemigroup of a quantale that preserves any existing joins and meets.*

3 Logic

We assume familiarity with the syntactic apparatus of first-order logic with infinite disjunctions and conjunctions. Sensitivity to the distinction between *large classes* and *sets* (small classes) is required, since collections of formulas may be large.

3.1 Formulas

Fix a denumerable list $v_0, \ldots, v_n, \ldots$ of individual variables and a set of predicate letters, with typical member P, that are k-ary for various $k < \omega$. These are used to define *atomic* formulas $P(v_{n_1}, \ldots, v_{n_k})$. A *preformula* is any expression generated from atomic formulas by using the binary connectives $\&, \to_l, \to_r$ and the quantifiers $\exists v_n, \forall v_n$, and by allowing the formation of the disjunction $\bigvee \Phi$ and conjunction $\bigwedge \Phi$ of any *set* Φ of formulas. A *formula* is a preformula that has only finitely many free variables. This constraint is a standard convention in infinitary logic, designed to avoid dealing with expressions that have too many free variables to be convertible into sentences by prefixing quantifiers [1, 4]. We confine our attention to formulas throughout.

The class of all formulas is large, so cannot be used in its entirety to build a Lindenbaum algebra as a quotient set (the requisite equivalence classes of formulas may themselves be large). The class $\operatorname{sub} \varphi$ of subformulas of a formula φ is defined in the usual way, e.g. if $\varphi = \bigvee \Phi$, then $\operatorname{sub} \varphi = \{\varphi\} \cup \bigcup_{\psi \in \Phi} \operatorname{sub} \psi$. Then if Ψ is any set of formulas, the class $\operatorname{sub} \Psi = \bigcup_{\psi \in \Psi} \operatorname{sub} \psi$ of all subformulas of members of Ψ is a set.

We adopt the usual conventions for variable-substitution, writing $\varphi(w/v)$ for the formula obtained by substituting w for all free occurrences of v in a suitable alphabetic variant of φ.

3.2 Quantal Models

These are structures $\mathfrak{A} = \langle \mathcal{Q}, D, V \rangle$ with $\mathcal{Q} = \langle Q, \leq, \bullet \rangle$ being a quantale, D a non-empty set of individuals, and V a function assigning to each k-ary predicate letter P a function $V(P) : D^k \to Q$. To interpret variables in the model we use D-*valuations*, which are sequences $\sigma = \langle \sigma_0, \ldots, \sigma_n, \ldots \rangle$ of elements of D, the idea being that σ assigns value σ_n to variable v_n. We write $\sigma(d/n)$ for the valuation obtained from σ by replacing σ_n by d. For each formula φ we specify a value $\|\varphi\|_\sigma^\mathfrak{A} \in Q$ for each valuation σ. This is defined inductively on the formation of φ, as follows:

- $\|P(v_{n_1}, \ldots, v_{n_k})\|_\sigma^\mathfrak{A} = V(P)(\sigma_{n_1}, \ldots, \sigma_{n_k})$
- $\|\varphi \& \psi\|_\sigma^\mathfrak{A} = \|\varphi\|_\sigma^\mathfrak{A} \bullet \|\psi\|_\sigma^\mathfrak{A}$
- $\|\varphi \to_l \psi\|_\sigma^\mathfrak{A} = \|\varphi\|_\sigma^\mathfrak{A} \Rightarrow_l \|\psi\|_\sigma^\mathfrak{A}, \qquad \|\varphi \to_r \psi\|_\sigma^\mathfrak{A} = \|\varphi\|_\sigma^\mathfrak{A} \Rightarrow_r \|\psi\|_\sigma^\mathfrak{A}$
- $\|\bigvee \Phi\|_\sigma^\mathfrak{A} = \sum_{\varphi \in \Phi} \|\varphi\|_\sigma^\mathfrak{A}$
- $\|\bigwedge \Phi\|_\sigma^\mathfrak{A} = \prod_{\varphi \in \Phi} \|\varphi\|_\sigma^\mathfrak{A}$
- $\|\exists v_n \varphi\|_\sigma^\mathfrak{A} = \sum_{d \in D} \|\varphi\|_{\sigma(d/n)}^\mathfrak{A}$
- $\|\forall v_n \varphi\|_\sigma^\mathfrak{A} = \prod_{d \in D} \|\varphi\|_{\sigma(d/n)}^\mathfrak{A}$.

We write $\varphi \models^\mathfrak{A} \psi$ if $\|\varphi\|_\sigma^\mathfrak{A} \leq \|\psi\|_\sigma^\mathfrak{A}$ for all D-valuations σ, and say that φ *semantically implies* ψ *over quantales*, written $\varphi \models^q \psi$, if $\varphi \models^\mathfrak{A} \psi$ for all quantal models $\mathfrak{A}$.

3.3 Proof Theory

A *sequent* is an expression $\varphi \vdash \psi$ with φ and ψ being formulas. Alternatively, a sequent may be thought of as an ordered pair of formulas, with the symbol $\vdash$ denoting a class of sequents, i.e. a binary relation between formulas.

Let $\vdash_q$ be the smallest class of sequents that includes all instances of the axiom schemas

- $\varphi \vdash \varphi$;
- $\varphi\&(\psi\&\rho) \vdash (\varphi\&\psi)\&\rho, \qquad (\varphi\&\psi)\&\rho \vdash \varphi\&(\psi\&\rho)$;
- $\varphi \vdash \bigvee \Phi$, if $\varphi \in \Phi$;
- $\varphi(w/v) \vdash \exists v \varphi$;
- $\bigwedge \Phi \vdash \varphi$, if $\varphi \in \Phi$;
- $\forall \varphi \vdash \varphi(w/v)$;
- $(\varphi \rightarrow_l \psi)\&\varphi \vdash \psi, \qquad \varphi\&(\varphi \rightarrow_r \psi) \vdash \psi$;

and is closed under the following rules:

- if $\varphi \vdash \psi$ and $\psi \vdash \rho$, then $\varphi \vdash \rho$;
- if $\varphi \vdash \psi$, then $\varphi\&\rho \vdash \psi\&\rho$ and $\rho\&\varphi \vdash \rho\&\psi$;
- if $\varphi \vdash \psi$ for all $\varphi \in \Phi$, then $\bigvee \Phi \vdash \psi$;
- if $\varphi \vdash \psi$, then $\exists v \varphi \vdash \psi$ provided v does not occur free in ψ.
- if $\psi \vdash \varphi$ for all $\varphi \in \Phi$, then $\psi \vdash \bigwedge \Phi$;
- if $\varphi \vdash \psi$, then $\varphi \vdash \forall v \psi$ provided v does not occur free in φ.
- if $\varphi\&\psi \vdash \rho$, then $\varphi \vdash \psi \rightarrow_l \rho$ and $\psi \vdash \varphi \rightarrow_r \rho$.

THEOREM 3 (Soundness). $\varphi \vdash_q \psi$ *implies* $\varphi \models_q \psi$.

Proof. For any quantal model $\mathfrak{A}$, the relation $\models^{\mathfrak{A}}$ includes all instances of the above axioms and is closed under the above rules, so it includes $\vdash_q$. Thus $\varphi \vdash_q \psi$ implies $\varphi \models^{\mathfrak{A}} \psi$ for all quantal models $\mathfrak{A}$. ∎

3.4 Lindenbaum Models of Fragments

A *fragment* is a set $\mathcal{F}$ of formulas that includes all atomic formulas; is closed under the binary connectives $\&, \rightarrow_l, \rightarrow_r$, and the quantifiers $\exists v_n, \forall v_n$; and is closed under subformulas and variable substitution. Any set Φ of formulas can be enlarged to a fragment: define $\mathcal{F}_0$ by adding all atomic formulas to Φ, and then inductively define $\mathcal{F}_{n+1}$ by closing $\mathcal{F}_n$ under the binary connectives and the quantifiers, and then closing under subformulas and variable-substitution. Then $\mathcal{F} = \bigcup_{n<\omega} \mathcal{F}_n$ is the smallest fragment including Φ.

Definition of the Lindenbaum quantal model $\mathfrak{A}^\mathcal{F}$ of a fragment begins with the standard construction. Let $\vdash$ be a relation satisfying all the above axioms and rules. The condition "$\varphi \vdash \psi$ and $\psi \vdash \varphi$" gives an equivalence relation on $\mathcal{F}$. Let $|\varphi|$ be the equivalence class of $\varphi \in \mathcal{F}$ and $S^\mathcal{F} = \{|\varphi| : \varphi \in \mathcal{F}\}$. Put $|\varphi| \leq |\psi|$ iff $\varphi \vdash \psi$; $|\varphi| \bullet |\psi| = |\varphi \& \psi|$, $|\varphi| \Rightarrow_l |\psi| = |\varphi \to_l \psi|$ and $|\varphi| \Rightarrow_r |\psi| = |\varphi \to_r \psi|$. The axioms and rules ensure that this yields a well-defined residuated posemigroup $\mathcal{S}^\mathcal{F}$ on $S^\mathcal{F}$ in which

$$|\exists v_n \varphi| = \sum_{p<\omega} |\varphi(v_p/v_n)|, \quad |\forall v_n \varphi| = \prod_{p<\omega} |\varphi(v_p/v_n)| \tag{3.1}$$

$$|\bigvee \Phi| = \sum_{\varphi \in \Phi} |\varphi|, \quad \text{when } \bigvee \Phi \in \mathcal{F} \tag{3.2}$$

$$|\bigwedge \Phi| = \prod_{\varphi \in \Phi} |\varphi| \quad \text{when } \bigwedge \Phi \in \mathcal{F}. \tag{3.3}$$

The proof of (3.1) is as for finitary first-order logic (e.g. [2, Lemma 3.4.1]), and depends on the fact that a formula has finitely many free variables.

Now put $\mathfrak{A}^\mathcal{F} = \langle \mathcal{Q}^\mathcal{F}, D, V \rangle$, where $\mathcal{Q}^\mathcal{F}$ is the quantale obtained from $\mathcal{S}^\mathcal{F}$ by Corollary 2 and having an embedding $f : \mathcal{S}^\mathcal{F} \to \mathcal{Q}^\mathcal{F}$; D is the set of all variables v_n; and $V(P)(v_{n_1}, \ldots, v_{n_k}) = f|P(v_{n_1}, \ldots, v_{n_k})|$. Then if σ is the D-valuation with $\sigma_n = v_n$, we get

$$\|P(v_{n_1}, \ldots, v_{n_k})\|_\sigma^{\mathfrak{A}^\mathcal{F}} = f|P(v_{n_1}, \ldots, v_{n_k})|.$$

We then extend this to show inductively that

$$\|\varphi\|_\sigma^{\mathfrak{A}^\mathcal{F}} = f|\varphi| \text{ for all } \varphi \in \mathcal{F}. \tag{3.4}$$

This uses the the definition of $\|\varphi\|_\sigma^\mathfrak{A}$, results (3.1)–(3.3), the fact that f preserves the residuated posemigroup operations and any joins and meets existing in $\mathcal{S}^\mathcal{F}$; and the general substitutional result that $\|\varphi(v_p/v_n)\|_\sigma^\mathfrak{A} = \|\varphi\|_{\sigma(\sigma_p/n)}^\mathfrak{A}$ (which holds of any σ in any model $\mathfrak{A}$).

THEOREM 4 (Completeness). *$\varphi \models_q \psi$ implies $\varphi \vdash_q \psi$.*

Proof. Let $\varphi \models_q \psi$, and take a fragment $\mathcal{F}$ containing φ and ψ. Construct $\mathfrak{A}^\mathcal{F}$ as above using the relation $\vdash_q$ to define $\mathcal{S}^\mathcal{F}$. Then $\varphi \models^{\mathfrak{A}^\mathcal{F}} \psi$, so $\|\varphi\|_\sigma \leq \|\psi\|_\sigma$ in $\mathfrak{A}^\mathcal{F}$ where $\sigma_n = v_n$. Hence $|\varphi| \leq |\psi|$ by (3.4) and the fact that f is order-invariant, so $\varphi \vdash_q \psi$. ∎

4 Covers and Model Structures

We now work with structures $\mathfrak{S} = \langle S, \triangleleft, \cdot, Cov \rangle$, where $\triangleleft$ is a *preorder* (i.e. reflexive and transitive relation); $\cdot$ is an associative operation that is $\triangleleft$-monotone in each argument; and Cov is a function assigning to each $x \in S$ a collection $Cov(x)$ of subsets of S, called the *covers of x*, or *x-covers*.

For $X, Y \subseteq S$, put $X \cdot Y = \{x \cdot y : x \in X \text{ and } y \in Y\}$, $x \cdot Y = \{x\} \cdot Y$ and $X \cdot y = X \cdot \{y\}$. Let $[X] = \{y \in S : (\exists x \in X) x \triangleleft y\}$ and $[x) = [\{x\}) = \{y : x \triangleleft y\}$. X is *increasing* if $[X) \subseteq X$, meaning that if $x \in X$ and $x \triangleleft y$, then $y \in X$.[2] In general, $[X)$ is the smallest increasing superset of X. We write $X \triangleleft Y$, and say that Y *refines* X, and that X is *refined by* Y, if $(\forall y \in Y)(\exists x \in X) x \triangleleft y$. Thus $X \triangleleft Y$ iff $Y \subseteq [X)$. A set X is *cover-closed* if, for all $x \in S$, $(\exists C \in Cov(x))(C \subseteq X)$ implies $x \in X$. A *c-filter* is a set that is increasing and cover-closed.

We call $\mathfrak{S}$ a *model structure*[3] if the following axioms hold for all $x \in S$:

cov1: there exists an x-cover $C \subseteq [x)$;

cov2: if $C \in Cov(x)$ and for all $y \in C$, $C_y \in Cov(y)$, then $\bigcup_{y \in C} C_y \in Cov(x)$.

cov3: if $x \triangleleft y$, then every x-cover can be refined to a y-cover: $(\forall C \in Cov(x))(\exists B \in Cov(y)) C \triangleleft B$.

cov4: if $C \in Cov(x)$ and $B \in Cov(y)$, then $C \cdot B$ can be refined to a $x \cdot y$-cover.

cov5: if there exists an x-cover refining $X \cdot Y$, then there exist x', y' with $x' \cdot y' \triangleleft x$; an x'-cover $X' \subseteq X$; and a y'-cover $Y' \subseteq Y$.

THEOREM 5. *If $\mathfrak{S}$ is a model structure, then $\mathcal{Q}^\mathfrak{S} = \langle S^\mathfrak{S}, \subseteq, \bullet \rangle$ is a quantale, where $S^\mathfrak{S}$ is the set of all c-filters of $\mathfrak{S}$ and $X \bullet Y = [X \cdot Y)$. Joins in $\mathcal{Q}^\mathfrak{S}$ are given by $\sum \mathcal{X} = \{x : (\exists C \in Cov(x)) C \subseteq \bigcup \mathcal{X}\}$ for all $\mathcal{X} \subseteq S^\mathfrak{S}$. The residuals of $\bullet$ are given by $X \Rightarrow_l Y = \{z \in S : z \cdot X \subseteq Y\}$ and $X \Rightarrow_r Y = \{z \in S : X \cdot z \subseteq Y\}$.*

Proof. This could be shown by direct set-theoretic reasoning, but more insight into the role of the (cov)-axioms is gained by constructing $\mathcal{Q}^\mathfrak{S}$ as the quantale of closed elements of a quantic nucleus. Put $\mathcal{Q}^\triangleleft = \langle S^\triangleleft, \subseteq, \bullet \rangle$ where $S^\triangleleft$ is the set of $\triangleleft$-increasing subsets of $\mathfrak{S}$ and $X \bullet Y = [X \cdot Y)$. It is readily seen that $\mathcal{Q}^\triangleleft$ is a quantale in which the join of $\mathcal{X} \subseteq S^\triangleleft$ is the set-union $\bigcup \mathcal{X}$ (and the meet is $\bigcap \mathcal{X}$). Note that the definition of $\mathcal{Q}^\triangleleft$ is independent of Cov.

Now define a $\subseteq$-monotonic function j_{Cov} on $S^\triangleleft$ by $j_{Cov} X = \{x \in S : (\exists C \in Cov(x)) C \subseteq X\}$. The axioms (cov1)–(cov4) then ensure that j_{Cov} is a quantic nucleus on $\mathcal{Q}^\triangleleft$, as follows.

First, (cov3) ensures that $j_{Cov} X$ is increasing when X is, for if $x \in j_{Cov} X$ and $x \triangleleft y$, then there exists $C \in Cov(x)$ with $C \subseteq X$, hence by (cov3) there exists $B \in Cov(y)$ with $B \subseteq [C) \subseteq [X) = X$, implying $y \in j_{Cov} X$.

Next, (cov1) ensures that j_{Cov} is inflationary, for if $x \in X \in S^\triangleleft$, then by (cov1) there exists $C \in Cov(x)$ with $C \subseteq [x) \subseteq X$, implying $x \in j_{Cov} X$. Thus $X \subseteq j_{Cov} X$.

---

[2] Some treatments of Kripke-Joyal semantics in posets for intuitionistic logic use decreasing sets rather than increasing ones, e.g. [3]. Here we follow the conventions of Kripke's original model theory, as well as of those using preordered groupoids to model other substructural logics [19, 18, 6, 7]. Formally, the distinction is no more than that between a preorder and its converse.

[3] Modal logicians would tend to call this a "frame", but we avoid this term since it has a different meaning in locale theory.

(cov2) ensures that $j_{Cov}j_{Cov}X \subseteq j_{Cov}X$, for if there exists $C \in Cov(x)$ with $C \subseteq j_{Cov}X$, then for all $y \in C$ there exists $C_y \in Cov(y)$ with $C_y \subseteq X$. Then by (cov2), $\bigcup_{y \in Y} C_y$ is an x-cover included in X, showing $x \in j_{Cov}X$.

Finally, (cov4) and (cov3) ensure that j_{Cov} satisfies (2.5). For if $z \in (j_{Cov}X) \bullet (j_{Cov}Y)$, then $x \cdot y \triangleleft z$ for some x, y such that there is an x-cover $C \subseteq X$ and a y-cover $B \subseteq Y$. By (cov4) there is some $x \cdot y$-cover $A \subseteq [C \cdot B]$, and then by (cov3) there is a z-cover $A' \subseteq [A] \subseteq [C \cdot B] \subseteq [X \cdot Y] = X \bullet Y$, showing that $z \in j_{Cov}(X \bullet Y)$. Hence $(j_{Cov}X) \bullet (j_{Cov}Y) \subseteq j_{Cov}(X \bullet Y)$.

Now for $X \in S^\triangleleft$ we have $j_{Cov}X = X$ iff $j_{Cov}X \subseteq X$ iff X is cover-closed. So the set of j_{Cov}-closed elements of $S^\triangleleft$ is just the set $S^\mathfrak{S}$ of c-filters of $\mathfrak{S}$. Thus by the theory explained in Section 2, $\mathcal{Q}^\mathfrak{S} = \langle S^\mathfrak{S}, \subseteq, \bullet_{j_{Cov}} \rangle$ is a quantale is which $\sum \mathcal{X} = j_{Cov} \bigcup \mathcal{X}$, as required by the statement of this Theorem. But (cov5) ensures that if X and Y are cover-closed, then so is $X \bullet Y$, for if there exists $C \in Cov(x)$ with $C \subseteq X \bullet Y$, then $X \cdot Y \triangleleft C$, so taking x', y', X', Y' as given by (cov5) we get $x' \in X$ as X is cover-closed, and likewise $y' \in Y$, hence $x \in [X \cdot Y]$ as required because $x' \cdot y' \triangleleft x$. Thus if X, Y are c-filters, then so is $X \bullet Y$, implying $X \bullet_{j_{Cov}} Y = j_{Cov}(X \bullet Y) = X \bullet Y$. So $\bullet_{j_{Cov}}$ is just $\bullet$ on $S^\mathfrak{S}$.

To show the residuals in $\mathcal{Q}^\mathfrak{S}$ are as stated, let $W = \{z : z \cdot X \subseteq Y\}$. Then in general $Z \cdot X \subseteq Y$ iff $Z \subseteq W$, so if Y is increasing, $Z \bullet X \subseteq Y$ iff $Z \subseteq W$. But if Y is increasing then so is W by $\triangleleft$-monotonicity of $\cdot$. Thus if $X, Y \in S^\triangleleft$, then since $W \in S^\triangleleft$ we must have $W = (X \Rightarrow_l Y)$ in $\mathcal{Q}^\triangleleft$. But the left residual of $\bullet_{j_{Cov}}$ in $\mathcal{Q}^\mathfrak{S}$ is just the restriction of the left residual of $\bullet$ in $\mathcal{Q}^\triangleleft$ to $S^\mathfrak{S}$, so if $X, Y \in S^\mathfrak{S}$, then $W = (X \Rightarrow_l Y)$ in $\mathcal{Q}^\mathfrak{S}$. Similarly for $X \Rightarrow_r Y$. ∎

The covering concept comes of course from topology, where an open cover of a set x is any collection C of open sets whose union includes x. A property holds *locally* of x if it holds of all members of some cover of x. For example, a function is *locally constant* on x if it is constant on each member of some x-cover. If we take S to be the set of all open subsets of some topological space and put $x \triangleleft y$ iff $y \subseteq x$, $x \cdot y = x \cap y$ and $C \in Cov(x)$ iff $x \subseteq \bigcup C$, then we obtain a model structure (in which "refines" has its usual meaning for topological covers). Indeed this construction works in any quantale, as we now show.

THEOREM 6. *Every quantale $\mathcal{Q}$ is isomorphic to the quantale $\mathcal{Q}^\mathfrak{S}$ of some model structure $\mathfrak{S}$.*

Proof. Let $\mathcal{Q} = \langle Q, \leq, \bullet \rangle$. Define $\mathfrak{S} = \langle S, \triangleleft, \cdot, Cov \rangle$ by putting $S = Q$; $x \triangleleft y$ iff $y \leq x$; $x \cdot y = x \bullet y$; and $C \in Cov(x)$ iff $x \leq \sum C$ for all $C \subseteq Q$. Then $\triangleleft$ is a preorder and $\cdot$ is associative and $\triangleleft$-monotone. Moreover, $[x] = (x\downarrow)$.

Thus a set X is increasing in $\mathfrak{S}$ iff it is $\leq$-*decreasing* in $\mathcal{Q}$, i.e. $x \in X$ implies $(x\downarrow) \subseteq X$. In particular, $x\downarrow$ is increasing, and is also cover-closed, for if $C \in Cov(y)$ and $C \subseteq (x\downarrow)$, then $y \leq \sum C \leq x$, hence $y \in (x\downarrow)$. Moreover, if X is any c-filter and we put $x = \sum X$, then $X \in Cov(x)$ and so $x \in X$ by cover-closure, hence $X = (x\downarrow)$ as X is $\leq$-decreasing.

Thus the map $x \mapsto (x{\downarrow})$ is a bijection between Q and the set $S^{\mathfrak{S}}$ of c-filters of $\mathfrak{S}$. This map is order-invariant: $x \leq y$ iff $(x{\downarrow}) \subseteq (y{\downarrow})$. But it is readily seen that $(x \bullet y){\downarrow} = [(x{\downarrow}) \cdot (y{\downarrow})]$ so altogether the map is an isomorphism between $\mathcal{Q}$ and the quantale $\mathcal{Q}^{\mathfrak{S}}$ of the previous Theorem. It remains to show $\mathfrak{S}$ is a model structure.

(cov1): Any C with $x \in C \subseteq [x)$ will do.

(cov2): If $x \leq \sum C$, and $(\forall y \in C)(y \leq \sum C_y)$, then $x \leq \sum_{y \in C}(\sum C_y) = \sum(\bigcup_{y \in C} C_y)$.

(cov3): If $C \in Cov(x)$ and $x \triangleleft y$, then $y \leq x \leq \sum C$, so just take $B = C$.

(cov4): If $C \in Cov(x)$ and $B \in Cov(y)$, then $x \cdot y \leq (\sum C) \cdot (\sum B) = \sum(C \cdot B)$, with the equality following from the distributive laws (2.1) and (2.2). Thus $C \cdot B$ is itself an $x \cdot y$-cover.

(cov5): Suppose there is an x-cover $C \subseteq [X \cdot Y)$. Put $X' = \{a \in X : (\exists b \in Y)(\exists c \in C) a \cdot b \triangleleft c\} \subseteq X$ and $Y' = \{b \in Y : (\exists a \in X)(\exists c \in C) a \cdot b \triangleleft c\} \subseteq Y$. Let $x' = \sum X'$ and $y' = \sum Y'$, so that $X' \in Cov(x')$ and $Y' \in Cov(y')$. It remains to show $x' \cdot y' \triangleleft x$. Now if $c \in C$, then $a \cdot b \triangleleft c$ for some $a \in X$ and $b \in Y$. Then $a \in X'$ and $b \in Y'$, so $c \leq a \cdot b \leq x' \cdot y'$. Hence $x \leq \sum C \leq x' \cdot y'$, as required. ∎

The axioms (cov1)–(cov5) are almost minimal requirements for $\mathcal{Q}^{\mathfrak{S}}$ to be a quantale in Theorem 5. (cov5) could be weakened as it is only needed when X and Y are c-filters. In that case we could add the requirement that every cover is an increasing set, since both Theorems 5 and 6 would hold under this requirement. As it stands, the model structure defined in Theorem 6 satisfies several other strengthenings and additional conditions. It has $[x) \in Cov(x)$ and $\{x\} \in Cov(x)$, each of which implies (cov1). For (cov3) it has the stronger property that if $x \triangleleft y$, then every x-cover is a y-cover. For (cov4) it has the conclusion that $C \cdot B$ is an $x \cdot y$ cover when $C \in Cov(x)$ and $B \in Cov(y)$. It even has the property that $Cov(x)$ is closed under supersets. But it does not have the property that *every x-cover is a subset of $[x)$*, which is a basic assumption in the definition of coverings on posets in locale theory [10, 3] and is fundamental to the categorical view of coverings in a Grothendieck topology on a category. In the $\mathfrak{S}$ of Theorem 6, this property would require that an x-cover have $x = \sum C$, rather than $x \leq \sum C$. This makes (cov3) problematic, although the other (cov)-axioms still hold. If $\mathcal{Q}$ is a *locale*, then (cov3) does hold in this case, for if $y \leq x = \sum C$, then the subset $\{y \sqcap c : c \in C\}$ of $[y)$ is a refinement of C whose join is y, by the distribution of the lattice meet $\sqcap$ over $\sum$. But that argument is not available in a general quantale.

5 Kripke-Joyal Semantics

A *structural model* for the language of Section 3 has the form $\mathcal{M} = \langle \mathfrak{S}, D, V \rangle$, where $\mathfrak{S}$ is a model structure; D is a set of individuals; and for each k-ary predicate letter P, $V(P)$ is a function assigning a c-filter of $\mathfrak{S}$ to each k-tuple of elements of D. In other words, $V(P) : D^k \to S^{\mathfrak{S}}$, where $S^{\mathfrak{S}}$ is the set of all c-filters of $\mathfrak{S}$. From such an $\mathcal{M}$ we immediately obtain the quantal model $\mathfrak{A}^{\mathcal{M}} = \langle \mathcal{Q}^{\mathfrak{S}}, D, V \rangle$, where $\mathcal{Q}^{\mathfrak{S}} = \langle S^{\mathfrak{S}}, \subseteq, \bullet \rangle$ is the quantale of Theorem 5. We define $\varphi \models^{\mathcal{M}} \psi$ to mean that $\varphi \models^{\mathfrak{A}^{\mathcal{M}}} \psi$. Also we write $\|\varphi\|_\sigma^{\mathcal{M}}$ for the value $\|\varphi\|_\sigma^{\mathfrak{A}^{\mathcal{M}}}$ given by the general definition of $\|\varphi\|_\sigma^{\mathfrak{A}}$ from Section 3. Thus $\|\varphi\|_\sigma^{\mathcal{M}}$ is a c-filter, and

$$\varphi \models^{\mathcal{M}} \psi \quad \text{iff} \quad \|\varphi\|_\sigma^{\mathcal{M}} \subseteq \|\psi\|_\sigma^{\mathcal{M}} \text{ for all } D\text{-valuations } \sigma. \tag{5.1}$$

In the converse direction, from any quantal model $\mathfrak{A} = \langle \mathcal{Q}, D, V \rangle$ we obtain the structural model $\mathcal{M}^{\mathfrak{A}} = \langle \mathfrak{S}, D, V^{\mathfrak{A}} \rangle$, where $\mathfrak{S}$ is the model structure of Theorem 6 for which there is an isomorphism of quantales $f : \mathcal{Q} \cong \mathcal{Q}^{\mathfrak{S}}$, and $V^{\mathcal{M}}(P) = f \circ V(P) : D^k \to S^{\mathfrak{S}}$.

Since the isomorphism f preserves all the operations involved in the definition of $\|\varphi\|_\sigma^{\mathfrak{A}}$, an inductive proof then shows that in general $f \circ \|\varphi\|_\sigma^{\mathfrak{A}} = \|\varphi\|_\sigma^{\mathcal{M}^{\mathfrak{A}}}$, and so $\|\varphi\|_\sigma^{\mathfrak{A}} \subseteq \|\psi\|_\sigma^{\mathfrak{A}}$ iff $\|\varphi\|_\sigma^{\mathcal{M}^{\mathfrak{A}}} \subseteq \|\psi\|_\sigma^{\mathcal{M}^{\mathfrak{A}}}$. Hence

$$\varphi \models^{\mathfrak{A}} \psi \quad \text{iff} \quad \varphi \models^{\mathcal{M}^{\mathfrak{A}}} \psi. \tag{5.2}$$

THEOREM 7 (Completeness for Structural Models). *For all formulas φ and ψ the following are equivalent:*

(1) $\varphi \vdash_q \psi$.

(2) $\varphi \models^{\mathcal{M}} \psi$ *for all structural models $\mathcal{M}$.*

Proof. (1) implies (2): by the Soundness Theorem 3, $\varphi \vdash_c \psi$ implies $\varphi \models^{\mathfrak{A}^{\mathcal{M}}} \psi$. (2) implies (1): if $\varphi \nvdash_q \psi$, then by the Completeness Theorem 4, $\varphi \nvDash^{\mathfrak{A}} \psi$ for some quantal model $\mathfrak{A}$. But then $\varphi \nvDash^{\mathcal{M}^{\mathfrak{A}}} \psi$ by (5.2). ∎

In a model structure $\mathcal{M} = \langle S, \triangleleft, \cdot, Cov, D, V \rangle$, a *satisfaction relation* can be defined by using the notation

$$\mathcal{M}, x \models \varphi[\sigma]$$

to mean that $x \in \|\varphi\|_\sigma^{\mathcal{M}}$. This can be read "$\varphi$ is true/satisfied in $\mathcal{M}$ at x under σ". Thus (5.1) becomes

$$\varphi \models^{\mathcal{M}} \psi \quad \text{iff} \quad \text{for all } x \text{ and } \sigma, \mathcal{M}, x \models \varphi[\sigma] \text{ implies } \mathcal{M}, x \models \psi[\sigma].$$

A purely model theoretic description of this satisfaction relation in a structural model can be derived by applying the description of the quantale operations of $\mathcal{Q}^{\mathfrak{S}}$ given in Theorem 5 to the general definition of $|\varphi\|_\sigma^{\mathfrak{A}}$. This is given below (with the symbol $\mathcal{M}$ suppressed as it is constant throughout). The cases of $\bigvee, \bigwedge$ and $\exists$ are just like the corresponding clauses in the Kripke-Joyal semantics of models in sheaf categories [11, Theorem VI.7.1].

$x \models P(v_{n_1}, \ldots, v_{n_k})[\sigma]$ iff $x \in V(P)(\sigma_{n_1}, \ldots, \sigma_{n_k})$.

$x \models \varphi \& \psi[\sigma]$ iff   for some y and z such that $y \cdot z \triangleleft x$, $y \models \varphi[\sigma]$ and $z \models \psi[\sigma]$.

$x \models \varphi \to_l \psi[\sigma]$ iff $y \models \varphi[\sigma]$ implies $x \cdot y \models \psi[\sigma]$.

$x \models \varphi \to_r \psi[\sigma]$ iff $y \models \varphi[\sigma]$ implies $y \cdot x \models \psi[\sigma]$.

$x \models \bigvee \Phi[\sigma]$ iff   there is an x-cover C such that for all $z \in C$, $z \models \varphi[\sigma]$ for some $\varphi \in \Phi$.

$x \models \bigwedge \Phi[\sigma]$ iff $x \models \varphi[\sigma]$ for all $\varphi \in \Phi$.

$x \models \exists v_n \varphi[\sigma]$ iff   there is an x-cover C such that for all $z \in C$, $z \models \varphi[\sigma(d/n)]$ for some $d \in D$.

$x \models \forall v_n \varphi[\sigma]$ iff $x \models \varphi[\sigma(d/n)]$ for all $d \in D$.

In place of the last clause, the sheaf semantics (and Kripke's intuitionistic semantics) typically has

$$x \models \forall v_n \varphi[\sigma] \quad \text{iff} \quad x \triangleleft y \text{ implies } y \models \varphi[\sigma(d/n)] \text{ for all } d \in D.$$

But this follows from the last clause because the c-filter $\|\varphi[\sigma(d/n)]\|_\sigma^{\mathcal{M}}$ is $\triangleleft$-increasing. The semantics of $\to_l$ is sometimes given in the form

$$x \models \varphi \to_l \psi[\sigma] \quad \text{iff} \quad (x \cdot y \triangleleft z \text{ and } y \models \varphi[\sigma]) \text{ implies } z \models \psi[\sigma],$$

but this follows from the above because $\|\psi\|_\sigma^{\mathcal{M}}$ is $\triangleleft$-increasing. Similarly for $\varphi \to_r \psi$.

If we take the *classical* disjunction of Φ to be a formula that is satisfied at x precisely when some member of Φ is satisfied at x, then the above criterion for $x \models \bigvee \Phi$ is that the classical disjunction of Φ is *locally* satisfied at x, i.e. satisfied throughout some cover of x. Similarly, the criterion for $x \models \exists v_n \varphi$ is that the classical existential quantification of φ is locally satisfied at x.

6   Modalities

We now take a few steps in the direction of modal logic over quantales. A *modal operator* on a quantal or other poset will be taken to be any unary function j that is monotone. This can be used to give algebraic semantics to a new unary connective (modality) ∇ by defining $\|\nabla \varphi\| = j\|\varphi\|$. Structurally ∇ can be interpreted by adding to the definition of a model structure $\mathfrak{S}$ a new binary relation $\prec$ on S, and requiring that in a model based on $\mathfrak{S}$,

$$\mathcal{M}, x \models \nabla \varphi[\sigma] \quad \text{iff} \quad \text{for some } y, \; x \prec y \text{ and } \mathcal{M}, y \models \varphi[\sigma]. \tag{6.1}$$

The relation $\prec$ induces the modal operator $j_\prec$ on $\langle \mathcal{P}S, \subseteq \rangle$ having

$$j_\prec X = \{x \in S : \exists y (x \prec y \in X)\}. \tag{6.2}$$

Then the definition $\|\nabla \varphi\|_\sigma^{\mathcal{M}} = j_\prec \|\varphi\|_\sigma^{\mathcal{M}}$ ensures that (6.1) holds. But we also want to ensure that $j_\prec X$ is a c-filter whenever X is. This can be achieved by requiring that a model structure satisfies the following two conditions

$$\text{if } x \triangleleft y \text{ and } x \prec z, \text{ then for some } w, z \triangleleft w \text{ and } y \prec w. \qquad (6.3)$$

cov7: if there exists an x-cover included in $j_{\prec} X$, then there exists a y with $x \prec y$, and a y-cover included in X.

(6.3) states that $(\triangleright \circ \prec) \subseteq (\prec \circ \triangleright)$, where $\triangleright$ is the converse of $\triangleleft$, and suffices to make $j_{\prec} X$ $\triangleleft$-increasing if X is. (cov7) makes $j_{\prec} X$ cover-closed if X is. Then the quantale $\mathcal{Q}^{\mathfrak{S}}$ of c-filters is closed under $j_{\prec}$.

In the opposite direction, starting with a modal operator j on a quantale $\mathcal{Q}$, define a relation $\prec_j$ on the structure $\mathfrak{S}$ of Theorem 6 having $\mathcal{Q} \cong \mathcal{Q}^{\mathfrak{S}}$, by putting $x \prec_j y$ iff $x \leq jy$. This in turn induces the operator $j_{\prec_j}$ on $\mathcal{Q}^{\mathfrak{S}}$.

THEOREM 8. $\prec_j$ satisfies (6.3) and (cov7), and the isomorphism $x \mapsto (x{\downarrow})$ from $\mathcal{Q}$ to $\mathcal{Q}^{\mathfrak{S}}$ preserves the operators j and $j_{\prec_j}$.

Proof. For (6.3), if $x \triangleleft y$ and $x \prec_j z$, then $y \leq x \leq jz$, so $y \prec_j z$, hence (6.3) holds in the strong form that we can take $w = z$.

For (cov7), suppose there exists $C \in Cov(x)$ with $C \subseteq j_{\prec_j} X$. Hence $x \leq \sum C$. Let $B = \{z \in X : \exists c \in C(c \prec_j z)\}$, and put $y = \sum B$. Then $B \in Cov(y)$ and $B \subseteq X$, so it remains to show $x \prec_j y$. But if $c \in C$, then $c \in j_{\prec_j} X$, so there exists z with $c \prec_j z \in X$. Then $c \leq jz$ and $z \in B$, so $z \leq y$, hence $jz \leq jy$ as j is monotone, and thus $c \leq jy$. Therefore $x \leq \sum C \leq jy$, implying $x \prec_j y$.

Preservation of the operators requires that $(jx){\downarrow} = j_{\prec_j}(x{\downarrow})$. But if $z \in (jx){\downarrow}$, then $z \prec_j x \in (x{\downarrow})$, so $z \in j_{\prec_j}(x{\downarrow})$ by (6.2). And if $z \in j_{\prec_j}(x{\downarrow})$, then $z \prec_j y \leq x$ for some y, so $z \leq jy \leq jx$, giving $z \in (jx){\downarrow}$. ∎

We can now consider correspondences between properties of j and properties of $\prec$ in a manner familiar from modal logic. If $\prec$ is reflexive, then $j_{\prec}$ is inflationary, and if $\prec$ is transitive, then $j_{\prec} j_{\prec} X \subseteq j_{\prec} X$. Hence if $\prec$ is a preorder, $j_{\prec}$ is a *closure operator* (inflationary and idempotent). Conversely, if j is an inflationary modal operator on a quantale, then $\prec_j$ is reflexive, and if $j \circ j \leq j$ (i.e. $\forall x (jjx \leq jx)$), then $\prec_j$ is transitive. Thus

a closure operator on a quantale can be represented as the operator $j_{\prec}$ induced on the quantale of c-filters of a model structure by a preorder relation $\prec$.

Recall from (2.5) that a quantic nucleus is a closure operator satisfying $jx \bullet jy \leq j(x \bullet y)$. For $j_{\prec}$ to satisfy (2.5) it suffices that

$$\text{if } x \cdot y \triangleleft z, x \prec x' \text{ and } y \prec y', \text{ then } \exists z'(z \prec z' \text{ and } x' \cdot y' \triangleleft z'). \qquad (6.4)$$

Conversely, if j is a quantic nucleus, then (6.4) holds in the strong form that we can replace its conclusion by $x' \cdot y' \triangleleft z$. For if $z \leq x \bullet y$, $x \leq jx'$

and $y \leq jy'$, then $z \leq jx' \bullet jy' \leq j(x' \bullet y')$. So any quantic nucleus can be represented as the operator induced by a preorder satisfying this strong form of (6.4).

To axiomatise the logic characterised by the class of quantal models with a modality, we just add the rule

$$\varphi \vdash \psi \text{ implies } \nabla\varphi \vdash \nabla\psi$$

to the proof theory of Section 3.3. Then putting $j|\varphi| = |\nabla\varphi|$ gives a well-defined modal operator on the Lindenbaum posemigroup $\mathcal{S}^{\mathcal{F}}$ of a fragment as in Section 3.4. Viewing $\mathcal{S}^{\mathcal{F}}$ as a subalgebra of the quantale $\mathcal{Q}^{\mathcal{F}}$, we can lift j to an operator j^+ on $\mathcal{Q}^{\mathcal{F}}$ by putting

$$j^+ x = \sum\{ja \in \mathcal{S}^{\mathcal{F}} : a \leq x\}.$$

j^+ agrees with j when restricted to $\mathcal{S}^{\mathcal{F}}$, and is monotone so can be represented as the operator induced by $\prec_{j^+}$.

If the axiom $\varphi \vdash \nabla\varphi$ is added, then j is inflationary, hence so is j^+ (proof: $x = \sum\{a \in \mathcal{S}^{\mathcal{F}} : a \leq x\} \leq \sum\{ja : a \leq x\}$). The axiom $\nabla\nabla\varphi \vdash \nabla\varphi$ enforces $j \circ j \leq j$, but this appears only to lift to j^+ if j^+ preserves joins, yielding $j^+ j^+ x = \sum\{jja : a \leq x\}$. On the other hand the nucleus condition (2.5) does lift to j^+, as shown by the following calculation, in which $a, b, c \in \mathcal{S}^{\mathcal{F}}$:

$$\left(\sum_{a \leq x} ja\right) \bullet \left(\sum_{b \leq y} jb\right) = \sum_{a \leq x, b \leq y} (ja \bullet jb) \leq \sum_{a \leq x, b \leq y} j(a \bullet b) \leq \sum_{c \leq x \bullet y} jc.$$

Here the equality is given by the quantale distribution laws (2.1), (2.2), the first inequality by (2.5) for j, and the second by $a \bullet b \leq x \bullet y$.

To give a completeness theorem for the logic of closure operators on quantales, a different lifting of j to $\mathcal{Q}^{\mathcal{F}}$ can be used, namely

$$j^{\#} x = \prod\{ja : a \in \mathcal{S}^{\mathcal{F}} \text{ and } x \leq ja\}.$$

Interestingly, this $j^{\#}$ is a closure operator for any j whatsoever, but its restriction to $\mathcal{S}^{\mathcal{F}}$ agrees with j (equivalently, the embedding of $\mathcal{S}^{\mathcal{F}}$ into $\mathcal{Q}^{\mathcal{F}}$ preserves j and $j^{\#}$) precisely when j itself is a closure operator, facts that the reader may like to verify.

On the other hand it does not appear that (2.5) lifts to $j^{\#}$ in general, so an axiomatisation of the modal logic of quantic nuclei on quantales awaits further investigation.

7 Conclusion and Further Work

The aim of this paper has been to extend the idea of Kripke-Joyal semantics to a generalisation of the intuitionistic context, and a minimal syntax has been used for this purpose. There are many possible additions and extensions that could be considered, and further questions that suggest themselves. On the syntactic side we have not considered the modelling of individual constants, function symbols, the equality predicate, or true and false propositional constants ⊤ and ⊥ (although for the latter we could use $\bigvee \emptyset$ and $\bigwedge \emptyset$). Semantically we used a kind of "constant domain" model, with a single set D forming the range of quantifiable variables. It may be of interest to explore the variable-domain approach of Kripke's intuitionistic models, in which each element x of a model structure has its own domain D_x that is used in evaluating quantifiers at x.

Proof-theoretically we used simple sequents with a single formula on each side of the symbol ⊢, so there is scope for discussion of something more like the usual Gentzen calculi. There are additional axioms required for the various kinds of quantale that have been considered. It is often assumed that a quantale is *unital*, i.e. there is an identity element for •, and this requires a new propositional constant E, and a distinguished c-filter in model structures to serve as $\|E\|^{\mathcal{M}}$ and be the identity element in $\mathcal{Q}^{\mathfrak{S}}$. Then there are quantales that are right-sided ($\varphi \& \top \vdash \varphi$), left-sided ($\top \& \varphi \vdash \varphi$), and idempotent ($\varphi \& \varphi \dashv\vdash \varphi$). The combination of axioms $\varphi \& \psi \vdash \varphi$, $\varphi \& \psi \vdash \psi$ and $\varphi \vdash \varphi \& \varphi$ force $a \bullet b$ to be the meet of a and b, hence $\|\varphi \& \psi\|^{\mathcal{M}} = \|\varphi\|^{\mathcal{M}} \cap \|\psi\|^{\mathcal{M}} = \|\varphi \wedge \psi\|^{\mathcal{M}}$.

The *involutary* quantales have an involution a^* that preserves joins and reverses the arguments of •, and amongst these the *Hilbert* quantales have an orthocomplement $a^\perp$ that obeys De Morgan's laws in relation to $\bigvee$ and $\bigwedge$. The enhancement of our model structures to interpret unary connectives corresponding to a^* and $a^\perp$ is a natural topic for investigation.

As to modal logic, a quantic nucleus j seems rather schizophrenic as a modal operator. Its properties as a closure operator suggest it should model a "diamond" modality with the bounded-existential modelling condition of (6.1), as we have done here. But in locales the nucleus property (2.5) implies that $j(a \bullet b)$ is the lattice meet of ja and jb, which is a property more reminiscent of the "box" modalities that have the bounded-universal semantics

$$\mathcal{M}, x \models \nabla \varphi[\sigma] \quad \text{iff} \quad \text{for all } y, \; x \prec y \text{ implies } \mathcal{M}, x \models \varphi[\sigma]. \tag{7.1}$$

Indeed in [8] we gave a Kripke semantics for the finitary modal logic of nuclei on locales, using this kind of semantics. So there is more to be explored here, including the general study of modalities fulfilling (7.1), regardless of whether they interpret quantic nuclei.

There are intimate connections between cover systems and quantic nuclei. From any preordered semigroup $\langle S, \triangleleft, \cdot \rangle$ we obtain a quantale $\langle S^\triangleleft, \subseteq, \bullet \rangle$ in which $S^\triangleleft$ is the set of ⊲-increasing subsets, $X \bullet Y = [X \cdot Y]$, and joins are set-unions. We saw in the proof of Theorem 5 that from any system Cov on S satisfying (cov1)–(cov4) we get a quantic nucleus j_{Cov} on $S^\triangleleft$ by putting $j_{Cov}(X) = \{x \in S : (\exists C \in Cov(x)) C \subseteq X\}$. But conversely, from any quantic nucleus j on $S^\triangleleft$ we get a system Cov_j satisfying (cov1)–(cov4) by putting $Cov_j(x) = \{C \in S^\triangleleft : x \in j(C)\}$. Then $j_{Cov_j} = j$. If $Cov(x)$ is always a subset of $S^\triangleleft$ closed under supersets in $S^\triangleleft$ (as indeed $Cov_j(x)$ is), then $Cov_{j_{Cov}} = Cov$. These relationships, and the general theory of cover systems for quantales, warrant further investigation.

Finally we raise the question of the separate study of quantal *geometric* formulas, which are those formed from atomic ones using only $\&$, $\bigvee$ and $\exists$. It would be of interest to axiomatise the class of sequents of geometric formulas that are valid in quantal models in the sense of (5.1). Our completeness method depended on the the presence of the residuals $\Rightarrow_l$ and $\Rightarrow_r$ in a Lindenbaum posemigroup $\mathcal{S}^{\mathcal{F}}$ to construct the quantale $\mathcal{Q}^{\mathcal{F}}$ by the method of Corollary 2. It is not clear that the approach would work for Lindenbaum algebras generated by a language without $\to_l$ and $\to_r$. It may be necessary to use more general sequents of lists of formulas to achieve this. An alternative approach to completeness might be to use the standard Henkin method to build "canonical" models whose points are sets of formulas with syntactic closure properties that mimic those of the "truth sets" $\{\varphi : \mathfrak{A}, x \models \varphi[\sigma]\}$ defined by points in models under valuations. Of course the construction of such canonical models would be of interest in general, and not just for geometric formulas.

BIBLIOGRAPHY

[1] Jon Barwise. *Admissible Sets and Structures*. Springer-Verlag, Berlin, Heidelberg, 1975.

[2] J. L. Bell and A. B. Slomson. *Models and Ultraproducts*. North-Holland, Amsterdam, 1969.

[3] John L. Bell. Cover schemes, frame-valued sets and their potential uses in spacetime physics. In Albert Reimer, editor, *Spacetime Physics Research Trends, Horizons in World Physics*, volume 248. Nova Science Publishers, 2006. Manuscript at http://publish.uwo.ca/~jbell.

[4] John L. Bell. Infinitary logic. In Edward N. Zalta, editor, *The Stanford Encyclopedia of Philosophy*.
http://plato.stanford.edu/archives/spr2006/entries/logic-infinitary/, Spring 2006.

[5] B. A. Davey and H. A. Priestley. *Introduction to Lattices and Order*. Cambridge University Press, 1990.

[6] Kosta Došen. Sequent systems and groupoid models II. *Studia Logica*, 48(1):41–65, 1989.

[7] Kosta Došen. A brief survey of frames for the Lambek calculus. *Zeitschrift für Mathematische Logik und Grundlagen der Mathematik*, 38:179–187, 1992.

[8] Robert Goldblatt. Grothendieck topology as geometric modality. *Zeitschrift für Mathematische Logik und Grundlagen der Mathematik*, 27:495–529, 1981. Reprinted in [9].

[9] Robert Goldblatt. *Mathematics of Modality*. CSLI Lecture Notes No. 43. CSLI Publications, Stanford, California, 1993. Distributed by Chicago University Press.

[10] P. T. Johnstone. *Stone Spaces*. Cambridge University Press, 1982.

[11] Saunders Mac Lane and Ieke Moerdijk. *Sheaves in Geometry and Logic : A First Introduction to Topos Theory*. Springer-Verlag, 1992.

[12] H. M. MacNeille. Partially ordered sets. *Transactions of the American Mathematical Society*, 42:416–460, 1937.

[13] Christopher J. Mulvey. &. *Rendiconti Circ. Mat. Palermo*, 12:99–104, 1986. Manuscript at www.maths.sussex.ac.uk/Staff/CJM.

[14] Christopher J. Mulvey and Joan Wick Pelletier. On the quantisation of points. *Journal of Pure and Applied Algebra*, 159:231–295, 2001.

[15] Christopher J. Mulvey and Joan Wick Pelletier. On the quantisation of spaces. *Journal of Pure and Applied Algebra*, 175:289–325, 2002.
[16] Christopher J. Mulvey and Pedro Resende. A noncommutative theory of Penrose tilings. *Internat. J. Theoret. Phys.*, 44(6):655–689, 2005.
[17] Susan B. Niefield and Kimmo I. Rosenthal. Constructing locales from quantales. *Mathematical Proceedings of the Cambridge Philosophical Society*, 104:215–234, 1988.
[18] Hiroakira Ono. Semantics for substructural logics. In Peter Schroeder-Heister and Kosta Došen, editors, *Substructural Logics*, pages 259–291. Oxford University Press, 1993.
[19] Hiroakira Ono and Yuichi Komori. Logics without the contraction rule. *The Journal of Symbolic Logic*, 50(1):169–201, 1985.
[20] Kimmo I. Rosenthal. *Quantales and Their Applications*, volume 234 of *Pitman Research Notes in Mathematics*. Longman Scientific & Technical, 1990.
[21] A. S. Troelstra. *Lectures on Linear Logic*. CSLI Lecture Notes No. 29. CSLI Publications, Stanford, California, 1992.

Robert Goldblatt
Centre for Logic, Language and Computation
Victoria University of Wellington, New Zealand
Rob.Goldblatt@vuw.ac.nz

A General Semantics for Quantified Modal Logic[1]

ROBERT GOLDBLATT AND EDWIN D. MARES

ABSTRACT. In [9] we developed a semantics for quantified relevant logic that uses general frames. In this paper, we adapt that model theory to treat quantified modal logics, giving a complete semantics to the quantified extensions, both with and without the Barcan formula, of every propositional modal logic S. If S is *canonical* our models are based on propositional frames that validate S. We employ frames in which not every set of worlds is an admissible proposition, and an alternative interpretation of the universal quantifier using greatest lower bounds in the lattice of admissible propositions. Our models have a fixed domain of individuals, even in the absence of the Barcan formula.

For systems with the Barcan formula it is possible to preserve the usual Tarskian reading of the quantifier, at the expensive of sometimes losing validity of S in the underlying propositional frames. We apply our results to a number of logics, including S4.2, S4M and KW, whose quantified extensions are incomplete for the standard semantics.

Keywords: Quantified modal logic, general frame, Tarskian, canonical logic, completeness, incompleteness, Barcan formula.

1 Introduction

A general semantics for quantified relevant logic is developed in [14]. In this paper, we adapt this to treat quantified modal logic, providing a complete semantics for the standard predicate extension QS of any propositional modal logic S, as well as for the logic QSB obtained by adding the Barcan formula to QS. In particular we provide complete semantics for modal predicate logics known to be incomplete for their usual possible worlds model theories. We employ *general* frames, in which not every set of worlds is admissible as a proposition, and use the set of admissible propositions to give an alternative interpretation of the universal quantifier $\forall$. Our results are applied to a number of logics

The idea behind the semantics for the universal quantifier is the following. Suppose that a formula A has at most the variable x free. The formula $\forall xA$ is true at a world a if and only if there is some proposition X such that X entails every instantiation of A and X obtains at a. This needs some explanation. A proposition on our theory is a set of worlds. A proposition

---

[1]Supported by grant 05-VUW-079 from the Marsden Fund of the Royal Society of New Zealand.

obtains at a world if that world is in that proposition. A proposition X entails a formula A if every world in X makes A true. More formally we have the following:

$$\mathcal{M}, a \models_f \forall x_n A \text{ iff there is a proposition } X \text{ such that } a \in X \text{ and}$$
$$X \subseteq |A|^{\mathcal{M}}_{f[j/n]} \text{ for all } j \in I.$$

A bit more explanation is in order. Here $\mathcal{M}$ is a model, a is a world, and f is an assignment to individual variables taking values in a domain I of individuals. The notation $f[j/n]$ refers to the function that is just like f except that it assigns the individual j to x_n. $|A|^{\mathcal{M}}_{f[j/n]}$ is the set of worlds at which formula A is true under $f[j/n]$, representing the proposition expressed by A when the variable x_n is instantiated to j. Note that the proposition X need not be $|\forall x_n A|^{\mathcal{M}}_f$ itself, so this is not the standard semantics for $\forall$.

The integrating of general propositions into our semantics for relevant logics can be justified by our attempt to obtain a theory of partial information. A general proposition is the information that informs one of the truth of a universal statement. But there are other philosophical projects that utilize general propositions. Recently David Armstrong has proposed his *truth-maker principle,* that every true statement is made true by some truth-maker (in the case of Armstrong, truth-makers are facts) (see [1] and [2]). Thus, to any true universal statement there must be some fact that corresponds to it – a general fact. In our parlance, a fact is just a true proposition. One way of understanding the model theory that we present in the current paper is as a modal theory of truth-makers that includes general propositions.

This paper shows that we can produce a complete semantics for any logic that results from adding the standard axioms and rules for quantification to a propositional modal logic. That there is a class of general frames over which these logics are complete is hardly surprising. General frames are essentially algebras in represented form, and every such logic has an algebraic semantics over which it is complete (i.e. the class consisting in just the Lindenbaum algebra of that logic). There is, however, something much more interesting about the present semantics. As we show, *for any* **canonical** *modal logic S, its quantified extension QS is complete over a class of general frames for which the underlying propositional frames are just the S-frames.* This means that our semantics for quantified logics just "sits gingerly on top of" the semantics for the corresponding canonical propositional logic.

We also explore some consequences of adding the Barcan formula to any of our logics QS. We show that the resulting system QSB is complete over a special class of our frames that we call "Tarskian", in which $\forall$ gets its classical reading of

$$\mathcal{M}, a \models_f \forall x_n A \quad \text{iff} \quad \text{for all } j \in I, \mathcal{M}, a \models_{f[j/n]} A.$$

But the price paid for this is that the underlying propositional frames of these Tarskian general frames may fail to validate S. On the other hand we also have another complete class of general frames for QSB which may be based on propositional frames for S while giving a non-Tarskian reading for ∀, *and yet still validating the Barcan formula*. This gives a new perspective on the semantics of the Barcan formula. Instead of seeing it as the axiom corresponding to constant domain models – indeed our models have a fixed domain I of individuals even for systems without the Barcan formula – we see it as the principle needed to ensure that we can confine ourselves to models that give the standard Tarskian interpretation of ∀. But at a price.

The range of possibilities here is illustrated by reference to models of a number of well-known logics whose quantified extensions are incomplete for the standard semantics, including S4.2, S4+ the McKinsey axiom, and the provability logic KW.

2 Logics

We use two languages in this paper. The first is a standard propositional modal language, $\mathcal{L}$, that includes an infinite set of propositional variables, the connectives ⊃, □, and ⊥, and parentheses. This language has the standard formation rules and the connectives ¬, ∧, ∨, ≡, and ◇ are defined in the usual way. We use lower case letters from the first part of the Greek alphabet as metavariables that range over formulas of $\mathcal{L}$.

The second language is a standard predicate language, $\mathcal{LQ}$, which contains a countable list of individual variables $(x_0, x_1, ...)$; some or all of: individual constants, predicate letters and function symbols; the same connectives as $\mathcal{L}$; and the universal quantifier ∀. The existential quantifier is defined in the usual way. We use upper case letters from the first part of the Roman alphabet as metavariables that range over formulas of $\mathcal{LQ}$.

Our paper concerns a class of normal propositional modal logics and their predicate extensions. Each of these propositional logics is a set of $\mathcal{L}$-formulas that includes the theorems of the logic K, that is, it contains all substitution instances of the theorems of the propositional calculus and all instances of the K-schema ($\Box(\alpha \supset \beta) \supset (\Box\alpha \supset \Box\beta)$), as well as being closed under the rules,

$$\text{MP} \; \frac{\vdash \alpha \supset \beta, \; \vdash \alpha}{\vdash \beta}, \qquad \text{N} \; \frac{\vdash \alpha}{\vdash \Box\alpha},$$

and the rule of uniform substitution.

Each of the predicate logics fulfils this same description but for $\mathcal{LQ}$-formulas, and includes all instances of the schema

(UI) $\quad \forall x A \supset A[\tau/x]$, where x is free for term τ in A,

and is also closed under the rule of restricted generalization[1], viz.,

---

[1] In [14], we called the relevant version of this rule "RIC", for the "rule of intensional confinement". This name was useful in that context in order to link it with the schema $\forall x(A \rightarrow B) \rightarrow (A \rightarrow \forall x B)$, where x is not free in A. This schema is usually called "confinement" and can be derived in the logic R from RIC. Moreover, we wished to distinguish this sort of confinement from the more "extensional" variety that we find in the schema $\forall x(A \vee B) \rightarrow (A \vee \forall x B)$, where x is not free in A. This latter schema is not derivable in our base system QR.

$$\text{RGen} \quad \frac{\vdash A \supset B}{\vdash A \supset \forall x B}, \text{ where } x \text{ is not free in } A.$$

Where S is a propositional modal logic as just defined, we call QS the smallest predicate logic that contains all the $\mathcal{L}Q$-substitution-instances of theorems of S.

We assume from the outset that $\mathcal{L}Q$ has infinitely many individual constants. For otherwise we could add such constants in a way that is conservative with respect to our finitary proof theory, as is well known.

We also assume that the set of propositional variables of $\mathcal{L}$ is at least as large as the set of $\mathcal{L}Q$-formulas. This will be used in the proof of Lemma 5 (see also the Appendix).

3 Frames and Models

As usual a frame for propositional modal logic is a pair $\mathcal{F} = \langle W, R \rangle$, where W is a non-empty set (of "worlds") and R is a binary relation on W. A *model* $\mathcal{M} = \langle W, R, v \rangle$ on $\mathcal{F}$ is given by a valuation v mapping each propositional variable to a subset of W. Each model determines a satisfaction relation $\mathcal{M}, a \models \alpha$, expressing truth/satisfaction of α at world a, as follows:

- $\mathcal{M}, a \models p$ iff $a \in v(p)$, for all propositional variables p;
- $\mathcal{M}, a \models \alpha \supset \beta$ iff $\mathcal{M}, a \not\models \alpha$ or $\mathcal{M}, a \models \beta$;
- $\mathcal{M}, a \not\models \bot$;
- $\mathcal{M}, a \models \Box \alpha$ iff $\forall b \in W(aRb$ implies $\mathcal{M}, b \models \alpha)$.

A formula A is *valid in* $\mathcal{M}$ if and only if it is satisfied at every world in $\mathcal{M}$, and *valid in* $\mathcal{F}$ if and only if it is valid in every model on $\mathcal{F}$. A logic S is *sound* over a class of frames if and only if every theorem of S is valid in that class of frames (i.e. valid in all members of the class). S is *complete* for a class when every formula valid in the class is an S-theorem; and *characterised* by a class when it both sound over and complete for that class. We say that a frame $\mathcal{F}$ is an *S-frame* if S is sound over $\{\mathcal{F}\}$.

Recall that $\langle W', R' \rangle$ is a *generated subframe* of $\langle W, R \rangle$ if it is a substructure of $\langle W, R \rangle$ that is R-closed in the sense that if $a \in W'$ and aRb, then $b \in W'$. In this case, any formula valid in $\langle W, R \rangle$ will be valid also in $\langle W', R' \rangle$.

A given frame $\langle W, R \rangle$ determines a function $[R]: \wp W \to \wp W$, where $\wp W$ is the powerset of W, having

$$[R]X = \{a \in W : \forall b \in W(aRb \text{ implies } b \in X)\}.$$

Then if $|\alpha|^{\mathcal{M}} = \{a : \mathcal{M}, a \models \alpha\}$ is the "truth set" of α defined by $\mathcal{M}$, we see that $[R]|\alpha|^{\mathcal{M}} = |\Box \alpha|^{\mathcal{M}}$. Also, if $X \Rightarrow Y = (W \backslash X) \cup Y$, then $|\alpha|^{\mathcal{M}} \Rightarrow |\beta|^{\mathcal{M}} = |\alpha \supset \beta|^{\mathcal{M}}$.

The operations $[R]$ and $\Rightarrow$ on $\wp W$ can be lifted to operations on functions of the form $I^\omega \to \wp W$, where I is any given set. If φ and ψ are two such functions, we define functions $[R]\varphi$ and $\varphi \Rightarrow \psi$ of the same form by putting $([R]\varphi)f = [R](\varphi f)$ and $(\varphi \Rightarrow \psi)f = (\varphi f \Rightarrow \psi f)$ for all $f \in I^\omega$.

These operations will be used in defining frames for predicate logics. For this purpose, fix a set $Prop \subseteq \wp W$. Then $Prop$ determines an operation $\sqcap : \wp\wp W \to \wp W$ defined, for each $S \subseteq \wp W$, by putting

$$\sqcap S = \bigcup \{X \in Prop : X \subseteq \bigcap S\}.$$

In general $\sqcap S \subseteq \bigcap S$, so $\sqcap S$ is a lower bound of S in the partially-ordered set $(\wp W, \subseteq)$. If $S \subseteq Prop$ and $\sqcap S \in Prop$, then $\sqcap S$ is the *greatest lower bound* of S in $(Prop, \subseteq)$. Moreover, if $\bigcap S \in Prop$, then $\sqcap S = \bigcap S$. In particular, this holds when $Prop = \wp W$. We emphasise that the definition of $\sqcap$ depends on the particular set $Prop$, and it may be that $\sqcap S \notin Prop$ for some $S \subseteq Prop$.

We use $\sqcap$ to define functions $\forall_n$ that interpret the quantifiers $\forall x_n$ for each $n \in \omega$. Where $f \in I^\omega$, we write $f[j/n]$ to mean the function g such that $g(m) = f(m)$ for all $m \neq n$ and $g(n) = j$. Now, for any $\varphi : I^\omega \longrightarrow \wp W$, we set

$$(\forall_n \varphi)f = \sqcap_{j \in I} \varphi(f[j/n]).$$

A *quantified general frame* (QG-frame) $\mathcal{F} = \langle W, R, I, Prop, PropFun \rangle$ is a structure in which W is a non-empty set, R is a binary relation on W, I is a non-empty set (of "individuals"), $Prop$ is a set of subsets of W, and $PropFun$ is a set of *propositional functions*, i.e. functions from I^ω to $Prop$, such that the following conditions hold:

CProp: $\emptyset \in Prop$, and if X and Y are in $Prop$, then so are $X \Rightarrow Y$ and $[R]X$. Hence $Prop$ is a modal algebra of subsets of W.

CFalse: The constant function $\varphi_\emptyset \in PropFun$, where $\varphi_\emptyset(f) = \emptyset$ for all $f \in I^\omega$.

CImp: If $\varphi, \psi \in PropFun$, then $\varphi \Rightarrow \psi \in PropFun$.

CMod: If $\varphi \in PropFun$, then $[R]\varphi \in PropFun$.

CAll: If $\varphi \in PropFun$, then $\forall_n \varphi \in PropFun$ for all $n \in \omega$.

A *valuation* for the language $\mathcal{L}Q$ on such a frame $\mathcal{F}$ is a function V that assigns to each individual constant c an element $V(c)$ of I, to each n-ary function symbol F a function $V(F) : I^n \to I$, and to each n-ary predicate letter P a function $V(P) : I^n \to Prop$. Put $\mathcal{M} = \langle \mathcal{F}, V \rangle$. Then each $\mathcal{L}Q$-term τ is interpreted by $\mathcal{M}$ as a function $\tau^\mathcal{M} : I^\omega \to I$, defined by putting $c^\mathcal{M} f = V(c)$ for each constant c; $(x_n)^\mathcal{M} f = fn$ for each variable x_n (so f functions as a *variable-assignment*); and inductively, $(F\tau_1 \ldots \tau_n)^\mathcal{M} f = V(F)(\tau_1^\mathcal{M} f, \ldots, \tau_n^\mathcal{M} f)$. Each *atomic* formula $P\tau_1 \ldots \tau_n$ gets interpreted as a function $|P\tau_1 \ldots \tau_n|^\mathcal{M} : I^\omega \to Prop$ defined by

$$|P\tau_1 \ldots \tau_n|^\mathcal{M}(f) = V(P)(\tau_1^\mathcal{M} f, \ldots, \tau_n^\mathcal{M} f).$$

The pair $\mathcal{M} = \langle \mathcal{F}, V \rangle$ is called a *model* on the QG-frame $\mathcal{F}$ if:

$|A|^{\mathcal{M}}$ belongs to *PropFun* for all *atomic* formulas A.

Each model has a truth/satisfaction relation $\mathcal{M}, a \models_f A$ between worlds $a \in W$, variable-assignments $f \in I^\omega$ and formulas A. This has associated truth sets $|A|_f^{\mathcal{M}} =_{df} \{b \in W : \mathcal{M}, b \models_f A\}$.[2] The inductive definition of $\mathcal{M}, a \models_f A$ is as follows:

- $\mathcal{M}, a \models_f P\tau_1 \ldots \tau_n$ if and only if $a \in |P\tau_1 \ldots \tau_n|^{\mathcal{M}}(f)$;
- $\mathcal{M}, a \models_f A \supset B$ iff $\mathcal{M}, a \not\models_f A$ or $\mathcal{M}, a \models_f B$;
- $\mathcal{M}, a \not\models_f \bot$;
- $\mathcal{M}, a \models_f \Box A$ iff $\forall b \in W(aRb$ implies $\mathcal{M}, b \models_f A)$.
- $\mathcal{M}, a \models_f \forall x A$ iff there is an $X \in Prop$ such that $X \subseteq \bigcap_{j \in I} |A|_{f[j/n]}^{\mathcal{M}}$ and $a \in X$.

The assignment $f \mapsto |A|_f^{\mathcal{M}}$ gives a function $|A|^{\mathcal{M}} : I^\omega \to \wp W$ for each formula A. These functions satisfy

- $|\bot|^{\mathcal{M}} = \varphi_\emptyset$;
- $|A \supset B|^{\mathcal{M}} = |A|^{\mathcal{M}} \Rightarrow |B|^{\mathcal{M}}$;
- $|\Box A|^{\mathcal{M}} = [R]|A|^{\mathcal{M}}$;
- $|\forall x_n A|^{\mathcal{M}} = \forall_n |A|^{\mathcal{M}}$.

These properties, and the closure properties of *PropFun*, ensure that we always have $|A|^{\mathcal{M}} \in PropFun$ and hence $|A|_f^{\mathcal{M}} \in Prop$, for any formula A. The satisfaction clause for $\forall x_n$ can be expressed as

$$|\forall x_n A|_f^{\mathcal{M}} = \prod_{j \in I} |A|_{f[j/n]}^{\mathcal{M}},$$

showing that $|\forall x_n A|_f^{\mathcal{M}}$ is the greatest lower bound of $\{|A|_{f[j/n]}^{\mathcal{M}} : j \in I\}$ in the partially ordered set $(Prop, \subseteq)$. This is the natural interpretation of $\forall$ in algebraic semantics. To reproduce the standard Tarskian semantics for $\forall$ we would need this greatest lower bound to be $\bigcap_{j \in I} |A|_{f[j/n]}^{\mathcal{M}}$. But this need not be so (see Section 6).

Taking $\exists x_n$ as abbreviating $\neg \forall x_n \neg$, we find that

$$|\exists x_n A|_f^{\mathcal{M}} = \bigsqcup_{j \in I} |A|_{f[j/n]}^{\mathcal{M}},$$

where $\bigsqcup$ is the operation defined, for $S \subseteq \wp W$, by

$$\bigsqcup S = \bigcap \{X \in Prop : \bigcup S \subseteq X\}.$$

---
[2] The symbols $\tau^{\mathcal{M}} f$ and $|A|_f^{\mathcal{M}}$ were written $Vf\tau$ and $|A|_{Vf}$ in [14].

For $S \subseteq Prop$, if $\bigsqcup S$ belongs to $Prop$ then it is the least upper bound of S in the partial-ordering $(Prop, \subseteq)$.

Now by a simple induction we can show that $|A|_f^{\mathcal{M}} = |A|_g^{\mathcal{M}}$ whenever f and g agree on all free variables of A [14, Lemma 4.4]. Hence if x_n is not free in A, $|A|_f^{\mathcal{M}} = |A|_{f[j/n]}^{\mathcal{M}}$ for all $j \in I$.

It can also be shown that if term τ is free for x_n in A, and $g = f[\tau^{\mathcal{M}} f/n]$, then $|A[\tau/x_n]|_f^{\mathcal{M}} = |A|_g^{\mathcal{M}} = |A|_{f[\tau^{\mathcal{M}} f/n]}^{\mathcal{M}}$ [14, Lemma 7.1].

LEMMA 1. *If $\mathcal{M}$ is any model on a QG-frame $\mathcal{F}$, then*

(i) *The K-schema $\Box(A \supset B) \supset (\Box A \supset \Box B)$ is valid in $\mathcal{M}$.*

(ii) *if A is valid in $\mathcal{M}$, then so is $\Box A$.*

(iii) *if $A \supset B$ and A are valid in $\mathcal{M}$, then so is B.*

(iv) *UI is valid is $\mathcal{M}$.*

(v) *if $A \supset B$ is valid in $\mathcal{M}$ and x is not free in A, then $A \supset \forall x B$ is valid in $\mathcal{M}$.*

Proof. (i)–(iii) are standard and straightforward. For (iv), if $\mathcal{M}, a \models_f \forall x_n A$, then the semantics of $\forall$ ensures that $\mathcal{M}, a \models_{f[j/n]} A$ for all $j \in I$. In particular $\mathcal{M}, a \models_{f[\tau^{\mathcal{M}} f/n]} A$, and hence $\mathcal{M}, a \models_f A[\tau/x_n]$ by the last observation before the Lemma.

For (v), if $A \supset B$ is valid in $\mathcal{M}$, then $|A|_g^{\mathcal{M}} \subseteq |B|_g^{\mathcal{M}}$ for any g. Thus if x_n is not free in A, for any f we get $|A|_f^{\mathcal{M}} = |A|_{f[j/n]}^{\mathcal{M}} \subseteq |B|_{f[j/n]}^{\mathcal{M}}$ for all $j \in I$. Hence if $\mathcal{M}, a \models_f A$, then $x \in |A|_f^{\mathcal{M}} \subseteq \bigcap_{j \in I} |B|_{f[j/n]}^{\mathcal{M}}$. Since $|A|_f^{\mathcal{M}} \in Prop$, this implies $\mathcal{M}, a \models_f \forall x_n B$. Thus $A \supset \forall x_n B$ is valid in $\mathcal{M}$. ∎

A QG-frame $\langle W, R, I, Prop, PropFun \rangle$ will be said to be *based on* the propositional frame $\langle W, R \rangle$. Where $\mathcal{C}$ is a class of propositional frames, $Q(\mathcal{C})$ is the class of QG-frames based on members of $\mathcal{C}$.

THEOREM 2. *If $\mathcal{C}$ is a class of S-frames then QS is sound over $Q(\mathcal{C})$.*

Proof. In view of Lemma 1, it is enough to show that every substitution instance of a theorem of S is valid over $Q(\mathcal{C})$. Suppose that α is an S-theorem and that $p_1, \ldots, p_n$ are all the propositional variables that occur in α. We show that $\alpha[B_1/p_1, \ldots, B_n/p_n]$ is valid over $Q(\mathcal{C})$ for any $\mathcal{LQ}$-formulas $B_1, \ldots, B_n$.

Let $\mathcal{M}$ be any model on a QG-frame based on some S-frame $\langle W, R \rangle \in \mathcal{C}$. Then α is valid in $\langle W, R \rangle$. Given $f \in I^{\omega}$, define a propositional model $\mathcal{M}_f = \langle W, R, v \rangle$ by putting $v(p_i) = |B_i|_f^{\mathcal{M}}$ for all $i \leq n$ (and $v(p)$ can be arbitrary otherwise). Thus

$$\mathcal{M}_f, a \models p_i \quad \text{iff} \quad \mathcal{M}, a \models_f B_i$$

for all $a \in W$ and $i \leq n$. A simple induction then shows that

$$\mathcal{M}_f, a \models \beta \quad \text{iff} \quad \mathcal{M}, a \models_f \beta[B_1/p_1, \ldots, B_n/p_n]$$

in general, where β is any $\mathcal{L}$-formula whose variables are among $p_1, \ldots, p_n$. In particular, when $\beta = \alpha$ we get $\mathcal{M}, a \models_f \alpha[B_1/p_1, \ldots, B_n/p_n]$ as required, since α is valid in $\mathcal{M}_f$. ∎

REMARK 3. The proof of this Theorem shows that if an $\mathcal{L}$-formula α is valid in a frame $\langle W, R \rangle$, then every $\mathcal{LQ}$-substitution instance of α is valid in every QG-frame based on $\langle W, R \rangle$. ∎

Historical Remarks

Our notion of QG-frame combines two prior notions. The first, concerning *Prop*, is the notion of a *general* frame for propositional modal logic, in which not every set of worlds is admissible as a proposition. Frames of the form $\langle W, R, Prop \rangle$ were introduced by S. K. Thomason [22], and their mathematical theory systematically developed in the first author's thesis [7].

The second precursor, concerning *PropFun*, is Halmos' notion of a *functional polyadic algebra* (see [12]). This is an algebra of "propositional" functions of the form $I^{\mathcal{V}} \to \mathbf{B}$ that is closed under operations corresponding to the standard connectives and quantifiers. Here $\mathbf{B}$ is a Boolean algebra, thought of as a collection of propositions, I is a domain of individuals, and $\mathcal{V}$ is a set of "variables". The operations corresponding to the quantifiers are defined using products (greatest lower bounds) and sums (least upper bounds) in $\mathbf{B}$. For QG-frames we have taken $\mathbf{B}$ to be a Boolean algebra of subsets of a Kripke frame $\langle W, R \rangle$, with the operation $[R]$ corresponding to $\Box$. But it is crucial to realise that even in this set-theoretic context, in which a binary product $X \sqcap Y$ is just the intersection $X \cap Y$, an infinite product $\bigsqcap S$ need not be the intersection $\bigcap S$, but in general is the operation that we have defined. This yields our alternative set-theoretic interpretation of the quantifier ∀.

Apparently it was Andrjez Mostowski [16] who introduced the method of interpreting an n-ary predicate as a lattice-valued function $I^n \to \mathbf{L}$, with the quantifiers interpreted by products and sums in $\mathbf{L}$. He took $\mathbf{L}$ to be a complete Brouwerian lattice for the application he was interested in and raised the question of whether his approach provided a complete semantics for intuitionistic logic. The idea was taken up by Rasiowa and Sikorski, who gave their famous algebraic proof of Gödel's completeness theorem for classical first-order logic [17], and systematically developed this method into an algebraic semantics for superintuitionistic predicate logics and for extensions of QS4 [18].

4 Canonical Frames and Models

In this section we remind the reader about certain facts concerning canonical models for propositional modal logics and give our construction of general canonical models for quantified logics.

Recall that for any logic L (propositional or quantified) a set Γ of formulas of the language of L is *L-consistent* if there are no formulas $A_1, ..., A_n \in \Gamma$ such that $(A_1 \wedge \cdots \wedge A_n) \supset \bot$ is an L-theorem. A *maximally L-consistent* set is one that is L-consistent and has no proper L-consistent extensions, or equivalently, is L-consistent and contains one of A and $\neg A$ for all formulas A. The intersection of the class of all maximally L-consistent sets is just the set of all L-theorems.

The *canonical frame* of a propositional modal logic S is $\mathcal{F}_S = \langle W_S, R_S \rangle$, where W_S is the set of maximally S-consistent sets and R_S is the binary relation on W_S having aR_Sb if and only if $\{\alpha : \Box\alpha \in a\} \subseteq b$. The *canonical S-model* $\mathcal{M}_S = \langle \mathcal{F}_S, v_S \rangle$ has the valuation v_S defined by $v_S(p) = \{a \in W_S : p \in a\}$. This satisfies the "Truth Lemma"

$$\mathcal{M}_S, a \models \alpha \quad \text{iff} \quad \alpha \in a,$$

from which it follows that a formula is valid in $\mathcal{M}_S$ iff it is an S-theorem. Thus S is characterised by the model $\mathcal{M}_S$, but not necessarily by the frame $\mathcal{F}_S$. While any formula valid in $\mathcal{F}_S$ is an S-theorem, the converse may not be true.

The *canonical QG-frame* of a quantified modal logic L is the structure

$$\mathcal{F}_L = \langle W_L, R_L, I_L, Prop_L, PropFun_L \rangle,$$

and the *canonical QG-model* is $\mathcal{M}_L = \langle \mathcal{F}_L, V_L \rangle$, where

- W_L is the set of maximally L-consistent sets of $\mathcal{L}\mathcal{Q}$-formulas;
- R_L is the binary relation on W_L defined as in the canonical logic for propositional modal logics;
- I_L is the set of closed terms of $\mathcal{L}\mathcal{Q}$;
- $||A||_L$ is the set of maximally L-consistent sets a such that $A \in a$;
- $Prop_L$ is the set $\{||A||_L : A \text{ is a closed formula of } \mathcal{L}\mathcal{Q}\}$;
- For any $f \in I_L^\omega$ and any formula A, $A^f = A[f0/x_0, \ldots, fn/x_n, \ldots] =$ the closed formula got by uniformly substituting the closed term fn for free occurrences of x_n in A;
- For each formula A, φ_A is the function from I_L^ω to $Prop_L$ such that $\varphi_A f = ||A^f||_L$;
- $PropFun_L$ is the set of functions φ_A for all formulas A;
- $V_L(c) = c$, for all individual constants c.
- For F an n-ary function symbol, $V_L(F)(\tau_1, \ldots, \tau_n) = F\tau_1 \ldots \tau_n$ for all closed terms $\tau_1, \ldots, \tau_n$.
- For P an n-ary predicate letter, $V_L(P)(\tau_1, \ldots, \tau_n) = ||P\tau_1 \ldots \tau_n||_L$.

Note that we do not require that the worlds in this frame/model are $\forall$-complete (see Section 7).

Many properties of $\mathcal{F}_L$ and $\mathcal{M}_L$ can be derived just as in [14]. In particular, as in Lemmas 9.1–9.4 of [14] we can show that the canonical frame $\mathcal{F}_L$ is in fact a QG-frame, that is, it satisfies CProp–CAll. This involves showing that $||A||_L \Rightarrow ||B||_L = ||A \supset B||_L$, $[R_L]||A||_L = ||\Box A||_L$, $\varphi_\emptyset = ||\bot||_L$, $\varphi_A \Rightarrow \varphi_B = \varphi_{A \supset B}$, $[R_L]\varphi_A = \varphi_{\Box A}$, $\forall_n \varphi_A = \varphi_{\forall x_n A}$.

The canonical model also satisfies a Truth Lemma in the form

(4.1) $\quad \mathcal{M}_L, a \models_f A \quad$ iff $\quad A^f \in a$.

Suppose that the variables that occur free in A are amongst $x_0, ..., x_n$, and that $c_0, ..., c_n$ are constants that do not occur in A (recall our assumption that $\mathcal{LQ}$ has infinitely many constants). We can then show that $A[c_0/x_0, ..., c_n/x_n]$ is an L-theorem if and only if A is an L-theorem. This implies that

(4.2) $\quad L \vdash A \quad$ iff $\quad$ for all $f \in I_L^\omega$, $L \vdash A^f$.

From (4.1) and (4.2) and the fact that $\bigcap W_L$ is the set of all L-theorems, it follows that a formula A is valid in $\mathcal{M}_L$ if and only if it is an L-theorem (see [14], Lemma 9.6 and Theorem 9.7). Consequently, any formula valid in $\mathcal{F}_L$ is an L-theorem, and so to show that $\mathcal{F}_L$ characterises L it suffices to show that it validates L.

An immediate application of this construction is a completeness theorem for any logic of the form QS:

THEOREM 4. *For any propositional modal logic S, the quantified logic QS is characterised by its canonical general frame $\mathcal{F}_{QS}$, and hence is complete for the class of all its validating quantified general frames.*

Proof. Completeness for $\mathcal{F}_{QS}$: if a QS-formula A is valid in $\mathcal{F}_{QS}$, then it is valid in the model $\mathcal{M}_{QS}$, and hence is a QS-theorem.

Soundness for $\mathcal{F}_{QS}$: it is enough to show that if $\mathcal{LQ}$-formula A is a substitution instance of a propositional S-theorem β, then A is valid in any model $\mathcal{M} = \langle \mathcal{F}_{QS}, V \rangle$ on $\mathcal{F}_{QS}$. Suppose A is obtained by uniformly substituting $\mathcal{LQ}$-formulas B_p for certain propositional variables p in β. Now the function $|B_p|^\mathcal{M}$ interpreting B_p in $\mathcal{M}$ belongs to $PropFun_{QS}$, and so is equal to φ_{A_p} for some $\mathcal{LQ}$-formula A_p. Hence $|B_p|^\mathcal{M}_f = ||A_p^f||_L = \{a : \mathcal{M}_{QS}, a \models_f A_p\}$ by the Truth Lemma.

A simple induction on $\mathcal{L}$-formulas α then shows that in general

$$\mathcal{M}, a \models_f \alpha[\ldots, B_p/p, \ldots] \quad \text{iff} \quad \mathcal{M}_{QS}, a \models_f \alpha[\ldots, A_p/p, \ldots].$$

But then when $\alpha = \beta$, $\beta[\ldots, A_p/p, \ldots]$ is an instance of an S-theorem, hence valid in $\mathcal{M}_{QS}$, so $A = \beta[\ldots, B_p/p, \ldots]$ is valid in $\mathcal{M}$. ∎

5 Completeness for Canonical Logics

Although $\mathcal{F}_{\mathrm{QS}}$ validates QS, its underlying propositional frame $\langle W_{\mathrm{QS}}, R_{\mathrm{QS}}\rangle$ need not validate S. There may be a propositional model $\langle W_{\mathrm{QS}}, R_{\mathrm{QS}}, v\rangle$ falsifying some S-theorem (this would require that $v(p) \notin \mathit{Prop}_{\mathrm{QS}}$ for some variable p). Examples will be given later.

We now show that this situation cannot arise if S is *canonical*, which means that it is valid in the canonical propositional frame $\mathcal{F}_{\mathrm{S}} = \langle W_{\mathrm{S}}, R_{\mathrm{S}}\rangle$, and hence is characterised by this frame. The class of canonical logics is wide: it includes every logic that is characterised by some first-order definable class of frames $\langle W, R\rangle$ [5], and many others besides [11, 10].

LEMMA 5. *If L is any quantified modal logic extending QS, then $\langle W_L, R_L\rangle$ is isomorphic to a generated subframe of $\mathcal{F}_S$.* ∎

A direct proof of this result appears in the Appendix. Here we give a brief explanation of it by invoking the duality between algebraic models and frames. The Lindenbaum algebra $\mathbf{A}^{\mathrm{S}}$ of S is a free S-algebra, freely generated by the (equivalence classes of) propositional variables. The Lindenbaum algebra $\mathbf{A}^{\mathrm{L}}$ of L is also an S-algebra, and is no bigger than the set of generators of $\mathbf{A}^{\mathrm{S}}$, by the assumption that there are at least as many variables in $\mathcal{L}$ as there are $\mathcal{L}Q$-formulas. Hence there is a surjective homomorphism $f : \mathbf{A}^{S} \twoheadrightarrow \mathbf{A}^{L}$. By duality, this induces an injective bounded morphism $f_+ : \mathbf{A}^{L}_+ \rightarrowtail \mathbf{A}^{S}_+$ in the reverse direction, between the canonical structures of these algebras. The image of f_+ is a generated subframe of $\mathbf{A}^{S}_+$ isomorphic to $\mathbf{A}^{L}_+$. But the points of $\mathbf{A}^{S}_+$ are the ultrafilters of $\mathbf{A}^{S}$, which can be identified with maximally S-consistent sets, and so $\mathbf{A}^{S}_+$ can be identified with $\langle W_{\mathrm{S}}, R_{\mathrm{S}}\rangle$. Similarly $\mathbf{A}^{L}_+$ can be identified with $\langle W_L, R_L\rangle$.

Since validity of formulas is preserved by generated subframes and isomorphism, Lemma 5 and Theorem 4 immediately give

THEOREM 6. *If S is a canonical propositional logic and L is a quantified logic extending QS, then $\langle W_L, R_L\rangle$ is an S-frame. In particular $\langle W_{\mathrm{QS}}, R_{\mathrm{QS}}\rangle$ is an S-frame and so QS is characterised by the class of all QG-frames whose underlying propositional frame validates S.* ∎

Now let Φ be any set of conditions on proposition frames such that (i) S is validated by all frames satisfying Φ, (ii) $\mathcal{F}_{\mathrm{S}}$ satisfies Φ, and (iii) Φ is preserved by generated subframes and isomorphism. Then S is canonical and $\langle W_{\mathrm{QS}}, R_{\mathrm{QS}}\rangle$ satisfies Φ. It follows from Theorems 2 and 6 that QS is characterised by the class of all QG-frames whose underlying propositional frame satisfies Φ.

For example, let S=S4M, the extension of S4 by the McKinsey axiom $\Box\Diamond A \supset \Diamond\Box A$. The S4M-frames are precisely those that are reflexive, transitive frames and *final* in the sense that every world has an accessible "endpoint":

$$\forall x \exists y (xRy \land \forall z(yRz \supset y = z)).$$

These conditions are preserved by generated subframes, and possessed by the canonical frame of S4M. It follows that $\langle W_{QS4M}, R_{QS4M}\rangle$ is reflexive, transitive and final, and that QS4M is characterised by the class of all quantified general frames based on S4M-frames.

The significance of this is that QS4M is incomplete for its standard quantificational semantics based on S4M-frames [13, p. 283].

6 Tarskian Frames

The *converse Barcan formula* $\Box \forall x A \supset \forall x \Box A$ is derivable as a schema in the proof theory of any system QS. In the standard semantics for quantified modal logics, this schema is usually validated by constraining models to have "increasing domains": each world a is assigned a domain Ia of individuals such that aRb implies $Ia \subseteq Ib$, and the satisfaction of a formula $\forall x A$ at world a is evaluated by having x range over the members of Ia. Our semantics also validates the converse Barcan formula, but is closer to what is usually called a *constant domain* semantics, in that quantified variables range over the one domain I of individuals relative to all worlds.

In the standard semantics, constant domain models validate the *Barcan formula*

(BF) $\forall x \Box A \supset \Box \forall x A$.

A standard constant domain model has the form $\mathcal{M} = \langle W, R, I, \nu \rangle$, where the valuation ν assigns to each constant an element of I, to each n-ary function letter an n-ary function on I, and to each n-ary predicate letter P and each element a of W an n-ary relation $\nu(P,a) \subseteq I^n$. The *standard satisfaction relation* $\mathcal{M}, a \models_f A$ is defined inductively, with the atomic case given by

$$\mathcal{M}, a \models_f P\tau_1\ldots\tau_n \quad \text{iff} \quad \langle \tau_1^{\mathcal{M}} f, \ldots, \tau_n^{\mathcal{M}} f \rangle \in \nu(P, a);$$

the propositional connectives treated as usual; and

(6.1) $\mathcal{M}, a \models_f \forall x_n A$ iff for all $j \in I$, $\mathcal{M}, a \models_{f[j/n]} A$.

Our general semantics does not validate the Barcan formula for most logics (QS5 is a notable exception), despite the fact that it has constant domains. To obtain a condition validating BF we say that QG-frame $\mathcal{F} = \langle W, R, I, Prop, PropFun \rangle$ is *Tarskian* if

$$\bigcap_{j \in I} \varphi(f[j/n]) \in Prop$$

for all $\varphi \in PropFun$, $n \in \omega$, and $f \in I^\omega$. When this holds, we have

$$\bigcap_{j \in I} \varphi(f[j/n]) = \prod_{j \in I} \varphi(f[j/n]) = (\forall_n \varphi) f,$$

so an equivalent definition of "Tarskian" is that in general

$$(\forall_n \varphi) f = \bigcap_{j \in I} \varphi(f[j/n]).$$

This condition implies that in any model on $\mathcal{F}$, the satisfaction clause for $\forall x_n A$ simplifies to the standard Tarskian clause (6.1), from which validity of BF follows readily.

Note that a QG-frame must be Tarskian if it has $Prop = \wp W$, since then $\bigcap S \in Prop$ for all $S \subseteq Prop$. Also a frame in which $Prop$ is finite must be Tarskian, since $Prop$ is closed under finite intersection. Finally, a frame must be Tarskian if I is finite, for then so is $\{\varphi(f[j/n]) : j \in I\}$.

If $\mathcal{M} = \langle \mathcal{F}, V \rangle$ is any model, in our sense, on a QG-frame $\mathcal{F}$, we obtain a standard model $\mathcal{M}^s = \langle W, R, I, \nu_\mathcal{M} \rangle$ by defining $\nu_\mathcal{M}$ to agree with V on constants and function letters, and putting

$$\nu_\mathcal{M}(P, a) = \{\langle j_1, \ldots, j_n \rangle : a \in V(P)(j_1, \ldots, j_n)\}.$$

Then we get $\tau^{\mathcal{M}^s} = \tau^\mathcal{M}$ for all $\mathcal{LQ}$-terms τ, because of the agreement of $\nu_\mathcal{M}$ and V on constants and function letters. Moreover, *if $\mathcal{F}$ is Tarskian*, then in general $\mathcal{M}, a \models_f A$ iff $\mathcal{M}^s, a \models_f A$, and so the two models validate the same formulas.

In the converse direction, given $\mathcal{M} = \langle W, R, I, \nu \rangle$ a standard constant domain model, we define the QG-model $\mathcal{M}^* = \langle W, R, I, \wp W, (\wp W)^I, V_\mathcal{M} \rangle$ by taking $V_\mathcal{M}$ to agree with ν on constants and function letters, and putting

$$V_\mathcal{M}(P)(j_1, \ldots, j_n) = \{a \in W : \langle j_1, \ldots, j_n \rangle \in \nu(P, a)\}.$$

Thus the underlying frame of $\mathcal{M}^*$ is the *full* frame based on $\langle W, R, I \rangle$, with $Prop$ containing every subset of W, and $PropFun$ every function $I \to \wp W$. This makes it immediately a QG-frame (i.e. CProp–CAll hold) *that is Tarskian*. Because of the agreement of $V_\mathcal{M}$ with ν we get $\tau^{\mathcal{M}^*} = \tau^\mathcal{M}$ for all terms τ. In general $\mathcal{M}^*, a \models_f A$ iff $\mathcal{M}, a \models_f A$, and so these two models also validate the same formulas.

Now if $\mathcal{M}^{*s} = \langle W, R, I, \nu_{\mathcal{M}^*} \rangle$ is the standard constant domain model constructed from the QG-model $\mathcal{M}^*$, then in fact $\nu_{\mathcal{M}^*} = \nu$, and so $\mathcal{M}^{*s} = \mathcal{M}$. Conversely, if $\mathcal{M}^{s*} = \langle W, R, I, \wp W, (\wp W)^I, V_{\mathcal{M}^s} \rangle$ is the Tarskian QG-model constructed from the standard model $\mathcal{M}^s = \langle W, R, I, \nu_\mathcal{M} \rangle$ associated with a QG-model $\mathcal{M} = \langle \mathcal{F}, V \rangle$, then $V_{\mathcal{M}^s} = V$. It does not follow however that $\mathcal{M}^{s*} = \mathcal{M}$, since the underlying frame $\mathcal{F}$ of $\mathcal{M}$ need not be full (i.e. maybe $Prop \neq \wp W$ in $\mathcal{F}$), whereas that of $\mathcal{M}^{s*}$ *is* full by definition. What we do get is that if $\mathcal{F}$ is Tarskian, then $\mathcal{M}, a \models_f A$ iff $\mathcal{M}^{s*}, a \models_f A$, so the satisfaction relations of the two models are identical.

The upshot of all this is an equivalence between the standard constant domain semantics and the semantics of our Tarskian general frames. In particular:

THEOREM 7. *Let $\mathcal{C}$ be a class of propositional frames of the form $\langle W, R \rangle$. Then for any $\mathcal{LQ}$-formula A, then following are equivalent.*

- A is valid in all standard constant domain models based on members of $\mathcal{C}$.

- A is valid in all Tarskian QG-frames based on members of $\mathcal{C}$.

- A is valid in all full QG-frames based on members of $\mathcal{C}$. ∎

7 Tarskian Canonical Frames

Let LB be the extension of quantified logic L by the Barcan formula BF. The canonical QG-frame $\mathcal{F}_{\text{LB}}$ for LB of Section 4 validates LB, but it need not be Tarskian. Thus the Tarskian condition is sufficient to ensure validity of BF, but it is not necessary.

To see this, we first show

LEMMA 8. *If L is any quantified logic containing the schema BF, then the canonical frame $\mathcal{F}_L$ validates BF.*

Proof. In any model $\mathcal{M}$ on $\mathcal{F}_L$, each formula A is interpreted as a function $|A|^{\mathcal{M}}$ that belongs to PropFun_L, and hence is equal to φ_B for some B. Thus to show that $\mathcal{M}$ validates $\forall x_n \Box A \supset \Box \forall x_n A$ it is enough to show that $\forall_n [R_L] \varphi_B(f) \subseteq [R_L] \forall_n \varphi_B(f)$ for all $f \in I_L^\omega$, i.e. that $\varphi_{\forall x_n \Box B}(f) \subseteq \varphi_{\Box \forall x_n B}(f)$. This means that $||(\forall x_n \Box B)^f||_L \subseteq ||(\Box \forall x_n B)^f||_L$. But that follows by properties of maximally L-consistent sets, since $(\forall x_n \Box B)^f \supset (\Box \forall x_n B)^f$ is an L-theorem derivable from BF. ∎

COROLLARY 9. *For any proposition modal logic S, QSB is characterised by its canonical frame $\mathcal{F}_{QSB}$.* ∎

Now consider the well-known logic S4.2, the extension of S4 by the schema $\Diamond \Box A \supset \Box \Diamond A$. S4.2 is a canonical logic whose frames are precisely those reflexive, transitive frames that are *convergent*, i.e.

$$\forall x, y, z (xRy \wedge xRz \supset \exists w (yRw \wedge zRw)).$$

Hence by Theorem 6, the canonical QG-frame for QS4.2B is based on an S4.2 frame. If this canonical frame were Tarskian, then QS4.2B would be characterised by the class of all Tarskian QG-frames based on S4.2-frames, and hence by Theorem 7 would be characterised by the class of all standard constant domain models based on S4.2-frames. But in fact QS4.2B is *incomplete* for the class of all standard constant domain models based on S4.2-frames [13, p. 271].[3]

Thus $\mathcal{F}_{\text{QS4.2B}}$ is a characterising frame for QS4.2B that is reflexive, transitive and convergent; validates BF; but is non-Tarskian. We have found a semantics for this standardly-incomplete logic that has the standard semantics for its propositional part, but an alternative interpretation of the quantifier $\forall$.

---

[3]In the notation of [13], QS is LPC+S, and QSB is S+BF.

Now for any logic L that includes the schema BF there is also a canonical model construction based on ∀-*complete* sets. A set Γ of formulas is ∀-complete if $\forall x A \in \Gamma$ whenever $A[\tau/x]$ for all closed terms τ. Let W_L^T be the set of all maximally L-consistent ∀-complete sets. The presence of infinitely many constants in $\mathcal{LQ}$ ensures that any L-consistent formula can be shown to belong to some member of W_L^T, and hence that the intersection of all members of W_L^T is just the set of L-theorems. A frame $\mathcal{F}_L^T$ and model $\mathcal{M}_L^T = \langle \mathcal{F}_L^T, V_L^T \rangle$ can be constructed just as for $\mathcal{F}_L$ and $\mathcal{M}_L$ but using W_L^T in place of W_L. The crucial role of BF here, as first shown in [21], is to allow a proof that for any $\Gamma \in W_L^T$,

$$\Box A \in \Gamma \quad \text{iff} \quad \text{for all } \Delta \in W_L^T, \, \Gamma R_L \Delta \text{ implies } A \in \Delta.$$

This can then be used to show that the Truth Lemma

$$\mathcal{M}_L^T, a \models_f A \quad \text{iff} \quad A^f \in a$$

holds in general, and hence that the formulas valid in $\mathcal{M}_L^T$ are just the L-theorems. But now the ∀-completeness of members of W_L^T ensures that if A has only x free, then $||\forall x A||_L = \bigcap \{||A[\tau/x]||_L : \tau \in I_L\}$. This implies that $\mathcal{F}_L^T$ is Tarskian (see Lemmas 9.3 and 9.4 of [14].) We call it the *Tarskian canonical frame* for L.

THEOREM 10. *For any propositional modal logic S, the logic QSB is characterised by its Tarskian canonical frame $\mathcal{F}_{QSB}^T$, and hence is complete for the class of all its validating Tarskian frames.*

Proof. As usual, any formula valid in $\mathcal{F}_{QSB}^T$ is valid in $\mathcal{M}_{QSB}^T$ and hence is a QSB-theorem. But $\mathcal{F}_{QSB}^T$ validates all instances of S-theorems by the same proof as for Theorem 4, and validates BF because it is Tarskian, so validates QSB. ∎

Corollary 9 and Theorem 10 now show that QSB has two characteristic canonical general frames: $\mathcal{F}_{QSB}$ and $\mathcal{F}_{QSB}^T$. If S is canonical, then $\mathcal{F}_{QSB}$ will be based on an S-frame but may not be Tarskian, as we saw with S4.2. Vice versa, $\mathcal{F}_{QSB}^T$ will be Tarskian, but may not be based on an S-frame. S4.2 shows this again: if $\mathcal{F}_{QS4.2B}^T$ were based on an S4.2 frame, then QS4.2B would be characterised by the class of all Tarskian QG-frames based on S4.2-frames, which we have already observed to be false.

Now $\mathcal{F}_{QS4.2B}^T$ is reflexive and transitive, by the usual application of S4-axioms, so the upshot of this discussion is that it cannot be convergent. It follows that $\langle W_{QS4.2B}^T, R_{QS4.2B}^T \rangle$ is not a generated subframe of the convergent frame underlying $\mathcal{F}_{QS4.2B}$.

8 Non-Canonical Cases

We have just seen that $\mathcal{F}_{QSB}^T$ need not be based on an S-frame even when S is canonical. In fact there is little gain when $\langle W_{QSB}^T, R_{QSB}^T \rangle$ *does* validate S, because in that case QSB is complete for Tarskian general frames based on S-frames, and hence by Theorem 7 is complete for the standard constant domain semantics based on S-frames. Our theory comes into its own in cases where QSB is incomplete for the standard semantics, by providing a complete semantics based on Tarskian general frames - albeit one in which $\langle W_{\text{QSB}}^T, R_{\text{QSB}}^T \rangle$ is not an S-frame.

There is one useful observation we can make about properties that are inherited by $\langle W_{\text{QSB}}^T, R_{\text{QSB}}^T \rangle$ from $\mathcal{F}_{\text{S}}$. $\langle W_{\text{QSB}}^T, R_{\text{QSB}}^T \rangle$ is a *substructure* of $\langle W_{\text{QSB}}, R_{\text{QSB}} \rangle$, in that R_{QSB}^T is the restriction of R_{QSB} to the subset W_{QSB}^T of W_{QSB}, and hence is isomorphic to a substructure of $\mathcal{F}_{\text{S}}$ (Lemma 5). If S includes the modal axiom(s) corresponding to a universal property, then that property will be possessed by the propositional frame underlying $\mathcal{F}_{\text{QSB}}^T$.

What if S is not canonical? Can QS or QSB still be complete for a class of quantified general frames based on S-frames? Of course we should not expect this if S itself is incomplete for S-frames. Take the case that S is the incomplete logic GH from [3]. Every frame for GH has transitive R so validates the axiom 4: $\Box p \to \Box\Box p$, but this axiom is not derivable in GH. Now if QGH or QGHB were complete for a class of quantified general frames based on GH-frames, then these structures would have transitive R and validate all $\mathcal{LQ}$-instances of 4, which would then be derivable in the relevant quantified logic, and in particular would be derivable in QGHB. But this is not so, since the GH-model of [3] falsifying 4 can readily be turned into a QGHB-model falsifying $\Box Px \to \Box\Box Px$, where P is a unary predicate letter.

This situation is general: if S is any incomplete propositional logic, then there is some $\mathcal{L}$-formula α that is valid in all S-frames but not an S-theorem. By [13, Corollary 14.6] there must then be some $\mathcal{LQ}$-substitition instance A of α that is not a QSB-theorem. But if $\langle W, R \rangle$ is an S-frame, then by our Remark 3, A is valid in every QG-frame based on $\langle W, R \rangle$. It follows, similarly to the previous paragraph, that neither QS nor QSB can be complete for any class of QG-frames based on S-frames.

But even when S is S-frame complete, the desired property can fail for QS and QSB. This is exemplified by the propositional modal logic KW, the smallest normal modal logic containing the schema $\Box(\Box\alpha \supset \alpha) \supset \Box\alpha$. The KW-frames are precisely those transitive frames that have no infinite R-chains $b_0 R b_1 R \cdots R b_n R \cdots$. KW is characterised by these frames, but is not canonical.

The logic QKW was shown in [15] not to be complete for its standard (expanding domain) semantics. We adapt that example to our present semantics, by considering the formula

$$(\ddagger) \quad \forall x \Box (Px \to \Diamond PFx) \to \forall x \neg \Diamond Px,$$

where P is a unary predicate letter, and F a unary function symbol. Inspection of the QKW-model defined in Theorem 2 of [15], based on a nonstandard model of arithmetic, shows that this model falsifies ($\ddagger$) with F interpreted as the successor function. Hence ($\ddagger$) is not a QKW-theorem.

But (‡) is valid in any quantified general frame $\mathcal{F}$ that is based on a KW-frame. For if not, there would be a model $\mathcal{M}$ on such an $\mathcal{F}$ having a point a at which (‡) was false. Then $\mathcal{M}, a \models \forall x \Box (Px \to \Diamond PFx)$ and $\mathcal{M}, a \not\models \forall x \neg \Diamond Px$. By our semantics for $\forall$, there must be some $X \in Prop$ with $a \in X$ and $X \subseteq |\Box (Pj \to \Diamond PFj)|^{\mathcal{M}}$ for all $j \in I$ (treating j as an individual constant). Since $\forall x \neg \Diamond Px$ is false at a, $X \not\subseteq |\neg \Diamond Pj^*|^{\mathcal{M}}$ for some j^*. Hence there is some world $b_0 \in X$ at which $b_0 \models \Diamond Pj^*$ in $\mathcal{M}$, so there is some b_1 such that $b_0 R b_1$ and $b_1 \models Pj^*$. Now suppose inductively that we have defined b_n ($n \geq 1$) such that $b_n \models PF^{n-1}j^*$ and $b_0 R b_n$. Here F^n is the n-th iteration of F, with $F^n j = F(F^{n-1}j)$ and $F^0 j = j$. Then

$$b_0 \in X \subseteq |\Box(PF^{n-1}j^* \to \Diamond PF^n j^*)|^{\mathcal{M}},$$

so $b_n \models PF^{n-1}j^* \to \Diamond PF^n j^*$. Hence there is some b_{n+1} with $b_n R b_{n+1}$ and $b_{n+1} \models PF^n j^*$. Then $b_0 R b_{n+1}$ by transitivity of R. This shows, by induction, that there is an infinite R-chain $b_0 R b_1 R \cdots R b_n R \cdots$ in $\mathcal{F}$, which contradicts the assumption that $\mathcal{F}$ is based on a KW-frame.

Thus QKW, while being characterised by $\mathcal{F}_{\text{QKW}}$ and by the class of its validating quantified general frames, is not complete for any such class of general frames that are based on KW-frames. Hence $\mathcal{F}_{\text{QKW}}$ is not based on a KW-frame (and so must contain an infinite R-chain).

The logic QKWB with the Barcan formula is, similarly, characterised by its Tarskian canonical frame $\mathcal{F}^T_{\text{QKWB}}$ and complete for its class of validating Tarskian frames, but is not complete for the class of Tarskian frames that are based on KW-frames. If it were, then it would be characterised by the class of all standard constant domain models based on KW-frames (Theorem 7). But the logic characterised by the standard constant domain models based on KW-frames was shown in [4] not to be recursively axiomatisable, so cannot be equal to QKWB.

9 Conclusion and Further Work

We have constructed a canonical general frame $\mathcal{F}_{\text{QS}}$ for every quantified modal logic of the form QS, and shown that if S is a canonical propositional logic, then the Kripke frame underlying $\mathcal{F}_{\text{QS}}$ validitates S. The semantics of $\forall$ in $\mathcal{F}_{\text{QS}}$ may be non-Tarskian.

We also shed new light on the Barcan formula, by revealing that its role is to ensure that the standard Tarskian semantics for $\forall$ can be provided. For systems QSB including BF there is a second characterising canonical general frame $\mathcal{F}^T_{\text{QSB}}$ in which $\forall$ gets the Tarskian semantics, but the underlying Kripke frame may not validate S. Thus there is some trade-off between these two desiderata.

There is much more to be done with our notion of quantified general frame. What about the case of a logic L that is not of the form QS or QSB? To show that $\mathcal{F}_L$ characterises such an L may require us to assume that L has a rule of substitution of general formulas in place of atomic formulas. Extensions of QS4 with this rule have been studied by Shehtman and

Skvortsov [19, 20], using a "Kripke metaframe" semantics based on functorial constructions. They construct a characterising canonical metaframe for their quantified S4-extensions of a canonical propositional logic, both with and without BF.

In this context we should also mention the work of Ghilardi [6] showing that if S is any canonical superintuitionistic propositional logic, then the the quantified extension of S has a characteristic canonical functor frame based on an S-frame. He also notes (p. 211) that his result does not fully apply to the functor semantics of quantified extensions of S4.

To give a general semantics for systems without the converse Barcan formula will require us to consider models with varying (world-dependent) domains of individuals, as indicated at the start of Section 6. Then there is the question of adding the equality symbol to our languages and addressing the whole morass of issues around the interpretation of existence and identity in intensional logics. We intend to take these matters up in future work.

Appendix

Here is a direct construction to prove Lemma 5. Let $p \mapsto p^*$ be a mapping of the set of propositional variables of $\mathcal{L}$ *onto* the set of $\mathcal{L}Q$-formulas. Such a mapping exists by the assumption the the former set is at least as large as the latter (but see the end for further discussion of this assumption).

Now for any $\mathcal{L}$-formula α, let α^* be the result of uniformly substituting p^* for p in α, for all p that occur in α. This operation commutes with the $\mathcal{L}$-connectives: $\bot^* = \bot$, $(\alpha \supset \beta)^* = \alpha^* \supset \beta^*$, $(\Box \alpha)^* = \Box(\alpha^*)$ etc. Moreover, if α is an S-theorem, then α^* is a QS-theorem by definition, and hence is an L-theorem.

Define $f : W_\mathrm{L} \to W_\mathrm{S}$ by putting $f(a) = \{\alpha : \alpha^* \in a\}$ for all maximally L-consistent sets a. Of course it has to be checked that this $f(a)$ is maximally S-consistent. First, $f(a)$ is S-consistent, for otherwise there would be $\alpha_1, \ldots, \alpha_n \in f(a)$ such that $(\alpha_1 \wedge \cdots \wedge \alpha_n) \supset \bot$ is an S-theorem, hence $(\alpha_1^* \wedge \cdots \wedge \alpha_n^*) \supset \bot$ is an L-theorem, contrary to the L-consistency of a. Second, a contains one of α^* and $\neg(\alpha^*) = (\neg \alpha)^*$ so $f(a)$ contains one of α and $\neg \alpha$, for all α, so $f(a)$ is negation complete, as required.

f is injective: if $a \neq b$ in W_L, then there is some formula $A = p^* \in a$ with $\neg A = (\neg p)^* \in b$, hence $p \in f(a)$ and $\neg p \in f(b)$, so $p \notin f(b)$ and $f(a) \neq f(b)$.

If $aR_\mathrm{L}b$, then $\Box \alpha \in f(a)$ implies $\Box(\alpha^*) = (\Box \alpha)^* \in a$, hence $\alpha^* \in b$ and $\alpha \in f(b)$; so $f(a)R_\mathrm{S}f(b)$.

Finally, to complete the proof that f is a bounded morphism, suppose $f(a)R_\mathrm{S}c$ in W_S. We have to show that $aR_\mathrm{L}b$ and $f(b) = c$ for some $b \in W_\mathrm{L}$. Put
$$b_0 = \{\alpha^* : \Box \alpha^* \in a\} \cup \{\beta^* : \beta \in c\}.$$

If b_0 were not L-consistent, then since the two sets that make up b_0 are each closed under finite conjunctions, there would be formulas α, β with $\Box \alpha^* \in a$ and $\beta \in c$, such that $\alpha^* \supset \neg \beta^*$ is an L-theorem. But then $\Box \alpha^* \supset \Box \neg \beta^*$ is an L-theorem, so belongs to a, implying that $\Box\neg\beta^* \in a$, hence $\Box\neg\beta \in f(a)$, and so $\neg b \in c$, contradicting the S-consistency of c.

Thus b_0 is L-consistent, and hence is included in some $b \in W_L$. Since $\{\alpha^* : \Box\alpha^* \in a\} \subseteq b$ we get $aR_L b$, and since $\{\beta^* : \beta \in c\} \subseteq b$ we get $c \subseteq f(b)$, and so $c = f(b)$ as required, by *maximal* S-consistency of c.

Altogether then, f is an injective bounded morphism from $\langle W_L, R_L \rangle$ into $\mathcal{F}_S$. The image of f is isomorphic to $\langle W_L, R_L \rangle$. But the image of a bounded morphism is always a generated subframe of its codomain. This proves Lemma 5.

In conclusion, we comment on the assumption that there are at least as many $\mathcal{L}$-variables as $\mathcal{L}Q$-formulas. While propositional modal languages are often based on a denumerable list of variables, we could assume a proper class $\{p_\lambda : \lambda \text{ is an ordinal}\}$ of such variables, generating a proper class Fma of formulas. For each infinite cardinal κ, let Fma^κ be the set of such formulas with variables from $\{p_\lambda : \lambda < \kappa\}$. A (normal propositional modal) logic S is a subclass of Fma defined as in Section 2. Then each $S^\kappa = S \cap Fma^\kappa$ is a logic in Fma^κ and has its own canonical frame $\mathcal{F}_S^\kappa$ (of size 2^κ). We say that S is *canonical* if it is valid in these frames $\mathcal{F}_S^\kappa$ for all κ.

Thus, given a predicate modal language $\mathcal{L}Q$ we can take any κ that is at least as large as the set of $\mathcal{L}Q$-formulas, and use S^κ as our version of S, and $\mathcal{F}_S^\kappa$ as the canonical frame having a generated subframe isomorphic to $\langle W_L, R_L \rangle$. In this sense our assumption is innocuous.

Note that a logic S in Fma is uniquely determined by S^ω: S is the only logic in Fma whose restriction to Fma^ω is S^ω, and likewise S^κ is the only logic in Fma^κ whose restriction to Fma^ω is S^ω. For further discussion of this, see [9].

BIBLIOGRAPHY

[1] D. M. Armstrong. *A World of States of Affairs*. Cambridge University Press, 1997.

[2] D. M. Armstrong. *Truth and Truth-Makers*. Cambridge University Press, 2004.

[3] George Boolos and Giovanni Sambin. An incomplete system of modal logic. *Journal of Philosophical Logic*, 14:351–358, 1985.

[4] M. J. Cresswell. Some incompletable modal predicate logics. *Logique et Analyse*, 160:321–334, 1997.

[5] Kit Fine. Some connections between elementary and modal logic. In Stig Kanger, editor, *Proceedings of the Third Scandinavian Logic Symposium*, pages 15–31. North-Holland, 1975.

[6] Silvio Ghilardi. Quantified extensions of canonical propositional intermediate logics. *Studia Logica*, 51(2):195–214, 1992.

[7] Robert Goldblatt. *Metamathematics of Modal Logic*. PhD thesis, Victoria University, Wellington, February 1974. Included in [8].

[8] Robert Goldblatt. *Mathematics of Modality*. CSLI Lecture Notes No. 43. CSLI Publications, Stanford, California, 1993. Distributed by Chicago University Press.

[9] Robert Goldblatt. Quasi-modal equivalence of canonical structures. *The Journal of Symbolic Logic*, 66:497–508, 2001.

[10] Robert Goldblatt, Ian Hodkinson, and Yde Venema. On canonical modal logics that are not elementarily determined. *Logique et Analyse*, 181:77—101, 2003. Published October 2004.

[11] Robert Goldblatt, Ian Hodkinson, and Yde Venema. Erdős graphs resolve Fine's canonicity problem. *The Bulletin of Symbolic Logic*, 10(2):186–208, June 2004.
[12] P. R. Halmos. *Algebraic Logic*. Chelsea, New York, 1962.
[13] G. E. Hughes and M. J. Cresswell. *A New Introduction to Modal Logic*. Routledge, 1996.
[14] Edwin D. Mares and Robert Goldblatt. An alternative semantics for quantified relevant logic. *The Journal of Symbolic Logic*, 71(1):163–187, 2006.
[15] Franco Montagna. The predicate modal logic of provability. *Notre Dame Journal of Formal Logic*, 25:179–189, 1984.
[16] Andrjez Mostowski. Proofs of non-deducibility in intuitionistic functional calculus. *The Journal of Symbolic Logic*, 13(4):204–207, 1948.
[17] H. Rasiowa and R. Sikorski. A proof of the completeness theorem of Gödel. *Fundamenta Mathematicae*, 37:193–200, 1950.
[18] H. Rasiowa and R. Sikorski. *The Mathematics of Metamathematics*. PWN–Polish Scientific Publishers, Warsaw, 1963.
[19] Valentin Shehtman and Dmitrij Skvortsov. Semantics of non-classical first-order predicate logics. In P. P. Petkov, editor, *Mathematical Logic*, pages 105–116. Plenum Press, 1990.
[20] D. P. Skvortsov and V. B. Shehtman. Maximal Kripke-type semantics for modal and superintuitionistic predicate logics. *Annals of Pure and Applied Logic*, 63:69–101, 1993.
[21] R. H. Thomason. Some completeness results for modal predicate calculi. In K. Lambert, editor, *Philosophical Problems in Logic*, pages 56–76. D. Reidel, 1970.
[22] S. K. Thomason. Semantic analysis of tense logic. *The Journal of Symbolic Logic*, 37:150–158, 1972.

Robert Goldblatt
Centre for Logic, Language and Computation
Victoria University of Wellington, New Zealand
Rob.Goldblatt@vuw.ac.nz

Edwin D. Mares
Centre for Logic, Language and Computation
Victoria University of Wellington, New Zealand
Edwin.Mares@vuw.ac.nz

A decidable modal logic that is finitely undecidable

Igor Gorbunov

ABSTRACT. The aim of this paper is to construct an example of a decidable modal logic above $K4$ such that the logic of its finite Kripke frames is undecidable.

1 Introduction

A theory is said to be *finitely decidable* if the theory of its finite models is decidable. In this paper we consider the following question: is it the case that if a theory is decidable then it is finitely decidable? That is, whether the following implication holds:

DECIDABILITY $\Rightarrow$ FINITE DECIDABILITY

For some theories, there is a correlation between decidability and finite decidability. For example, the implication above holds for the equational group theory, the equational ring theory, and some others (see, e.g., [5, 4, 2]). However, in general this implication does not hold. Examples of a decidable first-order theory and an equational theory that are finitely undecidable were given by Joohee Jeong [3]. The question whether the above implication holds for propositional (normal uni)modal logics was asked by Michael Zakharyaschev.

Here we give a 'negative' answer to this question by proving the following

THEOREM 1. *There exists a decidable normal unimodal logic L above $K4$ such that the logic of its finite frames is undecidable.*

2 Definitions

We deal with the standard modal language containing the connectives $\wedge$, $\vee$, $\rightarrow$, $\square$, and the propositional constant $\bot$ (falsehood). The connectives $\neg$ and $\lozenge$ are defined as abbreviations: $\neg \varphi = \varphi \rightarrow \bot$, $\lozenge \varphi = \neg \square \neg \varphi$. We also write $\square^+ \varphi$ for $\square \varphi \wedge \varphi$.

The following definitions are required for the proof of our theorem. A *Kripke frame* $\mathfrak{F}$ is a pair $\langle W, R \rangle$, where W is a non-empty set of worlds and R is a binary relation on W. A world y is *accessible from a world* x if the relation xRy holds. A set $G \subseteq W$ is *accessible from a world* w if every world of G is accessible from w. A frame $\mathfrak{F} = \langle W, R \rangle$ is called *transitive* if R is a transitive relation. A transitive frame is *rooted* if there exists a world d such that the set $W \setminus \{d\}$ is accessible from d. In this case the world d is called a *root* of the frame.

Let $\mathfrak{F} = \langle W, R \rangle$ be a frame and ν a valuation in $\mathfrak{F}$, i.e., $\nu(p) \subseteq W$, for every propositional variable p. Let x be a world in $\mathfrak{F}$. We define a *truth-relation* $x \models^\nu \varphi$, "φ is true at x under ν", by taking

$x \not\models^\nu \bot$;

$x \models^\nu p$ iff $x \in \nu(p)$, for every variable p;

$x \models^\nu \psi \wedge \chi$ iff $x \models^\nu \psi$ and $x \models^\nu \chi$;

$x \models^\nu \psi \vee \chi$ iff $x \models^\nu \psi$ or $x \models^\nu \chi$;

$x \models^\nu \psi \to \chi$ iff $x \models^\nu \psi$ implies $x \models^\nu \chi$;

$x \models^\nu \Box \psi$ iff $y \models^\nu \psi$ for all $y \in W$ such that xRy.

If $x \not\models^\nu \varphi$ then we say φ is *false* at x under ν.

All frames we consider below are assumed to be transitive. A world u is called a *dead end* if uRw for no $w \in W$. A set $\mathfrak{a} \subseteq W$ is called an *antichain* if, for any distinct worlds $w, v \in \mathfrak{a}$, we have $\neg wRv$ and $\neg vRw$. The cardinality of a set G is denoted by $\|G\|$. An antichain $\mathfrak{a}$ is called a *singleton* if $\|\mathfrak{a}\| = 1$. A set $\mathfrak{a} \subseteq W$ is called a *chain* if for every pair of worlds $w, v \in \mathfrak{a}$, $w \neq v$, we have wRv or vRw.

For a set $X \subseteq W$, let

- $(X){\uparrow} = \{w \in W \mid \exists x \in X, xRw\}$;
- $(X)\overline{\uparrow} = X \cup (X){\uparrow}$;
- $(X){\downarrow} = \{w \in W \mid \exists x \in X, wRx\}$;
- $(X)\overline{\downarrow} = X \cup (X){\downarrow}$.

We also write $x{\uparrow}$, $x\overline{\uparrow}$, and $x{\downarrow}$ instead of $\{x\}{\uparrow}$, $\{x\}\overline{\uparrow}$, and $\{x\}{\downarrow}$, respectively.

Denote by $\operatorname{Dom} f$ and $\operatorname{Im} f$ the domain and the range of a map f, respectively. Suppose that we have two frames $\mathfrak{F} = \langle W, R \rangle$ and $\mathfrak{G} = \langle V, S \rangle$. A partial map f from W onto V is called a *subreduction* (or a partial p-morphism) of $\mathfrak{F}$ to $\mathfrak{G}$ if, for any $x, y \in \operatorname{Dom} f$, the following conditions hold:

- xRy implies $f(x) S f(y)$;
- $f(x) S f(y)$ implies $\exists z \in \operatorname{Dom} f(xRz \wedge f(z) = f(y))$.

A subreduction f is called *cofinal* if, for every $z \in (\operatorname{Dom} f){\uparrow} \setminus \operatorname{Dom} f$, we have $z \in (\operatorname{Dom} f){\downarrow}$.

Let $\mathfrak{G}$ be a finite frame and $\mathfrak{D}$ a set of antichains in $\mathfrak{G}$ different from reflexive singleton. A subreduction f is said to satisfy the *closed domain condition* (CDC) for $\mathfrak{D}$ if $f(x{\uparrow}) = (\mathfrak{a})\overline{\uparrow}$ implies $x \in \operatorname{Dom} f$ for every $\mathfrak{a} \in D$ and every $x \in (\operatorname{Dom} f){\uparrow}$.

Let $\mathfrak{F} = \langle W, R \rangle$ be a finite rooted (transitive) frame, $W = \{w_0, \ldots, w_n\}$, with w_0 being the root. The *canonical formula* $\alpha(\mathfrak{F}, \mathfrak{D}, \bot)$ associated with $\mathfrak{F}$ and $\mathfrak{D}$ is defined (see [6] and [1], p. 310) by taking

$$\alpha(\mathfrak{F}, \mathfrak{D}, \bot) = \bigwedge_{w_i R w_j} \varphi_{ij} \wedge \bigwedge_{i=0}^{n} \varphi_i \wedge \bigwedge_{\mathfrak{a} \in \mathfrak{D}} \varphi_{\mathfrak{a}} \wedge \varphi_{\bot} \to p_0,$$

where

$$\varphi_{ij} = \Box^+(\Box p_j \to p_i),$$

$$\varphi_i = \Box^+((\bigwedge_{\neg a_i R a_k} \Box p_k \wedge \bigwedge_{j=0, j \neq i}^{n} p_j \to p_i) \to p_i),$$

$$\varphi_{\mathfrak{a}} = \Box^+(\bigwedge_{w_i \in W \setminus \mathfrak{a}\uparrow} \Box p_j \wedge \bigwedge_{i=0}^{n} p_i \to \bigvee_{w_j \in \mathfrak{a}} \Box p_j),$$

$$\varphi_{\bot} = \Box^+(\bigwedge_{i=0}^{n} \Box^+ p_i \to \bot).$$

We denote by $\mathfrak{D}^{\sharp}$ the set of all antichains in $\mathfrak{F}$ different from reflexive singletons. The formula $\alpha^{\sharp}(\mathfrak{F}, \bot) = \alpha(\mathfrak{F}, \mathfrak{D}^{\sharp}, \bot)$ is called the *frame formula for* $\mathfrak{F}$ (see [1], p. 312).

3 Logic L

In this section we construct a logic L satisfying the conditions of Theorem 1.

DEFINITION 2. Let $h : \omega \to \omega$ be some recursive function with a non-recursive range $Im\, h \subseteq \omega$. Let Y_{nm}, for $n, m \geq 0$, be the irreflexive transitive finite frame shown in Fig 1. Denote by Y the set of all such frames. (Note that $Y_{nm} = Y_{mn}$.)

Let la be the Gödel–Löb formula, i.e., $la = \Box(\Box p \to p) \to \Box p$. For any $n > 0$, denote by bw_n the formula

$$\bigwedge_{0 \leq i \leq n} \Diamond p_i \to \bigvee_{0 \leq i < j \leq n} (\Diamond(p_i \wedge (p_j \vee \Diamond p_j)) \vee \Diamond(p_j \wedge (p_i \vee \Diamond p_i))).$$

(The formula bw_n is called the *formula of width n*.)

Let $\Gamma = \{\alpha^{\sharp}(Y_{nm}, \bot) \mid h(n) \neq m, h(m) \neq n\}$. Now we can define the logic L we analyse in this paper.

DEFINITION 3. $L = K4 \oplus \{la, bw_2, \Box bw_1\} \oplus \Gamma$, where $\oplus$ assumes taking the closure under the standard inference rules of $K4$.

Since $bw_2 \in L$, it follows from Fine's theorem (see, e.g., [1], p. 354, p. 358) that L is Kripke complete.

DEFINITION 4. We say that a rooted frame $\mathfrak{F}$ is of *width n* if it contains an antichain of n worlds but no antichain of greater cardinality. A transitive rooted frame is called *pseudolinear* if its root is the only world from which

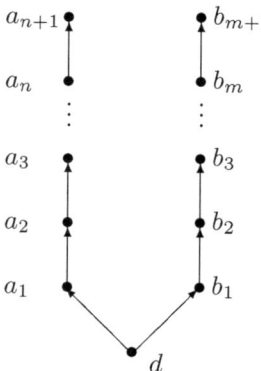

Figure 1

every antichain of the frame is accessible; in other words, in such a frame branching is possible only at the root.

The semantic meaning of the formulas la and bw_n is given in [1] (p. 80, p. 83): la is valid in those transitive Kripke frames that contain no infinite ascending chains, bw_n is valid in those rooted frames that are of width $\leq n$. For $\Box bw_1$, one can easily show that $\Box bw_1$ is valid in a frame $\mathfrak{F}$ iff $\mathfrak{F}$ is pseudolinear (the corresponding proof is similar to the one for bw_n presented in [1]).

Thus we have

LEMMA 5. *If $\mathfrak{F} \models L$, then $\mathfrak{F}$ is a transitive, irreflexive, pseudolinear frame of width ≤ 2 without infinite ascending chains.*

DEFINITION 6. Transitive, irreflexive, pseudolinear frames of width ≤ 2 without infinite ascending chains will be called *M-frames*. Denote by U the set of all M-frames with two dead ends.

Note that the set of all finite frames from U coincides with Y.

4 Technical results

Let us consider some properties of the frames in the sets M, U, Y and some properties of reductions of frames in $M \cup U \cup Y$ to frames in the same set.

LEMMA 7. *If $\mathfrak{F} = \langle W, R \rangle$ is a frame from U with root d, and $b_1, b_2 \in W$ are the dead ends of the frame, then*

- *the sets $C(b_i) = \{w \in W \mid wRb_i \vee w = b_i\}$, $i = 1, 2$, are maximal chains of $\mathfrak{F}$;*

- if $z \in W$ is not the root, then it belongs to exactly one chain in the frame;
- if $\mathfrak{c} \subseteq W$ is a chain of the frame $\mathfrak{F}$, then there exists a unique $i \in \{1,2\}$ such that $\mathfrak{c} \subseteq C(b_i)$;
- $\mathfrak{F}$ has only two maximal chains.

Proof. Let us show that $C(b_i) = \{w \in W \mid wRb_i \vee w = b_i\}$, $i = 1,2$, is a chain. Without loss of generality, let us assume that $i = 1$. Suppose that the set $\{z_1, z_2\}$, where $z_1, z_2 \in C(b_1)$, $z_1 \neq z_2$, is an antichain. Since the frame $\mathfrak{F}$ is pseudolinear and $z_1, z_2 \neq d$, we obtain that $\{z_1, z_2, b_2\}$ is an antichain. Now let us prove the maximality. Suppose that, for some $z \in W$ such that $z \notin C(b_1)$, the set $C(b_1) \cup \{z\}$ is a chain. Since $\neg zRb_1$, we have $b_1 R z$.

If $z \in W$ and $z \notin C(b_i)$, for any $i = 1,2$, then $\{b_1, b_2, z\}$ is an antichain. If $z \in W$ and $z \in C(b_i)$, for any $i = 1,2$, then $z = d$.

If $\mathfrak{c}$ is a chain in $\mathfrak{F}$ and $\mathfrak{c} \not\subseteq C(b_i)$, for any $i = 1,2$, then there exist two worlds $z_1, z_2 \in \mathfrak{c}$, such that $z_1 \notin C(b_1)$ and $z_2 \notin C(b_2)$ ($z_1, z_2 \neq d$). Therefore $z_1 \in C(b_2)$ and $z_2 \in C(b_1)$. If $z_1 R z_2$, then $z_1 R b_1$. If $z_2 R z_1$, then $z_2 R b_2$.

The last statement follows immediately from the previous one. ∎

By the definition of subreduction we obtain the following technical results.

LEMMA 8. Suppose $\mathfrak{F} = \langle W, R \rangle$ and $\mathfrak{G} = \langle V, S \rangle$ are frames, $\{u_1, \ldots, u_n\}$ is an antichain in the frame $\mathfrak{G}$, $f : \mathfrak{F} \to \mathfrak{G}$ is a subreduction of $\mathfrak{F}$ to $\mathfrak{G}$, and $x_1 \in f^{-1}(u_1), \ldots, x_n \in f^{-1}(u_n)$. Then $\{x_1, \ldots, x_n\}$ is an antichain in $\mathfrak{F}$.

Proof. Let $x_i \in f^{-1}(u_i)$, $x_j \in f^{-1}(u_j)$, $i \neq j$, and $x_i R x_j$. It follows from the definition of subreduction that $f(x_i) S f(x_j)$, and so $u_i S u_j$. ∎

COROLLARY 9. Suppose that $\mathfrak{F} = \langle W, R \rangle$ is an M-frame, $\mathfrak{G} = \langle V, S \rangle$ is an M-frame of width 2, and there exists a subreduction $f : \mathfrak{F} \to \mathfrak{G}$. Then $\mathfrak{F}$ is a frame of width 2.

LEMMA 10. Suppose $\mathfrak{F} = \langle W, R \rangle$ and $\mathfrak{G} = \langle V, S \rangle$ are frames, S is an irreflexive relation on V, $f : \mathfrak{F} \to \mathfrak{G}$ is a subreduction of $\mathfrak{F}$ to $\mathfrak{G}$, and $v \in V$. Then $f^{-1}(v)$ is an antichain.

Proof. Let $x, y \in f^{-1}(v)$, $x \neq y$, and xRy. By the definition of subreduction, we have $f(x) S f(y)$, and so vSv. ∎

LEMMA 11. Let $\mathfrak{F} = \langle W, R \rangle$ and $\mathfrak{G} = \langle V, S \rangle$ be frames in U and $f : \mathfrak{F} \to \mathfrak{G}$ be a subreduction of $\mathfrak{F}$ to $\mathfrak{G}$. If v is a root of the frame $\mathfrak{G}$, then $\|f^{-1}(v)\| = 1$ and $f^{-1}(v)$ is a root of the frame $\mathfrak{F}$.

Proof. Let $\{z_1, z_2\}$ be an antichain of $\mathfrak{G}$. If $x \in f^{-1}(v)$ and x is not the root $\mathfrak{F}$, then there exist worlds $y_1 \in f^{-1}(z_1)$ and $y_2 \in f^{-1}(z_2)$ such that xRy_1 and xRy_2. Since $\{y_1, y_2\}$ is an antichain, we obtain a contradiction with $\mathfrak{F}$ being pseudolinear. ∎

LEMMA 12. *If* $\mathfrak{F} = \langle W, R \rangle$ *and* $\mathfrak{G} = \langle V, S \rangle$ *are frames in* U *and* $f : \mathfrak{F} \to \mathfrak{G}$ *is a subreduction of* $\mathfrak{F}$ *to* $\mathfrak{G}$, *then for any world* $v \in V$, *we have* $\|f^{-1}(v)\| = 1$.

Proof. It is sufficient to consider the case when v is not a root of $\mathfrak{G}$. Suppose there are $x_1, x_2 \in W$ such that $x_1 \neq x_2$, $\{x_1, x_2\} \subseteq f^{-1}(v)$. Then $\{x_1, x_2\}$ is an antichain. Let b_1, b_2 be dead ends of the frame $\mathfrak{G}$ and $v \in C(b_1)$. Then $x_1, x_2 \in C(f^{-1}(b_1))$. Note that $f^{-1}(b_2) \neq \varnothing$. Let $y \in f^{-1}(b_2)$. Then $\{x_1, x_2, y\}$ is an antichain. It follows that $\mathfrak{F}$ is a frame of width 3. ∎

COROLLARY 13. *If* $\mathfrak{F} = \langle W, R \rangle$ *and* $\mathfrak{G} = \langle V, S \rangle$ *are frames in* U *and* $f : \mathfrak{F} \to \mathfrak{G}$ *is a subreduction of* $\mathfrak{F}$ *to* $\mathfrak{G}$, *then for any worlds* $w, v \in W$, *we have* $f(w)Sf(v)$ *iff* wRv.

Proof. Let $x = f(w)$, $y = f(v)$, and xSy. It follows from the definition of subreduction that $f(w)Sf(v)$ implies $\exists u \in \mathrm{Dom}\, f(wRu \land f(u) = f(v))$. If $u \neq v$ then $\|f^{-1}(y)\| > 1$. We obtain a contradiction with Lemma 12. Hence $u = v$, and so wRv. ∎

COROLLARY 14. *If* $\mathfrak{F} = \langle W, R \rangle$ *and* $\mathfrak{G} = \langle V, S \rangle$ *are frames from* U *and* $f : \mathfrak{F} \to \mathfrak{G}$ *is a subreduction of* $\mathfrak{F}$ *to* $\mathfrak{G}$, *then* f *is an isomorphism between* $\mathrm{Dom}\, f$ *and* V.

COROLLARY 15. *Suppose that* $\mathfrak{F} = \langle W, R \rangle$ *and* $\mathfrak{G} = \langle V, S \rangle$ *are frames from* U, *and* $f : \mathfrak{F} \to \mathfrak{G}$ *is a subreduction. Then we have:*

- *if* $\mathfrak{c} \subseteq W$ *is a chain such that* $f(\mathfrak{c}) \neq \varnothing$, *then* $f(\mathfrak{c})$ *is a chain of* $\mathfrak{G}$;

- *for any chain* $\mathfrak{d} \subseteq V$, *the set* $f^{-1}(\mathfrak{d})$ *is a chain of* $\mathfrak{F}$.

LEMMA 16. *Let* $\mathfrak{F} = \langle W, R \rangle$ *and* $\mathfrak{G} = \langle V, S \rangle$ *be frames and* $f : \mathfrak{F} \to \mathfrak{G}$ *be a cofinal subreduction of* $\mathfrak{F}$ *to* $\mathfrak{G}$. *If* $b \in V$ *is a dead end of* $\mathfrak{G}$ *then every world in* $f^{-1}(b)$ *is a dead end*.

Proof. If $x \in f^{-1}(b)$ and x is not a dead end, then there is $y \in W$ such that xRy. Therefore $y \in (\mathrm{Dom}\, f)\uparrow$. Now, by the definition of cofinal subreduction, we obtain the either $y \in \mathrm{Dom} f$ or there exists $z \in \mathrm{Dom}\, f$ such that $xRyRz$. Therefore, it follows that there is a world $w \in \{f(y), f(z)\}$ such that bSw. Hence b is not a dead end of the frame $\mathfrak{G}$. ∎

Since any frame from U has two dead ends, we obtain

COROLLARY 17. *If* $\mathfrak{F} = \langle W, R \rangle$ *is an* M-*frame*, $\mathfrak{G} = \langle V, S \rangle$ *is a frame from* U, $f : \mathfrak{F} \to \mathfrak{G}$ *is a cofinal subreduction, then* $\mathfrak{F} \in U$.

LEMMA 18. *Let* $\mathfrak{F} = \langle W, R \rangle$ *be a frame from* U, $\mathfrak{G} = \langle V, S \rangle \in Y$, *a world* $b_1 \in V$ *be a dead end, and* $\mathfrak{c} \subseteq W$ *be a maximal chain of* $\mathfrak{F}$. *Let* $f : \mathfrak{F} \to \mathfrak{G}$ *be a cofinal subreduction such that* $f^{-1}(C(b_1)) \subseteq \mathfrak{c}$. *Let* $\mathfrak{D}$ *be a set of antichains in* $\mathfrak{G}$ *and all singletons belong to* $\mathfrak{D}$. *If* $\|C(b_1)\| < \|\mathfrak{c}\|$, *then* f *does not satisfy* (CDC) *for* $\mathfrak{D}$.

Proof. Since f is one-to-one mapping, we have $\|f^{-1}(C(b_1))\| < \|\mathfrak{c}\|$. So there exists $x \in \mathfrak{c} \setminus f^{-1}(C(b_1))$ such that $x \notin Dom\, f$. Since x is not a dead end, we obtain that $x \in (Dom\, f)\downarrow$.

Let us denote by a_1 a dead end of $\mathfrak{F}$ such that $f(a_1) = b_1$. Note that $a_1 \in x\uparrow \cap\, Dom\, f$, and hence $x\uparrow \cap\, Dom\, f \neq \varnothing$. Since $\mathfrak{G}$ is a finite frame, it follows that there is a world $y \in x\uparrow \cap\, Dom\, f$ such that for any $z \in \mathfrak{c}$, the relation $xRzRy$ implies $z \notin Dom\, f$.

Let us prove that $f(x\uparrow) = f(y)\uparrow$. Since xRy, we have $f(y)\uparrow \subseteq f(x\uparrow)$. Suppose $w \neq f(y)$ and $w \in f(x\uparrow)$. Then there exists $z \in x\hat{}$, such that $f(z) = w$. Since $x\uparrow$ is a chain, we obtain yRz. This means that $f(y)Sw$. Therefore, we have $f(x\uparrow) = f(y)\uparrow$. Since $\{f(y)\} \in D$, we obtain a contradiction with (CDC) for $\mathfrak{D}$. ∎

LEMMA 19. *Let $\mathfrak{F} = \langle W, R \rangle$ be a frame in U, and worlds b_1, b_2 be the dead ends of $\mathfrak{F}$. If for some $n, m, \geq 0$ we have either $\|C(b_1)\| \neq n+1$ and $\|C(b_1)\| \neq m+1$, or $\|C(b_2)\| \neq n+1$ and $\|C(b_2)\| \neq m+1$, then there is no cofinal subreduction $f : \mathfrak{F} \to Y_{n,m}$ satisfying (CDC) for $\mathfrak{D}$, where $\mathfrak{D}$ is a set of antichains in the frame $Y_{n,m}$ and all singletons belong to $\mathfrak{D}$.*

Proof. Suppose that w_1, w_2 are dead ends of $Y_{n,m}$ and $\|C(b_1)\| = r$, $\|C(w_1)\| = n+1$, $\|C(w_2)\| = m+1$. Without loss of generality, we may assume that $r < n+1$, $r > m+1$. In this case we have $\|C(b_1)\| < \|C(w_1)\|$. By Corollary 14, there is no subreduction f such that $f(C(b_1)) \subseteq C(w_1)$, for otherwise we would have $\|C(b_1)\| > \|C(w_2)\|$. By taking into account Lemma 18, we obtain that a subreduction f with $f(C(b_1)) \subseteq C(w_2)$, does not satisfy (CDC) for $\mathfrak{D}$. ∎

LEMMA 20. *Let $\mathfrak{F} = \langle W, R \rangle$ be a frame from U, $\mathfrak{G} = \langle V, S \rangle \in Y$, $f : \mathfrak{F} \to \mathfrak{G}$ a subreduction. The subreduction f satisfies (CDC) for any antichain $\{u_1, u_2\}$ of the frame.*

Proof. Let us show that for every $x \in W$ if $f(x\uparrow) = \{u_1, u_2\}\uparrow$ then $x \in Dom\, f$. Suppose $f(y_1) = u_1$, $f(y_2) = u_2$. Then $y_1, y_2 \in x\hat{}$. It follows from Lemma 7 that $\{y_1, y_2\}$ is antichain. Hence x is a root of $\mathfrak{F}$. Using Lemma 11, we obtain $x \in Dom\, f$. ∎

From Lemma 19 and Lemma 20 we obtain the following

LEMMA 21. *Suppose that $\mathfrak{D}$ is a set of antichains in the frame $Y_{n,m}$ and that all singletons belong $\mathfrak{D}$. There exists a cofinal subreduction from Y_{rs} to Y_{nm} satisfying (CDC) for $\mathfrak{D}$ iff $r = n$ and $s = m$ or $r = m$ and $s = n$.*

5 Undecidability of L_{fin}

Let L_{fin} be the logic of all finite frames for the logic L defined above. Let us prove that L_{fin} is undecidable. For the proof we need the following formulas:

$$\gamma_{nm} = (\Diamond(\Box\bot \wedge p) \wedge \Diamond(\Box\bot \wedge \neg p)) \to$$
$$((\Diamond^{n+1}(\Box\bot \wedge p) \wedge \neg\Diamond^{n+2}(\Box\bot \wedge p)) \to$$
$$(\Diamond^{m+1}(\Box\bot \wedge \neg p) \wedge \neg\Diamond^{m+2}(\Box\bot \wedge \neg p))),$$

where n, m are arbitrary natural numbers.

LEMMA 22. *The formula γ_{nm} is refuted in a finite frame $\mathfrak{F} = \langle W, R \rangle$ for the logic L iff $\mathfrak{F}$ is isomorphic to Y_{rs}, where $n = s$ and $m \neq r$.*

Proof. ($\Rightarrow$) Suppose that, for some valuation ν, we have $w \not\models^\nu \gamma_{nm}$. Then $w \models^\nu \Diamond(\Box\bot \wedge p) \wedge \Diamond(\Box\bot \wedge \neg p)$. Thus two different dead ends of $\mathfrak{F}$ are accessible from the world w. It follows that $\mathfrak{F} \in U$ and the world w is a root of $\mathfrak{F}$. Since $\mathfrak{F}$ is finite, we obtain that $\mathfrak{F} \in Y$.

Denote by b_1, b_2 the dead ends in $\mathfrak{F}$. Suppose $b_1 \in \nu(p)$ and $b_2 \notin \nu(p)$. As $(\Diamond^{n+1}(\Box\bot \wedge p) \wedge \neg\Diamond^{n+2}(\Box\bot \wedge p)) \to (\Diamond^{m+1}(\Box\bot \wedge \neg p) \wedge \neg\Diamond^{m+2}(\Box\bot \wedge \neg p))$ is refuted at the root w, we obtain $w \models^\nu \Diamond^{n+1}(\Box\bot \wedge p) \wedge \neg\Diamond^{n+2}(\Box\bot \wedge p)$. Hence the world b_1 is accessible from the root in $n+1$ steps, that is, there is a maximal chain $x_1 R x_2 R \ldots R x_{n+2}$, where $x_1 = w$ and $x_{n+2} = b_1$. Otherwise we have $w \not\models^\nu \Diamond^{m+1}(\Box\bot \wedge \neg p) \wedge \neg\Diamond^{m+2}(\Box\bot \wedge \neg p)$. Therefore b_2 is not accessible in $m+1$ steps and $\|C(b_2)\| \neq m+2$.

($\Leftarrow$) This direction is trivial. ∎

Observe that the frame Y_{nm} is a frame for the logic L (this means that Y_{nm} is a frame for L_{fin} as well) iff $m = h(n)$. Thus we have

LEMMA 23. *We have $\gamma_{nh(n)} \notin L_{fin}$ iff $n \in Im\, h$, for any $n \geq 0$ such that $n \neq h(n)$ and $n \neq h^2(n)$.*

Proof. ($\Leftarrow$) Suppose $n \in Im\, h$. Then there exists $k \geq 0$ such that $h(k) = n$. Since $n \neq h(n)$ and $h^2(n) \neq n$, we obtain $k \neq n$ and $h(n) \neq k$. Thus the formula $\gamma_{nh(n)}$ is refuted in the frame Y_{kn}. Otherwise, since $h(k) = n$, we have that the frame Y_{kn} is a frame for the logic L_{fin}.

($\Rightarrow$) Suppose that $\gamma_{nh(n)} \notin L_{fin}$. Then there is a frame $\mathfrak{F}$ for L_{fin} such that $\mathfrak{F} \not\models \gamma_{nh(n)}$. Thus by Lemma 22 the frame $\mathfrak{F}$ is isomorphic to Y_{rs} with $r = n$ and $s \neq h(n)$. Otherwise, since Y_{rs} is a frame for L_{fin}, we have $s = h(r)$ or $r = h(s)$. Let $s = h(r)$, that is, $s = h(n)$. It follows that $r = h(s)$, and so $n = h(s)$ for any $s \geq 0$. Thus $n \in Im\, h$. ∎

COROLLARY 24. *The logic L_{fin} is undecidable.*

Proof. Suppose the logic L_{fin} is decidable. Let us show that in this case the set $Im\, h$ is recursive. Indeed, for any $n \geq 0$, we can easily verify the conditions $n = h(n)$ and $n = h^2(n)$, from which it follows that $n \in Im\, h$. If these conditions are not satisfied, then we can use the formulas $\gamma_{nh(n)}$: if $\gamma_{nh(n)} \notin L_{fin}$, then $n \in Im\, h$. ∎

6 Decidability of L

Let us prove that the logic L is decidable.

LEMMA 25. *Let $\mathfrak{F}$ be an M-frame and let $\mathfrak{G}$ be a frame satisfying one of the following conditions:*

- *it is of width ≥ 3;*
- *it contains a reflexive world;*
- *it is not pseudolinear.*

Then there is no subreduction of $\mathfrak{F}$ to $\mathfrak{G}$.

Proof. If the frame $\mathfrak{G}$ contains a three-element antichain, then by Lemma 8, its preimage must also contain an antichain of cardinality ≥ 3.

If $\mathfrak{G}$ contains a reflexive world, then its preimage is an infinite ascending chain.

If the frame $\mathfrak{G}$ contains an antichain which is accessible from some non-root world w, then the pseudolinearity of the $\mathfrak{F}$ implies that the preimage of w is a root of the frame $\mathfrak{F}$. ∎

LEMMA 26. *Let $\mathfrak{D}$ be a set of antichains in $Y_{n,m}$. If $\mathfrak{D}$ does not contain all non-root singletons, then there is an infinite frame of the logic L such that it is cofinally subreducible to $Y_{n,m}$ and the corresponding subreduction satisfies (CDC) for $\mathfrak{D}$.*

Proof. A maximal chain in the set of all chains of the form $\{a_1, \ldots, a_s, \ldots, a_k\}$ (or $\{b_1, \ldots, b_s, \ldots, b_k\}$) from a frame $Y_{n,m}$ (where s takes every value in interval $[1,k]$, $1 \leq k \leq l$, $l \in \{n,m\}$; see Fig. 1) such that $\{a_i\} \in \mathfrak{D}$ (respectively, $\{b_i\} \in \mathfrak{D}$), $1 \leq i \leq s$, is called a *rooted segment* for the pair $\langle Y_{n,m}, \mathfrak{D} \rangle$. A rooted segment is called a *full rooted segment* if $s = l + 1$.

Let us consider first the case when neither of the rooted segments is full and one of them contains s worlds and the other one contains r worlds. Denote by U_{sr} the infinite frame shown in Fig. 2.

By using Lemma 22, it is not difficult to prove that the frame U_{sr} is a frame for L. A cofinal subreduction f of U_{sr} to Y_{nm} satisfying (CDC) for $\mathfrak{D}$ is defined as follows:

- $f(x_i) = a_i, f(v_k) = a_{n+2-k}, 1 \leq k \leq n+1-s, 1 \leq i \leq s$;
- $f(y_j) = b_j, f(u_l) = b_{m+2-l}, 1 \leq l \leq m+1-r, 1 \leq j \leq r$;
- $f(d_1) = d$.

If one of the rooted segment is full (i.e., $s = n+1$), then consider the infinite frame for L shown in Fig. 3, which will be denoted by X_{nr}. The cofinal subreduction f of X_{nr} to Y_{nm} satisfying (CDC) for $\mathfrak{D}$ is defined as follows:

- $f(x_i) = a_i, 1 \leq i \leq n+1$;

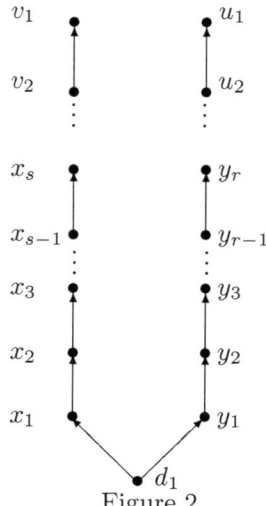

Figure 2

- $f(y_j) = b_j, f(u_l) = b_{m+2-l}, 1 \leq l \leq m+1-r, 1 \leq j \leq r$;
- $f(d_2) = d$.

Observe that $f(d_1) = d$, $f(d_2) = d$, $f(v_1) = a_{n+1}$, $f(u_1) = b_{m+1}$, and $f(x_s) = a_{n+1}$. Thus $w \in Dom\ f{\uparrow} \cap Dom\ f{\overline{\downarrow}}$, for any world w in U_{sr} or X_{nr}, i.e., f is cofinal. For (CDC), by Lemma 20, it is sufficient to observe that (CDC) holds for any one-element subset of $\mathfrak{D}$. Let $\{w\} \in \mathfrak{D}$. If w is in a rooted segment, then $f(z{\uparrow}) = w{\uparrow}$ implies $z = x_i$ or $z = y_j$, where $1 \leq i \leq s$, $1 \leq j \leq r$; in both cases $z \in Dom\ f$. If w is not in any rooted segment and $\{w\} \in \mathfrak{D}$ then $f(z{\uparrow}) = w{\uparrow}$ implies $z = v_k$ or $z = u_l$, where $1 \leq k < n+1-s$, $1 \leq l < m+1-r$; again $z \in Dom\ f$. ∎

LEMMA 27. *The logic L is decidable.*

Proof. Here we describe an algorithm deciding, for an arbitrary formula φ, whether it belongs to the logic L or not. It is proved in [6] (see also [1], p. 309) that there is an algorithm which for a given modal formula φ, returns a finite set of pairs $\{\langle \mathfrak{F}_1, \mathfrak{D}_1 \rangle, \ldots, \langle \mathfrak{F}_n, \mathfrak{D}_n \rangle\}$ (where $\mathfrak{F}_i$ is a finite rooted frame and $\mathfrak{D}_i$ is a set of antichains in it) such that, for any frame $\mathfrak{F}$, $\mathfrak{F} \not\models \varphi$ iff there is a cofinal subreduction $f : \mathfrak{F} \to \mathfrak{F}_i$, for some $1 \leq i \leq n$, satisfying (CDC) for $\mathfrak{D}_i$.

For every frame of the list $\mathfrak{F}_1 \ldots \mathfrak{F}_n$ we check whether this frame is a finite M-frame or not. If M-frames do not belong to the list, then for any frame $\mathfrak{F}$ for L and any frame $\mathfrak{F}_i$ of the list, there is no subreduction of $\mathfrak{F}$ to $\mathfrak{F}_i$. Hence $\varphi \in L$.

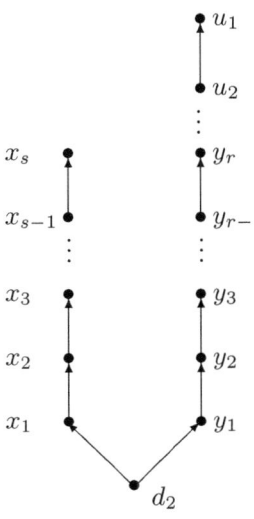

Figure 3

Suppose the list contains some M-frames. Denote them by $\mathfrak{F}_{\alpha_1} \ldots \mathfrak{F}_{\alpha_m}$, $m \leq n$. Suppose $\mathfrak{F}_{\alpha_i} \notin Y$, for some $1 \leq i \leq m$. Then $\mathfrak{F}_{\alpha_i}$ is frame for L, and so $\varphi \notin L$.

Let $\{\mathfrak{F}_{\alpha_1} \ldots \mathfrak{F}_{\alpha_m}\} \subseteq Y$. Then every frame $\mathfrak{F}_{\alpha_i}$ is equal to $Y_{n,m}$, for some pair (n,m). We can check that $m = h(n)$, for any such frame. If there is a pair (n,m) such that $m = h(n)$, then $Y_{n,m}$ is a frame of the logic L. Hence $\varphi \notin L$.

Suppose that the condition $m = h(n)$ does not hold for any frame of the list. Then by Lemma 20 and Lemma 26 it is enough to consider the rooted segments for the pairs $\langle \mathfrak{F}_{\alpha_i}, \mathfrak{D}_{\alpha_i} \rangle$ of the list. Since every set $\mathfrak{D}_{\alpha_i}$ of the list is finite, we can effectively check if the rooted segment corresponding to the pair $\langle \mathfrak{F}_{\alpha_i}, \mathfrak{D}_{\alpha_i} \rangle$ is full. If it is so for the pair $\langle \mathfrak{F}_{\alpha_i}, \mathfrak{D}_{\alpha_i} \rangle$ of the list, then any cofinal subreduction of $\mathfrak{F}_{\alpha_i}$ to some frame $\mathfrak{G}$ satisfies (CDC) for $\mathfrak{D}_{\alpha_i}$ only if $\mathfrak{G}$ is isomorphic to $\mathfrak{F}_{\alpha_i}$. Hence, if both rooted segments are full, for every pair $\langle \mathfrak{F}_{\alpha_i}, \mathfrak{D}_{\alpha_i} \rangle$ of the list, then $\varphi \in L$.

If it is not the case, then by Lemma 26, $\varphi \notin L$.

This completes our description of the decision algorithm. ∎

Theorem 1 is follows immediately from Corollary 24 and Lemma 27.

7 Acknowledgements

I am indebted to my Teacher Alexander Chagrov. I am also grateful to the referees for their constructive and friendly critique, to Mikhail Rybakov for his substantial technical help, and to Michael Zakharyaschev for his comments on the original version of this paper.

BIBLIOGRAPHY

[1] A. Chagrov and M. Zakharyaschev. *Modal Logic.* Oxford University Press, 1997.
[2] Yu. Ershov, I. Lavrov, A. Tatmanov, and Taitslin. Elementary theories. *Russian Mathematical Surveys*, 20(4):35–105, 1965.
[3] J. Jeong. A decidable variety that is finetely undecidable. *The Journal of Symbolic Logic*, 64:651–677, 1999.
[4] A. Zamyatin. Varieties of associative rings whose elementary theory is decidable. 17(4):996–999, 1976.
[5] A. Zamyatin. A non-abelian variety of group has an undecidable elementary theory. *Algebra and Logic*, 17:13–17, 1978.
[6] M. Zakharyaschev. Canonical formulas for $K4$. Part I: Basic results. *Journal of Symbolic Logic*, 57:1377–1402, 1992.

Igor Gorbunov
Department of Mathematics
Tver State University
Zhelyabova Street, 33
Tver
170000
Russia
i_gorbunov@mail.ru

Regarding Overlaps in 'Topologic'

BERNHARD HEINEMANN

ABSTRACT. We study a particular extension of Moss and Parikh's logic of knowledge and effort with regard to covering properties of spaces of sets. As is known, one can only specify properties of points close to the actual one by means of formulas of that logic, and this may be viewed as a deficiency. In this paper, we introduce a new modality describing overlaps, which is intended to rectify that shortcoming. By cooperating with knowledge the overlap operator in fact allows us to access remote points as well. This could be interesting in connection with a modal treatment of compactness. The outcome of the paper is a soundness and completeness theorem for the logic of set spaces including the overlap operator, the decidability of this logic, some interesting connections with the usual logic of knowledge, and a hybrid version of the logic of overlapping which appears in a sense more natural than previous hybridizations of the initial system.

1 Introduction

The starting point to this paper is Moss and Parikh's approach to reasoning about knowledge of some agent; see [3]. The bi-modal system invented there facilitates a fairly abstract description of procedures gaining knowledge. As a by-product, a certain *topological component* of knowledge was revealed. In fact, since knowledge is represented by the collection of all *knowledge states* of the agent, knowledge acquisition appears as a *shrinking procedure* regarding this space of sets. In this way, some particular concepts from topology like *closeness* or *neighbourhood* turn up naturally in connection with knowledge.

Moss and Parikh called one of their system topologic, and we adopt this naming here. We recall some features of the language underlying topologic. As it has just been indicated, formulas may contain two one-place operators: a modality K representing knowledge of the agent and another one, $\Box$, representing (computational) effort. The domains for evaluating formulas are triples $(X, \mathcal{O}, V)$ which we call *set spaces*, consisting of a non-empty set X of states, a set $\mathcal{O}$ of subsets of X representing the knowledge states of the agent, and a valuation V determining the states where the atomic propositions are true. The operator K then quantifies across any knowledge state $U \in \mathcal{O}$, whereas $\Box$ quantifies 'downward' across $\mathcal{O}$. That is, more knowledge, i.e., a closer proximity to states of 'complete' knowledge, can be achieved by *decreasing* (with respect to the set inclusion relation), and just this is modeled by $\Box$.

Some classes of set spaces including the ordinary topological ones, could be characterized by means of **topologic** subsequently; cf [5, 6, 10, 20]. However, more expressive power is needed to specify less basic properties of scenarios in which topology plays a part.

To this end, both a *temporal* and a *hybrid* version of **topologic** were introduced in [11] and [12], respectively, and further developed later on. Hybrid **topologic**, in particular, turned out to be quite a useful system; cf, eg, [15].

In this paper we nevertheless go 'back to the roots'. That is, we do neither change the original operators of **topologic** (as in the temporal setting) nor go beyond the purely modal framework (as it was done by using hybrid means of expression). Instead, we add a further 'spatial' modality, called the *overlap operator,* to the basic system and investigate the accompanying modal logic. (This is at least true up to the last but one section, where we hybridize that logic though.)

We now motivate the introduction of the new modality. Due to the semantics of **topologic**, formulas can only speak about properties of points close to the actual one; in other words, by evaluating a formula we cannot 'leave' the neighbourhood in question (see Definition 2 below). The overlap operator, however, will allow us to do exactly this. Thus we will have tools to hand for quantifying not only 'downward' but also 'diagonally' across $\mathcal{O}$. This may be viewed as a very first step towards a modal treatment of *coverings* of spaces. (Unfortunately, we have to confine ourselves to this first step here. The long-term goal is a treatment of *compactness,* actually, but we are still far away from this.)

It should be remarked that the investigation into overlapping is new for **topologic**, but not for modal logic in general. Overlapping is a very natural spatial concept and occurs, therefore, in many papers on modal logics of space. Standing in for all of them we mention two papers from different areas of application, [18] and [19].

The subsequent technical part of this paper is organized as follows. In the next section we integrate the overlap operator into the language of **topologic**. We give some examples of valid formulas of the extended language there, too, and we touch on the question of expressive power. Afterwards, in Section 3, we present an axiomatization of the new logic and prove its soundness and completeness with respect to the class of all set spaces. We also show in this section that the set of all theorems of the logic is decidable. In Section 4, we take a small step further towards coverings of set spaces by considering the reflexive and transitive closure of the relation which results from joining overlapping with moving inside sets. We prove (only) Kripke completeness and decidability in this case, and we comment on an interesting connection with the common logic of knowledge. (See the textbooks [4] or [17] for an exhaustive treatment regarding this.) We then discuss a hybrid extension of the logic of overlapping. This system once again demonstrates the power of names in the context of **topologic**. Moreover, it is also more natural (in a sense that will be made precise later on) than other approaches to this topic in the past (cf [13, 14]). Finally, we give a brief summary of the paper and point to further aspects and future research.

Apart from the appendix, which relies on several concepts and results from [3] and [14], respectively, the present paper is, taken by and large, self-contained.

2 Incorporating the overlap operator

In this section, the language of **topologic** is extended by a unary operator, O, describing overlaps of the actual neighbourhood situation. We first define the new syntax and semantics, and give some comments related to expressiveness afterwards.

Let PROP $= \{A, B, \ldots\}$ be a denumerable set of symbols called *proposition letters*. We define the set WFF of *well-formed formulas over* PROP by the rule

$$\alpha ::= A \mid \neg \alpha \mid \alpha \wedge \beta \mid \mathsf{K}\alpha \mid \Box \alpha \mid \mathsf{O}\alpha.$$

The operators K and $\Box$ represent *knowledge* and *effort*, respectively, as it is common for **topologic**. O is called the *overlap operator*. The missing boolean connectives $\top, \bot, \vee, \rightarrow, \leftrightarrow$ are treated as abbreviations. as needed. The duals of K, $\Box$ and O are denoted L, $\Diamond$ and P, respectively.

We now turn to semantics. First, we define the domains for interpreting formulas. For a given set X, we let $\mathcal{P}(X)$ denote the powerset of X.

DEFINITION 1 (Set frames and spaces).

1. Let $X \neq \emptyset$ be a set and $\mathcal{O} \subseteq \mathcal{P}(X)$ a set of subsets of X. Then $\mathcal{F} := (X, \mathcal{O})$ is called a *set frame*.

2. Let $\mathcal{F} = (X, \mathcal{O})$ be a set frame. The *set $\mathcal{N}_\mathcal{F}$ of neighbourhood situations* of $\mathcal{F}$ is defined by

$$\mathcal{N}_\mathcal{F} := \{(x, U) \mid x \in U \text{ and } U \in \mathcal{O}\}.$$

(Mostly, neighbourhood situations are written without brackets later on.)

3. Let $\mathcal{F}$ be as above. A mapping $V : \text{PROP} \longrightarrow \mathcal{P}(X)$ is called an $\mathcal{F}$-*valuation*.

4. A *set space* is a triple $\mathcal{M} := (X, \mathcal{O}, V)$, where $\mathcal{F} := (X, \mathcal{O})$ is a set frame and V an $\mathcal{F}$-valuation; $\mathcal{M}$ is then called *based on* $\mathcal{F}$.

The relation of satisfaction between neighbourhood situations and formulas, is now defined with regard to set spaces.

DEFINITION 2 (Satisfaction and validity). Let $\mathcal{M} = (X, \mathcal{O}, V)$ be a set space.

1. Let x, U be a neighbourhood situation of $\mathcal{F} = (X, \mathcal{O})$. Then

$$\begin{aligned}
x, U \models_\mathcal{M} A &: \iff x \in V(A) \\
x, U \models_\mathcal{M} \neg \alpha &: \iff x, U \not\models_\mathcal{M} \alpha \\
x, U \models_\mathcal{M} \alpha \wedge \beta &: \iff x, U \models_\mathcal{M} \alpha \text{ and } x, U \models_\mathcal{M} \beta \\
x, U \models_\mathcal{M} \mathsf{K}\alpha &: \iff \forall y \in U : y, U \models_\mathcal{M} \alpha \\
x, U \models_\mathcal{M} \Box\alpha &: \iff \forall U' \in \mathcal{O} : (x \in U' \subseteq U \Rightarrow x, U' \models_\mathcal{M} \alpha) \\
x, U \models_\mathcal{M} \mathsf{O}\alpha &: \iff \forall U' \in \mathcal{O} : (x \in U' \Rightarrow x, U' \models_\mathcal{M} \alpha),
\end{aligned}$$

where $A \in \text{PROP}$ and $\alpha, \beta \in \text{WFF}$. In case $x, U \models_\mathcal{M} \alpha$ is true we say that α *holds in* $\mathcal{M}$ *at* the neighbourhood situation x, U.

2. A formula α is called *valid in* $\mathcal{M}$ (written '$\mathcal{M} \models \alpha$'), iff it holds in $\mathcal{M}$ at every neighbourhood situation of the frame $\mathcal{M}$ is based on.

The following comments on this definition may be appropriate. The meaning of proposition letters is independent of neighbourhoods by definition, thus 'stable' with respect to $\Box$ and O. This fact is reflected by a special axiom below. – Intuitively, shrinking is a special case of overlapping. This is captured by the exact definition as well. – Note that the semantics of the overlap operator 'dismisses' the set component of the actual neighbourhood situation. Because of that, this operator turns out to be a very good-natured modality; see below.[1] – The general case of overlapping can be obtained by an iterated application of O and K; see Section 4.

Now, we look for the basic set space validities. The formula schema

$$\mathsf{K}\Box\alpha \to \Box\mathsf{K}\alpha$$

is the typical validity of **topologic**. This schema was called the *Cross Axiom* in the paper [3] and plays a key role in the completeness and the decidability proof for that logic. The Cross Axiom describes the interaction between knowledge and effort. It is now natural to ask whether knowledge and overlapping cooperate in a similar manner. The reader can easily convince himself or herself that the answer to this question is 'no'. On the other hand, the effort and the overlap operator are obviously related as displayed by item 1 of the following proposition. Besides that we state that O satisfies all the S5 laws.

PROPOSITION 3. *Let* $\mathcal{M}$ *be any set space and* $\alpha \in \text{WFF}$ *a formula. Then*

1. $\mathcal{M} \models \mathsf{O}\alpha \to \Box\alpha$,

2. $\mathcal{M} \models \mathsf{O}\alpha \to \alpha \wedge \mathsf{OO}\alpha$,

3. $\mathcal{M} \models \alpha \to \mathsf{OP}\alpha$.

---

[1] One of the anonymous referees pointed to the kinship between the overlap operator and the *bi-persistent sentences* of **topologic** (cf [3], Sec. 3.1). In fact, every formula $\mathsf{O}\alpha$ proves to be bi-persistent. However, the overlap operator is not definable in **topologic**.

The proof of Proposition 3 is straightforward from Definition 2 and, therefore, omitted.

Finally in this section, we ask for interesting properties that can be expressed with the aid of the new language. We give a simple, but instructive example.

DEFINITION 4. Let $\mathcal{F} = (X, \mathcal{O})$ be a set frame. $\mathcal{F}$ is called *directed*, iff for all $x \in X$ and any $U, U' \in \mathcal{O}$ that overlap at x there exists some $\tilde{U} \in \mathcal{O}$ such that $x \in \tilde{U} \subseteq U \cap U'$.

Directed frames were studied in [20], where it was shown among other things that the class of all directed frames is not finitely axiomatizable in the language of topologic. (As to a treatment of this class by means of hybrid logic and comments on the arising differences, cf [15].) We now show that the new language is strong enough to capture directedness.

PROPOSITION 5. *Let $\mathcal{F} = (X, \mathcal{O})$ be a set frame. Then, $\mathcal{F}$ is directed iff for all set spaces $\mathcal{M}$ based on $\mathcal{F}$ the formula schema*

$$\mathsf{P}\Box\alpha \to \Diamond\alpha$$

is valid in $\mathcal{M}$.

Proof. We only prove the sufficiency of the condition. To this end, suppose that $\mathcal{F}$ is not directed. Then there are $x \in X$ and $U, U' \in \mathcal{O}$ such that $x \in U \cap U'$, but for all $\tilde{U} \in \mathcal{O}$ containing x we have that $\tilde{U} \not\subseteq U$ or $\tilde{U} \not\subseteq U'$. Now, take a proposition letter $A \in \text{PROP}$ and any $\mathcal{F}$-valuation V satisfying $V(A) = U'$. Let $\mathcal{M}$ be the resulting set space and $\alpha := \mathsf{K}A$. Then $x, U' \models_\mathcal{M} \Box\alpha$, hence $x, U \models_\mathcal{M} \mathsf{P}\Box\alpha$. But $x, U \not\models_\mathcal{M} \Diamond\alpha$ as every $\tilde{U} \in \mathcal{O}$ such that $x \in \tilde{U} \subseteq U$ contains a point $y \notin U'$ which, therefore, falsifies A. Thus $\mathcal{M} \not\models \mathsf{P}\Box\alpha \to \Diamond\alpha$ for some formula α. ∎

Note that the modal correspondent to weak directedness, $\Diamond\Box\alpha \to \Box\Diamond\alpha$, can be derived from $\mathsf{P}\Box\alpha \to \Diamond\alpha$ by using the S5 axioms for O.

We will briefly come back to the topic just touched on in the penultimate section of this paper.

3 Completeness and decidability

Our starting point to this section is the system of axioms for topologic from [3]. We then add the schemata which the new modality is involved in. We prove the soundness and completeness of the resulting list with respect to the class of all set spaces. Later on in this section, we also show that the new logic is decidable.

One schema from the just mentioned axiomatization of topologic is missing in the following block since it is subsumed by the later Axiom 9.

1. All instances of propositional tautologies.

2. $\mathsf{K}(\alpha \to \beta) \to (\mathsf{K}\alpha \to \mathsf{K}\beta)$

3. $K\alpha \to (\alpha \land KK\alpha)$

4. $L\alpha \to KL\alpha$

5. $\Box(\alpha \to \beta) \to (\Box\alpha \to \Box\beta)$

6. $\Box\alpha \to \alpha \land (\Box\Box\alpha)$

7. $K\Box\alpha \to \Box K\alpha$,

where $A \in \text{PROP}$ and $\alpha, \beta \in \text{WFF}$. In this way, it is expressed that for every Kripke model validating these axioms

- the accessibility relation $\xrightarrow{K}$ belonging to the knowledge operator is an equivalence,

- the accessibility relation $\xrightarrow{\Box}$ belonging to the effort operator is reflexive and transitive, and

- the relations for knowledge and effort commute as described by the Cross Axiom (schema 7 above).

The next group of axioms concerns the overlap operator and its interaction with the effort operator $\Box$; cf Proposition 3.

8. $O(\alpha \to \beta) \to (O\alpha \to O\beta)$

9. $(A \to OA) \land (PA \to A)$

10. $O\alpha \to OO\alpha$

11. $\alpha \to OP\alpha$

12. $O\alpha \to \Box\alpha$,

where $A \in \text{PROP}$ and $\alpha, \beta \in \text{WFF}$. – Note that two of the expected schemata are missing in this list, viz $O\alpha \to \alpha$ (see Proposition 3.2) and $(A \to \Box A) \land (\Diamond A \to A)$. This is because all instances of these schemata are derivable by means of the calculus we are going to define right now.

We obtain a logical system from the above list of axioms by adding the standard proof rules of modal logic, i.e., *modus ponens* and *necessitation with respect to each modality*. We call the system ET, indicating *extended topologic*.

DEFINITION 6 (The logic). Let ET be the smallest set of formulas containing the axiom schemata 1 – 12 and closed under application of the following rule schemata:

(MODUS PONENS) $\dfrac{\alpha \to \beta, \alpha}{\beta}$ (Δ–NECESSITATION) $\dfrac{\alpha}{\Delta\alpha}$,

where $\alpha, \beta \in \text{WFF}$ and $\Delta \in \{K, \Box, O\}$.

The following result is the first of the main issues of this paper.

THEOREM 7 (Soundness and completeness). *A formula $\alpha \in$ WFF is valid in all set spaces, iff it is ET–derivable.*

Proof. (Sketch.) The soundness part of Theorem 7 is easy to prove. Thus we need not go into that in more detail. The proof of completeness is both an adaption to the new system and a suitable extension of the corresponding proof for topologic; cf [3], Sec. 2.2. Only the part concerning the modifications is given detailedly.

Let $\alpha \in$ WFF be not ET–derivable. We attain a set space falsifying α by an infinite 'three-dimensional' step-by-step construction. In each step, an approximation to the final model is defined. In order to ensure that this 'limit structure' behaves as desired, several requirements on the intermediate models have to be kept under control.

Let $\mathcal{C}$ be the set of all maximal ET–consistent sets of formulas and $\xrightarrow{K}$, $\xrightarrow{\square}$, and $\xrightarrow{O}$ the accessibility relations on $\mathcal{C}$ induced by the modalities K, $\square$ and O, respectively. Suppose that $\Gamma_0 \in \mathcal{C}$ is to be realized (i.e., Γ_0 contains $\neg\alpha$). We choose a denumerably infinite set of points, Y, fix an element $x_0 \in Y$, and construct inductively a sequence of quadruples (X_n, P_n, i_n, t_n) such that $P_0 = \{\bot\}$ and, for every $n \in \mathbb{N}$,

1. $X_n \subseteq Y$ is a finite set containing x_0,

2. P_n is a finite disjoint union of finite trees,[2]

3. $i_n : P_n \longrightarrow \mathcal{P}(X_n)$ is an injective function from P_n into the set of all non-singleton subsets of X_n such that $p \leq q \iff i_n(p) \supseteq i_n(q)$ holds for all $p, q \in P_n$, and

4. $t_n : X_n \times P_n \longrightarrow \mathcal{C}$ is a *partial* function assigning a maximal ET–consistent set to some pairs from $X_n \times P_n$ such that, whenever $x, y \in X_n$ and $p, q \in P_n$, then

    (a) $t_n(x, p)$ is defined iff $x \in i_n(p)$; in this case it holds that

    i. if $y \in i_n(p)$, then $t_n(x, p) \xrightarrow{K} t_n(y, p)$,

    ii. if $q \geq p$, then $t_n(x, p) \xrightarrow{\square} t_n(x, q)$ and

    iii. if $x \in i_n(q)$, then $t_n(x, p) \xrightarrow{O} t_n(x, q)$,

    (b) $t_n(x_0, \bot) = \Gamma_0$.

We now explain to what extent the intermediate structures (X_n, P_n, i_n, t_n) approximate the desired model. Actually, it can be guaranteed that, for all $n \in \mathbb{N}$,

5. $X_n \subseteq X_{n+1}$,

---

[2] The induced partial order on P_n is denoted $\leq$.

6. P_{n+1} is an *end extension* of P_n (i.e., a sup-structure of P_n such that no element of $P_{n+1} \setminus P_n$ is strictly smaller than some element of P_n),

7. $i_{n+1}(p) \cap X_n = i_n(p)$ for all $p \in P_n$, and

8. $t_{n+1}|_{X_n \times P_n} = t_n$.

Furthermore, the construction complies with the following requirements on existential formulas. For all $n \in \mathbb{N}$,

9. if $\mathsf{L}\beta \in t_n(x,p)$, then there are $k \in \mathbb{N}$, $n < k$, and $y \in i_k(p)$ such that $\beta \in t_k(y,p)$,

10. if $\Diamond\beta \in t_n(x,p)$, then there are $k \in \mathbb{N}$, $n < k$, and $q \in P_k$ such that $p \leq q$ and $\beta \in t_k(x,q)$, and

11. if $\mathsf{P}\beta \in t_n(x,p)$, then there are $k \in \mathbb{N}$, $n < k$, and $q \in P_k$ such that $\beta \in t_k(x,q)$.

Let us assume for the moment that the construction has been carried out successfully, meeting all these requirements. Let (X, P, i, t) be the *limit* of the structures (X_n, P_n, i_n, t_n), i.e.,

- $X = \bigcup\limits_{n \in \mathbb{N}} X_n$,

- $P = \bigcup\limits_{n \in \mathbb{N}} P_n$,

- i is given by $i(p) = \bigcup\limits_{m \geq n} i_m(p)$, where n is the smallest number l such that $p \in P_l$, and

- t is given by $t(x,p) := t_n(x,p)$, where n is the smallest number l such that $t_l(x,p)$ is defined.

Now let $\mathcal{O} := \mathrm{Im}(i)$ and $\mathcal{F} := (X, \mathcal{O})$. We define an $\mathcal{F}$–valuation V by

$$V(A) := \{x \in X \mid A \in t(x,p) \text{ for some minimal element } p \in P\},$$

for all $A \in \mathrm{PROP}$. Then, $\mathcal{M} := (X, \mathcal{O}, V)$ is obviously a set space. Moreover, the following properties can (among other things) easily be deduced from the above requirements:

12. for all $x \in X$ and $p \in P$ we have that $t(x,p)$ is defined iff $x \in i(p)$; in this case it holds that

    (a) if $q \in P$ and $x \in i(q)$, then $t(x,p) \xrightarrow{\circ} t(x,q)$, and

    (b) if $\beta \in \mathrm{WFF}$ and $\mathsf{O}\beta \in t(x,p)$, then there is some $q \in P$ such that $\beta \in t(x,q)$;

13. $t(x_0, \bot) = \Gamma_0$, where x_0 and Γ_0 are as above.

With that, the following *Truth Lemma* can be proved by induction on the structure of formulas.

LEMMA 8 (Truth Lemma). *For all formulas $\beta \in$ WFF and neighbourhood situations $(x, i(p)) \in \mathcal{N}_\mathcal{F}$, we have that $(x, i(p)) \models_\mathcal{M} \beta \iff \beta \in t(x, p))$.*

Proof. We consider only the case that $\beta = \mathsf{P}\gamma$. First, assume that $x, i(p) \models_\mathcal{M} \mathsf{P}\gamma$. (Note for later purposes that this means, in particular, that $x \in i(p)$, i.e., $t(x, p)$ is defined.) Then, by Definition 2 and the definition of $\mathcal{F}$ there exists $q \in P$ such that $x \in i(q)$ and $x, i(q) \models_\mathcal{M} \gamma$. The induction hypothesis now gives us $\gamma \in t(x, q)$. We then conclude $t(x, p) \xrightarrow{\mathsf{O}} t(x, q)$ with the aid of item 12 (a). Therefore, $\mathsf{P}\gamma \in t(x, p)$. – The converse is proven similarly, but item 12 (b) has to be used in this case instead of 12 (a). ∎

Letting now $x := x_0$, $p := \bot$ and $\beta := \neg\alpha$, then Theorem 7 follows immediately from Lemma 8 and item 13.

It remains to define (X_n, P_n, i_n, t_n), for all $n \in \mathbb{N}$. The case $n = 0$ is easy. In the induction step, some existential formula contained in some maximal ET–consistent set $t_n(x, p)$, where $x \in X_n$ and $p \in P_n$, must be realized in a way meeting all the above requirements. Again, we confine ourselves to the case of the overlap operator.

So, let $\mathsf{P}\beta \in t_n(x, p)$. We choose some element $y \in Y$ not yet processed, and we take some $q \notin P_n$. Then we let $X_{n+1} := X_n \cup \{y\}$, $P_{n+1} := P_n \cup \{q\}$, and

$$i_{n+1}(p') := \begin{cases} i_n(p') & \text{if } p' \in P_n \\ \{x, y\} & \text{if } p' = q, \end{cases}$$

for all $p' \in P_{n+1}$; furthermore, we let q be incomparable with any element of P_n. Finally, we define t_{n+1} as follows. Let Γ be a maximal ET–consistent set such that $t_n(x, p) \xrightarrow{\mathsf{O}} \Gamma$ and $\beta \in \Gamma$. Then,

$$t_{n+1}(x', p') := \begin{cases} t_n(x', p') & \text{if } x' \in X_n \text{ and } p' \in P_n \\ \Gamma & \text{if } (x' = x \text{ or } x' = y) \text{ and } p' = q \\ \text{undefined} & \text{otherwise,} \end{cases}$$

for all $x' \in X_{n+1}$ and $p' \in P_{n+1}$. This completes the definition of the approximating structure in the induction step, and we must now check that properties 1 – 8 and 11 remain valid (9 and 10 are irrelevant to the present case). Apart from 4 all items are more or less obvious from the construction. (Note, however, that we need that every subset has at least two elements in order to establish property 3 for i_{n+1}.) Thus only the verification of property 4 is left. As 4 (b) is also obvious we concentrate on 4 (a). First, it follows from the definition of t_{n+1} that the condition on the domain of this function is satisfied. Then, (i) is clear from the definition of t_{n+1} and the validity of this condition for n. (Note that $\xrightarrow{\mathsf{K}}$ is a reflexive relation.) And (ii) too is valid for corresponding reasons ($\xrightarrow{\square}$ is reflexive as well). Only for (iii) some arguments are needed. So, assume first that

$x \in i_{n+1}(p')$ and let $q' \in P_{n+1}$ be such that $x \in i_{n+1}(q')$ (x from above). We must then show that (∗) $t_{n+1}(x, p') \xrightarrow{\circ} t_{n+1}(x, q')$. If $p', q' \in P_n$, then this follows from the induction hypothesis. If $p', q' \notin P_n$, then (∗) is valid since $\xrightarrow{\circ}$ is a reflexive relation. If $p' \in P_n$ and $q' \notin P_n$, then we obtain $t_{n+1}(x, p') \xrightarrow{\circ} t_{n+1}(x, p)$ from the induction hypothesis, and as we also have $t_{n+1}(x, p) \xrightarrow{\circ} t_{n+1}(x, q)$, the assertion follows in this case because of the transitivity of the relation $\xrightarrow{\circ}$. Finally, if $p' \notin P_n$ and $q' \in P_n$, the symmetry of $\xrightarrow{\circ}$ is used in addition. – Second, if $x' \in X_{n+1}$ is a point different from x, then either $x' = y$ holds from which we conclude $p' = q' = q$ and, therefore, the validity of (∗), or $x' \in X_n$ thus $p', q' \in P_n$, and (∗) follows from the induction hypothesis. In this way, property 4 is proved for t_{n+1}.

In order to ensure that *all* possible cases are eventually exhausted, processing has to be suitably scheduled with regard to each of the modalities involved. This can be done by means of appropriate enumerations. Concerning this and the construction in case of a modality of topologic, the reader is referred to the paper [3] for further details. However, it should be mentioned here that Axiom 12 is additionally needed to establish property 4 (a) (iii) in the induction step for $L\beta$ and $\Diamond\beta$, respectively. – This completes the proof sketch of Theorem 7. ∎

We now show that the set of all ET–theorems is decidable. As the logic does not satisfy the finite model property with respect to the class of all set spaces (cf [3], Sec. 1.3), we make a little detour in order to obtain this result. We introduce a certain class of auxiliary Kripke models and prove that ET (is sound and complete and) satisfies the finite model property with respect to *this* class of models. This gives us the desired decidability result in a standard manner.

In the following, R and K, S and $\Box$, and T and O, respectively, correspond to each other.

DEFINITION 9 (ECA–model). Let $M := (W, R, S, T, V)$ be a trimodal model, where $R, S, T \subseteq W \times W$ are binary relations and V is a valuation. Then, M is called an *extended cross axiom model* (or, in short, an *ECA–model*) iff the following conditions are satisfied:[3]

1. R and T are equivalence relations, and S is reflexive and transitive,

2. $S \subseteq T$,

3. $S \circ R \subseteq R \circ S$, and

4. for all $w, w' \in W$ and $A \in \text{PROP}$: $(w \in V(A) \iff w' \in V(A))$, if $w\,T\,w'$.

---

[3]The term 'cross axiom model' was introduced in [3]; see Sec. 2.3 there.

Note that every set space induces a semantically equivalent ECA–model in a rather straightforward manner.

It is easy to see that all the axioms considered in the first part of this section are sound with respect to the class of all ECA–models. Furthermore, the canonical model of ET is an example of an ECA–model. This gives us the following theorem.

THEOREM 10 (Kripke completeness I). *The system* ET *is sound and complete with respect to the class of all ECA–models.*

We use the method of *filtration* in order to prove the finite model property of ET with respect to ECA–models. For a given ET–consistent formula $\alpha \in$ WFF, we define a filter set $\Sigma \subseteq$ WFF as follows. We first let

$$\Sigma_0 := \text{sf}(\alpha) \cup \{\neg\beta \mid \beta \in \text{sf}(\alpha)\} \cup \{\text{O}A \mid A \in \text{PROP} \cap \text{sf}(\alpha)\},$$

where $\text{sf}(\alpha)$ designates the set of all subformulas of α. Second, we form the closure of Σ_0 under finite conjunctions of pairwise distinct elements of Σ_0. Third, we close under single applications of the operator L. And finally, we form the set of all subformulas of elements of the set obtained last. Let Σ denote the resulting set of formulas.

We now consider the respective *smallest* filtrations of the accessibility relations $\xrightarrow{\text{K}}$ and $\xrightarrow{\square}$ of the canonical model, and the *symmetric and transitive* filtration of the relation $\xrightarrow{\text{O}}$; cf [7], Sec. 1.4. Let $M := (W, R, S, T, V)$ be the corresponding filtration of a suitably generated submodel of the canonical model, for which the valuation V assigns the empty set to all proposition letters not in Σ. Then we have the following lemma.

LEMMA 11. *The structure M is a finite ECA–model of which the size depends computably on the length of α.*

Proof. Most of it is clear from the definitions and the proof of [3], Theorem 2.11. Only items 2 and 4 of Definition 9 have to be checked.

For item 2, let Γ, Θ be two points of the canonical model such that $[\Gamma] \, S \, [\Theta]$, where the brackets [...] indicate the respective classes. Then there are $\Gamma' \in [\Gamma]$ and $\Theta' \in [\Theta]$ such that $\Gamma' \xrightarrow{\square} \Theta'$. Since $\xrightarrow{\square} \subseteq \xrightarrow{\text{O}}$ holds because of Axiom 12, we conclude $\Gamma' \xrightarrow{\text{O}} \Theta'$ from that. Due to the first filtration condition (marked (F1) in [7], Sec. 1.4) it follows that $[\Gamma] \, T \, [\Theta]$, as desired.

For item 4, let Γ, Θ be as above and assume now that $[\Gamma] \, T \, [\Theta]$. Moreover, let $A \in \Sigma$ and $[\Gamma] \in V(A)$. It follows that $A \in \Gamma$. From this we infer $\text{O}A \in \Gamma$ with the aid of Axiom 9. Due to the definition of Σ_0 we obtain $\text{O}A \in [\Gamma]$. The usual filtration lemma then gives us $M, [\Gamma] \models \text{O}A$. Consequently, $M, [\Theta] \models A$, i.e., $[\Theta] \in V(A)$. The desired assertion now follows for symmetry reasons. ∎

For later purposes, note that the completeness theorem of topologic is used in [3] for proving the transitivity of the relation S.

Since the model M from Lemma 11 realizes α, the claimed decidability result follows readily.

THEOREM 12 (Decidability). *The set of all* ET*-theorems is decidable.*

4 Towards coverings

We now introduce an 'iterated' overlap operator, C, for set spaces and touch on the question how coverings can be dealt with by means of this modality. We then show that C is strongly related to the *common knowledge operator* of the usual logic of knowledge (see [4], Sec. 3.3). This yields, in particular, that interpreting C in ECA–models leads to a finitely axiomatizable and decidable logic. Finally in this section, we mention a connection between our logic and reasoning about knowledge of agents in so-called *systems with perfect recall;* cf [4], Sec. 4.4 and 8.2.

DEFINITION 13 (Semantics of C). Let $\mathcal{M} = (X, \mathcal{O}, V)$ be a set space, $\alpha \in$ WFF a formula and x, U a neighbourhood situation. Then,

$$x, U \models_{\mathcal{M}} \mathsf{C}\alpha : \iff \begin{cases} \forall n \in \mathbb{N} \text{ and } y \in X : \text{ if } U_0, \ldots, U_n \in \mathcal{O} \text{ satisfy} \\ U = U_0, \ U_i \cap U_{i+1} \neq \emptyset \text{ for } i = 0, \ldots, n-1, \\ \text{and } y \in U_n, \text{ then } y, U_n \models_{\mathcal{M}} \alpha. \end{cases}$$

Thus all the formulas of the form

$$\underbrace{\mathsf{KO}\ldots\mathsf{KO}}_{n \text{ times}}\alpha$$

(for some $n \in \mathbb{N}$) must be true at x, U in order to make $\mathsf{C}\alpha$ true at the neighbourhood situation x, U. It follows that the relation belonging to C amounts to the reflexive and transitive closure of the union of the relations belonging to K and O in ECA–models; see below.

What about coverings? – Well, suppose at the moment that we have the *global modality* A to hand (cf [1], Sec. 7.1). Then, the formula $\mathsf{C}\alpha \to \mathsf{A}\alpha$ obviously describes the fact that the set system $\mathcal{O}$ of a given set frame $(X, \mathcal{O})$ represents a 'connected covering' of the domain X of the frame. (In other words, every neighbourhood situation is reachable from the actual one via a finite sequence of overlapping sets.)

Unfortunately, we do not have nice meta-theorems for the logic of set spaces regarding the richer language up to now. The situation changes, however, if we restrict attention to ECA–models.

DEFINITION 14 (Enriched ECA–model). A tuple $M := (W, R, S, T, Q, V)$ is called an *enriched ECA–model*, iff (W, R, S, T, V) is an ECA–model and $Q \subseteq W \times W$ the reflexive and transitive closure of the relation $R \cup T$.

Again, every set space induces a semantically equivalent enriched ECA–model; see the remark right after Definition 9.

Adding now the axiom schema

$$\mathsf{C}\alpha \to \mathsf{K}(\alpha \wedge \mathsf{C}\alpha) \wedge \mathsf{O}(\alpha \wedge \mathsf{C}\alpha)$$

and the rule schema

$$\frac{\alpha \to \mathsf{K}(\alpha \wedge \beta) \wedge \mathsf{O}(\alpha \wedge \beta)}{\alpha \to \mathsf{C}\beta}$$

to the system from Section 3, yields a logic we designate ETC. It turns out that ETC is sound and complete, and satisfies the finite model property, with respect to the class of all enriched ECA–models.

THEOREM 15 (Kripke completeness II). *The system* ETC *is sound and complete with respect to the class of all finite enriched ECA–models.*

Proof. Soundness is again easy. For completeness, we start off with the structure $M = (W, R, S, T, V)$ considered in Lemma 11. First note that both the new axiom and the rule schema are precisely those for common knowledge in case two agents are involved in the group; cf [4], Sec 3.3. From this we get in a standard manner that $Q := (R \cup T)^*$ is a filtration of the accessibility relation $\xrightarrow{C}$, belonging to the operator C, of the canonical model of ETC. Note that only the *definability lemma* (cf [7], 9.7), the new axiom and rule schemata, respectively, and the conditions defining a filtration are used for that.

However, we cannot yet conclude that the theorem is true since we do not have completeness of the system ETC with respect to set spaces; cf the remark following the proof of Lemma 11. Nevertheless, there is a way out. We have to change the filtration of $\xrightarrow{\Box}$ and consider the *transitive closure of the smallest filtration* (cf [2], p 141 f) instead of the smallest one. Then, the main result of [16] assures us that the structure $M' := (W, R, S, T, Q, V)$ is really a finite enriched ECA–model. The desired completeness result now follows easily. ∎

The following corollary is immediate from Theorem 15.

COROLLARY 16. *The set of all* ETC*–theorems is decidable.*

We conclude this section with an additional remark on the relation of the systems ET(C) to the usual logic of knowledge. We first state that the system ET contains the logic of knowledge of two agents having perfect recall as a sublogic. In fact, the Cross Axiom may be viewed as the schema corresponding to perfect recall in our context; cf the discussion in [4], Sec. 8.2. And the Cross Axiom holds for the second 'knower' as well since

$$(*) \quad \mathsf{O}\Box\alpha \to \Box\mathsf{O}\alpha$$

is ET–derivable. Calling the logic where Axiom 9 is skipped and Axiom 12 is replaced with $(*)$, therefore, L^{pr}, we have that $\mathsf{L}^{pr} \subseteq \mathsf{ET}$. The same is true if common knowledge is included: $\mathsf{LC}^{pr} \subseteq \mathsf{ETC}$. The question now arises whether LC^{pr} is decidable. Unfortunately, we cannot answer this question positively from our previous techniques and results. Rather on the contrary, undecidability seems more likely.[4]

---

[4]The question just raised is very natural in view of [9], Corollary 3.6, and the remark

5 Hybridization

Concluding the technical part of the paper, we now *hybridize* the logic of overlapping. Although it is true that we cannot solve the problems left over from the preceding section in this way (because of the first-order nature of hybrid logic), this method is advantageous to us in several respects. First, we obtain much more expressive power concerning properties of relations, and this can already be achieved by simply adding suitable sets of nominals to the ground language. Second, we are able to do without the global modality, which was mainly needed for technical reasons in previous hybridizations of topologic; cf [14, 15]. Actually, the global modality appears somewhat artificial in this context since its semantics lacks to be typical of spatial structures. Third, other special purpose hybrid connectives like the 'in'–operators utilized in [15] become definable in the new language. And finally, all the meta-theorems proved in the papers just cited remain valid for the new system.

We now verify all these points more or (mostly) less detailedly. For a start, we define the extended language. As we already indicated above, we merely add two sets of nominals to the language underlying ET. If the denotation of a nominal is non-empty, then it is to be either a unique state or a distinguished set of states. Let $N_{stat} = \{i, j, \ldots\}$ and $N_{sets} = \{I, J, \ldots\}$ be the corresponding sets of symbols, which we call *names of states* and *names of sets*, respectively.

DEFINITION 17 (Set spaces with names). Let $\mathcal{F} = (X, \mathcal{O})$ be a set frame such that $X, \emptyset \in \mathcal{O}$.

1. A *hybrid $\mathcal{F}$-valuation* is a mapping

   $$V : \text{PROP} \cup N_{stat} \cup N_{sets} \longrightarrow \mathcal{P}(X)$$

   such that

   (a) $V(i)$ is either $\emptyset$ or a singleton subset of X for every $i \in N_{stat}$,
   (b) $V(A) \in \mathcal{O}$ for every $A \in N_{sets}$.

2. A *set space with names* (or, in short, an *SSN*) is a triple $(X, \mathcal{O}, V)$, where $\mathcal{F} = (X, \mathcal{O})$ is a set frame and V a hybrid $\mathcal{F}$-valuation.

Note that X and the empty set are now included in $\mathcal{O}$. The inclusion of X is mainly because of technical reasons. Actually, a smooth proof of Theorem 19 is facilitated in this way. The inclusion of $\emptyset$ takes into account that nominals may have an empty denotation. This is appropriate for our purposes, but not common in standard hybrid logic.

---

at the end of Sec. 5 there, respectively. That corollary tells us that the usual logic of knowledge of two agents with perfect recall and common knowledge is undecidable. And in the remark, the authors suspect that the complexity of the logic of knowledge where $\square$ is the only temporal modality will significantly decrease. Now note that we have a reduction, induced by $\text{O}\alpha \mapsto \text{O}\square\alpha$, of the satisfiability problem of ETC to the satisfiability problem of LC^{pr}.

DEFINITION 18 (Satisfaction for nominals). Let $\mathcal{M} := (X, \mathcal{O}, V)$ be an SSN and x, U a neighbourhood situation of $(X, \mathcal{O})$. Then

$$x, U \models_{\mathcal{M}} i \quad :\Longleftrightarrow \quad x \in V(i)$$
$$x, U \models_{\mathcal{M}} I \quad :\Longleftrightarrow \quad V(I) = U,$$

for all $i \in \mathrm{N}_{stat}$ and $I \in \mathrm{N}_{sets}$.

In Proposition 5, we gave an example of a frame property that can be captured with the aid of the overlap operator. Here is another one: *weak connectedness of neighbourhood situations*. This means that for any two overlapping sets, one of them is contained in the other. An appropriate correspondent reads

$$(\#) \quad (\alpha \to \Diamond \beta) \vee \mathsf{O}(\beta \to \Diamond \alpha).$$

(By the way, the modal analogue $\Box(\Box \alpha \to \beta) \vee \Box(\Box \beta \to \alpha)$ is in turn derivable from (#) in the system ET.) However, certain desirable frame properties are *not* expressible in the language from Section 2, eg, that

(P)   any two overlapping sets are contained in a common superset

(which is obviously true for all frames $(X, \mathcal{O})$ SSNs are based on since $X \in \mathcal{O}$; cf Definition 17). This deficiency can be remedied by using nominals. Eg, the just mentioned property is captured by the formula schema

$$(\&) \quad I \wedge \mathsf{P} J \to \mathsf{P}(\Diamond I \wedge \Diamond J),$$

where $I, J \in \mathrm{N}_{sets}$.

This example is really important because it leads us to a sound and complete axiomatization of *hybrid* ET without using the global modality. In fact, a suitable variant of the schema (&) serves as a substitute for Axiom 15 from [14] which likewise corresponds to (P) and is crucial to the completeness proof there. Now, letting ETH denote the augmentation of ET by the appropriately modified axioms and rules for names in hybrid topologic, cf [14] (or [15]), we obtain the following theorem.[5]

THEOREM 19 (Hybrid completeness). ETH *is sound and complete with respect to the class of all set spaces with names.*

Proof. (Sketch.) The proof of Theorem 19 goes along the lines of the proof of Theorem 3.2 from [14], but some modifications are required during the construction of the model refuting a given non-derivable formula. Again, we start off with the canonical model (of ETH this time). Let $\mathcal{C}$, $\xrightarrow{\mathsf{K}}$, $\xrightarrow{\Box}$, $\xrightarrow{\mathsf{O}}$, and Γ_0, be according to the previous proof.

As it was already indicated above, we will make decisive use of the following axioms:

---

[5] Note that the global modality is involved in another schema for names, viz Axiom 14 from [14]. This has to be handled as follows: replace A with K$\Box$ therein. Furthermore, note that the difference between A and O concerning axioms, lies in the schemata 9 and 12 from Section 3.

(A1) $i \wedge I \wedge \mathsf{P}(j \wedge J) \to \mathsf{P}(\Diamond(i \wedge I) \wedge \mathsf{L}\Diamond(j \wedge J))$

(A2) $\mathsf{A}(I \wedge \alpha \to \mathsf{L}\beta) \vee \mathsf{A}(I \wedge \beta \to \mathsf{L}\alpha)$,

where $i \in \mathrm{N}_{stat}$, $I \in \mathrm{N}_{sets}$ and $\alpha, \beta \in \mathrm{WFF}$. Note that (A1) is related to the fact that $X \in \mathcal{O}$ for every SSN $(X, \mathcal{O}, V)$.

Moreover, the ENRICHMENT rules are not only different, but also more complicated compared to those from [14]. Consider the following set of strings:

$$L := (\mathsf{PL})^* \cup (\mathsf{PL})^*\mathsf{P} \cup (\mathsf{LP})^* \cup (\mathsf{LP})^*\mathsf{L}.$$

Then, for all $W \in L$ and $\nabla \in \{\mathsf{L}, \Diamond, \mathsf{P}\}$ we have an ETH–rule schema called W_∇–ENRICHMENT which reads

$$\frac{W(i \wedge I \wedge \nabla(j \wedge J \wedge \alpha)) \to \beta}{W(i \wedge I \wedge \nabla\alpha) \to \beta},$$

where $\alpha, \beta \in \mathrm{WFF}$, $i, j \in \mathrm{N}_{stat}$, $I, J \in \mathrm{N}_{sets}$, and j, J are new each time.

We then call a maximal consistent set Γ of formulas

- *named* iff Γ contains some $i \in \mathrm{N}_{stat}$ and some $I \in \mathrm{N}_{sets}$, and

- *enriched* iff for all $W \in L$ and every $\nabla \in \{\mathsf{L}, \Diamond, \mathsf{P}\}$ we have that $W(i \wedge I \wedge \nabla\alpha) \in \Gamma$ implies $W(i \wedge I \wedge \nabla(j \wedge J \wedge \alpha)) \in \Gamma$, for some $j \in \mathrm{N}_{stat}$ and $J \in \mathrm{N}_{sets}$.

Our first task now is extending Γ_0 to a named and enriched maximal consistent set Γ' in the language which has been increased by enough new constants. This can be done by means of a suitable *Extended Lindenbaum Lemma* in the usual way (i.e., by using, in particular, the above rules); cf [1], 7.25, and [14], Lemma 3.3.

Let $\mathcal{M}_0$ be the submodel, generated by Γ', of the canonical model of ETH in the extended language. Moreover, let $\mathcal{M}'$ be the substructure of $\mathcal{M}_0$ of which the domain, D, consists of all points that are named. $\mathcal{M}'$ is the model we want to operate in. We focus only on one of the points enabling us to do that successfully here, viz *uniqueness of the interpretation of set names*.

LEMMA 20. *Let I be a set name. Assume that $\Gamma, \Gamma' \in D$ both contain I. Then $\Gamma \xrightarrow{\mathsf{K}} \Gamma'$.*

Proof. Assume on the contrary that there are two points $\Gamma, \Gamma' \in D$ such that $I \in \Gamma \cap \Gamma'$, but not $\Gamma \xrightarrow{\mathsf{K}} \Gamma'$. Then there is some $n \in \mathbb{N}$ and a sequence $\Gamma^0, \Gamma^1, \ldots, \Gamma^n$ of elements of D such that $\Gamma^0 = \Gamma$, $\Gamma^n = \Gamma'$, and Γ^i, Γ^{i+1} are connected by either $\xrightarrow{\mathsf{K}}$ or $\xrightarrow{\mathsf{O}}$ for all $i = 0 \ldots n - 1$. This is true because $\mathcal{M}_0$ is generated and each Γ^i can in fact be chosen as named (which is proved with the aid of our special notion of enrichment). The assertion of the lemma now follows by an iterated application of the schema (A1), since the desired contradiction is then yielded by the schema (A2). ∎

The subsequent *Existence Lemma* can be proved with the aid of Lemma 20 (among other things).

LEMMA 21. *Assume that $\Gamma \in D$ contains the formula $\nabla \alpha$ (where ∇ is as above). Then there exists some $\Theta \in D$ satisfying $\Gamma \xrightarrow{\nabla} \Theta$ and $\alpha \in \Theta$.*

Lemma 21 is fundamental to the further progress of the proof of Theorem 19, which now can be continued essentially as in [14], Sec. 3. ∎

The proof of the following theorem is even more similar to the corresponding one from [14] (Theorem 4.4 there) and can be omitted here therefore.

THEOREM 22 (Hybrid decidability). *The set of all ETH–derivable formulas is decidable.*

It is no accident that we succeeded in axiomatizing ETH without the global modality because the latter is definable in set spaces with names, actually through

$$A\alpha :\equiv K\Box O\alpha.$$

Note that we again made use of the fact that $X \in \mathcal{O}$ for that (see above). Then, however, the usual hybrid *satisfaction operators* $@_{...}$, cf [1], Sec. 7.3, can be defined as well. This can be seen as follows. The formulas of the form $i \wedge I$, where $i \in N_{stat}$ and $I \in N_{sets}$, are taken as names of neighbourhood situations. With that, the satisfaction operator associated with such a name reads

$$@_{i \wedge I}\alpha :\equiv E(i \wedge I \wedge \alpha),$$

where $\alpha \in$ WFF and E is the dual of A.

And a final point should be stressed here: in SSNs, O is really *stronger* than the global modality. This is witnessed by the fact that even the 'in'–operators considered in the paper [15] become definable. (This is not the case with that system.) For convenience, we remind the reader of the semantics of these operators. Let $\mathcal{M} = (X, \mathcal{O}, V)$ be an SSN and x, U a neighbourhood situation of $(X, \mathcal{O})$. Then

$$x, U \models_{\mathcal{M}} [\epsilon_I]\alpha :\iff \text{if } x \in V(I), \text{ then } x, V(I) \models_{\mathcal{M}} \alpha,$$

where $I \in N_{sets}$ and $\alpha \in$ WFF. The 'in'–operators are, therefore, named variants of the overlap operator; cf Definition 2. Thus it is not surprising that they can be defined:

$$[\epsilon_I]\alpha :\equiv O(I \to \alpha).$$

It follows that we obtain all the results from the paper [15] for the present system as well, in particular, that the (hybrid) logic of directed spaces (see Definition 4) is decidable; cf Theorem 20 and Corollary 21 there. (The decidability problem for the logic of directed spaces was left open in the paper [20]; cf the remark right after Definition 4.) And one can certainly expect similar results for other extensions of ETH by additional axioms, which are due to the applications one has in mind.

All in all, we may say that ETH seems to be the 'right' hybrid logic for set spaces with names.

6 Concluding remarks

In the paper, we first added an operator describing overlapping to Moss and Parikh's bi-modal language of set spaces. We then proved the soundness and completeness as well as the decidability of the arising logic, which we called **extended topologic**, **ET**. Subsequent to that, we discussed further-reaching modifications of **ET**. On the one hand we focussed on *coverings* of set spaces and their relation to common knowledge (system **ETC**), and on the other hand we dealt with a natural hybridization that turned out to be very neat in several respects (system **ETH**). In both cases we obtained some interesting results indicating the adequacy of our approach.

Some topics could only be touched on in the paper. This concerns the topological aspects we mentioned above as well as the connection between **ET(C)** and the usual logic of knowledge. In particular, a certain epistemic interpretation of the overlap operator suggested by one of the anonymous referees will have to be closer examined.

At this point we may also take a different view of **topologic**: consider the bi-modal system with knowledge and overlapping the basic one and build extensions of *this* system as needed. (Note that this system syntactically coincides with the usual logic of knowledge of two agents (apart from the rigidity condition for proposition letters; see Axiom 9 above).) Then, eg, introducing a modality complementary to the effort operator, i.e., quantifying 'upward' across $\mathcal{O}$, is very natural (among other things). Such an operator corresponds to *systems with no learning* in the usual logic of knowledge; cf [9] or [8].

Our final remark concerns complexity. Unfortunately, we do not have any satisfactory results regarding this (apart from the one proved in [11]). In particular, it seems to be difficult to establish reasonable upper bounds. Thus future research into the complexity of **topologic** and its extensions is urgently needed.

BIBLIOGRAPHY

[1] Patrick Blackburn, Maarten de Rijke, and Yde Venema. *Modal Logic*, volume 53 of *Cambridge Tracts in Theoretical Computer Science*. Cambridge University Press, Cambridge, 2001.

[2] Alexander Chagrov and Michael Zakharyaschev. *Modal Logic*, volume 35 of *Oxford Logic Guides*. Clarendon Press, Oxford, 1997.

[3] Andrew Dabrowski, Lawrence S. Moss, and Rohit Parikh. Topological reasoning and the logic of knowledge. *Annals of Pure and Applied Logic*, 78:73–110, 1996.

[4] Ronald Fagin, Joseph Y. Halpern, Yoram Moses, and Moshe Y. Vardi. *Reasoning about Knowledge*. MIT Press, Cambridge, MA, 1995.

[5] Konstantinos Georgatos. Knowledge theoretic properties of topological spaces. In M. Masuch and L. Pólos, editors, *Knowledge Representation and Uncertainty, Logic at Work*, volume 808 of *Lecture Notes in Artificial Intelligence*, pages 147–159. Springer, 1994.

[6] Konstantinos Georgatos. Knowledge on treelike spaces. *Studia Logica*, 59:271–301, 1997.

[7] Robert Goldblatt. *Logics of Time and Computation*, volume 7 of *CSLI Lecture Notes*. Center for the Study of Language and Information, Stanford, CA, second edition, 1992.

[8] Joseph Y. Halpern, Ron van der Meyden, and Moshe Y. Vardi. Complete Axiomatizations for Reasoning about Knowledge and Time. *SIAM Journal on Computing*, 33:674–703, 2004.
[9] Joseph Y. Halpern and Moshe Y. Vardi. The complexity of reasoning about knowledge and time. I. Lower bounds. *Journal of Computer and System Sciences*, 38:195–237, 1989.
[10] Bernhard Heinemann. Topological Modal Logics Satisfying Finite Chain Conditions. *Notre Dame Journal of Formal Logic*, 39(3):406–421, 1998.
[11] Bernhard Heinemann. Topological nexttime logic. In M. Kracht, M. de Rijke, H. Wansing, and M. Zakharyaschev, editors, *Advances in Modal Logic 1*, volume 87 of *CSLI Publications*, pages 99–113, Stanford, CA, 1998. Kluwer.
[12] Bernhard Heinemann. Extended canonicity of certain topological properties of set spaces. In M. Vardi and A. Voronkov, editors, *Logic for Programming, Artificial Intelligence, and Reasoning*, volume 2850 of *Lecture Notes in Artificial Intelligence*, pages 135–149, Berlin, 2003. Springer.
[13] Bernhard Heinemann. The hybrid logic of linear set spaces. *Logic Journal of the IGPL*, 12(3):181–198, 2004.
[14] Bernhard Heinemann. A hybrid logic for reasoning about knowledge and topology, September 2005. See URL http://www.informatik.fernuni-hagen.de/thi1/ber.ps.
[15] Bernhard Heinemann. A two-sorted hybrid logic including guarded jumps. In R. Schmidt, I. Pratt-Hartmann, M. Reynolds, and H. Wansing, editors, *Advances in Modal Logic*, volume 5, pages 73–92, London, 2005. King's College Publications.
[16] Gisela Krommes. A new proof of decidability for the modal logic of subset spaces. In B. ten Cate, editor, *Proceedings of the Eighth ESSLLI Student Session*, pages 137–147, Vienna, Austria, August 2003.
[17] John-Jules Ch. Meyer and Wiebe van der Hoek. *Epistemic Logic for AI and Computer Science*, volume 41 of *Cambridge Tracts in Theoretical Computer Science*. Cambridge University Press, Cambridge, 1995.
[18] Werner Nutt. On the translation of qualitative spatial reasoning problems into modal logics. In W. Burgard, T. Christaller, and A. B. Cremers, editors, *Proceedings 23rd Annual German Conference on Artificial Intelligence, KI-99*, volume 1701 of *Lecture Notes in Artificial Intelligence*, pages 113–125, Berlin, 1999. Springer.
[19] Ilya Shapirovsky and Valentin Shetman. Modal logics of regions and Minkowski spacetime. *Journal of Logic and Computation*, 15(4):559–574, 2005.
[20] M. Angela Weiss and Rohit Parikh. Completeness of certain bimodal logics for subset spaces. *Studia Logica*, 71:1–30, 2002.

Bernhard Heinemann
FernUniversität in Hagen
58084 Hagen, Germany
E-mail: Bernhard.Heinemann@fernuni-hagen.de

Weaker-to-Stronger Translational Embeddings in Modal Logic

LLOYD HUMBERSTONE

ABSTRACT. After some general remarks (Section 1), we consider whether a translation popular from earlier discussions of deontic and doxastic logic (reviewed in Section 2) faithfully embeds the normal modal logic **KD** into one of its proper extensions. This would contrast with the more familiar such modal-to-modal embeddings in which a logic is embedded either into one which is strictly weaker or one which is incomparable with it in respect of strength. The translation in question turns out, somewhat surprisingly, not to deliver the desired result (Section 3) and simply in the interests of illustrating the possibility of weaker-to-stronger embeddings amongst the normal modal logics, Section 4 provides an example of a translation which does succeed in embedding **K** (though **KD** could also play this role) into one of its (normal) proper extensions.

Keywords: Translational embeddings, normal modal logics.

1 Introduction

The immediate stimulus for the present study came from some remarks by one of the author's students, Rohan French. We had been discussing the striking paper [2], in which Béziau finds the following situation paradoxical: we have two logics, one stronger than the other (in the sense of proving everything the latter proves, and more besides) while at the same time the stronger logic can be faithfully embedded by a translation into the weaker logic. After all, this embedding locates a copy (under the translation) of the whole of the stronger logic within the weaker – arguably undermining the latter's claim to be weaker. Béziau presents a particularly simple example of this phenomenon in [2], as well as recalling several further instances from the literature. (Still further examples, as well as a discussion of the paradoxicality reaction, may be found in [12].) The following aspect of Béziau's discussion struck French as strange. In *all* the familiar examples of such translational embeddings, a stronger logic is embedded into a weaker one, rather than – what we should expect if Béziau's surprised reaction were appropriate – a weaker logic into a stronger one. (See note 9 for a consideration ignored here.) Indeed, French further reasoned, it would be embedding in this latter direction that would be problematic, since as a general rule a stronger logic collapses more distinctions than a weaker logic, which should be expected to cause trouble for the faithfulness of the supposed translation. A special case of this kind of reasoning was used in the proof of Theorem

3.1 of [12], where it is argued that the consequence relation associated with the $\{\to, \neg\}$-fragment of intuitionistic logic cannot be embedded by (what we call below) a definitional translation into the consequence relation of classical logic on the grounds that for the first of these consequence relations there are, to within equivalence, six non-equivalent formulas in one variable, while for the latter there are only four. (The argument is spelt out in more detail below.) What had not, however, occurred to the author, was that analogous considerations might turn out to threaten *any* translation purporting to embed a weaker logic into a stronger.[1] This was French's suggestion, made in the full knowledge that only a qualified form of it could be expected to survive in view of the following. While it is typically the case that a deductively stronger logic makes fewer discriminations than a weaker one (than a proper sublogic, that is), this is not invariably the case; several types of exception to the general rule are provided in [13].

An explanation is due of some of the terminology in use here. Suppose we think of logics as consequence relations $\vdash_1$ and $\vdash_2$, and regard a *translation* as a function τ from the language of $\vdash_1$ to the language of $\vdash_2$ which is said to embed $\vdash_1$ into $\vdash_2$ *faithfully* when for all formulas $A_1, \ldots, A_n, B$ of the former language, we have:

(1.1) $A_1, \ldots, A_n \vdash_1 B$ if and only if $\tau(A_1), \ldots, \tau(A_n) \vdash_2 \tau(B)$.

The word 'faithfully' here records the 'if' direction of this condition and since we always impose this restriction, except for occasional emphasis, we simply say that the translation τ embeds $\vdash_1$ into $\vdash_2$ when (1.1) is satisfied. These consequence relations are called, respectively, the *source* and the *target* of the embedding in this case. Adapting terminology from previous work on such embeddings,[2] we say in addition that τ is *compositional* when for each k-ary primitive connective $\#$ of the language of $\vdash_1$ there is a formula $C(p_1, \ldots, p_k)$ of the language of $\vdash_2$ in which precisely the indicated propositional variables (or 'sentence letters') occur – and here we consider only sentential logics[3] – with the property that for all formulas $B_1, \ldots, B_k$ of the

---

[1][12], p. 162, quotes Ryszard Wójcicki as having said of two logics, one weaker than the other, that *nevertheless* the second can be definitionally embedded in the first, and envisages (p. 163) the following reaction: "There is no 'nevertheless' about it – the increase in expressive power [of the weaker over the stronger logic] is just a manifestation of the general phenomenon that deductively weaker logics exhibit greater discriminatory strength...". [12] then remarks that this response will be seen to fail in a later section of the paper, where (p. 175) what we find is an example of a pair of logics, the stronger of which cannot be embedded in the weaker. The question of current interest, however, is not whether a stronger logic can always be embedded in a weaker one, but rather whether a weaker logic can ever be embedded in a stronger one.

[2]See [11], p. 441, and [12], p. 147, for references. Embeddability – and especially definitional embeddability (introduced presently) – is variously referred to in this literature as interpretability, definability, and reconstructability.

[3]We presume also that the same countably infinite set of such variables $\{p_i\}_{i \in Nat}$ is a set of free generators for the languages of all logics, considered as algebras with their possibly differing sets of primitive connectives as fundamental operations. We generally write p, q, r for p_1, p_2, p_3 respectively. If $A = A(p_1, \ldots, p_k)$ is a formula in which the exhibited variables occur, then $A(B_1, \ldots, B_k)$ is the result of uniformly substituting B_i

language of $\vdash_1$, $\tau(\#(B_1,\ldots,B_k)) = C(\tau(B_1),\ldots,\tau(B_k))$. τ is *definitional* if it is compositional and we also have $\tau(p_i) = p_i$ for every propositional variable p_i. A certain kind of non-compositional embedding is also familiar from the literature, in which there is a single formula $C(p)$ of the language of $\vdash_2$ such that for any formula B of the language of $\vdash_1$, $\tau(B) = C(B)$. Essentially this amounts to applying a single derived connective from the language of $\vdash_2$ to a formula from the language of $\vdash_2$, and for it to count as a translation we must have the language of $\vdash_1$ included in that of $\vdash_2$. A well-known example is Glivenko's embedding of $\vdash_{CL}$ into $\vdash_{IL}$, where these are the usual consequence relations of classical and intuitionistic propositional logic respectively, and (1.1) holds with $\vdash_{CL} = \vdash_1$, $\vdash_{IL} = \vdash_2$ taking $\tau(B)$ as $\neg\neg B$. (So C is $\neg\neg p$.) In what follows we are concerned mainly with compositional embeddings, and in particular, with the definitional case.

Letting $\vdash_{IL_\to^\neg}$ and $\vdash_{CL_\to^\neg}$ be the restrictions of $\vdash_{IL}$ and $\vdash_{CL}$ to the language with connectives $\to, \neg$, let us elaborate upon the point made above (from [12]) that $\vdash_{IL_\to^\neg}$ cannot be embedded by a definitional translation into $\vdash_{CL_\to^\neg}$ because the latter makes too many distinctions. More specifically there are six equivalence classes of formulas in the single propositional variable p for $\vdash_{IL_\to^\neg}$ but only four in the case of $\vdash_{CL_\to^\neg}$. In general with n distinct such equivalence classes for a consequence relation $\vdash$ there will be $n(n-1)/2$ claims of the form $A_i \dashv\vdash A_j$ with $i < j$, where $A_1,\ldots,A_n$ is an enumeration comprising one representative from each equivalence class. (1.1) implies, for the current choice of $\vdash_1$, $\vdash_2$, and any envisaged τ:

(1.2) $\quad A_i \dashv\vdash_{IL_\to^\neg} A_j$ if and only if $\tau(A_i) \dashv\vdash_{CL_\to^\neg} \tau(A_j)$,

where the A_i, A_j, are chosen with $i < j$ (to avoid double-counting) from amongst representatives of the six different $\vdash_{IL_\to^\neg}$-equivalence classes of formulas in which the only variable to appear is p. There are thus 15 ($= (6\times 5)/2$) such $[A_i \dashv\vdash_{IL_\to^\neg} A_j]$-claims, all of which are false. Now passing from the left to the right hand side of (1.2), we have the equivalence claims $\tau(A_i) \dashv\vdash_{CL_\to^\neg} \tau(A_j)$, of which, since there are only four $\vdash_{CL_\to^\neg}$-equivalence classes to choose from, at most 6 ($= (4 \times 3)/2$) can be true – and perhaps even fewer, since we have not said that τ is surjective (to within equivalence). Thus (1.2) must fail in its 'if' direction for at least 9 ($= 15 - 6$) choices of $\langle i,j\rangle$.[4]

Very little of the hypothesis that τ is a definitional translational embedding from is actually used here in refuting that hypothesis. An examination of the argument establishes, in particular, that there is no τ satisfying the 'if' direction of (1.1) with the property that for all formulas A, $\tau(A)$ is constructed only from variables appearing in A, with $\vdash_1$ as $\vdash_{IL_\to^\neg}$ and $\vdash_2$ as $\vdash_{CL_\to^\neg}$. And analogous non-existence results for other weaker-into-stronger embeddings in this vicinity follow by similar reasoning. With full $\vdash_{IL}$ there are even more (namely denumerably many) one-variable formulas, making the shortage of matching distinctions on the $\vdash_{CL}$ side even worse. The same

---

for p_i ($i = 1,\ldots,k$) in A.
[4]See Theorem 3.1 of [12] and Section 5 of [11] for related observations.

style of argument as has just been given works for the pure ¬-fragments of these consequence relations, as well as for their →-fragments (though in the latter case one must consider instead the two-variable → fragments, since $\vdash_{CL}$ and $\vdash_{IL}$ agree on the one-variable purely implicational formulas, so there is no weaker/stronger situation here).

The above argument about classical and intuitionistic logic ignores the possibility that formulas might be equivalent, each being a consequence of the other, by the consequence relation $\vdash$ at issue, and yet discriminated between ('discernible', 'non-synonymous',...) by $\vdash$ in because of the existence of a contexts in which the replacement of one formula by the other does not yield similarly equivalent results. This is a possibility not arising for $\vdash_{IL}, \vdash_{CL}$, and their fragments, but in general it would need to be taken into account in defining the – shall we say – discrimination relation $D_\vdash$ of a consequence relation $\vdash$.[5] Suppose we have a set $\mathbb{C}$ of consequence relations on the same language in which the anomalies reviewed in [13] do not arise, and the "strengthening the logic reduces its discriminations" motto applies: i.e., for all $\vdash_0, \vdash_1 \in \mathbb{C}$, we have $\vdash_0 \subsetneq \vdash_1 \Rightarrow D_{\vdash_1} \subsetneq D_{\vdash_0}$. Does this imply that whenever $\vdash_0 \subsetneq \vdash_1$ (with $\vdash_0, \vdash_1 \in \mathbb{C}$), $\vdash_0$ cannot be definitionally embedded into $\vdash_1$? In fact the earlier argument about the implication-negation fragments of intuitionistic and classical logic does not generalize to give this conclusion. That argument was basically an appeal to the pigeonhole principle: we can't match up the equivalence classes of formulas in one variable in the two logics because there are more such equivalence classes in the intuitionistic case than in the classical case. And what 'more' means in such contexts is that the set of such classes has a greater cardinality: it does not just mean "proper superset".[6] But the latter is all we have from the current hypothesis about stronger logics losing discriminations: $D_{\vdash_1} \subsetneq D_{\vdash_0}$. (A fuller discussion would relativize these D relations further, to the number of variables in the formulas concerned.) If, for example, the two logics both have infinitely many equivalence classes of n-variable formulas, then even if some of these equivalence classes are merged as we pass from the weaker to the stronger logic, a definitional translation which is the identity map except for some n-ary connective, may yet succeed in embedding the weaker logic into the stronger one.

Rather than spend further time on diagnostic remarks, we treat French's idea instead as a challenge: exhibit a concrete example of such an embedding. We pursue this in the arena of modal logic, taking the above n as 1. In Sections 2 and 3, we have a historical look at what come close to being examples, and at logic inspired by them (**KDH**) of considerable interest in its own right. In Section 4 a problem with using that logic as the target of the proposed embedding is overcome, and the translation is changed to

---

[5]That is, we define: $D_\vdash(A, B)$ iff for some formula $C(p)$ of the language of $\vdash$, we have either $C(A) \nvdash C(B)$ or $C(B) \nvdash C(A)$. Here we do *not* require that $C(p)$ have p for its only propositional variable. (Here we are considering how to make allowance for non-congruential – also called "non-selfextensional" – consequence relations.)

[6]Interestingly, 'Galileo's Paradox' is also mentioned in Béziau [2], though not quite in the present connection.

something of less independent interest but of great service in meeting the challenge just described.

2 Turning to Modal Logic

We will consider modal logics for simplicity as sets of formulas rather than as consequence relations, and as all cast in the same language with any functionally complete set of primitive boolean connectives (taken as fixed throughout, and preferably containing the nullary truth connective ⊤) together with the 1-ary connective □ (◇A being defined as ¬□¬A). A modal logic **S** is any set of formulas in this language which contains all truth-functional tautologies and is closed under uniform substitution (of arbitrary formulas for propositional variables) and Modus Ponens. A (not necessarily primitive) 1-ary connective # is *normal* in **S** if $A \to B \in$ **S** implies #$A \to$ #$B \in$ **S** (all A, B), (#$p \wedge$ #q) $\to$ #($p \wedge q$) $\in$ **S**, and #⊤ $\in$ **S**. Finally, a modal logic **S** is *normal* if □ is normal in **S**. The reasoning for defining the notion of a normal modal logic in this way will be evident as we proceed. Its upshot coincides with that of the usual definition – though for some purposes it is not convenient to build in the condition of closure under uniform substitution that has been imposed here. In future rather than writing "$A \in$ **S**" we write "⊢$_\mathbf{S}$ A", and describe A as provable in **S**. (The above definition of normality evidently gives rise to an axiomatization of the smallest normal modal logic **K**, and likewise with other logics isolated, so provability here amounts to the existence of a proof from the axiomatization in question. Familiarity with reasonably standard names of various modal logics and modal formulas is assumed in what follows.) Given this use of "⊢", the definitions of embedding etc., from the preceding section can be transferred to the present setting but with $n = 0$ in (1.1).[7] Several non-compositional embeddings in the ('one-off prefixing') style of Glivenko's, mentioned in that section, are known: for example –

- ⊢$_\mathbf{S5}$ A if and only if ⊢$_\mathbf{S4}$ ◇□A (Matsumoto [19]);
- ⊢$_\mathbf{S5}$ A if and only if ⊢$_\mathbf{K5}$ □A (Chellas [6]).

Of interest to us here, however, are compositional – indeed definitional – embeddings, and since we are discussing modal logics, the languages of the source and target logics for these embeddings coincide. (The contrast here is with modal translations of intuitionistic, intermediate and substructural logics. See [4], [9]; these are typically compositional though not definitional translations.) Matters are further simplified if we restrict attention to those translations τ which translate the boolean connectives by themselves, in

---

[7]We could of course pursue the modal discussion in terms of consequence relations also, for which purpose the most useful consequence relation to associate with **S** would be the 'local' consequence relation holding between a set Γ of formulas and a formula A when a conjunction of formulas from Γ provably implies A in **S**. There seems little advantage in doing this, though. The discernibility relations $D_\mathbf{S}$ relate A to B if for some formula $C(p)$, ⊢$_\mathbf{S}$ $C(A)$ while ⊬$_\mathbf{S}$ $C(B)$ or vice versa, which in turn holds for normal modal logics **S** just in case ⊬$_\mathbf{S}$ $A \leftrightarrow B$.

the sense that for all primitive connectives # (of arity k) other than $\Box$, we have $\tau(\#(B_1,\ldots,B_k)) = \#(\tau(B_1),\ldots,\tau(B_k))$. A familiar example of a definitional embedding along these lines is given by the translation τ for which $\tau(\Box B) = \tau(B) \wedge \Box\tau(B)$, which embeds **KT** into **K**. This example is discussed in [20], where the question is raised of a similar translation in the reverse direction, embedding **K** into **KT**. This last question was answered negatively in [12], p. 174, with the observation that no translation which translates a formula into one in the same propositional variables and translates ∨ (disjunction) by itself can embed a Halldén-incomplete logic into a logic which is Halldén-complete.[8] This may seem a somewhat parochial feature of the relationship between **K** and **KT** in view of the considerations of Section 1: the more glaring aspect of this relationship is that **K** $\subsetneq$ **KT** so the envisaged embedding would be in the contentious "weaker into stronger" direction. Note incidentally that each of the two non-compositional examples listed above again embeds a stronger logic into a weaker one. Evidently we shall need to look elsewhere if we are to find a weaker-to-stronger embedding between modal logics.[9]

Where else to look, then? We can usefully cast our minds back to 1959 for an early pointer, with the publication of Dawson [8], in which the problem of representing a suitable deontic logic within the expressive resources of a suitable alethic modal logic.[10] Here the main features of suitability are that in the alethic case, we should have the formula **T** (= $\Box p \to p$) provable, whereas in the deontic case we should not, asking only for the provability of the weaker **D** (i.e., $\Box p \to \Diamond p$, or equivalently, if we are working amongst the normal modal logics, $\Diamond\top$), and perhaps also the formula called **U** in [17]: $\Box(\Box p \to p)$. Thus our deontic $\Box$ operator (often written "O" in such discussions) can't just be the alethic $\Box$ operator; but Dawson noticed that if we took the deontic $\Box$ as, instead, the alethic $\Diamond\Box$, and took **S4.2** as our alethic modal logic, the normal extension of **S4** by the formula **G**, as it sometimes known (after Geach), then a more suitable deontic logic emerges. We retain the conventional label **S4.2** rather than the explicit name **S4G** (or indeed, more explicitly still, **KT4G**) for this logic, in the interests of familiarity:

G $\Diamond\Box p \to \Box\Diamond p$

then the prospects are rosier, since while the unwanted "**T**-for-$\Diamond\Box$" formula remains unprovable, the prefix $\Diamond\Box$ is normal in **S4.2** (though not in **S4**) – cf. note 15 – and **G** itself is the required deontic weakening "**D**-for-$\Diamond\Box$"

---

[8] For another application of this fact, see note 19 of [15].
[9] Nothing in this formulation should make us ignore the third kind of case: embedding one logic into another which is neither stronger nor weaker than it. It is easy to find modal examples illustrating this possibility. For instance take the logic determined by the class of all frames when truth at a point in a model is defined in such a way that $\Box A$ is taken to be true at a point just in case A is not true at any accessible point. This modal logic – not a normal modal logic, to be sure – is $\subseteq$-incomparable with **K**, but can be embedded in **K** by the translation of $\Box A$ as $\Box\neg A$. We will see an example of the same phenomenon amongst the normal modal logics in the last paragraph of this section.
[10] Variations on Dawson's theme may be found in Åqvist [1] and Kielkopf [16].

("*Ought* implies *May*") of that unwanted principle. (Note that when '□' is translated into '◇□', '◇' becomes '□◇'.)[11]

Twenty years later, this same alethic/deontic issue came up as an epistemic/doxastic issue, with the problem of representing a suitable doxastic □ in terms of an epistemic □. Again we may want **D** for the former (belief consistency as part of our idealization), and perhaps **U** as well – *cf.* note 42 in [14] – but we certainly don't want **T**, and Wolfgang Lenzen, in [18],[12] again reached for **S4.2** as a suitable epistemic logic, and the suggestion that the belief operator (our doxastic □) be defined in terms of the epistemic □ (often written as K in this context), namely as ◇□ again. Lenzen

---

[11]**T** here is more than we need: ◇□ is easily seen to be normal in any extension of **KD4.2**. In particular, its normality in **KD4.3** was noted – though not described in these terms – in Rescher and Urquhart [22], pp. 135–137, where it is, in effect, observed that the embedding which replaces □ with ◇□ embeds **KD5** into **KD4.3**, in the sense of 'embed' not requiring faithfulness, and it is conjectured that the embedding is indeed faithful. Byrd [3], p. 592, tries to establish this conjecture by appeal to a result from Segerberg [24] to the effect that the only consistent proper extensions of **KD5** (called *D5* in [3]) are **S5**, **KD5Alt**$_n$, and **S5Alt**$_n$ ($n \in Nat$); but there is no such result to be found in the pages cited by Byrd from [24], and it would not be correct since most conspicuously missing from this list is **KD45** (together with its **Alt**$_n$ extensions limiting the numbers of points accessible to a given point). The unprovability of 4 in **KD5** shows in fact that Rescher and Urquhart's conjecture is false, since its translation is a theorem of **KD4.3** (indeed of **KD4**). (Whether, revised so as to have **KD45** as a source logic for the translation, the conjecture would be correct, the author does not know.) It should be pointed out that in the works cited, none of this terminology is employed. Instead, Rescher and Urquhart observe that for the (future-directed) tense logic of linear unending time if we interpret □ as what in Prior's notation would be written as FG, the logic **KD5** (for □) is sound, and conjecture that it is complete. (Soundness and completeness here are to be understood as "with respect to the class of (forward) serial, transitive, and (forward) connected frames".) Byrd took himself to be establishing that this logic is indeed complete for the intended interpretation of □ ("eventual permanence") over these frames, and point just made would be put by saying that □$p \to$ □□p is valid but not provable, contradicting the supposed completeness result.

[12]Not being a reader of German, I owe my knowledge of the contents of this paper to correspondence from Lenzen in 1982; the parallel with Dawson occurred to me only somewhat later. A significant difference between the two – which does not bear on the issue of embeddings – is that Lenzen is also concerned with a plausible *mixed* epistemic-doxastic logic in which, to use the conventional notations for the □ operators in this setting, BA is defined to be $\neg K \neg KA$. Thus there are questions about the interaction of knowledge and belief to consider. The most interesting interaction effect is a 'Cartesian' principle: $Bp \to BKp$, reflecting the fact that belief here is taken to be of the 'fully confident' variety. Dawson, on the other hand, was not concerned in [8] with developing a plausible combined alethic-deontic logic – except on the second last page of [8], where he ventures into this territory as a kind of afterthought. There is, for instance, no need to defend the analogue of the mixed epistemic-doxastic principle just cited: that whatever ought to be the case ought necessarily to be the case. Another way of putting this point is to say that whereas Lenzen is interested in the definitional extension of an epistemic logic by the definition of B given above, Dawson (or an idealized version of Dawson, ignoring the afterthought) is interested only in the image under the correlated τ – in the present instance, that replacing □ with ◇□ – of a given alethic logic. (Note that it is because of the correlation between these embeddings and explicit definitions that the former are called definitional embeddings; when used as explicit definitions, however, it is mandatory that the □ in the *definiens* and the □ in the *definiendum* be notationally distinguished, as with the K-versus-B notation earlier.)

further observed that the (definitional) translation which simply replaces $\Box$ with $\Diamond\Box$ throughout a formula embeds **KD45** (a suitably idealized doxastic logic – sometimes also referred to as 'deontic **S5**'[13] into **S4.2** (a candidate epistemic logic).[14]

Although the above description represents the construction of a deontic or doxastic logic on the basis of an alethic or epistemic logic as a process of weakening (weakening **T** to **D**), when we look more closely, we can see that there is also some strengthening going on. This is evident from the fact that the source of the embedding just mentioned is **KD45**, which because of the '**5**' is not after all a sublogic of its target, **S4.2**. So we are not quite in possession of an example of the phenomenon of interest: a weaker-to-stronger definitional embedding. (This example may be added to that in note 9 to illustrate the case of incomparable source and target logics, this time from amongst the normal modal logics.) Still fixing on the translation that replaces $\Box$ by $\Diamond\Box$ we could try correcting matters by either strengthening the target logic – say, to **S5**, since the source logic proves **5** – or by weakening the source logic, to something in which **5** in particular, is not provable. The former avenue is in fact not open because the fact that $\nvdash_{\mathbf{KD45}} \Box p \to p$ would require $\nvdash_{\mathbf{S5}} \Diamond\Box p \to p$. So in the following section we pursue the latter possibility.

3 The Logic KDH

The most important feature of **S4.2** in explaining its appearance as the target logic for the embedding mentioned in the previous section is not the leading axiom **G** itself, but the fact that in concert with the rest of **S4**, this logic includes amongst its theorems the formula we shall call **H**, simplifying some given in in Chellas [7] to two of the formulas listed here;[15] we use the label indiscriminately for any of the following four following evidently interdeducible forms:

---

[13]This same label has been used for several different logics. Even as early as 1971, Rescher and Urquhart [22], p. 136 (n. 14 and text to which the note is appended), cite three such logics.

[14]And further still, Lenzen notes that this remains correct if the target is replaced by the stronger logic **S4.4**, for which case there is also a translation in the reverse direction. This aspect of Lenzen [18] is recalled in Pelletier and Urquhart [21]. Interestingly, in the final paragraph of [8], Dawson had also mentioned **S4.4**, though without using that name.

[15]The labels in question are "(**H**$^{++}$)" and"(**H**$^{++}\Diamond$)". We have taken the liberty of removing the "++" superscript, even though another formula on Chellas's list, in Exercise 4.56 ([7], p. 144) has this label (i.e., **H** *tout court*), and – contrary to the claim there made – the result of adding the present formula to **S4** is not **S4.3** but **S4.2**. The result of adding the formula which *Chellas* calls **H** (following, e.g., [17], p. 69, and named after Hintikka) is indeed **S4.3**, and the list of formulas in the exercise falls into two parts, some of which give **S4.2** and others of which give **S4.3**, when added to **S4**. The needed correction seems not to be amongst the several which have been made in subsequent reprintings of [7] – at least, this is so for the 1995 printing. Yet another (unrelated) formula is called **H** in Hughes and Cresswell [10], p. 160; likewise in the cases of Segerberg [24], p. 148, and Chagrov and Zakharyaschev [5], Exercise 5.21, p. 158.

H $(\Diamond\Box p \wedge \Diamond\Box q) \to \Diamond\Box(p \wedge q)$ $(\Diamond\Box p \wedge \Diamond\Box q) \to \Diamond(\Box p \wedge \Box q)$
$\Box\Diamond(p \vee q) \to (\Box\Diamond p \vee \Box\Diamond q)$ $\Box(\Diamond p \vee \Diamond q) \to (\Box\Diamond p \vee \Box\Diamond q)$.

While **H** and **G** yield the same normal modal logic (namely **S4.2**) as additions to **S4**, for purposes of extending **K** (and **KD**) it is the effects of **H** that we need. (**G** is easily seen to be provable in **KDH**.)[16]

It will also be convenient to isolate the formulas $\mathbf{H}_n$, though this time we mention only the analogues of the first formula listed under **H**:

$\mathbf{H}_n$ $(\Diamond\Box p_1 \wedge \ldots \wedge \Diamond\Box p_n) \to \Diamond\Box(p_1 \wedge \ldots \wedge p_n)$.

it is easy enough to see that any normal modal logic proving $\mathbf{H}_2$ (= **H**) proves $\mathbf{H}_n$ for all $n \geq 1$. Since the converse is even more easily seen, we have $\mathbf{KH}_m = \mathbf{KH}_n$ for any $m, n \geq 2$. And $\mathbf{KH}_1 = \mathbf{K}$.[17] It also easy to check that **H** modally defines the class of frames $\langle W, R\rangle$ satisfying a condition[18] we shall call H, taking $\langle W, R\rangle$ as a first-order structure:

H $\forall w, x, y((Rwx \wedge Rwy) \to \exists u(Rwu \wedge \forall v(Ruv \to (Rxv \wedge Ryv))))$

and a familiar canonical model argument shows that **KH** is determined by (= sound and complete w.r.t.) the class of frames satisfying H. The key step, recapitulated in the Appendix for the sake of a contrast with a variant mentioned immediately below, involves showing that the canonical frame for any (consistent) normal modal logic extending **KH** satisfies H. For present purposes what is of greater interest is that the canonical frame for any such logic satisfies a stronger condition, H$^+$, from which it follows that **KH** is determined not only by the class of frames satisfying H but also by the narrower class of frames satisfying H$^+$. (See the Appendix for a proof of this.)

H$^+$ $\forall w[\exists u(Rwu) \to \exists u(Rwu \wedge \forall x(Rwx \to \forall v(Ruv \to Rxv)))]$

Why are we taking all this interest in the axiom **H**? Not because of the logic **KH**, but because of the logic **KDH**, in fact. Recall that for a primitive or derived 1-ary connective # to be normal in **S** we require for all formulas A, B: that (i) whenever $\vdash_\mathbf{S} A \to B$ we have $\vdash_\mathbf{S} \#A \to \#B$, (ii) $\vdash_\mathbf{S} (\#A \wedge \#B) \to \#(A \wedge B)$, and (iii) $\vdash_\mathbf{S} \#\top$. Now in any normal modal logic condition (i) is satisfied by any affirmative modality, and hence by the choice of # as $\Diamond\Box$, and condition (ii) is equivalent, given the requirement of closure under uniform substitution, to the provability of **H**; but condition (iii) for the present choice of # is not satisfied by **KH**, since $\not\vdash_\mathbf{KH} \Diamond\Box\top$. Thus we need to consider an extension of **KH**, and since in any normal modal logic $\Diamond\Box\top$ and $\Diamond\top$ are equivalent (because $\Box\top$ and $\top$ are), the extension we need is **KDH** – alias the smallest normal modal logic containing $\mathbf{H}_n$ for all all $n \geq 0$ (as we saw in note 17).

---

[16] The interdeducibility of **G** and **H** over **S4** is shown on p. 75 of [8]; this is part of the point of Chellas's exercise, mentioned in the previous footnote.

[17] What about $\mathbf{H}_0$? Understanding the conjunction with no conjuncts as $\top$, the antecedent can be ignored, leaving us with $\Diamond\Box\top$, which is, as noted below, equivalent to **D**.

[18] That is, **H** is valid on all and only frames satisfying this condition.

The considerations just rehearsed show not just that $\Diamond\Box$ is normal in **KDH**, but that **KDH** is the *smallest* normal modal logic in which $\Diamond\Box$ is normal. This means that all the behaviour demanded of $\Box$ for the logic to count as normal is displayed by $\Diamond\Box$ in **KDH** but in no weaker logic. But what has still not been ruled out is that $\Diamond\Box$ might, considered as a single operator, exhibit other logical behaviour beyond that demanded by normality (i.e., exhibited by $\Box$ in **K**). Ruling that out would constitute establishing, that the Dawson–Lenzen translation – call it $\tau_{\Diamond\Box}$ – which uniformly replaces $\Box$ with $\Diamond\Box$ does indeed embed **K** into **KDH**. We can see without much difficulty that this can't be quite right, since any extension of **KD** proves $\Box\Diamond\top$ and this is $\tau_{\Diamond\Box}(\Diamond\top)$ while $\Diamond\top$ is not **K**-provable. Thus, we need to replace **K** as a candidate source logic by **KD**, and adjust the conjecture to: $\tau_{\Diamond\Box}$ embeds **KD** into **KDH**. Spelling this out explicitly:

KDH CONJECTURE: For any formula A: $\vdash_{\mathbf{KD}} A \Leftrightarrow \vdash_{\mathbf{KDH}} \tau_{\Diamond\Box}(A)$.

The $\Rightarrow$ direction follows from the normality of $\Diamond\Box$ in **KDH** and the **KDH**-provability of $\tau_{\Diamond\Box}(\mathbf{D})$. This amounts to a proof of the $\Rightarrow$-claim by induction on the length of (a shortest) proof of A in **KD**. Before addressing the converse direction, we pause to ponder one aspect of the current direction.

The normality of $\Diamond\Box$ in **KDH** suggests that it should be interpretable by universal quantification over accessible points for some suitable accessibility relation. Making this explicit gives an alternative proof of the $\Rightarrow$ direction of the **KDH** conjecture, but seems to be of some interest in its own right, as one would like to see this implicit binary relation exhibited. As a preliminary step, we need a general observation concerning frames satisfying a strengthening of the condition H$^+$ above, which we call DH$^+$.

DH$^+$ $\quad \forall w \exists u (Rwu \wedge \forall x (Rwx \to \forall v (Ruv \to Rxv)))$

Note that DH$^+$ is equivalent to the conjunction of H$^+$, given earlier, with the 'seriality' condition D $= \forall w \exists z (Rwz)$.

For any w, call any point u as promised by DH$^+$ a *special successor* of w. Thus a special successor of w is an R-successor of w which, for any R-successor x of w, bears the relation R only to such points as x bears the relation R to. (The original condition H$^+$ says only that any point with a successor has a special successor.) To fix notation: we regard a model as a structure $\langle W, R, V \rangle$, where $\langle W, R \rangle$ is the frame of the model – $W \neq \varnothing$ and $R \subseteq W \times W$ – and V is a function mapping propositional variables to subsets of W. Truth at a point in such a model is then taken as defined in the standard way for arbitrary formulas, and notated with the aid of "$\models$".

PROPOSITION 1. *Let* $\mathcal{M} = \langle W, R, V \rangle$ *be any model whose frame* $\langle W, R \rangle$ *satisfies* DH$^+$. *Thus every* $w \in W$ *has a special successor in the sense just defined. Arbitrarily choose one such special successor* w^* *for each* $w \in W$ *and define a new model* $\mathcal{M}^* = \langle W, S, V \rangle$ *where* Sxy *just in case* Rx^*y. *Then for every formula* A: *for all* $w \in W$ *we have* $\mathcal{M}^* \models_w A$ *if and only if* $\mathcal{M} \models_w \tau_{\Diamond\Box}(A)$.

Proof. By induction on the complexity of A. The only case of interest arises in the induction step, and concerns the case in which A is $\Box B$ for some formula B. Since $\tau_{\Diamond\Box}(\Box B)$ is $\Diamond\Box(\tau_{\Diamond\Box}(B))$, what we have to show is that:

$$\mathcal{M}^* \models_w \Box B \text{ if and only if } \mathcal{M} \models_w \Diamond\Box(\tau_{\Diamond\Box}(B)).$$

For the 'only if' direction, suppose that $\mathcal{M}^* \models_w \Box B$. Then for all $v \in W$ such that Swv, we have $\mathcal{M}^* \models_v B$. This means that for all v such that Rw^*v, $\mathcal{M}^* \models_v B$. Suppose, for a contradiction that $\mathcal{M} \not\models_w \Diamond\Box(\tau_{\Diamond\Box}(B))$. In particular then, $\mathcal{M} \not\models_{w^*} \Box(\tau_{\Diamond\Box}(B))$, so for some $x \in W$ for which Rw^*x, $\mathcal{M} \not\models_x \tau_{\Diamond\Box}(B)$ and thus by the inductive hypothesis, $\mathcal{M}^* \not\models_x B$. Now we have our contradiction with the earlier conclusion that for all v such that Rw^*v, $\mathcal{M}^* \models_v B$.

For the 'if' direction, suppose that (i) $\mathcal{M} \models_w \Diamond\Box(\tau_{\Diamond\Box}(B))$ but (ii) $\mathcal{M}^* \not\models_w \Box B$. From (ii) we infer that for some $v \in W$, Swv, which is to say: Rw^*v, and $\mathcal{M}^* \not\models_v B$. From (i), there is $x \in W$ with Rwx and $\mathcal{M} \models_x \Box(\tau_{\Diamond\Box}(B))$. Since w^* is a special successor of w and Rw^*v and Rwx, we must have Rxv (as $\mathcal{M}$ satisfies DH^+); so, since $\mathcal{M} \models_x \Box(\tau_{\Diamond\Box}(B))$, $\mathcal{M} \models_v \tau_{\Diamond\Box}(B)$. The inductive hypothesis then gives: $\mathcal{M}^* \models_v B$, contradicting what was inferred from (ii) above. ∎

The S appearing here is the 'hidden' accessibility relation which provides a model-theoretic explanation of the normality of $\Diamond\Box$ in **KDH**. To use this observation to derive the (already settled) "⇒" direction of the **KDH** Conjecture, suppose that $\nvdash_{\mathbf{KDH}} \tau_{\Diamond\Box}(A)$. Thus A is false as some point w the canonical model for **KDH**, which satisfies the condition DH^+ (see Appendix). Thus the construction used in Proposition 1 gives another model (with the same universe) in which A itself is false at w. Further, this new model has a serial frame. Therefore, by the soundness of **KD** w.r.t. the class of such frames, $\nvdash_{\mathbf{KD}} A$.

Let us return to the ⇐ direction of the **KDH** Conjecture above. By analogy with the case of Proposition 1, one would like a recipe for turning any serial model into a model for **KDH**, perhaps adding additional points (to serve as special successors) in such a way that whenever a point in the first model falsifies a formula A, a corresponding point in the second model falsifies $\tau_{\Diamond\Box}(A)$. That would guarantee that whenever **KDH** proves the $\tau_{\Diamond\Box}$-translation of a formula, the formula itself is **KD**-provable. But, perhaps surprisingly in view of the fact, emphasized above, that **KDH** is the *weakest* normal modal logic in which $\Diamond\Box$ is normal, it is too strong for any such recipe to be found:

PROPOSITION 2. *The* ⇐ *direction of the* **KDH** *Conjecture is false.*

Proof. It suffices to provide a counterexample – a formula A for which $\vdash_{\mathbf{KDH}} \tau_{\Diamond\Box}(A)$ while $\nvdash_{\mathbf{KD}} A$. We do so by selecting $A = (\Diamond\Diamond p \wedge \Diamond\Diamond q) \to \Diamond(\Diamond p \wedge \Diamond q)$. There is no difficulty in falsifying this formula at a point in a model on a serial frame, so it is not **KD**-provable. But its $\tau_{\Diamond\Box}$-translation is

KDH-provable, as the following considerations show. Begin with the upper right-hand form of the **H** axiom above:

$$(\Diamond\Box p \wedge \Diamond\Box q) \to \Diamond(\Box p \wedge \Box q).$$

By the normality of $\Box$, we conclude that the following is provable:

$$(\Box\Diamond\Box p \wedge \Box\Diamond\Box q) \to \Box\Diamond(\Box p \wedge \Box q).$$

Finally, substituting $\Diamond p$ for p and $\Diamond q$ for q:

$$(\Box\Diamond\Box\Diamond p \wedge \Box\Diamond\Box\Diamond q) \to \Box\Diamond(\Box\Diamond p \wedge \Box\Diamond q).$$

We have here arrived at $\tau_{\Diamond\Box}(A)$ for the choice of A indicated above. ∎

Earlier, we noted that the source logic for the conjectured embedding had to be boosted from **K** to **KD** because $\tau_{\Diamond\Box}(\mathbf{D})$ was **KDH**-provable, which was no problem for the general prospect of a weaker-to-stronger embedding, since $\mathbf{KD} \subsetneq \mathbf{KDH}$. The above example does not allow for a similar 'minor adjustment', as the formula $(\Diamond\Diamond p \wedge \Diamond\Diamond q) \to \Diamond(\Diamond p \wedge \Diamond q)$ is not provable in **KDH**. Because of this negative finding, we conclude our exploration of the translation $\tau_{\Diamond\Box}$ at this point, and despite the interest, in view of the observations of Dawson and Lenzen, of this translation turn our attention to finding a weaker-to-stronger embedding (via a definitional translation) from whatever quarter we may, setting aside such questions as whether the translation concerned is of independent interest.

4 A Successful Weaker-to-Stronger Embedding

We will make use of the idea, mentioned before Proposition 1, of a model construction adding a new layer of points between points and those accessible to them in some given model, so that what took one step by the accessibility relation now takes two steps by the new relation. This was problematic in the case where we wanted the new model to be a model for **KDH**, because such conditions as H and H$^+$ had to be satisfied by the new model – which Proposition 1 shows is a requirement that could not be satisfied. We will now consider the translation $\tau_{\Box\Box}$ which replaces every $\Box$ by $\Box\Box$. The latter is normal in every normal modal logic, allowing us to bypass the difficulties (just alluded to) raised by **H**, whose presence was required in the target logic for $\Diamond\Box$ to be normal. We start with a model $\mathcal{M} = \langle W, R, V \rangle$ and construct what we shall call $\mathcal{M}^+$. (To visualize this process, it is simplest to imagine that the frame of $\mathcal{M}$ is a tree, restriction to which cases will be sufficient for our applications of the idea, in view of [23].) For each element of W pick a new element $\bar{w}$ where $\bar{w} \notin W$ and for $w, v \in W$, $\bar{w} \neq \bar{v}$ whenever $w \neq v$. Now define $\mathcal{M}^+$ to be $\langle W^+, S, V^+ \rangle$ where:

- $W^+ = W \cup \{\bar{w} \mid w \in W\}$;

- for $x, y \in W^+$, Sxy if and only if for some $w \in W$ we have $x = w$ and $y = \bar{w}$ or else $x = \bar{w}$, $y \in W$ and Rwy;

- V^+ is any assignment of truth-sets to atomic formulas whose restriction to W is V. (I.e., for all $w \in W$, $w \in V^+(p_i) \Leftrightarrow w \in V(p_i)$; this means we have not uniquely specified $\mathcal{M}^+$, but may take it to be any model meeting the constraints given.)

LEMMA 3. *Let $\mathcal{M} = \langle W, R, V \rangle$ be any model and take $\mathcal{M}^+$ to be the model $\langle W^+, S, V^+ \rangle$ obtained by the construction just described. Then for all formulas A, and all $w \in W$ we have $\mathcal{M} \models_w A$ iff $\mathcal{M}^+ \models_w \tau_{\Box\Box}(A)$.*

Proof. As for Proposition 1, this is shown by induction on the complexity of A and the only inductive case of interest is again that of $A = \Box B$, which is much simpler here as Rwv (for $w, v \in W$) iff $Sw\bar{w}$ and $S\bar{w}v$, which in turn holds iff $S^2 wv$. ∎

Lemma 3 allows us to draw the not very exciting conclusion that $\tau_{\Box\Box}$ embeds **K** into itself. The '⇒' half of this conclusion:

For all A, $\vdash_{\mathbf{K}} A \Leftrightarrow \vdash_{\mathbf{K}} \tau_{\Box\Box}(A)$,

is something already mentioned in passing, that $\Box\Box$ is normal in any normal modal logic, and so in particular in **K**, which itself, like the corresponding claim for $\Diamond\Box$ in our discussion of **KDH** in the preceding section, can be shown either by an induction on the length of proofs in an axiomatization (in the present instance, of **K**), or model-theoretically, as in the proof of Proposition 1: in the present case this amounts to the observation that the relative product R^2 of a relation R with itself, is just another binary relation – which can be used to interpret $\Box$. The '⇐' half of the inset claim, presented contrapositively, invites us to consider a point in a model $\mathcal{M}$ at which A is false (by the completeness of **K** w.r.t. the class of all frames), from which Lemma 3 allows us to infer that $\tau_{\Box\Box}(A)$ is false at the corresponding point in the model $\mathcal{M}^+$, and is therefore not **K**-provable, this time by the soundness of **K** w.r.t. the class of all frames.

To get some mileage out of Lemma 3 for where we have been heading, however, we need to observe that the frames of the derived models $\mathcal{M}^+$ have some specific structural features that can be exploited to validate modal formulas not valid on all frames. Specifically, note that each point in the universe W^+ of such a frame is either one of the original elements w of W, in which case it bears the new accessibility relation S to at most one point, namely the 'new' point $\bar{w}$, or else it is itself one of these new ("barred") points, in which case all its S-successors in turns bear the relation S to at most one point. Thus the following formula, for which we adopt an *ad hoc* notation simply in the interest of brevity:

X $(\Diamond p \to \Box p) \vee \Box(\Diamond q \to \Box q)$

is valid at each point in the frame, as the first first disjunct is valid at the surviving 'old' points and the second at the new points.[19] Where **KX** is the smallest normal modal logic containing the above formula **X**, we can combine this 'soundness' observation with a completeness result provided straightforwardly by the canonical model method to record the following, stated with a reversion to the 'R' notation for an arbitrary accessibility relation and also write $R(w)$ for $\{z \in W | Rwz\}$, etc. (and $|Z|$ is the cardinality of a set Z):

PROPOSITION 4. **KX** *is determined by the class of frames* $\langle W, R \rangle$ *in which for all* $x \in W$ *either* $|(R(x))| \leq 1$ *or for all* $y \in R(x), |R(y)| \leq 1$.

All we need for present purposes is the soundness half of this result, though, already evident in the observation that **X** is valid on the frames of the models $\mathcal{M}^+$.

THEOREM 5. *The translation* $\tau_{\Box\Box}$ *embeds* **K** *(faithfully) into* **KX**.

Proof. We have already seen that $\tau_{\Box\Box}(A)$ is **K**-provable – so *a fortiori* it is **KX**-provable – for any **K**-provable A. On the other hand if $\nvdash_{\mathbf{K}} A$, then for some model $\mathcal{M}$ and some point w in that model, $\mathcal{M} \nvDash_w A$ so by Lemma 3, $\mathcal{M}^+ \nvDash_w \tau_{\Box\Box}(A)$, and since $\mathcal{M}^+$ is a model for **KX**, $\nvdash_{\mathbf{KX}} \tau_{\Box\Box}(A)$. ∎

Note incidentally, *à propos* of the observation in Section 1 about the unembeddability of Halldén-incomplete logics, such as **K**, into Halldén-complete logics, that **KX** is evidently Halldén-incomplete, in view of the 'Halldén-unreasonable' disjunction **X**. (Use Proposition 4 to check that neither disjunct is **KX**-provable.)

The same considerations show that $\tau_{\Box\Box}$ embeds **KD** into **KDX**, because the transformation $\mathcal{M} \leadsto \mathcal{M}^+$ preserves seriality. (For the analogue of Proposition 4 in the case of **KDX**, replace the occurrences of "$\leq$" with "$=$"; the "$\to$"s in **X** can also be equivalently replaced with "$\leftrightarrow$"s in the presence of **D**.) It would be interesting to know what range is of normal modal logics for which an argument along these lines can be given.

It is also worth recording the fact that the transformation from $\mathcal{M}$ to $\mathcal{M}^+$ also allows us to prove the analogue of Lemma 3 with $\tau_{\Box\Box}$ replaced by the translation that occupied us in Section 3, $\tau_{\Diamond\Box}$, and that for all $w \in W$, $\mathcal{M}^+ \vDash_w A$ for all A for which $\vdash_{\mathbf{KDH}} A$. But $\mathcal{M}^+$ is not a model for **KDH** because the 'new points' $\bar{w}$ in W^+ need not verify **H**, and any effort to fix this by ensuring that every point has a special successor then prevents the $\tau_{\Diamond\Box}$-version of Lemma 3 from going through.

Finally, returning to the theme of Section 1, the fact that $\mathbf{KX} \supsetneq \mathbf{K}$ does not prevent the translational embedding via $\tau_{\Box\Box}$ from succeeding even though, in the notation of note 7, $D_{\mathbf{KX}} \subsetneq D_{\mathbf{K}}$, since every distinction in **K** between formulas A and B is matched by a corresponding distinction in **KX** between $\tau_{\Box\Box}(A)$ and $\tau_{\Box\Box}(B)$.

---

[19] In calling a formula *valid at* a point in a frame, we mean that it is true at that point in any model on that frame.

Appendix

We begin by showing, as claimed in Section 3, that the canonical frame for any consistent normal modal logic extending **KH** satisfies the condition there called H, repeated below for convenience. The purpose of including this standard style of argument here is to provide a contrast with the more interesting completeness result given later (involving the the condition H$^+$).

H $\qquad \forall w, x, y((Rwx \wedge Rwy) \to \exists u(Rwu \wedge \forall v(Ruv \to (Rxv \wedge Ryv))))$

Let $\langle W, R \rangle$ be the canonical frame of any consistent normal modal logic in which **H** is provable. We need only show that this frame satisfies the condition H. It has to be shown how, given $w, x, y \in W$, one may find a $u \in W$ behaving as required: a $u \in R(w)$ with $R(u) \subseteq R(x) \cap R(y)$. We construct such a u as any maximally consistent (w.r.t. the logic in question) superset of the following set, once the set's own consistency has been established; it is clear enough that this will guarantee that u behaves as required:

$$\{A \mid \Box A \in w\} \cup \{\Box B \mid \Box B \in x\} \cup \{\Box C \mid \Box C \in y\}.$$

By familiar steps, we get that if this set were not consistent then we should have ("⊢" indicating provability in the logic in question):

$\vdash (A_1 \wedge \ldots \wedge A_k) \to \neg(\Box B_1 \wedge \ldots \wedge \Box B_m \wedge \Box C_1 \wedge \ldots \wedge \Box B_n)$

where the A's, B's and C's are drawn from the three terms of the union inset above; further routine steps follow:

$\vdash (A_1 \wedge \ldots \wedge A_k) \to \neg(\Box B_1 \wedge \ldots \wedge \Box B_m \wedge \Box C_1 \wedge \ldots \wedge \Box C_n)$

$\vdash (A_1 \wedge \ldots \wedge A_k) \to \neg(\Box(B_1 \wedge \ldots \wedge B_m) \wedge \Box(C_1 \wedge \ldots \wedge C_n))$

$\vdash (A_1 \wedge \ldots \wedge A_k) \to ((\Diamond\neg(B_1 \wedge \ldots \wedge B_m) \vee \Diamond\neg(C_1 \wedge \ldots \wedge C_n)),$

$\vdash (\Box A_1 \wedge \ldots \wedge \Box A_k) \to \Box(\Diamond\neg(B_1 \wedge \ldots \wedge B_m) \vee \Diamond\neg(C_1 \wedge \ldots \wedge C_n)),$

at which point we invoke the an appropriate substitution instance of the axiom **H** to conclude that

$\vdash (\Box A_1 \wedge \ldots \wedge \Box A_k) \to (\Box\Diamond\neg(B_1 \wedge \ldots \wedge B_m) \vee \Box\Diamond\neg(C_1 \wedge \ldots \wedge C_n)).$

So, since each $\Box A_i \in w$ forces one or other disjunct of the consequent into w, thereby forcing either $\Diamond\neg(B_1 \wedge \ldots \wedge B_m)$ into $x \in R(w)$, contradicting the fact that each $\Box B_j \in x$, or $\Diamond\neg(C_1 \wedge \ldots \wedge C_m)$ into $y \in R(w)$, this time contradicting the provenance of the C's.

Another completeness result for **KH** was mentioned in Section 3, namely that this logic is determined by the class of frames satisfying:

H$^+$ $\qquad \forall w(R(w) \neq \varnothing \to \exists u \in R(w) \forall x \in R(w) \,.\, R(u) \subseteq R(x)).$

We show this, too, by looking at the canonical frame $\langle W, R \rangle$ of any consistent normal extension of **KH**. (The condition is abbreviated here with the '$R(w)$' notation.)

We must show how to find, for any $w \in W$ with a successor, a special successor (to use Section 3's terminology): a u as promised by the consequent of H$^+$. So suppose $R(w) \neq \varnothing$ To obtain the desired $u \in W$, we take any maximal consistent extension of the set $\{A \mid \Box A \in w\} \cup \{\Box B \mid \Diamond\Box B \in w\}$ – leaving the reader to check that this suffices for u to serve as a special successor to w – assuring ourselves of the consistency of this set (relative to the logic in question) by the following reasoning. Its inconsistency would mean (with $\vdash$ as above) that for some A_i ($i = 1, \ldots, k$) with $\Box A_i \in w$ and B_j ($j = 1, \ldots, n$) with $\Diamond\Box B_j \in w$:

$$\vdash (A_1 \wedge \ldots \wedge A_k) \to \neg(\Box B_1 \wedge \ldots \wedge \Box B_n) \tag{1}$$

and so by the normality of $\Box$ and a little De Morgan on the consequent:

$$\vdash (\Box A_1 \wedge \ldots \wedge \Box A_k) \to \Box(\Diamond\neg B_1 \vee \ldots \vee \Diamond\neg B_n). \tag{2}$$

Since the antecedent here is in w, so is the consequent. But by $\mathbf{H}_n$, or more accurately, a variant of the dual of $\mathbf{H}_n$ as given in Section 3:

$$\vdash \Box(\Diamond\neg B_1 \vee \ldots \vee \Diamond\neg B_n) \to (\Box\Diamond\neg B_1 \vee \ldots \vee \Box\Diamond\neg B_n). \tag{3}$$

Thus one of the disjuncts $\Box\Diamond\neg B_j$ of the consequent must belong to w, contradicting the choice of the various B_j as formulas for which $\Diamond\Box B_j \in w$. What, however, of the possibility that there are no such B_j? This would mean that the negation of the antecedent in (1) was provable, which would contradict the fact that each $\Box A_i \in w$ and $R(w) \neq \varnothing$.

As a simple corollary, we have that the canonical frame for any consistent normal modal extension of **KDH** satisfies the condition DH$^+$ from Section 3 (H without its antecedent), and thus that **KDH** itself is determined by the class of frames meeting this condition. Also, we note that since whenever a class of frames contains the canonical frame for some logic but not every frame for the logic, that class is not modally definable, it follows that the class of frames satisfying H$^+$ is not modally definable.

There is a point of some theoretical interest here arising over the extent to which a condition (+), in general *stronger* than a corresponding condition (−), is *equivalent* to it when considered specifically as a condition on canonical frames for normal modal logics. (What follows is a tentative and provisional attempt to characterize the phenomenon of interest; a more general formulation may be desirable, and the use just made of the phrase "in general" turns out to fall short of meaning "universally". This will become evident presently, *à propos* of **KH**, and it suggests that these remarks will need to be corrected in other ways too.) We consider the first-order language with dyadic predicate letter R as sole non-logical primitive. Where $\Phi(w, x, u)$ is a formula of this language in which at most the exhibited variables are free – though possibly containing further quantifiers and their associated bound variables – let $\Phi_n^\wedge(w, x, u)$ be the result of replacing any subformula Rxv (v any variable distinct from x[20]) in $\Phi(w, x, u)$ by

---

[20]The author has not considered the case of subformulas Rxx, or, for that matter, $x = v$.

$Rx_1v \wedge \ldots \wedge Rx_nv$, with an analogous replacement in the case of Rvx. Then the $(-)$ and $(+)$ conditions induced by Φ are the following:

$(-)$ For all $n \in$ Nat: $\forall w \forall x_1 \ldots \forall x_n \exists u \Phi_n^{\wedge}(w, x, u)$

$(+)$ $\forall w \exists u \forall x \Phi(w, x, u)$.

(In $(-)$, the $n = 0$ case is to be understood as for the modal object language: see note 17.) For the case of **KDH**, $(+)$ is DH$^+$, here rewritten into the above format (with the key quantifier prefix preceding a suitable Φ):

$$\forall w \exists u \forall x [Rwu \wedge (Rwx \to \forall y (Ruy \to Rxy))]$$

and the corresponding $(-)$ condition is the combined requirement that for each $n \in$ Nat:

$$\forall w \forall x_1 \ldots \forall x_n \exists u [Rwu \wedge (\bigwedge_{i=1}^{n} Rwx_i \to \forall y (Ruy \to \bigwedge_{i=1}^{n} Rx_iy))].$$

If the canonical frame for a normal modal logic satisfies the latter – (in general) weaker – condition then the logic must prove $\mathbf{H}_n$ for all $n \in$ Nat, which in turn guarantees that the frame satisfies the former – stronger – condition. (For a variation inspired by the case of **KH** the $(+)$ condition would be

$$\forall w \exists u \forall x [Rwx \to (Rwu \wedge \forall y (Ruy \to Rxy))],$$

which is equivalent to our condition H$^+$, which is a "canonical consequence" of the corresponding $(-)$ condition, requiring for all n:

$$\forall w \forall x_1 \ldots \forall x_n \exists u [\bigwedge_{i=1}^{n} Rwx_i \to (Rwu \wedge \forall y (Ruy \to \bigwedge_{i=1}^{n} Rx_iy))].$$

By contrast with the **KDH** weak condition, the $n \geq 1$ cases of this condition do not imply seriality, and the $n = 1$ case is a logical truth, matching the vacuity noted in Section 3 of $\mathbf{H}_1$; further, as in the discussion there, the m and n cases are equivalent whenever $m, n \geq 2$. This $(-)$ condition can accordingly be simplified to consist of the requirements given by the $n = 0$ and $n = 2$ cases, the latter being our condition H. But the former implies seriality, which, as noted, does not follow from the present $(+)$ condition, contrary to our stage-setting remarks characterizing the $(-)$ condition as weaker than – implied by – the stronger $(+)$ condition. This anomaly shows that the author's current grasp of the general situation leaves much to be desired.)

A simpler example, both in reducing the complexity of Φ and in not involving the anomaly just parenthetically alluded to, of the $(-)$ form implying the $(+)$ form over canonical frames is the following, which may already be familiar to the reader. Here the initial $\forall w$ is absent (or vacuous) and $\Phi(w, x, u)$ is just Rux. The $(-)$ condition then says that any finitely many points have a common R-predecessor while the $(+)$ condition says that some point bears the relation R to every point.[21] While in general the former condition is weaker than the latter, restricting attention to canonical frames, the "$(-) \Rightarrow (+)$" implication holds, since if the canonical frame for (normal modal) **S** satisfies $(-)$, then **S** enjoys the Rule of Disjunction, as it is called in [17], p. 44, from which one may conclude that the canonical frame for **S** satisfies the $(+)$ condition.

A case intermediate in complexity is provided by the following generalized form of **G** (from Section 3), namely the schema:

G$_n$ $\quad (\Diamond\Box p_1 \wedge \ldots \wedge \Diamond\Box p_n) \to \Box\Diamond(p_1 \wedge \ldots \wedge p_n)$.

For a particular choice of n, the formula **G**$_n$ – which differs from **H**$_n$ in having "$\Box\Diamond$" in place of "$\Diamond\Box$" in the consequent – is valid at precisely those points any $n+1$ successors of which themselves have a common successor; in speaking of the schema **G**$_n$, we have in mind the simultaneous presence of all such formulas. Their provability is guaranteed for a normal modal logic whose canonical frame satisfies the $(-)$ condition we get by taking $\Phi(w, x, u)$ as $Rwx \to Rxu$, which in turn suffices for the canonical frame to satisfy the corresponding $(+)$ condition, according to which

$$\forall w \exists u \forall x (Rwx \to Rxu).[22]$$

It would be interesting to know what the range of conditions Φ is for which this '$\exists$-fronting' $(-)$-to-$(+)$ transition goes through (as well as clarifying the situation described parenthetically as anomalous *à propos* of the **KH**-inspired $(+)$ and $(-)$ conditions). The whole issue is presumably closely connected with other compactness-induced phenomena in modal logic such as Fine's modal saturation property and the Goldblatt–Thomason–van Benthem ultrafilter extension construction, but exactly how this connection goes is not clear to the author – though it may well have been explored in the literature of the subject somewhere.

Acknowledgments. I am grateful to Allen Hazen and to Bryn Humberstone for their comments on the material in Section 4 of this paper, and to Rohan French for providing the challenge in response to which it was written, as well as to several careful referees for their corrections.

Note added in press. After this paper was completed, my attention was drawn to [25], which provides an elegant and highly informative discussion of the embedding possibilities amongst various (not necessarily normal) modal logics, though without special attention to our weaker/stronger theme.

---

[21] Such formulations call to mind a second-order representation of the $(-)$ conditions, with monadic second-order variables ranging over finite sets. Instead of the above $\Phi_n^\wedge(w, x, u)$ formulas we would have $\Phi^\forall$, say, in which subformulas Rxv are replaced by $\forall x(Xx \to Rxv)$ – and analogously for Rvx – and the $(-)$ form is given by: $\forall w \forall X \exists u \Phi^\forall(w, x, u)$. This may be more suggestive for sorting out the present topic.

[22] To obtain u given w in the canonical frame, take $u \supseteq \{A \mid \Diamond\Box A \in w\}$. The consistency of this set follows from the fact that a normal modal logic proving all instances of **G**$_n$ is closed under the rule: *From* $A_1 \vee \ldots \vee A_k$ *to* $\Box\Diamond A_1 \vee \ldots \vee \Box\Diamond A_k$. (In fact the admissibility of this rule for a given k is equivalent to the provability of the particular instance **G**$_{k-1}$.)

BIBLIOGRAPHY

[1] Lennart Åqvist, 'On Dawson Models for Deontic Logic', *Logique et Analyse* **7** (1964), 14–21.
[2] Jean-Yves Béziau, 'Classical Negation Can Be Expressed by One of its Halves', *Logic Journal of the IGPL* **7** (1999), 145–151.
[3] Michael Byrd, 'Eventual Permanence', *Notre Dame Journal of Formal Logic* **21** (1980), 591–601.
[4] A. V. Chagrov and M. Zakharyaschev, 'Modal Companions of Intermediate Propositional Logics', *Studia Logica* **51** (1992), 49–82.
[5] A. V. Chagrov and M. Zakharyaschev, *Modal Logic*, Clarendon Press, Oxford 1997.
[6] B. F. Chellas, 'Modalities in Normal Systems Containing the S5 Axiom', pp. 261–265 in C. Diamond and J. Teichman (eds.), *Intention and Intentionality: Essays in Honour of G. E. M. Anscombe*, Harvester Press, Brighton 1979.
[7] B. F. Chellas, *Modal Logic: An Introduction*, Cambridge University Press, Cambridge 1980.
[8] E. E. Dawson, 'A Model for Modal Logic', *Analysis* **19** (1959), 73–78.
[9] K. Došen, 'Modal Translations in Substructural Logics', *Journal of Philosophical Logic* **21** (1992), 283–336.
[10] G. E. Hughes and M. J. Cresswell, *A New Introduction to Modal Logic*, Routledge, London 1996.
[11] Lloyd Humberstone, 'Contra-Classical Logics', *Australasian Journal of Philosophy* **78** (2000), 437–474.
[12] Lloyd Humberstone, 'Béziau's Translation Paradox', *Theoria* **71** (2005). 138–181.
[13] Lloyd Humberstone, 'Logical Discrimination', pp. 207–228 in J.-Y. Béziau (ed.) *Logica Universalis: Towards a General Theory of Logic*, Birkhäuser, Basel 2005.
[14] Lloyd Humberstone, 'Modality', Chapter 20 (pp. 534–614) in Frank Jackson and Michael Smith (eds.), *The Oxford Handbook of Contemporary Philosophy*, Oxford University Press 2005.
[15] Lloyd Humberstone, 'Modal Logic for Other-World Agnostics: Neutrality and Halldén Incompleteness', forthcoming, *Journal of Philosophical Logic*.
[16] Charles F. Kielkopf, 'K1 as a Dawson Modelling of A. R. Anderson's Sense of "Ought"', *Notre Dame Journal of Formal Logic* **15** (1974), 402–410.
[17] E. J. Lemmon and D. S. Scott, *An Introduction to Modal Logic*, ed. K. Segerberg, American Philosophical Quarterly Monograph Series, Basil Blackwell, Oxford 1977. (Originally circulated 1966.)
[18] Wolfgang Lenzen, 'Epistemologische Betrachtungen zu [S4, S5]', *Erkenntnis* **14** (1979), 33–56.
[19] K. Matsumoto, 'Reduction Theorem in Lewis' Sentential Calculi', *Math. Japonicae* **3** (1955), 133–135.
[20] F. J. Pelletier, 'Six Problems in "Translational Equivalence"', *Logique et Analyse* **27** (1984), 423–434.
[21] F. J. Pelletier and A. Urquhart, 'Synonymous Logics', *Journal of Philosophical Logic* **32** (2003) 259–285.
[22] N. Rescher and A. Urquhart, *Temporal Logic*, Springer-Verlag, New York 1971.
[23] H. Sahlqvist, 'Completeness and Correspondence in the First and Second Order Semantics for Modal Logic', pp. 110–143 in S. Kanger (ed.), *Procs. Third Scandinavian Logic Symposium, Uppsala 1973*, North-Holland, Amsterdam 1975.
[24] K. Segerberg *An Essay in Classical Modal Logic*, Filosofiska Studier, Uppsala 1971.
[25] Evgeni E. Zolin, 'Embeddings of Propositional Monomodal Logics', *Logic Journal of the IGPL* **8** (2000), 861–882.

Lloyd Humberstone
Department of Philosophy, Monash University, Vic. 3800, Australia
Lloyd.Humberstone@arts.monash.edu.au

Dynamic topological logics over spaces with continuous functions

B. KONEV, R. KONTCHAKOV, F. WOLTER AND

M. ZAKHARYASCHEV

ABSTRACT. Dynamic topological logics are combinations of topological and temporal modal logics that are used for reasoning about dynamical systems consisting of a topological space and a continuous function on it. Here we partially solve a major open problem in the field by showing (by reduction of the ω-reachability problem for lossy channel systems) that the dynamic topological logic over arbitrary topological spaces as well as those over Euclidean spaces $\mathbb{R}^n$, for each $n \geq 1$, are undecidable. Actually, we prove this result for the natural and expressive fragment of the full dynamic topological language where the topological operators cannot be applied to formulas containing the temporal eventuality. Using Kruskal's tree theorem we also show that the formulas of this fragment that are valid in arbitrary topological spaces with continuous functions are recursively enumerable, which is not the case for spaces with homeomorphisms.

Keywords: dynamic topological logic, spatial logic, temporal logic.

1 Introduction

Dynamic topological logics were introduced in [10, 11, 13, 2] with the aim of logical modelling of and reasoning about the asymptotic properties of iterations of continuous functions on topological spaces, that is, orbits $\{f(x), f^2(x), \dots\}$ of points x in a topological space under a continuous function f. The full dynamic topological language $\mathcal{DTL}$ of these logics can be regarded as a natural combination of the standard modal language of $\mathcal{S}4$ interpreted over topological spaces and the propositional temporal language of $\mathcal{LTL}$ interpreted over $\langle \mathbb{N}, < \rangle$. For instance, the $\mathcal{DTL}$-formulas

$$\bigcirc \mathbf{I}\varphi \quad \text{and} \quad \mathbf{I}\square_F \varphi \qquad (1)$$

are interpreted in a dynamical system $\langle \langle T, \mathbb{I} \rangle, f \rangle$ (where $\mathbb{I}$ is the interior operator on the space T and f is a continuous function on $\langle T, \mathbb{I} \rangle$) under a valuation $\mathfrak{V}$ as, respectively, the sets

$$f^{-1}\big(\mathbb{I}\mathfrak{V}(\varphi)\big) \quad \text{and} \quad \mathbb{I}\bigcap_{n=1}^{\infty} f^{-n}(\mathfrak{V}(\varphi))$$

(we denote the modal box of $\mathcal{S}4$ by **I**). The available body of knowledge in this area can be roughly classified as follows.

The fragment $\mathcal{DTL}^\circ$ of $\mathcal{DTL}$ with sole temporal operator $\circ$:

- The sets of $\mathcal{DTL}^\circ$-formulas that are valid in dynamic topological systems (DTSs, for short) with *homeomorphisms* based on (i) arbitrary topological spaces, (ii) Aleksandrov spaces, (iii) the Euclidean spaces $\mathbb{R}^n$, for each $n \geq 1$, coincide, enjoy the finite model property (fmp), are finitely axiomatisable and decidable [2, 13, 12, 8].

- The sets of $\mathcal{DTL}^\circ$-formulas that are valid in DTSs (with *continuous functions*, not only homeomorphisms) based on (i) topological spaces and (ii) Aleksandrov spaces coincide, enjoy the fmp, are finitely axiomatisable and decidable [2, 12, 5]. A challenging open problem here is to investigate the set of $\mathcal{DTL}^\circ$-formulas that are valid in $\mathbb{R}$.

The fragment $\mathcal{DTL}_0$ of $\mathcal{DTL}$ where no temporal operator can occur in the scope of a topological operator: The logic in this language over arbitrary topological spaces and continuous functions was axiomatised in [12]. It is easy to see, in fact, that this logic is decidable and coincides with the corresponding logics over Aleksandrov spaces and $\mathbb{R}^n$, $n \geq 1$, no matter whether the functions are continuous or homeomorphisms only.

Full $\mathcal{DTL}$:

- The sets of $\mathcal{DTL}$-formulas that are valid in DTSs with *homeomorphisms* based on (i) topological spaces, (ii) Aleksandrov spaces, (iii) $\mathbb{R}^n$, for each $n \geq 1$, are all pairwise distinct and are not recursively enumerable [8]. It is of interest to note that the logics over (i)–(iii) with *finitely many iterations* of homeomorphisms coincide, but are still not recursively enumerable [8].

- Only two results have been known for the logics of DTSs with *continuous functions*. First, the logics over arbitrary and Aleksandrov spaces with *finitely many iterations* coincide and are decidable, but not in primitive recursive time [7]. Second, the logic over Aleksandrov spaces turns out to be undecidable [9]. The major open problem has been to investigate the computational properties (decidability, axiomatisability, etc.) of the logic over $\mathbb{R}^n$ and arbitrary topological spaces.

In this paper we present a partial solution to this open problem by showing that the logics in question are undecidable.

One key observation that has led to this result was the following fact discovered in [6] (and probably elsewhere) in a somewhat different context: the fragment $\mathcal{DTL}_1$ of $\mathcal{DTL}$, where we are not allowed to apply the topological operators to formulas containing $\Box_F$ or $\Diamond_F$ (as in the latter formula in (1)) but can have $\circ$ in the scope of **I**, is still expressive enough to encode various undecidable problems. But, on the other hand, it is not sufficiently strong to distinguish between arbitrary and Aleksandrov topological spaces. Clearly, $\mathcal{DTL}_1$ extends both $\mathcal{DTL}^\circ$ and $\mathcal{DTL}_0$. It follows from [8] that the sets of $\mathcal{DTL}_1$-formulas valid in (i) topological spaces, (ii) Aleksandrov spaces,

(iii) $\mathbb{R}^n$, for each $n \geq 1$, with *homeomorphisms* are still not recursively enumerable (even for systems with finitely many iterations).

The first main result of this paper is that the undecidable ω-reachability problem for lossy channel systems can be reduced, via $\mathcal{DTL}_1$-formulas, to satisfiability in Aleksandrov (and so arbitrary topological) spaces. On the other hand, using Kruskal's tree theorem, we show that the set of valid $\mathcal{DTL}_1$-formulas is recursively enumerable (we actually conjecture that it is finitely axiomatisable). However, it is still an open problem whether the set of all valid $\mathcal{DTL}$-formulas is axiomatisable.

The second key idea is that the technique of embedding Aleksandrov spaces into $\mathbb{R}$ from [3] can be used to show that, given an arbitrary $\mathcal{DTL}_1$-formula φ, one can construct a formula $\varphi^{\mathbb{R}}$, the *relativisation* of φ, such that φ is satisfiable if and only if $\varphi^{\mathbb{R}}$ is satisfiable in a DTS based on $\mathbb{R}$. Therefore, the ω-reachability problem is in fact reduced to the satisfiability of relativised $\mathcal{DTL}_1$-formulas in Euclidean spaces. For the reader's convenience we summarise the results discussed above in the table below where merged cells mean that the corresponding logics coincide:

language	functions	spaces		
		arbitrary	Aleksandrov	$\mathbb{R}^n$
$\mathcal{DTL}^{\bigcirc}$	continuous	fmp, finite axiom. [2, 12, 5]		?
	homeomorphisms	fmp, finite axiom. [2, 13, 12, 8]		
$\mathcal{DTL}_0$	continuous	decidable, finite axiom. (partly in [12])		
	homeomorphisms			
$\mathcal{DTL}_1$	continuous	undecidable, but r.e.		undecidable
	homeomorphisms	non-r.e. [8]		non-r.e. [8]
$\mathcal{DTL}_1$ / $\mathcal{DTL}$	continuous/ finite iterations	decidable in non-prim. recursive time [7]		?
	homeomorphisms/ finite iterations	non-r.e. [8]		
$\mathcal{DTL}$	continuous	undecidable	undecidable	undecidable
	homeomorphisms	non-r.e. [8]	non-r.e. [8]	non-r.e. [8]

2 $\mathcal{DTL}$

The full language $\mathcal{DTL}$ of *dynamic topological logic* is constructed in the usual way from a countably infinite set $\{p_1, p_2, \dots\}$ of *spatial variables*, the Booleans $\neg$, $\wedge$ (and their standard derivatives $\vee$, $\to$, etc.), the modal (or rather topological) operators $\mathbf{I}$ and $\mathbf{C}$, and the temporal operators $\bigcirc$ (next-time), $\square_F$ (always in the future) and $\Diamond_F$ (eventually). More precisely, $\mathcal{DTL}$-*formulas* are given by the following definition:

$$\varphi ::= p_i \mid \neg\varphi \mid \varphi_1 \wedge \varphi_2 \mid \mathbf{I}\varphi \mid \bigcirc\varphi \mid \square_F\varphi.$$

We use $\mathbf{C}\varphi$ and $\Diamond_F\varphi$ as abbreviations for $\neg\mathbf{I}\neg\varphi$ and $\neg\square_F\neg\varphi$, respectively.

A *dynamic topological structure* (DTS, for short) is a pair of the form $\mathfrak{F} = \langle \mathfrak{T}, f \rangle$, where $\mathfrak{T} = \langle T, \mathbb{I} \rangle$ is a topological space with the interior operator

I, and $f\colon T \to T$ is a continuous function on $\mathfrak{T}$, that is, $f^{-1}(A)$ is open whenever $A \subseteq T$ is open (alternatively, $f^{-1}(\mathbb{I}A) \subseteq \mathbb{I}f^{-1}(A)$, for all $A \subseteq T$).

A *dynamic topological model* (DTM, for short) is a pair $\mathfrak{M} = \langle \mathfrak{F}, \mathfrak{V} \rangle$, where $\mathfrak{F}$ is a DTS and $\mathfrak{V}$ a *valuation* assigning to each variable p_i a subset $\mathfrak{V}(p_i)$ of T. The *truth-relation* $(\mathfrak{M}, w) \models \varphi$, for $w \in T$, is defined inductively as follows:

- $(\mathfrak{M}, w) \models p_i$ iff $w \in \mathfrak{V}(p_i)$,
- $(\mathfrak{M}, w) \models \neg \varphi$ iff $(\mathfrak{M}, w) \not\models \varphi$,
- $(\mathfrak{M}, w) \models \varphi_1 \wedge \varphi_2$ iff $(\mathfrak{M}, w) \models \varphi_1$ and $(\mathfrak{M}, w) \models \varphi_2$,
- $(\mathfrak{M}, w) \models \mathbf{I}\varphi$ iff $w \in \mathbb{I}\{v \in T \mid (\mathfrak{M}, v) \models \varphi\}$,
- $(\mathfrak{M}, w) \models \bigcirc \varphi$ iff $(\mathfrak{M}, f(w)) \models \varphi$,
- $(\mathfrak{M}, w) \models \square_F \varphi$ iff $(\mathfrak{M}, f^n(w)) \models \varphi$ for all $n > 0$.

For a class $\mathcal{K}$ of DTSs, we denote by $\mathsf{Log}\,\mathcal{K}$, the *dynamic topological logic of* $\mathcal{K}$, the set of all $\mathcal{DTL}$-formulas φ such that $(\mathfrak{M}, w) \models \varphi$ for every DTM $\mathfrak{M}$ based on a DTS from $\mathcal{K}$ and every point w in $\mathfrak{M}$. If we are only interested in the restriction of $\mathsf{Log}\,\mathcal{K}$ to a certain fragment $\mathcal{DTL}_i$ of $\mathcal{DTL}$, then we write $\mathsf{Log}_i\,\mathcal{K}$ for $\mathsf{Log}\,\mathcal{K} \cap \mathcal{DTL}_i$.

In this paper, we deal with (i) the class $\mathcal{T}$ of DTSs based on arbitrary topological spaces, (ii) the classes $\mathcal{R}^n$ of DTSs based on the Euclidean $\mathbb{R}^n$, for $n \geq 1$, and (iii) the class $\mathcal{A}$ of DTSs based on Aleksandrov spaces. If we restrict these classes to DTSs with homeomorphisms then we write $\mathcal{T}_h$, $\mathcal{R}_h^n$, and $\mathcal{A}_h$, respectively.

We remind the reader that every quasi-order $\mathfrak{G} = \langle W, R \rangle$ (R is a reflexive and transitive relation on W) gives rise to the topological space $\mathfrak{T}_\mathfrak{G}$ over W consisting of all R-closed subsets of W. In other words, the interior operator $\mathbb{I}_\mathfrak{G}$ on $\mathfrak{T}_\mathfrak{G}$ can be defined by taking, for every $X \subseteq W$,

$$\mathbb{I}_\mathfrak{G} X = \{w \in X \mid \forall v \in W \ (wRv \to v \in X)\}.$$

Such spaces are known as *Aleksandrov spaces*. Alternatively they can be defined as topological spaces where arbitrary (not only finite) intersections of open sets are open; for details see [1, 4]. Clearly, for $\mathfrak{M} = \langle \langle \mathfrak{T}_\mathfrak{G}, f \rangle, \mathfrak{V} \rangle$,

$(\mathfrak{M}, w) \models \mathbf{I}\varphi$ iff $(\mathfrak{M}, v) \models \varphi$ for every $v \in W$ with wRv.

It should be also clear that a function $f\colon W \to W$ is *continuous* on $\mathfrak{T}_\mathfrak{G}$ iff

$$wRv \quad \text{implies} \quad f(w)Rf(v),$$

for all $w, v \in W$. Such a function f is called *monotone*. A bijection f is a *homeomorphism* on $\mathfrak{T}_\mathfrak{G}$ iff both the implication above and its converse hold.

3 $\mathcal{DTL}_1$

Let us consider now the sublanguage $\mathcal{DTL}_1$ of $\mathcal{DTL}$ where neither $\Box_F$ nor $\Diamond_F$ can occur in the scope of **I** ($\bigcirc$ is allowed); more precisely, $\mathcal{DTL}_1$-formulas φ are defined as follows:

$$\psi ::= p_i \mid \neg\psi \mid \psi_1 \wedge \psi_2 \mid \mathbf{I}\psi \mid \bigcirc\psi \qquad (2)$$

$$\varphi ::= \psi \mid \neg\varphi \mid \varphi_1 \wedge \varphi_2 \mid \bigcirc\varphi \mid \Box_F\varphi. \qquad (3)$$

This language turns out to be quite interesting and useful. Note first that the formulas used in the proof of [8, Theorem 9] belong to $\mathcal{DTL}_1$, and so:

THEOREM 1. *None of the logics* $\mathsf{Log}_1 \mathcal{T}_h$, $\mathsf{Log}_1 \mathcal{A}_h$, $\mathsf{Log}_1 \mathcal{R}_h^n$, *for* $n \geq 1$, *is recursively enumerable.*

On the other hand, $\mathcal{DTL}_1$ does not distinguish between arbitrary topological and Aleksandrov spaces and, moreover, enjoys a kind of 'local finite model property.' These features of $\mathcal{DTL}_1$ are formulated and proved in Lemmas 2, 5, and 6. They will be heavily used later on in this paper.

LEMMA 2. *Every satisfiable $\mathcal{DTL}_1$-formula φ is satisfiable in a DTM based on an Aleksandrov topological space. Moreover, one can choose a DTM $\mathfrak{M}_\mathfrak{G}$ satisfying φ such that $\mathfrak{M}_\mathfrak{G}$ is based on the Aleksandrov space $\mathfrak{T}_\mathfrak{G}$ induced by a quasi-order $\mathfrak{G} = \langle W, R \rangle$ where,*

(max) *for all $x \in W$ and all $\mathcal{DTL}_1$-formulas ψ with $(\mathfrak{M}_\mathfrak{G}, x) \models \psi$, the set $A_{x,\psi} = \{y \in W \mid xRy \text{ and } (\mathfrak{M}_\mathfrak{G}, y) \models \psi\}$ contains an R-maximal point (i.e., a point z such that if zRz' for some $z' \in A_{x,\psi}$ then $z'Rz$).*

To illustrate the idea of the proof, suppose that φ is satisfied in a DTM $\mathfrak{M}$ based on a topological space $\mathfrak{T}$. We consider the DTM $\mathfrak{M}_\mathfrak{G}$ based on the Aleksandrov space $\mathfrak{T}_\mathfrak{G}$ of all ultrafilters over $\mathfrak{T}$ and show by induction on the formula structure that φ is satisfied in $\mathfrak{M}$ iff it is satisfied in $\mathfrak{M}_\mathfrak{G}$. There is a subtlety in the proof though: the inductive step for subformulas of the form (2) goes through for *all* ultrafilters whereas its counterpart for subformulas of the form (3) for *principal* ultrafilters only. It also follows from the proof that the claim of Lemma 2 is true for DTMs with homeomorphisms (cf. Lemma B.1 from [6]).

COROLLARY 3. $\mathsf{Log}_1 \mathcal{T} = \mathsf{Log}_1 \mathcal{A}$ *and* $\mathsf{Log}_1 \mathcal{T}_h = \mathsf{Log}_1 \mathcal{A}_h$.

It is known [15] that $\mathsf{Log}_1 \mathcal{T} \subsetneq \mathsf{Log}_1 \mathcal{R}^n$, $n \geq 1$, while the question whether the equality $\mathsf{Log}_1 \mathcal{T}_h = \mathsf{Log}_1 \mathcal{R}_h^n$ holds remains open.

REMARK 4. Note that neither equality in the corollary above holds for full $\mathcal{DTL}$. For instance, $\varphi = \Box_F \mathbf{I}p \to \mathbf{I}\Box_F p$ is valid in all DTMs based on Aleksandrov spaces (as infinite intersections of open sets are open there). However, $\neg\varphi$ can be satisfied in a DTM $\mathfrak{M}$ based on $\mathbb{R}$: if one takes the continuous function $f(x) = 2x$ and $\mathfrak{V}(p) = (-1, 1)$ then $\bigcap_{k=1}^{\infty} f^{-k}(\mathbb{I}_\mathbb{R}\mathfrak{V}(p)) = \{0\}$ but $\mathbb{I}_\mathbb{R} \bigcap_{k=1}^{\infty} f^{-k}(\mathfrak{V}(p)) = \emptyset$, and so $(\mathfrak{M}, 0) \models \neg\varphi$.

The following lemma shows that, for $\mathcal{DTL}_1$, DTMs with continuous functions are closely connected to expanding domain product models introduced in [7]:

LEMMA 5. *Let φ be a $\mathcal{DTL}_1$-formula and $\mathfrak{M}$ a DTM based on an Aleksandrov space and satisfying* (max) *such that $(\mathfrak{M}, x_0) \models \varphi$ for some x_0. Then there exist a DTM $\mathfrak{M}^{ep}$ based on the Aleksandrov space induced by a quasi-order $\langle W^{ep}, R^{ep} \rangle$ and a continuous function f^{ep} on it such that $(\mathfrak{M}^{ep}, r_0) \models \varphi$ for some $r_0 \in W^{ep}$ and*

(fin) $\langle W^{ep}, R^{ep} \rangle$ *is the disjoint union of finite quasi-orders $\langle W_n^{ep}, R_n^{ep} \rangle$, for $n \geq 0$;*

(root) *for $n \geq 0$, $r_{n+1} = f^{ep}(r_n)$ and $W_n^{ep} = \{y \in W^{ep} \mid r_n R^{ep} y\}$;*

(inj) f^{ep} *is injective;*

(size) $|W_n^{ep}| \leq ((1 + \ell(\varphi))!)^{n+1}$, *for $n \geq 0$, where $\ell(\varphi)$ is the length of φ.*

The proof of this lemma is similar to that of [7, Lemma 2.2].

Before proceeding to Lemma 6, we remind the reader that, for a quasi-order R on W, a set $C \subseteq W$ is called a *cluster* in $\langle W, R \rangle$ if, for some $x \in W$, $C = \{y \in W \mid xRy \ \& \ yRx\}$; in this case we also say that C is the cluster *generated by x* and denote it by $C(x)$. A *tree of clusters* is a rooted quasi-order $\langle W, R \rangle$ such that, for all $x, y, z \in W$, if xRz and yRz, then xRy or yRx. A cluster $C(y)$ is said to be an *immediate strict successor* of a cluster $C(x)$ if xRy, $C(x) \neq C(y)$ and whenever $xRzRy$ then either $C(z) = C(x)$ or $C(z) = C(y)$.

LEMMA 6. *Let φ be a $\mathcal{DTL}_1$-formula satisfiable in a DTM based on an Aleksandrov space and meeting conditions* (fin), (root), (inj). *Then there exist a recursive function $F(\varphi, n)$ and a DTM $\mathfrak{M}^{ex}$ based on the Aleksandrov space induced by a quasi-order $\langle W^{ex}, R^{ex} \rangle$ and a continuous function f^{ex} on it such that $(\mathfrak{M}^{ex}, r^{ex}) \models \varphi$ for some $r^{ex} \in W^{ex}$, $\mathfrak{M}^{ex}$ satisfies* (inj) *and*

(tree) $\langle W^{ex}, R^{ex} \rangle$ *is the disjoint union of finite trees of clusters $\langle W_n^{ex}, R_n^{ex} \rangle$, for $n \geq 0$;*

(top) *for $n \geq 0$, $y \in W_n^{ex}$, $x \in W_{n+1}^{ex}$ with $xR^{ex} f^{ex}(y)$, there is $z \in W_n^{ex}$ such that $zR^{ex}y$ and $f^{ex}(z) = x$; in particular, for all $x, y \in W^{ex}$, $f^{ex}(x) R^{ex} f^{ex}(y)$ implies $xR^{ex}y$;*

(size') $|W_n^{ex}| \leq F(\varphi, n)$.

Proof. Let $(\mathfrak{M}, r) \models \varphi$, where $\mathfrak{M} = \langle \langle \mathfrak{T}, f \rangle, \mathfrak{V} \rangle$, $\mathfrak{T}$ is induced by $\langle W, R \rangle$ and $\mathfrak{M}$ satisfies (fin), (root), (inj). The DTM $\mathfrak{M}^{ex}$ satisfying the conditions of the lemma will be constructed by a kind of 'unravelling' $\mathfrak{M}$.

A sequence $\vec{C} = (C_0, \ldots, C_n)$, $n \geq 0$, is called a *cluster sequence over* $\langle W, R \rangle$ iff the C_i are clusters in $\langle W, R \rangle$ such that C_{i+1} is an immediate strict successor of C_i. Define a partial order $\leq$ on cluster sequences by taking

$$(C_0, \ldots, C_n) \leq (C'_0, \ldots, C'_m) \quad \text{iff} \quad n \leq m \text{ and } C_i = C'_i, \text{ for } 0 \leq i \leq n.$$

For a point $x \in W$, let $T(x)$ be the set of all pairs $\boldsymbol{y} = \langle \vec{C}, y \rangle$, where $\vec{C} = (C_0, \ldots, C_n)$ is a cluster sequence in $\langle W, R \rangle$ such that $C_0 = C(x)$ and $C_n = C(y)$. Such pairs are ordered by the relation

$$\langle \vec{C}, y \rangle \leq \langle \vec{C}', y' \rangle \quad \text{iff} \quad \vec{C} \leq \vec{C}'.$$

Clearly, $\langle T(x), \leq \rangle$ is a tree of clusters, and every y with xRy is represented by at least one point $\boldsymbol{y}$ in this tree. Moreover, for every $\boldsymbol{y} = \langle \vec{C}, y \rangle$ and $z \in C(y)$, $T(x)$ contains $\boldsymbol{z} = \langle \vec{C}, z \rangle$ and both $\boldsymbol{y}$ and $\boldsymbol{z}$ are in the same cluster in $\langle T(x), \leq \rangle$. It should be noted that if $\langle W, R \rangle$ is a tree of clusters then, for every $x \in W$, $\langle T(x), \leq \rangle$ is isomorphic to the quasi-order $\langle W_x, R_x \rangle$ generated by x.

The new model $\mathfrak{M}^{ex}$ will be based on the Aleksandrov space induced by a quasi-order on sequences of pairs of the form $\langle \vec{C}, y \rangle$. We define it inductively: let $W_0^{ex} = \{(\boldsymbol{y}) \mid \boldsymbol{y} \in T(r)\}$ (the set of sequences of length 1) and

$$R_0^{ex} = \{\, ((\boldsymbol{y}), (\boldsymbol{y}')) \in W_0^{ex} \times W_0^{ex} \mid \boldsymbol{y} \leq \boldsymbol{y}' \,\}.$$

Suppose that $\langle W_k^{ex}, R_k^{ex} \rangle$, $k \geq 0$, has already been defined. Let W_{k+1}^{ex} be the union of two sets:

$$\overline{W}_k^{ex} = \{(\boldsymbol{y_0}, \ldots, \boldsymbol{y_k}, f(\boldsymbol{y_k})) \mid (\boldsymbol{y_0}, \ldots, \boldsymbol{y_k}) \in W_k^{ex}\},$$
$$X_{k+1}^{ex} = \bigcup_{(\boldsymbol{y_0}, \ldots, \boldsymbol{y_k}) \in W_k^{ex}} \{(\boldsymbol{y_0}, \ldots, \boldsymbol{y_k}, \boldsymbol{y}) \mid \boldsymbol{y_k} = \langle \vec{C}_k, y_k \rangle \text{ and } \boldsymbol{y} \in T(f(y_k))\},$$

where we write $f(\boldsymbol{y_k})$ for $\langle \vec{C}_k, f(y_k) \rangle$ whenever $\boldsymbol{y_k} = \langle \vec{C}_k, y_k \rangle$. Note that W_{k+1}^{ex} contains sequences of length $k+2$ only and, for every sequence $(\boldsymbol{y_0}, \ldots, \boldsymbol{y_k}) \in W_k^{ex}$ with $\boldsymbol{y_k} = \langle \vec{C}_k, y_k \rangle$, y_k indicates the element of W this sequence represents.

Let R_k^{ex} be the transitive closure of the union of the following relations:

$$\{((\boldsymbol{y_0}, \ldots, \boldsymbol{y_k}, f(\boldsymbol{y_k})), (\boldsymbol{y'_0}, \ldots, \boldsymbol{y'_k}, f(\boldsymbol{y'_k}))) \in \overline{W}_k^{ex} \times \overline{W}_k^{ex} \mid$$
$$(\boldsymbol{y_0}, \ldots, \boldsymbol{y_k}) R_k^{ex} (\boldsymbol{y'_0}, \ldots, \boldsymbol{y'_k})\},$$
$$\{((\boldsymbol{y_0}, \ldots, \boldsymbol{y_k}, f(\boldsymbol{y_k})), (\boldsymbol{y_0}, \ldots, \boldsymbol{y_k}, \boldsymbol{y})) \in \overline{W}_k^{ex} \times X_{k+1}^{ex} \mid \boldsymbol{y} \in T(f(y_k))\},$$
$$\{((\boldsymbol{y_0}, \ldots, \boldsymbol{y_k}, \boldsymbol{y}), (\boldsymbol{y_0}, \ldots, \boldsymbol{y_k}, \boldsymbol{y}')) \in X_{k+1}^{ex} \times X_{k+1}^{ex} \mid \boldsymbol{y} \leq \boldsymbol{y}'\}.$$

Informally, the first relation above states that $\overline{W}_k^{ex}$ is an isomorphic copy of W_k^{ex}; the other two relations 'attach' a tree to every point $(\boldsymbol{y_0}, \ldots, \boldsymbol{y_k}, f(\boldsymbol{y_k}))$

of $\overline{W}_k^{ex}$ such that the attachment is an isomorphic copy of the tree generated by $f(y_k)$ in W. It is readily checked that $\langle R_{k+1}^{ex}, W_{k+1}^{ex}\rangle$ is a tree of clusters.

The W_k^{ex} are pairwise disjoint. Let $W^{ex} = \bigcup_{k=0}^{\infty} W_k^{ex}$, $R^{ex} = \bigcup_{k=0}^{\infty} R_k^{ex}$, $\mathfrak{T}^{ex}$ be the Aleksandrov space induced by $\langle W^{ex}, R^{ex}\rangle$. Define f^{ex} by taking

$$f^{ex}((\boldsymbol{y_0}, \ldots, \boldsymbol{y_k})) = (\boldsymbol{y_0}, \ldots, \boldsymbol{y_k}, f(\boldsymbol{y_k})),$$

for all $(\boldsymbol{y_0}, \ldots, \boldsymbol{y_k}) \in W^{ex}$. Clearly, f^{ex} is monotone, injective and satisfies **(top)** (in fact, it is a homeomorphism between each pair W_k^{ex} and $\overline{W}_k^{ex}$). Finally, define a valuation $\mathfrak{V}^{ex}$ by taking, for every variable p_i,

$$\mathfrak{V}^{ex}(p_i) = \{(\boldsymbol{y_0}, \ldots, \boldsymbol{y_k}) \in W^{ex} \mid \boldsymbol{y_k} = \langle \vec{C}_k, y_k\rangle \text{ and } y_k \in \mathfrak{V}(p_i)\}.$$

Let $\mathfrak{M}^{ex} = \langle\langle \mathfrak{T}^{ex}, f^{ex}\rangle, \mathfrak{V}^{ex}\rangle$ and $r^{ex} = \langle C(r), r\rangle$. It is not difficult to see that $(\mathfrak{M}^{ex}, r^{ex}) \models \varphi$. ∎

We use Lemma 6 and Kruskal's tree theorem to prove, in a way similar to the proof of [9, Theorem 4], the following (cf. Theorem 1):

THEOREM 7. $\mathsf{Log}_1 \mathcal{T}$ *is recursively enumerable.*

We present here only a sketch of the proof. To apply Kruskal's tree theorem, we require a representation of the models of Lemma 6 as sequences of labelled trees. This can be achieved in a straightforward way using Hintikka-type structures of the following form. For a $\mathcal{DTL}_1$-formula φ, let $cl\varphi$ be the set of all subformulas of φ and their negations. A *full φ-type* t is Boolean-closed subset of $cl\varphi$, that is,

- $\neg\psi \in t$ iff $\psi \notin t$, for every subformula ψ of φ,
- $\psi_1 \wedge \psi_2 \in t$ iff $\psi_1, \psi_2 \in t$, for every subformula $\psi_1 \wedge \psi_2$ of φ.

A subset s of a φ-type t is a *(simple) φ-type* if it does not contain formulas with occurrences of $\square_F$ but satisfies the conditions above for all subformulas of φ not containing $\square_F$.

A *quasistate* $\mathfrak{S} = \langle W, R, l\rangle$ is a finite transitive *irreflexive* tree $\langle W, R\rangle$ with a labelling function l which assigns

- a set $l(x)$ of simple φ-types to every $x \in W$ that is not the root of $\langle W, R\rangle$ (this set is supposed to encode the cluster represented by x),
- a set $l(r) = \{l^1(r)\} \cup l^2(r)$, where $l^1(r)$ is a full φ-type and $l^2(r)$ is a set of simple φ-types, to the root r of $\langle W, R\rangle$

such that the following condition holds for all $\mathbf{I}\psi \in \varphi$, $x \in W$, and $t \in l(x)$: we have $\mathbf{I}\psi \in t$ iff $\psi \in s$ for every $s \in l(y)$ with $x = y$ or xRy.

A sequence $\mathfrak{S}_n = \langle W_n, R_n, l_n\rangle$, $n < \omega$, of quasistates is called a *quasimodel* for φ if there are injective maps $f_n : W_n \to W_{n+1}$ such that

(1) for all $x, y \in W_n$, xR_ny iff $f_n(x)R_{n+1}f_n(y)$,

(2) $|\mathfrak{S}_n| \leq F(\varphi, n)$, where $F(\varphi, n)$ is the recursive function from Lemma 6,

(3) $f(r_n) = r_{n+1}$ for the roots r_n and r_{n+1} of $\langle W_n, R_n \rangle$ and $\langle W_{n+1}, R_{n+1} \rangle$, respectively,

(4) $\varphi \in l_0^1(r_0)$,

(5) if $f_n(x) = y$, then for every $t \in l_n(x)$ there exists $t' \in l_{n+1}(y)$ such that for every $\bigcirc \psi \in cl\varphi$, $\bigcirc \psi \in l_n(x)$ iff $\psi \in l_{n+1}(y)$; if $x = r_n$ and $t = l_n^1(r_n)$, then we can choose $t' = l_{n+1}^1(r_{n+1})$,

(6) for every $\square_F \psi \in cl\varphi$, $\neg \square_F \psi \in l_n^1(r_n)$ iff there exists $m > n$ such that $\neg \psi \in l_m^1(r_m)$; if m is minimal with this property, then we say that m *realises the eventuality* $\neg \square_F \psi$.

The proof of the following lemma is easy and left to the reader.

LEMMA 8. *A $\mathcal{DTL}_1$-formula φ is satisfiable in a DTM iff there exists a quasimodel for φ.*

We remind the reader of the following weak version of *Kruskal's tree theorem*. An injective function $h : \langle W_1, R_1, l_1 \rangle \to \langle W_2, R_2, l_2 \rangle$ between labelled trees is called an *embedding* if, for all $x, y \in W_1$, xR_1y iff $h(x)R_2h(y)$, and $l_1(x) = l_2(h(x))$. We assume that the labels are taken from some finite set.

THEOREM 9 (Kruskal). *For any sequence $\mathfrak{S}_n$, $n < \omega$, of labelled trees, there exists $i < j$ such that $\mathfrak{S}_i$ is embeddable into $\mathfrak{S}_j$.*

Consider now a quasimodel $\mathfrak{S}_n$, $n < \omega$, for φ. Clearly, if there are $i < j$ such that $\mathfrak{S}_i$ is embeddable into $\mathfrak{S}_j$, then the sequence

$$\mathfrak{S}_0, \ldots, \mathfrak{S}_{i-1}, \mathfrak{S}_j, \mathfrak{S}_{j+1} \ldots$$

is a quasimodel for φ as well. Moreover, we can 'prune' the quasistates $\mathfrak{S}_j, \mathfrak{S}_{j+1} \ldots$ in such a way that the bound from (2) above still holds.

Now we can recursively enumerate the $\mathcal{DTL}_1$-formulas φ which are not satisfiable as follows: for every φ enumerate all finite sequences $\mathfrak{S}_0, \ldots, \mathfrak{S}_m$ such that conditions (1)–(5) above and the following modification of condition (6) hold:

- For every $\square_F \psi \in cl\varphi$ and every $n < m$, we have $\neg \square_F \psi \in l_n^1(r_n)$ iff $\neg \square_F \psi \in l_m^1(r_m)$ or there exists k, $n < k \le m$, such that k realises the eventuality $\neg \square_F \psi$.

- Let $0 = k_0 < k_1 < k_2 \cdots < k_n \le m$ be the minimal numbers such that every $\neg \square_F \psi \in l_{k_i}^1(r_{k_i})$ is realised until k_{i+1}. If no $\neg \square_F \psi$ exists in $l_{k_i}^1(r_{k_i})$, then set $k_{i+1} = k_i + 1$. Then, if there are m_1, m_2 with $k_i < m_1 < m_2 < k_{i+1}$ or $k_n < m_1 < m_2 < m$ such that $\mathfrak{S}_{m_1}$ is embeddable into $\mathfrak{S}_{m_2}$, then there exists a $\neg \square_F \psi \in l_{k_1}^1(r_{k_1})$ which is realised somewhere in the interval $[m_1, m_2)$.

Now suppose that the enumeration of such sequences does not terminate for φ (i.e., infinite many such sequences are generated). Then, by König's lemma, there exists an infinite sequence whose finite initial segments satisfy the conditions above. But then, by Kruskal's tree theorem, this sequence is a quasimodel for φ, and so φ is satisfiable. Conversely, if the enumeration of such finite sequences terminates (i.e., there are only finitely many sequences), then there is no quasimodel for φ. Hence φ is not satisfiable. It follows that the non-satisfiable formulas (and therefore the valid formulas) are recursively enumerable.

4 Embedding DTMs into the real line

THEOREM 10. *A $\mathcal{DTL}_1$-formula φ is satisfiable iff $\varphi^{\mathbb{R}}$, the relativisation of φ, is satisfiable in the DTS $\langle\langle\mathbb{R}, \mathbb{I}_{\mathbb{R}}\rangle, x \mapsto x+1\rangle$, where*[1]

$$\varphi^{\mathbb{R}} = dom \wedge \Box_F^+ \mathbf{I}(dom \to \bigcirc dom) \wedge \varphi^{dom},$$

dom is a fresh variable, and φ^{dom} is the result of replacing every occurrence of a subformula of the form $\mathbf{I}\psi$ in φ with $\mathbf{I}(dom \to \psi)$.

Proof. Suppose that $\varphi^{\mathbb{R}}$ is satisfied in $\langle\langle\mathbb{R}, \mathbb{I}_{\mathbb{R}}\rangle, x \mapsto x+1\rangle$. Then, by Lemma 2, $\varphi^{\mathbb{R}}$ is satisfied in a DTM $\mathfrak{M}$ based on an Aleksandrov space. It is readily seen that φ is satisfied in the DTM $\mathfrak{M}'$ obtained from $\mathfrak{M}$ by removing all those points where *dom* is false.

Conversely, suppose that φ is satisfiable. Let $\mathfrak{M}^{ex} = \langle\langle\mathfrak{T}^{ex}, f^{ex}\rangle, \mathfrak{V}^{ex}\rangle$ be the model provided by Lemma 6 and satisfying φ. Our plan is (1) to extend $\mathfrak{M}^{ex}$ to a DTM $\widehat{\mathfrak{M}}$ that satisfies $\varphi^{\mathbb{R}}$ and is based on ω-trees of clusters of finite depth, and (2) to embed $\widehat{\mathfrak{M}}$ into a model based on $\langle\langle\mathbb{R}, \mathbb{I}_{\mathbb{R}}\rangle, x \mapsto x+1\rangle$.

We remind the reader that a tree of clusters $\langle W, R\rangle$ is *of depth n* if n is the length of the longest sequence $C(x_1), \ldots, C(x_n)$ of clusters in $\langle W, R\rangle$ such that $C(x_{i+1})$ is an immediate strict successor of $C(x_i)$. A cluster C is called *final* if it has no strict successor. A tree of clusters is called an ω-tree if every non-final cluster has ω distinct strict immediate successors.

(1) Fix some $k \geq 0$. By **(tree)**, $\langle W_k^{ex}, R_k^{ex}\rangle$ is a finite tree of clusters. We construct an ω-tree of clusters $\mathfrak{G}_k = \langle \widehat{W}_k, \widehat{R}_k\rangle$ by attaching an ω-tree of depth 2 to every cluster in $\langle W_k^{ex}, R_k^{ex}\rangle$: let $\widehat{W}_k$ be the union of W_k^{ex} with $\{(C(x), n) \mid x \in W_k^{ex},\ n \in \mathbb{N}\}$ and $\widehat{R}_k$ the transitive and reflexive closure of

$$R_k^{ex} \cup \{(x, (C(x), n)) \mid x \in W_k^{ex},\ n \in \mathbb{N}\}.$$

Note that the $\widehat{W}_k$ are pairwise disjoint. Let $\widehat{W} = \bigcup_{k=0}^{\infty} \widehat{W}_k$, $\widehat{R} = \bigcup_{k=0}^{\infty} \widehat{R}_k$, and let $\widehat{\mathfrak{T}}$ be the Aleksandrov space induced by $\mathfrak{G} = \langle \widehat{W}, \widehat{R}\rangle$.

Define a function $\widehat{f} \colon \widehat{W} \to \widehat{W}$ by taking, for all $x \in W^{ex}$ and $n \in \mathbb{N}$,

$$\widehat{f}(x) = f^{ex}(x),$$

$$\widehat{f}((C(x), n)) = \begin{cases} y_n, & \text{if } n < m, \\ (C(f^{ex}(x)), n-m), & \text{otherwise,} \end{cases}$$

---

[1] We use $\Box_F^+ \psi$ as an abbreviation for $\psi \wedge \Box_F \psi$.

where $\{y_0, \ldots, y_{m-1}\} \subseteq W^{ex}_{k+1} \setminus \{f(z) \mid z \in W^{ex}_k\}$ and $C(y_0), \ldots, C(y_{m-1})$ are all the distinct strict immediate successors of $C(f^{ex}(x))$.

Clearly, $\widehat{f}$ is a well-defined function on $\widehat{W}$. Informally, (i) each $x \in W^{ex}_k$ is mapped onto its image $f^{ex}(x)$; (ii) for each $x \in W^{ex}_k$, its first m successors of the form $(C(x), n)$ from $\widehat{W}_k \setminus W^{ex}_k$ are mapped onto the roots of the m distinct trees that are attached to $C(f^{ex}(x))$ at step $k+1$ and the remaining successors of this sort are simply renumbered. Since every $\widehat{\mathfrak{G}}_{k+1}$ contains an isomorphic copy of $\widehat{\mathfrak{G}}_k$ and in view of (**top**), the function $\widehat{f}$ is monotone.

Clearly, $\varphi^{\mathbb{R}}$ is satisfied in the DTM $\widehat{\mathfrak{M}}$ under the valuation $\widehat{\mathfrak{V}}$ defined by taking $\widehat{\mathfrak{V}}(dom) = W^{ex}$ and $\widehat{\mathfrak{V}}(p_i) = \mathfrak{V}^{ex}(p_i)$, for every variable p_i in φ.

(2) Next, we require the following generalisation of a result from [3]:

LEMMA 11. *For every open interval $I \subseteq \mathbb{R}$ and every ω-tree of clusters $\mathfrak{G} = \langle W, R \rangle$ of finite depth, there is a surjective open and continuous function $f^{\mathfrak{G}}_I : I \to \mathfrak{T}_{\mathfrak{G}}$. Moreover, for every initial ω-subtree $\mathfrak{G}' = \langle W', R' \rangle$ of $\mathfrak{G}$ (where each cluster of $\mathfrak{G}'$ either inherits all strict immediate successors from $\mathfrak{G}$ or none of them) and every point $z \in I$,*

(cons) *if $f^{\mathfrak{G}'}_I(z)$ is not in a final cluster of $\mathfrak{G}'$, then $f^{\mathfrak{G}}_I(z) = f^{\mathfrak{G}'}_I(z)$;*

(final) *if $f^{\mathfrak{G}'}_I(z)$ is in a final cluster of $\mathfrak{G}'$, then there is an open interval I_z with $z \in I_z$ such that $f^{\mathfrak{G}}_I(I_z) = \{y \in W \mid f^{\mathfrak{G}'}_I(z) R y\}$.*

The above result can be proved by extending [3, Theorem 16] so that in [3, Lemma 15], the enumeration of all countably many immediate strict successors is chosen in such a way that each of them occurs infinitely often in the enumeration.

Thus, Lemma 11 provides us with a surjective open and continuous function $f^{\widehat{\mathfrak{G}}_k}_{(0,1)} : (0,1) \to \widehat{\mathfrak{G}}_k$, for each $k \geq 0$. Let $D = \bigcup_{k=0}^{\infty}(k, k+1) \subseteq \mathbb{R}$. Define a function $g : D \to \widehat{\mathfrak{G}}$ by taking:

$$g(k+z) = f^{\widehat{\mathfrak{G}}_k}_{(0,1)}(z), \qquad \text{for all } k \in \mathbb{N} \text{ and } z \in (0,1).$$

The valuation $\mathfrak{V}_{\mathbb{R}}$ is defined by taking $\mathfrak{V}_{\mathbb{R}}(dom) = \{z \in D \mid g(z) \in \widehat{\mathfrak{V}}(dom)\}$ and $\mathfrak{V}_{\mathbb{R}}(p_i) = \{z \in D \mid g(z) \in \widehat{\mathfrak{V}}(p_i)\}$, for every variable p_i of φ.

First, we show that, for all $z \in \mathbb{R}$,

$$\text{if } z \in \mathfrak{V}_{\mathbb{R}}(dom) \quad \text{then} \quad (z+1) \in \mathfrak{V}_{\mathbb{R}}(dom) \text{ and } \widehat{f}(g(z)) = g(z+1). \quad (4)$$

Indeed, suppose that $z \in \mathfrak{V}_{\mathbb{R}}(dom)$. Then $g(z) \in \widehat{\mathfrak{V}}(dom)$ and therefore, $g(z) \in W^{ex}_k$, for some $k \geq 0$. Then $\widehat{f}(g(z)) \in W^{ex}_{k+1} \subseteq \widehat{\mathfrak{V}}(dom)$. Moreover, by (**cons**), $\widehat{f}(g(z)) = g(z+1)$. Hence, $z+1 \in \mathfrak{V}_{\mathbb{R}}(dom)$. Note that it is essential that $g(z)$ is not in a final cluster of $\widehat{\mathfrak{G}}_k$; and this is precisely the reason why at step **(1)** we needed to extend $\langle W^{ex}, R^{ex} \rangle$ to $\langle \widehat{W}, \widehat{R} \rangle$ by ω-trees of depth 2 (for each final cluster of each $\widehat{\mathfrak{G}}_k$ the attached ω-tree of depth 2 provides countably many 'placeholders' for the trees that will be attached to this cluster in $W_{k+1}^{ex}, W_{k+2}^{ex}, \ldots$; note also that these placeholders do not belong to the 'domain').

Now, we prove by induction on the construction of subformulas ψ of φ^{dom} that, for every $z \in \mathfrak{V}_\mathbb{R}(dom)$,[2]

$$g(z) \in \widehat{\mathfrak{V}}(\psi) \quad \text{iff} \quad z \in \mathfrak{V}_\mathbb{R}(\psi). \tag{5}$$

The case of variables follows immediately from the definition and the cases of the Booleans are trivial.

Case $\psi = \mathbf{I}(dom \to \psi')$. Since $z \in \mathfrak{V}_\mathbb{R}(dom)$, there is $k \in \mathbb{N}$ such that $z \in (k, k+1)$. Then

$z \in \mathfrak{V}_\mathbb{R}(\mathbf{I}(dom \to \psi'))$ iff [def. of open set]

there is an open $U \subseteq (k, k+1)$ such that
$z \in U$ and $U \subseteq \mathfrak{V}_\mathbb{R}(dom \to \psi')$ iff [definition]

there is an open $U \subseteq (k, k+1)$ such that
$z \in U$ and $U \cap \mathfrak{V}_\mathbb{R}(dom) \subseteq \mathfrak{V}_\mathbb{R}(\psi')$ iff [IH]

there is an open $U \subseteq (k, k+1)$ such that
$z \in U$ and $g(U) \cap \widehat{\mathfrak{V}}(dom) \subseteq \widehat{\mathfrak{V}}(\psi')$ iff $[V = g(U)$
                                                            g open and cont.]

there is an open $V \subseteq \widehat{W_k}$ such that
$g(z) \in V$ and $V \cap \widehat{\mathfrak{V}}(dom) \subseteq \widehat{\mathfrak{V}}(\psi')$ iff [definition]

$g(z) \in \widehat{\mathfrak{V}}(\mathbf{I}(dom \to \psi'))$

Case $\psi = \bigcirc \psi'$. Then

$z \in \mathfrak{V}_\mathbb{R}(\bigcirc \psi')$ iff $z+1 \in \mathfrak{V}_\mathbb{R}(\psi')$ iff   [IH]

                                 $g(z+1) \in \widehat{\mathfrak{V}}(\psi')$ iff   [(4)]

                                 $\widehat{f}(g(z)) \in \widehat{\mathfrak{V}}(\psi')$ iff $g(z) \in \widehat{\mathfrak{V}}(\bigcirc \psi')$.

Case $\psi = \Box_F \psi'$ is considered in the same way.

By (4) and (5), $\varphi^\mathbb{R}$ is satisfied in $\langle \langle \langle \mathbb{R}, \mathbb{I}_\mathbb{R} \rangle, x \mapsto x+1 \rangle, \mathfrak{V}_\mathbb{R} \rangle$. ∎

COROLLARY 12. *A $\mathcal{DTL}_1$-formula φ is satisfiable iff $\varphi^\mathbb{R}$ is satisfiable in the DTS $\langle \langle \mathbb{R}^n, \mathbb{I}_{\mathbb{R}^n} \rangle, (x_1, \ldots, x_{n-1}, x_n) \mapsto (x_1, \ldots, x_{n-1}, x_n + 1) \rangle, n \geq 1$.*

It should be noted that although the relativisation $\varphi^\mathbb{R}$ is satisfied in a DTS based on an Aleksandrov space iff φ is satisfied in some (in general, different) DTS based on an Aleksandrov space, the analogous statement fails for the Euclidean spaces: φ may not be satisfiable in a DTS based on $\mathbb{R}$ even if its relativisation $\varphi^\mathbb{R}$ is satisfiable (e.g., a counterexample in [15]). The explanation lies in the variable *dom* which 'carves' out of $\mathbb{R}$ a subspace that topologically resembles an Aleksandrov space.

---

[2] For a DTM $\mathfrak{M} = \langle \mathfrak{F}, \mathfrak{V} \rangle$, we denote by $\mathfrak{V}(\varphi)$ the set $\{w \in \mathfrak{F} \mid (\mathfrak{M}, w) \models \varphi\}$.

5 Undecidability

As usual, for a finite alphabet Σ, the set of all finite words over Σ (including the empty word ϵ) is denoted by Σ^* (also called Σ-words). Given a word $w = a_1 \cdot a_2 \cdot \dots \cdot a_n$, we denote its length n by $|w|$ and refer to its elements using the bracket notation: $w(i) = a_i$, for $1 \le i \le n$. For a pair of words u and w, we say that u is a *subword* of w and write $u \sqsubseteq_h w$ if there is a strictly monotone map $h \colon \{1, \ldots, |u|\} \to \{1, \ldots, |w|\}$ such that $u(i) = w(h(i))$, for every i, $1 \le i \le |u|$. We also write $u \sqsubseteq w$ if $u \sqsubseteq_h w$ for some h.

A *single channel system* is a triple $S = \langle Q, \Sigma, \Delta \rangle$, where $Q = \{q_1, \ldots, q_n\}$ is a set of *control states*, $\Sigma = \{a_1, \ldots, a_m\}$ is an alphabet of *messages* and $\Delta \subseteq Q \times \{?, !\} \times \Sigma \times Q$ is a set of *transitions*.

A *configuration* of S is a pair $\gamma = \langle q, w \rangle$, where $q \in Q$ and $w \in \Sigma^*$. Define a *lossy relation* and two types of *perfect transition relations for* S between configurations by taking:

$$(1) \quad \langle q, w \rangle \to_\ell \langle q, w' \rangle \quad \text{iff} \quad w' \sqsubseteq w,$$

$$(\mathbf{s}) \quad \langle q, w \rangle \xrightarrow{\langle q, !, a, q' \rangle}_p \langle q', a \cdot w \rangle, \qquad (\mathbf{r}) \quad \langle q, w \cdot a \rangle \xrightarrow{\langle q, ?, a, q' \rangle}_p \langle q', w \rangle.$$

(Here **l** stands for 'loose', **s** for 'send' and **r** for 'receive.') Finally, define *lossy transition relations for* S as the following compositions:

$$(\mathbf{ls}) \quad \langle q, w \rangle \xrightarrow{\langle q, !, a, q' \rangle}_\ell \langle q', w' \rangle \quad \text{iff} \quad \langle q, w \rangle \xrightarrow{\langle q, !, a, q' \rangle}_p \langle q', a \cdot w \rangle \to_\ell \langle q', w' \rangle,$$

$$(\mathbf{lr}) \langle q, w \cdot a \rangle \xrightarrow{\langle q, ?, a, q' \rangle}_\ell \langle q', w' \rangle \quad \text{iff} \quad \langle q, w \cdot a \rangle \xrightarrow{\langle q, ?, a, q' \rangle}_p \langle q', w \rangle \to_\ell \langle q', w' \rangle.$$

For a binary relation $\to$ on configurations of S, a sequence $\gamma_0, \ldots, \gamma_m$ of configurations of S is called a $\to$-*computation of* S if $\gamma_k \to \gamma_{k+1}$, for $k < m$.

The undecidable 'master problem' that will be used to prove the undecidability of the logics under consideration is the ω-*reachability problem for channel systems*: given a channel system $S = \langle Q, \Sigma, \Delta \rangle$, two states $q_0, q_{rec} \in Q$, and a relation $\to$ in the interval

$$\left(\bigcup_{\delta_k \in \Delta} \xrightarrow{\delta_k}_p \right) \subseteq \to \subseteq \left(\bigcup_{\delta_k \in \Delta} \xrightarrow{\delta_k}_\ell \right) \tag{6}$$

decide whether for every $n \in \mathbb{N}$ there exists a $\to$-computation of S starting with $\langle q_0, \epsilon \rangle$ and reaching q_{rec} at least n times. The following lemma can be proved by a reduction of the undecidable boundedness problem for lossy channel systems [14]. The reduction was suggested by Ph. Schnoebelen:

LEMMA 13. *The ω-reachability problem is undecidable.*

We are now in a position to prove the main result of the paper:

THEOREM 14. *None of* $\mathsf{Log}_1 \mathcal{T}$ *and* $\mathsf{Log}_1 \mathcal{R}^n$, *for* $n \ge 1$, *is decidable.*

Proof. By Corollary 12, it suffices to consider (the complement of) $\mathsf{Log}_1 \mathcal{T}$. Given a single channel system $S = \langle Q, \Sigma, \Delta \rangle$ and states $q_0, q_{rec} \in Q$, we construct a $\mathcal{DTL}_1$-formula $\varphi_{S, q_0, q_{rec}}$ such that

($\Rightarrow$) if $\varphi_{S,q_0,q_{rec}}$ is satisfiable then, for every $n \in \mathbb{N}$, there is a *lossy* computation starting with $\langle q_0, \epsilon \rangle$ and reaching q_{rec} at least n times;

($\Leftarrow$) if, for every $n \in \mathbb{N}$, there is a *perfect* computation starting with $\langle q_0, \epsilon \rangle$ and reaching q_{rec} at least n times then $\varphi_{S,q_0,q_{rec}}$ is satisfiable in a DTM.

By Lemma 13 this will imply that $\mathsf{Log}_1 \mathcal{T}$ is undecidable.

The formula $\varphi_{S,q_0,q_{rec}}$ is the conjunction of four formulas σ, θ_{S,q_0}, τ_S and $\rho_{S,q_{rec}}$ and is constructed in four steps $(\boldsymbol{\sigma})$, $(\boldsymbol{\theta})$, $(\boldsymbol{\tau})$ and $(\boldsymbol{\rho})$, respectively.

($\Rightarrow$) To explain the meaning of each of the four conjuncts, let us assume that $(\mathfrak{M}, x_0) \models \varphi_{S,q_0,q_{rec}}$, for some DTM $\mathfrak{M} = \langle \langle \mathfrak{T}, f \rangle, \mathfrak{V} \rangle$, where $\mathfrak{T}$ is the Aleksandrov topological space induced by a quasi-order $\langle W, R \rangle$.

For every $n \geq 0$, let $W_n = \{y \in W \mid f^n(x_0)Ry\}$ and $R_n = R \cap (W_n \times W_n)$. Clearly, the $\langle W_n, R_n \rangle$ are quasi-orders. By Lemma 5, we may assume that the W_n are *finite* and f is injective.

($\boldsymbol{\sigma}$) First we introduce a new operator $\mathbf{S}$ (to be interpreted as an almost 'irreflexive' diamond in $\langle W, R \rangle$). Namely, we fix some fresh variable s and put, for every $\mathcal{DTL}^\circ$-formula ψ,

$$\mathbf{S}\psi \;=\; (s \wedge \mathbf{C}(\neg s \wedge \mathbf{C}\psi)) \;\vee\; (\neg s \wedge \mathbf{C}(s \wedge \mathbf{C}\psi)).$$

Define new relations $\overline{R}_n$ on W_n, $n \geq 0$, by taking, for all $x, y \in W_n$,

$$x\overline{R}_n y \quad \text{iff} \quad \exists z \in W_n \; \bigl(xR_n z R_n y \;\text{ and }\; (\mathfrak{M}, x) \models s \Leftrightarrow (\mathfrak{M}, z) \models \neg s\bigr).$$

It can be checked that $\overline{R}_n$ is transitive, $\overline{R}_n \subseteq R_n$ and, for every $x \in W_n$,

$$(\mathfrak{M}, x) \models \mathbf{S}\psi \quad \text{iff} \quad \text{there is } y \in W_n \text{ such that } x\overline{R}_n y \text{ and } (\mathfrak{M}, y) \models \psi.$$

Let

$$\sigma \;=\; \Box_F^+ \mathbf{I}(s \leftrightarrow \bigcirc s). \tag{7}$$

By the monotonicity of f, if $(\mathfrak{M}, x_0) \models \sigma$ then, for all $x, y \in W_n$,

$$\text{if } \; x\overline{R}_n y \; \text{ then } \; f(x)\overline{R}_{n+1}f(y).$$

($\boldsymbol{\theta}$) Next, we encode infinitely many computations along the orbit of x_0. Each computation is encoded backwards (from its end to the beginning), and the variable m delimits computations in the sense that each computation is encoded between two consecutive occurrences of m. For every transition $\delta_k \in \Delta$, we introduce a fresh variable $\mathrm{tr}\delta_k$ and, for every state $q_i \in Q$, a variable q_i.[3] Let θ_{S,q_0} be the conjunction of the following formulas:

$$\bigwedge_{\delta_i \neq \delta_j} \Box_F^+ \neg(\mathrm{tr}\delta_i \wedge \mathrm{tr}\delta_j) \quad \wedge \quad \bigwedge_{q_i \neq q_j} \Box_F^+ \neg(q_i \wedge q_j), \tag{8}$$

$$\bigwedge_{\delta_k = \langle q_i, *, a, q_j \rangle \in \Delta} \Box_F^+ (\mathrm{tr}\delta_k \to q_j \wedge \bigcirc q_i), \tag{9}$$

$$\Box_F^+ \big(\neg m \to \bigvee_{\delta_i \in \Delta} \mathrm{tr}\delta_i \big), \tag{10}$$

$$\Box_F^+ (m \to q_0), \tag{11}$$

$$m \wedge \Box_F^+ \Diamond_F m, \tag{12}$$

where $*$ denotes either of $\{?, !\}$. It follows that if $(\mathfrak{M}, x_0) \models \theta_{S,q_0}$ then there is an infinite sequence of natural numbers $0 = M_0 < M_1 < \ldots$ such that

$$(\mathfrak{M}, f^{M_n}(x_0)) \models m,$$

for every $n \geq 0$. Let $N_n = M_n - (M_{n-1} + 1)$, for $n > 0$. Clearly, $N_n \geq 1$, for $n > 0$. Let $N_0 = 0$. For every $n \geq 0$, there are also unique sequences

$$q_{i_0^n}, \ldots, q_{i_{N_n}^n} \quad \text{and} \quad \delta_{k_0^n}, \ldots, \delta_{k_{N_n-1}^n}$$

of, respectively, states from Q and transitions from Δ such that

- $(\mathfrak{M}, f^{M_n - j}(x_0)) \models q_{i_j^n}$, for each $0 \leq j \leq N_n$,
- $(\mathfrak{M}, f^{M_n - (j+1)}(x_0)) \models \mathrm{tr}\delta_{k_j^n}$, for each $0 \leq j < N_n$,
- $\delta_{k_j^n} = \langle q_{i_j^n}, *, a, q_{i_{j+1}^n} \rangle$ and $q_{i_0^n} = q_0$.

(τ) Our next step is to encode Σ-words in the 'topological dimension.' For every message $a_i \in \Sigma$, we introduce a fresh variable a_i and let $\lambda = \bigvee_{a_i \in \Sigma} a_i$. Intuitively, Σ-words are encoded in $\overline{R}_k$-connected points of W_k. Denote by τ_S the conjunction of the following formulas:

$$\bigwedge_{a_i \neq a_j} \Box_F^+ \mathbf{I} \neg (a_i \wedge a_j), \tag{13}$$

$$\Box_F^+ \mathbf{I}(\lambda \to \mathbf{I}\lambda), \tag{14}$$

$$\bigwedge_{a_i \in \Sigma} \Box_F^+ \mathbf{I}(a_i \to \bigcirc(\lambda \to a_i)), \tag{15}$$

$$\Box_F^+ (m \to \mathbf{I} \neg \lambda), \tag{16}$$

$$\bigwedge_{\delta_k = \langle q, !, a, q' \rangle \in \Delta} \Box_F^+ \big(\mathrm{tr}\delta_k \to \mathbf{I}(\lambda \to \neg \mathbf{S} \bigcirc \neg \lambda) \wedge \mathbf{I}(\lambda \wedge \bigcirc \neg \lambda \to a) \big), \tag{17}$$

$$\bigwedge_{\delta_k = \langle q, ?, a, q' \rangle \in \Delta} \Box_F^+ \big(\mathrm{tr}\delta_k \to \mathbf{I}(\lambda \to \bigcirc \lambda) \wedge \mathbf{I}(\neg \mathbf{S}\lambda \to \bigcirc \mathbf{S}\lambda) \wedge \bigcirc \mathbf{I}(\neg \mathbf{S}\lambda \to a) \big). \tag{18}$$

---

[3] Note the different font we use to denote these variables: their values will be needed only along the orbit of x_0 (in contrast to s and other variables below).

CLAIM 15. Let $y \in W_n$ be such that $(\mathfrak{M}, y) \models \lambda$. Then $y\overline{R}_n y$ does not hold.

Proof of claim. Suppose that $y\overline{R}_n y$. By (16), $(\mathfrak{M}, f^n(x_0)) \models \neg \mathsf{m}$ and thus, by (10), $(\mathfrak{M}, f^n(x_0)) \models \mathrm{tr}\delta_k$ for some $\delta_k \in \Delta$. Consider two cases: (a) if $\delta_k = \langle q, ?, a, q' \rangle$ then $(\mathfrak{M}, f(y)) \models \lambda$, by the first conjunct of (18); (b) if $\delta_k = \langle q, !, a, q' \rangle$ then, as $y\overline{R}_n y$, we have, by (17), $(\mathfrak{M}, f(y)) \models \lambda$ again.

By monotonicity of f, we obtain $f(y) \in W_{n+1}$ and $f(y)\overline{R}_{n+1} f(y)$; therefore, one can repeat the above step for $f(y)$. However, this process cannot continue indefinitely since there is $M_j > n$ such that $(\mathfrak{M}, f^{M_j}(x_0)) \models \mathsf{m}$, and so, by (16), $(\mathfrak{M}, y) \models \neg \lambda$ for all $y \in W_{M_j}$. ∎

Say that a finite sequence $\vec{y} = (y_1, y_2, \ldots, y_k)$ of elements of W_n with $y_1 \overline{R}_n y_2 \overline{R}_n \ldots$ *carries the Σ-word* w and write $\mathit{val}(\vec{y}) = w$ if $(\mathfrak{M}, y_i) \models w(i)$, for all $1 \leq i \leq |w|$. Note that, by (14), if $(\mathfrak{M}, y) \models \lambda$ for some $y \in W_n$ then, for every $z \in W_n$ with $y\overline{R}_n z$, we will also have $(\mathfrak{M}, z) \models \lambda$. Let $\mathit{val}(\vec{y}) = \epsilon$ if $\vec{y}$ is the empty sequence. We say that a sequence $\vec{y}\,' = (y_1', \ldots, y_r')$ with $y_i'\overline{R}_n y_{i+1}'$ is an *extension* of $\vec{y} = (y_1, \ldots, y_k)$ with $y_i \overline{R}_n y_{i+1}$ if there is a strictly monotone function $h \colon \{1, \ldots, k\} \to \{1, \ldots, r\}$ (i.e., $h(i) > h(j)$, for $i > j$) such that $y_i = y_{h(i)}'$, for $1 \leq i \leq k$. A sequence $\vec{y} = (y_1, \ldots, y_k)$ with $y_i \overline{R}_n y_{i+1}$ is said to be *maximal carrying a Σ-word* if no extension of $\vec{y}$ carries a Σ-word.

Suppose $(\mathfrak{M}, f^n(x_0)) \models \neg \mathsf{m}$. Then, by (10), $(\mathfrak{M}, f^n(x_0)) \models \mathrm{tr}\delta_k$ for some transition $\delta_k \in \Delta$. Let $\vec{y}\,' = (y_1, \ldots, y_l)$ with $y_i \overline{R}_n y_{i+1}$ be a maximal sequence carrying a Σ-word and $\mathit{val}(\vec{y}\,') = d_1 \ldots d_l$. First, consider the case when $\vec{y}\,'$ is empty. This means that $(\mathfrak{M}, y) \models \neg \lambda$ for all $y \in W_n$. Take any maximal sequence $\vec{y}$ in W_{n+1} carrying a Σ-word. Clearly, $\vec{y}$ can be regarded as an extension of $f(\vec{y}\,') = \epsilon$ and, as ϵ is a subword of any word, we have $\langle q, \mathit{val}(\vec{y}) \rangle \xrightarrow{\delta_k}_\ell \langle q', \epsilon \rangle$, for any transition $\delta_k = \langle q, *, a, q' \rangle \in \Delta$ (this corresponds to a situation when everything is lost after the transition).

Suppose now that $\vec{y}\,'$ is not empty. Let $f(\vec{y}\,') = (f(y_1), \ldots, f(y_l))$. Note that points of this sequence are all distinct since f is injective. Consider the cases of sending and receiving messages:

Case $\delta_k = \langle q, !, a, q' \rangle$. By the first conjunct of (17), only three subcases are possible:

(0) There is no m, $1 \leq m \leq l$, such that $(\mathfrak{M}, y_m) \models \bigcirc \lambda$. Then, by the first conjunct of (17), $\vec{y}\,' = (y_1)$ and, by its last conjunct, $\mathit{val}(\vec{y}\,') = a$. Take any maximal sequence $\vec{y}$ in W_{n+1} carrying a Σ-word. As $\vec{y}$ does not contain $f(y_1)$, it can trivially be regarded as an extension of $f(\vec{y}\,')$. Thus $\langle q, \mathit{val}(\vec{y}) \rangle \xrightarrow{\delta_k}_\ell \langle q', a \rangle$ (everything but the sent message is lost).

(1) Let $(\mathfrak{M}, y_1) \models \bigcirc \lambda$. Then $(\mathfrak{M}, f(y_1)) \models \lambda$ and, by (15), $\mathit{val}(\vec{y}\,') = \mathit{val}(f(\vec{y}\,'))$. Take any maximal extension $\vec{y}$ of $f(\vec{y}\,')$ carrying a Σ-word. Then $\mathit{val}(\vec{y}) \sqsupseteq \mathit{val}(\vec{y}\,')$ and $\langle q, \mathit{val}(\vec{y}) \rangle \xrightarrow{\delta_k}_\ell \langle q', \mathit{val}(\vec{y}\,') \rangle$ (the sent message is lost; other messages are possibly lost but not completely).

(2) Let $(\mathfrak{M}, y_1) \models \bigcirc \neg \lambda$ but $(\mathfrak{M}, y_2) \models \bigcirc \lambda$. By the last conjunct of (17), we have $(\mathfrak{M}, y_1) \models a$ and, by (15), $val(\vec{y}') = a \cdot val(f(\vec{y}'))$. Take any maximal extension $\vec{y}$ of $f(\vec{y}')$ carrying a Σ-word. Then we have $\langle q, val(\vec{y}) \rangle \xrightarrow{\delta_k}_\ell \langle q', val(\vec{y}') \rangle$ (the sent message is not lost; other messages are possibly lost but not completely).

Case $\delta_k = \langle q, ?, a, q' \rangle$. By the maximality of $\vec{y}$ and Claim 15, $(\mathfrak{M}, y_l) \models \lambda \wedge \neg \mathbf{S}\lambda$. Then, by the first and second conjuncts of (18), $(\mathfrak{M}, f(y_l)) \models \lambda \wedge \mathbf{S}\lambda$. By Claim 15, the set $\{z \in W_{n+1} \mid f(y_l) \overline{R}_{n+1} z, (\mathfrak{M}, z) \models \lambda\}$ contains no infinite ascending $\overline{R}_{n+1}$-chain, and so we can find some $y \in W_{n+1}$ such that $f(y_l) \overline{R}_{n+1} y$ and $(\mathfrak{M}, y) \models \lambda \wedge \neg \mathbf{S}\lambda$. By the last conjunct of (18), we have $(\mathfrak{M}, y) \models a$. As, by the first conjunct of (18) and (15), $val(f(\vec{y}')) = val(\vec{y}')$, we obtain $val(f(\vec{y}') \cdot y) = val(\vec{y}') \cdot a$. Finally, take any maximal extension $\vec{y}$ of $f(\vec{y}') \cdot y$ carrying a Σ-word. Clearly, $\langle q, val(\vec{y}) \rangle \xrightarrow{\delta_k}_\ell \langle q', val(\vec{y}') \rangle$.

Now, given $n \geq 1$, we can find a lossy computation. Let $(q_{i_0^n}, \ldots, q_{i_{N_n}^n})$ and $(\delta_{k_0^n}, \ldots, \delta_{k_{N_n-1}^n})$ be the unique sequences of states and transitions, respectively, defined by the model $\mathfrak{M}$ as described in step $(\boldsymbol{\theta})$. We start from the tail: let $j = N_n$ and take any maximal sequence $\vec{y}_j$ in $W_{M_n - j}$ carrying a Σ-word. We know from the considerations that there is a maximal sequence $\vec{y}_{j-1}$ in $W_{M_n - (j-1)}$ carrying a Σ-word such that

$$\langle q_{i_{j-1}^n}, val(\vec{y}_{j-1}) \rangle \xrightarrow{\delta_{k_j^n}}_\ell \langle q_{i_j^n}, val(\vec{y}_j) \rangle.$$

By repeating this procedure sufficiently many times, we arrive at $\langle q_0, \epsilon \rangle$, where the n-th computation starts (here we use (11) and (16)).

$(\boldsymbol{\rho})$ At the final step we enforce the n-th computation to visit the state q_{rec} at least n times. We require two fresh variables *light* and *on*. Let $\rho_{S, q_{rec}}$ be the conjunction of the following formulas:

$$light \wedge \Box_F(\mathsf{m} \rightarrow \mathbf{I}(light \rightarrow \bigcirc \mathbf{S} light)), \tag{19}$$

$$\Box_F^+ \mathbf{I}(light \rightarrow \bigcirc light), \tag{20}$$

$$\Box_F^+(\mathsf{m} \rightarrow \bigcirc \mathbf{I}(light \rightarrow on)), \tag{21}$$

$$\Box_F^+(\mathbf{C}(light \wedge on \wedge \bigcirc \neg on) \rightarrow \mathsf{q}_{rec}), \tag{22}$$

$$\Box_F^+ \mathbf{I}((light \wedge on \wedge \bigcirc \neg on) \rightarrow \neg \mathbf{S}(light \wedge on \wedge \bigcirc \neg on)), \tag{23}$$

$$\Box_F^+(\mathsf{m} \rightarrow \mathbf{I}(light \rightarrow \neg on)). \tag{24}$$

The above conjunction works as follows. By (19), with every start of a computation (i.e., whenever m is true) a fresh point x is picked up and the variable *light* is true at it; we say in this case that a new light is created. By (20), *light* is also true at all images $f^i(x)$ of x. At the next iteration, by (21), all lights are switched on, i.e., *on* is true at every point with *light*. By (22), whenever a light is switched off the current state of the computation must be q_{rec}. Finally, by (24), all lights must be switched off at the next occurrence of m. It should also be noted that, by (19), the root x_0 is always a light and whenever m is true (for $j > 0$) a fresh light is created as an $\overline{R}_j$-successor for every existing light. Therefore, after k occurrences of m there will be an $\overline{R}_j$-sequence of lights of length $\geq k$. As, by (23), along every $\overline{R}_j$-sequence only one light may be switched off at a time and, by (23), this may only happen in the state q_{rec}, the variable m cannot become true sooner than the computation visits q_{rec} at least k times. This completes the encoding of the ω-reachability problem.

($\Leftarrow$) For the converse direction, suppose that, for every $n \geq 1$, we have a *perfect* computation of S starting with $\langle q_0, \epsilon \rangle$ and reaching q_{rec} at least n times. Let N_n be the number of transitions in the n-th computation; the 0-th computation is considered to be empty and $N_0 = 0$. For $n > 0$, let $M_n = \sum_{k=1}^{n}(N_k + 1)$ and $M_0 = 0$ (for technical reasons, $M_{-1} = -1$).

Fix $n \geq 0$ and consider the n-th computation:

$$\langle q_0, \epsilon \rangle \xrightarrow{\delta_{k_1^n}}_p \langle q_{i_1^n}, w_1^n \rangle \xrightarrow{\delta_{k_2^n}}_p \langle q_{i_2^n}, w_2^n \rangle \xrightarrow{\delta_{k_3^n}}_p \cdots \xrightarrow{\delta_{k_{N_n}^n}}_p \langle q_{i_{N_n}^n}, w_{N_n}^n \rangle.$$

Let H_0^n be the number of send (!) transitions in the above sequence. Clearly, the number of messages that can be held in the channel during the above computation is bounded by H_0^n.

We inductively define the sequence (L_j^n, H_j^n), $0 \leq j \leq N_n$ (w_j^n will be written between L_j^n and H_j^n). Let $L_0^n = H_0^n$. Suppose (L_{j-1}^n, H_{j-1}^n) is defined. Then

$$(L_j^n, H_j^n) = \begin{cases} (L_{j-1}^n, H_{j-1}^n - 1), & \text{if } \delta_{i_j^n} = \langle q_{i_{j-1}^n}, ?, a, q_{i_j^n} \rangle, \\ (L_{j-1}^n - 1, H_{j-1}^n), & \text{if } \delta_{i_j^n} = \langle q_{i_{j-1}^n}, !, a, q_{i_j^n} \rangle. \end{cases}$$

We are in a position now to define a satisfying model for $\varphi_{S,q_0,q_{rec}}$ based on the computations above. First, for each computation number $n \in \mathbb{N}$ and a step j in it, $0 \leq j \leq N_n$, let

$$U_j^n = \{0, 1, \ldots, H_j^n\} \times \{0, 1\}$$

be the chunk of the model required to encode this step of the computation: one extra point (0) is needed for a light that is created for this computation; moreover, points are duplicated to make the strict modality **S** work properly. Note that U_0^n is the largest among them and includes all of the U_j^n.

Next, for every $m \in \mathbb{N}$, let

$$W_m = \{m\} \times \left[\bigcup_{n=0}^{M_n < m} (\{n\} \times U_0^n) \cup (\{n_0\} \times U_{M_{n_0} - m}^{n_0}) \right],$$

where n_0 is such that $M_{n_0-1} < m \leq M_{n_0}$ (i.e., n_0 is the number of the computation at iteration m and $j = M_{n_0} - m$ is the step number inside that computation). Note that the computations are encoded backwards.

Let R_m be the lexicographic order on W_m, i.e.,

$(m, n, k, q) R_m (m, n', k', q')$ iff $n < n'$ or
$\qquad \qquad \qquad \qquad \qquad (n = n'$ and $(k < k'$ or $(k = k'$ and $q \leq q'))$,

for all quadruples $(m, n, k, q), (m, n', k', q') \in W_m$ (n and n' are for computation numbers, k and k' for elements to write words on, and q and q' range $\{0, 1\}$ for duplication). Note that $r_m = (m, 0, 0, 0)$ is the root of $\langle W_m, R_m \rangle$. Let $W = \bigcup_{m=0}^{\infty} W_m$, $R = \bigcup_{m=0}^{\infty} R_m$ and let $\mathfrak{T}$ be the Aleksandrov space induced by $\langle W, R \rangle$. Define f by taking, for all $(m, n, k, q) \in W$:

$$f((m, n, k, q)) = (m + 1, n, k, q).$$

Clearly, f is monotone on $\langle W, R \rangle$. Finally, $\mathfrak{V}$ is defined by taking

$\mathfrak{V}(s) = \{(m, n, k, 0) \in W\}$,
$\mathfrak{V}(\mathsf{m}) = \{(m, 0, 0, 0) \in W \mid \exists n \in \mathbb{N} \quad C(m; n, 0)\}$,
$\mathfrak{V}(\mathsf{tr}\delta_k) = \{(m, 0, 0, 0) \in W \mid \exists n, j \in \mathbb{N} \quad C(m; n, j + 1) \text{ and } \delta_{k_j^n} = \delta_k\}$,
$\mathfrak{V}(\mathsf{q}_i) = \{(m, 0, 0, 0) \in W \mid \exists n, j \in \mathbb{N} \quad C(m; n, j) \text{ and } q_{i_j^n} = q_i\}$,
$\mathfrak{V}(a_i) = \{(m, n, k, q) \in W \mid \exists j \in \mathbb{N} \quad C(m; n, j) \text{ and } w_j^n(L_j^n + k) = a_i\}$,
$\mathfrak{V}(light) = \{(m, n, 0, q) \in W\}$,
$\mathfrak{V}(on) = \{(m, n, 0, q) \in W \mid \exists j \in \mathbb{N} \quad C(m; n, j) \text{ and}$
$\qquad q_{rec}$ occurs $> n$ times among $q_{i_{N_n}^n}, \ldots, q_{i_j^n}\}$,

where the predicate $C(m; n, j)$ is true iff $m = M_n - j$ and $0 \leq j \leq N_n$. It can be readily verified that $(\langle \langle \mathfrak{T}, f \rangle, \mathfrak{V} \rangle, r_0) \models \varphi_{S, q_0, q_{rec}}$. ∎

As a consequence we obtain the following

THEOREM 16. *None of* Log $\mathcal{T}$, Log $\mathcal{A}$, Log $\mathcal{R}^n$, *for* $n \geq 1$, *is decidable.*

Acknowledgements. The work on this paper was partially supported by U.K. EPSRC grants no. GR/S63175/02, GR/S61973/02, GR/S63182/01, GR/S61966/01.

BIBLIOGRAPHY

[1] P. S. Alexandroff. Diskrete Räume. *Matematicheskii Sbornik*, 2 (44):501–518, 1937.
[2] S. Artemov, J. Davoren, and A. Nerode. Modal logics and topological semantics for hybrid systems. Technical Report MSI 97-05, Cornell University, 1997.
[3] G. Bezhanishvili and M. Gehrke. Completeness of **S4** with respect to the real line: revisited. *Annals of Pure and Applied Logic*, 131(1–3):287–301, 2005.
[4] N. Bourbaki. *General Topology, Part 1*. Hermann, Paris and Addison-Wesley, 1966.
[5] J. Davoren. *Modal logics for continuous dynamics*. PhD thesis, Cornell University, 1998.
[6] D. Gabelaia, R. Kontchakov, A. Kurucz, F. Wolter, and M. Zakharyaschev. Combining spatial and temporal logics: expressiveness vs. complexity. *Journal of Artificial Intelligence Research*, 23:167–243, 2005.
[7] D. Gabelaia, A. Kurucz, F. Wolter, and M. Zakharyaschev. Non-primitive recursive decidability of products of modal logics with expanding domains. *Annals of Pure and Applied Logic*, 142(1–3):245–268, 2006.
[8] B. Konev, R. Kontchakov, F. Wolter, and M. Zakharyaschev. On dynamic topological and metric logics. *Studia Logica*, 2006. In print.

[9] B. Konev, F. Wolter, and M. Zakharyaschev. Temporal logics over transitive states. In R. Nieuwenhuis, editor, *CADE*, volume 3632 of *Lecture Notes in Computer Science*, pages 182–203. Springer, 2005.
[10] P. Kremer. Temporal logic over **S4**: an axiomatizable fragment of dynamic topological logic. *Bulletin of Symbolic Logic*, 3:375–376, 1997.
[11] P. Kremer and G. Mints. Dynamic topological logic. *Bulletin of Symbolic Logic*, 3:371–372, 1997.
[12] P. Kremer and G. Mints. Dynamic topological logic. *Annals of Pure and Applied Logic*, 131(1–3):133–158, 2005.
[13] P. Kremer, G. Mints, and V. Rybakov. Axiomatizing the next-interior fragment of dynamic topological logic. *Bulletin of Symbolic Logic*, 3:376–377, 1997.
[14] R. Mayr. Undecidable problems in unreliable computations. *Theoretical Computer Science*, 1–3(297):337–354, 2003.
[15] S. Slavnov. Two counterexamples in the logic of dynamic topological systems. Technical Report TR–2003015, Cornell University, 2003.

Boris Konev, Frank Wolter
Department of Computer Science
The University of Liverpool
Liverpool L69 3BX, U.K.
{b.konev, frank}@csc.liv.ac.uk

Roman Kontchakov, Michael Zakharyaschev
School of Computer Science and Information Systems
Birkbeck College
Malet Street, London WC1E 7HX, U.K.
{roman, michael}@dcs.bbk.ac.uk

Topological Modal Logics with Difference Modality

ANDREY KUDINOV

ABSTRACT. We consider propositional modal logic with two modal operators □ and [≠]. In topological semantics □ is interpreted as an interior operator and [≠] as difference. We show that some important topological properties are expressible in this language. In addition, we present a few logics and proofs of f.m.p. and of completeness theorems.

1 Introduction.

This paper deals with the topological semantics of modal logic The study of topological semantics of modal logic was started in 1944 by McKinsey and Tarski [12]. Recently, this topic has been attracting more attention partly due to applications in AI (cf. [2] and [8]). Reading the modal box as an interior operator one can easily show that logic of all topological spaces is **S4**. In addition, McKinsey and Tarski proved that **S4** is also the complete logic of the reals, Cantor space and indeed of any metric separable space without isolated points (for a new proof of this fact see [1]). Therefore, all these spaces are modally equivalent, hence many natural properties of topological spaces such as connectedness, density-in-itself and T_1 are undefinable. For more information on spatial logics and spatial reasoning see [1, 9, 15, 16].

There are two ways of enriching the definability of a language: to change semantics or to extend the language. According to the first way, Esakia in [6] and Shehtman in [16] considered the derivational logic (more recent paper on this [3]). According to the other way, we can add the universal modality. In this new language we can express connectedness (cf. [15]).

In this paper, however, we add the difference modality (or modality of inequality) [≠], interpreted as "true everywhere except here". Using the difference modality was suggested by several people independently (in [10] for one). This modality and its interpretation in Kripke frames were studied more deeply in [14]. It has been shown that the difference modality greatly increases the expressive power of a language (cf. [11, 14]). The expressive power of this language in topological spaces has been studied by Gabelaia in [9]; the author presented axioms that defines T_1 and T_0 spaces. Added to the topological modal logic, the difference modality allows us to express topological properties that were unreachable before. The topological properties mentioned in the end of the first paragraph become definable. The universal modality is expressible as well in the following way: $[\forall]A = [\neq]A \wedge A$.

Here we also introduce three logics: **S4D**, **S4DS**, and **S4DT$_1$S**. We prove their f.m.p. and the following completeness theorems: **S4D** is complete with respect to all topological spaces (Theorem 31), **S4DS** is complete with respect to all dense-in-themselves topological spaces (Theorem 33), and **S4DT$_1$S** is complete with respect to any zero-dimensional dense-in-itself metric space (Theorem 36).

2 Definitions and basic notions.

Let us introduce some notations the reader will meet in this paper. Assume that B is a set, $R, R' \subseteq B \times B$ are relations on B, then

$$R \restriction_A = R \cap (A \times A), \text{ for any } A \subseteq B;$$
$$Id_B = \{(x,x) \mid x \in B\};$$
$$R^+ = R \cup Id_B;$$
$$R \circ R' = \{(x,z) \mid \exists y \, (xRy \,\&\, yR'z)\};$$
$$R^1 = R, \; R^n = R^{n-1} \circ R;$$
$$R^* = \bigcup_{n=1}^{\infty} R^n.$$

In this paper, we study propositional modal logics with two modal operators, $\Box$ and $[\neq]$. A formula is defined as follows:

$$\phi ::= p \mid \bot \mid \phi \to \phi \mid \Box \phi \mid [\neq]\phi.$$

The standard classic logic operators ($\lor, \land, \neg, \top, \equiv$) are expressed in terms of $\to$ and $\bot$. The dual modal operators $\Diamond, \langle \neq \rangle$ are defined in the usual way as $\Diamond A = \neg\Box\neg A$, $\langle \neq \rangle A = \neg[\neq]\neg A$ respectively. $[\forall] A$ stands for $[\neq]A \land A$.

DEFINITION 1. A *bimodal logic* (or *a logic*, for short) is a set of modal formulas closed under Substitution $\left(\frac{A(p_i)}{A(B)}\right)$, Modus Ponens $\left(\frac{A, A \to B}{B}\right)$ and two Generalization rules $\left(\frac{A}{\Box A}, \frac{A}{[\neq]A}\right)$; containing all classical tautologies and the following axioms

$$\Box(p \to q) \to (\Box p \to \Box q),$$
$$[\neq](p \to q) \to ([\neq]p \to [\neq]q).$$

K$_2$ denotes *the minimal bimodal logic*.

Let L be a logic and let Γ be a set of formulas, then $L + \Gamma$ denotes the minimal logic containing L and Γ. If $\Gamma = \{A\}$, then we write $L + A$ rather then $L + \{A\}$.

In this paper, however, we consider a few additional axioms:

(B_D^-) $p \to [\neq]\langle\neq\rangle p$
(4_D^-) $(p \land [\neq]p) \to [\neq][\neq]p$
$(T_\Box)$ $\Box p \to p$
$(4_\Box)$ $\Box p \to \Box\Box p$
$(D_\Box)$ $[\forall]p \to \Box p$
(AT_1) $[\neq]p \to [\neq]\Box p$
(DS) $[\neq]p \to \Diamond p$

The first two axioms are for $[\neq]$ and they are from the paper by de Rijke [14]. These axioms correspond to some basic properties of inequality: symmetry and pseudo-transitivity[1] respectively.

The next two axioms are axioms for **S4**. These axioms have well-known correspondence to the properties of topological interior operator: $\mathbf{I}Y \subseteq Y$ and $\mathbf{I}Y \subseteq \mathbf{II}Y$ respectively (where Y is an arbitrary set). We denote the interior and the closure operators by $\mathbf{I}$ and $\mathbf{C}$ respectively.

Axiom $(D_\Box)$ is needed to connect $\Box$ and $[\neq]$ and to make sure that $[\forall]$ is the universal modality.

The meaning of the next two axioms will be explained later.

In this paper we study the following three logics:

$$\mathbf{S4D} = \mathbf{K_2} + \{B_D, 4_D^-, D_\Box, T_\Box, 4_\Box\},$$
$$\mathbf{S4DS} = \mathbf{S4D} + DS,$$
$$\mathbf{S4DT_1S} = \mathbf{S4DS} + AT_1.$$

3 Topological models.

Let us define topological models.

DEFINITION 2. A *topological model* is a pair $(\mathfrak{X}, \theta)$, where $\mathfrak{X}$ is a topological space and θ is a function assigning to each proposition letter p a subset $\theta(p)$ of $\mathfrak{X}$. The function θ is called *a valuation*.

DEFINITION 3. The truth of a formula at a point of a topological model is defined by induction:

(i) $\mathfrak{X}, \theta, x \models p$ iff $x \in \theta(p)$
(ii) $\mathfrak{X}, \theta, x \not\models \bot$
(iii) $\mathfrak{X}, \theta, x \models \phi \to \psi$ iff $\mathfrak{X}, \theta, x \not\models \phi$ or $\mathfrak{X}, \theta, x \models \psi$
(iv) $\mathfrak{X}, \theta, x \models \Box \phi$ iff   there is a neighborhood U of x such that for any $y \in U$ $\mathfrak{X}, \theta, y \models \phi$
(v) $\mathfrak{X}, \theta, x \models [\neq]\phi$ iff $\mathfrak{X}, \theta, y \models \phi$ for any $y \neq x$

If U is a subset of $\mathfrak{X}$, then $\mathfrak{X}, \theta, U \models A$ denotes that $\mathfrak{X}, \theta, x \models A$ for any $x \in U$. A formula A is called *valid* in a topological space $\mathfrak{X}$ (notation: $\mathfrak{X} \models A$), if it is true at any point under any valuation. Also in notation $\mathfrak{X}, \theta, x \models A$ we will omit the space and/or the valuation, if it is clear what space and/or valuation we consider.

DEFINITION 4. The *D-logic* of a class of topological spaces $\mathcal{T}$ (in notation $L_D(\mathcal{T})$) is the set of all formulas that are valid in all topological spaces from $\mathcal{T}$.

Let us describe the classes of topological spaces axiomatized by (AT_1), (DS).

DEFINITION 5. A T_1-*space* is a topological space such that all its one-element subsets are closed.

---
[1] In this paper relation R is pseudo-transitive iff R^+ is transitive. In some papers this property is called weakly transitive (cf. [6, 3])

As we mentioned in introduction there is an axiom that defines T_1 spaces in [9], but it has a little bit different form. And due to the next lemma they are equivalent on topological spaces.

LEMMA 6. *Let $\mathfrak{X}$ be a topological space then $\mathfrak{X} \models AT_1$ iff $\mathfrak{X}$ is a T_1-space.*

Proof. ($\Rightarrow$) Ad absurdum. Suppose there exists $x \in \mathfrak{X}$ such that $\{x\}$ is not closed. Hence $\mathfrak{X} - \{x\} \neq \mathbf{I}(\mathfrak{X} - \{x\})$. Let $U = \mathfrak{X} - \{x\}$. There exists

(3.1) $y \in U - \mathbf{I}U$

We take a valuation θ in $\mathfrak{X}$ such that $\theta(p) = U$. Then $x \models [\neq]p$ but $x \models AT_1$; hence $x \models [\neq]\Box p$. Since $y \neq x$, we have $y \models \Box p$, which means that y together with some its neighborhood is in U. This contradicts to (3.1).

($\Leftarrow$) Assume that $\mathfrak{X}$ is a T_1-space. Let $\mathfrak{X}, \theta, x \models [\neq]p$ then $\theta(p) \supseteq \mathfrak{X} - \{x\}$. We need to prove that $x \models [\neq]\Box p$. It means that for all $y \in \mathfrak{X} - \{x\}$ $y \models \Box p$. Take any $y \in \mathfrak{X} - \{x\}$. Since $\mathfrak{X} - \{x\}$ is open, there exists an open $U \ni y$ and $U \subseteq \mathfrak{X} - \{x\}$. So $U \models p$, then $y \models \Box p$, hence $x \models [\neq]\Box p$. ∎

DEFINITION 7. Let $\mathfrak{X}$ be a topological space. A point $x \in \mathfrak{X}$ is called *isolated*, if $\{x\}$ is open. $\mathfrak{X}$ is called *dense-in-itself*, if it has no isolated points.

LEMMA 8. *Let $\mathfrak{X}$ be a topological space then $\mathfrak{X} \models DS$ iff $\mathfrak{X}$ is dense-in-itself.*

Proof. ($\Rightarrow$) Ad absurdum. Assume that $\mathfrak{X}$ is not dense-in-itself and $x \in \mathfrak{X}$ is isolated.

Let us take a valuation θ in $\mathfrak{X}$ such that $\theta(p) = \mathfrak{X} - \{x\}$; then $x \models [\neq]p$. Since $\{x\}$ is open and $x \models \neg p$, it follows that $x \models \Box \neg p$ or equivalently, $x \models \neg \Diamond p$. This contradicts to the axiom (DS).

($\Leftarrow$) Assume that $\mathfrak{X}$ is dense-in-itself and $(\mathfrak{X}, \theta), x \models [\neq]p$; then there are two cases:
(i) $\theta(p) = \mathfrak{X}$, in this case it is obvious that $(\mathfrak{X}, \theta), x \models \Diamond p$;
(ii) $\theta(p) = \mathfrak{X} - \{x\}$, then $(\mathfrak{X}, \theta), x \models \Diamond p$ since $x \in \mathbf{C}(\mathfrak{X} - \{x\}) = \mathfrak{X}$. ∎

4 Kripke frames and models.

Kripke frames and models are well-known basic notions of modal logic (cf. [4] and [5]).

DEFINITION 9. A *Kripke frame* is a tuple $F = (W, R_1, \ldots R_n)$ such that

(i) W is a non-empty set,
(ii) R_i for $i = 1 \ldots n$ are binary relations on W.

In this paper however, we consider Kripke frames with one or two relations only. The first is denoted as R and the second (if it is present) — as R_D.

DEFINITION 10. A *Kripke model* is a pair $\mathcal{M} = (F, \theta)$, where F is a frame and θ is a valuation (a function from the set of all proposition letters to the set of all subsets of W).

$\mathcal{M}, x \models A$ denotes that formula A is true in model $\mathcal{M}$ at point x; $\mathcal{M} \models A$ denotes that A is true at all points of model $\mathcal{M}$; $F \models A$ denotes that $(F, \theta), x \models A$ for all valuations θ and all points $x \in W$; $F, x \models A$ denotes that $(F, \theta), x \models A$ for all valuations θ. For a subset $U \subseteq W$ $\mathcal{M}, U \models A$ denotes that for any $x \in U$ $(\mathcal{M}, x \models A)$.

DEFINITION 11. The *logic* of a class of frames $\mathcal{F}$ (in notation $L(\mathcal{F})$) is the set of all formulas that are valid in all frames from $\mathcal{T}$. For a single frame F, $L(F)$ stands for $L(\{F\})$.

DEFINITION 12. A frame F is called *a Λ-frame* for a modal logic Λ, if $\Lambda \subseteq L(F)$.

DEFINITION 13. A *p-morphism* from a Kripke frame $F = (W, R, R_D)$ onto a Kripke frame $F' = (W', R', R'_D)$ is a map $f : W \to W'$ satisfying the following conditions:

1. f is surjective;

2. $\forall x \forall y (xRy \Rightarrow f(x)R'f(y))$ and the same for R_D and R'_D;

3. $\forall x \forall z (f(x)R'z \Rightarrow \exists y(xRy \& f(y) = z))$ and the same for R_D and R'_D.

In notation: $f : F \twoheadrightarrow F'$.

DEFINITION 14. By cone F^x we will understand the frame

$$(W^x, R \upharpoonright_{W^x}, R_D \upharpoonright_{W^x}),$$

where $W^x = (R \cup R_D)^+(x)$. If for some x $F = F^x$ then F called *rooted*.

The following two lemmas are well-known (cf. [4] and [5]).

LEMMA 15. *Let $F = (W, \ldots)$ be a Kripke frame, then*

$$L(F) = \bigcap \{L(F^x) \mid x \in W\}.$$

LEMMA 16. *(p-morphism Lemma)* $f : F \twoheadrightarrow F'$ *implies* $L(F) \subseteq L(F')$.

In this paper we consider only **S4D**-frames. The axioms $B_D, 4_D^-, D_\square, T_\square, 4_\square$ put constraints on relations R and R_D. So from now on we assume that all Kripke frames satisfy the following conditions:

- R is reflexive (axiom $T_\square$) and transitive ($4_\square$),
- R_D is symmetric (B_D)
- R_D is pseudo-transitive (4_D^-),
- $R \subseteq R_D \cup Id_W$ ($D_\square$).

Note that we can further assume that $R_D \cup Id = W \times W$, because according to Lemma 15 we can consider only generated subframes.

Now let us see what formulas AT_1 and DS mean in a Kripke frame.

Let $F = (W, R, R_D)$ be a **S4D**-frame, then $Top(F) = Top(W, R)$ denotes the topological space on the set W with the topology $\{R(V) \mid V \subseteq W\}$. For formulas with the difference modality, validity in F and $Top(F)$ may not be equivalent. This is because R_D may not be the real inequality relation.

DEFINITION 17. Let R be a transitive reflexive relation on W. Then $x \in W$ is called R-*minimal* (respectively R-*maximal*), if for any y, yRx (respectively xRy) implies $x = y$.

DEFINITION 18. Let $F = (W, R, R_D)$ be an **S4D**-frame; we say that F is a T_1-*frame* (or has the T_1-*property*), if all R_D-irreflexive points are R-minimal.

LEMMA 19. *Let* $F = (W, R, R_D)$ *be* **S4D**-*frame. Then* $F \models AT_1$ *iff* F *is a* T_1-*frame.*

Proof. ($\Rightarrow$) Suppose $F \models AT_1$ and there exists an R-non-minimal and R_D-irreflexive point in F. To be more specific, let x and y be two different points such that $\neg xR_Dx$ and yRx. Take a valuation θ such that $\theta(p) = W - \{x\}$. Then $x \models [\neq]p$ and $x \models \neg p$, thus $y \models \Diamond \neg p$. Since $x \neq y$ and yRx, we have xR_Dy, $x \models \langle \neq \rangle \Diamond \neg p$. Hence $x \models \neg [\neq] \Box p$. This contradicts $x \models AT_1$.

($\Leftarrow$) Assume that F is a T_1-frame and for some valuation for F we have $x \models [\neq]p$. Let us show that $x \models [\neq] \Box p$. As we mentioned above, generated subframes preserve validity, so we can assume that $F = F^x$ hence $R_D(x) \cup \{x\} = W$. There are two possibilities:

1) xR_Dx. Then $y \models p$ for any $y \in W$, hence for all $y \in W$ we have $y \models \Box p$; so $x \models [\neq] \Box p$.

2) $\neg xR_Dx$. Then $y \models p$ for every $y \neq x$. By assumption, $y \neq x$, yRz implies $z \neq x$, hence $z \models p$. So for any $y \neq x$ $y \models \Box p$; hence $x \models [\neq] \Box p$. ■

DEFINITION 20. Let $F = (W, R, R_D)$ be an **S4D**-frame; we say that F is a DS-*frame*, if every R_D-irreflexive point has an R-successor (called just a *successor* further on).

LEMMA 21. *Let* $F = (W, R, R_D)$ *be an* **S4D**-*frame. Then* $F \models DS$ *iff* F *is a DS-frame.*

Proof. ($\Rightarrow$) Suppose $F \models DS$ and there exists an R_D-irreflexive point x without successors. We take a valuation θ such that $\theta(p) = W - \{x\}$; then $x \models [\neq]p$ but $x \not\models \Diamond p$. This contradicts $F \models DS$.

($\Leftarrow$) Suppose that every R_D-irreflexive point in F has a successor. Let us prove that for any $x \in W$ $x \models DS$. Suppose $(F, \theta), x \models [\neq]p$, then there are two cases: (i) x is R_D-reflexive; then $\theta(p) = W$, and so $x \models \Diamond p$ since R is reflexive; (ii) x is R_D-irreflexive, then $\theta(p) \supseteq W - \{x\}$, and by our assumption, there exists $y \neq x$ such that xRy; hence $y \models p$ and $x \models \Diamond p$. ■

5 Kripke completeness and finite model property.

All our axioms are Sahlqvist formulas. So we easily obtain Kripke completeness for logics **S4D, S4DS, S4DT$_1$S**.

Following the common way of proving f.m.p. we use filtration (cf. [5] and [4]).

DEFINITION 22. Let $M = (F, \theta)$ be a Kripke model, where $F = (W, R, R_D)$ is a Kripke frame and Ψ is a set of formulas closed under subformulas. Let $\approx_\Psi$ be the equivalence relation on the elements of W defined as follows:

$$w \approx_\Psi v \text{ iff for all } \phi \text{ in } \Psi : (M, w \models \psi \text{ iff } M, v \models \phi).$$

By $[w]$ we denote the equivalence class of w. Suppose $M' = (F', \theta')$ and $F' = (W', R', R'_D)$ such that

1. $W' = W_\Psi = \{[w] \mid w \in W\}$.

2. If wRv then $[w]R'[v]$ (and similarly for R_D),

3. If $[w]R'[v]$ then for all $\Box\phi \in \Psi$; $M, w \models \Box\phi$ only if $M, v \models \phi$ (and similarly for R_D and $[\neq]$).

4. $\theta'(p) = \{[w] \mid M, w \models p\}$, for all atomic symbols p in Ψ.

Then M' is called a filtration of M through Ψ.

LEMMA 23. (Filtration Lemma) *Let M' be a filtration of M through Ψ, then for any $x \in M_1$ and for any $\psi \in \Psi$*

$$M, w \models \psi \iff M', [w] \models \psi.$$

LEMMA 24. *Let F_1 be an **S4D**-frame, $M_1 = (F_1, \theta_1)$ a model, Ψ a finite set of formulas closed under subformulas. Then there exists a filtration M_2 of M_1 through Ψ, such that $M_2 = (F_2, \theta_2)$ and F_2 is an **S4D**-frame.*

Proof. Let $M' = (W_\Psi, R', R'_D, \theta')$ be the minimal filtration of M. The minimal filtration is well-known (cf. [5] or [12]). Briefly, $[x]R'[y]$ iff there exist $x' \in [x]$ and $y' \in [y]$ such that $x'R_1y'$, and the same for R'_D and R_{D1}.

Let R_2 be the transitive closure of R':

$$R_2 = R'^* = \bigcup_{n \geq 1} R'^n;$$

and let R_{D2} be the pseudo-transitive closure of R'_D:

$$R_{D2} = R'^*_D - (Id - R'_D).$$

Note that the only difference between the pseudo-transitive and the transitive closure is that the irreflexive points remain irreflexive.

One can easily see that the reflexivity of R' is inherited by R_2, and the reflexivity of R' follows from the reflexivity of R_1. In the same way the symmetry of R_{D1} implies the symmetry of R_{D2}. The transitivity of R_2 and the pseudo-transitivity of R_{D2} are provided by construction. Next, we can easily show that $R' \subseteq R'_D \cup Id$; hence $R_2 \subseteq R_{D2} \cup Id$ holds.

To complete the proof, we have to show that the relations R_2 and R_{D2} satisfy the definition of filtration. Since Filtration Lemma for the minimal filtration and its transitive closure in transitive logics are well-known (cf. [4]), we will only check R_{D2}.

1. For arbitrary $w, v \in W_1$ assume $vR_{D1}w$, let us prove that $[v]R_{D2}[w]$. If $[v] \neq [w]$, the proof is the same as for the transitive closure. So assume $[w] = [v]$; then $[v]R'_D[w]$, and so $[v]R_{D2}[w]$.

2. Assume $[v]R_{D2}[w]$, let us prove that for all $[\neq]\psi \in \Psi$; $M_1, v \models [\neq]\psi$ only if $M_1, w \models \psi$. If $[v] \neq [w]$, then the proof is the same as for the transitive closure. If $[w] = [v]$, then from $[v]R_{D2}[v]$ follows $[v]R'_D[v]$. But R'_D was already filtration.

So we obtain a filtration that reduces M_1 to a finite model over an **S4D**-frame. ∎

THEOREM 25. *Let L be one of the logics: **S4D**, **S4DS**, **S4DT$_1$S**. Then L has the finite model property.*

Proof. Assume that A is a formula such that A is not in L. Hence, A is refuted in some generated submodel $M_1 = (W_1, R_1, R_{D1}, \theta)$ of the canonical model of logic L. Note that since M_1 is a generated submodel, $R_{D1} \cup Id_{W_1}$ is the universal relation.

Let Ψ be the set of all subformulas of formula A. By Lemma 24 there exists model $M_2 = (F_2, \theta_2)$ such that F_2 is a **S4D**-frame and M_2 is a filtration of M_1 through Ψ.

Since M_2 is a filtration, A is refuted in M_2. So it remains to prove (if needed) the T_1–property and the DS-property for F_2.

Let us prove that axiom AT_1 is valid in frame F_2. By Lemma 19, it is sufficient to prove that for any η such that $\neg \eta R_{D2}\eta$ there does not exist ψ such that $\psi \neq \eta$ and $\psi R_2 \eta$.

Assume the contrary, i.e. there exists a point $\psi \neq \eta$ such that $\psi R_2 \eta$. Then consider their inverse images: $[x] = \eta$ & $[y] = \psi$. By construction of R_2 we obtain
$$y \approx_\Psi y_0 R_1 z_0 \approx_\Psi y_1 \ldots y_k R_1 z_k \approx_\Psi x$$

Since $y \not\approx_\Psi x$, we can take maximal l such that $y_l \not\approx_\Psi z_l$. By transitivity of $\approx_\Psi$ we conclude that $z_l \approx_\Psi x$ and $[z_l] = [x] = \eta$. Assume that $z_l \neq x$, since $R_{D1} \cup Id$ is the universal relation $z_l R_{D1} x$, hence $[z_l] R_{D2}[x]$; which contradicts $\neg \eta R_{D2}\eta$.

So $y_l \neq x$ & $y_l R_1 x (= z_l)$, at the same time reflexivity is preserved under filtration; hence $\neg x R_{D1} x$. So we came to a contradiction, because the generated subframe $F_1 = (W_1, R_1, R_{D1})$ of the canonical model has the T_1–property.

Now let $[x]$ be an R_{D2}-irreflexive point. Hence, x is also an R_{D1}-irreflexive point, so for some y, xR_1y and $xR_{D1}y$, then $[x]R_2[y]$ and $[x]R_{D2}[y]$, so $[y] \neq [x]$; hence $[x]$ is not maximal. ∎

6 Topological completeness.

Let us define analogue of p-morphism for maps from topological space onto finite **S4D**-frame.

DEFINITION 26. Let $\mathfrak{X}$ be a topological space and let $F = (W, R, R_D)$ be a finite Kripke frame. A function $f : \mathfrak{X} \to F$ is called a *cd-p-morphism*, if it is surjective and satisfies the following two conditions

(6.1) $\mathbf{C}f^{-1}(w) = f^{-1}(R^{-1}(w))$,

(6.2) $R_D^{-1}(f^{-1}(w)) = f^{-1}(R_D^{-1}(w))$,

where $R_D =$ " $\neq$ " in $\mathfrak{X}$ (in particular $R_D^{-1}(\{x\}) = \mathfrak{X} - \{x\}$). In notation $f : \mathfrak{X} \xrightarrow{cd} F$.

Note that since f is surjective, (6.2) is equivalent to the following: if w is R_D-irreflexive then $f^{-1}(w)$ is one-element.

LEMMA 27. *If F is a finite Kripke frame, $\mathfrak{X}$ is a topological space and $f : \mathfrak{X} \xrightarrow{cd} F$ then $L_D(\mathfrak{X}) \subseteq L(F)$.*

Proof. Note that f is a *cd-p*-morphism and $\mathbf{C}$ distributes over finite[2] unions. So for $U \subseteq W$ we have

(6.3) $$f^{-1}(R^{-1}(U)) = f^{-1}(\bigcup_{w \in U} R^{-1}(w)) = \bigcup_{w \in U} f^{-1}(R^{-1}(w)) \stackrel{(6.1)}{=}$$
$$= \bigcup_{w \in U} \mathbf{C}f^{-1}(w) = \mathbf{C}f^{-1}(U)$$

In other terms, f is an interior map between topological spaces $\mathfrak{X}$ and $Top(F)$.

Similarly

(6.4) $$f^{-1}(R_D^{-1}(U)) = f^{-1}(\bigcup_{w \in U} R_D^{-1}(w)) = \bigcup_{w \in U} f^{-1}(R_D^{-1}(w)) \stackrel{(6.2)}{=}$$
$$= \bigcup_{w \in U} R_D^{-1} f^{-1}(w) = R_D^{-1} f^{-1}(U).$$

Now let θ be an arbitrary valuation on the frame F. Take a valuation Θ on $\mathfrak{X}$ such that $\Theta(p) = f^{-1}(\theta(p))$. Then a standard inductive argument shows that for any formula ϕ

(6.5) $\Theta(\phi) = f^{-1}(\theta(\phi))$,

---
[2] $\mathbf{C}$ does not distribute over infinite unions so finiteness of F is essential.

where $\theta(\phi) = \{v \,|\, (F,\theta), v \models \phi\}$ and $\Theta(\phi) = \{x \,|\, (\mathfrak{X}, \Theta), x \models \phi\}$.

For this proof we rewrite all formulas using $\Diamond$ and $\langle \neq \rangle$ (rather then $\Box$ or $[\neq]$).

There are only two nontrivial cases:

i) $\phi \equiv \Diamond \psi$. Then

$$f^{-1}(\theta(\Diamond \psi)) = f^{-1}(R^{-1}(\theta(\psi))) \stackrel{(6.3)}{=} \mathbf{C} f^{-1}(\theta(\psi)) \stackrel{induction}{=} \mathbf{C}\Theta(\psi) = \Theta(\Diamond \psi).$$

ii) $\phi \equiv \langle \neq \rangle \psi$. Then

$$f^{-1}(\theta(\langle \neq \rangle \psi)) = f^{-1}(R_D^{-1}(\theta(\psi))) \stackrel{(6.4)}{=} R_D^{-1} f^{-1}(\theta(\psi)) \stackrel{induction}{=} R_D^{-1} \Theta(\psi)$$
$$= \Theta(\langle \neq \rangle \psi).$$

Now if $\phi \notin \mathbf{L}(F)$, there exists a valuation θ such that $\theta(\phi) \neq W$. By (6.5) $\Theta(\phi) = f^{-1}(\theta(\phi))$, and so $\Theta(\phi) \neq \mathfrak{X}$ since f is subjective. Thus $\phi \notin \mathbf{L}(\mathfrak{X})$. ∎

The following proposition uses ideas from [14, 6]

PROPOSITION 28. *Let $F = (W, R, R_D)$ be a **S4D**-Kripke frame, $R_D \cup Id_W = W \times W$. There exists **S4D**-Kripke frame $F' = (W', R', R_D')$, such that $F' \twoheadrightarrow F$ and $x' R_D' y'$ iff $x' \neq y'$.*

Proof. Let us put $W^0 = \{x \in W | x R_D x\}$ and $W^\times = W - W^0$. Then

(6.6) $\quad W' = W^\times \cup W^0 \times \{0, 1\}$

Let us define the function $f : F' \to F$ such that

$$f(x') = \begin{cases} x, & \text{if } x' = x \in W^\times; \\ x, & \text{if } x' = (x, i); \end{cases}$$

and the relation R':

$$x' R' y' \iff f(x') R f(y')$$

Let us prove that f is a p-morphism.

1. Obviously f is surjective.

2. Assume that $x' R' y'$ then by definition of R' $f(x') R f(y')$. Assume that $x' R_D' y'$ (or $x' \neq y'$), $f(x') = x$ and $f(y') = y$. If $x \neq y$ then $x R_D y$. If $x = y$ then $x' = (x, 0)$ and $y' = (x, 1)$ (or vice verse); using (6.6) we conclude that $x R_D x = y$.

3. Assume that $f(x') R y$. If $y \in W^0$ then $y' = (y, 0)$ or $y' = y$ otherwise. Easy to see that $f(y') = y$ and $x' R' y'$. Assume that $f(x') R_D y$. Case when $f(x') \neq y$ is obvious so let $f(x') = y$. It means that $y \in W^0$ and $x' = (y, i)$. So we put $y' = (y, (i+1) \bmod 2)$ and this will do. ∎

COROLLARY 29. *Let $\mathcal{C}$ be the class **S4D**-frames of the form $F = (W, R, \neq)$ then **S4D** is complete with respect to $\mathcal{C}$.*

It is easy to show that for any **S4D**-frame $F = (W, R, \neq)$

$$Top(F) \xrightarrow{cd} F$$

but we can prove a stronger statement:

LEMMA 30. *Let (F, θ) be a Kripke model then for any formula A and $x \in W$*

$$F, \theta, x \models A \iff Top(F), \theta, x \models A$$

Proof. By induction on the complexity of A. The only case that is not trivial or classical is when $A = [\neq]B$.

$$F, \theta, x \models [\neq]B \iff \forall y \, (y \neq x \Rightarrow F, \theta, y \models B)$$

but by induction it holds iff

$$\forall y \, (y \neq x \Rightarrow Top(F), \theta, y \models B) \iff Top(F), \theta, x \models [\neq]B$$

∎

THEOREM 31. ***S4D** is the D-logic of all topological spaces.*

Proof. Let A be a formula that is not in **S4D**. Then by Corollary 29 there exists a Kripke frame $F = (W, R, \neq)$ such that $F \not\models A$. By Lemma 30 we obtain $Top(F) \not\models A$. ∎

PROPOSITION 32. *Let $F = (W, R, \neq)$ be a DS-frame, then $Top(F)$ is a dense-in-itself topological space.*

Proof. In $Top(F)$ the least open neighborhood of point x is $R(x)$. Since F is a DS-frame, $R(x) - \{x\} \neq \varnothing$; hence $Top(F)$ is dense-in-itself. ∎

THEOREM 33. ***S4DS** is logic of all dense-in-itself topological spaces.*

Proof. From Theorem 25 we know that **S4DS** is complete with respect to all finite DS-frames. Now we can apply Proposition 32 and Lemma 30. ∎

If a logic contains the axiom (AT_1) then we cannot use the above methods. Indeed if $F = (W, R, \neq)$ and $F \models AT_1$, then $R = Id_W$. The logic of such frames will be the logic of isolated points. So we need to find more sophisticated ways.

Recall a few definitions.

DEFINITION 34. *A non-empty topological space $\mathfrak{X}$ is called zero-dimensional if clopen sets constitute its open base.*

DEFINITION 35. A pair (X, ρ) called metric space if X is a set and ρ is a function from $X \times X$ onto $\mathbb{R}$, such that $\rho(x, y) \geq 0$, $\rho(x, y) = 0$ iff $x = y$, $\rho(x, y) = \rho(y, x)$, and $\rho(x, y) + \rho(y, z) \geq \rho(x, z)$.

On metric space can be defined natural topology based on open balls: $\{y \mid \rho(x, y) < r\}$.

THEOREM 36. **S4DT$_1$S** *is compete with respect to any zero-dimensional dense-in-itself metric space.*

Proof. Let $\mathfrak{X}$ be a zero-dimensional dense-in-itself metric space and ρ be the distance in it; $O(x, r)$ denotes the open ball $\{y \in \mathfrak{X} \mid \rho(x, y) < r\}$.

We know from Theorem 25 that **S4DT$_1$S** is complete with respect to all finite DS-T_1-frames. If we prove that for an arbitrary finite DS-T_1-frame $F = (W, R, R_D)$,

$$\mathfrak{X} \xrightarrow{cd} F$$

then we prove the theorem.

We use induction on the size of F. Consider three cases.

Case I. $W = R(w_0)$, $R_D = W \times W$ for some w_0. Since **S4** is complete with respect to $\mathfrak{X}$(cf. [1]) and $F^- = (W, R)$ is an **S4**-frame, there exists a continuous function $f : \mathfrak{X} \to Top(F^-)$. It is easy to check that $f : \mathfrak{X} \xrightarrow{cd} F$.

Case II. $W = R(w_0)$, $R_D = W \times W - (w_0, w_0)$. Since W is finite let us enumerate all points in W starting with w_0: $W = \{w_0, w_1, w_2, \ldots w_n\}$. Any generated subframe F^{w_i} for $i > 0$ satisfies case I.

Take an arbitrary point x_0 and clopen sets $Y_0, Y_1, \ldots$ such that

$$\{x_0\} \subset \ldots \subset Y_n \subset \ldots \subset Y_1 \subset Y_0 = \mathfrak{X}$$

and

$$Y_n \subseteq O(x_0, \frac{1}{n})$$

for every $n > 0$. We can do it because $\mathfrak{X}$ is zero-dimensional (cf. [13]).

Since $Y_n \subseteq O(x_0, \frac{1}{n})$, it follows that

$$\bigcap_n Y_n = \{x_0\}$$

and further we obtain

$$\mathfrak{X} - \{x_0\} = \mathfrak{X} - \bigcap_n Y_n = \bigsqcup_n \mathfrak{X}_n,$$

where $\mathfrak{X}_n = Y_n - Y_{n+1}$. Each set $\mathfrak{X}_n$ is open, metric, dense-in-itself, and zero-dimensional.

For any open neighborhood U of x_0 there exists n such that $O(x_0, \frac{1}{n}) \subset U$, it follows that $Y_n \subset U$, hence for all $i \geq n$ $\mathfrak{X}_i \subset U$.

So, by induction, for any $j > 0$ there exists

$$f_j : \mathfrak{X}_j \xrightarrow{cd} F^{w_k}, \text{ where } (k-1) \equiv j \pmod{n}$$

Now consider
$$f(x) = \begin{cases} w_0, & \text{if } x = x_0; \\ f_j(x), & \text{if } x \in \mathfrak{X}_j. \end{cases}$$

Let us prove that $f : \mathfrak{X} \xrightarrow{cd} F$.

First, we note that f is surjective.

Second, we check (6.1). Assume that $y \in \mathfrak{X}_j$ then:
$$y \in \mathbf{C}f^{-1}(w) \implies y \in \mathbf{C}f_j^{-1}(w) = f_j^{-1}(R^{-1}(w)) \subseteq f^{-1}(R^{-1}(w));$$

and the other way around:
$$y \in f^{-1}(R^{-1}(w)) \implies f_j^{-1}(R^{-1}(w)) = y \in \mathbf{C}f_j^{-1}(w) \subseteq \mathbf{C}f^{-1}(w).$$

Now assume that $y = x_0$. For any $w \in W$, $w_0 \in R^{-1}(w)$; hence
$$x_0 \in f^{-1}(R^{-1}(w)).$$

On the other hand, for some i, $w = w_i$ and for any open neighborhood of x_0, there exists m such that $(i-1) \equiv m \pmod{n}$ and $U \supset \mathfrak{X}_m \supset f_m^{-1}(w_i)$. In other words, x_0 is a limit point for $f^{-1}(w_i)$, hence $x_0 \in \mathbf{C}f^{-1}(w_i)$.

Third, we check (6.2). Since $f^{-1}(w_0)$ is a one-element set, (6.2) holds.

Case III. Everything else. Let us take all R-minimal R-clusters of F and from each one of them we choose an arbitrary point. So we get the following set: $\{v_1, v_2, \ldots, v_k\}$. Standard unravelling arguments show that
$$F' = F^{v_1} \sqcup F^{v_2} \sqcup \ldots \sqcup F^{v_n} \twoheadrightarrow F$$

So we need to show that $\mathfrak{X} \xrightarrow{cd} F'$.

Since F is a **S4DT$_1$S**-frame, each F^{v_i} satisfies case I or case II.

Since $\mathfrak{X}$ is zero-dimensional, we can present $\mathfrak{X}$ as the disjunctive union of clopen subsets:
$$\mathfrak{X} = \mathfrak{X}_1 \sqcup \ldots \sqcup \mathfrak{X}_{k-1} \sqcup \mathfrak{X}_k.$$

By induction we have
$$f_1 : \mathfrak{X}_1 \xrightarrow{cd} F^{v_1},$$
$$\ldots$$
$$f_{k-1} : \mathfrak{X}_{k-1} \xrightarrow{cd} F^{v_{k-1}},$$
$$f_k : \mathfrak{X}_k \xrightarrow{cd} F^{v_k}$$

It is easy to show that $f = f_1 \sqcup \ldots \sqcup f_k$ (if $x \in \mathfrak{X}_i$ then $f(x) = f_i(x)$) is a cd-p-morphism. ∎

The immediate and obvious corollary of this theorem is that **S4DT$_1$S** is complete with respect to all dense-in-themselves T_1 spaces.

7 Conclusions and open problems.

The language with difference modality shows much more expressive power than the basic topological language, and even more than the basic language with universal modality. We can express density-in-itself, T_1, and connectedness in it. Moreover the axiom

$$(AE_1) \qquad [\neq]p \wedge \neg p \wedge \Box(p \to \Box q \vee \Box \neg q) \to \Box(p \to q) \vee \Box(p \to \neg q)$$

distinguishes $\mathbb{R}$ from $\mathbb{R}^2$ (cf. [7]). It was proved that logic $\mathbf{S4DT_1S}$ + (AE_1) + "connectedness" is complete with respect to $\mathbb{R}^n$, $n \geq 2$ (the full proof is to be published). We still do not know the D-logic of $\mathbb{R}$ and whether $\mathbf{S4D} + (AT_1)$ is complete with respect to all T_1 spaces.

BIBLIOGRAPHY

[1] M. Aiello, J. van Benthem, G. Bezhanishvili. *Reasoning about Space: the Modal Way.* Technical Report PP-2001-18, University of Amsterdam, 2001.

[2] B. Bennett, A. G. Cohn, F. Wolter and M. Zakharyaschev. *Multi-Dimensional Modal Logic as a Framework for Spatio-Temporal Reasoning.* Applied Intelligence, 17 (3), pp 239–251, 2002.

[3] G. Bezhanishvili, L. Esakia, D. Gabelaia *Some Results on Modal Axiomatization and Definability for Topological Spaces.* Studia Logica, Vol. 81, No. 3. (December 2005), 325–355.

[4] P. Blackburn, M. de Rijke, Y. Venema. *Modal Logic.* Cambridge University Press, 2001.

[5] A.V.Chagrov, M.V.Zakharyaschev *Modal Logic.* Oxford University Press, 1996.

[6] L. Esakia. *Weak transitivity — a restitution.* Logical investigations, v.8, p.244–245. Moscow, Nauka, 2001.(in Russian)

[7] A. Kudinov *Difference modality in topological spaces.* Algebraic and Topological Methods in Non-classical Logics II, Barcelona, Abstracts, p.50–51, 2005.

[8] D. Gabbay, A. Kurucz, F. Wolter, M. Zakharyaschev. *Many-dimensional modal logics. Theory and applications.* Elsevier, 2003.

[9] D.Gabelaia. *Modal Definability in Topology.* Master's thesis, ILLC, University of Amsterdam, 2001.

[10] G. Gargov, S. Passy, T. Tinchev. *Modal environment for Boolean speculations.* Mathematical logic and its applications, Plenum Press, New York, 1987, 253–263.

[11] V. Goranko. *Modal definability in enriched languages.* Notre Dame Journal of Formal Logic, vol.31, 1990, 81–105.

[12] J.C.C. McKinsey, A. Tarski. *The algebra of topology.* Annals of Mathematics, v.45(1944), 141–191.

[13] H. Rasiowa, R. Sikorski. *The Mathematics of Metamathematics.* Warsaw, 1963.

[14] M. de Rijke. *The Modal Logic of Inequality.* Journal of Symbolic Logic, v.57 (1992), 566–584.

[15] V. Shehtman *"Everywhere" and "Here".* Journal of Applied Non-classical Logics, v.9 (1999), No 2/3, 369–380.

[16] V. Shehtman. *Derived sets in Euclidean spaces and modal logic.* ITLI Prepublication Series, X-90-05, University of Amsterdam, 1990.

Andrey Kudinov
Department of Mathematical Logic and Theory of Algorithms
Faculty of Mechanics and Mathematics Moscow State University
Leninskie gory, Moscow, 119992 RUSSIA
kudinov.andrey@gmail.com

Isomorphism via translation
TADEUSZ LITAK

ABSTRACT. We observe that the known fact that the difference logic and the hybrid logic with universal modality have the same expressive power on Kripke frames can be strengthened for a far wider class of general frames. This observation, together with a general completeness result and some algebraic theory of closure operators, is used to show that lattices of difference logics and of hybrid logics are isomorphic.

Keywords: closure operators, difference operator, discrete frames, hybrid logic, universal modality, atomic algebras

Gargov and Goranko proved in [13] that languages of difference logic $\mathcal{ML}(D, \diamond)$ and hybrid logic with universal modality $\mathcal{H}(E, \diamond)$ are equivalent with respect to frame definability. This observation was improved upon by Areces [1] who showed how to define a polynomial translation. It was suggested by Patrick Blackburn (personal communication) that these results should be strengthened to show something more. Namely, one would expect the existence of an isomorphism between the lattice of difference logics and the lattice of hybrid logics. We are going to show that weakly atomic frames — duals of atomic algebras — provide a natural tool to attack this problem. The apparatus behind the isomorphism proof is the standard algebraic theory of closure operators. In addition, weakly atomic frames also allow to generalize the correspondence results to the topological setting. To avoid notational complications, we work with the unimodal language, but virtually nothing hinges on it: all results transfer to the polymodal case.

The main results of this paper are Corollaries 17 and 18, Theorem 26 together with Corollaries 29 and 30. Theorem 6 is also of some independent interest. Corollary 18 allows for immediate transfers of known results on topological definability from $\mathcal{H}(E, \diamond)$ to $\mathcal{ML}(D, \diamond)$ and back. A recent example: Sustretov [19] has obtained a Goldblatt-Thomason-style characterization of topo-definability in $\mathcal{H}(E, \diamond)$, which by our result must be also a characterization of topo-definability for $\mathcal{ML}(D, \diamond)$.[1] In the converse direction, Kudinov [17] announced an axiomatization of the $\mathbf{K}_\mathsf{D}^{\mathbf{Name}}$-logic of Euclidean spaces of dimension at least 2. Hybrid translations of these axioms must then yield an axiomatization of the $\mathbf{K}_\mathcal{H}^{\mathbf{Name}}$-logic of these spaces.

The idea that translations in logics can be used to prove that certain lattices of logics are isomorphic occurred already in Kracht and Wolter's [16]

---
[1] Thanks to Balder ten Cate for pointing out this example.

improvement of Thomason's translation from polymodal logics to a subclass of unimodal logics. Actually, in their survey work [15] on modal translations and simulations, the authors mention the original result of Gargov and Goranko. However, they do not discuss the possibility of lifting it to an isomorphism between lattices of logics or otherwise put in on equal footing with other translations discussed in that paper. What they say, instead, is that *both nominals and the difference [operator] are rather nonstandard devices which work fine on Kripke structures but present special problems for generalized frames* [15]. One of the main purposes of this note is to show that those special problems are not impossible to overcome and it is possible to treat both formalisms in a general mathematical framework. More generally, the presence of *non-orthodox* (or *non-structural*) rules is not necessarily an impenetrable barrier for algebraic methods. There is more to universal algebra than varieties, quasi-varieties, structural rules and structural closure operators. Finally, our results make clear that the theory of closure operators can and should be applied to translations and embeddings between *classes* of logics.

The author wishes to thank Nick Galatos, who showed him how to simplify the formulation of Lemma 19 and how to relate it to existing results, to Patrick Blackburn for the inspiration and to the anonymous referees whose comments led to significant re-organization of the paper. Most of all, however, thanks are due to Balder ten Cate for his emails and comments on earlier versions of this work.

1 Languages and semantics

1.1 Weakly atomic frames

We are going to consider two formalisms extending the one of basic modal logic. The first is *the hybrid language with the universal modality*, obtained by extending the basic modal language with an infinite set of nominals $NOM = \{i, j, \ldots\}$ and a new modal operator E. The formulas of this language are generated by the following recursive definition:

$$\phi ::= \top \mid p \mid i \mid \neg\phi \mid \phi \wedge \psi \mid \Diamond\phi \mid \mathsf{E}\phi,$$

where p is a proposition letter and i is a nominal. It is usually assumed that the set of nominals NOM, as well as the set of proposition letters $PROP$, is countable. $VAR := PROP \cup NOM$. The second extension arises by adding *the difference operator* D, i.e., its syntax is:

$$\phi ::= \top \mid p \mid \neg\phi \mid \phi \wedge \psi \mid \Diamond\phi \mid \mathsf{D}\phi.$$

The set of hybrid formulas is denoted by $\mathcal{H}(\mathsf{E}, \Diamond)$. The set of difference formulas is denoted by $\mathcal{ML}(\mathsf{D}, \Diamond)$. To improve readability, we drop references to VAR and $PROP$. For every formula ϕ, $Sub(\phi)$ denotes the set of its subformulas. The remaining connectives — $\bot, \vee, \rightarrow, \Box, \mathsf{A}$ — are defined as abbreviations in a standard way. Define also $\overline{\mathsf{D}}p := \neg\mathsf{D}\neg p$, $\mathsf{O}p := p \wedge \neg\mathsf{D}p$.

DEFINITION 1. *A weakly atomic frame* is a structure of the form $\mathfrak{F} := \langle W, R, \mathbb{A} \rangle$, where $R \subseteq W \times W$ and $\mathbb{A}$ is a family of subsets of W closed under the Boolean operations, the operator $\Diamond_R X := \{w \in W \mid \exists x \in X.wRx\}$ s.t. for every non-empty $P \in \mathbb{A}$ there is $x \in P$ s.t. $\{x\} \in \mathbb{A}$. If $\{x\} \in \mathbb{A}$, x is called an *admissible element*. The set of admissible elements of W is denoted as At$\mathbb{A}$. Similarly, members of $\mathbb{A}$ are called *admissible subsets*.

Thus, weakly atomic frames are those where every subset contains an admissible element. A *propositional valuation* is one assigning members of $\mathbb{A}$ to propositional variables, a *nominal valuation* is one assigning admissible members of W to nominal variables, a *(total) valuation* for hybrid logic is one which is both nominal and propositional. In the case of difference operator, a valuation is simply a propositional valuation. A *model* $\mathfrak{M}$ is a pair $\langle \mathfrak{F}, \mathfrak{V} \rangle$, consisting of a weakly atomic frame and a valuation in it. Depending on the kind of valuation, it is propositional or (total) hybrid model (we do not consider purely nominal models). For $x \in W$ and $c \in NOM \cup PROP$, we write $\mathfrak{M}, x \vDash c$ if $c \in PROP$ and $x \in \mathfrak{V}(c)$ or $c \in NOM$ and $x = \mathfrak{V}(c)$. Clauses for booleans and the modal operator are standard. For universal modality, the clause is $\mathfrak{M}, x \vDash \mathsf{E}\phi$ if $\exists w \in W.\mathfrak{M}, w \vDash \phi$. For difference modality, the clause is $\mathfrak{M}, x \vDash \mathsf{D}\phi$ if $\exists w \neq x.\mathfrak{M}, w \vDash \phi$. We write $\mathfrak{F}, \mathfrak{V} \vDash \phi$ if ϕ holds under $\mathfrak{V}$ at all points in W. If ϕ is a hybrid formula and $\mathfrak{V}$ is a propositional valuation, we write $\mathfrak{F}, \mathfrak{V} \vDash \phi$ if $\mathfrak{F}, \mathfrak{V}' \vDash \phi$ for every total valuation $\mathfrak{V}'$ whose propositional component coincides with $\mathfrak{V}$. We write $\mathfrak{F}, x \vDash \phi$ if ϕ holds at x under every valuation. Finally, $\mathfrak{F} \vDash \phi$ means that $\mathfrak{F}, x \vDash \phi$ for every $x \in W$.

Say that a model $\langle \mathfrak{F}, \mathfrak{V} \rangle$ is *weakly named* for $\mathcal{H}(\mathsf{E}, \Diamond)$ if for every formula ϕ there is a nominal i s.t. $\mathfrak{V}(i) \in \mathfrak{V}(\phi)$. A model is *weakly named* for $\mathcal{ML}(\mathsf{D}, \Diamond)$ if for every formula ϕ there is a variable p s.t. $\emptyset \neq \mathfrak{V}(p \land \neg \mathsf{D}p) \subseteq \mathfrak{V}(\phi)$. General frames associated with weakly named models — i.e., frames where admissible subsets of W are exactly those which are values of some formula under $\mathfrak{V}$ — are weakly atomic.

Important subclasses of atomic frames are:

- *discrete frames*: frames where every singleton is admissible;

- *full frames* or *Kripke frames*: frames where every set is admissible, i.e., $\mathbb{A} = 2^W$. In such a case, we may simply drop $\mathbb{A}$ from the signature. This is how Kripke frames are usually defined. The only non-logical constant of *first-order correspondence language* for Kripke frames is a binary constant corresponding to the accessibility relation.

If for a given family of subsets $X \subseteq 2^W$ there is $A \in \mathbb{A}$ s.t. $\bigcup X \subseteq A$ and for every $B \in \mathbb{A}$, $\bigcup X \subseteq Z$ implies $A \subseteq Z$, we call A *the supremum of* X and denote it as $\bigvee X$; observe it is not necessarily equal to $\bigcup X$.

LEMMA 2. *If* $\mathfrak{F} = \langle W, R, \mathbb{A} \rangle$ *is a weakly atomic frame, then for every admissible subset* $A \in \mathbb{A}$, $A = \bigvee A \downarrow_{\mathsf{At}}$, *where* $A \downarrow_{\mathsf{At}} := \{\{x\} \mid x \in \mathsf{At}\mathbb{A} \cap A\}$.

Proof. It is clear that $\bigcup A \downarrow_{At} \subseteq A$. Now assume $\bigcup A \downarrow_{At} \subseteq B$ and $A \not\subseteq B$. Then $A \wedge \neg B$ is a non-empty admissible subset. Hence, by weak atomicity it has to contain an element whose singleton is in $A \downarrow_{At}$ — a contradiction. ∎

The reader probably recognized this lemma as a thinly disguised version of a proof that in an atomic algebra every element can be represented as the supremum of a family of atoms. The next subsection makes this connection explicit.

1.2 Connection with algebra

This subsection is addressed to readers interested in algebra, hence we do not define basic notions of duality theory appearing here: such readers are likely to know them anyways. It is not hard to recognize that weakly atomic frames are atomic modal algebras in disguise. The condition of weak atomicity readily implies that the algebra of admissible sets is atomic. Conversely, assume that the algebra is atomic and take the descriptive frame corresponding to it. Every admissible subset contains a principal ultrafilter and singletons of principal ultrafilters are admissible: hence, the frame is weakly atomic. Thus, we can obtain a full-blown duality between descriptive weakly atomic frames and atomic algebras as a restriction of standard duality between modal algebras and descriptive frames, as discussed in Chapter 5 of Blackburn et al. [2]. The only reason why we used the name *weakly atomic frames* instead of simply *atomic frames* is that the latter was sometimes used for discrete frames.

1.3 Neighborhood frames and topological spaces

The fact that we can identify so-called (normal) neighborhood frames (or Scott-Montague semantics) with a certain subclass of weakly atomic frames follows readily from the duality theory developed by Došen [8] and the above discussion. To make the paper more self-contained, let us describe it in more detail. Let us say that a weakly atomic frame $\langle W, R, \mathbb{A} \rangle$ is *set-theoretical* if for every $X \subseteq \mathbb{A}$ there exists $\bigvee X \in \mathbb{A}$. In other words, the family of admissible sets is lattice-complete. It is thus straightforward to prove that descriptive set-theoretical frames are exactly duals of atomic and complete modal algebras. Every Kripke frame is set-theoretical, but the converse does not hold for weakly atomic frames. However, a discrete frame is set-theoretical iff it is a Kripke frame.

There is another way of representing atomic and complete modal algebras: as *Scott-Montague semantics* or *normal neighborhood frames*. Such a structure consists of a family W and a function f assigning to every element of W a filter over W. Recall that a filter is a nonempty family of sets X satisfying $A, B \in X$ iff $A \cap B \in X$. $f(x)$ is called *the family of neighborhoods of x*. The dual algebra of a neighborhood frame is the powerset algebra of W together with the operator $\Box_f A = \{x \in W \mid A \in f(x)\}$.

Given a set-theoretical frame $\mathfrak{F} := \langle W, R, \mathbb{A} \rangle$, define the neighborhood frame associated with $\mathfrak{F}$ as $\mathfrak{F}_\sqcup := \langle \mathsf{At}\mathbb{A}, f_R \rangle$, where

$$f_R(x) := \{X \subseteq \mathsf{At}\mathbb{A} \mid x \in \Box_R \bigvee X \downarrow_{\mathsf{At}\mathbb{A}}\}.$$

Observe that for a non-admissible X this may be a larger set than $\{X \subseteq \mathsf{At}\mathbb{A} \mid x \in \Box_R X\}$: a definition of f_R using this smaller set would not work as it should. It is an instructive exercise to find a suitable counterexample. Conversely, for every neighborhood frame $\mathfrak{G}$ we can take the descriptive frame corresponding to its dual algebra to be the corresponding set-theoretical frame $\mathfrak{G}^\sqcap$.

FACT 3. $\mathfrak{F}_\sqcup$ is a neighborhood frame, $\mathfrak{G}^\sqcap$ is a set-theoretical frame, $\mathfrak{G} \simeq (\mathfrak{G}^\sqcap)_\sqcup$ and if $\mathfrak{F}$ is a descriptive set-theoretical frame, $\mathfrak{F} \simeq (\mathfrak{F}_\sqcup)^\sqcap$.

To see how this idea can be lifted to a category-theoretical equivalence, check Došen [8]. Topological spaces can be identified with neighborhood frames s.t. for every $X \in f(x)$ (1) $x \in X$ and (2) $\Box_f X \in f(x)$.

FACT 4. For any neighborhood frame $\mathfrak{G}$, t.f.a.e.

1. $\mathfrak{G}$ satisfies (1) and (2) above.

2. $\mathfrak{G}$ is a neighborhood base in the topological sense.

3. $\mathfrak{G}$ is a **S4** frame.

4. the accessibility relation of $\mathfrak{G}^\sqcap$ is a quasi-order.

2 Axiomatizations and completeness

2.1 The hybrid language

Axiomatization of $\mathbf{K}^{\mathbf{Name}}_\mathcal{H}$ — basic hybrid logic with the non-standard $\mathbf{Name}_\mathcal{H}$ rule — is given in Table 1. A $\mathbf{K}^{\mathbf{Name}}_\mathcal{H}$-*logic* is any set of formulas containing all the axioms of $\mathbf{K}^{\mathbf{Name}}_\mathcal{H}$ and closed under all its rules. For every $\Gamma \subseteq \mathcal{H}(\mathsf{E}, \Diamond)$, $\mathbf{K}^{\mathbf{Name}}_\mathcal{H}\Gamma$ denotes the smallest logic containing Γ.

LEMMA 5. *Every* $\mathbf{K}^{\mathbf{Name}}_\mathcal{H}$-*logic* Λ *is closed under the rule* $\mathbf{Name}^+_\mathcal{H}$:

   from $\mathsf{A}(i \to \phi)$ *deduce* $\mathsf{A}\phi$, *for* $i \notin Sub(\phi)$

Proof.

   1: $\mathsf{A}(i \to \phi)$ (assumption)

   2: $i \to \phi$ (by the dual of $\mathbf{Ref}_\mathsf{E}$ and MP)

   3: ϕ (by $\mathbf{Name}_\mathcal{H}$, as $i \notin Sub(\phi)$)

   4: $\mathsf{A}\phi$ (by $\mathbf{Nec}_\mathsf{A}$)

■

Axioms and rules for $\mathbf{K}_{\mathcal{H}}^{\mathbf{Name}}$	
CT	ϕ, for all classical tautologies ϕ
K	$\Box(p \to q) \to \Box p \to \Box q$
K$_\mathsf{A}$	$\mathsf{A}(p \to q) \to \mathsf{A}p \to \mathsf{A}q$
Ref$_\mathsf{E}$	$p \to \mathsf{E}p$
Trans$_\mathsf{E}$	$\mathsf{EE}p \to \mathsf{E}p$
Sym$_\mathsf{E}$	$p \to \mathsf{AE}p$
Incl$_\Diamond$	$\Diamond p \to \mathsf{E}p$
Incl$_i$	$\mathsf{E}i$
Nom	$\mathsf{E}(i \wedge p) \to \mathsf{A}(i \to p)$
MP	From $\phi \to \psi$ and ϕ deduce ψ
Nec	From ϕ deduce $\Box \phi$
Nec$_\mathsf{A}$	From ϕ deduce $\mathsf{A}\phi$
Subst	From ϕ deduce $\phi\sigma$, where σ is a substitution that uniformly replaces proposition letters by formulas and nominals by nominals
Name$_\mathcal{H}$	From $i \to \phi$ deduce ϕ, for $i \notin \phi$.
Additional rule of $\mathbf{K}_{\mathcal{H}}^{\mathbf{BG}}$	
BG$_\mathcal{H}$	From $\mathsf{E}(i \wedge \Diamond j) \to \mathsf{E}(j \wedge \phi)$ deduce $\mathsf{E}(i \wedge \Box \phi)$, for $i \neq j$ and $j \notin Sub(\phi)$

Table 1. Axiomatization for the hybrid language

Axioms and rules for $\mathbf{K}_{\mathsf{D}}^{\mathbf{Name}}$	
CT	ϕ, for all classical tautologies ϕ
K	$\Box(p \to q) \to \Box p \to \Box q$
K$_{\overline{\mathsf{D}}}$	$\overline{\mathsf{D}}(p \to q) \to \overline{\mathsf{D}}p \to \overline{\mathsf{D}}q$
WTrans$_\mathsf{D}$	$\mathsf{D}^2 p \to p \vee \mathsf{D}p$
Sym$_\mathsf{D}$	$p \to \overline{\mathsf{D}}\mathsf{D}p$
Incl$_\mathsf{D}$	$\Diamond p \to p \vee \mathsf{D}p$
MP	From $\phi \to \psi$ and ϕ deduce ψ
Nec	From ϕ deduce $\Box \phi$
Nec$_{\overline{\mathsf{D}}}$	From ϕ deduce $\overline{\mathsf{D}}\phi$
Subst	From ϕ deduce $\vdash \phi\sigma$, where σ is an arbitrary substitution
Name$_\mathsf{D}$	From $\mathsf{O}p \to \phi$ deduce ϕ, for any $p \notin Sub(\phi)$
Additional rule of $\mathbf{K}_{\mathsf{D}}^{\mathbf{BG}}$	
BG$_\mathcal{H}$	From $\mathsf{E}(\mathsf{O}p \wedge \Diamond \mathsf{O}q) \to \mathsf{E}(q \wedge \phi)$ deduce $\mathsf{E}(\mathsf{O}p \wedge \Box \phi)$, for $p \neq q$ and $q \notin Sub(\phi)$

Table 2. Axiomatization for the difference language

We write $\Gamma \vDash \gamma$ if for every weakly atomic frame $\mathfrak{F}$, if $\mathfrak{F} \vDash \Gamma$ then $\mathfrak{F} \vDash \gamma$. More generally, for any class K' of weakly atomic frames, we write $\Gamma \vDash_{K'} \gamma$ if for every $\mathfrak{F} \in K'$, $\mathfrak{F} \vDash \Gamma$ implies $\mathfrak{F} \vDash \gamma$. We say that a hybrid logic Γ is *atomically complete* if for every $\gamma \in \mathcal{H}(\mathsf{E}, \Diamond)$, $\Gamma \vDash \gamma$ iff $\gamma \in \mathbf{K}_{\mathcal{H}}^{\mathbf{Name}}\Gamma$. Definition of K'-completeness is analogous, with $\vDash$ replaced by $\vDash_{K'}$. We also say that a set of formulas Γ is *atomically Λ-consistent* if $\bot$ cannot be deduced from Γ by means of theorems of Λ, **MP**, **Name**$_\mathcal{H}$ and **Name**$_\mathcal{H}^+$. Observe that we don't allow the use of **Nec**$_A$ rule here, so we cannot use Lemma 5 and eliminate **Name**$_\mathcal{H}^+$ from this definition.

THEOREM 6 (Atomic completeness for hybrid logics). *Every $\mathbf{K}_{\mathcal{H}}^{\mathbf{Name}}$-logic Λ is atomically complete.*

Proof. (sketch) It is enough to show that every $\mathbf{K}_{\mathcal{H}}^{\mathbf{Name}}$-logic Λ is complete with respect to weakly named models. Extend every atomically Λ-consistent set of formulas Γ to *a weakly distinguishing MCS* Γ^+ — i.e., a set of formulas s.t.

- Γ is closed under all theorems of Λ and MP,
- for every ϕ, either ϕ or its negation belongs to Γ^+, but not both,
- there is a nominal $i \in \Gamma^+$ and
- for every ϕ, $\mathsf{E}\phi \in \Gamma^+$ only if $\mathsf{E}(i \wedge \phi) \in \Gamma^+$, for some $i \notin Sub(\phi)$.

The third requirement can be met because of **Name**$_\mathcal{H}$-consistency of Γ, the fourth — because of **Name**$_\mathcal{H}^+$-consistency of Γ. Compare this strategy to Lindenbaum-style lemmas in Gargov et al. [14], [13], de Rijke [18] or ten Cate, Litak [4]. Weakly distinguishing MCS's could be also called *weakly pasted MCS's* or *MCS's pasted for* E-*modality*, to show both similarities and differences with *distinguishing* or *pasted* MCS's used in these papers.

Our model is then built out of all MCS's Δ s.t. for every $\phi \in \Delta$, $\mathsf{E}\phi \in \Gamma^+$. Observe that — as opposed to proofs in papers mentioned above — we *don't* assume that every MCS in the model is a weakly distinguishing one and hence named (i.e., contains a nominal). The accessibility relation $R_\Diamond$ is then defined as usual in canonical models: $\Delta_1 R_\Diamond \Delta_2$ iff for every $\Box\phi \in \Delta_1$, $\phi \in \Delta_2$. The model thus obtained is weakly named and the general frame associated with it is a weakly atomic frame for the logic in question. ∎

This strategy then is a mixture of the standard canonical model technique and the hybrid technique of surjectively named models [13], which gives rise to discrete frames. The relationship with "mainstream" hybrid techniques and rules is discussed further in Section 2.3.

2.2 The difference language

Axiomatization of $\mathbf{K}_{\mathsf{D}}^{\mathbf{Name}}$-logic — difference logic with the non-standard **Name**$_\mathsf{D}$ rule — is given in Table 2. Note that the universal modality is definable in this system: in the case of difference operator, $\mathsf{E}\phi$ is defined as an abbreviation of $\phi \vee \mathsf{D}\phi$. The definition of atomic completeness is the same as in the hybrid case and we can prove

THEOREM 7 (Atomic completeness for difference logics). *Every difference logic is atomically complete.*

Proof. (sketch) Essentially the same as Theorem 6. The role of names for the points in the weakly named model construction is performed by formulas $\mathsf{O}p$. The only point one has to take care of is that R_D is really irreflexive, as the canonical model construction for difference logic — as opposed to the technique of *distinguishing sets* [21] — does not preclude that for some Δ, $\Delta R_\mathsf{D} \Delta$. Nevertheless, weak namedness implies that for every ϕ and every Δ, $\mathsf{E}\phi \in \Delta$ iff $\mathsf{E}(\mathsf{O}p \wedge \phi) \in \Delta$ for some p. If $\mathsf{O}p \in \Delta$, then Δ must be R_D-irreflexive. Hence, the variant of the canonical model obtained by deleting all pairs $\langle x,x \rangle$ from the interpretation of R_D validates exactly the same formulas. ∎

2.3 The role of non-standard rules

The axiomatizations used above are weaker than those used in Gargov and Goranko [13], Gargov et al. [14], Venema [21], ten Cate and the present author [4] and other papers on hybrid and difference logic. The difference lies not in the choice of axioms, but in non-standard rules. We are using only **Name**$_\mathcal{H}$ for $\mathcal{H}(\mathsf{E}, \Diamond)$ and **Name**$_\mathsf{D}$ for $\mathcal{ML}(\mathsf{D}, \Diamond)$. The Bulgarian logicians used a rule scheme called *COV* for $\mathcal{H}(\mathsf{E}, \Diamond)$. **Name**$_\mathcal{H}$ only is not enough to derive all instances of *COV*. Venema [21] made an analogous observation concerning **Name**$_\mathsf{D}$, which he denoted as *IR*, and replaced it with a (set of) rule(s) IR_D^*. Blackburn et al. [2] used both **Name**$_\mathcal{H}$ and a rule called *PASTE*. Our counterparts of these stronger rules are **BG**$_\mathcal{H}$ (Table 1) for $\mathcal{H}(\mathsf{E}, \Diamond)$ and **BG**$_\mathsf{D}$ (Table 2) for $\mathcal{ML}(\mathsf{D}, \Diamond)$. It can be proven that — as we have the universal modality in the language — both in the hybrid case and in the difference case all these strengthenings are equivalent. $\mathbf{K}_\mathcal{H}^{\mathbf{Name}}$-logics closed under **BG**$_\mathcal{H}$ are called $\mathbf{K}_\mathcal{H}^{\mathbf{BG}}$-logics, $\mathbf{K}_\mathsf{D}^{\mathbf{BG}}$-logics are defined analogously.

Every $\mathbf{K}_\mathcal{H}^{\mathbf{BG}}$-logic is complete with respect to *surjectively named models* (Theorem 5.4 in [13]), i.e., models where every point is named by a nominal. Analogously, every $\mathbf{K}_\mathsf{D}^{\mathbf{BG}}$-logic is complete with respect to models where every point is named by $\mathsf{O}p$. This gives us the following

THEOREM 8. *Every $\mathbf{K}_\mathcal{H}^{\mathbf{BG}}$-logic and every $\mathbf{K}_\mathsf{D}^{\mathbf{BG}}$-logic is complete with respect to discrete frames.*

How and why are these stronger rules non-conservative? It was observed first by Gabbay [10] and later restated in Venema [21] or Gargov et al. [13] that **Name**$_\mathcal{H}$ (**Name**$_\mathsf{D}$) is enough if $\Diamond$ is conjugated.

FACT 9. *If Λ is $\mathbf{K}_\mathcal{H}^{\mathbf{Name}}$-logic ($\mathbf{K}_\mathsf{D}^{\mathbf{Name}}$-logic) containing $p \to \Box \Diamond p$, then Λ is also closed under* **BG**$_\mathcal{H}$ (**BG**$_\mathsf{D}$). *Consequently, Λ is complete with respect to discrete frames.*

As every other result in the paper, this observation can be easily generalized to the polymodal context, e.g., for tense logics. Nevertheless, in general atomic completeness does not imply di-completeness. The logic of the so-called van Benthem frame [20] is atomically complete but not di-complete. Here, let us consider a more natural example taken from ten Cate and the present author [5]. That paper contains a more thorough discussion of non-standard rules in the topological context.

FACT 10. The $\mathbf{BG}_{\mathcal{H}}$ ($\mathbf{BG}_\mathsf{D}$) rule does not preserve validity on the real line, and, indeed, on any non-discrete T_1 space.

2.4 Properties of logics

Let us sum up by compiling a list of some standard properties one would like to be preserved and/or reflected by mappings between lattices of logics.

Decidability, finite axiomatizability. Definitions are standard.

Di-completeness, neighborhood completeness, Kripke completeness, finite model property. Substitute a suitable K' in definition of K'-completeness.

Elementary generation. Completeness with respect to a first-order definable class of Kripke frames.

Elementarity. The class of Kripke frames validating theorems of the logic is first-order definable.

At-persistence, di-persistence. If a weakly atomic (discrete) frame validates Λ, its underlying Kripke frame validates Λ as well.

Sahlqvist property. We present the notion of a Sahlqvist formula along the lines of Venema [21]. Let $c_1, c_2, c_3 \ldots$ be syntactic metavariables ranging over VAR, let $\blacklozenge_1, \blacklozenge_2, \blacklozenge_3, \ldots$ be syntactic metavariables ranging over arbitrary combinations of $\{\Diamond, \mathsf{E}\}$ in the hybrid case ($\{\Diamond, \mathsf{D}\}$ in the in the difference case) and define $\blacksquare_1, \blacksquare_2, \blacksquare_3 \ldots$ dually, i.e., a syntactic metavariable ranging over words in $\{\Box, \overline{\mathsf{D}}\}$ in the difference case ($\{\Box, \mathsf{A}\}$ in the hybrid case). A *strongly positive* formula is a conjunction of formulas of the form $\blacksquare c_i$. A formula is *positive* (*negative*) if every c_i occurs under an even (odd) number of negation symbols. A formula is *untied* if it obtained from strongly positive and negative formulas by applying only $\wedge$ and $\blacklozenge_1, \ldots \blacklozenge_r$. Formulas of the form $UNTIED \to POS$ (i.e., where antecedent is untied and consequent is positive) are called *Sahlqvist formulas*. Logics axiomatizable with Sahlqvist formulas are called Sahlqvist logics.

3 From $\mathcal{ML}(\mathsf{D}, \Diamond)$ to $\mathcal{H}(\mathsf{E}, \Diamond)$ and back

This section is based on the ideas of Gargov and Goranko [13] and Areces [1, Chapter 7].

DEFINITION 11 (Translation from $\mathcal{ML}(\mathsf{D}, \Diamond)$ to $\mathcal{H}(\mathsf{E}, \Diamond)$).

Fix any recursive $1-1$ mapping $f : \mathcal{ML}(\mathsf{D}, \Diamond) \mapsto \mathbb{N}$ s.t. $\mathbb{N} - f[\mathbb{N}]$ is an infinite recursive set and any recursive $1-1$ mapping $g : \mathbb{N} \mapsto \mathbb{N} - f[\mathbb{N}]$. Let $\tau' : \mathcal{ML}(\mathsf{D}, \Diamond) \mapsto \mathcal{ML}(\mathsf{D}, \Diamond)$ and $\tau : \mathcal{ML}(\mathsf{D}, \Diamond) \mapsto \mathcal{ML}(\mathsf{D}, \Diamond)$ be preprocessing functions s.t. τ' replaces every propositional variable p_n in ψ by $p_{g(n)}$ and $\tau(\psi)$ replaces all occurrences of subformulas of the form $\mathsf{D}\phi$

in $\tau'(\psi)$ by $p_{f(\mathsf{D}\tau(\phi))}$ by induction on the number of nested D operators. Denote $h(\phi) := f(\mathsf{D}\tau(\phi))$. Define

$$\sigma(\psi) := \tau(\psi) \wedge \bigwedge_{\mathsf{D}\phi \in Sub(\tau'(\psi))} (\mathsf{A}p_{h(\phi)} \vee \mathsf{A}\neg p_{h(\phi)} \vee (\mathsf{A}(p_{h(\phi)} \leftrightarrow \neg i_{h(\phi)}) \wedge \mathsf{E}p_{h(\phi)})) \wedge$$
$$(\mathsf{A}p_{h(\phi)} \to \mathsf{E}(\tau(\phi) \wedge i_{h(\phi)}) \wedge \mathsf{E}(\tau(\phi) \wedge \neg i_{h(\phi)})) \wedge$$
$$(\mathsf{A}\neg p_{h(\phi)} \to \mathsf{A}\neg \tau(\phi)) \wedge$$
$$((\mathsf{A}(p_{h(\phi)} \leftrightarrow \neg i_{h(\phi)}) \wedge \mathsf{E}p_{h(\phi)}) \to \mathsf{A}(\tau(\phi) \leftrightarrow i_{h(\phi)})).$$

As opposed to Areces [1, Chapter 7], we tried to avoid adding new variables to the language: we want to keep the language fixed, hence slightly more involved formulation. Nevertheless, the translation is in fact the same.

THEOREM 12. *For every weakly atomic frame $\mathfrak{F}$ and every $\psi \in \mathcal{ML}(\mathsf{D}, \Diamond)$, $\mathfrak{F} \vDash \psi$ iff $\mathfrak{F} \vDash \sigma(\psi)$.*

Proof. (sketch) First, observe that $\mathfrak{F} \vDash \psi$ iff $\mathfrak{F} \vDash \tau'(\psi)$, as the logic of an arbitrary frame is closed under substitution. Let $\mathfrak{V}$ be a propositional valuation in $\mathfrak{F}$. Take $\mathfrak{V}'$ to be any total valuation s.t. $\mathfrak{V}'$ agrees with $\mathfrak{V}$ on variables with indices from $g[\mathbb{N}]$, $\mathfrak{V}'(p_{h(\phi)}) = \mathfrak{V}(\mathsf{D}\tau(\phi))$, $\mathfrak{V}'(i_{h(\phi)})$ is some admissible singleton in $\mathfrak{V}(\tau(\phi))$ if this set is non-empty (here we use weak atomicity) and arbitrary otherwise. Clearly, $\mathfrak{V}(\tau'(\psi)) = \mathfrak{V}'(\sigma(\psi))$. Conversely, for every total hybrid model $\mathfrak{M} := \langle \mathfrak{F}, \mathfrak{V} \rangle$ and $\psi \in \mathcal{ML}(\mathsf{D}, \Diamond)$, if $\mathfrak{V}(\sigma(\psi)) \neq \emptyset$, then for every $\mathsf{D}\phi \in Sub(\tau'(\psi))$, $\mathfrak{V}(p_{h(\phi)}) = \mathfrak{V}(\mathsf{D}\tau(\phi))$ and hence $\mathfrak{V}(\sigma(\psi)) = \mathfrak{V}(\tau'(\psi))$. ∎

The converse direction is quite simple. As we already saw, by means of the difference operator, we can explicitly force a variable to serve as a name.

DEFINITION 13 (Translation from $\mathcal{H}(\mathsf{E}, \Diamond)$ to $\mathcal{ML}(\mathsf{D}, \Diamond)$).
Choose any $1-1$ and onto recursive mapping $\theta : PROP \cup NOM \mapsto PROP$ and extend it inductively to all formulas $\phi \in \mathcal{H}(\mathsf{E}, \Diamond)$. Let $\pi(\phi) := \bigwedge_{i \in NOM \cap Sub(\phi)} \mathsf{E}\mathsf{O}\theta(i) \to \theta(\phi)$. In addition, for every hybrid model $\mathfrak{M} = \langle W, R, \mathfrak{V} \rangle$ define $\mathfrak{M}_{\neq} := \langle W, R, \mathfrak{V}_{\neq} \rangle$, where $\mathfrak{V}_{\neq}(p) := \mathfrak{V}(\theta^{-1}(p))$ for each $p \in PROP$.

THEOREM 14. *For every weakly atomic frame $\mathfrak{F}$ and every $\phi \in \mathcal{H}(\mathsf{E}, \Diamond)$, $\mathfrak{F} \vDash \phi$ iff $\mathfrak{F} \vDash \pi(\phi)$.*

Proof. (sketch) Let $\mathfrak{F} := \langle W, R, \mathbb{A} \rangle$ be a weakly atomic frame, $x \in W$ and $\phi \in \mathcal{H}(\mathsf{E}, \Diamond)$. Then

$$\mathfrak{F}, x \vDash \phi \text{ iff } \mathfrak{F}, x \vDash \pi(\phi)$$

and thus $\mathfrak{F} \vDash \phi$ iff $\mathfrak{F} \vDash \pi(\phi)$. The proof of the above fact is based on two claims whose proofs can be adopted from Gargov and Goranko [13]:

CLAIM 15. *For every hybrid valuation $\mathfrak{V}$, $\mathfrak{V}(\phi) = \mathfrak{V}_{\neq}(\pi(\phi))$.*

CLAIM 16. For every propositional valuation $\mathfrak{V}$, $x \notin \mathfrak{V}(\pi(\phi))$ only if there is a hybrid valuation $\mathfrak{V}_\phi$ s.t. $\mathfrak{V}_\phi(\phi) = \mathfrak{V}(\theta(\phi)) = \mathfrak{V}(\pi(\phi))$.

Proof. (of claim, sketch) Fix an admissible $w \in W$. Define

$$\mathfrak{V}_\phi(q) = \begin{cases} \mathfrak{V}(\theta(q)) & : \theta(q) \in Sub(\pi(\phi)), \\ \{w\} & : \theta(q) \notin Sub(\pi(\phi)). \end{cases}$$

∎

∎

COROLLARY 17. $\mathcal{ML}(\mathsf{D}, \Diamond)$ and $\mathcal{H}(\mathsf{E}, \Diamond)$ *are equally expressive with respect to weakly atomic frames.*

This in turn using the observations of Section 1 gives us the following

COROLLARY 18. $\mathcal{ML}(\mathsf{D}, \Diamond)$ and $\mathcal{H}(\mathsf{E}, \Diamond)$ *are equally expressive with respect to discrete frames, normal neighborhood frames and topological spaces.*

4 Isomorphism between lattices of logics

4.1 Equivalence of closure operators

Ever since the early work of Tarski and the Polish school in the 1930's, it became clear that the study of logics should be intimately connected with the study of *closure operators*. Let us recall (cf. [7]) that given an arbitrary set X and a function $C : 2^X \mapsto 2^X$, we say that C is *a closure operator* on X if the following three conditions are satisfied for every $A, B \in 2^X$:

C1. $A \subseteq C(A)$,

C2. $A \subseteq B$ implies $C(A) \subseteq C(B)$,

C3. $C(C(A)) = C(A)$.

For convenience, a pair $\langle X, C \rangle$ consisting of a set and a closure operator on it will be called *a closure space*. A logic is often identified with a deductive consequence operator, which is indeed a closure operator on the set of formulas. The problem with this approach is that distinct consequence operators can often generate the same set of theorems. That is, the notion of *a theory* (deductively closed set of sentences) may differ even if the set of *tautologies* (deductive closure of the empty set) is the same.

Here, we take a more Hilbert-style approach: we identify logics with sets of formulas. But it doesn't mean that the theory of consequence operators is of no use for us. Because of its generality, it works well in contexts, where other algebraic techniques may pose certain problems, such as those mentioned in the introduction. There is nothing which prevents us from studying in this manner deductive consequence *with the substitution rule* and also *with non-orthodox rules*. Set of formulas which are closed under the substitution rule are *logics* rather than theories. It is a standard fact that for any closure operator C, C-closed sets form a complete lattice, where arbitrary meets coincide with set-theoretical intersections (cf. [7, Chapter 2]). Recall also that an element a of a complete lattice L is called *compact* if for every $A \subseteq L$, $a \leq \bigvee A$ implies the existence of $B \subseteq_{fin} A$ s.t. $a \leq \bigvee B$. If $\langle A, C \rangle$ is a closure space, C is *algebraic* if for every X, $C(X) = \bigcup \{C(Y) \mid Y \subseteq_{fin} X\}$. For algebraic closure operators, compact closed sets are those of the form $C(Y)$ for a finite Y.

LEMMA 19 (Isomorphism of lattices of closed elements). *Let* $\mathfrak{X} := \langle X, C_X \rangle$ *and* $\mathfrak{Y} := \langle Y, C_Y \rangle$ *be two closure spaces and assume there are mappings* $\Sigma : X \mapsto Y$, $\Pi : Y \mapsto X$ *s.t. for every* $\{a\}, A \subseteq X$, $\{b\}, B \subseteq Y$:

I1. $a \in C_X(A)$ *iff* $\Sigma(a) \in C_Y(\Sigma[A])$,

I2. $b \in C_Y(B)$ *iff* $b \in C_Y(\Sigma\Pi[B])$.

Then the lattices of C_X *and* C_Y*-closed sets are isomorphic by* $\bar{\Sigma}(A) := C_Y(\Sigma[A])$. *Moreover,* $\bar{\Pi}(B) := C_X(\Pi[B])$ *is the converse isomorphism, i.e.,* $B = \bar{\Sigma}(\bar{\Pi}(B))$ *for any* C_Y*-closed* B. *For algebraic closure operators, this mapping preserves and reflects compactness; that is, A is compact iff $\bar{\Sigma}$ is.*

Proof. For the sake of readability, we omit almost all parentheses, both round and square. It is straightforward to see that $\bar{\Sigma}$ is well-defined and preserves order. To see that it also reflects the order and hence is $1-1$, assume A_1, A_2 are C_X-closed, $a \in C_X A_1$ and $C_Y \Sigma C_X A_1 \subseteq C_Y \Sigma C_X A_2$. Then by I1, $\Sigma a \in C_Y \Sigma C_X A_1$ and by assumption and another application of I1, it gives us $a \in C_X A_2$.

To prove the mapping is onto and the 'moreover' part, take any C_Y-closed B and let $A := C_X \Pi B$. Then $b \in C_Y \Sigma C_X \Pi B$ iff (C1–C3) $C_Y b \subseteq C_Y \Sigma C_X \Pi B$ iff (I2) $C_Y \Sigma \Pi b \subseteq C_Y \Sigma C_X \Pi B$ iff (C1–C3) $\Sigma \Pi b \in C_Y \Sigma C_X \Pi B$ iff (I1 and C3) $\Pi b \in C_X \Pi B$ iff (I1) $\Sigma \Pi b \in C_Y \Sigma \Pi B$ iff (I2) $\Sigma \Pi b \in C_Y B$ iff (C1–C3) $C_Y \Sigma \Pi b \subseteq C_Y B$ iff (I2) $C_Y b \subseteq C_Y B$ iff (C1–C3) $b \in C_Y B$. Thus, $\bar{\Sigma} A = C_Y B = B$ and surjectivity follows. Preservation and anti-preservation of compactness is straightforward. ∎

Cf. Blok, Jónsson [3, Theorem 3.7] or Galatos, Tsinakis [12] for more general results of this kind.

4.2 Isomorphism via translation

Lemma 19 allows us to obtain a lattice isomorphism result as soon as we have the following ingredients:

- a class of frames or algebras K;

- two languages L_1 and L_2 and two closure operators C_1 and C_2 on them — the sets of formulas closed under C_1 (C_2) are called L_1-logics (L_2-logics);

- a proof that every L_1-logic and every L_2-logic is complete with respect to a subclass of K;

- two translations $F_1 : L_1 \mapsto L_2$ and $F_2 : L_2 \mapsto L_1$ s.t. every $\mathfrak{A} \in K$ satisfies $\phi \in L_1$ iff it satisfies $F_1(\phi)$ and $\mathfrak{A}$ satisfies $\psi \in L_2$ iff it satisfies $F_2(\psi)$.

We will be also able to prove that this isomorphism preserves and reflects many desirable properties, such as finite axiomatizability or completeness with respect to some well-behaved subclass of K. The idea is clear, but in order to prove it formally we need to introduce some definitions in the spirit of Abstract Algebraic Logic.

DEFINITION 20 (Logical family). $\langle F, C_\vdash, K, \vDash_K \rangle$ is called *a logical family* if

- F is an arbitrary *set of formulas*. We assume here that this set is recursive.

- $C_\vdash$ is an algebraic closure operator on F. $C_\vdash$-closed sets are called *logics*. Compact $C_\vdash$-closed sets are called *finitely axiomatizable logics*.

- K is an arbitrary class of structures (frames, algebras, topological spaces ...) called *a semantics*.

- $\vDash_K \subseteq K \times F$ is called *a validity relation*. If $\mathfrak{A} \vDash \phi$, we say ϕ *holds* in $\mathfrak{A}$.

This definition is so general that it has to be unsatisfying. For a start, it does not say anything about the relationship between $\vdash$ and $\vDash$. For $\Gamma \subseteq F$ and $K' \subseteq K$, denote *closure operators induced on F by K'* as

$$C_{K'}(\Gamma) := \{\phi \in F \mid \forall \mathfrak{A} \in K'(\forall \gamma \in \Gamma. \mathfrak{A} \vDash \gamma \Rightarrow \mathfrak{A} \vDash \phi)\}.$$

It is straightforward to see $C_{K'}$ is a closure operator; we say that $C_{K'}(\Gamma)$ is the *K'-closure of* Γ. If $\Gamma = C_{K'}(\Gamma)$, we say Γ is *a K'-complete logic*. Conversely, for any $\Gamma \subseteq F$, we can define $Mod(\Gamma) := \{\mathfrak{A} \in K \mid \mathfrak{A} \vDash \Gamma\}$. Thus, Γ is a K'-complete logic iff $\Gamma = C_{K' \cap Mod(\Gamma)}(\Gamma)$.

DEFINITION 21 (Soundness, completeness, complete family).
Let $\langle F, C_\vdash, K, \vDash_K \rangle$ be a logical family. If $C_\vdash(\Gamma) \subseteq C_K(\Gamma)$ for every $\Gamma \subseteq F$, K is *a sound semantics*. If the converse inclusion holds, K is called *a complete semantics*. A logical family with sound and complete semantics is called *a complete family*.

FACT 22. In every complete family, $C_\vdash$-closed elements are exactly C_K-complete logics.

DEFINITION 23 (Persistence and relative soundness).
Assume $\mathcal{F}_1 = \langle F_1, C_1, K, \vDash_1 \rangle$, $K' \subseteq K$ and $\nu : K' \mapsto K'$ is a mapping s.t.

- restriction of ν to $\nu[K']$ is the identity mapping;

- for every $\mathfrak{A} \in K'$ and every $\phi \in F$, $\nu\mathfrak{A} \vDash \phi$ implies $\mathfrak{A} \vDash \phi$.

$\nu\mathfrak{A}$ is called then *the underlying ν-frame of $\mathfrak{A}$*. ν itself is called *a carrier mapping*. We say that $\Gamma \subseteq F$ is K'-persistent (relative to ν) if $K' \ni \mathfrak{A} \vDash \Gamma$ implies $\nu\mathfrak{A} \vDash \Gamma$. Also, assume $\mathcal{F}_2 := \langle F_2, C_2, \nu[K'], \vDash_2 \rangle$, i.e., the semantics of $\mathfrak{F}_2$ is the range of ν. We say that $\mathcal{F}_1$-logic Γ is *sound relative to $\mathcal{F}_2$* if there is $\Gamma' \subseteq F_2$ s.t. $Mod_{\mathcal{F}_1}(\Gamma) \cap \nu[K'] = Mod_{\mathcal{F}_2}(\Gamma')$.

A paradigm example of such a ν is the mapping assigning to a weakly atomic frame its underlying Kripke frame. We allowed the case when the domain of ν is smaller than K itself to cover the case of di-persistence too. The notion of relative soundness generalizes the notion of elementarity. To see how, take $\mathfrak{F}_2$ to be the family of first-order theories in the frame correspondence language.

To sum up: we saw how to generalize notions such as topo-completeness, di-completeness, Kripke completeness, finite model property, elementary generation (K'-completeness, with K' replaced by respective class of frames), at-persistence, di-persistence (K'-persistence), and elementarity (relative soundness). Now let us see how to generalize the idea of translations preserving and reflecting validity — and how to use this notion to prove the existence of isomorphism preserving and reflecting the above-defined properties.

DEFINITION 24 (Equivalent families).
Let $\mathcal{F}_1 := \langle F_1, C_1, K, \vDash_1 \rangle$, $\mathcal{F}_2 := \langle F_2, C_2, K, \vDash_2 \rangle$ be two complete logical families sharing the same class of structures as semantics and let f_1, f_2 be a pair of functions s.t.

T1. $f_1 : F_1 \mapsto F_2$ and $f_2 : F_2 \mapsto F_1$;

T2. for every $\mathfrak{A} \in K$ and every $\phi_1 \in F_1$, $\mathfrak{A} \vDash_1 \phi_1$ iff $\mathfrak{A} \vDash_2 f_1(\phi_1)$;

T3. for every $\mathfrak{A} \in K$ and every $\phi_2 \in F_2$, $\mathfrak{A} \vDash_2 \phi_2$ iff $\mathfrak{A} \vDash_1 f_2(\phi_2)$.

We say then $\mathcal{F}_1$ and $\mathcal{F}_2$ are K-equivalent by $\langle f_1, f_2 \rangle$.

LEMMA 25. *Assume $\mathcal{F}_1 := \langle F_1, C_1, K, \vDash_1 \rangle$ and $\mathcal{F}_2 := \langle F_2, C_2, K, \vDash_2 \rangle$ are K-equivalent by $\langle f_1, f_2 \rangle$ and let $K' \subseteq K$. Then $\langle F_1, C_{1K'}, K', \vDash_1 \rangle$ and $\mathcal{F}_2 := \langle F_2, C_{2K'}, K', \vDash_2 \rangle$ are K'-equivalent by $\langle f_1, f_2 \rangle$, where $C_{1K'}$ and $C_{2K'}$ are closure operators induced by K' on F_1 and F_2, respectively.*

THEOREM 26. *Assume $\mathcal{F}_1 := \langle F_1, C_1, K, \vDash_1 \rangle$ and $\mathcal{F}_2 := \langle F_2, C_2, K, \vDash_2 \rangle$ are K-equivalent by $\langle f_1, f_2 \rangle$. For $\Gamma \subseteq F_1$, define $F(\Gamma) = C_2(f_1[\Gamma])$. This mapping restricted to C_1-closed sets is an isomorphism onto the lattice of C_2-closed sets, which preserves and reflects*

- *finite axiomatizability;*

- *for every $K' \subseteq K$, K'-completeness. Moreover, lattices of K'-complete $\mathcal{F}_1$-logics and $\mathcal{F}_2$-logics are isomorphic too;*

- *for any carrier mapping $\nu : K' \mapsto K'$, K'-persistence and for any logical family $\mathcal{G}$ whose semantics is $\nu[K']$, relative $\mathcal{G}$-soundness;*

- *if both f_1 and f_2 are effectively defined, the isomorphism preserves and reflects decidability. If, moreover, both are computable in polynomial time, the isomorphism preserves complexity up to a polynomial.*

Proof. The existence of such an isomorphism follows directly from Lemma 19. All we have to do is to check that the following two conditions hold for arbitrary $\{\delta\} \cup \Delta \subseteq F_1$, $\{\gamma\} \cup \{\Gamma\} \subseteq F_2$:

CLAIM 27. $\delta \in C_1(\Delta)$ iff $f_1(\delta) \in C_2(f_1[\Delta])$.

Proof. (of claim) $\delta \in C_1(\Delta)$
 iff $\Delta \vDash_{\mathcal{F}_1} \delta$ ($\mathcal{F}_1$ is a complete family)
 iff for every $\mathfrak{F} \in K$, $\mathfrak{F} \vDash_1 \Delta$ implies $\mathfrak{F} \vDash_1 \delta$
 iff for every $\mathfrak{F}$, $\mathfrak{F} \vDash_2 f_1[\Delta]$ implies $\mathfrak{F} \vDash_2 f_1(\delta)$ (T2)
 iff $f_1(\delta) \in C_2(f_1[\Delta])$ ($\mathcal{F}_2$ is a complete family). ∎

CLAIM 28. $\gamma \in C_2(f_2[f_1[\Gamma]])$ iff $\gamma \in C_2(\Gamma)$.

Proof. (of claim) $\gamma \in C_2(f_1[f_2[\Gamma]])$
 iff for every $\mathfrak{F} \in K$, $\mathfrak{F} \vDash_2 f_1[f_2[\Gamma]]$ implies $\mathfrak{F} \vDash_2 \gamma$ ($\mathcal{F}_2$ is a complete family)
 iff $\mathfrak{F} \vDash_2 \Gamma$ implies $\mathfrak{F} \vDash_2 \gamma$ (by T2 and T3)
 iff $\gamma \in C_2(\Gamma)$ ($\mathcal{F}_2$ is a complete family). ∎

Preservation and antipreservation of

- finite axiomatizability: follows from preservation and antipreservation of compactness;

- decidability: assume $\Delta = C_1(\Delta)$ is a decidable $\mathcal{F}_1$-logic. The problem whether $\gamma \in F_2$ is in $F(\Delta)$ reduces then to checking if $f_2(\gamma)$ is in Δ. So, if f_2 is computable, computability is preserved and if f_1 is computable, computability is reflected. Reasoning for complexity is analogous;

- K'-completeness: is a consequence of Lemma 25;

- K'-persistence:
    $K' \ni \mathfrak{A} \vDash \Gamma$ implies $\nu\mathfrak{A} \vDash \Gamma$
 is equivalent to
    $K' \ni \mathfrak{A} \vDash f_1(\Gamma)$ implies $\nu\mathfrak{A} \vDash f_1(\Gamma)$;

- relative soundness: because $Mod_{\mathcal{F}_1}(\Gamma) \cap \nu[K'] = Mod_{\mathcal{F}_2}(f_1[\Gamma]) \cap \nu[K']$.

∎

COROLLARY 29. *The lattices of* $\mathbf{K}_{\mathcal{H}}^{\mathbf{Name}}$-*logics and* $\mathbf{K}_{\mathsf{D}}^{\mathbf{Name}}$-*logics are isomorphic. This isomorphism preserves and reflects di-completeness, topo-completeness, Kripke completeness, finite model property, elementary generation, elementarity, at-persistence, di-persistence, finite axiomatizability and decidability.*

COROLLARY 30. *The lattices of* $\mathbf{K}_{\mathcal{H}}^{\mathbf{BG}}$-*logics and* $\mathbf{K}_{\mathsf{D}}^{\mathbf{BG}}$-*logics are isomorphic. This isomorphism preserves and reflects Kripke completeness, finite model property, elementary generation, elementarity, di-persistence, finite axiomatizability and decidability.*

4.3 The Sahlqvist problem

One more property whose preservation under modal translations is desirable is the property of being Sahlqvist. Its preservation and/or reflection are usually discussed while introducing translations and interpretations in modal logic; cf. [15]. Nevertheless, as observed by Conradie et al. [6], the problem with the syntactic definition of the Sahlqvist property is that it *is extremely fragile as it does not withstand even simple boolean transformations, or even substitutions changing the polarity of propositional variables.* And so, even if $\phi \in \mathcal{H}(\mathsf{E}, \Diamond)$ is Sahlqvist, $\pi(\phi)$ can, strictly speaking, fail to be one. The antecedent $\bigwedge_{i \in NOM \cap Sub(\phi)} \mathsf{E}\Diamond\theta(i)$ is indeed an untied formula, but the consequent, which is $\theta(\phi)$ itself, is not necessarily positive. However, it is of course a matter of trivial boolean pre-processing to show that if ϕ is a Sahlqvist formula of the form $\alpha \to \beta$, then $\pi(\phi)$ can be taken to be $\bigwedge_{i \in NOM \cap Sub(\phi)} \mathsf{E}\Diamond\theta(i) \wedge \theta(\alpha) \to \theta(\beta)$, and this is again a Sahlqvist formula. Therefore, it is safe to say that π preserves the property of being Sahlqvist as well. Moreover, it is clear that this slightly modified version of π (i.e. with the clause that formulas whose main connective is implication are translated as described above) yields a Sahlqvist formula *iff* the input was a Sahlqvist formula. Thus, it is justified to say that this translation *preserves and reflects* the property of being Sahlqvist. Things, however, look different if one takes σ as a starting point. It is enough to glance at the definition of σ to see there is no straightforward way of ensuring that this translation preserves the Sahlqvist property.

5 Concluding remarks

5.1 Further applications

There is nothing in the isomorphism proof from Section 4 which crucially depends on the nature of weakly atomic frames, hybrid logic or the difference operator. Therefore, our note is meant as a methodological suggestion: a proof that two languages are equivalent with respect to expressivity over certain class K of frames, algebras, spaces etc. yields automatically that lattices of K-complete logics in both languages are isomorphic. If K is large enough to provide a general completeness proof, it proves the isomorphism of lattices of *all* logics in both languages. Moreover, this isomorphism preserves many desirable properties. We feel this method can find other applications. First idea: using results of Gabbay et al. [11], one can try to apply ideas presented here to the correspondence between lattice of modal logics with Stavi connectives and the lattice of weak second-order theories (in the sense of van Benthem [20]) of linear orders. It would be also interesting to find other examples of this kind in the modal realm or elsewhere.

Besides, it should be possible to relate the isomorphism-via-translation techniques to the theory of residuation and Galois connections. It was not necessary to study this connection in depth for our present purposes, but anyone aiming to develop a more general mathematical theory of translations between classes of logics along the lines of Section 4 should investigate this option seriously. A good starting point is Erne et al. [9]

5.2 Life without rules

Coming back to the correspondence between $\mathbf{K}_\mathcal{H}^{\text{Name}}$-logics and $\mathbf{K}_\mathsf{D}^{\text{Name}}$-logics, the reader may wonder now what was the role of non-standard rules in the isomorphism proof. As we have shown, general completeness and isomorphism results can be proven both with and without the paste-like rules ($\mathbf{BG}_\mathcal{H}$ and $\mathbf{BG}_\mathsf{D}$, respectively). They are geared towards discrete frames — algebraically, they force complete additivity of corresponding algebras. It means that behaviour of $\Diamond$ on all sets is completely determined by its behaviour on admissible individuals. There is, however, nothing in the translation which prevents semantics of more topological character. In these semantics, constraints on $\Diamond$ imposed by paste-like rules are not natural.

But the situation with $\mathbf{Name}_\mathcal{H}$ and $\mathbf{Name}_\mathsf{D}$ seems to be different. These rules apparently capture something very fundamental about the nature of the difference operator and nominals. Let us consider briefly what could happen if we delete these rules. Of course, we could not take weak namedness and atomicity for granted anymore. So, in the hybrid case the set of admissible singletons could become arbitrarily small. And in the $\mathcal{ML}(\mathsf{D}, \Diamond)$ case, as was already noted, these rules are also necessary to ensure that we can restrict our attention to frames where R_D is irreflexive. So, the idea of translation based on non-standard semantics for such weak deductive systems would be to treat exactly the set of those points for which R_D is irreflexive as the set of admissible singletons. However, the axiom $\mathbf{Incl}_i$ poses an immediate problem. In the $\mathcal{H}(\mathsf{E}, \Diamond)$-case, it forces non-emptiness of the collection of admissible singletons. In the $\mathcal{ML}(\mathsf{D}, \Diamond)$-case, this would correspond to the requirement that every frame contains at least one point on which R_D is irreflexive. But this condition cannot be forced on non-standard semantics by any modal formula.

Balder ten Cate suggested two ways out of this predicament. One was to retain a very weak form of $\mathbf{Name}_\mathsf{D}$ for $\mathcal{ML}(\mathsf{D}, \Diamond)$, with $\mathsf{O}p$ replaced by $\mathsf{EO}p$. Another was to remove the problematic axiom from the hybrid axiomatization. The present author is not happy with either choice. The first one, while sacrificing a nice completeness result, would fail to achieve the main goal of eliminating non-standard rules from both languages. The second option feels, if anything, even worse. It would remove not only the axiom whose roots in hybrid logics can be traced back to Prior, but also the underlying fundamental idea: that nominals should behave like genuine individual names and hence be true not just at *at most* one point, but at *exactly* one point. Still more unacceptably, **Incl**$_i$ would not make these pseudo-semantics specialize to standard semantics in hybrid logic: Ei would define an *empty* class of frames

And it is doubtful anyways that either of bad solutions would restore the isomorphism result. An interested reader may investigate this question. The present author feels contended with the conclusion that every general (meta-)theory of logics with individual names has to take the non-orthodox rules seriously.

BIBLIOGRAPHY

[1] C. Areces. *Logic Engineering. The Case of Description and Hybrid Logics.* PhD thesis, Institute for Logic, Language and Computation, University of Amsterdam, 2000. ILLC Dissertation Series 2000-5.

[2] P. Blackburn, M. de Rijke, and Y. Venema. *Modal logic.* Cambridge University Press, Cambridge, UK, 2001.

[3] W. Blok and B. Jónsson. Equivalence of consequence operations. *Studia Logica*, submitted.

[4] B. ten Cate and T. Litak. The importance of being discrete. Under submission.

[5] B. ten Cate and T. Litak. Topological perspective on the hybrid proof rules. In *International Workshop on Hybrid Logic 2006*, ENTCS. Elsevier, to appear.

[6] W. Conradie, V. Goranko, and D. Vakarelov. Elementary canonical formulae: a survey on syntactic, algorithmic, and model-theoretic aspects. In *Advances in Modal Logic 5*. King's College, 2005.

[7] B.A. Davey and H.A. Priestley. *Introduction to Lattices and Order.* Cambridge University Press, 1990.

[8] K. Došen. Duality between modal algebras and neighbourhood frames. *Studia Logica*, 48:219–234, 1989.

[9] M. Erne, J. Koslowski, A. Melton, and G. Strecker. A primer on Galois connections. In A. R. Todd, editor, *Papers on general topology and applications*. Madison, WI, 1991.

[10] D. Gabbay. An irreflexivity lemma with applications to axiomatization of conditions on tense frames. In U. Monnich, editor, *Aspects of Philosophical Logic*, pages 67–89. Reidel, 1981.

[11] D. Gabbay, A. Pnueli, S. Shelah, and J. Stavi. On the temporal analysis of fairness. In *POPL '80: Proceedings of the 7th ACM SIGPLAN-SIGACT symposium on Principles of programming languages*, pages 163–173, New York, NY, USA, 1980. ACM Press.

[12] N. Galatos and C. Tsinakis. Equivalence of consequence relations: a categorical and order-theoretic perspective. In preparation.

[13] G. Gargov and V. Goranko. Modal logic with names. *Journal of Philosophical Logic*, 22:607–636, 1993.

[14] G. Gargov, S. Passy, and T. Tinchev. Modal environment for Boolean speculations. In D. Skordev, editor, *Mathematical Logic and its Applications. Proceedings of the Summer School and Conference dedicated to the 80th Anniversary of Kurt Gödel (Druzhba, 1986)*, pages 253–263. Plenum Press, 1987.

[15] M. Kracht and F. Wolter. Simulation and transfer results in modal logic - a survey. *Studia Logica*, 59:229–259, 1997.

[16] M. Kracht and F. Wolter. Normal modal logics can simulate all others. *Journal of Symbolic Logic*, 64:99–138, 1999.
[17] A. Kudinov. Difference modality in topological spaces. A talk at Algebraic and Topological Methods in Non-Classical Logics II, June 15-18, 2005, Barcelona, Spain. Abstract available at http://atlas-conferences.com/c/a/p/u/76.htm.
[18] M. de Rijke. The modal logic of inequality. *Journal of Symbolic Logic*, 57:566–584, 1992.
[19] D. Sustretov. Hybrid definability in topological spaces. In *Proceedings of ESSLLI'2005 Student Session*, 2005.
[20] J.F.A.K. van Benthem. Syntactic aspects of modal incompleteness theorems. *Theoria*, 45:63–77, 1979.
[21] Y. Venema. Derivation rules as anti-axioms in modal logic. *Journal of Symbolic Logic*, 58:1003–1034, 1993.

Tadeusz Litak
School of Information Science, JAIST
Asahidai 1-1, Nomi-shi, Ishikawa-ken
923-1292 JAPAN
litak@jaist.ac.jp

Quantification over names and modalities

Eric Martin

ABSTRACT. Making quantification and modalities interact properly has raised numerous issues that have been tackled in various ways. We propose an alternative approach with quantifiers ranging over names (closed terms) rather than individuals. To be general enough and not restricted to dealing exclusively with Henkin or Herbrand structures, we distinguish between the language used to describe a structure, and the language used to talk about a structure. We assume that the former is rich enough to provide designators for all individuals, while the latter might be a subset of the former, allowing one to make some individuals "unspeakable of," as is needed for the developments of classical logic. These considerations are applied to Kripke semantics, and classical results—the Löwenheim Skolem and compactness theorems—are revisited. With quantification over names, the results are more general and the proofs simpler, being based on restrictions rather than on extensions of the original vocabulary.

1 Introduction

Providing a suitable semantics for first-order modal logic has proved to be a challenging enterprise. The issues have been extensively discussed in [2], and a solution proposed. The crux of the matter is how to let quantifiers interact with modalities. In a formula of the form $\exists x \Box \psi(x)$, some object o is selected in the current possible world M, and asserted to have property ψ in all possible worlds accessible from M. But o might not exist in some of those worlds. In the constant domain semantics with rigid designators, the matter is simple. It is also simple with the generalization provided by intuitionistic Kripke frames, where the domain of a possible world that is accessible from M has to contain the domain of M (see [7] for a recent use of these frames). But for arbitrary domains and interpretations of the function symbols that vary from possible world to possible world, the issues are numerous. To resolve these issues, distinctions between existence of objects and designation of objects have been carefully unravelled ([2]), or an 'actually' operator has been introduced ([11]). In this paper we take another path, and suggest that quantification could range over names rather than individuals. This is not a new proposal. It goes back to the work of Ruth Barcan-Marcus [8], and has been embraced or investigated by many researchers. See for instance [1, 3, 4, 5, 6] for various developments. But when quantification over names is used, it is assumed that the domains of the intended interpretations are such that all individuals have names, or even that every individual has a unique name. This assumption is suitable in some contexts, like when dealing with the Herbrand semantics of logic programs, but inappropriate for classical semantics.

If we consider canonical models (structures all of whose individuals interpret a closed term), that arise from Henkin's proof of the completeness of first-order logic (see *e.g.* [10]), it is clear that models where all objects have a name play a natural role even in the realm of classical logic. They are obtained by *enriching* the initial vocabulary with enough constants to provide witnesses for existential formulas. Why not proceed in the opposite way, by starting with a language that is rich enough to denote all individuals in any intended interpretation, and then *restricting* the vocabulary so that some individuals cannot be named, if required? In other words, distinguish between the language used to describe a structure and the language used to talk about a structure. Canonical models are particularly interesting for another reason: they identify individuals with the set of all names for this individual. This concept is clear enough when only one canonical structure is involved, but what if many canonical structures are considered at once? Then we might encounter sets of names that are distinct but have a nonempty intersection. The implications of this observation suggest to consider, besides the usual mathematical structures, more abstract, 'linguistic' structures that consist of a set of constraints to denote objects and interpret predicates, and use the more abstract structures as the components of a Kripke frame.

We proceed as follows. In Section 2 we develop these considerations in more detail. In Section 3, we provide the formal machinery and express the natural relationships between mathematical structures, built from sets of individuals, and abstract structures, that generalize Herbrand structures and can be thought of as sets of linguistic constraints in the use of a set of names. In Section 4 we let abstract structures be the components of a Kripke frame, yielding abstract Kripke frames. We then put classical Kripke frames in relation with abstract Kripke frames, and propose a semantics based on this relationship. In Section 5 we show how the Löwenheim–Skolem and compactness results can be reformulated in the context of abstract Kripke frames. We conclude in Section 6.

2 Two kinds of structures

From now on $\mathcal{V}$ denotes a *vocabulary*, that is, a set of function and predicate symbols. For reasons that will soon become clear, we depart from the usual practice of including only nonlogical predicate symbols in $\mathcal{V}$, and we require that the distinguished predicate symbol = be a member of $\mathcal{V}$ when dealing with a language with equality, and that = not be a member of $\mathcal{V}$ when dealing with a language without equality. In this section, a *formula* over $\mathcal{V}$ can refer to a member of $\mathcal{L}_{\omega\omega}(\mathcal{V})$ (the set of first-order formulas over $\mathcal{V}$), or to a member of $\mathcal{L}_{\omega_1\omega}(\mathcal{V})$ (the extension of $\mathcal{L}_{\omega\omega}(\mathcal{V})$ that accepts countable disjunctions or conjunctions of statements), the latter set playing a role in the second part of the paper. Recall that a mathematical structure M over

$\mathcal{V}$ is a pair, with the first element of the pair being the domain of M, denoted $|M|$, and the second element of the pair being the set of interpretations in M of the function and predicate symbols in $\mathcal{V}$.

The usual proof of the completeness theorem for classical first-order logic is based on the construction (Henkin's construction) of a special mathematical structure whose domain is a partition of the set of closed terms over some extension V of the original vocabulary, with the members of a given individual being viewed as the possible names for that individual; such a mathematical structure, which is said to be *Henkin*, can then be simply represented as a pair consisting of a partition of the set of closed terms over V and a set of closed atoms over V. The *Herbrand structures* considered in Logic programming are a particular kind of mathematical structure over a vocabulary V that contains at least one constant, where every individual interprets a unique closed term, viewed as the unique name of that individual, and is usually identified with it; a Herbrand structure can then be simply represented as a set of closed atoms over V (this set being referred to as the *Herbrand base*), and identified with a Henkin structure whose individuals are singletons.

Take a partition U of the set of closed terms over $\mathcal{V}$ and a set of B of closed atoms over $\mathcal{V}$. As we have reminded, the pair (U, B) is usually identified with a (Henkin) mathematical structure, but we can take a different perspective. Rather than identifying (U, B) with a mathematical representation of some form of reality, we can view (U, B) as a purely syntactic, linguistic object, a blueprint from which mathematical representations of some forms of reality can be obtained. More precisely, given a vocabulary $\mathcal{V}^\star \subseteq \mathcal{V}$, the pair (U, B) can act as a blueprint for any mathematical representation of some form of reality, or for any mathematical structure M over $\mathcal{V}^\star$ that is *described* by (U, B) in the following sense:

- any individual ι of M is given at least one name, *i.e*, is associated with at least one closed term over $\mathcal{V}$, with a unique member of U being the set of names given to ι;

- a closed term t over $\mathcal{V}^\star$ is one of the names given to the unique individual of M that interprets t;

- a closed term of the form $f(t_1, \ldots, t_n)$ is such that if individuals $\iota_1, \ldots, \iota_n$ have respective names $t_1, \ldots, t_n$, then $f(t_1, \ldots, t_n)$ is a name of the individual $f^M(\iota_1, \ldots, \iota_n)$;

- a closed atom over $\mathcal{V}^\star$ of the form $p(t_1, \ldots, t_n)$ belongs to B iff the individuals of M denoted by $t_1, \ldots, t_n$ are in relation p^M in M.

From this perspective, a closed formula φ over $\mathcal{V}$ cannot be true in (U, B)—that notion is meaningless if (U, B) does not mathematically represent some form of reality, but just offers a blueprint for such a representation: φ can only be believed or *accepted* in (U, B), in the sense that for any reality and for any representation of that reality in the form of a mathematical structure M over $\mathcal{V}$, if M is described by (U,B) then φ is true in M. We will refer to (U,B), viewed as we have just described, as an *abstract structure* over $\mathcal{V}$. The distinction is futile in classical first-order logic, and in Logic programming for the particular case of Herbrand mathematical structures. We claim that it is beneficial to make the distinction in Kripke semantics. This is because in Kripke semantics, the basic notion of truth involves a whole set of structures rather than a unique one. To make this point clearer, assume that the only function symbols in $\mathcal{V}$ are four constants a,b,c,d, and consider the following sets:

$$U_1 = \{\{a,b\},\{c,d\}\} \quad U_2 = \{\{a\},\{b\},\{c,d\}\} \quad U_3 = \{\{a\},\{b,c\},\{d\}\}.$$

First consider three mathematical structures M_1, M_2 and M_3 over $\mathcal{V}$ with respective domains U_1, U_2 and U_3, such that for all $M \in \{M_1, M_2, M_3\}$, $\iota \in |M|$ and $x \in \{a,b,c,d\}$, ι is the interpretation of x in M iff $x \in \iota$. So M_1, M_2 and M_3 are Henkin mathematical structures over $\mathcal{V}$. The view that the individual $\{a,b\}$ of M_1 has two names a and b, the individual $\{a\}$ of M_2 has one name a, etc., is not problematic as long as M_1, M_2 and M_3 are considered independently. But consider M_1, M_2 and M_3 together. Since the sets $\{a,b\}$, $\{a\}$ and $\{b\}$ are three distinct mathematical objects, they are different individuals. How can we then have the absolute view that a and b are the names of $\{a,b\}$, and also that a is the name of $\{a\}$ and b the name of $\{b\}$? How can different objects share common names? How can two constants be the two names of some individual, and also the respective names of two distinct individuals? It is conceptually much simpler to assume that a and b are not the names of the individual $\{a,b\}$ of M_1, that a is not the name of the individual $\{a\}$ of M_2, etc: as far as the activity of naming individuals is concerned, the individuals $\{a,b\}$ and $\{a\}$ are no more special than the numbers 2 and π: all are mathematical objects. It is conceptually much simpler to assume that in some contexts, some agent names the individual $\{a,b\}$ of M_1 as either a or b, in some contexts, some agent names the individual $\{a\}$ of M_2 as a, and in some contexts, some agent names the individual $\{b\}$ of M_3 as b. So the point is that it is fine to consider M_1, M_2 and M_3 independently and view $\{a,b\}$ and $\{c,d\}$, or $\{a\}$, $\{b\}$ and $\{c,d\}$, or $\{a\}$, $\{b,c\}$ and $\{d\}$ both as individuals and sets of names. But it is difficult to keep that absolute view and consider M_1, M_2 and M_3 together, without relativizing and externalizing the notion of "being the name of an individual." This is precisely what can be achieved with the notion of an *abstract structure* describing a mathematical structure, as we now explain.

When dealing with abstract structures over $\mathcal{V}$, we conceive of the members of $\mathcal{V}$ as names of functions and relations. But when dealing with mathematical structures over $\mathcal{V}$, we conceive of the members of $\mathcal{V}$ as just placeholders to refer to unnamed functions and relations. This means that "$\{c,d\}$ is the interpretation of c in M_1" only says that some nullary function in M_1 exists and maps $\emptyset$ to the individual $\{c,d\}$. To better express our intentions, we could use distinct kinds of vocabularies to define mathematical structures and abstract structures. The vocabularies involved in the definition of mathematical structures could be sets of symbols like

$$\{-_0^{f,0}, -_1^{f,0}, -_2^{f,2}, -_3^{p,1}, -_4^{p,1}\}$$

with $-_0^{f,0}$ and $-_1^{f,0}$ being placeholders for two nullary functions, $-_2^{f,2}$ being a placeholder for a binary function, and $-_3^{p,1}$ and $-_4^{p,1}$ being two placeholders for two unary predicates. These placeholders, being devoid of any meaning, would make it easier to conceptualize that we are only pointing to mathematical objects—functions, relations—but not naming them. Vocabularies of the form that we commonly use would only be involved in the definition of abstract structures, to name from the outside the mathematical objects that make up the domain of a mathematical structure.

3 Abstract structures

Now for the formal details. The notion of abstract structure is defined in two steps:[1]

DEFINITION 1. A *labeling partition over* $\mathcal{V}$ is a partition X of the set of closed terms over $\mathcal{V}$ with the following properties.

- If $\mathcal{V}$ does not contain equality then all members of X are singletons.

- Let $n \in \mathbb{N}$, n-ary function symbol f in $\mathcal{V}$, and closed terms $t_1, \ldots, t_n$, $t'_1, \ldots, t'_n$ over $\mathcal{V}$ be such that for all non-null $i \leq n$, t_i and t'_i belong to the same member of X. Then $f(t_1, \ldots, t_n)$ and $f(t'_1, \ldots, t'_n)$ also belong to the same member of X.

The first clause in Definition 1 allows one a smooth treatment of languages with or without equality. Note that if $\mathcal{V}$ contains no constant then $\emptyset$ is the unique labeling partition over $\mathcal{V}$.

DEFINITION 2. An *abstract structure* (*over* $\mathcal{V}$) is a pair $\mathfrak{M}$ consisting of:

- a labeling partition over $\mathcal{V}$, called the *labeling partition of* $\mathfrak{M}$, whose members are referred to as *labeling classes* (*of* $\mathfrak{M}$);

- a set of closed atoms over $\mathcal{V}$, called the *acceptance set of* $\mathfrak{M}$, having the property that for all $n \in \mathbb{N}$, n-ary predicate symbols p in $\mathcal{V}$ distinct from equality, and closed terms $t_1, \ldots, t_n, t'_1, \ldots, t'_n$ over $\mathcal{V}$, if

    - $p(t_1, \ldots, t_n)$ belongs to the acceptance set of $\mathfrak{M}$, and
    - for all nonull $i \leq n$, t_i and t'_i belong to the same labeling class

    then $p(t'_1, \ldots, t'_n)$ belongs to the acceptance set of $\mathfrak{M}$.

So given an abstract structure $\mathfrak{M}$, $\mathfrak{M}(0)$ is the labeling partition of $\mathfrak{M}$ and $\mathfrak{M}(1)$ is the acceptance set of $\mathfrak{M}$.

---
[1] $\mathbb{N}$ denotes the set of natural numbers.

The correspondence between mathematical structures and abstract structures is defined as expected.

DEFINITION 3. Let an abstract structure $\mathfrak{M}$ over $\mathcal{V}$ be given.

Let a vocabulary $\mathcal{V}^\star \subseteq \mathcal{V}$ and a mathematical structure M over $\mathcal{V}^\star$ be given. A *homomorphism from* $\mathfrak{M}$ *into* M is a surjective mapping h from $\bigcup \mathfrak{M}(0)$ into the domain of M, with the following properties.

- For all closed terms t_1, t_2 over $\mathcal{V}$, $h(t_1) = h(t_2)$ iff t_1 and t_2 belong to the same member of $\mathfrak{M}(0)$.
- For all $n \in \mathbb{N}$, n-ary function symbols f in $\mathcal{V}^\star$, and closed terms $t_1, \ldots, t_n$ over $\mathcal{V}$, $h(f(t_1, \ldots, t_n)) = f^M(h(t_1), \ldots, h(t_n))$.
- For all $n \in \mathbb{N}$, n-ary predicate symbols p in $\mathcal{V}^\star$ distinct from $=$, and closed terms $t_1, \ldots, t_n$ over $\mathcal{V}$, $p(t_1, \ldots, t_n)$ belongs to $\mathfrak{M}(1)$ iff $(h(t_1), \ldots, h(t_n))$ belongs to p^M.

Given a vocabulary $\mathcal{V}^\star$ and a mathematical structure M over $\mathcal{V}^\star$, we say that $\mathfrak{M}$ *describes* M iff $\mathcal{V}^\star \subseteq \mathcal{V}$ and a homomorphism from $\mathfrak{M}$ into M exists.

For instance, assume that $\mathcal{V}$ is defined as in Section 2, and recall the mathematical structures M_1, M_2 and M_3 and the sets U_1, U_2 and U_3 defined in Section 2. Consider three abstract structures $\mathfrak{M}_1$, $\mathfrak{M}_2$ and $\mathfrak{M}_3$ over $\mathcal{V}$ of respective form (U_1, B_1), (U_2, B_2) and (U_3, B_3). Then for all $i \in \{1, 2, 3\}$, M_i could be described by $\mathfrak{M}_i$, but it could not be described by $\mathfrak{M}_j$ for $j \in \{1, 2, 3\} \setminus \{i\}$ (whether M_i is actually described by $\mathfrak{M}_i$ depends of course on B_i and on how the predicate symbols in $\mathcal{V}$ are interpreted in M_i). Each of $\mathfrak{M}_1$, $\mathfrak{M}_2$ and $\mathfrak{M}_3$ could describe other mathematical structures over $\mathcal{V}$. For instance, $\mathfrak{M}_1$ could describe a mathematical structure over $\mathcal{V}$ whose domain is U_1 where both a and b are interpreted by $\{c, d\}$ and both c and d are interpreted by $\{a, b\}$; $\mathfrak{M}_2$ could describe a mathematical structure over $\mathcal{V}$ whose domain is U_3 where a is interpreted by $\{a\}$, b is interpreted by $\{b, c\}$ and both c and d are interpreted by $\{d\}$; $\mathfrak{M}_1$ could describe a mathematical structure over $\mathcal{V}$ whose domain is $\{2, 4\}$, where both a and b are interpreted by 2 and both c and d are interpreted by 4. Moreover, each of $\mathfrak{M}_1$, $\mathfrak{M}_2$ and $\mathfrak{M}_3$ could describe mathematical structures over a vocabulary strictly included in $\mathcal{V}$. For instance, let $\mathcal{V}^\star$ be the union of $\{a, b, d\}$ with the set of predicate symbols in $\mathcal{V}$. Then $\mathfrak{M}_2$ could describe a mathematical structure over $\mathcal{V}^\star$ whose domain is U_2 where a is interpreted by $\{a\}$, b is interpreted by $\{b\}$, and d is interpreted by $\{c, d\}$; also, $\mathfrak{M}_2$ could describe a mathematical structure over $\mathcal{V}^\star$ whose domain is U_3 where a is interpreted by $\{a\}$, b is interpreted by $\{b, c\}$, and d is interpreted by $\{d\}$. More generally, an abstract structure over $\mathcal{V}$ can describe many mathematical structures over a vocabulary that is included in $\mathcal{V}$, and a mathematical structure over a vocabulary that is included in $\mathcal{V}$ can be described by many abstract structures over $\mathcal{V}$.

The next property states that all mathematical structures can be represented by abstract structures. Though not every individual of a mathematical structure M over $\mathcal{V}$ has a name built from the function symbols in $\mathcal{V}$, every individual of M can still be assumed to have a name that is possibly "hidden,", *i.e.*, not expressible with $\mathcal{V}$. Distinguishing between the vocabulary that describes a mathematical structure and the vocabulary that is used to make formulas about it is the key to defining the semantics of classical logic on the basis of abstract structures rather than on the basis of mathematical structures.

PROPERTY 4. Assume that $\mathcal{V}$ contains equality. Let a mathematical structure M over $\mathcal{V}$ be given. Let F be a set of function symbols not in $\mathcal{V}$ such that the set of closed terms over $\mathcal{V} \cup F$ of the form $f(t_1, \ldots, t_n)$, where $n \in \mathbb{N}$ and f is an n-ary function symbol in F of arity n, has cardinality at least equal to the cardinality of M. Then M is described by at least one abstract structure over $\mathcal{V} \cup F$.

COROLLARY 5. *Assume that $\mathcal{V}$ contains equality. Let a mathematical structure M over $\mathcal{V}$ be given. Let F be a set of constants not in $\mathcal{V}$ of cardinality at least equal to the cardinality of M. Then M is described by at least one abstract structure over $\mathcal{V} \cup F$.*

The previous property and corollary do not generalize to vocabularies that do not contain equality. This is because if $\mathcal{V}$ does not contain equality, then for every mathematical structure M over $\mathcal{V}$ that is described by an abstract structure over $\mathcal{V}$, two distinct closed terms over $\mathcal{V}$ have to be interpreted in M by two distinct individuals. This is at no real loss of generality, and it is technically convenient.

Given an abstract structure $\mathfrak{M}$ over $\mathcal{V}$ and a closed member φ of $\mathcal{L}_{\omega_1\omega}(\mathcal{V})$, we can define the notion $\mathfrak{M} \models \varphi$ either directly from Definition 2, or indirectly from the relationship between formal and mathematical structures. Directly:

DEFINITION 6. Let an abstract structure $\mathfrak{M}$ over $\mathcal{V}$ be given. Given a closed member φ of $\mathcal{L}_{\omega_1\omega}(\mathcal{V})$, the notion "$\varphi$ is accepted in $\mathfrak{M}$," or "$\mathfrak{M}$ is a model of φ," denoted $\mathfrak{M} \models \varphi$, is inductively defined as follows.

- Suppose that $\mathcal{V}$ contains equality, and let closed terms t_1, t_2 over $\mathcal{V}$ be given. Then:
    - $\mathfrak{M} \models t_1 = t_2$ if t_1 are t_2 belong to the same labeling class of $\mathfrak{M}$;
    - $\mathfrak{M} \models t_1 \neq t_2$ if t_1 are t_2 belong to distinct labeling classes of $\mathfrak{M}$.
- Let $n \in \mathbb{N}$, an n-ary predicate symbol p in $\mathcal{V}$, and closed terms $t_1, \ldots, t_n$ over $\mathcal{V}$ be given. Then:
    - $\mathfrak{M} \models p(t_1, \ldots, t_n)$ if $p(t_1, \ldots, t_n) \in \mathfrak{M}(1)$;
    - $\mathfrak{M} \models \neg p(t_1, \ldots, t_n)$ if $p(t_1, \ldots, t_n) \notin \mathfrak{M}(1)$.
- Let a set X of closed members of $\mathcal{L}_{\omega_1\omega}(\mathcal{V})$ be given. Then:
    - $\mathfrak{M} \models \bigvee X$ iff there exists $\psi \in X$ such that $\mathfrak{M} \models \psi$;
    - $\mathfrak{M} \models \bigwedge X$ iff for all $\psi \in X$, $\mathfrak{M} \models \psi$.

- Let $\psi \in \mathcal{L}_{\omega_1\omega}(\mathcal{V})$ and variable x be such that $\mathrm{fv}(\psi) = \{x\}$. Then:
  - $\mathfrak{M} \models \exists x\psi$ iff $\mathfrak{M} \models \psi[t/x]$ for some closed term t over $\mathcal{V}$;
  - $\mathfrak{M} \models \forall x\psi$ iff $\mathfrak{M} \models \psi[t/x]$ for all closed terms t over $\mathcal{V}$.

Indirectly:

PROPERTY 7. *For all abstract structures $\mathfrak{M}$ over $\mathcal{V}$, vocabularies $\mathcal{V}^\star \subseteq \mathcal{V}$ and closed members φ of $\mathcal{L}_{\omega_1\omega}(\mathcal{V}^\star)$, the following conditions are equivalent.*

- $\mathfrak{M} \models \varphi$.
- $\mathfrak{M}$ describes some mathematical structure M over $\mathcal{V}^\star$ with $M \models \varphi$.
- For all mathematical structures M over $\mathcal{V}^\star$ that $\mathfrak{M}$ describes, $M \models \varphi$.

We write $\mathfrak{M} \not\models \varphi$ if φ is not accepted in $\mathfrak{M}$. Also, we will use the notation $\mathfrak{M} \models T$ and $\mathfrak{M} \not\models T$ for subsets T of $\mathcal{L}_{\omega_1\omega}(\mathcal{V})$.

Recall that if $\mathcal{V}$ does not contain equality then the labeling classes of a abstract structure over $\mathcal{V}$ are singletons. This constraint does not have any implication on the notion of a formula being accepted in an abstract structure:

LEMMA 8. *Assume that $\mathcal{V}$ does not contain equality. Let an abstract structure $\mathfrak{M}$ over $\mathcal{V} \cup \{=\}$ and a closed $\varphi \in \mathcal{L}_{\omega_1\omega}(\mathcal{V})$ be given. Then φ is accepted in $\mathfrak{M}$ iff φ is accepted in the (unique) abstract structure over $\mathcal{V}$ that has the same acceptance set as $\mathfrak{M}$.*

COROLLARY 9. *Assume that $\mathcal{V}$ does not contain equality, and let φ be a closed member of $\mathcal{L}_{\omega_1\omega}(\mathcal{V})$. Then φ is accepted in an abstract structure over $\mathcal{V}$ iff φ is accepted in an abstract structure over $\mathcal{V} \cup \{=\}$.*

The choice of "accepted" in Definition 6 is more convenient than dictated by any epistemic considerations. It emphasizes that the relationship between a formula and an abstract structure can be kept at the linguistic level, with no need to refer to a mathematical model of some form of reality. At a formal level, the differences and connections between abstract structures and mathematical structures are straightforward. The formalization that supports these differences and connections would remain a rather sterile exercise if it was not utilized in a modal framework, where in contrast to the classical framework of first-order logic, structures are considered in relation to each other. The key point is in the range of the variables. Variables range over individuals w.r.t. mathematical structures, but they should be viewed as ranging over names (not labeling classes!), w.r.t. abstract structures. Let a closed member of $\mathcal{L}_{\omega_1\omega}(\mathcal{V})$ of the form $\exists x\varphi$ be given. If M is a mathematical structure over $\mathcal{V}$, the intended interpretation of $M \models \exists x\varphi$ is that one of M's individuals has property φ. If $\mathfrak{M}$ is an abstract structure over $\mathcal{V}$, the intended interpretation of $\mathfrak{M} \models \exists x\varphi$ is that for some name (some closed term over $\mathcal{V}$) t, the formula $\varphi[t/x]$ is accepted in $\mathfrak{M}$, meaning that for all mathematical structures M over $\mathcal{V}$, if M is described by $\mathfrak{M}$ then the individual of M which is named t by $\mathfrak{M}$, has property φ. This is crucially important when dealing simultaneously with mathematical structures that have arbitrary domains: a variable that denotes a particular individual u in a particular mathematical structure is comparable to a pointer that has been assigned the value u when viewed from a mathematical structure whose domain contains u, but it is comparable to a void pointer when viewed from any other mathematical structure. On the other hand, a variable that denotes a particular name of a particular abstract structure $\mathfrak{M}$ can point to any individual of any mathematical structure M that can be described by $\mathfrak{M}$: it suffices to let $\mathfrak{M}$ properly describe M.

Most logical notions can be defined on the basis of abstract structures; formally, mathematical structures are not really needed. Mathematical structures can be viewed as lower level interpretations, closer to reality. More precisely, mathematical structures provide a way to reinterpret the notion that a closed member of $\mathcal{L}_{\omega_1\omega}(\mathcal{V})$ is accepted in an abstract structure over $\mathcal{V}$, in terms of the classical notion of truth.

4 Abstract Kripke frames

Abstract structures provide a real benefit in a modal-theoretic context. It suffices to let abstract structures rather than mathematical structures play the role of possible worlds in Kripke semantics. More precisely:

DEFINITION 10. An *abstract Kripke frame over* $\mathcal{V}$ is a pair (K, R) where K is a set of abstract structures over $\mathcal{V}$, and R is a binary relation over K whose domain is K.

An abstract Kripke frame (K, R) is said to have *constant labeling* iff all members of K have the same labeling partition.

We impose the weak restriction that the domain of R is equal to K in order to simplify the formal developments, but that assumption is otherwise inessential. Of course, the notion of an abstract Kripke frame having a constant labeling is the counterpart to the notion of a Kripke frame having constant domain.

To be precise enough, let us define the language we will be using in the rest of the paper.

DEFINITION 11. The set $\mathcal{L}^{\text{mod}}_{\omega_1\omega}(\mathcal{V})$ of (*infinitary*) *modal formulas over* $\mathcal{V}$ is inductively defined as follows.

- All atoms over $\mathcal{V}$ and their negations belong to $\mathcal{L}^{\text{mod}}_{\omega_1\omega}(\mathcal{V})$.

- For all countable $X \subseteq \mathcal{L}^{\text{mod}}_{\omega_1\omega}(\mathcal{V})$, $\bigvee X$ and $\bigwedge X$ belong to $\mathcal{L}^{\text{mod}}_{\omega_1\omega}(\mathcal{V})$.

- For all variables x and for all $\varphi \in \mathcal{L}^{\text{mod}}_{\omega_1\omega}(\mathcal{V})$, $\exists x\varphi$ and $\forall x\varphi$ belong to $\mathcal{L}^{\text{mod}}_{\omega_1\omega}(\mathcal{V})$.

- For all $\varphi \in \mathcal{L}^{\text{mod}}_{\omega_1\omega}(\mathcal{V})$, $\Diamond\varphi$ and $\Box\varphi$ both belong to $\mathcal{L}^{\text{mod}}_{\omega_1\omega}(\mathcal{V})$.

If disjunctions and conjunctions are applied to finite sets only, then the resulting set of (finite) modal formulas is denoted $\mathcal{L}^{\text{mod}}_{\omega\omega}(\mathcal{V})$.

In the definition of both $\mathcal{L}_{\omega_1\omega}(\mathcal{V})$ and $\mathcal{L}_{\omega\omega}(\mathcal{V})$, negation is assumed to be applicable to atomic formulas only, which amounts to imposing a negation normal form (to negate arbitrary modal formulas, it suffices to inductively define an operator $\sim$ with $\sim\Box\varphi = \Diamond\sim\varphi$, $\sim\Diamond\varphi = \Box\sim\varphi$, $\sim\exists x\varphi = \forall x\sim\varphi$, $\sim\forall x\varphi = \exists x\sim\varphi$, $\sim\neg\varphi = \varphi$, and for atomic φ's, $\sim\varphi = \neg\varphi$). The usual syntax on finite formulas is introduced as abbreviations: $\varphi_1 \vee \varphi_2$ for $\bigvee\{\varphi_1, \varphi_2\}$, $\varphi_1 \wedge \varphi_2$ for $\bigwedge\{\varphi_1, \varphi_2\}$, etc. These decisions on the primitive syntax result in no loss in generality, but simplify the formal developments. The set of all variables that occur free in a modal formula φ is denoted $\mathrm{fv}(\varphi)$.

Given an abstract Kripke frame $\mathcal{K} = (K, R)$ and a member $\mathfrak{M}$ of K, the notion of a closed modal formula φ being *accepted in* $(\mathcal{K}, \mathfrak{M})$, or of $(\mathcal{K}, \mathfrak{M})$ being a *model* of φ, denoted $(\mathcal{K}, \mathfrak{M}) \Vdash \varphi$, is defined in the usual way from Definition 6, the key clauses still being those that deal with the quantifiers. Of course, we have to accept to use the same vocabulary to define all the abstract structures involved in a given abstract Kripke frame.

We see the benefit of variables ranging over closed terms rather than individuals, and of Kripke frames defined from abstract structures rather than from mathematical structures. We avoid some issues that have to be faced in order to interpret in the context of a (nonabstract) Kripke frame any closed formula that contains a nonclosed subformula of the form $\Diamond\psi$ or $\Box\psi$. In particular, both following issues are avoided, where $\mathcal{K} = (K, R)$ denotes a (nonabstract) Kripke frame and M a member of K.

First issue: A natural intended interpretation of $\exists x \Diamond \psi$ in $(\mathcal{K}, M)$ is that some individual ι in the domain of M has property $\Diamond \psi$ in $(\mathcal{K}, M)$. But for all $N \in K$ with $R(M, N)$, the domain of N does not have to include the domain of M, hence might not contain ι.

Second issue: Assume that the domain of N is included in the domain of M for every member N of K with $R(M, N)$. Then the previous issue disappears. Now let ι in the domain of M be such that ι has property $\Diamond \psi$ of x in $(\mathcal{K}, M)$, and as a consequence, ι has property ψ of x in $(\mathcal{K}, N)$, for some member N of K with $R(M, N)$. Suppose that $\mathcal{V}$ contains a constant c, and that ι is the interpretation of c in M. A natural syntactic operation is the substitution operation that transforms $\Diamond \psi$ into $\Diamond \psi[c/x]$. But $\Diamond \psi[c/x]$ does not necessarily mean that the individual denoted by c in N has property ψ of x in $(\mathcal{K}, N)$. Indeed, it might be that c does not denote ι in N. In that case, substituting in $\Diamond \psi$ all free occurrences of x by c "twists" the meaning of $\Diamond \psi$.

Let us relate abstract Kripke frames to mathematical (classical) Kripke frames. We can generalize Definition 3 as follows.

DEFINITION 12. Let an abstract Kripke frame $\mathcal{K} = (K, R)$ over $\mathcal{V}$, a vocabulary $\mathcal{V}^\star \subseteq \mathcal{V}$, and a mathematical Kripke frame $\mathcal{K}^\star = (K^\star, R^\star)$ over $\mathcal{V}^\star$ be given.

A *homomorphism from* $\mathcal{K}$ *into* $\mathcal{K}^*$ is a bijective mapping h from K into K^* with the following properties.

- For all $\mathfrak{M} \in K$, $\mathfrak{M}$ describes $h(\mathfrak{M})$.
- For all $\mathfrak{M}, \mathfrak{N} \in K$, $R^*(h(\mathfrak{M}), h(\mathfrak{N}))$ holds iff $R(\mathfrak{M}, \mathfrak{N})$ holds.

A *full Kripke frame (based on $\mathcal{K}$)* is a pair of the form (K^*, R^*, h) where h is a homomorphism from $\mathcal{K}$ into $\mathcal{K}^*$.

Note that Property 7 does not generalize from structures to full Kripke frames. The semantics of modal logic can be defined w.r.t. a mathematical Kripke frame and one of its linguistic blueprints, with different linguistic blueprints for the same mathematical Kripke frame possibly yielding different logical notions. More precisely:

DEFINITION 13. Let an abstract Kripke frame $\mathcal{K}$ over $\mathcal{V}$, a vocabulary $\mathcal{V}^* \subseteq \mathcal{V}$, a mathematical Kripke frame $\mathcal{K}^* = (K^*, R^*)$ over $\mathcal{V}^*$, and a homomorphism h from $\mathcal{K}$ into $\mathcal{K}^*$ be given. Let $M \in K^*$ and $\varphi \in \mathcal{L}^{\text{mod}}_{\omega_1\omega}(\mathcal{V}^*)$ be given. We write $(\mathcal{K}^*, h, M) \Vdash \varphi$ if $(K, h^{-1}(M)) \Vdash \varphi$.

Let us illustrate the previous definition with a few examples. We can write that for all $M \in K^*$,

(†) $(\mathcal{K}^*, h, M) \Vdash \forall x \Box \exists y (x = y)$

because for all $\mathfrak{M} \in K$, $(\mathfrak{M}, \mathcal{K})$ is a model of $\forall x \Box \exists y (x = y)$. In other words, we do not read (†) as: for all $N \in K^*$ with $R^*(M, N)$, the domain of M is included in the domain of N. We read (†) as: w.r.t. the chosen linguistic blueprint $\mathcal{K}$ for K^*, every name that can be used to make statements about M can also be used to make statements about any $N \in K^*$ with $R^*(M, N)$ (actually, about any $N \in K^*$). Note that with this example, the linguistic blueprint for $\mathcal{K}^*$ can be chosen arbitrarily. Now assume that K^* consists of two mathematical structures M and N over $\mathcal{V}^*$, with the domain of N being strictly larger than the domain of M. Then

(‡) $(K^*, h, M) \Vdash \exists x \exists y ((x = y) \wedge \Diamond (x \neq y))$,

in the sense that if $\mathfrak{M}$ is the (unique) member of K that describes M then $(\mathfrak{M}, \mathcal{K})$ is a model of $\exists x \exists y ((x = y) \wedge \Diamond (x \neq y))$. In other words, we read (‡) as: w.r.t. the chosen linguistic blueprint $\mathcal{K}$ for $\mathcal{K}^*$, there exists two names that denote the same individual in M but different individuals in N. Note that with this example again, the linguistic blueprint for $\mathcal{K}^*$ can be chosen arbitrarily. Finally, assume that $\mathcal{V}^*$ contains no constant and K^* consists of two mathematical structures M and N over $\mathcal{V}^*$, with the domain of N being at most as large as the domain of M. Then

$(K^*, h, M) \Vdash \exists x \exists y ((x = y) \wedge \Diamond (x \neq y)).$

for some choices of h, whereas

$(K^*, h, M) \nVdash \exists x \exists y ((x = y) \wedge \Diamond (x \neq y)).$

for other choices of h. Indeed, there exists a linguistic blueprint for $\mathcal{K}^\star$ in two names denote the same individual in M but distinct individuals in N, and there exists a linguistic blueprint for $\mathcal{K}^\star$ in which no two names denote the same individual in M but distinct individuals in N.

To conclude, with quantification over names, we can develop modal logic either on the basis of abstract Kripke frames, or on the basis of mathematical (classical) Kripke frames together with a homomorphism from an abstract Kripke frame into the latter, that represents a set of underlying linguistic blueprints—with different sets of linguistic blueprints possibly resulting in different semantics.

5 Classical results revisited

The classical statement of the Löwenheim–Skolem theorem is: given a set X of closed finite formulas over $\mathcal{V}$, if all members of X are true in a mathematical structure over $\mathcal{V}$ then all members of X are true in a mathematical structure over $\mathcal{V}$ whose domain is countable. If we develop a logical framework on the basis of abstract structures rather than on the basis of mathematical structures, the Löwenheim–Skolem theorem needs to be reformulated. A natural reformulation is: given a set X of closed finite formulas over $\mathcal{V}$, if all members of X are accepted in an abstract structure over $\mathcal{V}$, then all members of X are accepted in an abstract structure over $\mathcal{V}^\star$, for some countable subset $\mathcal{V}^\star$ of $\mathcal{V}$ that contains all symbols which occur in φ. This claim is indeed correct, and will be a particular case of one of the main results to be proved in this section.

The Löwenheim–Skolem theorem has been generalized to $\mathcal{L}_{\omega_1\omega}(\mathcal{V})$. Let X be a set of closed infinitary formulas over $\mathcal{V}$. It is not the case that if all members of X are true in a mathematical structure over $\mathcal{V}$, then all members of X are necessarily true in a mathematical structure over $\mathcal{V}$ whose domain is countable. But $\mathcal{L}_{\omega_1\omega}(\mathcal{V})$ still enjoys a weak form of the Löwenheim–Skolem theorem in the sense that if X is countable, then the previous claim becomes correct. In the realm of abstract structures, this translates to: if X is a countable set of closed infinitary formulas over $\mathcal{V}$ and if all members of X are accepted in an abstract structure over $\mathcal{V}$, then all members of X are necessarily accepted in an abstract structure over $\mathcal{V}^\star$ for some countable subset $\mathcal{V}^\star$ of $\mathcal{V}$ that contains all symbols which occur in X. This will also be a particular case of one of the main results to be proved in this section. Of course, if X is countable then $\bigwedge X$ belongs to $\mathcal{L}_{\omega_1\omega}(\mathcal{V})$, and the claims about $\mathcal{L}_{\omega_1\omega}(\mathcal{V})$ are better expressed in terms of closed infinitary formulas rather than in terms of countable sets of closed infinitary formulas.

In this section, we show how to express the Löwenheim–Skolem theorem for $\mathcal{L}_{\omega_1\omega}^{\mathrm{mod}}(\mathcal{V})$ and both the class of all abstract Kripke frames over $\mathcal{V}$ and the class of all abstract Kripke frames over $\mathcal{V}$ that have constant labeling. Countability can then also be imposed on the number of abstract structures that make up the abstract Kripke frame. This revisiting of the Löwenheim–Skolem theorem will allow one to easily derive the corresponding versions of the compactness theorem for $\mathcal{L}_{\omega\omega}^{\mathrm{mod}}(\mathcal{V})$. These results can be derived from a common set of arguments, based on a variation of the usual argument based on the classical notions of tableaux and Hintikka sets. One of the interesting aspects of the revisited proof is that we need to restrict vocabularies rather than restricting domains. This means that Henkin's construction has to be changed. Rather than *enriching* the underlying vocabulary with a countable set of constants, suitable to define a mathematical structure over the extended vocabulary where every individual has a name, and providing potential witnesses for all existential formulas, we will *reduce* the underlying vocabulary, still making sure that enough function symbols remain so that potential witnesses for all existential formulas can be provided. This is achieved in the lemma below, based on the following definition that uses a form of *term subsumption*.

DEFINITION 14. Let a $\varphi \in \mathcal{L}_{\omega_1\omega}^{\mathrm{mod}}(\mathcal{V})$ and a closed term t over $\mathcal{V}$ be given. We say that t is *unrestricted in* φ iff for all terms t' over $\mathcal{V}$ and for all substitutions ν, if t' occurs in φ and $t = t'\nu$ then t' is a variable.

Intuitively, if t is unrestricted in φ then φ cannot make a statement that is specific to t: if φ is used to make a statement about t then for any closed term t', φ can also be used to make the same statement about t'. For instance, assume that $\mathcal{V}$ consists of a constant a, a unary function symbol s, a binary function symbol g, and a unary predicate symbol P.

- No closed term over $\mathcal{V}$ is unrestricted in

$$P(g(a,y)) \wedge P(g(g(x,y), g(x, s(s(s(x))))))$$

as every closed term over $\mathcal{V}$ is an instance of a, $s(x)$, or $g(x,y)$.

- a is the only closed term over $\mathcal{V}$ that is unrestricted in

$$P(x) \wedge P(s(s(y))) \wedge P(g(y,x))$$

as x and y are the only terms t that occur in the formula above and such that a is an instance of t.

- No closed term over $\mathcal{V}$ is unrestricted in

$$\bigwedge \{P(g(g(a,x),x_1)), P(g(g(s(x),x_0),x_1)),$$
$$P(g(g(g(a,x),x_1),x_2)), P(g(g(g(s(x),x_0),x_1),x_2))\ldots\}$$

as can be easily verified.

- Infinitely many closed terms over $\mathcal{V}$ are unrestricted in

$$P(s(s(a))) \wedge P(g(x,y)),$$

for instance, all terms of the form $s^n(a)$ with n (the number of successive applications of s) at least equal to 3.

- Infinitely many closed terms over $\mathcal{V}$ are unrestricted in

$$P(g(a,y)) \wedge P(g(s(x),y)) \wedge P(g(x,x)),$$

for instance, all terms of the form $g(g(a,a), s^n(a))$.

LEMMA 15. *Let a modal formula φ over $\mathcal{V}$, a variable x, an abstract Kripke frame $\mathcal{K} = (K, R)$ over $\mathcal{V}$, and a member $\mathfrak{M}$ of K be such that $\mathrm{fv}(\varphi) = \{x\}$ and $(\mathcal{K}, \mathfrak{M}) \Vdash \exists x \varphi$. Assume that infinitely many closed terms over $\mathcal{V}$ are unrestricted in φ. Let a closed term t over $\mathcal{V}$ be unrestricted in φ. Then there exists an abstract Kripke frame $\mathcal{K}^\star = (K^\star, R^\star)$ over $\mathcal{V}$ and a member $\mathfrak{M}^\star$ of $K^\star$ with the following properties.*

- $(\mathcal{K}^\star, \mathfrak{M}^\star) \Vdash \varphi[t/x]$.

- *If $\mathcal{K}$ has constant labeling then $\mathcal{K}^\star$ has constant labeling.*

Proof. Let X denote the set of closed terms over $\mathcal{V}$ that are unrestricted in φ. Let a closed term v over $\mathcal{V}$ be such that $(\mathcal{K}, \mathfrak{M}) \Vdash \varphi[v/x]$. Let $u \in X$ be given. Let h be a bijection from $X \setminus \{u\}$ into X. Let $h^\star$ be the unique total mapping from the set of closed terms over $\mathcal{V}$ into the set of closed terms over $\mathcal{V}$ that satisfies the following conditions.

- For all $t \in X \setminus \{u\}$, $h^\star(t) = h(t)$.

- $h^\star(u) = v$.

- For all $n \in \mathbb{N}$, n-ary function symbols f in $\mathcal{V}$, and closed terms $t_1, \ldots, t_n$ over $\mathcal{V}$, if $f(t_1, \ldots, t_n) \notin X$ then $h^\star(f(t_1, \ldots, t_n))$ is equal to $f(h^\star(t_1), \ldots, h^\star(t_n))$.

Now for all $\mathfrak{N} \in K$, let $\mathfrak{N}^\star$ be the unique structure over $\mathcal{V}$ that satisfies the following conditions.

- Assume that $\mathcal{V}$ contains equality. Let closed terms t_1, t_2 over $\mathcal{V}$ be given. Then t_1 and t_2 belong to the same member of $\mathfrak{N}^\star(0)$ iff $h^\star(t_1)$ and $h^\star(t_2)$ belong to the same member of $\mathfrak{N}(0)$.

- Let an $n \in \mathbb{N}$, an n-ary predicate symbol p over $\mathcal{V}$ distinct from equality, and closed terms $t_1, \ldots, t_n$ over $\mathcal{V}$ be given. Then $p(t_1, \ldots, t_n)$ belongs to $\mathfrak{N}^\star(1)$ iff $p(h^\star(t_1), \ldots, h^\star(t_n))$ belongs to $\mathfrak{N}(1)$.

Put $K^\star = \{\mathfrak{N}^\star \mid \mathfrak{N} \in K\}$. Let $R^\star$ be the accessibility relation on $K^\star$ such that for all $\mathfrak{S}_1, \mathfrak{S}_2 \in K^\star$, $R^\star(\mathfrak{S}_1, \mathfrak{S}_2)$ holds iff there exists $\mathfrak{N}_1, \mathfrak{N}_2 \in K$ such that $\mathfrak{S}_1 = \mathfrak{N}_1^\star$, $\mathfrak{S}_2 = \mathfrak{N}_2^\star$, and $R(\mathfrak{N}_1, \mathfrak{N}_2)$ holds. It is easy to verify that $\mathcal{K}^\star$ has constant labeling if $\mathcal{K}$ has constant labeling. To complete the proof of the proposition, it suffices to show the following.

(∗) Let a subformula ψ of φ, $n \in \mathbb{N}$, distinct variables $x_1, \ldots, x_n$, closed terms $t_1, \ldots, t_n$ over $\mathcal{V}$, and a member $\mathfrak{M}$ of K be such that $\mathrm{fv}(\psi) = \{x_1, \ldots, x_n\}$. Then the formula $\psi[t_1/x_1, \ldots, t_n/x_n]$ is accepted in $(\mathcal{K}, \mathfrak{M})$ iff the formula $\psi[h^\star(t_1)/x_1, \ldots, h^\star(t_n)/x_n]$ is accepted in $(\mathcal{K}^\star, \mathfrak{M}^\star)$.

Let $m, n \in \mathbb{N}$, m-ary predicate symbol p in $\mathcal{V}$, distinct variables $x_1, \ldots, x_n$, closed terms $t_1, \ldots, t_n$, and terms $t'_1, \ldots, t'_m$ be such that $p(t'_1, \ldots, t'_m)$ is a subformula of φ and $x_1, \ldots, x_n$ are the variables that occur in at least one of $t'_1, \ldots, t'_m$. Then

$$p(h^\star(t'_1[t_1/x_1, \ldots, t_n/x_n]), \ldots, h^\star(t'_m[t_1/x_1, \ldots, t_n/x_n])) =$$
$$p(t'_1[h^\star(t_1)/x_1, \ldots, h^\star(t_n)/x_n], \ldots, t'_m[h^\star(t_1)/x_1, \ldots, h^\star(t_n)/x_n]).$$

Since
$$p(t'_1, \ldots, t'_m)[t_1/x_1, \ldots, t_n/x_n] =$$
$$p(t'_1[t_1/x_1, \ldots, t_n/x_n], \ldots, t'_m[t_1/x_1, \ldots, t_n/x_n])$$

and
$$p(t'_1, \ldots, t'_m)[h^\star(t_1)/x_1, \ldots, h^\star(t_n)/x_n] =$$
$$p(t'_1[h^\star(t_1)/x_1, \ldots, h^\star(t_n)/x_n], \ldots, t'_m[h^\star(t_1)/x_1, \ldots, h^\star(t_n)/x_n])$$

it follows that for all members $\mathfrak{M}$ of K, the following are equivalent.

- $p(t'_1, \ldots, t'_m)[t_1/x_1, \ldots, t_n/x_n]$ is accepted in $\mathfrak{M}$;
- $p(t'_1, \ldots, t'_m)[h^\star(t_1)/x_1, \ldots, h^\star(t_n)/x_n]$ is accepted in $\mathfrak{M}^\star$.

With the fact that $h^\star$ is onto the set of closed terms over $\mathcal{V}$, this implies that for all $\mathfrak{N}_1, \mathfrak{N}_2 \in K$, $R^\star(\mathfrak{N}_1^\star, \mathfrak{N}_2^\star)$ holds iff $R(\mathfrak{N}_1, \mathfrak{N}_2)$ holds. Using the fact that $h^\star$ is onto the set of closed terms over $\mathcal{V}$ again, it is then easy to verify $(*)$ by induction on the complexity of the subformulas of φ. ∎

The classical definition of Hintikka sets is easily extended to the syntax of $\mathcal{L}^{\mathrm{mod}}_{\omega_1\omega}(\mathcal{V})$. Very standard techniques yield technical details, definitions, lemmas and propositions—provided in the appendix—that allow one to derive the following form of the Löwenheim–Skolem and compactness theorems.

PROPOSITION 16. *Let $\mathcal{C}$ be either the set of abstract Kripke frames over $\mathcal{V}$, or the set of abstract Kripke frames over $\mathcal{V}$ that have constant labeling.*

1. *Let a closed $\varphi \in \mathcal{L}^{\mathrm{mod}}_{\omega_1\omega}(\mathcal{V})$ be satisfiable in $\mathcal{C}$. Then there exists a countable subset $\mathcal{V}^\star$ of $\mathcal{V}$, an abstract Kripke frame $\mathcal{K} = (K, R)$ over $\mathcal{V}^\star$, and a member $\mathfrak{M}$ of K such that $\varphi \in \mathcal{L}^{\mathrm{mod}}_{\omega_1\omega}(\mathcal{V}^\star)$, $(\mathcal{K}, \mathfrak{M}) \Vdash \varphi$, and:*

   - *$\mathcal{K}$ has constant labeling if all members of $\mathcal{C}$ have constant labeling;*
   - *K is countable.*

2. *For all countable $T \subseteq \mathcal{L}^{\mathrm{mod}}_{\omega\omega}(\mathcal{V})$, if all finite subsets of T are consistent with $\mathcal{C}$ and if infinitely many closed terms over $\mathcal{V}$ are unrestricted in $\bigwedge T$, then T has a model in $\mathcal{C}$.*

Proof. See appendix. ∎

Note that the second clause of the previous proposition, the compactness result, applied for instance to $\mathcal{V}$ contain some nonnullary function symbol, and T not containing any occurrence of that symbol. More generally, the compactness result holds for any subset of $\mathcal{L}^{\mathrm{mod}}_{\omega\omega}(\mathcal{V})$ that are consistent with an arbitrary, independent interpretation of countably many closed terms.

6 Conclusion

We have shown that quantifying over names rather than individuals results in an elegant framework, that does not have to be restricted to semantics based on Henkin or Herbrand structures, thanks to the distinction between the language used to describe a structure, and the language used to talk about a structure. Classical first-order and modal first-order logics can be developed on the basis of this framework, and allow one to avoid the issues raised by combining quantifiers and modalities.

The class of structures involved in the semantics of classical logic is a proper class, not a set. By the Löwenheim–Skolem theorem, this class can be restricted to the class of structures of cardinality at most equal to some ordinal λ, equal to ω if the underlying vocabulary is countable, but that class is still not a set. It is sometimes desirable to deal with sets of structures rather than proper classes. For instance, in [9], a topology is defined over the class of intended interpretations; this requires the class of intended interpretations to be a set. Using the Löwenheim–Skolem theorem and selecting 'representative structures' for all structures of cardinality λ at most, it is possible to indeed select a set of intended interpretations and still be faithful to the semantics defined on the basis of the class of all structures over the underlying vocabulary. The question then arises: how to choose "representative structures"? A natural idea is to use canonical structures, obtained by enriching the vocabulary and making sure that every individual can be identified with the nonempty set of all its names. Then taking the enriched vocabulary as primitive, and considering the original vocabulary as a restriction of the latter, yields the notions that have been presented in this paper. We hope that abstract Kripke frames and quantification over names is not only an effective formal machinery, but can also be supported by epistemic arguments.

Appendix

We start with a couple of technical definitions. They use a member ϵ of $\{0, 1\}$, whose purpose is to indicate whether we are interested in the class of all abstract Kripke frames over $\mathcal{V}$ ($\epsilon = 0$) or in the class of all abstract Kripke frames over $\mathcal{V}$ that have constant labeling ($\epsilon = 1$). Let $\mathbb{N}^\star$ denote the set of finite sequences of members of $\mathbb{N}$. We first define a notion of Hintikka set for $(\mathcal{V}, \epsilon)$. Intuitively, such a set $\mathfrak{H}$ will determine, for all members σ of $\mathbb{N}^\star$, an abstract Hintikka frame over $\mathcal{V}$ and a particular abstract structure in this frame, associated with σ. Requiring $\mathfrak{H}$ to contain a pair of the form (σ, ψ) expresses that ψ should be accepted in the abstract Hintikka frame and structure that are determined by $\mathfrak{H}$ and associated with σ.

DEFINITION 17. Let $\epsilon \in \{0, 1\}$ be given. A *Hintikka set for* $(\mathcal{V}, \epsilon)$ is any set $\mathfrak{H}$ consisting of pairs of the form (σ, φ), where σ is a member of $\mathbb{N}^\star$ and φ a closed member of $\mathcal{L}^{\mathrm{mod}}_{\omega_1\omega}(\mathcal{V})$, with the following properties.[2]

---

[2] Given two finite sequences σ and τ, we let $\sigma \star \tau$ denote the concatenation of σ and τ, and write $\sigma \star t$ in case τ is a singleton (t) that is not itself represented as a sequence.

1. $\mathfrak{H}$ contains no pair of the form $(\sigma, \bigvee \emptyset)$, and $\mathfrak{H}$ does not contain two pairs of the form (σ, ξ) and $(\sigma, \neg \xi)$.

2. If $\mathfrak{H}$ contains a pair of the form $(\sigma, \bigvee X)$ with $X \neq \emptyset$ then $\mathfrak{H}$ contains (σ, ψ) for some $\psi \in X$, and if $\mathfrak{H}$ contains a pair of the form $(\sigma, \bigwedge X)$ then $\mathfrak{H}$ contains (σ, ψ) for all $\psi \in X$.

3. If $\mathfrak{H}$ contains a pair of the form $(\sigma, \exists x \psi)$ then $\mathfrak{H}$ contains $(\sigma, \psi[t/x])$ for some closed term t over $\mathcal{V}$, and if $\mathfrak{H}$ contains a pair of the form $(\sigma, \forall x \psi)$ then $\mathfrak{H}$ contains $(\sigma, \psi[t/x])$ for all closed terms t over $\mathcal{V}$.

4. If $\mathfrak{H}$ contains a pair of the form $(\sigma, \Diamond \psi)$ then $\mathfrak{H}$ contains $(\sigma \star i, \psi)$ for some $i \in \mathbb{N}$, and if $\mathfrak{H}$ contains a pair of the form $(\sigma, \Box \psi)$ then $\mathfrak{H}$ contains $(\sigma \star i, \psi)$ for all $i \in \mathbb{N}$. $(\sigma, t_1 = t_2) \in \mathfrak{H}$ then $(\sigma, t_1 = t_2)$ $(\sigma, t_1 = t_3) \in \mathfrak{H}$. $(\sigma, p(t'_1, \ldots, t'_n)) \in \mathfrak{H}$;

5. If $\mathcal{V}$ contains equality then clauses that deal with it are added (details omitted).

6. Suppose that $\epsilon = 1$. Then for all $\sigma, \tau \in \mathbb{N}^*$ and for all closed equalities or inequalities ψ over $\mathcal{V}$, $(\sigma, \psi) \in \mathfrak{H}$ iff $(\tau, \psi) \in \mathfrak{H}$.

Let $\epsilon \in \{0, 1\}$ and a Hintikka set $\mathfrak{H}$ for $(\mathcal{V}, \epsilon)$ be given. Then for all members σ of $\mathbb{N}^*$, there exists a unique abstract structure over $\mathcal{V}$ and a unique abstract Kripke frame over $\mathcal{V}$ that are determined by $\mathfrak{H}$ and associated with σ, as formalized in the next definition.

DEFINITION 18. Let $\epsilon \in \{0, 1\}$ and a Hintikka set $\mathfrak{H}$ for $(\mathcal{V}, \epsilon)$ be given. For all $\sigma \in \mathbb{N}^*$, let $\mathfrak{M}_\mathfrak{H}^\sigma$ be the (unique) abstract structure over $\mathcal{V}$ with the following properties.

- Two distinct closed terms t_1 and t_2 over $\mathcal{V}$ belong to the same labeling class of $\mathfrak{M}_\mathfrak{H}^\sigma$ iff $\mathcal{V}$ contains equality and $(\sigma, t_1 = t_2) \in \mathfrak{H}$.

- Let $n \in \mathbb{N}$, an n-ary predicate symbol p in $\mathcal{V}$ distinct from equality, and closed terms $t_1, \ldots, t_n$ over $\mathcal{V}$ be given. Then $p(t_1, \ldots, t_n)$ belongs to $\mathfrak{M}_\mathfrak{H}^\sigma(1)$ iff $(\sigma, p(t_1, \ldots, t_n))$ belongs to $\mathfrak{H}$.

We define for all $\sigma \in \mathbb{N}^*$ an abstract Kripke frame $\mathcal{K}_\mathfrak{H} = (K_\mathfrak{H}^\sigma, R_\mathfrak{H}^\sigma)$ by:

- $K_\mathfrak{H}^\sigma = \{\mathfrak{M}_\mathfrak{H}^\tau \mid \tau \in \mathbb{N}^*, \sigma \subseteq \tau\}$;

- for all $\tau_1, \tau_2 \in \mathbb{N}^*$ that extend σ, $R_\mathfrak{H}^\sigma(\mathfrak{M}_\mathfrak{H}^{\tau_1}, \mathfrak{M}_\mathfrak{H}^{\tau_2})$ holds iff $\tau_1 \subseteq \tau_2$.

The following property directly follows from Definitions 17 and 18:

PROPERTY 19. Let $\epsilon \in \{0, 1\}$ and a Hintikka set $\mathfrak{H}$ for $(\mathcal{V}, \epsilon)$ be given. For all $\sigma \in \mathbb{N}^*$, the following holds.

1. $K_\mathfrak{H}^\sigma$ is countable.

2. Suppose that $\epsilon = 1$. Then for all $\tau \in \mathbb{N}^*$, $\mathfrak{M}_\mathfrak{H}^\sigma(0) = \mathfrak{M}_\mathfrak{H}^\tau(0)$.

3. For all closed $\varphi \in \mathcal{L}_{\omega_1\omega}^{\mathrm{mod}}(\mathcal{V})$, if $(\sigma, \varphi) \in \mathfrak{H}$ then $(\mathcal{K}_\mathfrak{H}, \mathfrak{M}_\mathfrak{H}^\sigma) \Vdash \varphi$.

To apply Property 19, it suffices to set ϵ to 0 or 1, choose a closed member φ of $\mathcal{L}_{\omega_1\omega}^{\mathrm{mod}}(\mathcal{V})$, and exhibit a Hintikka set $\mathfrak{H}$ for $(\mathcal{V}, \epsilon)$ that contains $((), \varphi)$. The set $\mathfrak{H}$ is obtained by constructing a tableau $\mathcal{T}$ for $(\mathcal{V}, \varphi, \epsilon)$, as defined in Definition 21 with the help of the technical notions introduced in Definition 20. The fundamental relationship between $\mathcal{T}$ and $\mathfrak{H}$ is then expressed in Property 22.

DEFINITION 20. Let $\epsilon \in \{0, 1\}$, $\sigma \in \mathbb{N}^\star$, and a closed $\varphi \in \mathcal{L}_{\omega_1\omega}^{\mathrm{mod}}(\mathcal{V})$ be given. Let ρ be a finite sequence of pairs of the form (τ, ψ), where τ is a member of $\mathbb{N}^\star$ and ψ a closed member of $\mathcal{L}_{\omega_1\omega}^{\mathrm{mod}}(\mathcal{V})$. We say that (σ, φ) is an F-sprout of ρ iff one of the following holds.

1. ρ contains two pairs of the form (σ, ξ) and $(\sigma, \neg\xi)$, and φ is $\bigvee \emptyset$.

2. ρ contains a pair of the form $(\sigma, \bigwedge X)$ with $\varphi \in X$.

3. ρ contains a pair of the form $(\sigma, \forall x \xi)$ and φ is of the form $\xi[t/x]$ for some closed term t over $\mathcal{V}$.

4. ρ contains a pair of the form $(\tau, \Box\varphi)$ and σ is of the form $\tau \star i$.

5. $\epsilon = 1$, φ is an equality or an inequality, and there exists $\tau \in \mathbb{N}^\star$ such that ρ contains an occurrence of (τ, φ).

6. If $\mathcal{V}$ contains equality then extra cases that deal with it are added (details omitted). and ρ contains occurrences of both $(\sigma, t_1 = t_2)$ and occurrences of

DEFINITION 21. Let $\epsilon \in \{0, 1\}$ and a closed $\varphi \in \mathcal{L}_{\omega_1\omega}^{\mathrm{mod}}(\mathcal{V})$ be given. We call *tableau for* $(\mathcal{V}, \varphi, \epsilon)$ any tree $\mathcal{T}$ over the set of pairs of the form (σ, ψ), where σ is a member of $\mathbb{N}^\star$ and ψ a closed member of $\mathcal{L}_{\omega_1\omega}^{\mathrm{mod}}(\mathcal{V})$, with the following properties.

- $((), \varphi)$ is the only child of $()$ in $\mathcal{T}$.

- For all $\rho \in \mathcal{T}$, if the last member of ρ is of the form $(\sigma, \bigvee X)$ with $X \neq \emptyset$, then $\{\rho \star ((\sigma, \psi)) \mid \psi \in X\}$ is the set of children of ρ in $\mathcal{T}$.

- Let $\rho \in \mathcal{T}$ be such that the last member of ρ is of the form $(\sigma, \exists x \psi)$. Let X be the set of closed terms over $\mathcal{V}$ that are unrestricted in all the modal formulas that occur in ρ.

  – If X is finite then $\{\rho \star ((\sigma, \psi[t/x])) \mid t$ is a closed term t over $\mathcal{V}\}$ is the set of children of ρ in $\mathcal{T}$.

  – If X is infinite then ρ has a unique child which is of the form $\rho \star ((\sigma, \psi[t/x]))$ for some $t \in X$.

- For all $\rho \in \mathcal{T}$, if the last member of ρ is of the form $(\sigma, \bigvee \emptyset)$ then ρ has no child in $\mathcal{T}$.

- For all $\rho \in \mathcal{T}$, if the last member of ρ is of the form $(\sigma, \Diamond\psi)$, then ρ has a unique child which is of the form $\rho \star ((\sigma \star 2i + 1, \psi))$ for some $i \in \mathbb{N}$ such that $2i + 1$ does not occur in ρ.

- For all nonempty $\rho \in \mathcal{T}$, if the last member of ρ is not of the form $(\sigma, \bigvee X)$, $(\sigma, \exists x \psi)$ or $(\sigma, \Diamond \psi)$, then ρ has a unique child in $\mathcal{T}$, which is either $((), \varphi)$ or one of ρ's F-sprouts.

- For all infinite branches B in $\mathcal{T}$ and for all strict initial segments ρ of B, all F-sprouts of ρ occur in B.

Note that for all branches B in $\mathcal{T}$, either B is infinite and contains no occurrence of the form $(\sigma, \bigvee \emptyset)$, or B is of the form $\rho \star ((\sigma, \bigvee \emptyset))$.

PROPERTY 22. *Let $\epsilon \in \{0,1\}$, a closed $\varphi \in \mathcal{L}^{\mathrm{mod}}_{\omega_1\omega}(\mathcal{V})$, and a tableau for $(\mathcal{V}, \varphi, \epsilon)$ be given. If B is an infinite branch in $\mathcal{T}$ then the set of members of B is a Hintikka set for $(\mathcal{V}, \epsilon)$.*

Let $\epsilon \in \{0,1\}$, a closed $\varphi \in \mathcal{L}^{\mathrm{mod}}_{\omega_1\omega}(\mathcal{V})$, and a tableau $\mathcal{T}$ for $(\mathcal{V}, \varphi, \epsilon)$ be given. Then either $\mathcal{T}$ has an infinite branch, or $\mathcal{T}$ is infinite but has no infinite branch, or $\mathcal{T}$ is finite. The first case has been exploited in Property 22. The next proposition considers the third case, which is the key to proving compactness properties.

PROPOSITION 23. *Suppose that $\mathcal{V}$ is countable. Let $\epsilon \in \{0,1\}$ and a set T of closed members of $\mathcal{L}^{\mathrm{mod}}_{\omega\omega}(\mathcal{V})$ be such that:*

- *infinitely many closed terms over $\mathcal{V}$ are unrestricted in $\bigwedge T$;*

- *some tableau for $(\mathcal{V}, \bigwedge T, F)$ has no infinite branch.*

Then there exists a finite $D \subseteq T$ and a finite tableau for $(\mathcal{V}, \bigwedge D, F)$.

Proof. Let $\mathcal{T}$ be a tableau for $(\mathcal{V}, \bigwedge T, \epsilon)$ with no infinite branch. Since all members of T are finite, all disjunctions in $\bigwedge T$ are finite. Since infinitely many closed terms over $\mathcal{V}$ are unrestricted in $\bigwedge T$, all member of $\mathcal{T}$ that end in pair of the form $(\sigma, \exists x \psi)$ have a unique child. Hence $\mathcal{T}$ is finitely branching. It follows from König's lemma that $\mathcal{T}$ is finite. Let D denote the (finite) set of members ψ of T such that (σ, ψ) occurs in $\mathcal{T}$ for some $\sigma \in \mathbb{N}^\star$. Let $\mathcal{T}'$ be the tree obtained from $\mathcal{T}$ by replacing all occurrences of $\bigwedge T$ in $\mathcal{T}$ by $\bigwedge D$. It is immediately verified that $\mathcal{T}'$ is a tableau for $(\mathcal{V}, \bigwedge D, \epsilon)$. ∎

The next lemma follows immediately from clause 2. in Property 19.

LEMMA 24. *For all Hintikka sets $\mathfrak{H}$ for $(\mathcal{V}, 1)$, $K^{()}_{\mathfrak{H}}$ has constant labeling.*

Putting together the previous results, we can now derive Proposition 16. Let $\mathcal{C}$ be either the set of all abstract Kripke frames over $\mathcal{V}$, or the set of all abstract Kripke frames over $\mathcal{V}$ that have constant labeling. Let a closed $\varphi \in \mathcal{L}^{\mathrm{mod}}_{\omega_1\omega}(\mathcal{V})$ be given. Let A be the (countable) set of members of $\mathcal{V}$ that occur in φ, and let S be the set of function symbols in $\mathcal{V}$ that do not occur in φ. If S is countable then put $\mathcal{V}^\star = A \cup S$. Otherwise let $\mathcal{V}^\star$ be the union of A with a countably infinite subset of S. Clearly, if infinitely many closed terms over $\mathcal{V}$ are unrestricted in φ then infinitely many closed terms over $\mathcal{V}^*$ are unrestricted in φ. Put $\epsilon = 0$ if $\mathcal{C}$ is the set of all abstract Kripke frames over $\mathcal{V}$, and $\epsilon = 1$ otherwise. It follows immediately from Lemma 15 that if φ is satisfiable in $\mathcal{C}$ then there exists a tableau for $(\mathcal{V}^*, \varphi, \epsilon)$ with an infinite branch. The first part of the Proposition 16 follows from the first and third clauses in Property 19, Property 22, and Lemma 24 by taking for φ a closed member of $\mathcal{L}^{\mathrm{mod}}_{\omega_1\omega}(\mathcal{V})$ that is satisfiable in $\mathcal{C}$. The second part of Proposition 16 follows from Property 22 and Proposition 23 by taking for φ a closed member of $\mathcal{L}^{\mathrm{mod}}_{\omega_1\omega}(\mathcal{V})$ of the form $\bigwedge T$ for some $T \subseteq \mathcal{L}^{\mathrm{mod}}_{\omega\omega}(\mathcal{V})$ that has no model in $\mathcal{C}$.

BIBLIOGRAPHY

[1] G. Corsi. A unified completeness theorem for quantified modal logic. *The Journal of Mathematical Logic*, 67:(4):1483–1510, 2002.

[2] M. Fitting and R. L. Mendelsohn. *First-order modal logic*. Kluwer Academic Publishers, 1999.

[3] J. W. Garson. Unifying quantified modal logic. *Journal of Philosophical Logic*, 34:621–649, 2005.

[4] A. R. Haas. An epistemic logic with quantification over names. *Computational Intelligence*, 11:460–497, 1995.

[5] M. Kaminski and G. Rey. Revisiting quantification in autoepistemic logic. *ACM Transactions on Computational Logic*, 3:(4):542–561, 2002.

[6] K. Konolige. Autoepistemic logic. In C. Hogger D. Gabbay and J. Robinson, editors, *Handbook of Logic in Artificial Intelligence and Logic Programming*, volume 3, pages 217–295. Oxford University Press, 1994.

[7] R. Kontchakov, A. Kurucz, and M. Zakharyaschev. Undecidability of first-order intuitionistic and modal logics with two variables. *The Bulletin of Symbolic Logic*, 11:(3):428–438, 2005.

[8] R. C. Barcan Marcus. Modalities and intensional languages. *Synthese*, 13:(4):303–322, 1961.

[9] E. Martin, A. Sharma, and F. Stephan. Unifying logic, topology and learning in parametric logic. *Theoretical Computer Science, Special Issue Algorithmic Learning Theory (ALT 2002)*, 350:(1):103–124, 2006.

[10] J. R. Shoenfield. *Mathematical logic*. Addison-Wesley Publishing Company, 1967.

[11] Y. Stephanou. First-order modal logic with an 'actually' operator. *Notre Dame Journal of Formal Logic*, 46:(4):381–405, 2005.

Eric Martin
School of Computer Science and Engineering,
The University of New South Wales, and National ICT Australia[3]
UNSW Sydney NSW 2052, Australia
emartin@cse.unsw.edu.au

---

[3]National ICT Australia is funded by the Australian Government's Department of Communications, Information Technology and the Arts and the Australian Research Council through Backing Australia's Ability and the ICT Centre of Excellence Program.

On the Deterministic Horn Fragment of Test-free PDL

LINH ANH NGUYEN

ABSTRACT. We study the deterministic Horn fragment of test-free propositional dynamic logic (PDL$^{(0)}$). This fragment adopts the restriction that, in bodies of program clauses and goals, special universal modal operators which are a kind of combination of $\Box$ and $\Diamond$ are used instead of $\Box$. The fragment contains deterministic positive logic programs and deterministic negative clauses, whose negations form serial positive formulae. A least Kripke model for a deterministic positive logic program in PDL$^{(0)}$ may not exist, because PDL$^{(0)}$ is a non-serial modal logic. In this work, we present an algorithm that, given a deterministic positive logic program P in PDL$^{(0)}$, constructs a least pseudo-model of P. A pseudo-model is similar to a Kripke model except that it contains two sets of accessibility relations, one for dealing with existential modal operators and the other for dealing with universal modal operators. A least pseudo-model M of P has the property that, for every serial positive formula φ, $P \models \varphi$ iff $M \models \varphi$. Furthermore, checking whether $M \models \varphi$ is solvable in polynomial time in the sizes of M and φ. Our algorithm runs in exponential time and returns a pseudo-model with size $2^{O(n^2)}$. We give a deterministic positive logic program in PDL$^{(0)}$ such that every pseudo-model characterizing it must have size $2^{\Omega(n)}$.

KEYWORDS: PDL, finite automata, Horn logic, minimal models.

1 Introduction

Modal logic programming extends classical logic programming with modalities. There are two approaches in modal logic programming: the translation approach [3, 10] and the direct approach [1, 2, 8, 9]. In the translation approach, both the functional translation method of Debart et al. [3] and the semi-functional translation method of Nonnengart [10] assume that the base modal logic is serial, i.e. it contains axiom $\Diamond_i\top$ for every modal index i. Using the direct approach for modal logic programming, Balbiani et al. [1] considered only serial modal logics, and in our previous works [7, 8, 9], we considered only serial or almost serial modal logics. In [2] Baldoni et al. studied modal logic programming using the direct approach for grammar logics, which are non-serial normal modal logics with axioms of the form $[t_1]\ldots[t_n]\varphi \to [s_1]\ldots[s_m]\varphi$. However, they considered only modal logic programs without existential modal operators.

Seriality has thus played an important role in the theory of modal logic programming. It is an essential assumption for the functional and semi-functional translation methods. For the direct approach, let us consider the following program in the modal logic K:

$$\Box p \leftarrow$$
$$q \leftarrow \Diamond p$$
$$s \leftarrow \Box r$$

The problem is whether there exists a world accessible from the actual world. If there exists then $\Box p$ implies $\Diamond p$, which then implies q. If there does not then $\Box r$ holds and implies s. The program is thus "nondeterministic" because the accessibility relation is not serial. In the above program, $\Box r$ does not follow from the program, but it may unwantedly become true when there are no worlds accessible from the actual world. To overcome this problem, instead of the program clause $s \leftarrow \Box r$ we can use $s \leftarrow \boxplus r$, where $\boxplus$ has the semantics defined by $\boxplus \varphi \equiv (\Box \varphi \wedge \Diamond \varphi)$ or $\boxplus \varphi \equiv (\Box \varphi \wedge \Diamond \top)$. One can say that allowing modal operators like $\boxplus$ is not a "solution" for dealing with non-seriality because $\boxplus$ contains "seriality" itself. Our justifications for using modal operators like $\boxplus$ are as follows:

- If the base modal logic is deliberately chosen to be K, then adopting $\boxplus$ is an appropriate solution. Note that we will still allow $\Box$ to appear in contexts and heads of program clauses.

- While seriality is a natural assumption in some applications, e.g. to state that knowledge and belief are consistent, it cannot be assumed in some cases. For example, if a is an action, we may not want a to be always admissible. That is, we may not want to adopt axiom $\langle a \rangle \top$, and in that case $[a]\varphi$ does not imply $\langle a \rangle \varphi$.

- Finally, program clauses like $s \leftarrow \boxplus r$ are more acceptable than $s \leftarrow \Box r$. In $s \leftarrow \boxplus r$, the premise $\boxplus r$ guarantees that $\Box r$ actually follows from the program, while in $s \leftarrow \Box r$, the premise $\Box r$ may "accidentally" be true.

In this work, we study the deterministic Horn fragment of test-free propositional dynamic logic (PDL$^{(0)}$), which is a non-serial modal logic and can express K_n. This fragment adopts the restrictions that: i) in bodies of program clauses and goals, modal operators like $\boxplus$ are used instead of universal modal operators like $\Box$; ii) implicit disjunction such as $\langle \pi \cup \pi' \rangle$ and $\langle \pi^* \rangle$ is not allowed in heads of program clauses. The deterministic Horn fragment of PDL$^{(0)}$ contains deterministic positive logic programs and deterministic negative clauses. Negation of a deterministic negative clause is a serial positive formula. In general, a serial positive formula is a positive formula which may contains modal operators like $\boxplus$ but not modal operators like $\Box$.

Constructing a least Kripke model for a given positive modal logic program in a serial propositional modal logic L is a useful starting point for developing semantics for positive modal logic programs in the corresponding first-order modal logic L. We have demonstrated this in [7, 8] for the basic serial monomodal logics KD, T, KDB, B, $KD4$, $S4$, $KD5$, $KD45$, $S5$.

In [7], we presented algorithms that, given a positive propositional modal logic program P, construct a least L-model of P, where L is one of the listed modal logics. As a continuation of [7], in [8] we have developed the least model semantics, fixpoint semantics, and SLD-resolution calculi for positive modal logic programs in the corresponding first-order modal logics L.

In $PDL^{(0)}$, a least Kripke model of a deterministic positive logic program P may not exist.[1] However, we can talk about least "pseudo-models" of P. A pseudo-model is similar to a Kripke model except that it contains two sets of accessibility relations, one for dealing with existential modal operators and the other for dealing with universal modal operators. Informally, M is a least pseudo-model of P if it satisfies P and for every pseudo-model M' of P, M is less than or equal to M'. A least pseudo-model M of P has the property that, for every serial positive formula φ, $P \models \varphi$ iff $M \models \varphi$. Furthermore, checking whether $M \models \varphi$ is solvable in polynomial time in the sizes of M and φ.

In this work, we present an algorithm that, given a deterministic positive logic program P, constructs a least pseudo-model of P. Our algorithm runs in exponential time and returns a pseudo-model with size $2^{O(n^2)}$. We give a deterministic positive logic program in $PDL^{(0)}$ such that every pseudo-model characterizing it must have size $2^{\Omega(n)}$.

From the view of the theory of complexity and expressiveness, the deterministic Horn fragment of $PDL^{(0)}$ does not have interesting properties. However, this fragment and our method for it are useful for the following reasons:

1. If a knowledge base is represented by a deterministic positive logic program P and the given query is a serial positive formula φ, then having a least pseudo-model M of P, checking whether $P \models \varphi$ can be reduced to checking whether $M \models \varphi$. This method is especially useful when the knowledge base rarely changes.

2. Our method for answering whether $P \models \varphi$ by constructing a least pseudo-model of P is bottom-up. It does not create choice points, while the traditional tableaux method for this problem would intensively use the "or" splitting rule and we know that a wrong choice when exploring tableaux, e.g. one near the root of the search tree, would cost much. Hence, our bottom-up method is more efficient than the traditional tableaux method, even though both the methods can give an algorithm with EXPTIME complexity.

3. The deterministic Horn fragment of $PDL^{(0)}$ eliminates nondeterminism (of $PDL^{(0)}$). How much important is this property? In [6], Hus-

---
[1] In [7], we showed that the logic program $\{\Box p\}$ does not have any least K-model. This implies that the deterministic positive logic program $\{[\sigma]p\}$ does not have any least Kripke model in $PDL^{(0)}$. The only reason is the non-seriality of $PDL^{(0)}$. It can be shown that adding the seriality axiom $[\pi]\varphi \rightarrow \langle\pi\rangle\varphi$ to $PDL^{(0)}$ causes that every positive logic program without implicit disjunctions $\langle\pi' \cup \pi''\rangle$ and $\langle\pi'^*\rangle$ in heads of program clauses has a finite least model in the resulting logic.

tadt et al. proved that the data complexity of query answering in the Horn fragment Horn-$\mathcal{SHIQ}$ of the description logic $\mathcal{SHIQ}$ is in PTIME, while in the full description logic $\mathcal{SHIQ}$ it is complete in coNP.[2] Here, we can also prove that the data complexity of query answering in the deterministic Horn fragment of the PDL$^{(0)}$-like description logic is in PTIME (see Section 6 for more details). This is an interesting property for practical applications. Also note that our deterministic Horn fragment of PDL$^{(0)}$ is more relaxed than the Horn-$\mathcal{SHIQ}$ fragment in the aspect that the constructor $\forall R.C$ (a $[\pi]$-like constructor) is disallowed in bodies of program clauses and queries of Horn-$\mathcal{SHIQ}$, while we allow $[\pi]$ in the form $[\pi]_\diamond$ to appear in bodies of program clauses and queries of the deterministic Horn fragment of PDL$^{(0)}$.

4. As mentioned earlier, we can extend the method of this work for dealing with logic programming in PDL$^{(0)}$.

Our algorithm uses formulae with automaton-modal operators, which are similar to formulae of automaton propositional dynamic logic (APDL) [5]. In [4], Goré and Nguyen also used such formulae for developing analytic tableau calculi with the superformula property for regular grammar logics. In both [4] and this work, formulae with automaton-modal operators are used to record the potentiality inherited from predecessor worlds, which guarantees that when a world w is created from u, the content of w can be computed from the content of u. This technique plays an essential role in constructing finite models.

The rest of this paper is structured as follows. In Section 2, we define PDL$^{(0)}$, the deterministic Horn fragment of PDL$^{(0)}$, automaton-modal operators, pseudo-models, and introduce an ordering of pseudo-models. In Section 3, we present our algorithm. In Section 4, we give characterizations of least pseudo-models of deterministic positive logic programs in PDL$^{(0)}$. In Section 5, we study the lower bound of sizes of such pseudo-models. Further work and concluding remarks are given in Section 6.

2 Preliminaries

2.1 Test-free Propositional Dynamic Logic

The language of test-free propositional dynamic logic (PDL$^{(0)}$) is built from two disjoint sets: Π_0 is a countable set of atomic programs and Φ_0 is a countable set of atomic propositions. We use σ to denote an element of Π_0 and p to denote an element of Φ_0. Programs and formulae are recursively defined using the BNF grammar below:

$$\Pi \ni \pi ::= \sigma \mid \pi \cup \pi \mid \pi;\pi \mid \pi^*$$
$$\Phi \ni \varphi ::= \top \mid p \mid \neg\varphi \mid \varphi \wedge \varphi \mid \varphi \vee \varphi \mid \varphi \rightarrow \varphi \mid [\pi]\varphi \mid \langle\pi\rangle\varphi$$

---

[2] When measuring the data complexity, the TBox of the considered knowledge base is treated as the intensional part, while the ABox is treated as the extensional part.

(The version PDL with test contains also the construction φ? as a program.[3]) We use π, α, β to denote elements of Π. A word $\sigma_1 \ldots \sigma_k$ over alphabet Π_0 will also be treated as the program $\sigma_1; \ldots; \sigma_k$. An operator $[\pi]$ is called a *universal* modal operator, while $\langle \pi \rangle$ is called an *existential* modal operator.

A program of PDL$^{(0)}$ is a regular expression over the alphabet Π_0. Such an expression π generates a regular language $\mathcal{L}(\pi)$ specified as follows: $\mathcal{L}(\sigma) = \{\sigma\}$, $\mathcal{L}(\pi \cup \pi') = \mathcal{L}(\pi) \cup \mathcal{L}(\pi')$, $\mathcal{L}(\pi; \pi') = \mathcal{L}(\pi).\mathcal{L}(\pi')$, and $\mathcal{L}(\pi^*) = (\mathcal{L}(\pi))^*$, where if L and M are sets of words then $L.M = \{\alpha\beta \mid \alpha \in L, \beta \in M\}$ and $L^* = \bigcup_{n \geq 0} L^n$ with $L^0 = \{\varepsilon\}$ and $L^{n+1} = L.L^n$, where ε denotes the empty word.

The semantics of PDL$^{(0)}$ comes from the semantics of modal logic. The structures over which programs and formulae of PDL$^{(0)}$ are interpreted are called *Kripke structures*. A *Kripke structure*, also called a (Kripke) model, is a tuple $M = \langle W, \tau, (R_\sigma)_{\sigma \in \Pi_0}, h \rangle$, where W is a set of states, $\tau \in W$ is the current state, R_σ for $\sigma \in \Pi_0$ is a binary relation on W (representing the set of input/output pairs of states of the program σ), and h is a function mapping states to sets of atomic propositions. For $w \in W$, the set of atomic propositions "true" at w is $h(w)$.

Given a Kripke model $M = \langle W, \tau, (R_\sigma)_{\sigma \in \Pi_0}, h \rangle$, define $R_{\pi \cup \pi'} = R_\pi \cup R_{\pi'}$, $R_{\pi;\pi'} = R_\pi \circ R_{\pi'}$, $R_{\pi^*} = R_\pi^*$, where $R_\pi^* = \bigcup_{n \geq 0} R_\pi^n$ with $R_\pi^0 = \{(w, w) \mid w \in W\}$ and $R_\pi^{n+1} = R_\pi \circ R_\pi^n$. It is easily seen that for $\pi \in \Pi$, $R_\pi = \bigcup_{\alpha \in \mathcal{L}(\pi)} R_\alpha$.

Given a Kripke model $M = \langle W, \tau, (R_\sigma)_{\sigma \in \Pi_0}, h \rangle$, a state $w \in W$, and a formula φ, the satisfaction relation $M, w \models \varphi$ is defined as usual for the classical connectives and that:

$$M, w \models p \quad \text{iff} \quad p \in h(w)$$
$$M, w \models [\pi]\varphi \quad \text{iff} \quad \forall v \in W. R_\pi(w, v) \text{ implies } M, v \models \varphi$$
$$M, w \models \langle \pi \rangle \varphi \quad \text{iff} \quad \exists v \in W. R_\pi(w, v) \text{ and } M, v \models \varphi$$

We say that M *satisfies* φ and φ is *true* in M, written $M \models \varphi$, if $M, \tau \models \varphi$. Let Γ be a formula set. We write $M \models \Gamma$ to denote that $M \models \varphi$ for every $\varphi \in \Gamma$. If $M \models \Gamma$ then we call M a *model* of Γ and say that Γ is *satisfiable*. We write $\Gamma \models \varphi$ to denote that every model of Γ satisfies φ.

2.2 The Deterministic Horn Fragment of PDL$^{(0)}$

We extend the primitive language with universal modal operators $[\pi]_\diamond$, which have the same role as the modal operator ⊠ discussed in the Introduction. The semantics of $[\pi]_\diamond$ is defined as follows: $M, w \models [\pi]_\diamond \varphi$ iff $M, w \models [\pi]\varphi$ and for every $\alpha, \beta \in \Pi_0^*$, $\sigma \in \Pi_0$, if $\alpha\sigma\beta \in \mathcal{L}(\pi)$ then $M, w \models [\alpha]\langle \sigma \rangle \top$.

Note that $[\sigma]_\diamond \varphi \equiv [\sigma]\varphi \land \langle \sigma \rangle \top$ like the case of ⊠, but in general we do not have that $[\pi]_\diamond \varphi \equiv [\pi]\varphi \land \langle \pi \rangle \top$. Informally, $\langle \pi \rangle \top$ means that there exists a run of π (with the stop property), while the additional condition of $[\pi]_\diamond$

---
[3] The semantics of φ? w.r.t. a Kripke model $M = \langle W, \tau, (R_\sigma)_{\sigma \in \Pi_0}, h \rangle$ is specified by $R_{\varphi?} = \{(w, w) \mid M, w \models \varphi\}$.

(w.r.t. $[\pi]$) means that every partial run of π is not blocked. Also note that formulae of the form $[\pi]_\diamond \varphi$ are expressible in PDL$^{(0)}$.[4]

A *positive formula* is a formula (in the extended language) without the connectives $\to$ and $\neg$. A *serial positive formula* is a positive formula which may contain modal operators $\langle \pi \rangle$ and $[\pi]_\diamond$ but not $[\pi]$ (and $[A]$ defined later for a finite automaton A).

A *deterministic Horn formula* in PDL$^{(0)}$ is a formula of one of the forms:

- $\top$ or an atomic proposition;
- $\neg \varphi$ or $\varphi \to \psi$, where φ is a serial positive formula and ψ is a deterministic Horn formula;
- $\varphi \wedge \psi$, where φ and ψ are deterministic Horn formulae;
- $[\pi]\varphi$ or $[\pi]_\diamond \varphi$, where φ is a deterministic Horn formula;
- $\langle \pi \rangle \varphi$, where φ is a deterministic Horn formula and π is a program without $\cup$ and $*$.

A *deterministic program clause* in PDL$^{(0)}$ is a formula of one of the forms:

$$B_1 \wedge \ldots \wedge B_n \to A$$
$$[\pi](B_1 \wedge \ldots \wedge B_n \to A)$$

and a *deterministic negative clause* in PDL$^{(0)}$ is a formula of one of the forms:

$$B_1 \wedge \ldots \wedge B_n \to \bot$$
$$[\pi](B_1 \wedge \ldots \wedge B_n \to \bot)$$

where $n \geq 0$, $B_1, \ldots, B_n$ are formulae of the form p, $[\alpha]_\diamond p$, $[\alpha]_\diamond \top$, $\langle \alpha \rangle p$, or $\langle \alpha \rangle \top$, A is a formula of the form p, $[\alpha]p$, $\langle \beta \rangle p$, or $\langle \beta \rangle \top$, where β is a program without $\cup$ and $*$, and $\bot$ denotes $\neg \top$.

A *deterministic Horn clause* in PDL$^{(0)}$ is either a deterministic program clause or a deterministic negative clause. (This notion is used for Proposition 1 and Corollary 2 given below.)

A *deterministic positive logic program* in PDL$^{(0)}$ is a finite set of deterministic program clauses.

PROPOSITION 1. *Every set of deterministic Horn formulae can be transformed into a set of deterministic Horn clauses preserving satisfiability.*

Sketch of the proof Apply the technique of [7], which is based on replacing a complex formula by a fresh atomic proposition and adding a formula defining that atomic proposition. For example, $[\pi]([\pi']_\diamond \varphi \to \psi)$, where φ is not $\top$ or an atomic proposition, is replaced by $[\pi]([\pi']_\diamond p \to \psi)$ and $[\pi;\pi'](\varphi \to p)$, where p is a fresh atom proposition. ∎

---

[4] Let $A = \langle \Pi_0, Q, q_I, \delta, F \rangle$ be a deterministic finite automaton equivalent to π. Let $S \subseteq Q \times \Pi_0$ be the set of pairs (q, σ) such that $\delta(q, \sigma)$ is productive in A. For $(q, \sigma) \in S$, let $\pi_{q,\sigma}$ be the regular expression equivalent to the automaton $\langle \Pi_0, Q, q_I, \delta, \{q\} \rangle$. Then $[\pi]_\diamond \varphi$ is expressible in PDL$^{(0)}$ as $[\pi]\varphi \wedge \bigwedge_{(q,\sigma) \in S}[\pi_{q,\sigma}]\langle \sigma \rangle \top$.

COROLLARY 2. *Every finite set Γ of deterministic Horn formulae can be transformed into a deterministic positive logic program P and a serial positive formula φ such that Γ is unsatisfiable iff $P \models \varphi$.*

For the proof, just note that $[\pi](B_1 \wedge \ldots \wedge B_n \rightarrow \bot) \equiv \neg\langle\pi\rangle(B_1 \wedge \ldots \wedge B_n)$.

2.3 Automaton-Modal Operators

Recall that a *finite automaton* A is a tuple $\langle \Sigma, Q, I, \delta, F \rangle$, where Σ is the alphabet, Q is a finite set of states, $I \subseteq Q$ is the set of initial states, $\delta \subseteq Q \times \Sigma \times Q$ is the transition relation, and $F \subseteq Q$ is the set of accepting states. A *run* of A on a word $\sigma_1 \ldots \sigma_k$ is a finite sequence of states $q_0, q_1, \ldots, q_k$ such that $q_0 \in I$ and $\delta(q_{i-1}, \sigma_i, q_i)$ holds for every $1 \leq i \leq k$. It is an *accepting run* if $q_k \in F$. We say that A *accepts* word w if there exists an accepting run of A on w. The set of all words accepted/recognized by A is denoted by $\mathcal{L}(A)$.

If A is a finite automaton then we call $[A]$ a (universal) *automaton-modal operator*. If A is a finite automaton and φ is a formula in the primitive language, i.e. $\varphi \in \Phi$, then we call $[A]\varphi$ a formula in the extended language.

The semantics of formulae with automaton-modal operators are defined as follows: $M, w_0 \models [A]\varphi$ if $M, w_k \models \varphi$ for every path $w_0 R_{\sigma_1} w_1 \ldots w_{k-1} R_{\sigma_k} w_k$ with $k \geq 0$ and $\sigma_1 \ldots \sigma_k \in \mathcal{L}(A)$.

It is well known that every regular expression π is equivalent to a finite automaton A (with the same alphabet) in the sense that $\mathcal{L}(\pi) = \mathcal{L}(A)$. It is easy to see that if π is equivalent to A then $M, w \models [\pi]\varphi$ iff $M, w \models [A]\varphi$. For every regular expression π, let $A_\pi = \langle \Pi_0, Q_\pi, I_\pi, \delta_\pi, F_\pi \rangle$ be a fixed finite automaton equivalent to π. A_π can be constructed from π in linear time. We assume that every state of Q_π is reachable from some state of I_π via a path using the transition relation δ_π.

If A is a finite automaton and Q is a subset of the states of A, then by (A, Q) we denote the finite automaton obtained from A by using Q as the set of initial states, and we will write $[A, Q]$ for the automaton-modal operator $[(A, Q)]$.

For a finite automaton $A = \langle \Pi_0, Q_A, I_A, \delta_A, F_A \rangle$ and $\alpha \in \Pi_0^*$, define

$$\delta_A(Q, \sigma) = \{q' \mid \exists q \in Q.(q, \sigma, q') \in \delta_A\},$$
$$\widetilde{\delta_A}(Q, \varepsilon) = Q,$$
$$\widetilde{\delta_A}(Q, \alpha\sigma) = \delta_A(\widetilde{\delta_A}(Q, \alpha), \sigma).$$

We have that $M, w_0 \models [A, Q]\varphi$ iff $M, w_k \models \varphi$ for every path $w_0 R_{\sigma_1} w_1 \ldots w_{k-1} R_{\sigma_k} w_k$ with $k \geq 0$ and $\widetilde{\delta_A}(Q, \sigma_1 \ldots \sigma_k) \cap F_A \neq \emptyset$.

2.4 Pseudo-models for Dealing with Non-seriality

A *pseudo-model* is a tuple $M = \langle W, \tau, (R_\sigma)_{\sigma \in \Pi_0}, (S_\sigma)_{\sigma \in \Pi_0}, h \rangle$, which is similar to a model except that for every $\sigma \in \Pi_0$, there are two accessibility relations R_σ and S_σ. We require that $R_\sigma \subseteq S_\sigma$ for every $\sigma \in \Pi_0$. The accessibility relations R_σ, resp. S_σ, are used to deal with existential, resp. universal, modal operators.

Given a pseudo-model $M = \langle W, \tau, (R_\sigma)_{\sigma \in \Pi_0}, (S_\sigma)_{\sigma \in \Pi_0}, h \rangle$, for $\pi \in \Pi$, define S_π analogously as for R_π; and for $w \in W$, define the relation $M, w \models \varphi$ in the usual way for classical connectives and:

- $M, w \models \langle \pi \rangle \varphi$ iff there exists $v \in W$ such that $R_\pi(w, v)$ and $M, v \models \varphi$;

- $M, w \models [\pi]\varphi$ iff for every $v \in W$, $S_\pi(w, v)$ implies $M, v \models \varphi$;

- $M, w \models [\pi]_\diamond \varphi$ iff $M, w \models [\pi]\varphi$ and for every $\alpha, \beta \in \Pi_0^*$, $\sigma \in \Pi_0$, if $\alpha \sigma \beta \in \mathcal{L}(\pi)$ then $M, w \models [\alpha]\langle \sigma \rangle \top$;

- $M, w \models [A]\varphi$ iff $M, w_k \models \varphi$ for every path $w_0 S_{\sigma_1} w_1 \ldots w_{k-1} S_{\sigma_k} w_k$ with $k \geq 0$ and $\sigma_1 \ldots \sigma_k \in \mathcal{L}(A)$.

Other related definitions remain unchanged.
If $M \models \Gamma$ then we say that M is a pseudo-model of Γ.
Every model is also a pseudo-model, with $S_\sigma = R_\sigma$ for every $\sigma \in \Pi_0$.

PROPOSITION 3. *The problem of checking $M \models \varphi$ for a given pseudo-model M and a given formula φ is solvable in polynomial time in the sizes of M and φ.*

This proposition can be proved by induction on the construction of φ. To deal with modal operators, we can run corresponding automata along paths in M.

2.5 Ordering Pseudo-models

In [7] we introduced an ordering between Kripke models. In this subsection, we provide an analogue for ordering pseudo-models. A pseudo-model M is said to be *less than* or *equal to* a pseudo-model M', write $M \leq M'$, if for every positive formula φ (in the extended language with $[\pi]_\diamond$ and $[A]$), if $M \models \varphi$ then $M' \models \varphi$. This relation $\leq$ is a pre-order.[5] We write $M \equiv M'$ to denote that $M \leq M'$ and $M' \leq M$.

A pseudo-model M is a *least pseudo-model* of a deterministic positive logic program P if $M \models P$ and $M \leq M'$ for every pseudo-model M' of P.

Let $M = \langle W, \tau, (R_\sigma)_{\sigma \in \Pi_0}, (S_\sigma)_{\sigma \in \Pi_0}, h \rangle$ and $M' = \langle W', \tau', (R'_\sigma)_{\sigma \in \Pi_0}, (S'_\sigma)_{\sigma \in \Pi_0}, h' \rangle$ be pseudo-models. We say that M is *less than or equal to* M' w.r.t. a binary relation $r \subseteq W \times W'$, and write $M \leq_r M'$, if the following conditions hold:

1. $r(\tau, \tau')$

2. $\forall \sigma \in \Pi_0 \ \forall x, x', y \ R_\sigma(x, y) \wedge r(x, x') \to \exists y' \ R'_\sigma(x', y') \wedge r(y, y')$

3. $\forall \sigma \in \Pi_0 \ \forall x, x', y' \ S'_\sigma(x', y') \wedge r(x, x') \to \exists y \ S_\sigma(x, y) \wedge r(y, y')$

4. $\forall x, x' \ r(x, x') \to h(x) \subseteq h(x')$

---
[5]i.e. a reflexive and transitive binary relation

In the above definition, the first three conditions state that r is a forward-backward bisimulation of the frames of M and M'.[6] Intuitively, $r(x, x')$ states that the state x is less than or equal to x'.

LEMMA 4. *If $M \leq_r M'$ then $M \leq M'$.*

Proof. Let $M = \langle W, \tau, (R_\sigma)_{\sigma \in \Pi_0}, (S_\sigma)_{\sigma \in \Pi_0}, h \rangle$ and $M' = \langle W', \tau', (R'_\sigma)_{\sigma \in \Pi_0}, (S'_\sigma)_{\sigma \in \Pi_0}, h' \rangle$ be pseudo-models and suppose that $M \leq_r M'$. We prove that for every positive formula φ and every u, u' such that $r(u, u')$, if $M, u \models \varphi$ then $M', u' \models \varphi$. We do this by induction on the construction of φ. Suppose that $r(u, u')$ holds and $M, u \models \varphi$.

The cases when $\varphi = p$ or $\varphi = \psi \wedge \zeta$ or $\varphi = \psi \vee \zeta$ are trivial.

Case $\varphi = \langle \pi \rangle \psi$: Since $M, u \models \langle \pi \rangle \psi$, there exists $v \in W$ such that $R_\pi(u, v)$ holds and $M, v \models \psi$. There exist $\sigma_1, \ldots, \sigma_k \in \Pi_0$ such that $\sigma_1 \ldots \sigma_k \in \mathcal{L}(\pi)$ and $R_{\sigma_1;\ldots;\sigma_k}(u, v)$ holds. Let $u_0 = u, u_1, \ldots, u_{k-1}, u_k = v$ be states such that $R_{\sigma_i}(u_{i-1}, u_i)$ holds for $1 \leq i \leq k$. Let $u'_0 = u'$. For every $1 \leq i \leq k$, by the condition 2, there exists a state $u'_i \in W'$ such that $R'_{\sigma_i}(u'_{i-1}, u'_i)$ and $r(u_i, u'_i)$ hold. Hence, $R'_\pi(u', v')$ and $r(v, v')$ hold for $v' = u'_k$. Since $r(v, v')$ holds and $M, v \models \psi$, by the inductive assumption, $M', v' \models \psi$. Hence $M', u' \models \langle \pi \rangle \psi$.

Case $\varphi = [\pi]\psi$: Let v' be an arbitrary state of W' such that $S'_\pi(u', v')$ holds. There exist $\sigma_1, \ldots, \sigma_k \in \Pi_0$ such that $\sigma_1 \ldots \sigma_k \in \mathcal{L}(\pi)$ and $S'_{\sigma_1;\ldots;\sigma_k}(u', v')$ holds. Let $u'_0 = u', u'_1, \ldots, u'_{k-1}, u'_k = v'$ be states such that $S'_{\sigma_i}(u'_{i-1}, u'_i)$ holds for $1 \leq i \leq k$. Let $u_0 = u$. For every $1 \leq i \leq k$, by the condition 3, there exists a state $u_i \in W$ such that $S_{\sigma_i}(u_{i-1}, u_i)$ and $r(u_i, u'_i)$ hold. Hence, $S_\pi(u, v)$ and $r(v, v')$ hold for $v = u_k$. Since $S_\pi(u, v)$ holds and $M, u \models [\pi]\psi$, we have that $M, v \models \psi$. Since $r(v, v')$ holds and $M, v \models \psi$, by the inductive assumption, $M', v' \models \psi$. Hence $M', u' \models [\pi]\psi$.

The case $\varphi = [A]\psi$ is similar to the case $\varphi = [\pi]\psi$.

The proof for the case $\varphi = [\pi]_\diamond \psi$ is a combination of the proofs of the case $\varphi = [\pi]\psi$ and the case $\varphi = \langle \pi \rangle \psi$. ∎

3 Constructing Finite Least Pseudo-models

In this section, we present an algorithm that, given a deterministic positive logic program P in $\text{PDL}^{(0)}$, constructs a finite least pseudo-model of P.

Let X be a set of formulae, which may contain modal operators of the form $[\pi]_\diamond$ or $[A, Q]$. The saturation of X, denoted by $\mathsf{Sat}(X)$, is defined to be the least extension of X such that:

- $\top \in \mathsf{Sat}(X)$,

- if $\langle \pi; \pi' \rangle \varphi \in \mathsf{Sat}(X)$ then $\langle \pi \rangle \langle \pi' \rangle \varphi \in \mathsf{Sat}(X)$,

- if $[\pi]\varphi \in \mathsf{Sat}(X)$ then $[A_\pi, I_\pi]\varphi \in \mathsf{Sat}(X)$,

- if $[A_\pi, Q]\varphi \in \mathsf{Sat}(X)$ and $Q \cap F_\pi \neq \emptyset$ then $\varphi \in \mathsf{Sat}(X)$.

---

[6]The condition 2 corresponds to the forward direction, while the condition 3 corresponds to the backward direction.

The transfer of X through $\langle\sigma\rangle$, where $\sigma \in \Pi_0$, is defined as follows:

$$\mathsf{Trans}(X,\sigma) = \mathsf{Sat}(\{[A_\pi,\delta_\pi(Q,\sigma)]\varphi \mid [A_\pi,Q]\varphi \in X\}).$$

The compact form $\mathsf{CF}(X)$ of X is the least set of formulae obtained as follows:

- if $\varphi \in X$ and φ is not of the form $[A_\pi, Q]\varphi$ then $\varphi \in \mathsf{CF}(X)$,
- if $[A_\pi, Q]\varphi \in X$ and $Q_1, \ldots, Q_k$ are all the sets such that $[A_\pi, Q_i]\varphi \in X$ for $1 \leq i \leq k$, then $[A_\pi, Q_1 \cup \ldots \cup Q_k]\varphi \in \mathsf{CF}(X)$.

We use the following data structures:

- W : a set of states, where $\tau \in W$ is the current state.
- H : for every $w \in W$, $H(w)$ is a set of formulae called the content of w.
- $Next : W \times \{\langle\sigma\rangle\varphi \mid \sigma \in \Pi_0, \varphi \in \Phi\} \to W$, a partial function interpreted as follows: $Next(u, \langle\sigma\rangle\varphi) = v$ means $\langle\sigma\rangle\varphi \in H(u)$, $\varphi \in H(v)$, and $\langle\sigma\rangle\varphi$ is "realized" at u by going to v via R_σ.
- $Next_S : W \times \Pi_0 \to W$, where $Next_S(u, \sigma) = v$ implies $S_\sigma(u, v)$.

Using the above data structures, we define:

- h to be the restriction of H such that $h(u) = H(u) \cap \Phi_0$ for $u \in W$;
- R_σ, for $\sigma \in \Pi_0$, to be $\{(u, v) \mid Next(u, \langle\sigma\rangle\varphi) = v$ for some $\varphi\}$;
- S_σ, for $\sigma \in \Pi_0$, to be $R_\sigma \cup \{(u, v) \mid Next_S(u, \sigma) = v\}$;
- $M = \langle W, \tau, (R_\sigma)_{\sigma \in \Pi_0}, (S_\sigma)_{\sigma \in \Pi_0}, h\rangle$.

In the algorithm given below, we use the function $\mathsf{Find}(X)$ defined as follows: if there exists a state $u \in W$ with $H(u) = X$ then return u, else add a new state u to W with $H(u) = X$ and return u.

A pseudo-model of P is constructed by building a "pseudo-model graph" for P. At the beginning the pseudo-model graph contains only one state with content P. Then for every state u and every formula φ belonging to the content of u, if φ is not true at u then the algorithm makes a change to satisfy it. There are three main forms for φ: $[A_\pi, Q]\psi$, $(B_1 \wedge \ldots \wedge B_k \to A)$, and $\langle\sigma\rangle\psi$ (the form $[\pi]\psi$ is reduced to $[A_\pi, I_\pi]\psi$, and the form $\langle\pi\rangle\psi$ is reduced to $\langle\sigma\rangle\psi'$). For the case when φ is of the form $[A_\pi, Q]\psi$, for every $\sigma \in \Pi_0$, we would like to add $[A_\pi, \delta_\pi(Q, \sigma)]\psi$ to the content of every state w accessible from u via S_σ. But such an action may affect other states involved with w. So, instead of adding the formula to the content of w, we discard the connection $S_\sigma(u, w)$ and connect u via S_σ (in an appropriate way using $Next$ or $Next_S$) to a state w_* with an appropriate content, which is created if necessary. That is we use w_* to replace the role of w. For the case of $(B_1 \wedge \ldots \wedge B_k \to A)$, if all $B_1, \ldots, B_k$ are "certainly" true at u (the truth of $[\pi]_\circ p$ at u is checked in a special way) then we would like to add A to the content of u. But analogously as for the previous case, instead of modifying the content of u, we just redirect connections appropriately. States are cached and never deleted. For the case when φ is of the form $\langle \sigma \rangle \psi$, to satisfy φ at u, we connect u via R_σ to the state with content consisting of ψ and the formulae "inherited" from u via R_σ. To guarantee the constructed pseudo-model to be smallest, for every $u \in W$ and $\sigma \in \Pi_0$, we connect u via S_σ using $Next_S$ to the state with content inherited from u via S_σ. Such connections are also useful for checking the truth cf formulae of the form $[\pi]_\circ p$ in a state.

ALGORITHM 5.
Input: A deterministic positive logic program P in $\text{PDL}^{(0)}$.
Output: $M = \langle W, \tau, (R_\sigma)_{\sigma \in \Pi_0}, (S_\sigma)_{\sigma \in \Pi_0}, h \rangle$:
    a finite least pseudo-model of P.

1. $W := \{\tau\}$; $H(\tau) := \mathsf{CF}(\mathsf{Sat}(P))$;

2. for every $u \in W$ and every $\varphi \in H(u)$

   (a) case $\varphi = [A_\pi, Q]\psi$:
   for every $w \in W$ and $\sigma \in \Pi_0$ such that $S_\sigma(u,w)$ holds:

      i. $w_* := \mathsf{Find}(\mathsf{CF}(H(w) \cup \mathsf{Trans}(\{\varphi\}, \sigma)))$;

      ii. for every $\xi \in \Phi$,
   if $Next(u, \langle \sigma \rangle \xi) = w$ then $Next(u, \langle \sigma \rangle \xi) := w_*$;

      iii. if $Next_S(u, \sigma) = w$ then $Next_S(u, \sigma) := w_*$;

   (b) case $\varphi = (B_1 \wedge \ldots \wedge B_k \to A)$:
   if

     for every $1 \leq i \leq k$, $M, u \models B_i$ and if $B_i = [\pi]_\circ p$ then for every $w \in W$ and $\sigma \in \Pi_0$, if $S_\alpha(u, w)$ holds and $\alpha \sigma \beta \in \mathcal{L}(\pi)$ for some $\alpha, \beta \in \Pi_0^*$, then $Next_S(w, \sigma)$ is defined,

   then

      i. $u_* := \mathsf{Find}(\mathsf{CF}(H(u) \cup \mathsf{Sat}(\{A\})))$;

      ii. for every $v \in W$, $\sigma \in \Pi_0$, and $\psi \in \Phi$,
   if $Next(v, \langle \sigma \rangle \psi) = u$ then $Next(v, \langle \sigma \rangle \psi) := u_*$;

      iii. for every $v \in W$ and $\sigma \in \Pi_0$,
   if $Next_S(v, \sigma) = u$ then $Next_S(v, \sigma) := u_*$;

      iv. if $\tau = u$ then $\tau := u_*$;

   (c) case $\varphi = \langle \sigma \rangle \psi$:
   if $Next(u, \langle \sigma \rangle \psi)$ is not defined then
   $Next(u, \langle \sigma \rangle \psi) := \mathsf{Find}(\mathsf{CF}(\mathsf{Trans}(H(u), \sigma) \cup \mathsf{Sat}(\{\psi\})))$;

3. for every $u \in W$ and every $\sigma \in \Pi_0$,
if $Next_S(u, \sigma)$ is not defined then
$Next_S(u, \sigma) := \mathsf{Find}(\mathsf{CF}(\mathsf{Trans}(H(u), \sigma)))$;

4. while some change occurred, go to step 2.

PROPOSITION 6. *Algorithm 5 terminates in $2^{O(n^2)}$ steps and returns a pseudo-model with $2^{O(n^2)}$ states, where n is the size of P (i.e. the sum of the lengths of the clauses of P).*

Proof. For each $u \in W$ and $\varphi \in H(u)$, there are three cases: i) φ is a subformula of a clause of P, ii) φ is a formula of the form $[A_\pi, Q]\psi$ with $[\pi]\psi$ being a subformula of a clause of P, iii) φ is a formula of the form $\langle \pi \rangle \psi$, where ψ is a subformula of a clause of P and π is a subprogram occurring in P. There are less than n possible values for ψ and π, and less than 2^n possible values for Q. Hence, due to the compact form, there are no more than $2^{O(n^2)}$ possible values for $H(u)$. Since the states of W have different contents, the size of W is $2^{O(n^2)}$.

The if condition of the step 2b can be checked in $2^{O(n^2)}$ steps.

The steps 2c and 3 make a change no more than $2^{O(n^2)}.n.n = 2^{O(n^2)}$ times. For the steps 2a and 2b, note that the content of u_* (resp. w_*) is "bigger" than the content of u (resp. w). Hence $Next$ and $Next_S$ are modified by the steps 2a or 2b no more than $2^{O(n^2)}.n.n.2^{O(n^2)} = 2^{O(n^2)}$ times, and τ is modified no more than $2^{O(n^2)}$ times.

Therefore, Algorithm 5 terminates in $2^{O(n^2)}$ steps and returns a pseudo-model with $2^{O(n^2)}$ states, where each state is of size $O(n)$. ∎

LEMMA 7. *Let P be a deterministic positive logic program in $PDL^{(0)}$ and M the pseudo-model constructed by Algorithm 5 for P. Then $M \models P$.*

Proof. We will refer to the data structures used in Algorithm 5.

For $u, u' \in W$, we write $H(u) \leq H(u')$ to denote that, for every $\varphi \in H(u)$, either $\varphi \in H(u')$ or $\varphi = [A_\pi, Q]\psi$ and there exists $[A_\pi, Q']\psi \in H(u')$ with $Q' \supseteq Q$. Observe that, for every v, σ, ψ, if $Next(v, \langle \sigma \rangle \psi)$ or $Next_S(v, \sigma)$ changes its current value from u to u' then $H(u) \leq H(u')$.

To prove that $M \models P$, we show that for every $u \in W$ reachable from τ via a path using the accessibility relations $(S_\sigma)_{\sigma \in \Pi_0}$ and for every formula $\varphi \in H(u)$ without automaton-modal operators, $M, u \models \varphi$. We prove this by induction on the structure of φ.

Consider the case when $\varphi = [\pi]\psi$. Suppose that $S_\pi(u, w)$ holds. By the inductive assumption, it is sufficient to show that $\psi \in H(w)$. There exist $w_0, \ldots, w_k$ in W with $w_0 = u$, $w_k = w$, and $\sigma_1, \ldots, \sigma_k \in \Pi_0$ such that $\sigma_1 \ldots \sigma_k \in \mathcal{L}(\pi)$ and $S_{\sigma_i}(w_{i-1}, w_i)$ holds for $1 \leq i \leq k$. Since $[\pi]\psi \in H(w_0)$, we have $[A_\pi, Q]\psi \in H(w_0)$ for some $Q \supseteq I_\pi$. Hence, for every $1 \leq i \leq k$, there exists $[A_\pi, Q_i]\psi \in H(w_i)$ with $Q_i \supseteq \widetilde{\delta}_\pi(I_\pi, \sigma_1 \ldots \sigma_i)$. Since $\sigma_1 \ldots \sigma_k \in \mathcal{L}(\pi)$, $\widetilde{\delta}_\pi(I_\pi, \sigma_1 \ldots \sigma_k) \cap F_\pi \neq \emptyset$, and hence $Q_k \cap F_\pi \neq \emptyset$ and $\psi \in H(w)$.

Consider the case when $\varphi = (B_1 \wedge \ldots \wedge B_k \to A)$ and the steps 2(b)ii, 2(b)iii, 2(b)iv are executed. As no changes occur (at the end) and u is reachable from τ via a path using the accessibility relations $(S_\sigma)_{\sigma \in \Pi_0}$, we have that $u_* = u$. Thus, by the inductive assumption, $M, u \models A$, and hence $M, u \models \varphi$.

The case $\varphi = \langle\pi\rangle\psi$ is reduced to the case $\varphi = \langle\sigma\rangle\psi'$, which is trivial. ∎

LEMMA 8. *Let P be a deterministic positive logic program in $PDL^{(0)}$ and $M' = \langle W', \tau', (R'_\sigma)_{\sigma\in\Pi_0}, (S'_\sigma)_{\sigma\in\Pi_0}, h'\rangle$ be an arbitrary pseudo-model of P. Consider a moment after executing a numerated step in an execution of Algorithm 5 for P. Let $r = \{(x,x') \in W \times W' \mid M', x' \models H(x)\}$. Then the following conditions hold:*

- $r(\tau, \tau')$

- $\forall x, y, x', y', \sigma, \psi$
  $r(x,x') \wedge (Next(x, \langle\sigma\rangle\psi) = y) \wedge R'_\sigma(x',y') \wedge (M', y' \models \psi) \rightarrow r(y,y')$

- $\forall x, y, x', y', \sigma \;\; r(x,x') \wedge (Next_S(x,\sigma) = y) \wedge S'_\sigma(x',y') \rightarrow r(y,y')$

Proof. By induction on the number of executed steps.

The base case occurs after executing step 1 and the assertions clearly hold. Consider some latter step of the algorithm. As induction hypothesis, assume that the assertions hold before executing that step. Suppose that after executing the step we have r_2, W_2, H_2, $Next_2$, $Next_{S2}$, $R_{2,\sigma}$, $S_{2,\sigma}$ (for $\sigma \in \Pi_0$), and M_2 in the places of r, W, H, $Next$, $Next_S$, R_σ, S_σ, and M. We prove that:

- $r_2(\tau, \tau')$

- $\forall x, y, x', y', \sigma, \psi$
  $r_2(x,x') \wedge (Next_2(x, \langle\sigma\rangle\psi) = y) \wedge R'_\sigma(x',y') \wedge (M', y' \models \psi) \rightarrow r_2(y,y')$

- $\forall x, y, x', y', \sigma \;\; r_2(x,x') \wedge (Next_{S2}(x,\sigma) = y) \wedge S'_\sigma(x',y') \rightarrow r_2(y,y')$

It suffices to consider steps 2(a)ii, 2(a)iii, 2(b)ii-2(b)iv, 2c, and 3.

Consider the step 2(a)ii. It suffices to show that if $r(u,u') \wedge (Next(u, \langle\sigma\rangle\psi) = w) \wedge R'_\sigma(u',w') \wedge (M', w' \models \psi)$ then $M', w' \models H(w_*)$. Suppose that the premise holds. By the inductive assumption, $r(w,w')$ holds and $M', w' \models H(w)$. Since $r(u,u')$ holds and $[A_\pi, Q]\psi \in H(u)$, we have that $M', u' \models [A_\pi, Q]\psi$. Hence $M', w' \models [A_\pi, \delta_\pi(Q, \sigma)]\psi$ (since $R'_\sigma(u',w')$ holds). Hence $M', w' \models H(w_*)$.

Consider the step 2(a)iii. It suffices to show that if $r(u,u') \wedge (Next_S(u,\sigma) = w) \wedge S'_\sigma(u',w')$ holds then $M', w' \models H(w_*)$. This can be proved analogously as for the step 2(a)ii.

Consider the steps 2(b)ii-2(b)iv. Let u' be a state of W' such that $r(u,u')$ holds. It is sufficient to show that $r_2(u_*, u')$ holds. We need only to show that $M', u' \models A$. Since $r(u,u')$ holds, $M', u' \models (B_1 \wedge \ldots \wedge B_k \rightarrow A)$. Hence, it is sufficient to show that $M', u' \models B_i$ for every $1 \leq i \leq k$. Fix such an index i. There are three cases to consider:

- Case $B_i = p$: Since $M, u \models B_i$, we have that $p \in H(u)$. Since $r(u,u')$ holds, it follows that $M', u' \models B_i$.

- Case $B_i = \langle \pi \rangle p$: There exists $w \in W$ such that $R_\pi(u,w)$ holds and $p \in H(w)$. Thus, there exist $w_0, \ldots, w_k$ in W with $w_0 = u$, $w_k = w$, and $\sigma_1, \ldots, \sigma_k \in \Pi_0$ such that $\sigma_1 \ldots \sigma_k \in \mathcal{L}(\pi)$ and $R_{\sigma_i}(w_{i-1}, w_i)$ holds for $1 \leq i \leq k$. Let $\psi_1, \ldots, \psi_k$ be formulae such that $Next(w_{i-1}, \langle \sigma_i \rangle \psi_i) = w_i$ for $1 \leq i \leq k$. Let $w_0' = u$. Since $r(w_0, w_0')$ holds and $\langle \sigma_1 \rangle \psi_1 \in H(w_0)$, we have that $M', w_0' \models \langle \sigma_1 \rangle \psi_1$. Hence, there exists $w_1' \in W'$ such that $R'_{\sigma_1}(w_0', w_1')$ holds and $M', w_1' \models \psi_1$. By the inductive assumption, $r(w_1, w_1')$ holds. Analogously, there exist $w_2', \ldots, w_k' \in W'$ such that $R'_{\sigma_i}(w_{i-1}', w_i')$ and $r(w_i, w_i')$ hold for every $1 \leq i \leq k$. Thus $R'_\pi(w_0', w_k')$ holds. Since $p \in H(w)$, $w = w_k$, and $r(w_k, w_k')$ holds, we have $M', w_k' \models p$. It follows that $M', w_0' \models \langle \pi \rangle p$, which means $M', u' \models B_i$.

- Case $B_i = [\pi]_\diamond p$:
  – We first show that $M', u' \models [\pi]p$. Suppose that $S'_\pi(u', w')$ holds. There exist $\sigma_1, \ldots, \sigma_k \in \Pi_0$ and $w_0' = u', w_1', \ldots, w_k' \in W'$ such that $\sigma_1 \ldots \sigma_k \in \mathcal{L}(\pi)$ and $S'_{\sigma_i}(w_{i-1}', w_i')$ holds for every $1 \leq i \leq k$. Let $w_0 = u$ and $w_i = Next_S(w_{i-1}, \sigma_i)$ for $1 \leq i \leq k$. The "if" condition of step 2b guarantees the existence of the states w_i. By the inductive assumption, $r(w_i, w_i')$ holds for $1 \leq i \leq k$. Since $M, u \models [\pi]_\diamond p$, we have that $p \in H(w_k)$. Since $r(w_k, w_k')$ holds, it follows that $M', w_k' \models p$, which means $M', w' \models p$. This implies that $M', u' \models [\pi]p$.
  – Suppose that $\alpha \sigma \beta \in \mathcal{L}(\pi)$ and $S'_\alpha(u', w')$ holds, where $\alpha, \beta \in \Pi_0^*$ and $\sigma \in \Pi_0$. We show that $M', w' \models \langle \sigma \rangle \top$. Let $\alpha = \sigma_1 \ldots \sigma_k$. There exist $w_0', \ldots, w_k'$ such that $w_0' = u'$, $w_k' = w'$, $S'_{\sigma_i}(w_{i-1}', w_i')$ holds for every $1 \leq i \leq k$. Let $w_0 = u$ and $w_i = Next_S(w_{i-1}, \sigma_i)$ for $1 \leq i \leq k$. The "if" condition of step 2b guarantees the existence of the states w_i. By the inductive assumption, $r(w_i, w_i')$ holds for $1 \leq i \leq k$. Since $M, u \models [\pi]_\diamond p$, we have that $M, w_k \models \langle \sigma \rangle \top$. Hence there exists $v \in W$ such that $R_\sigma(w_k, v)$ holds. Thus $v = Next(w_k, \langle \sigma \rangle \zeta)$ for some ζ, and $\langle \sigma \rangle \zeta \in H(w_k)$. Since $r(w_k, w_k')$ holds, by the definition of r, $M', w_k' \models \langle \sigma \rangle \zeta$. Hence $M', w' \models \langle \sigma \rangle \top$.

Consider the step 2c. Let w denote the state $\mathsf{Find}(\mathsf{CF}(\mathsf{Trans}(H(u), \sigma) \cup \mathsf{Sat}(\{\psi\})))$. Suppose that $r(u, u')$ and $R'_\sigma(u', w')$ hold and $M', w' \models \psi$. It suffices to show that $M', w' \models H_2(w)$. Since $r(u, u')$ holds, $M', u' \models H(u)$. It follows that $M', w' \models \mathsf{Trans}(H(u), \sigma)$ (since $R'_\sigma(u', w')$ holds). Hence $M', w' \models H_2(w)$.

Consider the step 3. Let w denote the state $\mathsf{Find}(\mathsf{CF}(\mathsf{Trans}(H(u), \sigma)))$. Suppose that $r(u, u')$ and $S'_\sigma(u', w')$ hold. It suffices to show that $M', w' \models H_2(w)$. Since $r(u, u')$ holds, $M', u' \models H(u)$. It follows that $M', w' \models \mathsf{Trans}(H(u), \sigma)$ (since $S'_\sigma(u', w')$ holds). Hence $M', w' \models H_2(w)$. ∎

LEMMA 9. *Let P be a deterministic positive logic program in $PDL^{(0)}$, M be the pseudo-model constructed by Algorithm 5 for P, and $M' = \langle W', \tau',$*

$(R'_\sigma)_{\sigma \in \Pi_0}$, $(S'_\sigma)_{\sigma \in \Pi_0}$, $h'\rangle$ be an arbitrary pseudo-model of P. Then $M \leq M'$. In particular, if r is the relation defined as in Lemma 8, then $M \leq_r M'$.

Proof. We will refer to the data structures used in Algorithm 5. Let r be the relation specified in Lemma 8 for the end of an execution of Algorithm 5 for P. By definition, $\forall x, x'\ r(x,x') \rightarrow h(x) \subseteq h(x')$ is true. By Lemma 8, $r(\tau, \tau')$ holds.

We prove that $\forall \sigma, x, x', y\ R_\sigma(x,y) \wedge r(x,x') \rightarrow \exists y'\ R'_\sigma(x',y') \wedge r(y,y')$. Suppose that $R_\sigma(x,y)$ and $r(x,x')$ hold. There exists ζ such that $y = Next(x, \langle\sigma\rangle\zeta)$. We have that $\langle\sigma\rangle\zeta \in H(x)$. Since $r(x,x')$ holds, $M', x' \models \langle\sigma\rangle\zeta$. There thus exists $y' \in W'$ such that $R'_\sigma(x',y')$ holds and $M', y' \models \zeta$. By Lemma 8, $r(y,y')$ holds.

We now prove that $\forall \sigma, x, x', y'\ S'_\sigma(x',y') \wedge r(x,x') \rightarrow \exists y\ S_\sigma(x,y) \wedge r(y,y')$. Suppose that $S'_\sigma(x',y')$ and $r(x,x')$ hold. Let $y = Next_S(x,\sigma)$. By Lemma 8, $r(y,y')$ holds.

We have prove that $M \leq_r M'$. Therefore $M \leq M'$. ∎

THEOREM 10. *Let P be a deterministic positive logic program in $PDL^{(0)}$. The pseudo-model M constructed by Algorithm 5 for P is a least pseudo-model of P.*

This theorem follows from Lemmas 7 and 9.

4 Characterizations of Least Pseudo-models

In classical propositional logic, if M is the least model of a positive logic program P then for every positive formula φ, $P \models \varphi$ iff $M \models \varphi$. Similarly, in a basic serial monomodal logic L, if M is a least L-model of a positive modal logic program P then for every positive (modal) formula φ, $P \models_L \varphi$ iff $M \models \varphi$ (see [7]). In this section, we extend such an assertion for the deterministic Horn fragment of $PDL^{(0)}$. The main result says that if P is a deterministic positive logic program in $PDL^{(0)}$, M is a least pseudo-model of P, and φ is a serial positive formula, then $P \models \varphi$ iff $M \models \varphi$.

Given a pseudo-model $M = \langle W, \tau, (R_\sigma)_{\sigma \in \Pi_0}, (S_\sigma)_{\sigma \in \Pi_0}, h\rangle$, let $M' = \langle W, \tau, (R'_\sigma)_{\sigma \in \Pi_0}, (S'_\sigma)_{\sigma \in \Pi_0}, h\rangle$ be the pseudo-model specified as follows:

$$R'_\sigma = R_\sigma \cup \{(u,w) \mid S_\sigma(u,w) \text{ and } R_\sigma(u,w') \text{ hold for some } w'\},$$
$$S'_\sigma = S_\sigma \setminus \{(u,w) \mid w \in W \text{ and } R_\sigma(u,w') \text{ does not hold for any } w'\}.$$

Thus, $R'_\sigma = S'_\sigma$ for every $\sigma \in \Pi_0$, and M' can be treated as a Kripke model. We call M' the *model corresponding to* M.

LEMMA 11. *Let P be a deterministic positive logic program in $PDL^{(0)}$, M the pseudo-model constructed by Algorithm 5 for P, M' the model corresponding to M, and φ a serial positive formula. If $M' \models \varphi$ then $M \models \varphi$.*

Proof. We will refer to the data structures used by Algorithm 5 for P and M. Let $r = \{(x,x') \in W \times W \mid M, x' \models H(x)\}$. By Lemma 8, we have that:

(i) $r(\tau, \tau)$

(ii) $\forall x, y, x', y', \sigma, \psi$
$r(x, x') \wedge (Next(x, \langle \sigma \rangle \psi) = y) \wedge R_\sigma(x', y') \wedge (M, y' \models \psi) \to r(y, y')$

(iii) $\forall x, y, x', y', \sigma \ r(x, x') \wedge (Next_S(x, \sigma) = y) \wedge S_\sigma(x', y') \to r(y, y')$

and by Lemma 9, $M \leq_r M$.

Let $M' = \langle W, \tau, (R'_\sigma)_{\sigma \in \Pi_0}, (S'_\sigma)_{\sigma \in \Pi_0}, h \rangle$. It suffices to prove by induction on the construction of φ that, for every $x, x' \in W$, if $r(x, x')$ holds and $M', x \models \varphi$ then $M, x' \models \varphi$. Suppose that $r(x, x')$ holds and $M', x \models \varphi$.

- Case $\varphi = p$: Since $M', x \models \varphi$, we have that $p \in h(x)$. Since $M \leq_r M$ and $r(x, x')$ holds, we derive that $p \in h(x')$. Hence $M, x' \models p$.

- Case $\varphi = \psi \vee \zeta$ or $\varphi = \psi \wedge \zeta$ is trivial.

- Case $\varphi = [\pi]_\circ \psi$: Since $M', x \models [\pi]_\circ \psi$, for every $\alpha, \beta \in \Pi_0^*$, $\sigma \in \Pi_0$, if $\alpha \sigma \beta \in \mathcal{L}(\pi)$ and $S'_\alpha(x, y)$ holds, then $R'_\sigma(y, z')$ holds for some z', and hence $R_\sigma(y, z)$ holds for some z. It follows that, for every $\alpha, \beta \in \Pi_0^*$ and $\sigma \in \Pi_0$, if $\alpha \sigma \beta \in \mathcal{L}(\pi)$ and $S_\alpha(x, y)$ holds, then $R_\sigma(y, z)$ holds for some z and $\{z \mid S'_\sigma(y, z)\} = \{z \mid S_\sigma(y, z)\}$. This together with $M', x \models [\pi]_\circ \psi$ implies that $M, x \models [\pi]_\circ \psi$. Since $M \leq_r M$ and $r(x, x')$ holds, it follows that $M, x' \models [\pi]_\circ \psi$.

- Case $\varphi = \langle \pi \rangle \psi$: There exist $\sigma_1, \ldots, \sigma_k \in \Pi_0$ and $x_0 = x, x_1, \ldots, x_k \in W$ such that $\sigma_1 \ldots \sigma_k \in \mathcal{L}(\pi)$, $R'_{\sigma_i}(x_{i-1}, x_i)$ holds for $1 \leq i \leq k$, and $M', x_k \models \psi$. Let $x'_0 = x'$. For $1 \leq i \leq k$, choose x'_i as follows:

  – Case $R_{\sigma_i}(x_{i-1}, x_i)$ holds and $x_i = Next(x_{i-1}, \langle \sigma_i \rangle \zeta_i)$: We have that $\langle \sigma_i \rangle \zeta_i \in H(x_{i-1})$. By the proof of Lemma 7, $M, x_{i-1} \models \langle \sigma_i \rangle \zeta_i$. Since $M \leq_r M$ and $r(x_{i-1}, x'_{i-1})$ holds, it follows that $M, x'_{i-1} \models \langle \sigma_i \rangle \zeta_i$. Let x'_i be a state such that $R_{\sigma_i}(x'_{i-1}, x'_i)$ holds and $M, x'_i \models \zeta_i$. Thus, by (ii), $r(x_i, x'_i)$ holds.

  – Case $R_{\sigma_i}(x_{i-1}, x_i)$ does not hold and $x_i = Next_S(x_{i-1}, \sigma_i)$: There must exist ζ_i such that $Next(x_{i-1}, \langle \sigma_i \rangle \zeta_i)$ is defined. Choose x'_i as in the above subcase. Thus $R_{\sigma_i}(x'_{i-1}, x'_i)$ holds. By (iii), it follows that $r(x_i, x'_i)$ holds.

Since $r(x_k, x'_k)$ holds and $M', x_k \models \psi$, by the inductive assumption, $M, x'_k \models \psi$. Since $R_{\sigma_i}(x'_{i-1}, x'_i)$ holds for every $1 \leq i \leq k$, it follows that $M, x'_0 \models \langle \pi \rangle \psi$, which means $M, x' \models \varphi$. ∎

THEOREM 12. *Let P be a deterministic positive logic program in $PDL^{(0)}$, M a least pseudo-model of P, and φ a serial positive formula. Then $P \models \varphi$ iff $M \models \varphi$.*

Proof. Consider the "if" direction. Suppose that $M \models \varphi$. Let M' be an arbitrary Kripke model of P. As M' is also a pseudo-model, we have that $M \leq M'$. Hence $M' \models \varphi$. Therefore $P \models \varphi$.

Now consider the "only if" direction. Suppose that $P \models \varphi$. We can assume that $M = \langle W, \tau, (R_\sigma)_{\sigma \in \Pi_0}, (S_\sigma)_{\sigma \in \Pi_0}, h \rangle$ is the pseudo-model constructed by Algorithm 5 for P. Let $M' = \langle W, \tau, (R'_\sigma)_{\sigma \in \Pi_0}, (S'_\sigma)_{\sigma \in \Pi_0}, h \rangle$ be the model corresponding to M. It is sufficient to show that $M' \models P$, because this implies that $M' \models \varphi$, and by Lemma 11, $M \models \varphi$.

Let H be the data structure of Algorithm 5 which specifies the contents of states of M. To show that $M' \models P$, we prove by induction on the construction of φ that if $\varphi \in H(u)$ then $M', u \models \varphi$. Suppose that $\varphi \in H(u)$. By the proof of Lemma 7, $M, u \models \varphi$. Using this, the only non-trivial case is when φ is of the form $B_1 \wedge \ldots \wedge B_k \rightarrow A$. Suppose that $M', u \models B_1 \wedge \ldots \wedge B_k$. By Lemma 11, $M, u \models B_1 \wedge \ldots \wedge B_k$. Since $M, u \models \varphi$, it follows that $M, u \models A$. This implies that $M', u \models A$ (because $R_\sigma \subseteq R'_\sigma$ and $S'_\sigma \subseteq S_\sigma$ for every $\sigma \in \Pi_0$). Therefore $M', u \models \varphi$. This completes the proof. ∎

COROLLARY 13. *Let P be a deterministic positive logic program in $PDL^{(0)}$, M' the model corresponding to the pseudo-model constructed by Algorithm 5 for P, and φ a serial positive formula. Then $P \models \varphi$ iff $M' \models \varphi$.*

Proof. By the proof of the above theorem, $M' \models P$. Hence, $P \models \varphi$ implies $M' \models \varphi$. For the conversion, suppose that $M' \models \varphi$. Let M be the pseudo-model constructed by Algorithm 5 for P. By Lemma 11, $M \models \varphi$. Let M'' be an arbitrary model of P. Since M is a least pseudo-model of P, we have that $M \leq M''$, which implies $M'' \models \varphi$. Hence $P \models \varphi$. ∎

5 Lower Bound

Proposition 6 states that, given a deterministic positive logic program P in $PDL^{(0)}$, a finite least pseudo-model of P can be constructed in exponential time and it has an exponential size (in the worst case). In this section, we give an example showing that, in general, this estimation is tight.

Let M and M' be pseudo-models. Define that $M \leq_\diamond M'$ if for every serial positive formula φ, $M \models \varphi$ implies $M' \models \varphi$. Define that $M \equiv_\diamond M'$ if $M \leq_\diamond M'$ and $M' \leq_\diamond M$.

LEMMA 14. *Let $M = \langle W, \tau, (R_\sigma)_{\sigma \in \Pi_0}, (S_\sigma)_{\sigma \in \Pi_0}, h \rangle$ be a finite pseudo-model such that, for every $\sigma \in \Pi_0$, R_σ is a function, i.e. $\forall x \exists ! y R_\sigma(x, y)$. Let $M' = \langle W', \tau', (R'_\sigma)_{\sigma \in \Pi_0}, (S'_\sigma)_{\sigma \in \Pi_0}, h' \rangle$ be a finite pseudo-model such that $M' \equiv_\diamond M$. Then for every $w \in W$ reachable from τ via a path using $(R_\sigma)_{\sigma \in \Pi_0}$ there exists $w' \in W'$ such that $(M, w) \equiv_\diamond (M', w')$.*

Proof. We first show that, for every $\sigma \in \Pi_0$,

$$\forall x, x', y \; R_\sigma(x, y) \wedge ((M, x) \leq_\diamond (M', x')) \rightarrow \\ \exists y' \; R'_\sigma(x', y') \wedge ((M, y) \leq_\diamond (M', y'))$$

Suppose that $R_\sigma(x,y)$ and $(M,x) \leq_\diamond (M',x')$. We show that there exists $y' \in W'$ such that $R'_\sigma(x',y')$ and $(M,y) \leq_\diamond (M',y')$. Suppose oppositely that for every $y' \in W'$ such that $R'_\sigma(x',y')$, $(M,y) \leq_\diamond (M',y')$ does not hold, i.e. there exists a serial positive formula $\varphi_{y'}$ such that $M,y \models \varphi_{y'}$ but $M',y' \not\models \varphi_{y'}$. Let φ be the conjunction of all such $\varphi_{y'}$. We have $M,x \models \langle\sigma\rangle\varphi$, while $M',x' \not\models \langle\sigma\rangle\varphi$, which contradicts the assumption that $(M,x) \leq_\diamond (M',x')$.

Similarly, we also have that, for every $\sigma \in \Pi_0$,

$$\forall\, x',y',x\; R'_\sigma(x',y') \wedge ((M',x') \leq_\diamond (M,x)) \to$$
$$\exists y\; R_\sigma(x,y) \wedge ((M',y') \leq_\diamond (M,y))$$

We now prove the claim of the lemma. It suffices to prove by induction on k that if $R_{\sigma_1\ldots\sigma_k}(\tau, w_k)$ holds then there exists $w'_k \in W'$ such that $(M, w_k) \equiv_\diamond (M', w'_k)$. The base case $k = 0$ holds for $w'_0 = \tau'$. For the induction step, suppose that the hypothesis holds for k, and $R_{\sigma_{k+1}}(w_k, w_{k+1})$ holds for some $\sigma_{k+1} \in \Pi_0$. We show that there exists $w'_{k+1} \in W'$ such that $(M, w_{k+1}) \equiv_\diamond (M', w'_{k+1})$. Since $(M, w_k) \leq_\diamond (M', w'_k)$ and $R_{\sigma_{k+1}}(w_k, w_{k+1})$ holds, by (i), there exists $w'_{k+1} \in W'$ such that $R'_{\sigma_{k+1}}(w'_k, w'_{k+1})$ holds and $(M, w_{k+1}) \leq_\diamond (M', w'_{k+1})$. On the other hand, since $(M', w'_k) \leq_\diamond (M, w_k)$ and $R'_{\sigma_{k+1}}(w'_k, w'_{k+1})$ holds, by (ii), there exists $w''_{k+1} \in W$ such that $R_{\sigma_{k+1}}(w_k, w''_{k+1})$ holds and $(M', w'_{k+1}) \leq_\diamond (M, w''_{k+1})$. Since $R_{\sigma_{k+1}}$ is a function, $w''_{k+1} = w_{k+1}$, and hence $(M, w_{k+1}) \equiv_\diamond (M', w'_{k+1})$. ∎

PROPOSITION 15. *Let $\Pi_0 = \{a, b\}$ and $P = \{[(a \cup b)^*; a; (a \cup b)^{(n-1)}]p,$ $[(a \cup b)^*]\langle a\rangle\top, [(a \cup b)^*]\langle b\rangle\top\}$. If M' is a pseudo-model characterizing P w.r.t. serial positive formulae (i.e. for every serial positive formula φ, $P \models \varphi$ iff $M' \models \varphi$), then M' has size $2^{\Omega(n)}$.*

Proof. Let $M = \langle W, \tau, (R_\sigma)_{\sigma \in \Pi_0}, (S_\sigma)_{\sigma \in \Pi_0}, h\rangle$ be the pseudo-model constructed by Algorithm 5 for P. It is easy to see that M satisfies the conditions stated in Lemma 14.

Let $\pi = (a \cup b)^*; a; (a \cup b)^{(n-1)}$. For $\alpha, \beta \in \Pi_0^*$, define that $\alpha \sim \beta$ if for every $\gamma \in \Pi_0^*$, $\alpha\gamma \in \mathcal{L}(\pi)$ iff $\beta\gamma \in \mathcal{L}(\pi)$. The equivalence relation $\sim$ has exactly 2^n abstract classes. Let $\alpha, \beta \in \Sigma^*$ and $\alpha \not\sim \beta$. There exist w_α and w_β such that $R_\alpha(\tau, w_\alpha)$ and $R_\beta(\tau, w_\beta)$ hold. Since $\alpha \not\sim \beta$, there exists $\gamma \in \Pi_0^*$ such that exactly one of $\alpha\gamma$ and $\beta\gamma$ belongs to $\mathcal{L}(\pi)$. Thus, $[\gamma]_\diamond p$ is true at exactly one of the worlds w_α and w_β. Hence $(M, w_\alpha) \not\equiv_\diamond (M, w_\beta)$. This implies that M contains at least 2^n states, which are reachable from τ (via paths using $(R_\sigma)_{\sigma \in \Pi_0}$) and not equivalent to each other. By Lemma 14, it follows that M' has at least 2^n states. ∎

6 Further Work and Conclusions

Recall that if L is one of the basic propositional serial monomodal logics and P is a positive logic program in L then there exists a finite least L-model of P [7]. To obtain a similar result for PDL$^{(0)}$, we have restricted to the deterministic Horn fragment and used pseudo-models. The restriction is necessary to overcome the problem of nondeterminism caused by non-seriality, and pseudo-models are needed because that deterministic positive logic programs in PDL$^{(0)}$, e.g. $\{[\sigma]p\}$, do not always have least Kripke models. Pseudo-models satisfy the following expectations:

- A least pseudo-model M of a deterministic positive logic program P has the property that for every serial positive formula φ, $P \models \varphi$ iff $M \models \varphi$.

- Every deterministic positive logic program has a finite least pseudo-model.

- Given a pseudo-model M and a formula φ, the problem of checking $M \models \varphi$ is solvable in polynomial time in the sizes of M and φ.

The model M' corresponding to the least pseudo-model M constructed by Algorithm 5 for a deterministic positive logic program P also characterizes P w.r.t. serial positive formulae. However, M is more useful then M' in the aspect that, for every positive formula φ, $M \models \varphi$ implies $P \models \varphi$ (because M is less than or equal to every (pseudo-)model of P), while M' does not have such a property.

Our Algorithm 5 runs in exponential time and returns a pseudo-model with size $2^{O(n^2)}$. We have given a deterministic positive logic program such that every pseudo-model characterizing it w.r.t. serial positive formulae must have an exponential size. This does not imply that the (combined) complexity of checking satisfiability of deterministic Horn formulae is EXPTIME-complete. It is an open problem.

In the PDL$^{(0)}$-like description logic (PDL$^{(0)}$-Desc), programs of PDL$^{(0)}$ are used as role constructors. A TBox of that logic is a finite set of formulae of PDL$^{(0)}$, which are treated as global assumptions for all the states (but not as local assumptions of the current state τ). Apart from the TBox, a knowledge base in a description logic contains also an ABox, which is a set of facts of the form $p(a)$ or $R(a,b)$, where p is a "concept", a and b are "objects", and R is a "role name". In the terminology of PDL, p is an atomic proposition, a and b are states, and R can be assumed to be an atomic program. Note that objects in description logics correspond to states in PDL (and possible worlds in modal logics). The instance checking problem in PDL$^{(0)}$-Desc is stated as follows: given a TBox T and an ABox A of PDL$^{(0)}$-Desc, a concept C, and an object a, check whether a is an instance of C in every model of $T \cup A$ (i.e. whether $T \cup A \models C(a)$, where $\models$ reflects "global semantic consequence" in PDL$^{(0)}$-Desc). The data complexity of that problem is measured w.r.t. the size of A, while assuming that T, C, and a are fixed. Our claim is that if T is a deterministic positive logic program in PDL$^{(0)}$ then the data complexity of that problem is in PTIME. A formal proof of this will appear in an extension of this paper. The sketch is as follows:

- We construct a finite least pseudo-model M for the ABox A and the global assumptions T by starting with the graph corresponding to A

and proceeding in a similar way as Algorithm 5, except that T is added to the content of every state.

- Then $T \cup A \models C(a)$ iff $M, a \models C$.
- Since T is fixed, the size of M and the complexity of constructing M are bounded by a polynomial in the size of A.

Additionally, similarly to the extension [8] of [7], our method can be extended to develop declarative and procedural semantics for deterministic positive logic programs in first-order dynamic logic, which will be useful for logic programming about actions, time, belief, and knowledge. Goals to such programs are deterministic negative clauses. This is a line to combine modal and temporal logic programming and remains as a future work.

Acknowledgements
I would like to thank the anonymous reviewers for useful comments.

BIBLIOGRAPHY

[1] Ph. Balbiani, L. Fariñas del Cerro, and A. Herzig. Declarative semantics for modal logic programs. In *Proceedings of the 1988 International Conference on Fifth Generation Computer Systems*, pages 507–514. ICOT, 1988.

[2] M. Baldoni, L. Giordano, and A. Martelli. A framework for a modal logic programming. In *Joint International Conference and Symposium on Logic Programming*, pages 52–66. MIT Press, 1996.

[3] F. Debart, P. Enjalbert, and M. Lescot. Multimodal logic programming using equational and order-sorted logic. *Theoretical Comp. Science*, 105:141–166, 1992.

[4] R. Goré and L.A. Nguyen. A tableau system with automaton-labelled formulae for regular grammar logics. In B. Beckert, editor, *Proceedings of TABLEAUX 2005*, *LNAI 3702*, pages 138–152. Springer-Verlag, 2005.

[5] D. Harel, D. Kozen, and J. Tiuryn. *Dynamic Logic*. MIT Press, 2000.

[6] U. Hustadt, B. Motik, and U. Sattler. Data complexity of reasoning in very expressive description logics. In L.P. Kaelbling and A. Saffiotti, editors, *IJCAI*, pages 466–471. Professional Book Center, 2005.

[7] L.A. Nguyen. Constructing the least models for positive modal logic programs. *Fundamenta Informaticae*, 42(1):29–60, 2000.

[8] L.A. Nguyen. A fixpoint semantics and an SLD-resolution calculus for modal logic programs. *Fundamenta Informaticae*, 55(1):63–100, 2003.

[9] L.A. Nguyen. Multimodal logic programming. To appear in TCS, 2006.

[10] A. Nonnengart. How to use modalities and sorts in Prolog. In C. MacNish, D. Pearce, and L.M. Pereira, editors, *Proceedings of JELIA'94, LNCS 838*, pages 365–378. Springer, 1994.

Linh Anh Nguyen
Institute of Informatics, University of Warsaw
ul. Banacha 2, 02-097 Warsaw, Poland
nguyen@mimuw.edu.pl

Complexity of intuitionistic and Visser's basic and formal logics in finitely many variables

MIKHAIL RYBAKOV

ABSTRACT. The main result of the paper is that the decision problem for the variable free fragment of **BPL**, one-variable fragment of **FPL**, and two-variable fragment of **Int** is PSPACE-complete. Some relevant questions are also discussed.

Keywords: complexity, decision problem, propositional logic, intuitionistic logic, Visser's formal logic, non-classical logic, complexity function.

1 Introduction

It is well known that practically all standard propositional logics are decidable but the decision problem for them is usually very hard: even for classical propositional logic **Cl** only exponential time algorithms deciding it are known, and the decision (satisfiability) problem for **Cl** is NP-complete [10]. Complexity problems for non-classical (more exactly, superintuitionistic) logics were first explicitly mentioned by A. V. Kuznetsov in 1975, see [18]. One of the questions raised by A. V. Kuznetsov was the problem of polynomial approximability of intuitionistic propositional logic **Int** and some of its extensions. A. V. Kuznetsov showed [19] that if this problem is solved positively for **Int**, then **Int** and **Cl** are polynomially equivalent. As follows from [19], it is the same situation with some other non-classical propositional logics.

In 1977 R. E. Ladner proved that the decision problem for **K**, **T**, **S4** is PSPACE-complete [20]. In 1979 R. Statman proved that the decision problem for **Int** is also PSPACE-complete [33]. So, a positive solution to Kuznetsov's question would imply that NP = PSPACE.

Later Ladner's and Statman's ideas were applied by other authors to prove PSPACE-hardness for very wide classes of logics: for example, see [1, 4, 9, 11, 40].

Because the decision problem even for 'simple' logics — such as **K**, **T**, **K4**, **S4**, **S5**, **Int**, etc. — is quite hard, it is natural to consider some restrictions for these logics and the decision problem for corresponding fragments. We shall consider the situation when the propositional language contains only finitely many variables. The main reason for this restriction is that in applications we need only finitely many variables, and it is even sufficient to have a fixed number of variables.

Apparently, the question about complexity of non-classical propositional logics in finitely many variables arose immediately after Ladner's and Statman's results were published (both R. E. Ladner and R. Statman used infinitely many propositional variables in their proofs). As an open problem[1] it was put by A. V. Chagrov in 1980s at several conferences and published later in [9, Problem 18.4].

For any propositional logic L and any natural number n, by $L(n)$ we denote the least fragment of L containing all formulas constructed from variables $p_1, \ldots, p_n$. By $L(0)$ we denote the variable free fragment of L.

The question about complexity of the decision problem for $L(n)$ is quite well studied in the case of standard modal logics. So, E. Spaan [32] in fact proved PSPACE-hardness for **K**(1); later J. Y. Halpern [15] presented the same result for **K**(1), **T**(1), **S4**(1) (some inaccuracies in [15] in the case of **S4** were corrected in [8]). In [8] PSPACE-hardness and NP-hardness is proved for n-variable fragments of wide classes of normal modal logics; in particular, it is proved that the decision problem for **K**(0), **K4**(0), **S4**(1), **GL**(1), **Grz**(1) is PSPACE-complete (for **GL**(1) see also [36]).

REMARK 1. Of course, from [8] one can easily obtain similar results for non-normal modal logics too. In fact in [8] it is proved that *all sets of formulas* (in particular, all non-normal modal logics) between **K**(0) and **K4**(0), between **K**(1) and **GL**(1), between **K**(1) and **Grz**(1) have PSPACE-hard decision problem; the same is true for NP-hardness results presented in [8]. One can also extend results of [8] to some non-normal logics incomparable with **K**. For example, using the fact that for any formula φ,

$$\varphi \in \mathbf{S4} \iff \Box\Box\top \to \varphi \in \mathbf{S3},$$

we obtain PSPACE-hardness (and hence, PSPACE-completeness) of the decision problem for **S3**(1).

In this paper I present some related results for non-modal logics, which have an intuitionistic-like Kripke semantics. First of all I mean intuitionistic propositional logic **Int**, Visser's basic propositional logic **BPL**, and Visser's formal propositional logic[2] **FPL**.

Intuitionistic logic is well-known and quite well studied in both the propositional and predicate cases; because I cannot say the same about Visser's basic and formal logics, I give some references and some reasons for interest in these logics.

First of all, the reader can see for these logics the paper [38] of A. Visser, where **BPL** and **FPL** are defined (for predicate case of **BPL** see [25]). The

---

[1]More exactly, as the conjecture that **Int** in finitely many variables is decidable in polynomial time.

[2]The logic **FPL** has at least two names: *formal propositional logic* and *fixed point logic*. We have an 'interesting' coincidence: these names have the same abbreviation, i. e., **FPL**.

main logic in [38] is **FPL**, which is obtained by interpreting implication as formal provability (in Peano Arithmetic); **BPL** is used in [38] as a 'technical' logic.

Some advanced results about **BPL** and **FPL** the reader can find, for example, in [5, 6, 35, 39]. I note only that in [35], in particular, **BPL** is described as a Hilbert-type calculus (A. Visser presented it in [38] by a set of rule schemes) and that some 'undecidability' results for **FPL** were presented at the conference 'Computer Science Applications of Modal Logic' [7]. As for complexity results for these logics, I give several references below (see Remarks 2 and 3).

Both **BPL** and **FPL** are connected with the modal logics **K4** and **GL** respectively (note that **K4** may be also considered as a 'technical' logic: we are interested mainly in its extensions such as **GL**, **S4**, **Grz**, etc.) via the following Gödel translation T:

$$\begin{aligned} \mathsf{T}(\bot) &= \Box\bot; \\ \mathsf{T}(p) &= \Box p, \text{ where } p \text{ is a propositional variable}; \\ \mathsf{T}(\varphi \wedge \psi) &= \mathsf{T}(\varphi) \wedge \mathsf{T}(\psi); \\ \mathsf{T}(\varphi \vee \psi) &= \mathsf{T}(\varphi) \vee \mathsf{T}(\psi); \\ \mathsf{T}(\varphi \rightarrow \psi) &= \Box(\mathsf{T}(\varphi) \rightarrow \mathsf{T}(\psi)). \end{aligned}$$

Exactly, as it follows from [38], for every formula φ,

$$\begin{aligned} \varphi \in \mathbf{BPL} &\iff \mathsf{T}(\varphi) \in \mathbf{K4}; \\ \varphi \in \mathbf{FPL} &\iff \mathsf{T}(\varphi) \in \mathbf{GL}. \end{aligned}$$

It is also well known [9] that for every formula φ,

$$\varphi \in \mathbf{Int} \iff \mathsf{T}(\varphi) \in \mathbf{S4},$$

so, we may mean **BPL** and **FPL** as 'superintuitionistic' fragments of **K4** and **GL** respectively. Roughly speaking, in particular, we may mean the implication in **Int**, **BPL**, and **FPL** as the strong implication in **S4**, **K4**, and **GL**, correspondingly (but instead of variables and $\bot$ we use 'boxed' variables and 'boxed' $\bot$).

Thus, **Int**, **BPL**, and **FPL** are 'naturally' connected with well-known modal logics; moreover, for these modal logics we know about their complexity in finitely many variables [8]. At the same time there are some 'technical' problems[3] to (easily) extend the proofs in [8] for the cases of **Int**, **BPL**, and **FPL**: the connectives in these logics are not classical and we do not have 'pure' modalities in the language. There is also an observation of another kind: the decision problem for **S4**(1) is PSPACE-complete while the one for **Int**(1) is in P: cf. [21]. The last observation gives us a hope that the situation with the complexity of **Int**(n), **BPL**(n), and **FPL**(n) differs from the one for corresponding modal logics.

---

[3] The main problem is with negation. But I omit any details here: after all I am going to give a proof.

In this paper I am going to clarify the situation: unfortunately (or maybe, fortunately!) the decision problem for **Int**(2), **BPL**(0), and **FPL**(1) is PSPACE-complete. See Section 3, Theorem 4. This theorem has some corollaries about the complexity of wide classes of logics and their positive fragments in finitely many variables (including some extensions of **Int**); the corollaries are presented in Section 4. As an application of the construction in Theorem 4 we obtain exponential lower bounds on the complexity function for all PSPACE-hard fragments of logics considered in this paper; for details see Section 5.

REMARK 2. Most results in this article are presented in other papers — see [26, 27, 28, 29] — but they are in Russian (the exception [29] does not contain any proof), so, as I think, actually they are not known in modal logic society, and I am thankful to AiML for that I can present these results here.

2 Definitions

Our definitions for propositional language and Kripke semantics for it mainly correspond to [9] and [38]; for complexity problems the reader can consult [13, 14, 16, 22]. So, the reader can omit this section.

We shall consider formulas constructed from propositional variables $p_1, p_2, p_3, \ldots$ and constant $\bot$ with $\wedge$, $\vee$, $\rightarrow$, and brackets. We define $\neg$ and $\top$ as the usual abbreviations: $\neg \varphi = \varphi \rightarrow \bot$, $\top = \neg \bot$. We shall use the following standard conventions on representation of formulas: we assume $\neg$ to connect formulas more strongly than $\wedge$ and $\vee$, which in turn are stronger than $\rightarrow$, and omit those brackets that can be recovered according to this priority of the connectives.

Below we shall mainly consider positive formulas. A formula φ is called *positive* if φ is constructed from propositional variables with $\wedge$, $\vee$, and $\rightarrow$. For a logic (a set of formulas) L denote by L^+ the positive fragment of L.

Let us define intuitionistic-like Kripke semantics for this language (the difference from the usual definitions, as in, for example, [9], is in that some worlds in Kripke frames can be irreflexive).

An *intuitionistic Kripke frame* is a pair $\mathfrak{F} = \langle W, R \rangle$ consisting of a non-empty set W and a relation R on W, which is transitive and antisymmetric. The elements of W are called the *worlds* of the frame $\mathfrak{F}$, and the relation R is called the *accessibility relation*. If for some $w, w' \in W$ the relation wRw' holds, we say that w' *is accessible from* w.

A *valuation* in an intuitionistic frame $\mathfrak{F} = \langle W, R \rangle$ is a map v associating with each propositional variable p some subset $v(p)$ of W such that for every $w \in v(p)$ and every $w' \in W$, wRw' implies $w' \in v(p)$.

An *intuitionistic Kripke model* is a pair $\mathfrak{M} = \langle \mathfrak{F}, v \rangle$, where $\mathfrak{F}$ is an intuitionistic frame and v is a valuation in $\mathfrak{F}$.

Let $\mathfrak{M} = \langle \mathfrak{F}, v \rangle$ be a model and w be a world in the frame $\mathfrak{F} = \langle W, R \rangle$. By induction on the construction of formula φ we define a relation $(\mathfrak{M}, w) \models \varphi$:

$(\mathfrak{M}, w) \not\models \bot$;

$(\mathfrak{M}, w) \models p \iff w \in v(p)$, where p is a propositional variable;

$(\mathfrak{M}, w) \models \psi' \wedge \psi'' \iff (\mathfrak{M}, w) \models \psi'$ and $(\mathfrak{M}, w) \models \psi''$;

$(\mathfrak{M}, w) \models \psi' \vee \psi'' \iff (\mathfrak{M}, w) \models \psi'$ or $(\mathfrak{M}, w) \models \psi''$;

$(\mathfrak{M}, w) \models \psi' \to \psi'' \iff$ for every $w' \in W$ such that wRw', $(\mathfrak{M}, w') \not\models \psi'$ or $(\mathfrak{M}, w') \models \psi''$.

From the definition it follows that $(\mathfrak{M}, w) \models \top$ and

$(\mathfrak{M}, w) \models \neg \psi \iff$ for every $w' \in W$ such that wRw', $(\mathfrak{M}, w') \not\models \psi$,

for every $w \in W$.

If $(\mathfrak{M}, w) \models \varphi$ holds, we say that φ *is true at the world w in $\mathfrak{M}$*; otherwise that φ *is refuted at w*. We say that φ *is true in a model* $\mathfrak{M} = \langle \mathfrak{F}, v \rangle$ defined on a frame $\mathfrak{F} = \langle W, R \rangle$ if $(\mathfrak{M}, w) \models \varphi$, for every $w \in W$; if φ is true in $\mathfrak{M}$, we write $\mathfrak{M} \models \varphi$. We say that φ *is true in a frame* $\mathfrak{F} = \langle W, R \rangle$ if φ is true in every model based on $\mathfrak{F}$; if φ is true in $\mathfrak{F}$, we write $\mathfrak{F} \models \varphi$. We say that φ *is true at the world w in frame $\mathfrak{F}$* if $(\mathfrak{M}, w) \models \varphi$, for every model $\mathfrak{M}$ defined on $\mathfrak{F}$; if φ is true at the world w in frame $\mathfrak{F}$, we write $(\mathfrak{F}, w) \models \varphi$.

We define propositional intuitionistic logic **Int** to be the set of formulas that are true in every reflexive frame, Visser's basic propositional logic **BPL** to be the set of formulas that are true in every frame, and Visser's formal propositional logic **FPL** to be the set of formulas that are true in every irreflexive and finite frame.

Let us recall the definitions of the complexity classes P, NP, and PSPACE. We consider only algorithmic recognition problems, i.e., problems that can be formulated as questions '$x \in X$?', for some set X (as an example, the decision problem for a logic L looks as '$\varphi \in L$?'). The complexity classes P, NP, and PSPACE are the classes of problems that can be solved respectively by polynomial time deterministic algorithms, by polynomial time nondeterministic algorithms, and by polynomial space nondeterministic[4] algorithms.

We are mainly interested in PSPACE-hard and PSPACE-complete problems. A problem '$x \in X$?' is called PSPACE-*hard* if any problem '$y \in Y$?' in PSPACE is polynomially reducible to '$x \in X$?', i.e., if there exists a polynomial time function (algorithm) f such that for every y,

$$y \in Y \iff f(y) \in X.$$

A problem '$x \in X$?' is called PSPACE-*complete* if it belongs to PSPACE and is PSPACE-hard.

---

[4]We can replace 'nondeterministic' with 'deterministic' here; the equivalence between the definitions follows from [30].

3   Main Result

Recall (see the Introduction) that **Int**, **BPL**, and **FPL** are connected with **S4**, **K4**, and **GL**, correspondingly, via the Gödel translation T. It is known [8] that the decision problem for **K4**(0), **GL**(1), **S4**(1) is PSPACE-complete. It is also known that the decision problem for **GL**(0) is in P, and hence the one for **FPL**(0) is also in P. Finally, the decision problem for **Int**(1) is also in P: cf. [21]. Thus, there is a reason to consider the question of the complexity of **Int**(n) for $n \geq 2$, **BPL**(n) for $n \geq 0$, and **FPL**(n) for $n \geq 1$.

In 1985 A. V. Chagrov [4] proved PSPACE-hardness for many logics, in particular, for **Int**, **BPL**, and **FPL** in the language with implication only (for PSPACE-hardness of **BPL** and **FPL** see also [2]). Below we shall use more weak facts, which follow from [4]: the decision problem for **Int**$^+$, **BPL**$^+$, and **FPL**$^+$ is PSPACE-complete.

REMARK 3. In fact, below we can use the facts that **Int**, **BPL**, and **FPL** have PSPACE-complete decision problem for their implicative fragments, moreover there is a reason to do so: one connective in the language means for some of our proofs only one case in the induction step. But on the one hand, [4] is in Russian and unfortunately it is difficult to access even in Russia. On the other hand, PSPACE-completeness of **Int**$^+$, **BPL**$^+$, **FPL**$^+$ is sufficient for the aim of this article, and there is no problem to find corresponding proofs in English: for **Int**$^+$ it is given in [9, 33] and, which is important, one can easily modify the proof in [9] for the cases of **BPL**$^+$ and **FPL**$^+$ (it is sufficient to replace the last occurrences of q with $(q \rightarrow q) \rightarrow q$ in both formulas on page 560 in [9]); or see [2] (to obtain PSPACE-hardness for **BPL**$^+$ and **FPL**$^+$ one can easily eliminate $\bot$ in the corresponding proofs for **BPL** and **FPL** in [2]).

Our main aim is in to prove the following statement.

THEOREM 4. *The decision problem for* **Int**$^+$(2), **BPL**(0), *and* **FPL**(1) *is* PSPACE-*complete*.

PROOF. The fact that this problem is in PSPACE follows from the PSPACE-completeness of the decision problem for **Int**, **BPL**, and **FPL** (in the language with infinitely many variables). So, it is sufficient to prove PSPACE-hardness of this problem.

The corresponding proof for **BPL**(0) and **FPL**(1) is similar to the one for the logic **K4** in [8]. There are some differences for the case of **Int**$^+$(2) but the general idea is the same. Below I give a proof for **Int**$^+$(2) (it is more difficult in a 'technical' sense than the ones for **BPL**(0) and **FPL**(1)) and a detailed sketch of proof for the cases of **BPL**(0) and **FPL**(1).

We start with **Int**$^+$(2). Let p and q be two propositional variables. We define the following positive formulas constructed from p and q:

$$D_1 = p \to q; \quad D_2 = q \to p; \quad A_1^1 = A_1^0 \wedge A_2^0 \to B_1^0 \vee B_2^0;$$
$$D_3 = D_1 \wedge D_2 \to p \vee q; \quad\quad\quad A_2^1 = A_1^0 \wedge B_1^0 \to A_2^C \vee B_2^0;$$
$$A_1^0 = D_2 \to D_1 \vee D_3; \quad\quad\quad A_3^1 = A_1^0 \wedge B_2^0 \to A_2^C \vee B_1^0;$$
$$A_2^0 = D_3 \to D_1 \vee D_2; \quad\quad\quad B_1^1 = A_2^0 \wedge B_1^0 \to A_1^C \vee B_2^0;$$
$$B_1^0 = D_1 \to D_2 \vee D_3; \quad\quad\quad B_2^1 = A_2^0 \wedge B_2^0 \to A_1^C \vee B_1^0;$$
$$B_2^0 = A_1^0 \wedge A_2^0 \wedge B_1^0 \to D_1 \vee D_2 \vee D_3; \quad B_3^1 = B_1^0 \wedge B_2^0 \to A_1^0 \vee A_2^0.$$

To define other formulas we need, let us fix the following order $\prec$ on the set $\{(i,j) : i, j \geq 2\}$:

$$(i,j) \prec (i',j') \iff \text{either } i+j < i'+j', \text{ or } i+j = i'+j' \text{ and } i < i'.$$

Clearly, this order is strict and linear. So, we can enumerate all such pairs of natural numbers by means of some function, say $g(x)$, accordingly to this order, i.e., $g(1) = (2,2)$, $g(2) = (2,3)$, $g(3) = (3,2)$, $g(4) = (3,3)$, etc.

Suppose the formulas $A_1^k, \ldots, A_m^k, B_1^k, \ldots, B_m^k$ are already defined, for some $k \geq 1$. Let $i,j \in \{2,\ldots,m\}$ and let $(i,j) = g(s)$; then we put

$$A_s^{k+1} = A_1^k \to B_1^k \vee A_i^k \vee B_j^k; \quad B_s^{k+1} = B_1^k \to A_1^k \vee A_i^k \vee B_j^k.$$

We call A_i^k and B_i^k formulas of level k. By N_k denote the number of formulas A_i^k of level k. Clearly, N_k is also the number of formulas B_i^k of level k. It is easy to see that

$$N_0 = 2, \ N_1 = 3, \ N_{k+2} = (N_{k+1} - 1)^2.$$

For a formula A we denote by $|A|$ the length of A. Let us fix some number l such that

$$l > \max_{1 \leq i \leq 2} \{|A_i^0| + |B_i^0|\};$$

for example, we can take $l = 200$. It is easy to show that $|A_i^k| < 5^k \cdot l$ and $|B_i^k| < 5^k \cdot l$ (by induction on k).

It is not hard to see that there exists k_0 such that $5^k \cdot l < N_k$, for any $k \geq k_0$; one can check that as k_0 we can take any number greater than or equal to 6. Let $k_0 = 6$.

We construct a Kripke model refuting all formulas A_i^k and B_i^k such that, for every formula A_i^k and B_i^k, there exists a unique maximal world, at which this formula is refuted. Let

$$W = \{c_0, d_1, d_2, d_3\} \cup \{a_i^k, b_i^k : k \geq 0, \ 1 \leq i \leq N_k\}.$$

We call a_i^k and b_j^k worlds of level k. To define the accessibility relation R

on W, let

$$R_0 = \{\langle d_1, c_0\rangle, \langle d_2, c_0\rangle, \langle d_3, c_0\rangle, \langle a_1^0, d_1\rangle, \langle a_2^0, d_1\rangle, \langle b_2^0, d_1\rangle,$$
$$\langle a_2^0, d_2\rangle, \langle b_1^0, d_2\rangle, \langle b_2^0, d_2\rangle, \langle a_1^0, d_3\rangle, \langle b_1^0, d_3\rangle, \langle b_2^0, d_3\rangle\};$$
$$R_1^a = \{\langle a_1^1, b_1^0\rangle, \langle a_1^1, b_2^0\rangle, \langle a_2^1, a_2^0\rangle, \langle a_2^1, b_2^0\rangle, \langle a_3^1, a_2^0\rangle, \langle a_3^1, b_1^0\rangle\};$$
$$R_1^b = \{\langle b_1^1, a_1^0\rangle, \langle b_1^1, b_2^0\rangle, \langle b_2^1, a_1^0\rangle, \langle b_2^1, b_1^0\rangle, \langle b_3^1, a_1^0\rangle, \langle b_3^1, a_2^0\rangle\};$$
$$R_1 = R_1^a \cup R_1^b,$$

and let, for every $k \geq 1$,

$$R_{k+1}^a = \{\langle a_m^{k+1}, b_1^k\rangle, \langle a_m^{k+1}, a_i^k\rangle, \langle a_m^{k+1}, b_j^k\rangle : A_m^{k+1} = A_1^k \to B_1^k \vee A_i^k \vee B_j^k\};$$
$$R_{k+1}^b = \{\langle b_m^{k+1}, a_1^k\rangle, \langle b_m^{k+1}, a_i^k\rangle, \langle b_m^{k+1}, b_j^k\rangle : B_m^{k+1} = B_1^k \to A_1^k \vee A_i^k \vee B_j^k\};$$
$$R_{k+1} = R_{k+1}^a \cup R_{k+1}^b.$$

We put

$$R' = \bigcup_{k=1}^\infty R_k$$

and take as R the reflexive and transitive closure of R'. Let $\mathfrak{F} = \langle W, R\rangle$ and $\mathfrak{M} = \langle \mathfrak{F}, v\rangle$, where v is defined in the following way:

$$(\mathfrak{M}, w) \models p \iff w = c_0 \text{ or } w = d_1;$$
$$(\mathfrak{M}, w) \models q \iff w = c_0 \text{ or } w = d_2.$$

The model $\mathfrak{M}$ is depicted on Fig. 1.

LEMMA 5. *Let w be a world of $\mathfrak{M}$. Then*

$$(\mathfrak{M}, w) \not\models A_m^k \iff wRa_m^k;$$
$$(\mathfrak{M}, w) \not\models B_m^k \iff wRb_m^k.$$

PROOF proceeds by easy induction on k. To prove the basis of induction it is useful to check that for every world w in $\mathfrak{M}$ and for every $i \in \{1, 2, 3\}$,

$$(\mathfrak{M}, w) \not\models D_i \iff wRd_i.$$

□

Let φ be a positive intuitionistic formula and let $p_1, \ldots, p_n$ be all variables of φ. We denote by k the least natural number such that $|\varphi| < 5^k \cdot l$. Note that

$$N_{k+6} > 5^{k+6} \cdot l > 5^6 \cdot |\varphi| > |\varphi| > n,$$

and so the following formulas are well-defined for every $i \in \{1, \ldots, n\}$:

$$\alpha_i = A_i^{k+6} \vee B_i^{k+6}.$$

Let φ_α be the result of substituting the formulas $\alpha_1, \ldots, \alpha_n$ respectively for $p_1, \ldots, p_n$ in φ.

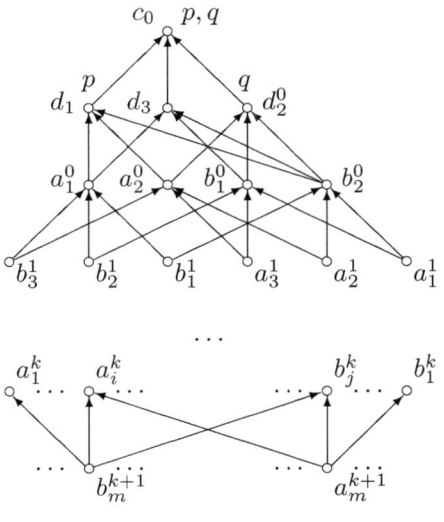

Figure 1. Model $\mathfrak{M}$

LEMMA 6. *There exists a number m such that for any φ, $|\varphi_\alpha| < m \cdot |\varphi|^2$.*

PROOF. From the fact that k is the least natural number satisfying the condition $|\varphi| < 5^k \cdot l$ it follows that $5^{k-1} \cdot l \leq |\varphi|$, and we obtain

$$5^{k+6} \cdot l \ \leq \ 5^7 \cdot |\varphi|.$$

Since $|A_i^{k+6}| < 5^{k+6} \cdot l$ and $|B_i^{k+6}| < 5^{k+6} \cdot l$, we have

$$|\alpha_i| \ < \ 2 \cdot 5^{k+6} \cdot l \ \leq \ 2 \cdot 5^7 \cdot |\varphi|.$$

Hence

$$|\varphi_\alpha| \ \leq \ |\varphi| \cdot \max_{1 \leq i \leq n} |\alpha_i| \ < \ 2 \cdot 5^7 \cdot |\varphi|^2,$$

i.e., we can take $m = 2 \cdot 5^7$. □

Using Lemma 6 one can show that the formula φ_α is computed from φ by a polynomial time algorithm.

LEMMA 7. *For every positive intuitionistic formula φ,*

$$\varphi \in \mathbf{Int} \ \iff \ \varphi_\alpha \in \mathbf{Int}.$$

PROOF. If $\varphi \in \mathbf{Int}$, then $\varphi_\alpha \in \mathbf{Int}$ because $\mathbf{Int}$ is closed under Substitution.

Let φ be a positive intuitionistic formula such that $\varphi \notin \mathbf{Int}$. Then there exists a model $\mathfrak{M}_\varphi = \langle \mathfrak{F}_\varphi, v_\varphi \rangle$ defined on reflexive, transitive, and antisymmetric frame $\mathfrak{F}_\varphi = \langle W_\varphi, R_\varphi \rangle$ such that $(\mathfrak{M}_\varphi, w_0) \not\models \varphi$, for some $w_0 \in W_\varphi$.

Let us construct an intuitionistic model refuting φ_α. Without a loss of generality we may assume that $W \cap W_\varphi$ is empty. We put $W^* = W \cup W_\varphi$. Let

$$R' = \{\langle w, a_i^{k+6}\rangle, \langle w, b_i^{k+6}\rangle : w \in W_\varphi, (\mathfrak{M}_\varphi, w) \not\models p_i, 1 \leq i \leq n\} \cup$$
$$\cup \{\langle w, a_{n+1}^{k+6}\rangle, \langle w, b_{n+1}^{k+6}\rangle : w \in W_\varphi\}.$$

Let R^* be the reflexive and transitive closure of the relation $R \cup R_\varphi \cup R'$ and let $\mathfrak{F}^* = \langle W^*, R^* \rangle$. We define the model $\mathfrak{M}^* = \langle \mathfrak{F}^*, v^* \rangle$ on $\mathfrak{F}^*$ by putting for every $w \in W^*$

$$(\mathfrak{M}^*, w) \models p \iff w = c_0 \text{ or } w = d_1;$$
$$(\mathfrak{M}^*, w) \models q \iff w = c_0 \text{ or } w = d_2.$$

For every subformula ψ of the formula φ, let ψ_α be the result of substituting the formulas $\alpha_1, \ldots, \alpha_n$ respectively for $p_1, \ldots, p_n$ in ψ. By induction on the construction of ψ let us show that for every $w \in W_\varphi$,

$$(\mathfrak{M}^*, w) \models \psi_\alpha \iff (\mathfrak{M}_\varphi, w) \models \psi.$$

Let $\psi = p_m$. If $(\mathfrak{M}_\varphi, w) \not\models p_m$, then a_m^{k+6} and b_m^{k+6} are accessible from w in $\mathfrak{M}^*$. Using Lemma 5, we see that $(\mathfrak{M}^*, a_m^{k+6}) \not\models A_m^{k+6}$ and $(\mathfrak{M}^*, b_m^{k+6}) \not\models B_m^{k+6}$, hence $(\mathfrak{M}^*, w) \not\models \alpha_m$, i.e., $(\mathfrak{M}^*, w) \not\models \psi_\alpha$.

Suppose $(\mathfrak{M}^*, w) \not\models \alpha_m$, for some world $w \in W_\varphi$. Let $g(m) = (i, j)$, i.e., $A_m^{k+6} = A_1^{k+5} \to B_1^{k+5} \vee A_i^{k+5} \vee B_j^{k+5}$, $B_m^{k+6} = B_1^{k+5} \to A_1^{k+5} \vee A_i^{k+5} \vee B_j^{k+5}$. Then there exist worlds w' and w'' in $\mathfrak{M}^*$ such that wR^*w', wR^*w'', and

$$(\mathfrak{M}^*, w') \models A_1^{k+5}, \quad (\mathfrak{M}^*, w') \not\models B_1^{k+5}, \quad (\mathfrak{M}^*, w') \not\models A_i^{k+5} \vee B_j^{k+5};$$
$$(\mathfrak{M}^*, w'') \not\models A_1^{k+5}, \quad (\mathfrak{M}^*, w'') \models B_1^{k+5}, \quad (\mathfrak{M}^*, w'') \not\models A_i^{k+5} \vee B_j^{k+5}.$$

Note that $w', w'' \in W$. Indeed, for every world $u \in W_\varphi$, we have $uR^*a_{n+1}^{k+6}$ and $uR^*b_{n+1}^{k+6}$. Since $(\mathfrak{M}^*, a_{n+1}^{k+6}) \not\models B_1^{k+5}$ and $(\mathfrak{M}^*, b_{n+1}^{k+6}) \not\models A_1^{k+5}$, we obtain $(\mathfrak{M}^*, u) \not\models A_1^{k+5}$, $(\mathfrak{M}^*, u) \not\models B_1^{k+5}$. But $(\mathfrak{M}^*, w') \models A_1^{k+5}$, $(\mathfrak{M}^*, w'') \models B_1^{k+5}$, thus $w', w'' \notin W_\varphi$, i.e., $w', w'' \in W$.

Since w' and w'' are R^*-accessible from a world of W_φ, then their levels are not greater than $k+6$. Besides, $(\mathfrak{M}^*, w') \not\models A_m^{k+6}$ and $(\mathfrak{M}^*, w'') \not\models B_m^{k+6}$, therefore the levels of w' and w'' are not less than $k+6$. So, w' and w'' are worlds of level $k+6$. But the unique world of level $k+6$ refuting A_m^{k+6} is a_m^{k+6} and the unique world of level $k+6$ refuting B_m^{k+6} is b_m^{k+6}, hence $w' = a_m^{k+6}$ and $w'' = b_m^{k+6}$.

Thus $wR^*a_m^{k+6}$. This means that there exists $u \in W_\varphi$ such that $wR_\varphi u$ and $uR'a_m^{k+6}$. But if $uR'a_m^{k+6}$, then, by the definition of $\mathfrak{F}^*$, we have both $uR'b_m^{k+6}$ and $(\mathfrak{M}_\varphi, u) \not\models p_m$. Since $wR_\varphi u$, we obtain $(\mathfrak{M}_\varphi, w) \not\models p_m$.

Let ψ' and ψ'' be subformulas of φ such that for every $w \in W_\varphi$,

$$(\mathfrak{M}^*, w) \models \psi'_\alpha \iff (\mathfrak{M}_\varphi, w) \models \psi';$$
$$(\mathfrak{M}^*, w) \models \psi''_\alpha \iff (\mathfrak{M}_\varphi, w) \models \psi'',$$

and let $\psi = \psi' \to \psi''$. If $(\mathfrak{M}_\varphi, w) \not\models \psi$, then there is a world w' in $\mathfrak{M}_\varphi$ accessible from w such that $(\mathfrak{M}_\varphi, w') \models \psi'$ and $(\mathfrak{M}_\varphi, w') \not\models \psi''$; but then $(\mathfrak{M}^*, w') \models \psi'_\alpha$ and $(\mathfrak{M}^*, w') \not\models \psi''_\alpha$, hence $(\mathfrak{M}^*, w) \not\models \psi_\alpha$.

Suppose for some $w \in W_\varphi$, $(\mathfrak{M}^*, w) \not\models \psi_\alpha$. Then there is $w' \in W^*$ such that wR^*w', $(\mathfrak{M}^*, w') \models \psi'_\alpha$, and $(\mathfrak{M}^*, w') \not\models \psi''_\alpha$. Note that the formulas $\alpha_1, \ldots, \alpha_n$ are true in every world $u \in W$ of level not greater than $k + 6$. Hence in all such worlds the formulas constructed from $\alpha_1, \ldots, \alpha_n$ using $\to$, $\wedge$, and $\vee$, must be true; in particular, ψ''_α must be true. From $(\mathfrak{M}^*, w') \not\models \psi''_\alpha$ it follows that $w' \notin W$, i.e., $w' \in W_\varphi$. By the inductive hypothesis, we obtain $(\mathfrak{M}_\varphi, w') \models \psi'$ and $(\mathfrak{M}_\varphi, w') \not\models \psi''$, hence $(\mathfrak{M}_\varphi, w) \not\models \psi$.

The cases $\psi = \psi' \wedge \psi''$ and $\psi = \psi' \vee \psi''$ are trivial.

As a result we obtain $(\mathfrak{M}^*, w_0) \not\models \varphi_\alpha$, and hence $\varphi_\alpha \notin \mathbf{Int}$. □

As a corollary we immediately obtain that the decision problem for $\mathbf{Int}^+(2)$ is PSPACE-complete.

The case with Visser's logics **BPL** and **FPL** is simpler. For the proof of PSPACE-hardness for $\mathbf{BPL}(0)$ and $\mathbf{FPL}(1)$ we define 'modality' $\square$ as the following abbreviation: $\square \psi = \top \to \psi$. Let also $\square^1 \psi = \square \psi$ and for any $n \geq 1$, let $\square^{n+1} \psi = \square \square^n \psi$.

Let us define formulas β_n and γ_n, for $n \geq 1$:

$$\beta_n = (\square^{n+2}\bot \to \square^{n+1}\bot) \to (\square^{n+1}\bot \to \square^n\bot) \vee \square^{n+2}\bot;$$
$$\gamma_n = \square^2 p \to (\square^{n+1}\bot \to \square^n\bot \vee p).$$

Let φ be a positive intuitionistic formula and let $p_1, \ldots, p_n$ be all variables of φ. We denote by φ_β the result of substituting the formulas $\beta_1, \ldots, \beta_n$ respectively for $p_1, \ldots, p_n$ in φ and by φ_γ the result of substituting the formulas $\gamma_1, \ldots, \gamma_n$ respectively for $p_1, \ldots, p_n$ in φ.

Clearly, φ_β and φ_γ are computed from φ by some polynomial time algorithms. Hence, to prove PSPACE-hardness for $\mathbf{BPL}(0)$ it is sufficient to prove the following statement.

LEMMA 8. *For every positive formula φ,*

$$\varphi \in \mathbf{BPL} \iff \varphi_\beta \in \mathbf{BPL}.$$

PROOF. If $\varphi \in \mathbf{BPL}$, then $\varphi_\beta \in \mathbf{BPL}$ because **BPL** is closed under Substitution.

Let $\varphi \notin \mathbf{BPL}$. We show that $\varphi_\alpha \notin \mathbf{BPL}$. There exists a model $\mathfrak{M} = \langle \mathfrak{F}, v \rangle$ on some transitive and antisymmetric Kripke frame $\mathfrak{F} = \langle W, R \rangle$ such that $(\mathfrak{M}, w_0) \not\models \varphi$, for some $w_0 \in W$.

Let $\mathfrak{F}_k = \langle W_k, R_k \rangle$ be the Kripke frame, depicted on Fig. 2 (b_k is a reflexive world, other are irreflexive, R_k is transitive), i.e.,

$$W_k = \{b_k, a_1^k, a_2^k, \ldots, a_{k+2}^k\};$$
$$wR_k w' \iff \text{either } w = w' = b_k,$$
$$\text{or } w = a_{k+2}^k, \ w' = b_k,$$
$$\text{or } w = a_i^k, \ w' = a_j^k, \ i > j.$$

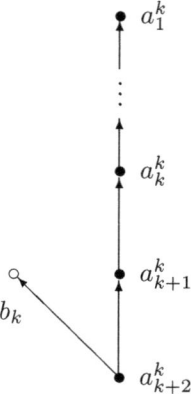

Figure 2. Frame $\mathfrak{F}_k$

Frame $\mathfrak{F}_k$ has the following property: if we add to it as a root some world w, then we obtain a frame $\mathfrak{F}'_k$ such that for every $m \geq 1$

$$(\mathfrak{F}'_k, w) \not\models \beta_m \iff k = m.$$

Using this observation it is easy to construct a frame of **BPL** refuting φ_β. Let

$$W^* = W \cup \bigcup_{k=1}^{n+1} W_k,$$

let R^* be the transitive closure of the relation

$$R \cup \bigcup_{k=1}^{n+1} R_k \cup \{\langle w, a_{k+2}^k\rangle : w \in W, (\mathfrak{M}, w) \not\models p_k\} \cup \{\langle w, a_{n+3}^{n+1}\rangle : w \in W\},$$

and let $\mathfrak{F}^* = \langle W^*, R^*\rangle$. In other worlds, for every variable p_k of φ and every world w in $\mathfrak{F}$ such that $(\mathfrak{M}, w) \not\models p_k$ we make $\mathfrak{F}_k$ to be accessible from w; and also for every world w in $\mathfrak{F}$ we make $\mathfrak{F}_{n+1}$ to be accessible from w.

For every subformula ψ of formula φ, let ψ_β be the result of substituting the formulas $\beta_1, \ldots, \beta_n$ respectively for $p_1, \ldots, p_n$ in ψ. By induction on construction of ψ one can show that for every $w \in W$

$$(\mathfrak{F}^*, w) \models \psi_\beta \iff (\mathfrak{M}, w) \models \psi.$$

Let $\psi = p_k$. If $(\mathfrak{M}, w) \not\models p_k$, then $wR^* a_{k+2}^k$, hence $(\mathfrak{F}^*, w) \not\models \beta_k$, i.e., $(\mathfrak{F}^*, w) \not\models \psi_\beta$.

Suppose $(\mathfrak{F}^*, w) \not\models \psi_\beta$, i.e., $(\mathfrak{F}^*, w) \not\models \beta_k$. Then there exists $w' \in W^*$ such that wR^*w',

$$(\mathfrak{F}^*, w') \models \Box^{k+2}\bot \to \Box^{k+1}\bot, \quad (\mathfrak{F}^*, w') \not\models (\Box^{k+1}\bot \to \Box^k\bot) \vee \Box^{k+2}\bot.$$

We claim $w' \notin W$. Indeed, suppose $w' \in W$. Then, by the definition of $\mathfrak{F}^*$, we have $w' R^* a_{n+3}^{n+1}$ (i.e., $\mathfrak{F}_{n+1}$ is accessible from w'). It is easy to see that $(\mathfrak{F}_{n+1}, a_{n+3}^{n+1}) \not\models \Box^{k+2}\bot \to \Box^{k+1}\bot$, hence $(\mathfrak{F}^*, a_{n+3}^{n+1}) \not\models \Box^{k+2}\bot \to \Box^{k+1}\bot$, and so $(\mathfrak{F}^*, w') \not\models \Box^{k+2}\bot \to \Box^{k+1}\bot$. We have a contradiction, therefore $w' \notin W$.

Since $w' \notin W$, we have $w' \in W_m$, for some $m \in \{1, \ldots, n+1\}$. One can easily check that for any $s \geq 1$,

$$(\mathfrak{F}_s, u) \not\models \beta_k \iff s = k \text{ and } u = a_{k+2}^k.$$

For our case it means that $w' = a_{k+2}^k$. By the construction of $\mathfrak{F}^*$, we see that $(\mathfrak{M}, w) \not\models p_k$.

The cases $\psi = \psi' \wedge \psi''$ and $\psi = \psi' \vee \psi''$ are trivial. The proof for the case $\psi = \psi' \to \psi''$ proceeds in the same way as in Lemma 7.

So, we obtain $(\mathfrak{F}^*, w_0) \not\models \varphi_\beta$, and hence $\varphi_\beta \notin \mathbf{BPL}$. □

So, the decision problem for $\mathbf{BPL}(0)$ is PSPACE-complete. To prove PSPACE-hardness for $\mathbf{FPL}(1)$ it is sufficient to prove

LEMMA 9. *For every positive formula* φ,

$$\varphi \in \mathbf{FPL} \iff \varphi_\gamma \in \mathbf{FPL}.$$

PROOF. The proof proceeds in the same way as for Lemma 8. Instead of frames $\mathfrak{F}_k$ one may take models $\mathfrak{M}_k$: the worlds in $\mathfrak{M}_k$ are numbers $1, \ldots, k+2$, the accessibility relation is the relation $>$, and the valuation for p is defined so that p is true in the worlds $1, \ldots, k$ and not true in $k+1$ and $k+2$. □

So, the decision problem for $\mathbf{FPL}(1)$ is PSPACE-complete, and Theorem 4 is proved. □

4 Corollaries

Of course, from Theorem 4 we immediately obtain

COROLLARY 10. *The decision problem for* $\mathbf{Int}(2)$ *is* PSPACE-*complete.*

From Theorem 4 one can obtain corollaries about complexity of the decision problem for n-variable fragments of wide classes of logics. I give some of them here.

First of all, with the help of the translation T we obtain some results in [8].

COROLLARY 11. [8]. *The decision problem for* $\mathbf{K4}(0)$, $\mathbf{GL}(1)$, $\mathbf{S4}(2)$ *is* PSPACE-*complete.*

REMARK 12. The logic $\mathbf{S4}$ has PSPACE-complete decision problem in the language with one variable: see [8].

Let us consider the logic of the weak law of the excluded middle $\mathbf{KC}$ (for details the reader can consult [9]), which is defined by adding to $\mathbf{Int}$ the corresponding axiom: $\mathbf{KC} = \mathbf{Int} + \neg p \vee \neg \neg p$.

The decision problem for **KC**$^+$ is PSPACE-complete. The proof of this fact is similar to the one for **Int**$^+$ [33]; in particular, to prove PSPACE-hardness of **KC**$^+$ it is sufficient to consider the same formulas.

V. A. Jankov [17] proved that, for every superintuitionistic logic L (i.e., an extension of **Int**),

$$L^+ = \mathbf{Int}^+ \iff L \subseteq \mathbf{KC}.$$

Hence from Theorem 4 we obtain the following statements.

COROLLARY 13. *The decision problem for* **KC**(2) *is* PSPACE*-complete.*

COROLLARY 14. *Let L be a logic such that* $\mathbf{Int} \subseteq L \subseteq \mathbf{KC}$. *Then the decision problem for $L(2)$ is* PSPACE*-hard.*

Corollary 14 has its own corollaries.

COROLLARY 15. *Let L be a logic such that $L \subset \mathbf{Cl}$ and it can be obtained by adding to* **Int** *a one-variable axiom. Then the decision problem for $L(2)$ is* PSPACE*-hard.*

PROOF immediately follows from [21] and Corollary 14. Indeed, using the Nishimura lattice [21], one can see that the greatest consistent extension of **Int** obtained by adding to it a one-variable axiom is **KC**. □

For the next corollary we need a definition: a formula φ is called *implication free* if φ is constructed from propositional variables with $\wedge$, $\vee$ and $\neg$.

COROLLARY 16. *Let L be a logic such that $L \subset \mathbf{Cl}$ and it can be obtained by adding to* **Int** *an implication free axiom. Then the decision problem for $L(2)$ is* PSPACE*-hard.*

PROOF. It is not hard to see that in this case $\mathbf{Int} + \varphi \subseteq \mathbf{KC}$ (see also [3]), and so the decision problem for the two-variable fragment of L is PSPACE-hard by Corollary 14. □

We consider two special logics: the Kreisel–Putnam logic **KP** and the Medvedev logic **ML**. The logic **KP** is defined by adding to **Int** the Kreisel–Putnam formula:

$$\mathbf{KP} = \mathbf{Int} + (\neg p \to q \vee r) \to (\neg p \to q) \vee (\neg p \to r).$$

The logic **ML** is an extension of **KP**. It can be defined semantically in the following way, see [9]. Let $n > 0$ and let W_n be the family of nonempty subsets of a set with n elements. Let $w R_n w'$ mean $w' \subseteq w$, for every $w, w' \in W_n$. The pair $\mathfrak{M}_n = \langle W_n, R_n \rangle$ is clearly a Kripke frame. Then

$$\mathbf{ML} = \{\varphi : \mathfrak{F}_n \models \varphi, \text{ for any } n > 0\}.$$

These two logics are rather complex: it is known that **KP** is decidable [12] but it is not known if the decision problem for **KP** belongs to PSPACE. The decidability of **ML** is still an open problem.

Since $\mathbf{ML}^+ = \mathbf{Int}^+$ [37], from [17] we obtain $\mathbf{ML} \subseteq \mathbf{KC}$, and hence we proved

COROLLARY 17. *The decision problem for logics* $\mathbf{KP}(2)$ *and* $\mathbf{ML}(2)$ *is PSPACE-hard.*

Note, that to prove PSPACE-hardness of **BPL**, **FPL**, **Int**, **KC** (in full language) we can use the same (positive) formulas (see [4]; one can also construct such formulas as a slight modification of the ones in [33] or [9], see Remark 3). More exactly, there exists a polynomial time algorithm f such that for every quantified Boolean formula φ,

$$\varphi \text{ is satisfiable} \iff f(\varphi) \notin L,$$

where $L \in \{\mathbf{BPL}, \mathbf{FPL}, \mathbf{Int}, \mathbf{KC}\}$, $f(\varphi)$ is a positive formula (for complexity of satisfiability problem for quantified Boolean formulas see [34]).

Using this observation we obtain the following statements.

COROLLARY 18. *Let L be a logic such that* $\mathbf{BPL} \subseteq L \subseteq \mathbf{KC}$. *Then the decision problem for* $L^+(2)$ *is PSPACE-hard.*

PROOF. Let $\varphi^* = f(\varphi)$, where f is the function mentioned above, φ is a quantified Boolean formula. If φ is not satisfiable, then $\varphi^* \in \mathbf{BPL}$, hence $\varphi_\alpha^* \in \mathbf{BPL}$ (because **BPL** is closed under Substitution), so $\varphi_\alpha^* \in L^+(2)$. If φ is satisfiable, then $\varphi^* \notin \mathbf{Int}$, hence, by Lemma 7, we have $\varphi_\alpha^* \notin \mathbf{Int}$, wherefore, by [17], $\varphi_\alpha^* \notin \mathbf{KC}$, and so $\varphi_\alpha^* \notin L^+(2)$. □

COROLLARY 19. *Let L be a logic such that* $\mathbf{BPL} \subseteq L \subseteq \mathbf{FPL}$. *Then the decision problem for $L(1)$ and $L^+(2)$ is PSPACE-hard.*

PROOF. It is easy to understand that the decision problem for $L(1)$ is PSPACE-hard. To prove PSPACE-hardness for $L^+(2)$ we may simply redefine the 'modality' $\square$: $\square\psi = (q \to q) \to \psi$, here q is a propositional variable different from p. All semantic properties of this 'new' $\square$ are the same as of the 'old' one, hence if in the definition of γ_n we use the 'new' $\square$ instead of the 'old' $\square$ and q instead of all other occurrences of $\bot$, then we obtain a positive version of φ_γ, and so, PSPACE-hardness for $L^+(2)$. □

Using the same argumentation and formulas β_n it is easy to show that the decision problem for $\mathbf{BPL}^+(1)$ is PSPACE-complete too.

COROLLARY 20. *The decision problem for the logic* $\mathbf{BPL}^+(1)$ *is PSPACE-complete.*

Note also that $\mathbf{BPL}^+(0)$ is trivial; in particular, it is decidable in polynomial time.

5 Complexity function

Recall that the complexity function $f_L(n)$ for a logic (a set of formulas) L is defined in the following way (see [9]):

$$f_L(n) = \max_{\substack{|\varphi| \leqslant n \\ \varphi \notin L}} \min_{\substack{\mathfrak{F} \models L \\ \mathfrak{F} \not\models \varphi}} |\mathfrak{F}|,$$

where $|\mathfrak{F}|$ is the cardinality of the frame $\mathfrak{F}$ and $|\varphi|$ is the length of the formula φ.

The complexity function for a logic L is a way to estimate the complexity of the decision problem for L. For example, it can provide us with bounds for some computational resources (for example, time resources) required for algorithms deciding L by searching for countermodels for tested formulas.

Moreover, the study of the complexity function for **Int** and for some other logics may be helpful in the search for an answer to Kuznetsov's question presented in [19] (see also [18]): is the problem '$\varphi \in \mathbf{Int}$?' polynomially reducible to the problem '$\varphi \in \mathbf{Cl}$?'? So, in fact A. V. Kuznetsov [19] showed that if the complexity function for **Int** is polynomial, then **Int** is polynomially equivalent to **Cl**. Note that **Int** here may be replaced with other logics — **BPL**, **FPL**, **K**, **T**, **K4**, etc. — which have PSPACE-hard decision problem and 'good' semantics. Note also that a positive answer to such question means that NP = PSPACE.

At the end of 1970s and later M. V. Zakharyaschev established for a lot of standard non-classical logics (in particular including **Int** and **KC**) that the complexity function for these logics is exponential, see [9, 23, 24], i.e., for some positive numbers k, m, r, s,

$$2^{k \cdot n^m} \leqslant f_L(n) \leqslant 2^{r \cdot n^s}.$$

So, the complexity function for these logics actually cannot be bounded by any polynomial function.

Zakharyaschev's results are in close connection with Ladner's and Statman's ones: using Zakharyaschev's formulas it is not hard to prove PSPACE-hardness of **Int**, **K**, **T**, **S4**, etc., and from Ladner's and Statman's constructions it is not hard to obtain that the complexity function for the corresponding logics is exponential.

In general we can observe a certain connection between the complexity function for a logic L and the complexity of L with respect to complexity classes such as NP and PSPACE. For example, known logics with NP-complete decision problem — **Cl**, **S5**, **S4.3**, etc. — have polynomial complexity function, known logics with PSPACE-complete decision problem — **Int**, **BPL**, **FPL**, **K**, **T**, etc. — have exponential complexity function; note also that the complexity function for every logic between **KP** and **ML** has the lower bound of order $2^{2^{k \cdot n}}$, for some $k > 0$ (for details see [9]).

All these results are obtained for logics with infinitely many variables in the language (I do not take into account our — with A. V. Chagrov —

article [8]), and it is natural to solve the problem about bounds of the complexity function for PSPACE-hard n-variable fragments of **Int**, **KC**, **BPL**, etc. Of course, it will be surprising if the complexity function for some of these fragments is polynomial, and it is really not the case: it has exponential lower bound in every case of the logics considered here. To prove it, it is sufficient to take corresponding Zakharyaschev's (or Statman's, or Chagrov's) formulas and substitute α_n's, or β_n's, or γ_n's in accordance with what logic we consider.

So, we have the following corollaries from the construction of Theorem 4.

COROLLARY 21. *Let a logic L be such that* $\mathbf{BPL}(1) \subseteq L(1) \subseteq \mathbf{FPL}(1)$. *Then the lower bound of the complexity function for $L(1)$ is exponential.*

COROLLARY 22. *Let a logic L be such that* $\mathbf{BPL}^+(2) \subseteq L(2) \subseteq \mathbf{FPL}^+(2)$, *or* $\mathbf{BPL}^+(2) \subseteq L(2) \subseteq \mathbf{Int}^+(2)$. *Then the lower bound of the complexity function for $L(2)$ is exponential.*

Taking into account results in [9] concerning the complexity function, in some cases we can specify the result.

COROLLARY 23. *The complexity function for the logics* $\mathbf{BPL}(0)$, $\mathbf{FPL}(1)$, $\mathbf{Int}(2)$, $\mathbf{KC}(2)$, $\mathbf{BPL}^+(1)$, $\mathbf{FPL}^+(2)$, $\mathbf{Int}^+(2)$ *is exponential.*

6 Acknowledgements

First of all, I am very grateful to my Teacher and scientific adviser Alexander Chagrov. I am also grateful to Valentin Shehtman and Philippe Balbiani: the basic results of this paper were presented and discussed on their seminars in Moscow and Toulouse. My special thanks to anonymous referees for their remarks and advises, which were very helpful to make the article more clear and understandable. In this connection I am also thankful to Aleksey Sosinskiy for his book [31]. I am thankful to Michael Zakharyaschev for his help in the presentation of my work.

BIBLIOGRAPHY

[1] P. Blackburn, M. de Rijke, and Y. Venema. *Modal Logic*. Cambridge University Press, 2001.
[2] F. Bou. Complexity of strict implication. *Advances in Modal Logic*, 5:1–16, 2005.
[3] E. Carpińska. On intermediate logics which can be axiomatized by means of implicationless formulas. *Reps Math. Logic*, (13):11–16, 1981.
[4] A. V. Chagrov. On the complexity of propositional logics. *Complexity Problems of Mathematical Logic*, pages 80–90, 1985. (In Russian).
[5] A. V. Chagrov. Formal propositional logic of A. Visser and its extensions. *Logical Investigations*, 10:204–211, 2003. (In Russian).
[6] A. V. Chagrov. Algorithmic problem of finitary semantical consequence for basic and formal logics of A. Visser. *Logical Investigations*, 11:282–289, 2004. (In Russian).
[7] A. V. Chagrov and L. A. Chagrova. Algorithmic problems in semantics of Visser's formal propositional logic. In *Computer Science Applications of Modal Logic. International Conference. September 5–9 2005*, pages 11–12, Moscow, 2005. Poncelet Laboratory of UMI 2615 and Independent University of Moscow.
[8] A. V. Chagrov and M. N. Rybakov. How many variables one needs to prove PSPACE-hardness of modal logics? *Advances in Modal Logic*, 4:71–81, 2003.

[9] A. V. Chagrov and M. V. Zakharyaschev. *Modal Logic*. Oxford University Press, 1997.
[10] S. A. Cook. The complexity of theorem-proving procedures. In *Proceedings of the Third Annual ACM Symposium on the Theory of Computation*, pages 151–158, 1971.
[11] R. Fagin, J. Y. Halpern, and M. Y. Vardi. What can machines know? On the properties of knowledge in distributed systems. *Jornal of the Association for Computing Machinery*, 39(2):328–376, 1992.
[12] D. M. Gabbay. The decidability of the Kreisel–Putnam system. *The Journal of Symbolic Logic*, 35(3):431–437, 1970.
[13] D. M. Gabbay, A. Kurucz, F. Wolter, and M. Zakharyaschev. *Many-Dimensional Modal Logics: Theory and Applications*. Studies in Logic and the Foundations of Mathematics, 148. Elsevier, North-Holland, 2003.
[14] M. R. Garey and D. S. Johnson. *Computers and Intractability, A Guide to the Theory of NP-completeness*. San Francisco, 1979.
[15] J. Y. Halpern. The effect of bounding the number of primitive propositions and the depth of nesting on the complexity of modal logic. *Artificial Intelligence*, 75(2):361–372, 1995.
[16] N. Immerman. *Descriptive Complexity*. Springer, 1999.
[17] V. A. Jankov. On the calculus of the weak law of excluded middle. *Mathematics of the USSR, Izvestia*, (5):1044–1051, 1968. (In Russian).
[18] A. V. Kuznetsov. On superintuitionistic logics. In *Proceedings of the International Congress of Mathematicians*, pages 243–249, Vancouver, 1975.
[19] A. V. Kuznetsov. On facilities for detection of non-derivability and non-expressibility. In *Logical Derivation*, pages 5–33. Nauka, Moscow, 1979. (In Russian).
[20] R. E. Ladner. The computational complexity of provability in systems of modal logic. *SIAM Journal on Computing*, 6:467–480, 1977.
[21] I. Nishimura. On formulas of the one variable in intuitionistic propositional calculus. *The Journal of Symbolic Logic*, 25:327–331, 1960.
[22] C. H. Papadimitriou. *Computational Complexity*. Addison–Wesley Publishing Company, 1995.
[23] S. V. Popov and M. V. Zakharyaschev. On the complexity of Kripke countermodels in intuitionistic propositional calculus. In *Proceedings of the 2nd Soviet–Finland Logic Colloquium*, pages 32–36, 1979. (In Russian).
[24] S. V. Popov and M. V. Zakharyaschev. On the cardinality of countermodels of intuitionistic calculus. Technical Report 45, Institute of Applied Mathematics, Russian Academy of Science, 1980. (In Russian).
[25] W. Ruitenburg. Basic predicate calculus. *Notre Dame Journal of Formal Logic*, 39(1):18–46, 1998.
[26] M. N. Rybakov. Complexity of the decision problem for basic and formal logics. *Logical Investigations*, 10:158–166, 2003. (In Russian).
[27] M. N. Rybakov. On complexity of the decision problem for basic and formal logics with finitely many variables in the language. In *Smirnov Readings. IV International Conference*, pages 49–50, Moscow, 2003. Publishing of Institute of Philosophy RAS.
[28] M. N. Rybakov. Embedding of intuitionistic logic into its two-variable fragment and complexity of this fragment. *Logical Investigations*, 11:247–261, 2004. (In Russian).
[29] M. N. Rybakov. Complexity of the two-variable fragment of intuitionistic propositional logic. In *Computer Science Applications of Modal Logic. International Conference. September 5–9 2005*, pages 35–36, Moscow, 2005. Poncelet Laboratory of UMI 2615 and Independent University of Moscow.
[30] W. J. Savitch. Relationships between nondeterministic and deterministic tape complexity. *Journal of Computer and System Science*, 4:177–192, 1970.
[31] A. B. Sosinskiy. *How to Write a Mathematical Article in English*. International Center "Sofus Li", Minsk, 1994. (In Russian).
[32] E. Spaan. *Complexity of Modal Logics*. PhD thesis, University of Amsterdam, Department of Mathematics and Computer Science, 1993.

[33] R. Statman. Intuitionistic propositional logic is polynomial-space complete. *Theoret. Comput. Sci.*, 9(1):67–72, 1979.
[34] L. Stockmeyer. Classifying the computational complexity of problems. *The Journal of Symbolic Logic*, 52(1):1–43, 1987.
[35] Y. Suzuki, F. Wolter, and M. Zakharyaschev. Speaking about transitive frames in propositional languages. *Journal of Logic, Language and Information* 7:317–339, 1998.
[36] V. Švejdar. The decision problem of provability logic with only one atom. *Arch. Math. Logic*, pages 1–6, 2003.
[37] M. Szatkowski. On fragments of medvedev's logic. *Studia Logica*, 40(1):39–54, 1981.
[38] A. Visser. A propositional logic with explicit fixed points. *Studia Logica*, 40:155–175, 1981.
[39] D. A. Viter. Basic logic and primitive-recursive realizability. *Logical Investigations*, 9:90–102, 2002. (In Russian).
[40] M. Zakharyaschev, F. Wolter, and A. Chagrov. Advanced modal logic. In D. M. Gabbay and F. Guenthner, editors, *Handbook of Philosophical Logic*, volume 3, pages 83–266. Kluwer Academic Publishers, 2nd edition, 2001.

Mikhail Rybakov

Tver State University

Departmant of Mathematics

Zhelyabova Street, 33

Tver

170000

Russia

m_rybakov@mail.ru

Downward-directed transitive frames with universal relations

Ilya Shapirovsky

ABSTRACT. In this paper we identify modal logics of some bimodal Kripke frames corresponding to geometrical structures. Each of these frames is a set of 'geometrical' objects with some natural accessibility relation plus the universal relation. For these logics we present finite axiom systems and prove completeness.

We also show that all these logics have the finite model property and are PSPACE-complete. To prove this, we show that under certain restrictions, adding the universal modality preserves 'good' properties of a monomodal logic.

Keywords: universal modality, relativistic temporal logics, regions, completeness, finite model property

1 Introduction

The subject of this paper is in the field of spatial and temporal modal logics. We focus mainly on relativistic time, and also consider some interval and regional structures.

Our basic temporal relations are *causal* ($\preceq$) and *chronological* ($\prec$) accessibility. Modal axiomatizations of real space and its domains ordered by $\preceq$ can be found in [3], [14]. Analogous results on $\prec$ were obtained in [12], [10]. In [15] it was noted that relativistic logics can be interpreted in interval or regional semantics. This approach was further developed in [13]. Thus at present we have quite a few monomodal logics axiomatizing relativistic spacetime, interval or regional structures; all these logics have the finite model property (for short, FMP) and they are known to be PSPACE-complete. But the expressive power of these systems is rather weak. On the other hand, richer spatial structures may have undecidable or even non recursively enumerable logics (cf. [6], [8], [13]). This brings up the following standard question: how to improve the expressivity preserving the FMP and complexity?

Consider the following property of relativistic time: any two points are accessible from another point, in other words, any two points have common past. This property (*downward-directedness*) holds in real space $\mathbb{R}^n$, but it may fail in its subsets, in particular, in the upper half-space $\mathbb{R}^{n-1} \times \{t \mid t > 0, t \in \mathbb{R}\}$. However, the monomodal logic of the upper half-space (ordered by $\prec$ or $\preceq$) is equal to the logic of the whole space. Using the universal modality, we can express downward-directedness with the modal formula $A^\downarrow = \exists p \wedge \exists q \to \exists(\Diamond p \wedge \Diamond q)$.

In this paper we axiomatize modal logics of some geometrical downward-directed frames expanded with universal relations, and show that the FMP and PSPACE-decidability are preserved.

The problem of enriching a modal language with the universal modality was first systematically investigated in [5]. Given a normal monomodal logic L, we consider the logic LU - the fusion of L and **S5** plus the *containment* axiom $\Diamond p \to \exists p$. Among others, it was proved in [5] that LU inherits strong Kripke completeness and compactness from L. However, some properties of L can be lost: [17] gives an example of a monomodal logic L with the FMP such that LU lacks the FMP; in [7], it was shown that there exists a decidable monomodal logic L such that LU is undecidable.

Nevertheless, we prove that adding the universal modality together with the axiom $A^\downarrow$ preserves properties of transitive logics such as Kripke-completeness, the finite model property, decidability within a certain polynomially closed complexity class.

2 Basic notions

In this paper we consider normal monomodal and bimodal propositional logics.

Let $FM(\Diamond)$ be the set of all formulas constructed using a countable set of propositional variables $PV = \{p_1, p_2, \ldots\}$, propositional constant $\bot$ (*false*), and connectives $\to$, $\Diamond$. To obtain $FM(\Diamond, \exists)$, we enrich the language with the unary modal operator $\exists$. We define $\Box$, $\forall$, $\neg$, $\vee$, $\wedge$, $\top$ in the usual way, in particular, $\Box A := \neg \Diamond \neg A$, $\forall A := \neg \exists \neg A$. Also let $\Diamond^+ A := \Diamond A \vee A$, $\Box^+ A := \Box A \wedge A$.

For a monomodal (bimodal) logic L and a formula $A \in FM(\Diamond)$ ($A \in FM(\Diamond, \exists)$), L + A denotes the smallest monomodal (bimodal) logic containing $L \cup \{A\}$. The notation $L \vdash A$ means $A \in L$.

$Sub(A)$ denotes the set of all subformulas of A; for $\odot \in \{\Diamond, \exists\}$, $Sub_\odot(A) := \{\odot A \mid \odot A \in Sub(A)\}$.

A *(Kripke) frame* is a tuple $(W, R_1, \ldots, R_n)$, where $W \neq \varnothing$, $R_i \subseteq W \times W$. In this paper we always assume that R_1 is *transitive*. We consider only monomodal or bimodal frames, i.e., $n \leq 2$.

A *(Kripke) model* $\mathfrak{M}$ over a frame $\mathfrak{F}$ is a pair $(\mathfrak{F}, \theta)$, where $\theta : PV \to 2^W$, 2^W denotes the power set of W. The truth of a formula in a model is defined in the standard way. In particular, for a model $\mathfrak{M}$ over a frame (W, R_1, R_2) and a formula $\exists A \in FM(\Diamond, \exists)$, we put for all $x \in W$

$$\mathfrak{M}, x \vDash \exists A \quad \text{iff} \quad \mathfrak{M}, y \vDash A \text{ for some } y \in R_2(x)$$

The notations $x \in \mathfrak{M}$, $x \in \mathfrak{F}$ mean $x \in W$. The notation $\mathfrak{M} \vDash A$ means that $\mathfrak{M}, x \vDash A$ for all $x \in \mathfrak{M}$. If for any model $\mathfrak{M}$ over $\mathfrak{F}$ we have $\mathfrak{M} \vDash A$ then we say that A is *valid* in $\mathfrak{F}$, $\mathfrak{F} \vDash A$ in symbols. For a class $\mathcal{F}$ of frames,

$\mathbf{L}(\mathcal{F})$ denotes the *modal logic determined by* $\mathcal{F}$, i.e., the set of all formulas that are valid in all frames from $\mathcal{F}$. For a single frame $\mathfrak{F}$, $\mathbf{L}(\mathfrak{F})$ abbreviates $\mathbf{L}(\{\mathfrak{F}\})$.

For a logic L, if $\mathbf{L}(\mathfrak{F}) \supseteq$ L, then we say that $\mathfrak{F}$ is an L-*frame*. A formula A is *satisfiable in a model* $\mathfrak{M}$ if for some $x \in \mathfrak{M}$ we have $\mathfrak{M}, x \vDash A$; A is *satisfiable at a point x in a frame* $\mathfrak{F}$ if $\mathfrak{M}, x \vDash A$ for some model $\mathfrak{M}$ over $\mathfrak{F}$. A is *satisfiable in a frame* $\mathfrak{F}$ if A is satisfiable at some $x \in \mathfrak{F}$. For a class $\mathcal{F}$ of frames, A is $\mathcal{F}$-*satisfiable* if A is satisfiable in some $\mathfrak{F} \in \mathcal{F}$. A is L-*satisfiable* if A is satisfiable in some L-frame.

Consider a monomodal frame (W, R). For $x \in W$ let $R(x) := \{y \mid xRy\}$. Id_W denotes the equality relation on W, and R^r denotes the reflexive closure of R, i.e., $R^r := R \cup Id_W$.

Consider some first-order properties of a relation R:

seriality:	$\forall x \exists y\ xRy$;
Church–Rosser property:	$\forall x \forall y_1 \forall y_2 \exists z (xRy_1 \wedge xRy_2 \to y_1 R z \wedge y_2 R z)$;
McKinsey property:	$\forall x \exists y \in R(x)\ R(y) = \{y\}$;
irreflexive McKinsey property:	$\forall x \exists y \in R^r(x)\ R(y) = \varnothing$;
2-density:	$\forall x \forall y_1 \forall y_2 \exists z(xRy_1 \wedge xRy_2 \to xRz \wedge zRy_1 \wedge zRy_2)$.

For a monomodal frame (W, R), we have the following correspondence between modal axioms and first-order properties (recall that we consider only transitive frames):

$A4 := \Diamond\Diamond p \to \Diamond p$	transitivity;
$AT := p \to \Diamond p$	reflexivity;
$AD := \Diamond \top$	seriality;
$A1 := \Box\Diamond p \to \Diamond\Box p$	McKinsey property;
$A2 := \Diamond\Box p \to \Box\Diamond p$	Church – Rosser property.

Put $A1^- := \Diamond^+ \Box \bot$, $Ad_2 := \Diamond p_1 \wedge \Diamond p_2 \to \Diamond(\Diamond p_1 \wedge \Diamond p_2)$.

The following proposition is straightforward:

PROPOSITION 1. *For a frame* $\mathfrak{F} = (W, R)$,

- $\mathfrak{F} \vDash A1^-$ *iff* $\mathfrak{F}$ *satisfies irreflexive McKinsey property;*

- $\mathfrak{F} \vDash Ad_2$ *iff* $\mathfrak{F}$ *is 2-dense iff for any* $n \geq 1$, $\mathfrak{F}$ *satisfies*

$$\forall x \forall y_1 \ldots \forall y_n (\{y_1, \ldots, y_n\} \subseteq R(x) \to \exists z \in R(x)\ \{y_1, \ldots, y_n\} \subseteq R(z)).$$

As usual, **K** denotes the smallest normal monomodal logic. Let

$$\mathbf{K4} := \mathbf{K} + A4, \qquad \mathbf{S4} := \mathbf{K4} + AT,$$
$$\mathbf{Cr} := \mathbf{K4} + Ad_2 + AD, \qquad \mathbf{CrB} := \mathbf{K4} + Ad_2 + A1^-.$$

For a logic L let L.1 := L + A1, L.2 := L + A2.

Consider a monomodal frame $\mathfrak{F} = (W, R)$. For $V \subseteq W$, let $R|V := R \cap (V \times V)$. For $x \in \mathfrak{F}$ let $\mathfrak{F}\langle x \rangle := (R^r(x), R|R^r(x))$. If for some $x \in \mathfrak{F}$ we have $\mathfrak{F} = \mathfrak{F}\langle x \rangle$, we say that $\mathfrak{F}$ is *rooted* (or *a cone*), and x is a *root* of $\mathfrak{F}$. We say that $x \in \mathfrak{F}$ is *minimal* in $\mathfrak{F}$, if yRx implies xRy for any $y \in \mathfrak{F}$. Let $\mathfrak{F}^u$ denote the frame with the additional *universal* relation: $\mathfrak{F}^u := (W, R, W \times W)$. For a model $\mathfrak{M} = (\mathfrak{F}, \theta)$ we put $\mathfrak{M}^u := (\mathfrak{F}^u, \theta)$. For a class $\mathcal{F}$ of monomodal frames we put $\mathcal{F}^u := \{\mathfrak{F}^u \mid \mathfrak{F} \in \mathcal{F}\}$.

Consider frames $\mathfrak{F} = (W, R_1, \ldots, R_n)$ and $\mathfrak{G} = (V, S_1, \ldots, S_n)$. Recall that a surjective map $f : W \to V$ is a *p-morphism* from $\mathfrak{F}$ onto $\mathfrak{G}$ (in notation, $f : \mathfrak{F} \twoheadrightarrow \mathfrak{G}$), if for any $x \in W$, $1 \leq i \leq n$, we have $f(R_i(x)) = S_i(f(x))$. The notation $\mathfrak{F} \twoheadrightarrow \mathfrak{G}$ means that there exists a p-morphism from $\mathfrak{F}$ onto $\mathfrak{G}$. Recall that $\mathfrak{F} \twoheadrightarrow \mathfrak{G}$ implies $\mathbf{L}(\mathfrak{F}) \subseteq \mathbf{L}(\mathfrak{G})$. For monomodal frames $\mathfrak{F}$ and $\mathfrak{G}$ the following two facts are trivial: if $\mathfrak{F}$ and $\mathfrak{G}$ are isomorphic then $\mathfrak{F}^u$ and $\mathfrak{G}^u$ are isomorphic; if $f : \mathfrak{F} \twoheadrightarrow \mathfrak{G}$ then $f : \mathfrak{F}^u \twoheadrightarrow \mathfrak{G}^u$, and so $\mathbf{L}(\mathfrak{F}^u) \subseteq \mathbf{L}(\mathfrak{G}^u)$.

The following syntactical introduction of the universal modality is due to [5]. For a monomodal logic L, let LU denote the smallest normal bimodal logic containing L and the formulas

$$A4_\exists = \exists\exists p \to \exists p, \quad AT_\exists = p \to \exists p, \quad AB_\exists = p \to \forall \exists p;$$
$$A_\subseteq = \Diamond p \to \exists p.$$

It follows that (W, R_1, R_2) is LU-frame iff $\mathfrak{F} = (W, R_1)$ is L-frame, R_2 is an equivalence relation on W, and $R_1 \subseteq R_2$.

3 Translation

Given a monomodal rooted frame $\mathfrak{F}$, it is possible to show that the satisfiability in $\mathfrak{F}^u$ can be reduced to the satisfiability in $\mathfrak{F}$. For this purpose we use the following construction, proposed in [7].

Consider a formula $A \in FM(\Diamond, \exists)$. In this section we assume that $PV(A) \subseteq \{p_1, \ldots, p_m\}$, $Sub(A) = \{A_1, \ldots, A_n\}$, and $Sub_\exists(A) = \{\exists A_{i_1}, \ldots, \exists A_{i_l}\}$.

Fix some variables $q_1, \ldots, q_l \notin PV(A)$. For any $B \in Sub(A)$ we define the formula $[B]$ as follows:

$[p] := p$, $[\bot] := \bot$, $[B_1 \to B_2] := [B_1] \to [B_2]$, $[\Diamond B] := \Diamond[B]$, $[\exists A_{i_j}] := q_j$.

LEMMA 2. *[7] Given a formula $A \in FM(\Diamond, \exists)$ and a model $\mathfrak{M} = ((W, R), \theta)$ such that:*

(3.1) *if $\mathfrak{M}^u \vDash \exists A_{i_j}$ then $\theta(q_j) = W$, otherwise $\theta(q_j) = \varnothing$.*

Then for any $y \in W$ and any $B \in Sub(A)$, we have

$$\mathfrak{M}^u, y \vDash B \iff \mathfrak{M}, y \vDash [B]$$

Proof. By induction on the length of the formula. We consider only the case when $B = \exists A_{i_j}$: $\mathfrak{M}^u, y \vDash \exists A_{i_j}$ iff for some $z \in W$ $\mathfrak{M}^u, z \vDash A_{i_j}$ iff $\theta(q_j) = W$ iff $\mathfrak{M}, y \vDash [\exists A_{i_j}]$ (recall that $[\exists A_{i_j}] = q_j$). ∎

We put

$$A_\exists := \bigwedge_{j=1}^{l} \left((\exists [A_{i_j}] \to \forall q_j) \wedge (\neg \exists [A_{i_j}] \to \forall \neg q_j) \right) \wedge [A]$$

(as usual, we put $\bigwedge_{i \in \varnothing} C_i := \top$)

LEMMA 3. *[7] A is satisfiable in $\mathfrak{F}^u$ iff $A_\exists$ is satisfiable in $\mathfrak{F}^u$.*

Proof. ($\Rightarrow$) For some $\mathfrak{M}_0 = (\mathfrak{F}, \theta_0)$, $x \in W$, we have $\mathfrak{M}_0^u, x \vDash A$. Since $q_j \notin PV(A)$, $1 \leq j \leq l$, there exists a model $\mathfrak{M} = (\mathfrak{F}, \theta)$ such that $\theta(p_i) = \theta_0(p_i)$, $1 \leq i \leq m$, and $\mathfrak{M}$ satisfies (3.1). Then $\mathfrak{M}^u, x \vDash A$, and by Lemma 2 $\mathfrak{M}^u, x \vDash A_\exists$.
($\Leftarrow$) For some $\mathfrak{M} = (\mathfrak{F}, \theta)$, $x \in W$, we have $\mathfrak{M}^u, x \vDash A_\exists$. By induction on the length of the formula one can see that for any $y \in \mathfrak{F}$ and any $B \in Sub(A)$, we have: $\mathfrak{M}^u, y \vDash B$ iff $\mathfrak{M}, y \vDash [B]$. Let us consider the case when $B = \exists A_{i_j}$: $\mathfrak{M}^u, y \vDash \exists A_{i_j}$ iff for some z $\mathfrak{M}^u, z \vDash A_{i_j}$ iff for some z $\mathfrak{M}, z \vDash [A_{i_j}]$ iff $\mathfrak{M}^u, y \vDash \exists [A_{i_j}]$ iff $\mathfrak{M}, y \vDash q_j$ (note that $\mathfrak{M}^u \vDash \forall q_j$ or $\mathfrak{M}^u \vDash \forall \neg q_j$).
Since $\mathfrak{M}, x \vDash [A]$, we have $\mathfrak{M}^u, x \vDash A$. ∎

Let us slightly modify the considered translation:

$$A_\diamond := \bigwedge_{j=1}^{l} \left((\diamond^+ [A_{i_j}] \to \square^+ q_j) \wedge (\neg \diamond^+ [A_{i_j}] \to \square^+ \neg q_j) \right) \wedge \diamond^+ [A]$$

Note that $A_\diamond \in FM(\diamond)$.

Consider a rooted frame $\mathfrak{F} = \mathfrak{F}\langle x \rangle$. Then for any model $\mathfrak{M}$ over $\mathfrak{F}$ and $B \in FM(\diamond)$ we have: $\mathfrak{M}^u \vDash \exists B$ iff $\mathfrak{M}, x \vDash \diamond^+ B$. Thus we obtain the following reformulation of Lemma 3:

LEMMA 4. *Let $\mathfrak{F}$ be a cone, and let x be a root of $\mathfrak{F}$. Then for any $A \in FM(\diamond, \exists)$, A is satisfiable in $\mathfrak{F}^u$ iff $A_\diamond$ is satisfiable at x in $\mathfrak{F}$.*

4 Downward-directed frames

In this section we prove transfer theorems for logics of downward-directed transitive frames.

A monomodal frame $\mathfrak{F} = (W, R)$ is *downward-directed* iff $\mathfrak{F}$ satisfies

$$\forall x \forall y \exists z (zRx \wedge zRy).$$

(W, R) is *weakly downward-directed* iff (W, R^r) is downward-directed.

Consider the following axioms:

$$A^{\Downarrow} := \exists p \wedge \exists q \to \exists (\diamond^+ p \wedge \diamond^+ q)$$
$$A^{\downarrow} := \exists p \wedge \exists q \to \exists (\diamond p \wedge \diamond q)$$

Straightforward arguments show

PROPOSITION 5. *For a frame* $\mathfrak{F} = (W, R)$,

- $\mathfrak{F}^u \vDash A^{\Downarrow}$ *iff* $\mathfrak{F}$ *is weakly downward-directed;*
- $\mathfrak{F}^u \vDash A^{\downarrow}$ *iff* $\mathfrak{F}$ *is downward-directed.*

Clearly, every rooted frame is weakly downward-directed, so $\mathfrak{F} = \mathfrak{F}\langle x \rangle$ implies $\mathfrak{F}^u \vDash A^{\Downarrow}$. Moreover, if x is reflexive then $\mathfrak{F}$ is downward-directed and $\mathfrak{F} \vDash A^{\downarrow}$. On the other hand, let x be minimal in $\mathfrak{F}$. Then $\mathfrak{F}^u \vDash A^{\Downarrow}$ implies $\mathfrak{F} = \mathfrak{F}\langle x \rangle$; if $\mathfrak{F}^u \vDash A^{\downarrow}$ then we also have that x is reflexive.

For a monomodal logic L, we put

$$\mathrm{LU}^{\downarrow} := \mathrm{LU} + A^{\downarrow}, \ \mathrm{LU}^{\Downarrow} := \mathrm{LU} + A^{\Downarrow}.$$

For a bimodal logic M, let $\mathfrak{F}^{\mathrm{M}} = (W^{\mathrm{M}}, R^{\mathrm{M}}, U^{\mathrm{M}})$ denote its canonical frame, $\mathfrak{M}^{\mathrm{M}}$ denote its canonical model (the notion of *canonicity* is defined in the standard way, see e.g. [1]). Note that if $\mathrm{M} \supseteq \mathbf{K}\mathrm{U}$ then $R^{\mathrm{M}} \subseteq U^{\mathrm{M}}$ and U^{M} is an equivalence relation on W^{M} [5]. One can see that $A^{\downarrow}$, $A^{\Downarrow}$ are Sahlqvist formulas, so $A^{\downarrow}$, $A^{\Downarrow}$ are canonical.

By induction on n it is not hard to check

PROPOSITION 6. *For* $n \geq 1$,

- $\mathbf{K4U}^{\Downarrow} \vdash \exists p_1 \wedge \ldots \wedge \exists p_n \to \exists (\Diamond^+ p_1 \wedge \ldots \wedge \Diamond^+ p_n)$.
- $\mathbf{K4U}^{\downarrow} \vdash \exists p_1 \wedge \ldots \wedge \exists p_n \to \exists (\Diamond p_1 \wedge \ldots \wedge \Diamond p_n)$.

The following lemma shows that the canonical frame for a logic $\mathrm{M} \supseteq \mathbf{K4U}^{\Downarrow}$ is a disjoint union of cones with universal relations:

LEMMA 7. *Consider a logic* $\mathrm{M} \supseteq \mathbf{K4U}$. *Let W be an equivalence class modulo* U^{M}, $R := R^{\mathrm{M}}|W$, $\mathfrak{F} := (W, R)$. *Then*

- *if* $\mathrm{M} \vdash A^{\Downarrow}$ *then $\mathfrak{F}$ has a root;*
- *if* $\mathrm{M} \vdash A^{\downarrow}$ *then $\mathfrak{F}$ has a reflexive root.*

Proof. Let $\Psi = \{\Diamond B \mid B \in y \text{ for some } y \in W\}$. If Ψ is M-consistent then by the Lindenbaum lemma there exists $x \in W^{\mathrm{M}}$ such that $x \supseteq \Psi$. Then by the construction, $W = R(x)$.

If $\mathrm{M} \vdash A^{\downarrow}$ then any finite subset of Ψ is M-consistent (see Proposition 6), so Ψ is M-consistent.

Suppose that $\mathrm{M} \vdash A^{\Downarrow}$ and Ψ is M-inconsistent. Then $\Phi := \{\Diamond B_1, \ldots, \Diamond B_n\}$ is M-inconsistent for some $\Diamond B_1, \ldots, \Diamond B_n \in \Psi$. By Proposition 6, there exists $x \in W$ such that $x \ni \bigwedge_{1 \leq i \leq n} \Diamond^+ B_i$. Then for any $x' \in W$, $x'Rx$ implies $x' \supset \Phi$. It follows that x is minimal in $\mathfrak{F}$. Since $\mathfrak{F} \vDash A^{\Downarrow}$, $\mathfrak{F} = \mathfrak{F}\langle x \rangle$. ∎

For a formula A, let $A^0 := \neg A$, $A^1 := A$.

THEOREM 8. *Let $\mathcal{F}$ be a class of rooted frames closed under taking cones. Then $\mathbf{L}(\mathcal{F}^u) = \mathbf{L}(\mathcal{F})\mathrm{U}^{\Downarrow}$.*

Proof. Put $\mathrm{L} := \mathbf{L}(\mathcal{F})$, $\mathrm{M} := \mathrm{L}\mathrm{U}^{\Downarrow}$. Since $\mathcal{F}^u$ consists of rooted LU-frames, we have $\mathbf{L}(\mathcal{F}^u) \supseteq \mathrm{M}$.

To prove the converse inclusion, consider an M-consistent formula A. Then $A \in y_0$ for some $y_0 \in W^{\mathrm{M}}$. Let $Sub_E(A) = \{B_1, \ldots, B_l\}$. Put $B := A \wedge \bigwedge_{1 \leq j \leq l} B_j^{v_j}$, where $v_1, \ldots, v_l$ are defined as follows: if $B_j \in y_0$ then $v_j := 1$, otherwise $v_j := 0$. Thus $B \in y_0$.

Note that for any M-consistent formula C, $q \notin PV(C)$, $v \in \{0,1\}$, we have $C \wedge \forall q^v$ is M-consistent. Indeed, suppose that $C \wedge \forall q^v$ is M-inconsistent, i.e., $\mathrm{M} \vdash \forall q^v \to \neg C$. Replacing q with $\bot$ or $\top$, we obtain $\mathrm{M} \vdash \forall \top \to \neg C$. Trivially, $\mathrm{M} \vdash \forall \top$, so $\mathrm{M} \vdash \neg C$: a contradiction.

Let $q_1, \ldots, q_l \notin PV(B)$. It follows that $B \wedge \bigwedge_{1 \leq j \leq l} \forall q_j^{v_j}$ is M-consistent, so for some $y \in W^{\mathrm{M}}$ we have $\{B, \forall q_1^{v_1}, \ldots, \forall q_l^{v_l}\} \subseteq y$.

Let $W := U^{\mathrm{M}}(y)$, and let $\mathfrak{M}^u = ((W, R, U), \theta)$ be the restriction $\mathfrak{M}^{\mathrm{M}}|W$, i.e., $R := R^{\mathrm{M}}|W$, $U := U^{\mathrm{M}}|W = W \times W$, and for any $p \in PV$, $\theta(p) := \{x \in W \mid p \in x\}$. As well as in $\mathfrak{M}^{\mathrm{M}}$, for any $z \in W$ and any $C \in FM(\Diamond, \exists)$ we have: $C \in x$ iff $\mathfrak{M}^u, x \vDash C$.

Put $\mathfrak{F} := (W, R)$. By Lemma 2, we have $A_{\exists} \in y$. By Lemma 7, $\mathfrak{F} = \mathfrak{F}\langle x \rangle$ for some $x \in W$. Then $A_{\Diamond} \in x$, thus $A_{\Diamond}$ is L-consistent.

Then $A_{\Diamond}$ is satisfiable at some $z \in \mathfrak{G}$, $\mathfrak{G} \in \mathcal{F}$, thus $A_{\Diamond}$ is satisfiable at the root of $\mathfrak{H} = \mathfrak{G}\langle z \rangle$, and by Lemma 4, A is satisfiable in $\mathfrak{H}^u$. Since $\mathcal{F}$ closed under taking cones, we have $\mathfrak{H} \in \mathcal{F}$ and so $\mathfrak{H}^u \in \mathcal{F}^u$. Thus A is $\mathcal{F}^u$-satisfiable.

It follows that $\mathbf{L}(\mathcal{F}^u) \subseteq \mathrm{M}$. ∎

COROLLARY 9. *For a monomodal transitive Kripke-complete logic* L,

- $\mathrm{LU}^{\Downarrow}$ *is Kripke-complete;*
- *if* L *has the FMP, then* $\mathrm{LU}^{\Downarrow}$ *has the FMP;*
- *for any $A \in \mathrm{FM}(\Diamond, \exists)$, A is $\mathrm{LU}^{\Downarrow}$-satisfiable iff $A_{\Diamond}$ is L-satisfiable;*
- L *and* $\mathrm{LU}^{\Downarrow}$ *are polynomially equivalent.*

Proof. Suppose $\mathrm{L} = \mathbf{L}(\mathcal{G})$. Put $\mathcal{F} = \{\mathfrak{G}\langle x \rangle \mid \mathfrak{G} \in \mathcal{G}, x \in \mathfrak{G}\}$. Then $\mathrm{L} = \mathbf{L}(\mathcal{F})$. By Theorem 8, $\mathrm{LU}^{\Downarrow} = \mathbf{L}(\mathcal{F}^u)$.

By Lemma 4, for any formula $A \in FM(\Diamond, \exists)$ we have:

A is $\mathcal{F}^u$-satisfiable iff $A_{\Diamond}$ is $\mathcal{F}$-satisfiable.

One can readily check that $A_{\Diamond}$ can be computed in time polynomial in the length of A, so $\mathrm{LU}^{\Downarrow}$ is polynomially reduced to L. Trivially, for any $B \in FM(\Diamond)$ we have: B is L-satisfiable iff B is $\mathrm{LU}^{\Downarrow}$-satisfiable, so L and $\mathrm{LU}^{\Downarrow}$ are polynomially equivalent. ∎

It is well-known that the logics **S4**, **S4.2**, **S4.1** have the FMP and are PSPACE-complete (see e.g. [2]). Clearly, if L ⊢ AT then LU$^\downarrow$ = LU$^\Downarrow$. Therefore we have

COROLLARY 10. *The logics* **S4**U$^\downarrow$, **S4.2**U$^\downarrow$, **S4.1**U$^\downarrow$ *have the FMP and are PSPACE-complete.*

Thus adding the universal modality together with the axiom $A^\Downarrow$ preserves Kripke completeness, the FMP, and the complexity of a transitive monomodal logic L. However, the situation with the axiom $A^\downarrow$ is more delicate: the following example shows that LU$^\downarrow$ may lack the FMP.

Consider the logic **GL** := **K** + $\Box(\Box p \to p) \to \Box p$. This logic is Kripke-complete, transitive and has the FMP: **GL** is complete with respect to the class of all finite strictly ordered trees (see e.g. [2]). The logic **GL**U$^\downarrow$ is consistent and has frames, for instance $\mathbf{L}((\mathbb{N}, >)^u) \supseteq \mathbf{GL}\mathrm{U}^\downarrow$. But it is not hard to see that this logic has no finite frames. (To simplify this example one can replace $A^\downarrow$ with the axiom $\exists p \to \exists \Diamond p$, expressing seriality of R^{-1}.)

Nevertheless, under some additional assumptions, LU$^\downarrow$ inherits the above mentioned properties of L.

For a monomodal frame $\mathfrak{F}$, let $\widetilde{\mathfrak{F}} := \mathfrak{C}_1 + \mathfrak{F}$, where $\mathfrak{C}_1$ is a reflexive singleton, + denotes the ordinal sum of frames.

For a formula $A \in FM(\Diamond, \exists)$, put

$$A_\Diamond^{refl} := A_\Diamond \wedge \bigwedge_{\Diamond B \in Sub(A_\Diamond)} (B \to \Diamond B)$$

LEMMA 11. *Suppose* $A_\Diamond^{refl}$ *is satisfiable at* y *in* $\mathfrak{F}$, $\mathfrak{G} := \widetilde{\mathfrak{F}\langle y \rangle}$. *Then* A *is satisfiable in* $\mathfrak{G}^u$.

Proof. For some model $\mathfrak{M} = (\mathfrak{F}, \eta)$, we have $\mathfrak{M}, y \vDash A_\Diamond^{refl}$. Let y_0 denote the root of $\mathfrak{G}$. Put $\mathfrak{N} := (\mathfrak{G}, \theta)$, where $\theta^{-1}(y_0) := \eta^{-1}(y)$, and $\theta^{-1}(z) := \eta^{-1}(z)$ for any $z \in \mathfrak{F}\langle y \rangle$.

By induction on the length of the formula one can readily check that $\mathfrak{N}, y_0 \vDash A_\Diamond$. By Lemma 4, A is satisfiable in $\mathfrak{G}^u$. ∎

THEOREM 12. *Let $\mathcal{F}$ be a class of frames such that:*
any $\mathfrak{F} \in \mathcal{F}$ has a reflexive root;
$\mathfrak{F} \in \mathcal{F}$, $y \in \mathfrak{F}$ implies $\widetilde{\mathfrak{F}\langle y \rangle} \in \mathcal{F}$.
Then $\mathbf{L}(\mathcal{F}^u) = \mathbf{L}(\mathcal{F})\mathrm{U}^\downarrow$.

Proof. Put L := $\mathbf{L}(\mathcal{F})$, M := LU$^\downarrow$. Given an M-consistent formula A, we have to show that A is $\mathcal{F}^u$-satisfiable.

Similarly to the proof of Theorem 8, for some reflexive $x \in W^M$ we have $A_\Diamond \in x$. So $A_\Diamond^{refl} \in x$, then $A_\Diamond^{refl}$ is L-consistent. Thus $A_\Diamond^{refl}$ is satisfiable at some $y \in \mathfrak{F}$, $\mathfrak{F} \in \mathcal{F}$. Let $\mathfrak{G} := \widetilde{\mathfrak{F}\langle y \rangle}$. By Lemma 11, A is satisfiable in $\mathfrak{G}^u$. Since $\mathfrak{G} \in \mathcal{F}$, A is $\mathcal{F}^u$-satisfiable.

It follows that $\mathbf{L}(\mathcal{F}^u) \subseteq$ M. Clearly, $\mathbf{L}(\mathcal{F}^u) \supseteq$ M, thus $\mathbf{L}(\mathcal{F}^u) =$ M. ∎

COROLLARY 13. *Let* L *be a monomodal transitive Kripke-complete logic such that for any cone* $\mathfrak{F}$, *if* $\mathbf{L}(\mathfrak{F}) \supseteq$ L *then* $\mathbf{L}(\widetilde{\mathfrak{F}}) \supseteq$ L. *Then*

- $\mathrm{LU}^\downarrow$ *is Kripke-complete;*
- *if* L *has the FMP, then* $\mathrm{LU}^\downarrow$ *has the FMP;*
- *for any* $A \in \mathrm{FM}(\Diamond, \exists)$, A *is* $\mathrm{LU}^\downarrow$-*satisfiable iff* $A_\Diamond^{refl}$ *is* L-*satisfiable;*
- L *and* $\mathrm{LU}^\downarrow$ *are polynomially equivalent.*

Proof. Let $\mathcal{G}$ be the class of all (finite) L-frames, and let $\mathcal{F}$ be the class of all (finite) cones from $\mathcal{G}$ with reflexive root. Then $L \subseteq \mathbf{L}(\mathcal{F})$.

Assume that $A \in FM(\Diamond)$ is L-satisfiable. $L = \mathbf{L}(\mathcal{G})$ implies that A is satisfiable at some $z \in \mathfrak{F}$, $\mathfrak{F} \in \mathcal{G}$. Trivially, A is satisfiable in $\widetilde{\mathfrak{F}\langle z \rangle} \in \mathcal{F}$. Thus $L = \mathbf{L}(\mathcal{F})$. Since $\mathcal{F}$ satisfies the conditions of Theorem 12, $\mathrm{LU}^\downarrow = \mathbf{L}(\mathcal{F}^u)$.

By Lemmas 4, 11, for any formula $A \in FM(\Diamond, \exists)$ we have:

$$A \text{ is } \mathcal{F}^u\text{-satisfiable iff } A_\Diamond^{refl} \text{ is } \mathcal{F}\text{-satisfiable.}$$

To complete the proof, note that the length of $A_\Diamond^{refl}$ is polynomial in the length of A. ∎

In [12], it was shown that the logics **Cr**, **Cr.2** have the FMP. The method proposed in [12] was used in [10] to prove the FMP of **CrB**. The complexity of 2-dense logics was studied in [11], where PSPACE-completeness of **Cr**, **Cr.2** was proved. A slight modification of this proof yields the PSPACE-completeness of **CrB**. By Proposition 1, if $L \in \{\mathbf{Cr}, \mathbf{Cr.2}, \mathbf{CrB}\}$ and $\mathfrak{F}$ is L-frame then $\widetilde{\mathfrak{F}}$ is L-frame. Therefore we have

COROLLARY 14. *The logics* $\mathbf{CrU}^\downarrow$, $\mathbf{Cr.2U}^\downarrow$, $\mathbf{CrBU}^\downarrow$ *have the FMP and are PSPACE-complete.*

5 Intervals, regions, and Minkowski spacetime

In this section we quote some results on modal axiomatization of relativistic spacetime and related interval and regional structures (for a detailed survey of this topic, see [13]).

5.1 Causal and chronological modalities

Let us recall the definition of *causal accessibility* $\preceq$ and *chronological accessibility* $\prec$ in Minkowski spacetime. For $(x_1, \ldots, x_n)$, $(y_1, \ldots, y_n) \in \mathbb{R}^n$, $n \geq 2$, we put:

$$(x_1, \ldots, x_n) \preceq (y_1, \ldots, y_n) \Leftrightarrow \sum_{i=1}^{n-1}(y_i - x_i)^2 \leq (x_n - y_n)^2 \ \& \ x_n \leq y_n,$$

$$(x_1, \ldots, x_n) \prec (y_1, \ldots, y_n) \Leftrightarrow \sum_{i=1}^{n-1}(y_i - x_i)^2 < (x_n - y_n)^2 \ \& \ x_n < y_n.$$

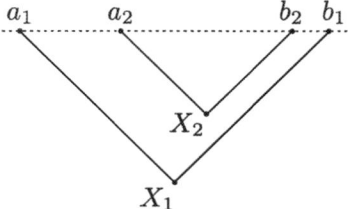

Figure 1.

For $D \subseteq \mathbb{R}^2$, $R \in \{\prec, \preceq\}$ let (D, R) abbreviate $(D, R|D)$. Put

$$\mathbb{R}^n_< := \{(x_1, \ldots, x_n) \in \mathbb{R}^n \mid x_n < 0\}, \quad \mathbb{R}^n_\leq := \{(x_1, \ldots, x_n) \in \mathbb{R}^n \mid x_n \leq 0\}.$$

The first results on modal axiomatization of relativistic relations are due to Goldblatt [3] and Shehtman [14], where modal logics of real space and its domains ordered by $\preceq$ were described. Analogous results on the relation $\prec$ were recently obtained in [12],[10].

THEOREM 15. *Let* $n \geq 2$.
$\mathbf{L}(\mathbb{R}^n, \preceq) = \mathbf{S4.2}$, $\mathbf{L}(\mathbb{R}^n_<, \preceq) = \mathbf{S4}$ *[3]*; $\mathbf{L}(\mathbb{R}^n_\leq, \preceq) = \mathbf{S4.1}$ *[14]*;
$\mathbf{L}(\mathbb{R}^n, \prec) = \mathbf{Cr.2}$, $\mathbf{L}(\mathbb{R}^n_<, \prec) = \mathbf{Cr}$ *[12]*; $\mathbf{L}(\mathbb{R}^n_\leq, \prec) = \mathbf{CrB}$ *[10]*.

5.2 Intervals and regions

Consider the sets of strict and non-strict intervals $\mathcal{I}$, $\mathcal{I}^*$ on the real line and the relations $\sqsupseteq$, $\sqsupset$:

$$\mathcal{I} := \{[a, b] \mid a, b \in \mathbb{R}, a < b\}, \quad \mathcal{I}^* := \{[a, b] \mid a, b \in \mathbb{R}, a \leq b\};$$
$$[a_1, b_1] \sqsupseteq [a_2, b_2] := a_1 \leq a_2 \text{ and } b_2 \leq b_1,$$
$$[a_1, b_1] \sqsupset [a_2, b_2] := a_1 < a_2 \text{ and } b_2 < b_1.$$

In [15], it was noted that there exists a simple isomorphism between the frames $(\mathcal{I}^*, \sqsupseteq, \sqsupset)$ and $(\mathbb{R}^2_\leq, \preceq, \prec)$, Fig 1.

PROPOSITION 16. *[15]*
$(\mathcal{I}, \sqsupseteq, \sqsupset) \cong (\mathbb{R}^2_<, \preceq, \prec)$, $(\mathcal{I}^*, \sqsupseteq, \sqsupset) \cong (\mathbb{R}^2_\leq, \preceq, \prec)$.

Thus the following fact is an immediate consequence of Theorem 15:

COROLLARY 17.
$\mathbf{L}(\mathcal{I}, \sqsupseteq) = \mathbf{S4}$, $\mathbf{L}(\mathcal{I}^*, \sqsupseteq) = \mathbf{S4.1}$;
$\mathbf{L}(\mathcal{I}, \sqsupset) = \mathbf{Cr}$; $\mathbf{L}(\mathcal{I}^*, \sqsupset) = \mathbf{CrB}$.

We consider the following sets as *regions*:

- *balls* in $\mathbb{R}^n$:

$$\mathcal{B}_n := \{B(X, r) \mid X \in \mathbb{R}^n, r > 0\}, \quad \mathcal{B}^*_n := \{B(X, r) \mid X \in \mathbb{R}^n, r \geq 0\},$$

where $B(X, r)$ is the closed ball with center X of radius r;

- *bricks* in $\mathbb{R}^n$:
$$\mathcal{R}_n := \{ \prod_{1 \leq i \leq n} [a_i, b_i] \mid [a_1, b_1], \ldots, [a_n, b_n] \in \mathcal{I} \},$$
$$\mathcal{R}_n^* := \{ \prod_{1 \leq i \leq n} [a_i, b_i] \mid [a_1, b_1], \ldots, [a_n, b_n] \in \mathcal{I}^* \};$$

- $\mathcal{CN}_n$ (respectively, $\mathcal{CV}_n$) is the set of all non-empty compact regular[1] sets with connected (respectively, convex) interior in $\mathbb{R}^n$. $\mathcal{CN}_n^*$ (respectively, $\mathcal{CV}_n^*$) is obtained from $\mathcal{CN}_n$ (respectively, $\mathcal{CV}_n$) by adding all singletons.

For a set $U \subseteq \mathbb{R}^n$, $\mathbf{I}U$ denotes its interior. Put
$$U \ni V := \mathbf{I}U \supseteq V \quad (U \text{ is a } \textit{non-tangential proper part of } V).$$

Note that if $W \in \{\mathcal{B}_1, \mathcal{CN}_1, \mathcal{CV}_1, \mathcal{R}_1\}$, then $(\mathcal{I}, \sqsupseteq, \sqsupset) = (W, \supseteq, \ni)$ and $(\mathcal{I}^*, \sqsupseteq, \sqsupset) = (W^*, \supseteq, \ni)$. Moreover, the following holds:

LEMMA 18. *[13] For $n \geq 1$, $W \in \{\mathcal{B}_n, \mathcal{CN}_n, \mathcal{CV}_n, \mathcal{R}_n\}$,*
$(W, \supseteq, \ni) \twoheadrightarrow (\mathcal{I}, \sqsupseteq, \sqsupset)$, $(W^*, \supseteq, \ni) \twoheadrightarrow (\mathcal{I}^*, \sqsupseteq, \sqsupset)$.

THEOREM 19. *[13] Let $W \in \{\mathcal{B}_n, \mathcal{CN}_n, \mathcal{CV}_n, \mathcal{R}_n\}$, $n \geq 1$. Then*
$\mathbf{L}(W, \supseteq) = \mathbf{S4}$, $\mathbf{L}(W^*, \supseteq) = \mathbf{S4.1}$;
$\mathbf{L}(W, \ni) = \mathbf{Cr}$, $\mathbf{L}(W^*, \ni) = \mathbf{CrB}$.

6 Main completeness results

Let $X, Y \in \mathbb{R}^n$. One can see that there exists a point $Z \in \mathbb{R}^n_{\leq}$ such that $Z \prec X$, $Z \prec Y$ (and therefore $Z \preceq X$, $Z \preceq Y$), and we have the following

LEMMA 20. *Let $\mathfrak{F} := (W, R)$, where $R \in \{\preceq, \prec\}$, $W \in \{\mathbb{R}^n, \mathbb{R}^n_{\leq}, \mathbb{R}^n_{<}\}$, $n \geq 2$. Then $\mathfrak{F}$ is downward-directed, and thus $\mathbf{L}(\mathfrak{F}^u) \supseteq \mathbf{L}(\mathfrak{F}) \mathbf{U}^{\downarrow}$.*

The following technical lemmas are needed for the sequel.

LEMMA 21. *Consider a frame $\mathfrak{F} = (W, R)$ and an infinite sequence of sets $W_0 \subseteq W_1 \subseteq \ldots$ such that $\bigcup_i W_i = W$. Put $\mathfrak{F}_i = (W_i, R|W_i)$ and suppose that for all $i \geq 0$, there exists a p-morphism $p_i : \mathfrak{F}_{i+1} \twoheadrightarrow \mathfrak{F}_i$, $p_i|W_i = Id_{W_i}$. Then $\mathfrak{F} \twoheadrightarrow \mathfrak{F}_0$.*

Proof. Put $f_i := p_i \cdot \ldots \cdot p_0$. Then $f_i : \mathfrak{F}_i \twoheadrightarrow \mathfrak{F}_0$, $i \geq 0$. One can see that $f := \bigcup_i f_i$ is the required p-morphism. ∎

LEMMA 22. *For $R \in \{\preceq, \prec\}$ we have:*
$(\mathbb{R}^2_{<}, R) \twoheadrightarrow ([-1, 1] \times \mathbb{R}_{<}, R)$, $(\mathbb{R}^2_{\leq}, R) \twoheadrightarrow ([-1, 1] \times \mathbb{R}_{\leq}, R)$.

Proof. For any $c \in \mathbb{R}$, we define the maps r_c, l_c as follows. For $x, t \in \mathbb{R}$, we put
$$r_c(x, t) := \begin{cases} (2c - x, t) & x \leq c \\ (x, t) & x > c \end{cases} \qquad l_c(x, t) := \begin{cases} (x, t) & x \leq c \\ (2c - x, t) & x > c \end{cases}$$

---
[1] Recall that *regular closed sets* are the closures of open sets

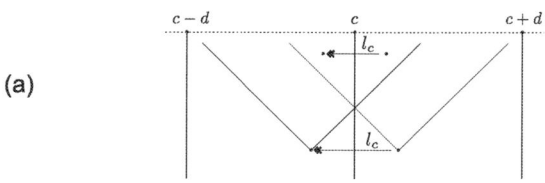

(a)

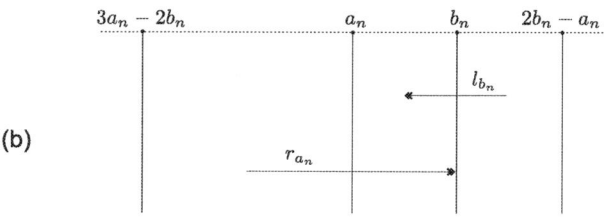

(b)

Figure 2.

It is not difficult to check that

$$r_c : ([c-d, c+d] \times \mathbb{R}_<, R) \twoheadrightarrow ([c, c+d] \times \mathbb{R}_<, R),$$

$$l_c : ([c-d, c+d] \times \mathbb{R}_<, R) \twoheadrightarrow ([c-d, c] \times \mathbb{R}_<, R),$$

for any $c, d \in \mathbb{R}$, $d \geq 0$ (see Fig 2,a).

Consider a sequence of segments $[a_0, b_0], [a_1, b_1], \ldots$, where $a_0 := -1$, $b_0 := 1$, $a_{n+1} := 3a_n - 2b_n$, $b_{n+1} := 2b_n - a_n$. Using maps l_{a_n}, r_{b_n}, we obtain

$$([a_{n+1}, b_{n+1}] \times \mathbb{R}_<, R) \twoheadrightarrow ([a_n, b_n] \times \mathbb{R}_<, R),$$

Fig 2,b. Clearly, a_n tends to $-\infty$, b_n tends to $+\infty$,[2] and therefore $\mathbb{R}_<^2 = \bigcup_i ([a_i, b_i] \times \mathbb{R}_<)$. By Lemma 21, $(\mathbb{R}_<^2, R) \twoheadrightarrow ([a_0, b_0] \times \mathbb{R}_<, R)$.

In complete analogy, $(\mathbb{R}_\leq^2, R) \twoheadrightarrow ([-1, 1] \times \mathbb{R}_\leq, R)$. ∎

For a relation $R \subseteq W \times W$, let $R^{\bowtie} := W \times W - (R \cup R^{-1} \cup Id_W)$.

LEMMA 23. *Consider a 2-dense frame $\mathfrak{F} = (W, R)$, $x \in W$, and suppose that the following holds:*

$$\forall y \in R^{\bowtie}(x) \; \exists z_y \; (R(z_y) = R(x) \cap R(y)).$$

Then for any $A \in \mathrm{FM}(\Diamond, \exists)$ we have: if $A^{\zeta efl}_\Diamond$ is satisfiable at x in $\mathfrak{F}$ then A is satisfiable in $\mathfrak{F}^u$.

Proof. Suppose that for some $\mathfrak{N} = (\mathfrak{F}\langle x \rangle, \eta)$ we have $\mathfrak{N}, x \vDash A^{\zeta efl}_\Diamond$.

Let $W_1 := R^r(x)$, $W_2 := R^{-1}(x) - W_1$, $W_3 := W - (W_1 \cup W_2) = R^{\bowtie}(x)$.

---

[2]By direct calculation, $a_n = \frac{1-4^{n+1}}{3}$, $b_n = \frac{2 \cdot 4^n + 1}{3}$

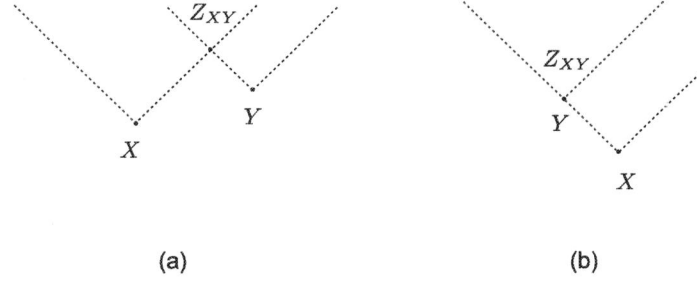

Figure 3.

For any $y \in W$ we define $y' \in W_1$ as follows. If $y \in W_1$, we put $y' := y$; if $y \in W_2$, we put $y' := x$. To define y' when $y \in W_3$, consider the set of formulas
$$\Psi_y := \{\Diamond B \in Sub(A_\Diamond) \mid \exists u \in R(z_y)\ \mathfrak{N}, u \vDash B\}.$$
Suppose $\Psi_y = \{\Diamond B_1, \ldots, \Diamond B_l\}$. Then $\mathfrak{N}, u_i \vDash B_i$ for some points $u_1, \ldots, u_l \in R(z_y)$. Since $\mathfrak{F}$ is 2-dense, by Proposition 1 we have $R(v) \supseteq \{u_1, \ldots, u_l\}$ for some $v \in R(z_y)$. We put $y' := v$.

Consider a model $\mathfrak{M} = (\mathfrak{F}, \theta)$ such that $\theta^{-1}(y) = \eta^{-1}(y')$ for any $y \in W$. We claim that for any $C \in Sub(A_\Diamond)$ and any $y \in W$,

(6.1) $\quad \mathfrak{M}, y \vDash C \iff \mathfrak{N}, y' \vDash C$

The proof is by induction on the length of the formula. Consider the only non-trivial case $C = \Diamond B$, $y \notin W_1$. Note that $R(y') \subseteq R(y)$, so $\mathfrak{N}, y' \vDash \Diamond B$ implies $\mathfrak{M}, y \vDash \Diamond B$ (by induction hypothesis). Conversely, assume that $\mathfrak{M}, y \vDash \Diamond B$. Thus $\mathfrak{M}, v \vDash B$ for some $v \in R(y)$, and by induction hypothesis $\mathfrak{N}, v' \vDash B$. There may be two options:

- $y \in W_2$. Then $y' = x$. Since $v' \in R^r(x)$, we have $\mathfrak{N}, x \vDash \Diamond^+ B$. Due to the definition of $A_\Diamond^{\xi efl}$, $\mathfrak{N}, x \vDash B \to \Diamond B$. Thus $\mathfrak{N}, x \vDash \Diamond B$.

- $y \in W_3$. Then $v \in W_1 - \{x\}$ or $v \in W_3$ (because $yR^\bowtie x$). In the former case $v' = v \in R(y) \cap R(x)$, so $v' \in R(z_y)$. If $v \in W_3$ then $R(z_v) \subseteq R(z_y)$, and since $v' \in R(z_v)$, $v' \in R(z_y)$. So, in either case, we have $v' \in R(z_y)$. Thus $\Diamond B \in \Psi_y$. Therefore for some $u \in R(y')$ we have $\mathfrak{N}, u \vDash B$, so $\mathfrak{N}, y' \vDash \Diamond B$.

Since $\mathfrak{N}, x \vDash A_\Diamond$, and due to (6.1), we obtain that $A_\exists$ is satisfiable in $\mathfrak{M}^u$. By Lemma 3, A is satisfiable in $\mathfrak{F}^u$. ∎

Observe that for any $X, Y \in \mathbb{R}^2$, there exists a unique point Z_{XY} such that $\prec(X) \cap \prec(Y) = \prec(Z_{XY})$, Fig. 3.

Now we are ready to prove the following key

LEMMA 24. *Let $\mathfrak{F} := (W, R)$, where $R \in \{\preceq, \prec\}$, $W \in \{\mathbb{R}^2, \mathbb{R}^2_\leq, \mathbb{R}^2_<\}$. Then $\mathbf{L}(\mathfrak{F}^u) = \mathbf{L}(\mathfrak{F})\mathrm{U}^\downarrow$, namely*

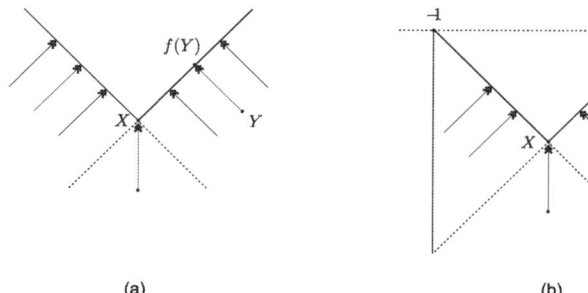

Figure 4.

$\mathbf{L}((\mathbb{R}^2, \preceq)^u) = \mathbf{S4.2U}^{\downarrow}$, $\mathbf{L}((\mathbb{R}^2_<, \preceq)^u) = \mathbf{S4U}^{\downarrow}$, $\mathbf{L}((\mathbb{R}^2_\leq, \preceq)^u) = \mathbf{S4.1U}^{\downarrow}$;
$\mathbf{L}((\mathbb{R}^2, \prec)^u) = \mathbf{Cr.2U}^{\downarrow}$, $\mathbf{L}((\mathbb{R}^2_<, \prec)^u) = \mathbf{CrU}^{\downarrow}$, $\mathbf{L}((\mathbb{R}^2_\leq, \prec)^u) = \mathbf{CrBU}^{\downarrow}$.

Proof. First, consider the frames ordered by $\preceq$.

Let $\mathfrak{F}$ be a cone in $(\mathbb{R}^2, \preceq)$, and let $\mathfrak{G}$ be a cone in $(\mathbb{R}^2_<, \preceq)$. Then $\mathbf{L}(\mathfrak{F}) = \mathbf{S4.2}$, $\mathbf{L}(\mathfrak{G}) = \mathbf{S4}$ [3],[14]. Put

$$\mathcal{F} := \{\mathfrak{F}\langle X\rangle \mid X \in \mathfrak{F}\}, \quad \mathcal{G} := \{\mathfrak{G}\langle X\rangle \mid X \in \mathfrak{G}\}.$$

By Theorem 8 we have: $\mathbf{L}(\mathcal{F}^u) = \mathbf{L}(\mathcal{F})\mathbf{U}^{\Downarrow}$, $\mathbf{L}(\mathcal{G}^u) = \mathbf{L}(\mathcal{G})\mathbf{U}^{\Downarrow}$. All frames in $\mathcal{F}$ (in $\mathcal{G}$) are isomorphic, thus all frames in $\mathcal{F}^u$ (in $\mathcal{G}^u$) are isomorphic. Thus

$$\mathbf{L}(\mathfrak{F}^u) = \mathbf{L}(\mathcal{F}^u) = \mathbf{L}(\mathcal{F})\mathbf{U}^{\Downarrow} = \mathbf{L}(\mathfrak{F})\mathbf{U}^{\Downarrow}.$$

Similarly, $\mathbf{L}(\mathfrak{G}^u) = \mathbf{L}(\mathfrak{G})\mathbf{U}^{\Downarrow}$. Since $\mathfrak{F}$ and $\mathfrak{G}$ are reflexive, we obtain $\mathbf{L}(\mathfrak{F}^u) = \mathbf{S4.2U}^{\downarrow}$, $\mathbf{L}(\mathfrak{G}^u) = \mathbf{S4U}^{\downarrow}$.

Let $\mathfrak{H}$ be a cone in $(\mathbb{R}^2_\leq, \preceq)$ whose root belongs to $\mathbb{R}^2_<$. By Theorem 8, we obtain $\mathbf{L}(\{\mathfrak{H}^u, \mathfrak{C}^u_1\}) = \mathbf{L}(\mathfrak{H})\mathbf{U}^{\Downarrow}$, where $\mathfrak{C}_1$ is a reflexive singleton. Trivially, $\mathbf{L}(\mathfrak{H}^u) \subseteq \mathbf{L}(\mathfrak{C}^u_1)$, so $\mathbf{L}(\mathfrak{H}^u) = \mathbf{L}(\mathfrak{H})\mathbf{U}^{\Downarrow}$. Since $\mathbf{L}(\mathfrak{H}) = \mathbf{S4.1}$ [14], we get $\mathbf{L}(\mathfrak{H}^u) = \mathbf{S4.1U}^{\downarrow}$.

Without loss of generality we may assume that the frames $\mathfrak{F}, \mathfrak{G}, \mathfrak{H}$ have the same root $X = (0, -1)$. Let us define the map f as follows: for any $Y \in \mathbb{R}^2$, put

$$f(Y) := \begin{cases} Y & X \preceq Y \\ X & Y \preceq X \\ Z_{XY} & \text{otherwise} \end{cases}$$

It is not difficult to see that $f : (\mathbb{R}^2, \preceq) \twoheadrightarrow \mathfrak{F}$ (Fig. 4,a), so $\mathbf{L}((\mathbb{R}^2, \preceq)^u) \subseteq \mathbf{S4.2U}^{\downarrow}$. We also have $f : ([-1, 1] \times \mathbb{R}_<, \preceq) \twoheadrightarrow \mathfrak{G}$, $f : ([-1, 1] \times \mathbb{R}_\leq, \preceq) \twoheadrightarrow \mathfrak{H}$ (Fig. 4,b), and by Lemma 22, $(\mathbb{R}^2_<, \preceq) \twoheadrightarrow \mathfrak{G}$ and $(\mathbb{R}^2_\leq, \preceq) \twoheadrightarrow \mathfrak{H}$. Thus $\mathbf{L}((\mathbb{R}^2_<, \preceq)^u) \subseteq \mathbf{S4U}^{\downarrow}$, $\mathbf{L}((\mathbb{R}^2_\leq, \preceq)^u) \subseteq \mathbf{S4.1U}^{\downarrow}$,

The converse inclusions hold by Lemma 20.

Now consider the frames ordered by $\prec$.

Suppose that $A \in FM(\diamondsuit, \exists)$ is $\mathbf{Cr.2U}^\downarrow$-satisfiable. By Corollary 13, $A\xi^{efl}$ is $\mathbf{Cr.2}$-satisfiable. Since $\mathbf{Cr.2} = \mathbf{L}(\mathbb{R}^2, \prec)$, $A\xi^{efl}$ is satisfiable in $(\mathbb{R}^2, \prec)$. By Lemma 23, A is satisfiable in $(\mathbb{R}^2, \prec)^u$. Thus $\mathbf{L}((\mathbb{R}^2, \prec)^u) \subseteq \mathbf{Cr.2U}^\downarrow$, and by Lemma 20, $\mathbf{L}((\mathbb{R}^2, \prec)^u) = \mathbf{Cr.2U}^\downarrow$.

Analogously, if A is $\mathbf{CrU}^\downarrow$-satisfiable then $A\xi^{efl}$ is satisfiable at some X in $(\mathbb{R}^2_<, \prec)$, and without loss of generality we may assume that $X = (0, -1)$. By Lemmas 20, 22, 23, we obtain $\mathbf{L}((\mathbb{R}^2_<, \prec)^u) = \mathbf{CrU}^\downarrow$.

Finally, consider a $\mathbf{CrBU}^\downarrow$-satisfiable formula A. Then $A\xi^{efl}$ is satisfiable in $(\mathbb{R}^2_\leq, \prec)$. It is easy to see that if $A\xi^{efl}$ is satisfiable at some non-serial point then $A\xi^{efl}$ is satisfiable at some $X \in \mathbb{R}^2_<$. Similarly to the previous cases, $\mathbf{L}((\mathbb{R}^2_\leq, \prec)^u) = \mathbf{CrBU}^\downarrow$. ∎

For $X = (x_1, \ldots, x_n) \in \mathbb{R}^n$, $n \geq 2$, put $p_n(X) := (x_1, x_n)$. Then $p_n : (\mathbb{R}^n, \preceq) \twoheadrightarrow (\mathbb{R}^2, \preceq)$, $p_n : (\mathbb{R}^n, \prec) \twoheadrightarrow (\mathbb{R}^2, \prec)$, [3]. So by Lemmas 20, 24, we have the following

THEOREM 25. *For* $n \geq 2$,
$\mathbf{L}((\mathbb{R}^n, \preceq)^u) = \mathbf{S4.2U}^\downarrow$;
$\mathbf{L}((\mathbb{R}^n, \prec)^u) = \mathbf{Cr.2U}^\downarrow$.

Lemma 24 together with Proposition 16 yields

PROPOSITION 26.
$\mathbf{L}((\mathcal{I}, \sqsupseteq)^u) = \mathbf{S4U}^\downarrow$, $\mathbf{L}((\mathcal{I}^*, \sqsupseteq)^u) = \mathbf{S4.1U}^\downarrow$;
$\mathbf{L}((\mathcal{I}, \sqsupset)^u) = \mathbf{CrU}^\downarrow$, $\mathbf{L}((\mathcal{I}^*, \sqsupset)^u) = \mathbf{CrBU}^\downarrow$.

Let $W \in \{\mathcal{B}_n, \mathcal{CN}_n, \mathcal{CV}_n, \mathcal{R}_n\}$, $R \in \{\supseteq, \ni\}$, $n \geq 1$. It is easy to see that $\mathfrak{F} := (W, R)$ is downward-directed, so $\mathbf{L}(\mathfrak{F}^u) \supseteq \mathbf{L}(\mathfrak{F})\mathbf{U}^\downarrow$; by Lemma 18, Proposition 26, and by Theorem 19, $\mathbf{L}(\mathfrak{F}^u) \subseteq \mathbf{L}(\mathfrak{F})\mathbf{U}^\downarrow$. Thus we obtain

THEOREM 27. *Let* $W \in \{\mathcal{B}_n, \mathcal{CN}_n, \mathcal{CV}_n, \mathcal{R}_n\}$, $n \geq 1$. *Then*
$\mathbf{L}((W, \supseteq)^u) = \mathbf{S4U}^\downarrow$, $\mathbf{L}((W^*, \supseteq)^u) = \mathbf{S4.1U}^\downarrow$;
$\mathbf{L}((W, \ni)^u) = \mathbf{CrU}^\downarrow$, $\mathbf{L}((W^*, \ni)^u) = \mathbf{CrBU}^\downarrow$.

7 Acknowledgements

The author is grateful to Prof. Valentin Shehtman for his help and also to the anonymous referees for their useful comments.

The work on this paper was supported by Poncelet Laboratory (UMI 2615 of CNRS and Independent University of Moscow) and by RFBR-NWO grant 047.011.2004.040.

BIBLIOGRAPHY

[1] P. Blackburn, M. de Rijke, and Y. Venema, *Modal logic*, Tracts in Theoretical Computer Science, Cambridge University Press, Cambridge, UK, 2001.

[2] A. Chagrov and M. Zakharyaschev, *Modal logic*, Oxford Logic Guides, vol. 35, Clarendon Press, Oxford, 1997.

[3] R. Goldblatt. *Diodorean modality in Minkowski spacetime*, Studia Logica, v. 39 (1980), 219-236.

[4] V. Goranko, A. Montanari, G. Sciavicco. *A roadmap of propositional interval temporal logics and duration calculi*, Journal of Applied Non-classical Logics, v. 14, No. 1-2, 2004, 11–56.

[5] V. Goranko, S. Passy. *Using the universal modality: gains and questions*, Journal of Logic and Computation. 1992; 2: 5-30.
[6] J. Halpern, Y. Shoham. *A propositional modal logic of time intervals*, In: Proceedings of LICS'86. IEEE Computer Society, 1986, 279–292.
[7] E. Hemaspaandra. *The price of universality*, Notre Dame Journal of Formal Logic 37(2): 174-203 (1996)
[8] C. Lutz and F. Wolter. *Modal logics of topological relations*, Advances in Modal Logic, pp. 249-263. Manchester. 2004
[9] A. Prior. *Past, present and future*, Clarendon Press, Oxford, 1967.
[10] I. Shapirovsky. *Modal logics of closed domains on Minkowski plane*, Journal of Applied Non-Classical Logics, to appear.
[11] I. Shapirovsky. *On PSPACE-decidability in transitive modal logics*, In: Advances in Modal Logic, Volume 5. R. Schmidt et al. (eds.). King's College Publications, 2005, 269-287
[12] I. Shapirovsky, V. Shehtman. *Chronological future modality in Minkowski spacetime*, Advances in Modal Logic, Volume 4, 437-459. King's College Publications, 2003.
[13] I. Shapirovsky, V. Shehtman. *Modal logics of regions and Minkowski spacetime*, Journal of Logic and Computation. 2005; 15: 559-574.
[14] V. Shehtman. *Modal logics of domains on the real plane*, Studia Logica, v. 42 (1983), 63-80.
[15] V. Shehtman. *On some two-dimensional modal logics*, In: 8th International Congress on Logic, Methodology, and Philosophy of Science, Moscow, Abstracts, v. 1, pp. 326–330, 1987.
[16] V. Shehtman. *"Everywhere" and "here"*, Journal of Applied Non-Classical Logics. 1999. V. 9, No. 2-3. P. 369-380.
[17] F. Wolter. *Solution to a problem of Goranko and Passy*, Journal of Logic and Computation. 1994 4(1):21-22.

Ilya Shapirovsky
Institute for Information Transmission Problems
Russian Academy of Sciences,
B.Karetny 19, Moscow, Russia, 101447
shapirov@mccme.ru

From topology to metric: modal logic and quantification in metric spaces

M. SHEREMET, D. TISHKOVSKY, F. WOLTER AND
M. ZAKHARYASCHEV

ABSTRACT. We propose a framework for comparing the expressive power and computational behaviour of modal logics designed for reasoning about qualitative aspects of metric spaces. Within this framework we can compare such well-known logics as **S4** (for the topology induced by the metric), **wK4** (for the derivation operator of the topology), variants of conditional logic, as well as logics of comparative similarity. One of the main problems for the new family of logics is to delimit the borders between 'decidable' and 'undecidable.' As a first step in this direction, we consider the modal logic with the operator 'closer to a set τ_0 than to a set τ_1' interpreted in metric spaces. This logic contains **S4** with the universal modality and corresponds to a very natural language within our framework. We prove that over arbitrary metric spaces this logic is EXPTIME-complete. Recall that over $\mathbb{R}$, $\mathbb{Q}$, and $\mathbb{Z}$, as well as their finite subspaces, this logic is undecidable.

Keywords: metric, topology, comparative similarity, conditional logic.

1 Introduction

Various 'modal-like' propositional logics have been introduced for reasoning about qualitative aspects of metric spaces. Recall, for example, the modal logic **S4** whose diamond and box are interpreted as the closure and interior operators of the induced topology [10], the modal logic **wK4** whose box is interpreted as the derivation operator on the topological space (see, e.g., [10, 2]), the extensions of these logics with the universal modality, conditional logic with a binary operator comparing distances between points [8], and the logics of comparative similarity [12]. In all of these cases, the truth conditions for the modal operators correspond to certain simple quantifier patterns of first-order logic:

- the interior $\mathbb{I}X$ of a set X is the set

$$\{w \mid \exists a \in \mathbb{R}^{>0} \, \forall v \, (d(w,v) < a \rightarrow v \in X)\},$$

- the universal box $\Box$ is defined by

$$\Box X \;=\; \{w \mid \forall a \in \mathbb{R}^{>0} \, \forall v \, (d(w,v) < a \rightarrow v \in X)\},$$

- the derived set ∂X of X is

$$\{w \mid \forall a \in \mathbb{R}^{>0} \, \exists v \, (0 < d(w,v) < a \land v \in X)\},$$

- the 'closer operator' $X \Subset Y$ for 'X is closer than Y' in the language $\mathcal{CSL}$ of comparative similarity or distances [12] is defined by

$$\{w \mid \exists a \in \mathbb{R}^{>0}(\exists v \in X(d(w,v) < a) \land \neg \exists v \in Y(d(w,v) < a))\},$$

- the interpretation of some variants of the conditional implication operator $X > Y$ [8] can be given by the formula (see, e.g., [1, 11])

$$(X \Subset (X \land \neg Y)) \lor \Box \neg X.$$

Of course, in the modal languages the quantifier patterns above are only implicit and therefore 'forgotten.' This probably explains why the relation between the first-order logic and different modal logics for metric spaces has received very little attention from the modal logic community; see [6] for a brief review of the available results.

In this paper, we first make the quantifier patterns above explicit by introducing 'modal' operators of the form $\exists^{<x}$, $\exists^{>x}$, $\exists^{=x}$, $\exists_{>0}^{\leq x}$ (and their duals $\forall^{<x}$, $\forall^{>x}$, etc.), where the variable x ranges over the *positive real numbers* and can be bound by the quantifiers $\forall x$ and $\exists x$. Intuitively, if x is assigned a value $a \in \mathbb{R}^{>0}$, then $\exists^{<x} X$ denotes the set of all points that are located at distance $< a$ from at least one point in X. In this language the intended meaning of the operators considered above can be represented in a clear and concise manner:

$$\mathbb{I}\tau = \exists x \forall^{<x} \tau, \tag{1.1}$$

$$\Box \tau = \forall x \forall^{<x} \tau, \tag{1.2}$$

$$\partial \tau = \forall x \exists_{>0}^{\leq x} \tau, \tag{1.3}$$

$$\tau_1 \Subset \tau_2 = \exists x (\exists^{<x} \tau_1 \land \neg \exists^{<x} \tau_2), \tag{1.4}$$

$$\tau_1 > \tau_2 = \exists x \left(\exists^{<x} \tau_1 \land \neg \exists^{<x} (\tau_1 \land \neg \tau_2) \right) \lor \Box \neg \tau_1. \tag{1.5}$$

We observe that the resulting modal language, called $\mathcal{QMS}$ (for qualitative metric system), has the same expressive power as the two-variable fragment of a certain two-sorted first-order language for metric spaces, and thereby obtain a first insight into the relation between first-order and (qualitative) modal languages for metric spaces. As follows from (1.1)–(1.3), the logics **wK4, S4** and $\mathbf{S4}_u$ (that is, **S4** enriched with the universal modalities) give rise to $\mathcal{CTL}$-like fragments of $\mathcal{QMS}$, with modal operators corresponding to the quantifier pattern 'a quantifier over the reals followed by a quantifier over the space.' These observations motivate the following general research programme:

Classify and investigate fragments of $\mathcal{QMS}$ with respect to their expressive power and computational behaviour.

In this paper, we contribute to this programme by investigating the $\mathcal{CTL}^+$-like extension of $\mathbf{S4}_u$ which allows quantifier patterns of the following form: a quantifier over the reals followed by a Boolean combination of quantifiers $\exists^{<x}$ over the space. A typical example is the closer operator $\Subset$ of $\mathcal{CSL}$.

First we show that this language has indeed the same expressive power over metric spaces as the modal logic $\mathcal{CSL}$ with sole operator $\leftmoon$ (although it might be exponentially more succinct). The computational properties of $\mathcal{CSL}$ over certain classes of metric spaces have already been investigated. We know from [12] that the satisfiability problem for $\mathcal{CSL}$ is EXPTIME-complete over metric spaces with the so-called *min-condition*:

$$d(X,Y) = \inf\{d(x,y) \mid x \in X, y \in Y\} = \min\{d(x,y) \mid x \in X, y \in Y\},$$

for all sets X and Y. Moreover, in this case the logic enjoys the finite model property. On the other hand, it has been shown in [12] that over (arbitrary or arbitrary finite subspaces of) $\mathbb{R}$, $\mathbb{Q}$, and $\mathbb{Z}$ the logic $\mathcal{CSL}$ can simulate arbitrary Diophantine equations, and so is undecidable. The major open and technically challenging problem has been to investigate the behaviour of $\mathcal{CSL}$ over arbitrary metric spaces, where already very simple formulas like $\neg(p \leftmoon \neg p) \sqcap \neg(\neg p \leftmoon p)$ require infinite converging sequences. Here we present a solution to this problem by proving that over arbitrary metric spaces $\mathcal{CSL}$ is still EXPTIME-complete.

$\mathcal{CSL}$ is closely related to certain conditional logics. In conditional logic, the min-condition above is often called the *limit assumption*, and spaces are not required to be symmetric. It has been shown in [12] that over possibly non-symmetric metric spaces with the min-condition, the closer operator $\leftmoon$ has the same expressive power as the conditional implication $>$. The resulting conditional logic is known as the logic of frames satisfying the normality, reflexivity, strict centering, uniformity and connectedness conditions [4]. In this paper, we do not consider distance spaces that are more general than metric spaces. However, the reader can easily modify the decidability proof given below for $\mathcal{CSL}$ over metric spaces in order to prove the decidability of $\mathcal{CSL}$ over distance spaces without symmetry and/or the triangle inequality.

2 The logic $\mathcal{QMS}$

In the examples above, we needed only one variable x over distances. In general, however, it is useful to have countably many *distance variables* $\{x_1, x_2, \ldots\}$ and, in order to represent *constraints* on relations between distances, an additional set Σ of formulas over these variables. As the distance variables range over $\mathbb{R}^{>0}$ we can take, for example,

- the set Σ_0 of inequalities $x_i < x_j$,
- the set Σ_1 of linear rational equalities $a_1 x_1 + \cdots + a_n x_n = a_{n+1}$,
- the set Σ_2 of linear rational inequalities $a_1 x_1 + \cdots + a_n x_n \leq a_{n+1}$.

Suppose now that we have such a set Σ of constraints. Let $\{p_1, p_2, \ldots\}$ be a countably infinite set of *set variables*. The $\mathcal{QMS}[\Sigma]$-*terms* are defined inductively as follows, where $\varkappa \in \Sigma$:

$$\tau ::= p_i \mid \varkappa \mid \neg \tau \mid \tau_1 \sqcap \tau_2 \mid$$
$$\exists x_i \tau \mid \exists^{<x_i} \tau \mid \exists^{=x_i} \tau \mid \exists^{>x_i} \tau \mid \exists^{<x_i}_{>x_j} \tau \mid \exists^{<x_i}_{>0} \tau.$$

The intended *metric models* for this language are structures of the form

$$\mathfrak{I} = (\mathfrak{D}, p_1^{\mathfrak{I}}, p_2^{\mathfrak{I}}, \ldots) \qquad (2.1)$$

where $\mathfrak{D} = (\Delta, d)$ is a *metric space* and the $p_i^{\mathfrak{I}}$ are subsets of Δ.

To interpret $\mathcal{QMS}[\Sigma]$-terms in metric models, we require *assignments* $\mathfrak{a}$ of *positive* real numbers $\mathfrak{a}(x_i) \in \mathbb{R}^{>0}$ to the distance variables x_i.[1] Given such an assignment $\mathfrak{a}$, we define the *extension* $p_i^{\mathfrak{I},\mathfrak{a}} \subseteq \Delta$ of a set variable p_i to be $p_i^{\mathfrak{I}}$. The *extension* $\varkappa^{\mathfrak{I},\mathfrak{a}} \in \{\emptyset, \Delta\}$ of $\varkappa \in \Sigma$ is defined by setting $\varkappa^{\mathfrak{I},\mathfrak{a}} = \Delta$ iff $(\mathbb{R}^{>0}, \mathfrak{a}) \models \varkappa$, and $\varkappa^{\mathfrak{I},\mathfrak{a}} = \emptyset$, otherwise.[2] The inductive definition of the *extension* $\tau^{\mathfrak{I},\mathfrak{a}} \subseteq \Delta$ of a $\mathcal{QMS}[\Sigma]$-term τ is now as usual for the Booleans and as follows for the remaining operators:

$$(\exists^{=x_i}\tau)^{\mathfrak{I},\mathfrak{a}} = (\exists^{=\mathfrak{a}(x_i)}\tau)^{\mathfrak{I}}, \qquad (\exists^{<x_i}\tau)^{\mathfrak{I},\mathfrak{a}} = (\exists^{<\mathfrak{a}(x_i)}\tau)^{\mathfrak{I}},$$

$$(\exists^{>x_i}\tau)^{\mathfrak{I},\mathfrak{a}} = (\exists^{>\mathfrak{a}(x_i)}\tau)^{\mathfrak{I}}, \qquad (\exists^{<x_i}_{>x_j}\tau)^{\mathfrak{I},\mathfrak{a}} = (\exists^{<\mathfrak{a}(x_i)}_{>\mathfrak{a}(x_j)}\tau)^{\mathfrak{I}},$$

$$(\exists x_i\,\tau)^{\mathfrak{I},\mathfrak{a}} = \bigcup \{\tau^{\mathfrak{I},\mathfrak{b}} \mid \mathfrak{b}(x_j) = \mathfrak{a}(x_j), \text{ for } x_j \neq x_i\},$$

where, for $a, b \in \mathbb{R}^{>0}$,

$$(\exists^{=a}\tau)^{\mathfrak{I}} = \{x \in \Delta \mid \exists y\,(d(x,y) = a \wedge y \in \tau^{\mathfrak{I}})\},$$
$$(\exists^{<a}\tau)^{\mathfrak{I}} = \{x \in \Delta \mid \exists y\,(d(x,y) < a \wedge y \in \tau^{\mathfrak{I}})\},$$
$$(\exists^{>a}\tau)^{\mathfrak{I}} = \{x \in \Delta \mid \exists y\,(d(x,y) > a \wedge y \in \tau^{\mathfrak{I}})\},$$
$$(\exists^{<b}_{>a}\tau)^{\mathfrak{I}} = \{x \in \Delta \mid \exists y\,(a < d(x,y) < b \wedge y \in \tau^{\mathfrak{I}})\}.$$

EXAMPLE 1. The sublanguage of $\mathcal{QMS}[\Sigma_2]$ with expressions of the form $\exists \vec{x}\,(\varkappa \wedge \tau)$, where $\varkappa$ is a conjunction of linear rational inequalities and τ is a $\mathcal{QMS}[\emptyset]$-term containing only the operators $\exists^{<x_i}$ and $\exists^{\leq x_i}$ and, additionally, quantifiers $\exists x_i$ only directly in front of $\exists^{<x_i}$ (as in the interior operator) has been investigated in [13]. In particular, it was proved that the satisfiability problem for this language is decidable.

To put this new language into a more familiar context, consider the following *two-sorted* first-order language $\mathcal{FM}[\Sigma]$. Its terms of sort $\mathbb{R}^{\geq 0}$ are the individual variables $x_1, x_2, \ldots$, and terms of sort *object* are the individual variables $w_1, w_2, \ldots$. The signature of $\mathcal{FM}$ also contains a countably infinite set $\{P_1, P_2, \ldots\}$ of unary predicates, binary predicates $<$ and $=$, and a binary function symbol d. The $\mathcal{FM}[\Sigma]$-*formulas* φ are defined inductively as follows, where $\varkappa \in \Sigma$:

$$\varphi ::= P_j(w_i) \mid \varkappa \mid d(w_i, w_j) < x_k \mid$$
$$x_k = 0 \mid \neg \varphi \mid \varphi_1 \wedge \varphi_2 \mid \exists x_i \varphi \mid \exists w_i \varphi.$$

---

[1] We quantify over *positive* real numbers rather than *non-negative* ones in order to obtain short and transparent definitions of standard topological operators; see (1.1). The expressiveness of the language does not depend on this assumption.

[2] It is straightforward to give a more conventional truth-definition for formulas in Σ by extending the language $\mathcal{QMS}$ with formulas and not regarding the members of Σ as terms. The semantics given here is a bit more concise.

$\mathcal{FM}[\Sigma]$ is interpreted in metric models $\mathfrak{J}$ of the form (2.1) with the help of two assignments $(\mathfrak{a}, \mathfrak{o})$, where $\mathfrak{a}$ assigns a non-negative real number to each x_i, while $\mathfrak{o}$ assigns an object from Δ to each w_i. The *satisfaction relation* $(\mathfrak{J}, \mathfrak{a}, \mathfrak{o}) \models \varphi$ is defined in the obvious way. Denote by $\mathcal{FM}_2[\Sigma]$ the fragment of $\mathcal{FM}[\Sigma]$ with only two variables of sort *object*.

The following expressive completeness result can be proved by an almost straightforward extension of the proof of Theorem 2.2 from [7] to quantifiers over distances. The succinctness result can be proved using the example and technique from [3]; see also [9].

THEOREM 2. *Let $\Sigma \supseteq \Sigma_0$. Then the language $\mathcal{QMS}[\Sigma]$ is expressively complete for the language $\mathcal{FM}_2[\Sigma]$ over metric models. More precisely, for each $\mathcal{QMS}[\Sigma]$-term τ, one can construct an $\mathcal{FM}_2[\Sigma]$-formula φ with a single free variable of sort object such that, for all metric models $\mathfrak{J}$ with assignments $\mathfrak{a}$ and all $o \in \Delta$,*

$$o \in \tau^{\mathfrak{J},\mathfrak{a}} \quad \textit{iff} \quad (\mathfrak{J}, \mathfrak{a}) \models \varphi[o], \qquad (2.2)$$

and conversely, for each $\mathcal{FM}_2[\Sigma]$-formula φ with exactly one free variable of sort object, there exists a $\mathcal{QMS}[\Sigma]$-term τ such that (2.2) holds for all metric models $\mathfrak{J}$ with assignments $\mathfrak{a}$ and all $o \in \Delta$.

$\mathcal{FM}_2[\Sigma]$ is, however, exponentially more succinct than $\mathcal{QMS}[\Sigma]$.

To classify and investigate $\mathcal{QMS}[\Sigma]$ and its various fragments can be regarded as an interesting and challenging research programme, with possible applications for reasoning about distances and similarity in various application domains. At this early stage, however, there are more open problems than answers. Here we mention just some of them (see also [6]). First, does the expressive completeness result above hold for the language $\mathcal{QMS}[\Sigma]$ with $\Sigma = \emptyset$? So far, the proof requires inequalities $x_i < x_j$ to be available in both languages. Second, the fragments of $\mathcal{QMS}[\Sigma]$ discussed above contain only the distance operators $\exists^{<x_i}$. Proofs of decidability results for those fragments often employ a certain *tree model property* (formulated in terms of tree metric spaces) as well as a technique that is close to standard unravelling (alias bisimulation). Is it possible to define a natural notion of bisimulation on metric spaces which could explain the 'good' behaviour of those fragments of $\mathcal{QMS}[\Sigma]$? Is there a natural characterisation of the fragment of $\mathcal{FM}[\Sigma]$ which is invariant under such bisimulations? Finally, an interesting problem is to find 'maximal' decidable fragments of $\mathcal{QMS}[\Sigma]$. It is not difficult to see using the technique of [7] that satisfiability of $\mathcal{QMS}[\Sigma_1]$-terms is undecidable. But we conjecture that much weaker fragments are undecidable already.

The contribution of this paper to the research programme above is an analysis of the computational behaviour and the expressive power of the fragment of $\mathcal{QMS}[\Sigma]$ known as the *logic of comparative similarity* [12].

3 The logic $\mathcal{CSL}$

Let us consider the $\mathcal{CTL}^+$-like fragment of $\mathcal{QMS}[\emptyset]$ where the role of branch quantifiers is played by $\exists x$ and $\forall x$, while instead of temporal operators we have the quantifiers $\exists^{<x}$ and $\forall^{<x}$. Thus, in this fragment we allow the quantifiers $\exists x$ to be applied to Boolean combinations of terms of the form $\exists^{<x}\tau$ and atoms p_i. More precisely, take a variable x and define the set $\mathcal{CLV}_{open}$ of *open terms* σ and the set $\mathcal{CLV}$ of *terms* τ by induction as follows:

$$\tau ::= p_i \mid \neg\tau \mid \tau_1 \sqcap \tau_2 \mid \exists x\sigma,$$
$$\sigma ::= \tau \mid \neg\sigma \mid \sigma_1 \sqcap \sigma_2 \mid \exists^{<x}\tau.$$

Another language we consider in this section is called $\mathcal{CSL}$ (which stands for the *logic of comparative similarity*) [12]. Its terms are defined by

$$\tau ::= p_i \mid \neg\tau \mid \tau_1 \sqcap \tau_2 \mid \tau_1 \Leftarrowtail \tau_2,$$

where $\Leftarrowtail$, the *closer operator*, is interpreted in a metric model $\mathfrak{J}$ of the form (2.1) as follows (this is equivalent to the definition of $\Leftarrowtail$ in the introduction):

$$(\tau_1 \Leftarrowtail \tau_2)^{\mathfrak{J}} = \{x \in \Delta \mid d(x, \tau_1^{\mathfrak{J}}) < d(x, \tau_2^{\mathfrak{J}})\}.$$

Here the *distance* $d(x, Y)$ *from a point* $x \in \Delta$ *to a subset* $Y \subseteq \Delta$ is defined as usual:

$$d(x, Y) = \begin{cases} \inf\{d(x, y) \mid y \in Y\}, & \text{if } Y \neq \emptyset \\ \infty, & \text{if } Y = \emptyset. \end{cases}$$

Two terms τ_1 and τ_2 are said to be *equivalent*, $\tau_1 \equiv \tau_2$ in symbols, if $\tau_1^{\mathfrak{J}} = \tau_2^{\mathfrak{J}}$ for every metric model $\mathfrak{J}$. It is not hard to see that

$$\tau_1 \Leftarrowtail \tau_2 \equiv \exists x \, (\exists^{<x}\tau_1 \sqcap \neg\exists^{<x}\tau_2).$$

Thus, $\mathcal{CSL}$ can be regarded as a sublanguage of $\mathcal{CLV}$. Despite its apparent simplicity, this language turns out to be quite expressive. In particular, $\top \Leftarrowtail \neg\tau$ is interpreted as the interior of τ in the topological space induced by the metric, and $\top \leftrightarrows \tau$ as the closure of τ, where $\top$ is the whole space $(p_1 \sqcup \neg p_1)$ and

$$\tau_1 \leftrightarrows \tau_2 = \neg(\tau_1 \Leftarrowtail \tau_2) \sqcap \neg(\tau_2 \Leftarrowtail \tau_1)$$

is the set of points located at the same distance from τ_1 and τ_2. The universal modalities can be also expressed via $\Leftarrowtail$:

$$\exists \tau \equiv (\tau \Leftarrowtail \bot) \qquad \text{and} \qquad \forall \tau \equiv \neg(\neg\tau \Leftarrowtail \bot)$$

(here $\bot$ stands for $p_1 \sqcap \neg p_1$). Thus, the logic $\mathcal{CSL}$ contains full $\mathbf{S4}_u$. We now show that actually $\mathcal{CSL}$ is as expressive as full $\mathcal{CLV}$.

THEOREM 3. *For every $\mathcal{CLV}$-term τ, there is a $\mathcal{CSL}$-term τ^* with $\tau \equiv \tau^*$.*

Proof. Observe first that, for all $\mathcal{CLV}$-terms $\tau_1, \ldots, \tau_n$ and ρ, we have

$$\bigsqcap_{i=1}^{n} \exists x (\exists^{<x} \tau_i \sqcap \neg \exists^{<x} \rho) \equiv \exists x (\bigsqcap_{i=1}^{n} \exists^{<x} \tau_i \sqcap \neg \exists^{<x} \rho).$$

(For a proof of this observation use the fact that distances are taken from the *linearly ordered* set $\mathbb{R}^{>0}$.)

Now the proof proceeds by induction on the construction of τ. The term τ is a Boolean combination of terms of the form $\exists x\sigma$ and atoms p_i. As both languages have the Boolean operators and the set variables p_i, it is sufficient to define the translation of a term of the form $\exists x\sigma$. We may assume that

$$\exists x\sigma = \exists x \bigsqcup_{i=1}^{n} \bigsqcap_{j=1}^{m_i} \rho_i^j,$$

where the ρ_i^j are negated or non-negated terms of the form $\exists^{<x}\tau$, p_i, or $\exists x\sigma'$. Clearly

$$\exists x\sigma \equiv \bigsqcup_{i=1}^{n} \exists x \bigsqcap_{j=1}^{m_i} \rho_i^j.$$

As $\neg \exists^{<x}\tau_1 \sqcap \neg \exists^{<x}\tau_2 \equiv \neg \exists^{<x}(\tau_1 \sqcup \tau_2)$, we obtain

$$\exists x\sigma \equiv \bigsqcup_{i=1}^{n} \exists x \bigsqcap_{j=1}^{k_i} (\exists^{<x}\tau_{ij} \sqcap \neg \exists^{<x}\tau_i \sqcap \beta_{ij}),$$

where the β_{ij} are negated or non-negated terms of the form $\exists x\sigma'$ or p_i. By the observation above, we then have

$$\exists x\sigma \equiv \bigsqcup_{i=1}^{n} \bigsqcap_{j=1}^{k_i} \left(\exists x(\exists^{<x}\tau_{ij} \sqcap \neg \exists^{<x}\tau_i) \sqcap \beta_{ij} \right).$$

By the induction hypotheses, we have translations τ_{ij}^*, τ_i^* and β_{ij}^* in $\mathcal{CSL}$. Then the translation we need can be obtained by taking

$$(\exists x\sigma)^* = \bigsqcup_{i=1}^{n} \bigsqcap_{j=1}^{k_i} ((\tau_{ij}^* \Leftarrow \tau_i^*) \sqcap \beta_{ij}^*),$$

which proves the theorem. ∎

Observe that the translation from $\mathcal{CLV}$ to $\mathcal{CSL}$ above introduces an exponential blow-up. The question whether $\mathcal{CLV}$ is indeed exponentially more succinct than $\mathcal{CSL}$ remains open.

4 Decidability of $\mathcal{CSL}$

A typical decidability proof for a modal (temporal, dynamic, etc.) logic L proceeds as follows. Given a formula φ, we take a proper 'closure' $\mathsf{cl}\,\varphi$ of the set $\mathsf{sub}\,\varphi$ of subformulas of φ, introduce a syntactical notion of a 'type' approximating those subsets of $\mathsf{cl}\,\varphi$ that can be realised in models for L, and then show how to construct an L-model for a given type t with $\varphi \in t$ by providing a 'witness type' for each $\Diamond \psi \in t$, that is, a type t' such that $\psi \in t'$ and $\chi \in t'$, for every $\Box \chi \in t$. This general scheme can be applied to $\mathcal{CSL}$ as well. As usual, however, the devil (or God?) is in the details.

Let us figure out first what a $\mathcal{CSL}$-type is. Throughout this section we assume that we are given a $\mathcal{CSL}$-term τ. Denote by $\mathsf{sub}\,\tau$ the set of subterms of τ. As we need to *compare* distances between types containing certain subterms of τ, we introduce the set

$$\mathsf{com}\,\tau \;=\; \{\bot, \top\} \cup \{\varphi \mid \varphi \sqsubseteq \psi \in \mathsf{sub}\,\tau \text{ or } \psi \sqsubseteq \varphi \in \mathsf{sub}\,\tau, \text{ for some } \psi\}.$$

Finally, we define $\mathsf{cl}\,\tau \supseteq \mathsf{sub}\,\tau$, the *closure* of $\mathsf{sub}\,\tau$, to be the smallest set of terms with the following properties:

- $\mathsf{cl}\,\tau$ is closed under single negations, and
- $\mathsf{cl}\,\tau$ contains $\varphi \sqsubseteq \psi$, for every $\varphi, \psi \in \mathsf{com}\,\tau$.

Clearly, the size of $\mathsf{cl}\,\tau$ is polynomial in the size $|\tau|$ of τ.

Suppose now that we have a metric model $\mathfrak{J}$ of the form (2.1) and a point $x \in \Delta^{\mathfrak{J}}$. Then the τ-*type* of x in $\mathfrak{J}$ is the set

$$t^{\mathfrak{J}}(x) \;=\; \{\varphi \in \mathsf{cl}\,\tau \mid x \in \varphi^{\mathfrak{J}}\}.$$

Clearly, this set is Boolean closed. Moreover, the model $\mathfrak{J}$ determines a natural linear quasi-order $\leq_{t^{\mathfrak{J}}(x)}$ on $\mathsf{com}\,\tau$: for all $\varphi, \psi \in \mathsf{com}\,\tau$, we have

$$\varphi \leq_{t^{\mathfrak{J}}(x)} \psi \quad \text{iff} \quad d(x, \varphi^{\mathfrak{J}}) \leq d(x, \psi^{\mathfrak{J}}) \quad \text{iff} \quad \neg(\psi \sqsubseteq \varphi) \in t^{\mathfrak{J}}(x).$$

Observe that

- φ is a $\leq_{t^{\mathfrak{J}}(x)}$-minimal element iff $d(x, \varphi^{\mathfrak{J}}) = 0$, i.e., $x \in \mathbf{C}\varphi^{\mathfrak{J}}$, and
- φ is a $\leq_{t^{\mathfrak{J}}(x)}$-maximal element iff $d(x, \varphi^{\mathfrak{J}}) = \infty$, i.e., $\varphi^{\mathfrak{J}} = \emptyset$.

This suggests the following syntactical approximation of the 'real' τ-types.

A subset t of $\mathsf{cl}\,\tau$ is said to be *Boolean closed* if $\top \in t$ and the following conditions are satisfied: (a) $\varphi \in t$ iff $\neg \varphi \notin t$, for all $\neg \varphi \in \mathsf{cl}\,\tau$, and (b) $\varphi \sqcap \psi \in t$ iff $\varphi \in t \wedge \psi \in t$, for all $\varphi \sqcap \psi \in \mathsf{cl}\,\tau$. With every Boolean closed $t \subseteq \mathsf{cl}\,\tau$ we associate the following binary relation $\leq_t$ on $\mathsf{com}\,\tau$:

$$\leq_t \;=\; \{(\varphi, \psi) \mid \neg(\psi \sqsubseteq \varphi) \in t\}.$$

Now, a τ-*type* (or simply a *type* if τ is understood) is a Boolean closed subset t of $\mathsf{cl}\,\tau$ such that

- $\leq_t$ is a linear quasi-order on $\mathsf{com}\,\tau$,
- all terms in $t \cap \mathsf{com}\,\tau$ are $\leq_t$-minimal elements,
- $\bot$ is a $\leq_t$-maximal element.

Denote by $<_t$ and $\simeq_t$ the strict linear order and the equivalence relation induced by $\leq_t$, respectively. It is easy to see that

$$<_t \;=\; \{(\varphi,\psi) \mid (\varphi \sqsubseteq \psi) \in t\}.$$

Let $\min t$ denote the set of $\leq_t$-minimal elements. It should be clear that there are at most exponentially ($2^{O(|\tau|^2)}$) many τ-types.

Recall that $\varphi \leftrightarrows \psi = \neg(\varphi \sqsubseteq \psi) \sqcap \neg(\psi \sqsubseteq \varphi)$. Clearly, $\top \leftrightarrows \varphi$ is actually equivalent to $\neg(\top \sqsubseteq \varphi)$, while $\bot \leftrightarrows \varphi$ is equivalent to $\neg(\varphi \sqsubseteq \bot)$.

Before we proceed to our next notion, let us consider an example explaining an essential difference between $\mathcal{CSL}$ and standard modal logics.

EXAMPLE 4. To satisfy the term $\neg p \sqcap (q \sqsubseteq \bot) \sqcap (p \sqsubseteq q) \sqcap (p \leftrightarrows \top)$

- we need, by the first conjunct, a point x from $\neg p$;
- by the second conjunct, we need a 'witness' y for $q \sqsubseteq \bot$, that is, a point y that belongs to q;
- by the third conjunct, we need a witness z for $p \sqsubseteq q$, that is, z belongs to p, neither x nor z are in q, and x should be closer to z than to y;
- finally, the fourth conjunct says that x must be 'infinitely close' to p, that is, we need an *infinite* sequence $\{z_i \mid i \in \omega\}$ of points from p converging to x. Note that, by the third conjunct, only a finite number of the z_i can be in q.

Thus, we require witnesses of two sorts: (i) those that are at some finite distance from a given point x, and (ii) those that represent infinite sequences converging to x (that the points of such a sequence can always be chosen to be of the same type follows from Lemma 5 below).

Two important facts should also be observed in connection with the example above. First, the concrete values of the distances $d(x,y)$ and $d(x,z)$ are of no importance at all; what really matters is that they should satisfy the inequalities $d(x,y) > d(x,z) > 0$. At the same time, the value $\lim_{i\to\infty} d(x,z_i)$ must be zero. The logic of comparative similarity *cannot* speak of any particular distance except 0.

The second important fact is that if a term requires some witness at a positive distance then a single witness—rather than an infinite sequence of witnesses as in (ii) above—is always enough.

LEMMA 5. *Let $\mathfrak{I}$ be a metric model.*

(1) *Suppose $x \in \Delta^{\mathfrak{I}}$ and $\varphi^{\mathfrak{I}} \neq \emptyset$ for some $\varphi \in \mathsf{com}\,\tau$. Then there is a type t with $\varphi \in t$ such that $d(x,\varphi^{\mathfrak{I}}) = d(x,t^{\mathfrak{I}})$, where $t^{\mathfrak{I}} = \{y \in \Delta^{\mathfrak{I}} \mid t^{\mathfrak{I}}(y) = t\}$. Moreover, the pair of types $(t^{\mathfrak{I}}(x), t)$ satisfies the following conditions:*

- *if $\top <_{t^{\mathfrak{I}}(x)} \varphi$ then $\varphi <_{t^{\mathfrak{I}}(x)} \psi$ implies $\varphi <_t \psi$, for all $\psi \in \mathsf{com}\,\tau$,*
- *if $\varphi \simeq_{t^{\mathfrak{I}}(x)} \top$ then $\chi <_{t^{\mathfrak{I}}(x)} \psi$ implies $\chi <_t \psi$, for all $\psi, \chi \in \mathsf{com}\,\tau$.*

(2) *For all $x,y \in \Delta^{\mathfrak{I}}$ and $\psi \in \mathsf{com}\,\tau$, we have $\psi <_{t^{\mathfrak{I}}(x)} \bot$ iff $\psi <_{t^{\mathfrak{I}}(y)} \bot$.*

Proof. To show (2), it is enough to observe that $\psi <_{t^{\mathfrak{J}}(x)} \bot$ iff $\psi^{\mathfrak{J}} \neq \emptyset$ iff $\psi <_{t^{\mathfrak{J}}(y)} \bot$, for all $x, y \in \Delta^{\mathfrak{J}}$ and $\psi \in \mathsf{com}\,\tau$.

Let us show (1) for the case $\varphi \simeq_{t^{\mathfrak{J}}(x)} \top$. Since $d(x, \varphi^{\mathfrak{J}}) = 0$, either $x \in \varphi^{\mathfrak{J}}$ or there is an infinite sequence $\{z_i \in \varphi^{\mathfrak{J}} \mid i \in \omega\}$ converging to x. In the former case we set $t = t^{\mathfrak{J}}(x)$. In the latter one, since the number of types is finite, there is an infinite subsequence $\{z_{i_j}\}$ of $\{z_i\}$ whose points are of the same type, and so we can set $t = t^{\mathfrak{J}}(z_{i_j})$.

Let $\chi <_{t^{\mathfrak{J}}(x)} \psi$ for some $\psi, \chi \in \mathsf{com}\,\tau$. Then $\varepsilon = d(x, \psi^{\mathfrak{J}}) - d(x, \chi^{\mathfrak{J}}) > 0$. Choose some $y \in t^{\mathfrak{J}}$ with $d(x, y) < \varepsilon/2$. By the triangle inequality we then have $d(y, \psi^{\mathfrak{J}}) - d(y, \chi^{\mathfrak{J}}) \geq \varepsilon - 2\,d(x, y) > 0$, which means $\chi <_t \psi$.

The case of $\top <_{t^{\mathfrak{J}}(x)} \varphi$ is considered analogously. ∎

We now define a notion of a φ-link using which we can provide witnesses for terms from a given type. Let s, t be types and $\varphi \in \mathsf{com}\,\tau$. Two cases are possible:

- Suppose that $\top <_s \varphi <_s \bot$. Then we say that the pair (s, t) is a φ-*link* (of types) if $\varphi \in t$ and, for all $\psi \in \mathsf{com}\,\tau$, we have

$$\psi <_s \bot \leftrightarrow \psi <_t \bot \quad \text{and} \quad \varphi <_s \psi \rightarrow \varphi <_t \psi$$

(note that $\varphi <_t \psi$ is equivalent here to $\top <_t \psi$, since $\varphi \in t$).

- Suppose that $\varphi \simeq_s \top$. Then we say that (s, t) is a φ-*link* (of types) if $\varphi \in t$ and, for all $\psi, \chi \in \mathsf{com}\,\tau$, we have

$$\psi <_s \bot \leftrightarrow \psi <_t \bot \quad \text{and} \quad \chi <_s \psi \rightarrow \chi <_t \psi$$

(note that the second implication is equivalent to $<_s\, \subseteq\, <_t$, and we have $\min t \subseteq \min s$).

In the latter case we will also say that (s, t) is a *short link*, while in the former the link will be called *long*. Clearly, a link (s, t) is short iff $<_s\, \subseteq\, <_t$.

Unfortunately, the notion of a link above does not take into account a possible interaction of two (or more) short links. To be more specific, consider the following situation. Suppose that t_0 is a type and $\varphi \notin t_0$, for some $\varphi \in \mathsf{com}\,\tau$ such that $\varphi \in \min t_0$ (i.e., $\varphi \simeq_{t_0} \top$). Then we need a short φ-link (t_0, t_1). Assume further that $\psi \notin t_1$, for some $\psi \in \mathsf{com}\,\tau$ with $\psi \in \min t_0$. This means that we also need a (long or short) ψ-link (t_1, t_2). But then, according to Lemma 6 below, (t_0, t_2) must be a short ψ-link, which by no means follows from the definition of a link.

LEMMA 6. *Let $\mathfrak{J}$ be a model, $x \in \Delta^{\mathfrak{J}}$, and let $\varphi, \psi \in \mathsf{com}\,\tau$ be such that $d(x, \varphi^{\mathfrak{J}}) = d(x, \psi^{\mathfrak{J}}) = 0$. Let s, t be types with $\varphi \in s$, $\psi \in t$, and let $S \subseteq s^{\mathfrak{J}}$ be such that $d(x, S) = d(x, \varphi^{\mathfrak{J}}) = 0$ and $d(y, t^{\mathfrak{J}}) = d(y, \psi^{\mathfrak{J}})$, for all $y \in S$. Then $d(x, t^{\mathfrak{J}}) = d(x, \psi^{\mathfrak{J}}) = 0$.*

Proof. Take an arbitrary $\varepsilon > 0$. As $d(x, S) = 0$, there is $y \in S$ with $d(x, y) < \varepsilon/2$. Then $d(y, t^{\mathfrak{J}}) = d(y, \psi^{\mathfrak{J}}) \leq d(y, x) + d(x, \psi^{\mathfrak{J}}) < \varepsilon/2$, and so $d(x, t^{\mathfrak{J}}) \leq d(x, y) + d(y, \psi^{\mathfrak{J}}) < \varepsilon$, i.e., $d(x, t^{\mathfrak{J}}) = 0$, as $\varepsilon > 0$ is arbitrary. ∎

Thus we should be careful when constructing sequences of links starting with a short one, in particular, we should remember some previous links in the sequence. Let us consider possible scenarios when we start with a short link (t_0, t_1).

1. Suppose that $<_{t_0} = <_{t_1}$ and we need a φ-link (t_1, t_2) for some $\varphi \in \text{com}\,\tau$. In this case the types t_0 and t_1 contain precisely the same terms of the form $\chi_1 \sqsubseteq \chi_2$ and can only differ in Boolean terms. It follows that (t_1, t_2) is a φ-link iff (t_0, t_2) is a φ-link. This means that the choice of t_2 does not depend on the link (t_0, t_1).

2. Suppose that $<_{t_0} \subsetneq <_{t_1}$ and we need a φ-link (t_1, t_2) for some $\varphi \in \text{com}\,\tau$. As we have $\min t_1 \subseteq \min t_0$, three cases are possible.

2.1: $\varphi \in \min t_1$. Then for any φ-link (t_1, t_2) we have $<_{t_0} \subset <_{t_1} \subseteq <_{t_2}$, and so (t_0, t_2) is also a short φ-link. Thus, no additional requirement should be imposed on (t_1, t_2).

2.2: $\varphi \in \min t_0 \setminus \min t_1$. In this case, when choosing a (long) φ-link (t_1, t_2), we must also ensure that (t_0, t_2) is a short φ-link.

2.3: $\varphi \notin \min t_0$, and so $\varphi \notin \min t_1$. In this case (t_0, t_1) does not have any influence on subsequent links at all.

3. Suppose that $<_{t_0} \subsetneq <_{t_1}$ and (t_1, t_2) is a φ-link, for $\varphi \in \min t_0 \setminus \min t_1$ (as in **2.2**), and so (t_0, t_2) is a short φ-link with $<_{t_0} \subseteq <_{t_2}$. Suppose also that we are looking for a ψ-link (t_2, t_3). As (t_1, t_2) is a long link, t_1 has no influence on the choice of t_3. However, (t_0, t_2) should be taken into account. We again have three cases.

3.1: $\psi \in \min t_2$. Then for any ψ-link (t_2, t_3) the pair (t_0, t_3) will automatically be a ψ-link.

3.2: $\varphi \in \min t_0 \setminus \min t_2$. Then, when choosing a long φ-link (t_2, t_3), we must also ensure that (t_0, t_3) is a short ψ-link.

3.3: $\varphi \notin \min t_0$. In this case no additional requirement is needed.

This analysis suggests the following definitions. A sequence $\mathbf{t} = (t_0, \ldots, t_n)$ of τ-types is called a *block* if we have the inclusions $<_{t_0} \subset \cdots \subset <_{t_{n-1}} \subseteq <_{t_n}$ (which means that all pairs $(t_0, t_1), \ldots, (t_{n-1}, t_n)$ are short links). We call t_n the *type* of $\mathbf{t}$, while $(t_0, \ldots, t_{n-1})$ is understood as its 'history' or 'heredity.' We say that $\mathbf{t}$ is *realised* in a model $\mathfrak{J}$ of the form (2.1) if there exist subsets $X_0 \subseteq t_0^{\mathfrak{J}}, \ldots, X_n \subseteq t_n^{\mathfrak{J}}$ such that $d(x_i, X_{i+1}) = 0$ for all $i < n$ and $x_i \in X_i$.

It is easy to see that the size of $\text{com}\,\tau$, and so the length of any block, is bounded by $|\tau|$. Therefore, the number of different blocks is at most exponential in $|\tau|$.

Now, for $\varphi \in \text{com}\,\tau$, we introduce a notion of a φ-*link of blocks*, which specialises the notion of a φ-link of types. Let $\mathbf{s}$ and $\mathbf{t}$ be blocks with $\mathbf{s} = (s_0, \ldots, s_m)$. Consider four cases.

- Suppose that $\varphi \notin \min s_0$. Then $(\mathbf{s}, \mathbf{t})$ is called a φ-*link* (of blocks) if $\mathbf{t} = (t)$ and (s_m, t) is a φ-link of types. In this case the long link (s_m, t) allows us to 'forget' everything that happened before t.

- Suppose that $\varphi \in \min s_{n-1} \setminus \min s_n$, for some $n \leq m$. Then $(\mathbf{s}, \mathbf{t})$ is a φ-*link* (of blocks) if $\mathbf{t} = (s_0, \ldots, s_{n-1}, t)$ and (s_m, t) is a φ-link of types. In this case (s_n, t) is a long link, while (s_{n-1}, t) is a short one, and so s_{n-1} and its 'heredity' should be kept.

- Suppose that $\varphi \in \min s_m \setminus s_m$ and $<_{s_{m-1}} = <_{s_m}$. Then $(\mathbf{s}, \mathbf{t})$ is a φ-*link* (of blocks) if $\mathbf{t} = (s_0, \ldots, s_{m-1}, t)$ and (s_m, t) is a φ-link of types. In this case s_{m-1} and s_m carry the same information on 'heredity' of t, so we can drop s_m.

- Suppose that $\varphi \in \min s_m \setminus s_m$ and $<_{s_{m-1}} \subset <_{s_m}$. Then $(\mathbf{s}, \mathbf{t})$ is a φ-*link* (of blocks) if $\mathbf{t} = (s_0, \ldots, s_m, t)$ and (s_m, t) is a φ-link of types.

Let D be a set of blocks and T the set of all types occurring in blocks from D. We call D a *diagram* if the following conditions hold:

$$\text{there exists } (t) \in D \text{ with } \tau \in t, \tag{4.1}$$

$$\text{for all } s, t \in T \text{ and } \varphi \in \operatorname{com} \tau, \text{ we have } \varphi <_s \bot \text{ iff } \varphi <_t \bot, \tag{4.2}$$

$$\text{for all } \mathbf{s} = (s_0, \ldots, s_n) \in D \text{ and } \varphi <_{s_n} \bot, \varphi \notin s_n, \text{ there exists } \mathbf{t} \in D \text{ such that } (\mathbf{s}, \mathbf{t}) \text{ is a } \varphi\text{-link}. \tag{4.3}$$

The rather abstract notions of block, link and diagram for the given term τ will become more transparent from the proofs of Lemmas 7 and 10.

LEMMA 7. *Let $\mathfrak{I}$ be a metric model where $\tau^{\mathfrak{I}} \neq \emptyset$, and let D be the set of blocks realised in $\mathfrak{I}$. Then D is a diagram.*

Before we turn to the proof of Lemma 7, we need to establish the following technical property.

LEMMA 8. *Suppose that a block $\mathbf{s} = (s_0, \ldots, s_m)$ is realised in $\mathfrak{I}$, and $X_0 \subseteq s_0^{\mathfrak{I}}, \ldots, X_m \subseteq s_m^{\mathfrak{I}}$ are such that $d(x_i, X_{i+1}) = 0$ for all $x_i \in X_i$ and $i < m$. Suppose also that $\varphi <_{s_m} \bot$ and $\varphi \notin s_m$. Then there exist a type t and subsets $Y_0 \subseteq X_0, \ldots, Y_m \subseteq X_m$ with the following properties:*

$$d(y, t^{\mathfrak{I}}) = d(y, \varphi^{\mathfrak{I}}), \quad \text{for all } y \in Y_m,$$

$$d(y, Y_{l+1}) = 0, \quad \text{for all } y \in Y_l \text{ and } l < m.$$

Proof. Choose an arbitrary $x_\lambda \in X_0$, where λ denotes the empty sequence. As $d(x_i, X_{i+1}) = 0$ for all $x_i \in X_i$ and $i < m$, we can choose $x_\alpha \in X_l$, for all $\alpha \in \omega^l$ and $l \leq m$, so that

$$x_\alpha = \lim_{i \to \infty} x_{(\alpha, i)}.$$

Now, for every $\alpha \in \omega^m$, there exists a type t_α such that $\varphi \in t_\alpha$ and $d(x_\alpha, \varphi^{\mathfrak{I}}) = d(x_\alpha, t_\alpha^{\mathfrak{I}})$. Since the number of types is finite, we obtain a partition $\omega^m = A_0 \cup \cdots \cup A_n$, where $t_\alpha = t_{\alpha'}$, for all $\alpha, \alpha' \in A_r$, $r \leq n$.

Say that a subset $A \subseteq \omega^m$ is *essential* if we have

$$(\exists^\infty a_0) \ldots (\exists^\infty a_{m-1}) \, (a_0, \ldots, a_{m-1}) \in A, \qquad (4.4)$$

where $\exists^\infty$ means 'there exist infinitely many.'

CLAIM 9. A_r is an essential subset of ω^m for some $r \leq n$.

Proof. We proceed by induction on m. Clearly, for $m = 1$ an essential subset is simply an infinite subset. So in this case Claim 9 is trivial.

Suppose that $m > 1$ and Claim 9 holds for $m-1$. For $A \subseteq \omega^m$ and $a \in \omega$, let $A|_a$ denote $\{\alpha \in \omega^{m-1} \mid (a, \alpha) \in A\}$. Then, for every $a \in \omega$, we have $\omega^{m-1} = A_0|_a \cup \cdots \cup A_n|_a$. Hence there exists $r(a) \leq n$ such that $A_{r(a)}|_a$ is essential in ω^{m-1}. As $r(a)$ has only a finite number of possible values, there exists $r < m$ such that $A_r|_a$ is essential in ω^{m-1} for infinitely many $a \in \omega$. This means that A_r is essential in ω^m. ∎

So let $A = A_r$ be an essential subset of ω^m for some $r \in \omega^m$. For every $l \leq m$, let

$$A_{(l)} = \{\alpha \in \omega^l \mid (\alpha, \beta) \in A \text{ for some } \beta \in \omega^{m-l}\}$$

(in particular, $A_{(m)} = X$ and $A_{(0)} = \{\lambda\}$). Then the following property is a straightforward consequence of the definition of essential sets:

for all $l < m$ and $\alpha \in A_{(l)}$, there are infinitely many $a \in \omega$ such that $(\alpha, a) \in A_{(l+1)}$. $\qquad (4.5)$

It remains to put $Y_l = \{x_\alpha \mid \alpha \in A_{(l)}\}$, for all $l \leq m$. ∎

We are now in a position to prove Lemma 7.

Proof. Clearly, D satisfies (4.1) and (4.2). Let us prove (4.3). Suppose that a block $\mathbf{s} = (s_0, \ldots, s_m)$ is realised in $\mathfrak{J}$ and $X_0 \subseteq s_0^{\mathfrak{J}}, \ldots, X_m \subseteq s_m^{\mathfrak{J}}$ are such that $d(x_i, X_{i+1}) = 0$ for all $i < m$ and all $x_i \in X_i$. Let $\varphi <_{s_m} -$, $\varphi \notin s_m$. By Lemma 8, there exist a type t and subsets $Y_0 \subseteq X_0, \ldots, Y_m \subseteq X_m$ with the following properties:

$$d(y, t^{\mathfrak{J}}) = d(y, \varphi^{\mathfrak{J}}), \quad \text{for all } y \in Y_m,$$

$$d(y, Y_{l+1}) = 0, \quad \text{for all } y \in Y_l \text{ and } l < m.$$

Note that the latter property implies that $d(y, Y_{l'}) = 0$, for all $y \in Y_l$ and $l < l' \leq m$. Four cases are now possible.

Case 1: $\varphi \notin \min s_0$. Then the block $\mathbf{t} = (t)$ is realised in $\mathfrak{J}$, because $t^{\mathfrak{J}} \neq \emptyset$, and $(\mathbf{s}, \mathbf{t})$ is a φ-link by construction.

Case 2: $\varphi \in \min s_{n-1} \setminus \min s_n$ for some $n \leq m$. Let us show that $\mathbf{t} = (s_0, \ldots, s_{n-1}, t)$ is a block realised in $\mathfrak{J}$. Take any $u \in Y_{n-1}$ and $v \in Y_m$. Then $d(u, t^{\mathfrak{J}}) \leq d(u, v) + d(v, t^{\mathfrak{J}}) = d(u, v) + d(v, \varphi^{\mathfrak{J}}) \leq d(u, v) + d(v, u) + d(u, \varphi^{\mathfrak{J}}) = 2\,d(u,v)$, and so $d(y, t^{\mathfrak{J}}) = 0$. We obtain $<_{s_{n-1}} \subseteq <_t$, since $u \in Y_{n-1} \subseteq s_{n-1}^{\mathfrak{J}}$. Thus $\mathbf{t}$ is a block. By considering the sets $Y_0, \ldots, Y_{n-1}, t^{\mathfrak{J}}$, we see that $\mathbf{t}$ is realised in $\mathfrak{J}$. Finally, $(\mathbf{s}, \mathbf{t})$ is a φ-link by construction.

Case 3: $\varphi \in \min s_m \setminus s_m$ and $<_{s_{m-1}} = <_{s_m}$. Similarly to the previous case we obtain that $d(u, t^{\mathfrak{J}}) = 0$ for all $u \in Y_{m-1}$ and therefore $<_{s_{m-1}} \subseteq <_t$. Thus, $\mathbf{t} = (s_0 \ldots, s_{m-1}, t)$ is a block, $\mathbf{t}$ is realised in $\mathfrak{J}$ (consider the sets $Y_0, \ldots, Y_{m-1}, t^{\mathfrak{J}}$), and $(\mathbf{s}, \mathbf{t})$ is a φ-link.

Case 4: $\varphi \in \min s_m \setminus s_m$ and $<_{s_{m-1}} \subset <_{s_m}$. Then $d(u, t^{\mathfrak{J}}) = d(u, \varphi^{\mathfrak{J}}) = 0$ for all $u \in Y_m$. Therefore $\mathbf{t} = (s_0 \ldots, s_m, t)$ is a block, $\mathbf{t}$ is realised in $\mathfrak{J}$ (consider the sets $Y_0, \ldots, Y_m, t^{\mathfrak{J}}$), and $(\mathbf{s}, \mathbf{t})$ is a φ-link. ■

LEMMA 10. *Let D be a diagram. Then there exists a model $\mathfrak{J}$ with $\tau^{\mathfrak{J}} \neq \emptyset$.*

Proof. Our first goal is to 'unravel' D into a certain tree that will serve as the underlying set of the model we need.

Let T be the set of all types from blocks in D. Let $\varphi_0, \ldots, \varphi_{k-1}$ be all different members of the set $\{\varphi \in \mathsf{com}\, \tau \mid \varphi <_t \bot \text{ for all } t \in T\}$. We are going to unravel D into a tree $\Delta \subseteq (\{0, \ldots, k-1\} \times \omega)^*$ together with three labelling functions $tp : \Delta \to T$, $bl : \Delta \to D$ and $hr : \Delta \to \Delta^*$ the intended meaning of which is as follows.

For all $\alpha \in \Delta$, $bl(\alpha)$ is some block in D of the type $tp(\alpha)$, and $tp(\alpha)$ should be the type of α in Δ after we turn Δ into a proper metric model. And if α is a child of β, i.e., $\alpha = \beta(i, j)$, for some $i < k$ and $j \in \omega$, then $(bl(\beta), bl(\alpha))$ should be a φ_i-link of blocks; in particular, we should have that $(tp(\alpha), tp(\beta))$ is a φ_i-link of types, $\varphi_i \in tp(\beta)$. Therefore, for $\alpha \in \Delta$, we set that $\alpha(i, 0)$ belongs to Δ iff φ_i does not belong to $tp(\alpha)$. Moreover, nodes $\alpha(i, j)$ with $j > 0$ are included into Δ iff φ_i belongs to $\min tp(\alpha)$, i.e., α should be a limit point of all the $\alpha(i, j)$. Finally, if $bl(\alpha) = (t_0, \ldots, t_n)$, then $hr(\alpha) = (\alpha_0, \ldots, \alpha_{n-1})$, where α_m is the node 'responsible' for the presence of t_m in $bl(\alpha)$.

We proceed by induction. First we choose some $(t_*) \in D$ with $\tau \in t_*$, and set

$$\lambda \in \Delta, \quad tp(\lambda) = t_*, \quad bl(\lambda) = (t_*), \quad hr(\lambda) = \lambda$$

(recall that λ denotes an empty sequence). Suppose now that $\alpha \in \Delta$ is constructed and $tp(\alpha), bl(\alpha), hr(\alpha)$ are defined, say,

$$tp(\alpha) = s_m, \quad bl(\alpha) = (s_0, \ldots, s_m), \quad hr(\alpha) = (\alpha_0, \ldots, \alpha_{m-1})$$

(i.e., $hr(\alpha) = \lambda$ when $m = 0$). Consider an arbitrary $i < k$. According to (4.3), there exists some $\mathbf{t} = (t_0, \ldots, t_n)$ in D such that $(bl(\alpha), \mathbf{t})$ is a φ_i-link (note that $n \leq m + 1$). Three cases are now possible.

$\varphi_i \in tp(\alpha)$. Then we set $\alpha(i, j) \notin \Delta$ for all $j \in \omega$.
$\varphi_i \notin \min tp(\alpha)$. Set $\alpha(i, 0) \in \Delta$ and $\alpha(i, j) \notin \Delta$ for all $j > 0$.
$\varphi_i \in \min tp(\alpha) \setminus tp(\alpha)$. Then we set $\alpha(i, j) \in \Delta$, for all $j \in \omega$.
Now, for all j with $\alpha(i, j) \in \Delta$ we define:

$$tp(\alpha(i, j)) = t_n, \quad bl(\alpha(i, j)) = \mathbf{t}, \quad hr(\alpha(i, j)) = (\alpha_0, \ldots, \alpha_{n-1}),$$

where α_m stands for α, if $n = m + 1$. Clearly, we have the following:

LEMMA 11. *Let $\alpha \in \Delta$ and $hr(\alpha) = (\alpha_0, \ldots, \alpha_{n-1})$, $bl(\alpha) = (t_0, \ldots, t_n)$. Then $hr(\alpha_m) = (\alpha_0, \ldots, \alpha_{m-1})$ and $bl(\alpha_m) = (t_0, \ldots, t_m)$, for all $m < n$.*

The next step is to convert Δ into a metric space $\mathfrak{D} = (\Delta, d^{\mathfrak{D}})$, i.e., to define a distance function $d^{\mathfrak{D}}$ on Δ. Recall that, by the construction of Δ, if $\varphi \in \mathrm{com}\,\tau$ and $\varphi \in tp(\beta)$ for some $\beta \in \Delta$, then every $\alpha \in \Delta$ with $\varphi \notin tp(\alpha)$ has a child β' with $\varphi \in tp(\beta')$. The main idea behind the construction of $d^{\mathfrak{D}}$ is to ensure that such a β' can always be chosen so that $d^{\mathfrak{D}}(\alpha, \beta') \leq d^{\mathfrak{D}}(\alpha, \beta)$. For this purpose we introduce a number of numerical parameters that will be defined by simultaneous induction on $\alpha \in \Delta$. These parameters are:

- The distance $d^{\alpha} = d^{\mathfrak{D}}(\alpha', \alpha)$, where α' is the parent of α.
- A sequence of numbers $c(\alpha)$ of the same length as $bl(\alpha)$. The distances d^{β}, for all children β of α, will form several slots within the interval $(0, 1)$, and $c(\alpha)$ stores some information on the boundaries of these slots.
- We use the following notation, for all $i < k$:

$$d_\alpha(i) = \begin{cases} 0, & \text{if } \varphi_i \in \min tp(\alpha), \\ d^{\alpha(i,0)}, & \text{if } \varphi_i \notin \min tp(\alpha). \end{cases}$$

- A 'sufficiently small' number $\varepsilon(\alpha)$ which is defined as follows. Suppose $c(\alpha) = (c_0, \ldots, c_n)$. Then

$$\varepsilon(\alpha) = \min\left(\{d_\alpha(i) - d_\alpha(j) \mid i, j < k,\ d_\alpha(i) > d_\alpha(j)\} \cup \right.$$
$$\left. \{c_m - d_\alpha(i) \mid m \leq n,\ i < k,\ c_m > d_\alpha(i)\}\right).$$

Roughly speaking, $\varepsilon(\alpha)$ measures the space available for 'splitting' the values $d_\alpha(i) = d_\alpha(j)$ with $i \neq j$.

We now list the principal conditions (4.6)–(4.11) that determine the choice of distances:

1) *For all $\gamma \in \Delta$ and $i, j < k$,*

$$d_\gamma(i) < d_\gamma(j) \ \leftrightarrow \ \varphi_i <_{tp(\gamma)} \varphi_j. \tag{4.6}$$

2) *Let $\gamma \in \Delta$ be such that $hr(\gamma) = \lambda$, $bl(\gamma) = (t)$, $c(\gamma) = (c)$. Then, for all $i < k$, $j \in \omega$,*

$$2c/3 \leq d_\gamma(i) < c, \quad \text{if } \varphi_i \notin \min t, \tag{4.7}$$
$$0 < d^{\gamma(i,j)} \leq \varepsilon(\gamma)/2, \quad \text{if } \varphi_i \in \min t \setminus t. \tag{4.8}$$

3) *Let $\gamma \in \Delta$ be such that $hr(\gamma) = (\gamma_0, \ldots, \gamma_{n-1})$, $bl(\gamma) = (t_0, \ldots, t_n)$, $c(\gamma) = (c_0, \ldots, c_n)$, where $n > 0$. Then, for all $i < k$, $j \in \omega$,*

$$d_{\gamma_{n-1}}(i) \leq d_\gamma(i) < d_{\gamma_{n-1}}(i) + c_n/3, \quad \text{if } \varphi_i \notin \min t_{n-1}, \tag{4.9}$$
$$2c_n/3 \leq d_\gamma(i) < c_n, \quad \text{if } \varphi_i \in \min t_{n-1} \setminus \min t_n, \tag{4.10}$$
$$0 < d^{\gamma(i,j)} \leq \varepsilon(\gamma)/2, \quad \text{if } \varphi_i \in \min t_n \setminus t_n, \tag{4.11}$$

And in the process of construction we will prove that the following property is satisfied as well:

LEMMA 12. *Let* $\gamma \in \Delta$ *and* $hr(\gamma) = (\gamma_0, \ldots, \gamma_{n-1})$, $bl(\gamma) = (t_0, \ldots, t_n)$, $c(\gamma) = (c_0, \ldots, c_n)$. *Then, for all* $m < n$, *we have*

$$c_{m+1} \leq \varepsilon(\gamma_m)/2, \quad c_{m+1} \leq c_m/2, \quad c(\gamma_m) = (c_0, \ldots, c_m). \tag{4.12}$$

Let us now turn to the construction. First, let $c(\lambda) = (1)$ and $d^\lambda = 2/3$ (the latter is defined simply for convenience). Suppose now that d^α and $c(\alpha) = (c_0, \ldots, c_n)$ are defined for some $\alpha \in \Delta$, condition (4.12) is satisfied for $\gamma = \alpha$, and conditions (4.6)–(4.12) are satisfied if γ is any ancestor of α. Let $hr(\alpha) = (\alpha_0, \ldots, \alpha_{n-1})$ and $bl(\alpha) = (t_0, \ldots, t_n)$. Two cases are possible.

Case 1: $n = 0$, i.e., $hr(\alpha) = \lambda$, $bl(\alpha) = (t_0)$, and $c(\alpha) = (c_0)$. Then, for all $i < k$ with $\varphi_i \notin \min t_0$, we can choose values $d_\alpha(i) = d^{\alpha(i,0)}$ in such a way that (4.6) and (4.7) are satisfied for $\gamma = \alpha$. Now $\varepsilon(\alpha)$ is defined, and we set $d^{\alpha(i,j)} = \varepsilon(\alpha)/(2j+2)$, for all $i < k$, $j \in \omega$ with $\varphi_i \in \min t_0 \setminus t_0$. Thus, (4.8) is satisfied as well for $\gamma = \alpha$, while (4.9)–(4.11) do not apply to this case. We further set

$$c(\alpha(i,0)) = (c_0/2), \quad \text{if } \varphi_i \notin \min t_0,$$
$$c(\alpha(i,j)) = (c_0, d^{\alpha(i,j)}), \quad \text{if } \varphi_i \in \min t_0 \setminus t_0,$$

for all $i < k$, $j \in \omega$. This makes (4.12) satisfied on the children of α (recall that $d^{\alpha(i,j)} \leq \varepsilon(\alpha)/2$ and $\varepsilon(\alpha) \leq c_n$ by definition).

Case 2: $n > 0$, i.e., $hr(\alpha)$ is a nonempty sequence. Since $(t_0, \ldots, t_n)$ is a block, we have $<_{t_{n-1}} \subseteq <_{t_n}$. And in view of (4.12) we have $c_n < d_{\alpha_{n-1}}(i)$ for all $i < k$ with $\varphi_i \notin \min t_{n-1}$. Therefore, for all $i < k$ with $\varphi_i \notin \min t_n$, we can choose values $d_\alpha(i) = d^{\alpha(i,0)}$ satisfying (4.6) and (4.9)–(4.10). Now $\varepsilon(\alpha)$ is defined, and we set $d^{\alpha(i,j)} = \varepsilon(\alpha)/(2j+2)$, for all $i < k$, $j \in \omega$ with $\varphi_i \in \min t_n \setminus t_n$. Thus, (4.11) is satisfied for $\gamma = \alpha$, while (4.7)–(4.8) do not apply to this case.

Further, consider any $i < k$ with $\varphi_i \notin t_n$. We then have several possibilities. First, let $\varphi_i \notin \min t_0$. Then $hr(\alpha(i,0)) = \lambda$, and we set $c(\alpha(i,0)) = (c_0/2)$. Clearly, (4.12) holds for $\gamma = \alpha(i,0)$.

Let $\varphi_i \in \min t_{m-1} \setminus \min t_m$ for some $1 \leq m \leq n$. Then $hr(\alpha(i,0)) = (\alpha_0, \ldots, \alpha_{m-1})$, and we set $c(\alpha(i,0)) = (c_0, \ldots, c_{m-1}, c_m/2)$. Now (4.12) holds for $\gamma = \alpha(i,0)$ in view of the induction hypothesis.

Let $\varphi_i \in \min t_n$ and $<_{t_{n-1}} = <_{t_n}$. Then $hr(\alpha(i,j)) = (\alpha_0, \ldots, \alpha_{n-1})$, and we set $c(\alpha(i,j)) = (c_0, \ldots, c_{n-1}, d^{\alpha(i,j)})$, for all $j \in \omega$. Again, (4.12) holds for $\gamma = \alpha(i,j)$ by the induction hypothesis.

Let finally $\varphi_i \in \min t_n$ and $<_{t_{n-1}} \subset <_{t_n}$. Then $hr(\alpha(i,j)) = (\alpha_0, \ldots, \alpha_n)$, and we set $c(\alpha(i,j)) = (c_0, \ldots, c_n, d^{\alpha(i,j)})$, for all $j \in \omega$. Recall that $d^{\alpha(i,j)} \leq \varepsilon(\alpha)/2$ and $\varepsilon(\alpha) \leq c_n$ by definition. Therefore (4.12) holds for $\gamma = \alpha(i,j)$ by the induction hypothesis.

Thus we define all the distances $d^\beta = d^{\mathfrak{D}}(\alpha, \beta)$, where α is a parent of β. Then we extend $d^{\mathfrak{D}}$ to the entire Δ by setting

$$d^{\mathfrak{D}}(\alpha,\alpha) = 0, \quad \text{for all } \alpha \in \Delta,$$

$$d^{\mathfrak{D}}(\beta,\alpha) = d^{\mathfrak{D}}(\alpha,\beta), \quad \text{if } \alpha \text{ is a parent of } \beta,$$

$$d^{\mathfrak{D}}(\alpha,\beta) = d^{\mathfrak{D}}(\alpha,\alpha_1) + \cdots + d^{\mathfrak{D}}(\alpha_n,\beta), \quad \text{if } \alpha, \alpha_1, \ldots, \alpha_n, \beta \text{ is the shortest path from } \alpha \text{ to } \beta.$$

This distance function satisfies the following properties:

LEMMA 13. Let $\alpha \in \Delta$ and $hr(\alpha) = (\alpha_0, \ldots, \alpha_{n-1})$, $bl(\alpha) = (t_0, \ldots, t_n)$, $c(\alpha) = (c_0, \ldots, c_n)$.

1) Let $m < n$. Then, for all $i < k$ with $\varphi_i \notin \min t_m$, we have

$$0 \leq d_\alpha(i) - d_{\alpha_m}(i) < (c_{m+1} + \cdots + c_n)/3 < 2c_{m+1}/3. \qquad (4.13)$$

2) For all $i < k$ and $1 \leq m \leq n$, we have

$$\begin{aligned} 2c_0/3 \leq d_\alpha(i) < c_0, & \quad \text{if } \varphi_i \notin \min t_0, \\ 2c_m/3 \leq d_\alpha(i) < c_m, & \quad \text{if } \varphi_i \in \min t_{m-1} \setminus \min t_m, \end{aligned} \qquad (4.14)$$

Proof. Let us prove (4.13) first. Note that, by (4.12), we have

$$c_{m+1} + \cdots + c_n \leq (1 + 1/2 + \cdots + 1/2^{n-m-1}) c_{m+1} < 2c_{m+1},$$

for any $m < n$. This proves the right-hand side inequality in (4.13). We then proceed by induction on $n - m$.

For $m = n - 1$, (4.13) follows directly from (4.9). Let now $m \leq n - 2$ and $\varphi_i \notin t_m$, for some $i < k$. By Lemma 11 and (4.12), we have $hr(\alpha_{n-1}) = (\alpha_0, \ldots, \alpha_{n-2})$, $bl(\alpha_{n-1}) = (t_0, \ldots, t_{n-1})$ and $c(\alpha_{n-1}) = (c_0, \ldots, c_{n-1})$. Therefore, by the induction hypothesis, we have

$$0 \leq d_{\alpha_{n-1}}(i) - d_{\alpha_m}(i) < (c_{m+1} + \cdots + c_{n-1})/3.$$

Combining this with (4.9) we obtain the required inequalities.

We now prove (4.14). Let $1 \leq m \leq n$ and $\varphi_i \in \min t_{m-1} \setminus \min t_m$, or $m = 0$ and $\varphi_i \notin \min t_0$. By Lemma 11 and (4.12) we have $hr(\alpha_m) = (\alpha_0, \ldots, \alpha_{m-1})$, $bl(\alpha_m) = (t_0, \ldots, t_m)$ and $c(\alpha_m) = (c_0, \ldots, c_m)$. Therefore, by (4.10), we have $2c_m/3 \leq d_{\alpha_m}(i) < c_m$, and moreover $d_{\alpha_m}(i) \leq c_m - \varepsilon(\alpha_m)$ by the definition of $\varepsilon(\alpha_m)$. But then, by applying (4.13) and (4.12), we obtain $2c_m/3 \leq d_\alpha(i) < c_m - \varepsilon(\alpha_m) + 2c_{m+1}/3 < c_m$. ∎

LEMMA 14. Let α be a parent of β in Δ. Then, for all $i < k$, we have:

$$|d_\alpha(i) - d_\beta(i)| \leq d^\beta.$$

Proof. Suppose that $hr(\alpha) = (\alpha_0, \ldots, \alpha_{n-1})$, $bl(\alpha) = (t_0, \ldots, t_n)$, $c(\alpha) = (c_0, \ldots, c_n)$, and $\beta = \alpha(j, l)$, $t = tp(\beta)$. Then (t_n, t) is a φ_j-link. Let $i < k$.

First, assume $\varphi_i \in \min t$. Then $d_\beta(i) = 0$, $\varphi_j \not<_t \varphi_i$, and so $\varphi_j \not<_{t_n} \varphi_i$. By (4.6), we obtain $d_\alpha(i) \leq d_\alpha(j)$, which implies $0 \leq d_\alpha(i) - d_\beta(i) \leq d^\beta$, since $d_\alpha(j) \leq d^\beta$.

Therefore we further assume that $\varphi_i \notin \min t$. Four cases are possible.

Case 1: $\varphi_j \notin \min t_0$. Then $\beta = \alpha(j,0)$, $c(\beta) = (c_0/2)$, and $d^\beta = d_\alpha(j) \in [2c_0/3, c_0)$, $d_\alpha(i) < c_0$, $d_\beta(i) \in [c_0/3, c_0/2)$. Hence we have $|d_\alpha(i) - d_\beta(i)| \leq c_0 - c_0/3 \leq d^\beta$.

Case 2: $\varphi_j \in \min t_{m-1} \setminus \min t_m$, for $1 \leq m \leq n$. Then $\beta = \alpha(j,0)$, $d^\beta = d_\alpha(j) \in [2c_m/3, c_m)$, and $hr(\beta) = (\alpha_0, \ldots, \alpha_{m-1})$, $bl(\beta) = (t_0, \ldots, t_{m-1}, t)$ $c(\beta) = (c_0, \ldots, c_{m-1}, c_m/2)$. Suppose first that $\varphi_i \in \min t_{m-1} \setminus t$. Then $d_\alpha(i) < c_m$ and $d_\beta(i) \in [c_m/3, c_m/2)$, i.e., $|d_\alpha(i) - d_\beta(i)| \leq c_m - c_m/3 \leq d^\beta$.

Suppose now that $\varphi_i \notin \min t_{m-1}$. Then $0 \leq d_\beta(i) - d_{\alpha_{m-1}}(i) < c_m/6$ by (4.9), and $0 \leq d_\alpha(i) - d_{\alpha_{m-1}}(i) < 2c_m/3$ by (4.13). Therefore, we have $|d_\alpha(i) - d_\beta(i)| \leq 2c_m/3 \leq d^\beta$.

Case 3: $\varphi_j \in \min t_n$ and $<_{t_{n-1}} = <_{t_n}$. Then $hr(\beta) = (\alpha_0, \ldots, \alpha_{n-1})$ and $c(\beta) = (c_0, \ldots, c_{n-1}, d^\beta)$. Suppose that $i \in \min t_n$. Then $d_\alpha(i) = 0$ and $d_\beta(i) < d^\beta$ by the construction. Hence $|d_\beta(i) - d_\alpha(i)| < d^\beta$.

Suppose now that $i \notin \min t_n$. Then we have $i \notin \min t_{n-1}$ and $d_\alpha(i), d_\beta(i) \in [d_{\alpha_{n-1}}(i), d_{\alpha_{n-1}}(i) + d^\beta/3)$ by (4.9). Hence $|d_\beta(i) - d_\alpha(i)| < d^\beta/3$.

Case 4: $\varphi_j \in \min t_{n-1}$ and $<_{t_{n-1}} = <_{t_n}$. Similar to the previous one. ∎

LEMMA 15. *Let $\alpha, \beta \in \Delta$ and $\varphi_i \in tp(\beta)$, for some $i < k$. Then*

$$d_\alpha(i) \leq d^\mathfrak{D}(\alpha, \beta).$$

Proof. First, note that $d_\beta(i) = 0$. Let $\alpha_0, \ldots, \alpha_n$ be the shortest path between α and β (i.e., $\alpha_0 = \alpha$, $\alpha_n = \beta$). We proceed by induction on n.

Let $n = 0$, i.e., $\alpha = \beta$. Then $d_\alpha(i) = d_\beta(i) = 0 = d^\mathfrak{D}(\alpha, \beta)$.

Let now $n \geq 1$. We have $d^\mathfrak{D}(\alpha, \beta) = d^\mathfrak{D}(\alpha, \alpha_1) + d^\mathfrak{D}(\alpha_1, \beta)$ by the definition of d^β. Then $|d_\alpha(i) - d_{\alpha_1}(i)| \leq d^\mathfrak{D}(\alpha, \alpha_1)$ by Lemma 14, and $|d_{\alpha_1}(i) - d_\beta(i)| \leq d^\mathfrak{D}(\alpha_1, \beta)$ by the induction hypothesis. Thus, we obtain $d_\alpha(i) = |d_\alpha(i) - d_\beta(i)| \leq d^\mathfrak{D}(\alpha, \beta)$, as required. ∎

Define now a model $\mathfrak{J} = (\mathfrak{D}, p_1^\mathfrak{J}, p_2^\mathfrak{J}, \ldots)$ by the following rule, for every atomic term p_i:
$$p_i^\mathfrak{J} = \{\alpha \in \Delta \mid p_i \in tp(\alpha)\}.$$

LEMMA 16. *For every $\alpha \in \Delta$ and $\varphi \in \mathsf{cl}\,\tau$, we have*

$$\alpha \in \varphi^\mathfrak{J} \quad \textit{iff} \quad \varphi \in tp(\alpha). \tag{4.15}$$

Proof. We proceed by induction on the construction of $\varphi \in \mathsf{cl}\,\tau$. If φ is an atomic term, then (4.15) holds by the definition of $\mathfrak{J}$. If $\varphi = \neg\psi_0$ or $\varphi = \psi_0 \sqcap \psi_1$, then (4.15) follows easily from the induction hypothesis.

So let now $\varphi = \psi_0 \sqsubseteq \psi_1$. Recall that D is the initial diagram, and T is the set of types occurring in blocks from D. Suppose that $\psi_0 \notin \{\varphi_0, \ldots, \varphi_{k-1}\}$. Then $\psi_0 \simeq_t \bot$, for all $t \in T$. Hence, by the definition of a type, $\psi_0 \notin t$ and $\psi_0 \sqsubseteq \psi_1 \notin t$, for all $t \in T$. On the one hand, we obtain by the induction hypothesis that $\psi_0^\mathfrak{J} = \emptyset$, and so $(\psi_0 \sqsubseteq \psi_1)^\mathfrak{J} = \emptyset$. On the other hand, we see that $\{\alpha \in \Delta \mid \psi_0 \sqsubseteq \psi_1 \in tp(\alpha)\} = \emptyset$. Thus, (4.15) is satisfied in this case.

We therefore assume from now on that $\psi_0 \in \{\varphi_0, \ldots, \varphi_{k-1}\}$. Then by the construction of Δ, we have, for every $\alpha \in \Delta$, that either $\psi_0 \in tp(\alpha)$, or $\psi_0 \in tp(\beta)$ for some child β of α. Hence $\psi_0^{\mathfrak{I}} = \{\alpha \in \Delta \mid \psi_0 \in tp(\alpha)\} \neq \emptyset$ by the induction hypothesis. Suppose now that $\psi_1 \notin \{\varphi_0, \ldots, \varphi_{k-1}\}$. Then $\psi_1^{\mathfrak{I}} = \emptyset$ similarly to the above, and $\psi_1 \simeq_t \bot$, hence $\psi_0 \rightleftharpoons \psi_1 \in t$, for all $t \in T$. We obtain that $\psi_0 \rightleftharpoons \psi_1^{\mathfrak{I}} = \Delta$, and $\{\alpha \in \Delta \mid \psi_0 \rightleftharpoons \psi_1 \in tp(\alpha)\} = \Delta$, i.e., (4.15) is satisfied in this case as well.

Suppose finally that $\psi_1 \in \{\varphi_0, \ldots, \varphi_{k-1}\}$, let $\psi_0 = \varphi_0$ and $\psi_1 = \varphi_1$ for concreteness. Then $\varphi_i^{\mathfrak{I}} = \{\alpha \in \Delta \mid \varphi_i \in tp(\alpha)\} \neq \emptyset$, $i = 0, 1$, similarly to the above. Therefore, by applying Lemma 15 and (4.6), we obtain for all $\alpha \in \Delta$: $\alpha \in (\varphi_0 \rightleftharpoons \varphi_1)^{\mathfrak{I}} \leftrightarrow d_\alpha(0) < d_\alpha(1) \leftrightarrow \varphi_0 \rightleftharpoons \varphi_1 \in tp(\alpha)$. ■

Thus, by Lemma 16, $\lambda \in \tau^{\mathfrak{I}}$ which completes the proof of Lemma 10. ■

We can now prove the main result of this paper.

THEOREM 17. *The satisfiability problem for $\mathcal{CSL}$-terms in metric models is* EXPTIME-*complete.*

Proof. The EXPTIME-hardness of this problem was shown in [12]. Let us prove the upper bound.

Given a term τ, let B be the set of all blocks for τ. As property (4.3) is clearly preserved under set unions, B contains the largest subset D' satisfying (4.3). It is not hard to see that D' can be constructed using the following elimination procedure (see, e.g., [5]).

Step 0: set $D_0 = B$.

Step $n + 1$: suppose that D_n is constructed. For each $\mathbf{s} \in D_n$ and each $\varphi \in \text{com } \tau$ such that $\varphi <_s \bot$ and $\varphi \notin s$, where s is the type of $\mathbf{s}$, we check whether there is a φ-link $(\mathbf{s}, \mathbf{t})$ for some $\mathbf{t} \in D_n$. If this is not the case then we set $D_{n+1} = D_n \setminus \mathbf{s}$ and go to the next step. Otherwise, we set $D' = D_n$.

As $|B| = 2^{O(|\tau|^2)}$, it should be clear that D' can be constructed in exponential time in $|\tau|$.

Suppose now that (4.1) does not hold for D'. Then obviously no diagram for τ exists, and so τ is not satisfiable by Lemma 7. So assume that $(t) \in D'$, $\tau \in t$ and consider the set

$$D = \{\mathbf{s}_n \mid ((t), \mathbf{s}_0), \ldots, (\mathbf{s}_{n-1}, \mathbf{s}_n) \text{ are } \varphi\text{-links, for some } \mathbf{s}_0, \ldots, \mathbf{s}_n \in D'\}.$$

Then D still satisfies (4.1) and (4.3), and by the definition of a link it satisfies (4.2) as well. Thus, D is a diagram and τ is satisfiable by Lemma 10.

It follows that τ is satisfiable iff D' satisfies (4.1), with the latter being verifiable in exponential time. ■

Acknowledgements. The work on this paper was partially supported by U.K. EPSRC grants no. GR/S63175/02, GR/S61973/02, GR/S63182/01, GR/S61966/01.

BIBLIOGRAPHY

[1] J. P. Delgrande. Preliminary considerations on the modelling of belief change operators by metric spaces. In *Proceedings of NMR*, pages 118–125, 2004.

[2] L. Esakia. Intuitionistic logic and modality via topology. *Annals of Pure and Applied Logic*, 127:155–170, 2004.

[3] K. Etessami, M. Vardi, and T. Wilke. First-order logic with two variables and unary temporal logic. In *Proceedings of the 12th IEEE Symp. Logic in Computer Science*, pages 228–235, 1997.

[4] N. Friedman and J. Halpern. On the complexity of conditional logics. In *Proceedings of KR'94*, pages 202–213, 1994.

[5] D. Harel, D. Kozen, and J. Tiuryn. *Dynamic Logic*. The MIT Press, 2000.

[6] A. Kurucz, F. Wolter, and M. Zakharyaschev. Modal logics for metric spaces: Open problems. In S. Artemov, H. Barringer, A. d'Avila Garcez, L. Lamb, and J. Woods, editors, *We Will Show Them! Essays in Honour of Dov Gabbay, Volume Two*, pages 193–208. College Publications, 2005.

[7] O. Kutz, H. Sturm, N.-Y. Suzuki, F. Wolter, and M. Zakharyaschev. Logics of metric spaces. *ACM Transactions on Computational Logic*, 4:260–294, 2003.

[8] D. Lewis. *Counterfactuals*. Blackwell, Oxford, 1973.

[9] C. Lutz, U. Sattler, and F. Wolter. Description logics and the two-variable fragment. In D. L. McGuiness, P. F. Pater-Schneider, C. Goble, and R. Möller, editors, *Proceedings of the 2001 International Workshop in Description Logics (DL-2001)*, pages 66–75, Stanford, California, USA, 2001.

[10] J.C.C. McKinsey and A. Tarski. The algebra of topology. *Annals of Mathematics*, 45:141–191, 1944.

[11] K. Schlechta. *Coherent Systems*. Elsevier, 2004.

[12] M. Sheremet, D. Tishkovsky, F. Wolter, and M. Zakharyaschev. Comparative similarity, tree automata, and Diophantine equations. In G. Sutcliffe and A. Voronkov, editors, *LPAR*, volume 3835 of *Lecture Notes in Computer Science*, pages 651–665. Springer, 2005.

[13] F. Wolter and M. Zakharyaschev. A logic for metric and topology. *Journal of Symbolic Logic*, 70:795–828, 2005.

Mikhail Sheremet,   Michael Zakharyaschev
School of Computer Science and Information Systems
Birkbeck College
Malet Street, London WC1E 7HX, U.K.
{mikhail, michael}@dcs.bbk.ac.uk

Dmitry Tishkovsky,   Frank Wolter
Department of Computer Science
The University of Liverpool
Liverpool L69 3BX, U.K.
{dmitry, frank}@csc.liv.ac.uk

Modality, Paraconsistency and Paracompleteness

RICARDO SOUSA SILVESTRE

ABSTRACT. We present here a modal logic that is both paraconsistent and paracomplete. The purpose of such endeavor is twofold. First we want to contribute to the understanding of the relations which some theoreticians have recently claimed that modal logic holds with paraconsistent and paracomplete logics. This will be achieved through the analysis of the formal features of our logic in comparison to traditional (normal) modal logic. We have made our logic to depart as less as possible from traditional modal logic, as well as provided some formal results such as a translation from one logic into another. Secondly, we try to show how situations which require a formal analysis of the notion of plausibility inside the context of what we called the skeptical and credulous approaches to induction might require a logic like the one we present here.

1 Introduction

From a point of view of truthfulness and falsehood, the principles of non-contradiction and excluded middle are the dual of each other. Roughly speaking, a paraconsistent logic [7] is a logic rejecting the principle of non-contradiction. Putting these two facts together, some theoreticians labeled the term "paracomplete logic" to designate those logics that do not satisfy the principle of excluded middle [15]. Intuitionist logic will obviously be classified as a paracomplete logic [2]. Following the same notational standard, logics that are both paraconsistent and paracomplete are called paranormal (or non-alethic) logics[1].

A more precise characterization of paraconsistent logic and paracomplete logic would go as follows. A paraconsistent negation is a unary operator that does not satisfy the principle of explosion (for any α, $\{\alpha, \neg\alpha\} \vdash \beta$ for any β) [2] and has enough properties to be called a negation; and a paracomplete negation is a unary operator that does not satisfy the excluded middle principle ($\vdash \alpha \lor \neg\alpha$) and has enough properties to be called a negation. A paraconsistent logic then is a logic having a paraconsistent negation, and a

---

[1] This "para" notation is all due to Miro Quesada [2].

[2] The need to consider the *ex-contradictio sequitur quod libet* principle instead of an internal principle of non contradiction is of course due to the non-equivalence between these two principles, but above all to the fact that the real novelty of paraconsistent logic is to be able to formalize inconsistent but non-trivial theories [6]

paracomplete logic is one having a paracomplete negation[3].

The relation between paraconsistent and paracomplete logic and modal logic is not new. After all, the first known system of paraconsistent logic, Jaskowski's discussive logic, was defined as a fragment of S5 [13]. Others have proposed semantics for paraconsistent and paracomplete logics which strongly resembles a Kripkean possible-world semantics [17, 18, 20]. More significant for us is the fact that intuitionistic negation can be interpreted (as it is commonly so) in terms of a certain translation into S4 that interprets this negation by $\Box\neg$. According to our tentative definition of paracompleteness, the existence of this derived negation inside S4 by itself allows us to classify S4 as a paracomplete logic. The same can be done to obtain a paraconsistent negation inside S4 or any other normal modal logic: just define $\sim$ as $\Diamond\neg$ and you will have a unary operator that does not satisfy the principle of explosion ($\sim \alpha, \alpha \vdash \beta$) and has enough properties to be called a negation [3]. This idea is of course not new: it has been deeply explored in between the mid-70s and the 80s [8, 23]. However, only recently we have seen the appearance of foundational works which try to relate in a more general way these two fields [1, 3, 4, 16], some of which with bold titles such as "S5 is a paraconsistent logic and so is first-order classical logic" [3] or "Nearly every normal modal logic is paranormal" [16].

At the heart of all this is the semantic structure of modal logic explored in the interpretation of $\Box$ and $\Diamond$. The fact that "α is possible" and "$\neg\alpha$ is possible" can coexist without trivializing the theory embodies a subtler sort of paraconsistency which some have named "hertian" [2] or conceptual [22] paraconsistency. Similarly, the fact that there might be a model where neither $\Box\neg\alpha$ nor $\Box\alpha$ are true entitles us to speak of a "hertian" or conceptual paracompleteness. That this conceptual paraconsistency, for example, cannot be turned into a truly or formal one by the use of $\sim$ can be easily checked. Even though regarding non-modal formulas $\sim$ will behave paraconsistently, with respect to modal formulas it is as classical as $\neg$: $\sim \Diamond\alpha, \Diamond\alpha \vdash \beta$, for any β, for example. (This or course might be different if we consider normal logics weaker than S5. Also, basing the truth-values of formulas of a non-modal language into a possible world semantics[4] might allow us to obtain a formal paraconsistent or paracomplete negation.)

But if the core of paraconsistency and paracompleteness of traditional modal logic (be it the conceptual or the formal one through $\sim$) is the behavior of $\Box$ and $\Diamond$ with respect to the negation, we can wonder: (1) What if

---

[3] All this should be seen as a rough characterization of paraconsistency and paracompleteness. Despite the vagueness of the positive side of our characterizations of paraconsistent and paracomplete negations, we think it fulfills its purpose of giving us something to start with. After all, one will hardly find something much superior in the literature, for the question of what is a paraconsistent (and paracomplete) negation is still something quite controversial. See for instance [2].

[4] Something like $V(p) = 1$ iff $v_w(p) = 1$ for all worlds w, where p is a propositional symbol, V is the valuation function which will give recursively the truth-value of every formula, and for each world w there is a function v_w that gives the truth-value of each propositional symbol in this world [5].

instead of a classical negation we had a truly paraconsistent and paracomplete negation in which $\neg \Diamond \alpha, \Diamond \alpha \nvdash \beta$, for some β, and $\nvdash \neg \Box \alpha \vee \Box \alpha$? (2) Will in this case our understanding of the mentioned connection between paraconsistent and paracomplete logic and modal logic somewhat improve? (3) If so, besides this theoretical interest, what other advantages, perhaps in terms of applications, the resulting logic would have? (4) And how much should it depart from traditional modal logic, or how much would it differ from the existing paraconsistent and paracomplete logics?

Our purpose in this paper is to contribute to the understanding of the relations that exist between paraconsistent and paracomplete logic and modal logic by trying to develop such a modal logic where $\Box$ and $\Diamond$ have a truly paracomplete and paraconsistent behavior, respectively. We will call such a logic paranormal modal logic. We acknowledge that this term might be misleading, for normal modal logic [5] have already an established meaning in modal logic and the "para" in paranormal does not apply to it. However, unfortunately the other term used to designate logics that are both paraconsistent and paracomplete, "non-alethic," also has a well established meaning in modal logic. In the lack of a better term then, we shall use the first one. In the next section we will set the basic assumption that will guide us in the construction of such a logic, partially answering question (4). In Section 3 we address (3) and try to say why it is worthy to have such a logic. In Section 4 we present the language and calculus of our logic and in Section 5 the semantics. Side by side to our exposition in these sections we will try to advance our answer to question (4) and perhaps also to (2). In Section 6 we present a formal comparison between paranormal modal logic and traditional normal modal logic, giving a big step, we believe, in answering (2), and finally, in Section 7, we try to summarize our contribution.

2 Paranormal Modal Logic

Here is the main prescription that will guide us in our task of developing our system: we want a paraconsistent and paracomplete logic in the way described in the previous section which departs the least possible from traditional modal logic. The rationale behind this is of course our foundational analysis: besides letting clear which formal and conceptual aspects we should give up and which ones we can retain when transforming modal logic into paranormal modal logic, this will make a technical comparison between the two logics easier. This principles of course the same as one of the conditions which according to da Costa any paraconsistent logic should fulfill [7], only now applied to classical modal logic instead of classical first-order logic. We will now proceed, based on this principle, to figure out which features of traditional modal logic we are willing to keep and which ones we should abandon.

First of all, due to the very task we set in the introduction, paranormal

---

[5] A normal modal logic is modal logic which has K ($\Box(\alpha \rightarrow \beta) \rightarrow (\Box \alpha\, rightarrow \Box \beta)$) as a valid principle and *modus ponens*, generalization (from α conclude $\Box \alpha$) and the rule of uniform substitution as valid inference rules [12].

modal logic should have two dual modal operators like $\Box$ and $\Diamond$. Just to avoid confusion, we shall use two different symbols -! and ?- in a post-fixed notation to represent our operators. ! should correspond to $\Box$ and ? to $\Diamond$. Another desirable feature is to have the same semantic structure used in traditional modal logic for the evaluation of formulas. Putting these things together, it seems that if we really want ! and ? to behave as much as possible like $\Box$ and $\Diamond$, respectively, then α? should be true iff α is true in at least one world, and α! iff α is true in all worlds [6]. Since our real concern is the negation, we should try to leave all else intact. Aside that, as far as possible we shall try to follow the standard presentation style of modal logics found in classical modal logic textbooks. We shall have a basic paranormal modal system, which we might call $K_?$, from which by adding new axioms, on the syntactic side, or by restricting the model structure, on the semantic one, we obtain several other paranormal modal systems, say, $T_?$, $D_?$, $S4_?$, $S5_?$, and so on.

About the distinguishing features of paranormal modal logic, ? and ! should have a paracomplete and a paraconsistent behavior, respectively. Given two formulae α and β, it may be that α? and $\neg(\alpha?)$ are both true, and that β! and $\neg(\beta!)$ are both false. This characterization of ? and ! reveals an odd behavior of $\neg$: when taken along with ?-marked formulas it is a paracomplete negation, but along with !-marked formulas it behaves like a paraconsistent negation. As a consequence of this, it is expected that many classical logical principles (in our case the excluded middle principle, the law of non-contradiction, the *reductio ad absurdum* principle, contraposition law, and others) shall not hold in paranormal modal logic. Regarding non-modal formulas, we can make $\neg$ behave exactly like classical negation, for our initial considerations do not require this sort of formulas to behave non-classically. Due to all this, it seems fair to classify our negation as a *paranormal modality-dependent negation*.

Despite these general considerations, there are many gaps to be filled. The crucial one seems to be how we will conceive a Kripkean semantic structure able to produce a model which satisfies both α? and $\neg(\alpha?)$, or that is such that neither α! nor $\neg(\alpha!)$ are satisfied. In order to avoid arbitrariness and give an initial answer to (3), we will try to limit the technical choices we have in the construction of paranormal modal logic by taking guidance of the intuitions contained in a possible application, to be presented in the next Section. After all, to have an intuitive justification was one of the conditions that according to Jaskowski a logic of contradictory systems should satisfy [13].

---

[6] In our informal use of the notions of possible world and satisfaction in this section and in the next one we will drop the reference to the notion of an accessibility and act as if the relation between the worlds were universal.

3 Why Paranormal Modal Logic? The Skeptical and Credulous Approaches to induction

The first answer we would be inclined to give to the question that begins the title of this section is that maybe answering (1) would be by itself a worthy task. After all, it was this sort of theoretical curiosity that led to the development of non-euclidean geometries and da Costa's calculi. In both cases, the applications came after. However, as we saw, there is a higher theoretical purpose involved in the creation of a paranormal modal logic. As we have said, the existence of such a logic might improve our understanding about the relations that exist between modal logic and paraconsistent and paracomplete logics. That this is so will be shown in the sections that will follow. Moreover, from the side of modal logic our work might be of some relevance. We have seen before different ways to base modal logic in some non-standard logic [9, 11, 14]. At this time we provide a minimal "traditional" modal logic based on a paraconsistent and paracomplete logic, that is to say, a paraconsistent and paracomplete modal logic which is as close as possible to traditional normal modal logic.

Nevertheless, and despite this, we do have an application which justifies the sort of modal logic presented here and which, we hope, provides an answer to question (3). This application is strongly connected with the main motivation for the construction of the systems of which paranormal modal logic is, say, a refined version [5, 17, 18]: the problem of dealing with contradictions that are sure to arise from the use of defeasible rules of inference [18]. As we have said, much of what we will set in this section will guide important choices to be made in the development of our logic, including our reading of the modal operators ! and ?.

The problem of inductive inconsistencies, extensively studied in both artificial intelligence and philosophy of science[7], strongly suggests the existence of at least two different approaches to the formal analysis of induction, which we shall call the *skeptical* and the *credulous* approaches to induction. They can be explained as follows. Consider a situation where some of our inductive rules of inference conflict with each other. In this case, there will be contradictory conclusions. One way to be able to keep reasoning classically about them is to keep all the conclusions we may obtain from our rules of inference without getting into a contradiction in different sets of formulas. We will call the deductive closure of each one of these sets an *inductive extension*. A skeptical approach to induction then is a position that rejects all sorts of contradictions and accepts as authentic inductive conclusions only those formulas that are in the intersection of all inductive extensions. It is a sort of rigid approach which requires quite a lot to accept something as a valid inductive conclusion. On the other hand, a credulous approach would take contradictions into consideration and accept the union of all inductive extensions, corresponding then to a tolerant approach which

---

[7]For some connections between the analysis of inductive inferences in artificial intelligence and philosophy of science see [21].

quite easily gives a formula the status of an authentic conclusion[8].

Now suppose that we decide, as it is common in philosophy of science, to distinguish inductive conclusions from ordinary formulas using some probability-like term like the term "plausibility." In this case we will have two different ways to appraise the plausibility of statements. We will say that α is plausible according to a skeptical approach iff α is true in all inductive extensions, and that α is plausible according to a credulous approach iff α is true in at least one extension. Here we see that the skeptical and credulous approaches work as evaluation functions which assess in different ways the truthfulness of plausible statements.

As one might have concluded, this picture can be represented with the aid of a modal framework. Let ! correspond to the skeptical plausibility and ? correspond to the credulous plausibility so that $\alpha!$ means that α is plausible according to the skeptical approach and $\alpha?$ that α is plausible according to the credulous approach. Let also each inductive extension correspond to a possible world, which we shall call from now on *plausible world*. In this case, we will have a quite familiar picture where $\alpha!$ is true iff α is true in all plausible worlds, and $\alpha?$ is true iff α is true in at least one plausible world. Given this, one might be wondering what is special about all that: traditional modal logic seems to be able to deal quite easily with the situation. Unfortunately this is not so, and the reason it is not so resides in the fact that having two dual ways to evaluate plausible formulas, a rigid and a tolerant one, makes negation to behave in a way that traditional logic cannot deal with.

According to the meaning we have set above, $\neg(\alpha!)$ and $\neg(\alpha?)$ will mean something like (I) "α is implausible according to a skeptical position" and (II) "α is implausible according to a credulous position," respectively. Note however that due to the qualification on the implausibility, which can be either skeptical or credulous, there is an ambiguity in the reading of these two sentences. This can be shown with the help of brackets. The first one, for example, can either mean that (i) it is not the case that [α is plausible according to a skeptical position] or that (ii) [it is not the case that α is plausible] according to a skeptical position. Clearly enough, the negation involved in (i) corresponds to the classical negation present in traditional modal logic: since "α is plausible according to a skeptical position" is true iff α is true in all worlds, (i) will be true if α is false in at least one world. That is not however the case with (ii). There, something very different is being said: it says that the whole sentence "it is not the case that α is plausible" is true according to a skeptical position. Now, how will this be evaluated in our plausible world framework?

If we stick to the idea behind the notion of a skeptical approach explained above, to evaluate "α is not plausible" according to a skeptical position means that we will be very strict in the matter of accepting "α is not

---

[8]The reader familiar with nonmonotonic literature should have already noticed the similarity of concepts. To a discussion of these two approaches from the standpoint of artificial intelligence see [22].

plausible" as true. In other words, we will require the maximum we can to give "α is not plausible" the value true or, going the other way round and taking true as 1 and false as 0, we will try the best we can to minimize the truth-value of "α is not plausible." I cannot think of anything else that could satisfy this except the situation where α is false in *all* plausible worlds. This would fairly express the idea behind [α being not plausible] according to a rigid, skeptical approach. Therefore, (ii) is true iff α is false in all plausible worlds.

Similarly for the credulous version of (ii): (ii') [it is not the case that α is plausible] according to a credulous position. (ii') involves being very tolerant in the matter of giving "α is not plausible" the value true or, going the other way round, it involves trying to maximize the truth-value of "α is not plausible." This of course means that being false in at least one world will suffice for us to attribute true to the sentence "α is not plausible." Therefore, (ii') is true iff α is false in *at least one* plausible world.

If one agrees on this analysis[9] as well as on the idea that the notion of implausibility is to be analyzed, represented or described in terms of the notion of plausibility, he will have to conclude that the ambiguity involved in (I) and (II) should (or at least might) be understood as primordially related to the negation, in such a way that a formal system like ours will have to have at least two negations, one for each reading of (I) and (II). In either case, both negations will evaluate $(\neg\alpha)!$ and $(\neg\alpha)?$ in the usual way, for there is no ambiguity involved here. $(\neg\alpha)!$ and $(\neg\alpha)?$ will be true, respectively, iff α is false in all plausible worlds and iff α is false in at least one plausible world. But this is exactly the evaluation, we have agreed above, that should be given to $\neg(\alpha!)$ and $\neg(\alpha?)$ in order to account for our second reading. Therefore, concerning our non-standard negation, we will have the strange relations expressed by the formulas $\neg(\alpha!) \to (\neg\alpha)!$ and $(\neg\alpha)? \to \neg(\alpha?)$.

The second important thing about this negation is that, since $\alpha!$ is true iff α is true in all worlds and $\neg(\alpha!)$ is true iff α is false in all worlds, it might happen that neither $\alpha!$ nor $\neg(\alpha!)$ are true, that is to say, a behavior exactly like the one a truly paracomplete negation should have and that initially motivated our theoretical investigation. Similarly, $\alpha?$ is true iff α is true in at least one world. Since $\neg(\alpha?)$ is true iff α is false in at least one world, then we will have that a model might satisfy both $\alpha?$ and $\neg(\alpha?)$, making $\neg$ correspond to what we have called a truly paraconsistent negation.

In this way, it seems that if we really stick to the meaning given to the skeptical and credulous positions, we will have to admit that in order to formalize a logic of skeptical and credulous plausibility inside a possible world framework we will have to develop something like a paranormal modal logic. From now on, we will make reference to this application to elucidate some key formal principles of our system.

---

[9]For the skeptic readers, a much more detailed argumentation is given in [22].

4 Language and Calculus

In this section we present the simplest paranormal modal system corresponding in traditional modal logic to system K. We call it $K_?$.

DEFINITION 1. Let $\Im$ be a propositional language. The paranormal modal language $\Im_?$ is defined as follows:
(i) If p is a propositional symbol of $\Im$, then p$\in \Im_?$;
(ii) If $\alpha, \beta \in \Im_?$, then $\neg \alpha, \alpha \to \beta, \alpha \wedge \beta, \alpha \vee \beta \in \Im_?$;
(iii) If $\alpha \in \Im_?$, then $\alpha!, \alpha? \in \Im_?$;
(iv) Nothing else belongs to $\Im_?$.

We call the set of all propositional symbols P. The logical symbols $\to$, $\wedge$ and $\vee$ are interpreted according to their usual meaning. $\neg$ will represent what we have named paranormal modality-dependent negation: regarding the elements of $\Im$, $\neg$ behaves classically, but in connection with !-marked and ?-marked formulae it behaves paracompletely and paraconsistently, respectively.

DEFINITION 2. Let $\alpha, \beta \in \Im_?$ be any formulae of $\Im_?$. We define the derived symbols $\leftrightarrow, \bot$ and $\sim$ as follows:
(i) $\alpha \leftrightarrow \beta =_{def} (\alpha \to \beta) \wedge (\beta \to \alpha)$;
(ii) $\bot =_{def}$ p $\wedge \neg$p, where p$\in$P is an arbitrary propositional symbol;
(iii) $\sim \alpha =_{def} \alpha \to \bot$.

The symbols $\leftrightarrow$ and $\bot$ are taken in accordance with their usual meaning. $\sim$ is a derived symbol intent to represent the classical negation. This is the negation we shall use to formalize the first reading of implausible sentences we saw in the last section.

DEFINITION 3. Let $\alpha \in \Im_?$ be a formula. We say that α is *?-free* iff ? does not occur in α and that α is *!-free* iff ! does not occur in α. If α is both ?-free and !-free we call it a ?!-free formula.

DEFINITION 4. The axiomatic of paranormal modal logic $K_?$ is as follows:
Positive Classical Axioms
P1: $\alpha \to (\beta \to \alpha)$
P2: $(\alpha \to (\beta \to \varphi)) \to ((\alpha \to \beta) \to (\alpha \to \varphi))$
P3: $\alpha \wedge \beta \to \alpha$
P4: $\alpha \wedge \beta \to \beta$
P5: $\alpha \to (\beta \to \alpha \wedge \beta)$
P6: $\alpha \to \alpha \vee \beta$
P7: $\beta \to \alpha \vee \beta$
P8: $(\alpha \to \beta) \to ((\varphi \to \beta) \to (\alpha \vee \varphi \to \beta))$
P9: $((\alpha \to \beta) \to \alpha) \to \alpha$
Paranormal Classical Axioms
A1: $(\alpha \to \beta) \to ((\alpha \to \neg \beta) \to \neg \alpha)$, wherein β is ?-free and α is !-free
A2: $\neg \alpha \to (\alpha \to \beta)$, wherein α is ?-free
A3: $\alpha \vee \neg \alpha$, wherein α is !-free
Non-Positive Additional Classical Axioms
N1: $\neg(\alpha \to \beta) \leftrightarrow (\alpha \wedge \neg \beta)$

N2: $\neg(\alpha \wedge \beta) \leftrightarrow (\neg\alpha \vee \neg\beta)$
N3: $\neg(\alpha \vee \beta) \leftrightarrow (\neg\alpha \wedge \neg\beta)$
N4: $\neg\neg\alpha \leftrightarrow \alpha$
Paranormal Modal Axioms
K1: $\alpha? \leftrightarrow\, \sim ((\sim \alpha)!)$
K2: $(\neg\alpha)! \leftrightarrow \neg(\alpha!)$
K3: $(\neg\alpha)? \leftrightarrow \neg(\alpha?)$
Modal Axioms
K?: $(\alpha \to \beta)! \to (\alpha! \to \beta!)$
Rules of Inference
MP: $\alpha, \alpha \to \beta / \beta$
N?: $\alpha/\alpha!$

The calculus of paranormal modal logic is a modal extension of positive classical logic (axioms P1-P9) in such a way as to consider a modality-dependent paranormal negation. The schemas of formula A1-A3 correspond to the negative axioms of classical logic restricted in such a way as to consider the paraconsistent and paracomplete behaviors of ? and !, respectively. Trivially enough, since these restrictions apply only to non-modal formulas, the non modal fragment of paranormal modal logic behaves exactly like classical logic. Axiom schemas N1-N4 are meant to restore the deductive power of paranormal modal logic weakened by the restrictions imposed on axioms A1-A3. The paranormal modal axioms K1-K3 set the basic properties of the modal operators ! and ?. K1 states that in connection with classical negation $\sim$, ? and ! are the dual operators of each other. K2 and K3 state, respectively, that the skeptical plausibility of $\neg\alpha$ is equivalent to the skeptical implausibility of α, and that the credulous plausibility of $\neg\alpha$ is equivalent to the credulous implausibility of α. K? is the paranormal equivalent to normal modal logic's axiom K and, finally, MP is *modus ponens* and N? is the paranormal rule of necessitation.

Due to K1, one might think that ? or ! could be introduced as definitions and that either K2 or K3 could be obtained as theorem. Unfortunately this is not so, the reason being the non-standard behavior of $\to$, which shall become clear when we present the semantics in the next section.

According to the best known contemporary taxonomy of paraconsistent logic [19], our logic belongs to the group of what has been labeled 'positive-plus' logics, that is, logics which, laying the blame for trivialization on the classical theory of negation, begin with a positive fragment of classical logic, and then add to it a weakened account of negation (which in our case is done by the restrictions imposed on axioms A1-A3.) We believe this style of attaining paraconsistency (and paracompleteness) is extremely suitable for our purposes in that it leaves explicit from an axiomatic point of view where paranormal modal logic effectively differs from normal modal logic. This difference can be seen very clearly if we consider that a fair axiomatic for K can be obtained from the above one by excluding P9, the non-positive additional classical axioms and the paranormal modal axioms, and removing the qualifications imposed on A1-A3[10]. This is also valid for all other normal modal systems, for in order to obtain, for example, T$_?$, S4$_?$ and S5$_?$ from K$_?$, we proceed in exactly the same way as we do to obtain T, S4 and S5 from K: add (T$_?$) $\alpha! \to \alpha$ to K$_?$ and you have system T$_?$; add (4$_?$) $\alpha! \to \alpha!!$ to T$_?$ and you have system S4$_?$; add (B$_?$) $\alpha \to \alpha?!$ to S4$_?$ and you have S5$_?$, and so on and so forth.

Letting A,B$\subseteq \Im_?$ be two sets of formulas, A representing the set of global premises and B the set of local ones, $\alpha \in \Im_?$ a formula and L$_?$ a paranormal modal logic, we define α's being L$_?$-deducted from A and B (in symbols: $A + B \vdash_{L?} \alpha$) in the standard way[11]. If $B = \emptyset$ we write $A \vdash_{L?} \alpha$ and if $A = B = \emptyset$, we write $\vdash_{L?} \alpha$, in which case we say that α is a L$_?$-theorem. In the case of K$_?$ for example, we say that α is K$_?$-deducted from A and B (in symbols: $A + B \vdash_{K?} \alpha$) and that α is a K$_?$-theorem (in symbols: $\vdash_{k?} \alpha$).

Theorems 5 and 6 below show some K$_?$-theorems which might give a flavor of the non-standard aspect of the calculus of paranormal modal logic.

THEOREM 5. *Some formulae of $\Im_?$ that satisfy one of the following schemas of formula are not K$_?$-theorems:*

$(\alpha \to \beta) \to (\alpha \to \neg\beta) \to \neg\alpha$; $\neg\alpha \to (\alpha \to \beta)$; $\neg\alpha \vee \alpha$;
$(\alpha \to \beta) \to (\neg\beta \to \neg\alpha)$; $(\neg\beta \to \neg\alpha) \to (\alpha \to \beta)$; $\neg(\alpha \wedge \neg\alpha)$;
$\neg\alpha \vee \beta \to (\alpha \to \beta)$; $(\alpha \to \beta) \to \neg\alpha \vee \beta$.

THEOREM 6. *Some formulae of $\Im_?$ that satisfy one of the following schemas of formula are not K$_?$-theorems:*

$\alpha? \to \neg((\neg\alpha)!)$; $\neg((\neg\alpha)!) \to \alpha?$; $\neg(\alpha!) \to (\neg\alpha)?$; $(\neg\alpha)? \to \neg(\alpha!)$;
$\alpha! \to \neg((\neg\alpha)?)$; $\neg((\neg\alpha)?) \to (\alpha!)$; $\neg(\alpha?) \to (\neg\alpha)!$; $(\neg\alpha)! \to \neg(\alpha?)$;

5 Semantics

In this section we present the semantics of paranormal modal logic. Again we tried to use a formal structure that resembles as most as possible traditional modal logic. The two definitions that follow define a semantic structure that is identical to the one we use to evaluate the truth-value of formulas in normal modal logic. The difference will be the function that will make use of this structure to recursively attribute true or false to all sorts of formulas, to be defined in Definition 9

DEFINITION 7. A frame F is a duple $\langle W, R \rangle$ where W is a non-empty set of entities called *(plausible) words* and R is a binary relation on W called *accessibility relation*.

DEFINITION 8. A model M is a triple $\langle W, R, v \rangle$ where $F = \langle W, R \rangle$ is a frame and v is function mapping pairs composed by elements of P and

---

[10]If we wish, we could only remove the qualifications on A1-A3 as well as K2 and K3 that we would still have an axiomatic for K, only a redundant one.

[11]From an axiomatic point of view, the difference between global and local premises is that the rule of necessitation can be applied only to those formulas derived exclusively from the set of global premises. For a textbook-style presentation of modal logic that uses global and local premises in the definition of the relations of deductibility and logical consequence see [10].

elements of W to truth-values 0 and 1. We say that the model M is *based on* F and that $w \in W$ is a world of M.

DEFINITION 9. The *max-min modal valuations* are functions Ω and $\mho$ which, given a model $M = \langle W, R, v \rangle$ and a world $w \in W$, map formulas of $\Im_?$ to truth-values 0 and 1. Ω and $\mho$ are defined as follows:

(i) $\Omega_{M,w}(p) = \mho_{M,w}(p) = 1$ iff $v_w(p) = 1$;
(ii) $\Omega_{M,w}(\neg \alpha) = 1$ iff $\mho_{M,w}(\alpha) = 0$;
(iii) $\mho_{M,w}(\neg \alpha) = 1$ iff $\Omega_{M,w}(\alpha) = 0$;
(iv) $\Omega_{M,w}(\alpha \to \beta) = 1$ iff $\Omega_{M,w}(\alpha) = 0$ or $\Omega_{M,w}(\beta) = 1$;
(v) $\mho_{M,w}(\alpha \to \beta) = 1$ iff $\Omega_{M,w}(\alpha) = 0$ or $\mho_{M,w}(\beta) = 1$;
(vi) $\Omega_{M,w}(\alpha \wedge \beta) = 1$ iff $\Omega_{M,w}(\alpha) = 1$ and $\Omega_{M,w}(\beta) = 1$;
(vii) $\mho_{M,w}(\alpha \wedge \beta) = 1$ iff $\mho_{M,w}(\alpha) = 1$ and $\mho_{M,w}(\beta) = 1$;
(viii) $\Omega_{M,w}(\alpha \vee \beta) = 1$ iff $\Omega_{M,w}(\alpha) = 1$ or $\Omega_{M,w}(\beta) = 1$;
(ix) $\mho_{M,w}(\alpha \vee \beta) = 1$ iff $\mho_{M,w}(\alpha) = 1$ or $\mho_{M,w}(\beta) = 1$;
(x) $\Omega_{M,w}(\alpha?) = 1$ iff, for some $w' \in W$ such that wRw', $\Omega_{M,w'}(\alpha) = 1$;
(xi) $\mho_{M,w}(\alpha?) = 1$ iff, for all $w' \in W$ such that wRw', $\mho_{M,w'}(\alpha) = 1$;
(xii) $\Omega_{M,w}(\alpha!) = 1$ iff, for all $w' \in W$ such that wRw', $\Omega_{M,w'}(\alpha) = 1$;
(xiii) $\mho_{M,w}(\alpha!) = 1$ iff, for some $w' \in W$ such that wRw', $\mho_{M,w'}(\alpha) = 1$;.

Ω and $\mho$ are functions that, depending on the modal operator attached to the formula, evaluate it either skeptically or credulously (if the formulas is a non-modal one, the evaluation is, we can say, neutral.) They represent therefore, depending on the formula being evaluated, our skeptical and credulous approaches to induction. In what follows we shall try to elucidate some key aspects of Ω and $\mho$[12].

Our first concern will be how, through (x)-(xiii), modal formulas are evaluated. While Ω, which is the function we shall use in the definition of logical consequence, gives the intended meaning of ! and ?, $\mho$ does the opposite and evaluates !-formulas credulously and ?-formulas skeptically. The need for this "anomalous" interpretation has to do with the way Ω and $\mho$ deal with the negation, which is materialized in items (ii) and (iii). The rationale behind these rules can be seen through a simple interpretation of the meaning of functions Ω and $\mho$. Consider $\neg(\alpha?)$. Since Ω will evaluate it credulously, $\mho$ evaluates $\alpha?$ skeptically. We can read $\Omega_{M,w}(\neg(\alpha?)) = 1$ as "according to a credulous position that tries to maximize the truthfulness of formulas, 'not [α is plausible]' is true." This can be said to be equivalent to "the task of maximizing the truthfulness of 'not [α is plausible]' was successful," which in its turn is the same as "the task of maximizing the falsehood of 'α is plausible' was successful." Now consider $\mho_{M,w}(\alpha?) = 0$. It can be read as "according to a position that tries to minimize the truthfulness of formulas, 'α is plausible' is not true," which in its turn is equivalent to "the task of minimizing the truth of 'α is plausible' failed," which is pretty the same as "the task of maximizing the falsehood of 'α

---

[12]It should be said that a full explanation of the intricacies of this unusual way to evaluate modal formulas would need much more space. For a full account of this and other aspects of paranormal modal logic, including a first-order formulation, see [22].

is plausible' was successful." That is why $\Omega_{M,w}(\neg\alpha) = 1$ iff $\mho_{M,w}(\alpha) = 0$. The same can be done to show that $\mho_{M,w}(\neg\alpha) = 1$ iff $\Omega_{M,w}(\alpha) = 0$. Thus we see why to evaluate $\neg(\alpha?)$ credulously and $\neg(\alpha!)$ skeptically we need to evaluate $\alpha?$ skeptically and $\alpha!$ credulously.

Our final comment concerns item (iv), which is asymmetric with respect to (v). It can be shown that making use of the simple idea of skeptical and credulous evaluation as described here, the correct form of (iv) would be $\Omega_{M,w}(\alpha \rightarrow \beta) = 1$ iff $\mho_{M,w}(\alpha) = 0$ or $\Omega_{M,w}(\beta) = 1$. From a purely conceptual point of view, $\alpha \rightarrow \beta$ is true according to a maximal position, for example, iff α is false according to a minimal position or β is true according to a maximal one. However, if we follow that, we would not be able to preserve *modus ponens*, since our consequence relation is defined over Ω[13] .

Alike to the axiomatic presented in the last section, there is a strong parallel between paranormal modal logic's semantics and normal modal logic's one. As we said, the notion of semantic model is exactly the same; what changes is the way to use the model to evaluate the truth-value of formulas. Roughly speaking, we just have, in addition to the usual way to evaluate modal formulas (which is, except for the asymmetry of (iv), captured by Ω), another function intended to reverse the position taken by Ω. This additional function is essential for the proper treatment of negation in (ii) and (iii), which is the key of the non-standard behavior of ! and ?. Aside that, other paranormal modal systems are obtained in the standard way by restricting the accessibility relation.

DEFINITION 10. Let $M = \langle W, R, v \rangle$ be a model, $w \in W$ a world of M and $\alpha \in \Im_?$ a formula: α *is satisfied by* M *at* w (in symbols: $M, w \Vdash \alpha$) iff $\Omega_{M,w}(\alpha) = 1$; α *is satisfied by* M (in symbols: $M \Vdash \alpha$) iff, for all $w' \in W$, $M, w' \Vdash \alpha$.

From the two notions above we define, in the customary way, the notion of logical consequence. Let $A, B \subseteq \Im_?$ be two sets of formulae, A representing the global premises and B the local ones, and $\alpha \in \Im_?$ a formula. If α is a logical consequence of A and B (all classes of models being considered) we say that α is a $K_?$-logical consequence of A (in symbols: $A + B \models_{k?} \alpha$). In the case $A = B = \emptyset$, we say that α is $K_?$-valid (in symbols: $\models_{k?} \alpha$).

THEOREM 11. $K_?$ *is sound and complete (in symbols: Let* $A, B \subseteq \Im_?$ *be two sets of formulas and* $\alpha \in \Im_?$ *a formula.* $A + B \vdash_{K?} \alpha$ *iff* $A + B \models_{K?} \alpha$).

6 Normal and Paranormal Modal Logic: K and $K_?$

In this section we show the results which, from a formal point of view, more strongly relate paranormal modal logic with normal modal logic. Let $\Im_\diamond$ be the language of propositional normal modal logic, $\diamond$ being taken either as a primitive or a derived symbol, and $\vdash_K$ and $\models_K$ the relations of deductibility and logical consequence of modal logic K.

---

[13]Although acknowledging some *ad hoc* flavor in this maneuver, we do not think this sole point represents a threat to our use of the skeptical and credulous approaches to induction to provide an intuition for our logic.

DEFINITION 12. We define the function $\Phi\colon \Im_\Diamond \to \Im_?$ as follows:
(i) $\Phi(p) = p$, where $p \in P$;
(ii) $\Phi(\neg\alpha) =\, \sim \Phi(\alpha)$;
(iii) $\Phi(\alpha \oplus \beta) = \Phi(\alpha) \oplus \Phi(\beta)$, where $\oplus \in \{\wedge, \vee, \to\}$;
(iv) $\Phi(\Diamond\alpha) = \Phi(\alpha)?$;
(v) $\Phi(\Box\alpha) = \Phi(\alpha)!$.

THEOREM 13. Let $\alpha \in \Im_?$. If $\vdash_K \alpha$, then $\vdash_{K?} \Phi(\alpha)$.

The meaning of theorem 13 is that, when taken in conjunction with $\sim$, paranormal modalities ! and ? are indistinguishable from normal modalities $\Box$ and $\Diamond$. This of course implies that the full expressive power of normal modal logic is contained in paranormal modal logic. Theorems 17-20 below show the stronger result that the inverse is also true.

DEFINITION 14. We define the functions Π and $\amalg$ of the form: $\Im_? \to \Im_\Diamond$ as follows:
(i) $\Pi(p) = \amalg(p) = p$;
(ii) $\Pi(\alpha?) = \Diamond \Pi(\alpha)$;
(iii) $\amalg(\alpha?) = \Box \amalg(\alpha)$;
(iv) $\Pi(\alpha!) = \Box\Pi(\alpha)$;
(v) $\amalg(\alpha!) = \Diamond \amalg(\alpha)$;
(vi) $\Pi(\neg\alpha) = \neg \amalg(\alpha)$;
(vii) $\amalg(\neg\alpha) = \neg\Pi(\alpha)$;
(viii) $\Pi(\alpha \oplus \beta) = \Pi(\alpha) \oplus \Pi(\beta)$, where $\oplus \in \{\wedge, \vee, \to\}$;
(ix) $\amalg(\alpha \oplus \beta) = \amalg(\alpha) \oplus \amalg(\beta)$, where $\oplus \in \{\wedge, \vee\}$;
(x) $\amalg(\alpha \to \beta) = \Pi(\alpha) \to \amalg(\beta)$.

DEFINITION 15. We define the functions Δ and ∇ of the form: $\Im_\Diamond \to \Im_?$ as follows:
(i) $\Delta(p) = \nabla(p) = p$;
(ii) $\Delta(\Diamond\alpha) = \Delta(\alpha)?$;
(iii) $\nabla(\Diamond\alpha) = \nabla(\alpha)!$;
(iv) $\Delta(\Box\alpha) = \Delta(\alpha)!$;
(v) $\nabla(\Box\alpha) = \nabla(\alpha)?$;
(vi) $\Delta(\neg\alpha) = \neg\nabla(\alpha)$;
(vii) $\nabla(\neg\alpha) = \neg\Delta(\alpha)$;
(viii) $\Delta(\alpha \oplus \beta) = \Delta(\alpha) \oplus \Delta(\beta)$, where $\oplus \in \{\wedge, \vee, \to\}$;
(ix) $\nabla(\alpha \oplus \beta) = \nabla(\alpha) \oplus \nabla(\beta)$, where $\oplus \in \{\wedge, \vee\}$;
(x) $\nabla(\alpha \to \beta) = \Delta(\alpha) \to \nabla(\beta)$.

DEFINITION 16. Let $A \subseteq \Im_?$ and $B \subseteq \Im_\Diamond$.
(i) $\Pi(A) = \{\Pi(\alpha) | \alpha \in A\}$;
(ii) $\amalg(A) = \{\amalg(\alpha) | \alpha \in A\}$;
(iii) $\Delta(B) = \{\Delta(\alpha) | \alpha \in B\}$;
(iv) $\nabla(B) = \{\nabla(\alpha) | \alpha \in B\}$.

THEOREM 17. Let $A, B \subseteq \Im_\Diamond$ and $\alpha \in \Im_\Diamond$. $A + B \vdash_K \alpha$ iff $\Delta(A) + \Delta(B) \vdash_{K?} \Delta(\alpha)$.

THEOREM 18. *Let* $A, B \subseteq \Im_?$ *and* $\alpha \in \Im_?$. $A + B \vdash_{K?} \alpha$ *iff* $\Pi(A) + \Pi(B) \vdash_K \Pi(\alpha)$.

THEOREM 19. *Let* $A, B \subseteq \Im_\Diamond$ *and* $\alpha \in \Im_\Diamond$. $A + B \models_K \alpha$ *iff* $\Delta(A) + \Delta(B) \models_{K?} \Delta(\alpha)$.

THEOREM 20. *Let* $A, B \subseteq \Im_?$ *and* $\alpha \in \Im_?$. $A + B \models_{K?} \alpha$ *iff* $\Pi(A) + \Pi(B) \models_K \Pi(\alpha)$.

Definitions 14 and 15 formalize a translation that comes naturally when we look at the semantics of normal and paranormal modal logics. Using them, theorems 17-20 state that, according to this translation, both normal and paranormal logics can be fully embedded inside each other. The implications of this are obvious. Since formulas resulting from the application of Δ can be seen as abbreviations inside $\Im_\Diamond$, it might be said that there is a formal paraconsistent and paracomplete inferential relation (in addition to a conceptual one) based on a truly paranormal modality-dependent negation inside normal modal logic. This we think strengths the thesis we have mentioned in Section 1 about normal modal logic being paranormal.

This result might give room for a sort of objection that questions the whole worthiness of our endeavor: if all the expressive and inferential power of paranormal modal logic is already contained in normal modal logic, what is the point of developing and studying it? Well, taking this objection seriously might allow us to refer to theorems 18 and 19 and claim that since all the expressive and inferential power of normal modal logic is contained in paranormal modal logic, why not to question instead the supremacy of normal modal logic? After all, given the above stated equivalence between the two systems, the only real reason for this supremacy would be the historical one that one logic was discovered, created, or whatever, before the other. Of course things are not so simple. In fact, the whole situation is very alike to one that might be created from the fact that it is possible to translate S5 into the fragment of monadic classical first-order logic with only one variable, and vice-versa [24].

I particularly prefer to stick to the formal apparatus available to us and think of two different but really strongly connected formal systems; so strongly connected that might be taken as different aspects of the same thing. From this perspective, I can only think of those funny toys we often see on cartoons which at one time look like a car and at other time look like a completely different object, say, an airplane. Despite the toy being, at one specific time, from the point of view of the child who plays with it at that time, only a car, all the stuffs needed for the airplane are already there, inside the car.

7 Conclusion

In this paper we presented what we have called paranormal modal logic with two main purposes in mind. The first, if the reader allows me to take the inverse order, was to formalize a logic of plausibility based on what we have called the skeptical and the credulous approaches to induction. The conceptual framework provided by this theory of induction was the main intuitive justification we used to rationalize some (somehow controversial) conceptual features of our system. Second, we intended to throw some light upon the relations that hold between modal logic and paraconsistent and paracomplete logic. We think we have succeeded in this task in that the simple fact of having a sound and complete paraconsistent and paracomplete modal logic which, from the formal point of view, is as close as possible to traditional modal logic already gives important hints about what in modal logic can be said to be connected to paraconsistency and paracompleteness and consequently might help in the development of a paranormal modal logic and what is incompatible and need to be given up in such endeavor. Also, and surely more surprising, we showed that paranormal and normal modal logics can from a representational and inferential point of view be fully embedded in each other.

Perhaps one might object to the above conclusions that we should have done a more general work, perhaps from a meta-level, instead of taking two specific modal systems. Well, despite having based our analysis on two systems, K and K?, the conclusions we drew are trivially applicable to the two classes of logical systems definable upon K and K?. Second, despite being "just one modal logic," we believe normal modal logic K and normal logics in general are very distinguished representatives of the class of modal logics, in that conclusions of the sort we got here about them are surely relevant for the whole field of modal logic.

BIBLIOGRAPHY

[1] Batens, Diderik (2002), On the Remarkable Correspondence between Paraconsistent Logics, Modal Logics and Ambiguity Logics. In Paraconsistency: The Logical Way to the Inconsistency, eds. W. A. Carnielli, M. E. Coniglio e I. D' Ottaviano: 445-454, New York: Marcel Dekker.
[2] Béziau, J. Y. (199), The future of paraconsistent logic. Logical Studies 2: 1-23.
[3] Béziau, J. Y. (2002), S5 is a paraconsistent logic and so is first-order classical logic. Logical Studies, 9:301-309.
[4] Béziau, J. Y. (2005), Paraconsistent logic from a modal viewpoint. Journal of Applied Logic, 3(1):7-14.
[5] Buchsbaum, Arthur R. and T. Pequeno (1993), Uma Família de Lógicas Paraconsistentes e/ou Paracompletas com Semânticas Recursivas. Coleção Documentos, Série Lógica e Teoria da Ciência 14, Universidade de São Paulo.
[6] da Costa, N. C. A. (1963), Calculs propositionnels pour les systèmes formels inconsistants. Comtes Rendus de l'Académie des Sciences de Paris 257: 3790-3793.
[7] da Costa, N. C. A. (1974), On the Theory of Inconsistent Formal Systems. Notre Dame Journal of Formal Logic 15.
[8] Dosen, K. (1986), Negation as a modal operator. Reports on Mathematical Logic 20:15-28.
[9] Fagin, Ronald and Y. H. Joseph and M. Y. Vardi (1996), A Nonstandard Approach to the Logical Omniscience Problem. Artificial Intelligence, 79: 203-240
[10] Fitting, Melvin (1993), Basic Modal Logic. In Handbook of Logic in Artificial Intelligence and Logic Programming, Vol. 1, Logical Foundations, eds. D. Gabbay and D. Hogger and J. Robinson, Oxford: Oxford University Press.
[11] Hintikka, J. (1975), Impossible possible worlds vindicated. Journal of Philosophical Logic 4: 475-484.

[12] Hughes, G. E and M. J. Cresswell (1996), A New Introduction to Modal Logic. New York: Routledge.
[13] Jaśkowski, S. (1948), A propositional calculus for inconsistent deductive systems (in Polish). Studia Societatis Scientiarum Torunensis, Sectio A, 5:57-77. Translated into English in Studia Logica, 24:143-157, 1967.
[14] Kripke, S. (1965), Semantical analysis of modal logic II: non-normal modal propositional calculi. In The theory of models, Vol. 9, eds. J. W. Addison, L. Henkin and A. Tarski: 206-220, Amsterdam: North-Holland.
[15] Loparíc, Andréa and N. da Costa (1984), Paraconsistency, Paracompletenes and Valuations. Logique et Analyse 106: 119-131.
[16] Marcos, J. (2005), Nearly every normal modal logic is paranormal. Logique et Analyse 48: 279-300.
[17] Martins, A. T. C and M. Pequeno and T. Pequeno (2002), A Multiple Worlds Semantics to a Paraconsistent Nonmonotonic Logic. In Paraconsistency: The Logical Way to the Inconsistency, eds. W. A. Carnielli, M. E. Coniglio e I. D' Ottaviano: 187-211, New York: Marcel Dekker.
[18] Pequeno, Tarcisio H. C. and Arthur R. Buchsbaum (1991), The Logic of Epistemic Inconsistency. In Principles of Knowledge Representation and Reasoning: Proceedings of Second International Conference, San Mateo: Morgan Kaufman.
[19] Priest, G. and R. Routley and J. Norman (eds.) Paraconsistent Logic: Essays on the Inconsistent. München: Philosophia Verlag, 1989.
[20] Rescher, Nicholas and Robert Brandom (1980), The Logic of Inconsistency: A study in nonstandard possible-world semantics and ontology. Basil Blackwell.
[21] Silvestre, Ricardo and T. Pequeno (2005), A Logic of Inductive Implication or AI Meets Philosophy of Science II. In Proceedings of the 18th Conference of the Canadian Society for the Computational Studies of Intelligence, (LNAI 3501), eds. B. Kégl and G. Lapalme (eds.): 232-243, Berlin-Heidelberg: Springer-Verlag.
[22] Silvestre, Ricardo (2005), Induction and Plausibility: A Formal Approach from the Standpoint of Artificial Intelligence. Ph.D. dissertation, University of Montreal, Montreal.
[23] Vakarelov, D. (1989), Consistency, completeness and negation. In Paraconsistent Logic: Essays on the inconsistent, eds. G. Priest, R. Sylvan and J. Norman, Philosophia Verlag: 328-363.
[24] Wajsberg, M (1933), Ein erweiteter Klassenkalkül. Monatshefte für Mathematik und Physik 40: 113-126.

Appendix: Sketch of the Proofs of Theorems

We will here just sketch the proofs of theorems 11, 17, 18, 19 and 20. In the spirit of the main purpose of the paper, we will base the proof of theorem 11 on theorems 17-20. Therefore the proof of theorems 17-20 shall come first. For a full account of the proofs see [22]. In order to prove theorems 17-20 some key lemmas are needed. Let P1,..., P11, K, NP, MP and N be the axiomatic for K in which the axioms P1, ..., P8 are as in definition 4, P9, P10 and P11 are the non-qualified axioms corresponding to A1, A2 and A3, respectively, K is $\Box(\alpha \to \beta) \to (\Box\alpha \to \Box\beta)$, NP is $\Diamond\alpha \leftrightarrow \neg\Box\alpha$, N is $\alpha/\Box\alpha$ and MP is *modus ponens*. Let also Ψ be the evaluation function of normal modal logic (one which functions like Ω) and $\Vdash_\Psi$ (and $\Vdash_\Omega$) the respective satisfaction relation.

LEMMA 21. *Let* $\alpha \in \Im_?$ *and* $\beta \in \Im_\Diamond$. $\Delta(\Pi(\alpha)) = \alpha$ *and* $\Pi(\Delta(\beta)) = \beta$. *Let* $A \subseteq \Im_?$ *and* $B \subseteq L_\Diamond$. $\Pi(\Delta(A)) = A$ *and* $\Pi(\Delta(B)) = B$.

LEMMA 22. *Let* $A, B \subseteq \Im_?$ *and* $\alpha \in \Im_?$. *If* $A + B \vdash_{K?} \alpha$ *then* $\Pi(A) + \Pi(B) \vdash_K \Pi(\alpha)$.

LEMMA 23. *Let* $A, B \subseteq \Im_\diamond$ *and* $\alpha \in \Im_\diamond$. *If* $A + B \vdash_K \alpha$ *then* $\Delta(A) + \Delta(B) \vdash_{K?} \Delta(\alpha)$.

LEMMA 24. *Let* $\alpha \in \Im_\diamond$ *be a formula,* $M = \langle W, R, v \rangle$ *a model and* $w \in W$ *a world of* M. $M, w \Vdash_\Psi \alpha$ *iff* $M, w \Vdash_\Omega \Delta(\alpha)$ *or, equivalently,* $\Psi_{M,w}(\alpha) = 1$ *iff* $\Omega_{M,w}(\Delta(\alpha)) = 1$.

LEMMA 25. *Let* $\alpha \in \Im_?$ *a formula,* M *be a model and* w *a world of* M. $M, w \Vdash_{\Omega?} \alpha$ *iff* $M, w \Vdash_\Psi \Pi(\alpha)$ *or, equivalently,* $\Omega_{M,w}(\alpha) = 1$ *iff* $\Psi_{M,w}(\Pi(\alpha)) = 1$.

Lemma 21 follows from the fact that Δ and Π are the inverse functions of each other. Proof of lemmas 22 and 23 are rather long. The key idea however is simple. In the case of lemma 23, we do the proof by induction on the size of the K-derivation of α from A and B. The base of induction is trivial for the case where $\alpha \in A \cup B$. In the case α is an axiom of K, we have to analyze the possibility of α's being an instance of each one of K's axiom schemas. For each one of these possibilities, we show that there is a K?-derivation of $\Delta(\alpha)$. Regarding K's classical axioms, the cases where α is an instance of P1-P8 are trivial. To show that if α is an instance of P9, P10, P11 then there is a K?-derivation of $\Delta(\alpha)$ requires much more labour (and space.) In order to deal with P9, for example, we are considering $\alpha \equiv (\varphi \to \beta) \to ((\varphi \to \neg\beta) \to \varphi)$. $\Delta(\alpha)$ then is $(\Delta\varphi \to \Delta\beta) \to ((\Delta\varphi \to \neg\nabla\beta) \to \neg\nabla\varphi)$. We then prove that there exists a K?-derivation of $(\Delta\varphi \to \Delta\beta) \to ((\Delta\varphi \to \neg\nabla\beta) \to \neg\nabla\varphi)$. In order to do that, we have to do another proof by induction, now on the size of φ. K and NP are quite easy. In the step of induction we consider K-derivations of α from A and B of size $n > 1$ and suppose that for K-derivations of sizes smaller than n the result holds. That is to say, if $A + B \vdash_K \varphi$ and the size of the derivation of φ from A and B is smaller than n, then $\Delta(A) + \Delta(B) \vdash_{K?} \Delta(\varphi)$. Then we just prove that with this supposition, either considering MP or N the result will hold.

The proof of lemma 22 follows the same idea, just that at this time we have to consider K?-derivations and Π and II.

The proof of lemma 24 is done by induction on the size of α. The basic case where $\alpha \equiv p$ is trivial. In the hypothesis of induction we take an arbitrary formula α and suppose that the result holds for all formulae φ of size $m < n$, where n is α's size. We then prove that, if this is the case, the result also holds for α. This is done by considering all possible forms α may have. The only situation which poses some difficulty is the case where $\alpha \equiv \neg\varphi$. For all others, the proof is trivial. In this difficult case, $\Delta\alpha \equiv \Delta(\neg\varphi) \equiv \neg\nabla\varphi$. We have then to prove that $\Psi_{M,w}(\neg\varphi) = 1$ iff $\Omega_{M,w}(\neg\nabla\varphi) = 1$, which is done by induction on the size of φ. For the hypothesis of induction (the basis is trivial), we suppose that the result holds for formulae of size smaller than φ's size. We then show that, if this supposition holds, independently of the form of φ, the general result that $\Psi_{M,w}(\neg\varphi) = 1$ iff $\Omega_{M,w}(\neg\nabla\varphi) = 1$ also holds. As usual, we consider all forms φ may have. The proof of lemma 25 is almost identical to lemma 24's. All we have to do is to properly erase the occurrences of Δ and consider function Π along with Ψ.

With these lemmas at hand, we can proceed to the proofs of theorems 17-20.

Theorem 17. Let $A, B \subseteq \Im_\diamond$ and $\alpha \in \Im_\diamond$. $A + B \vdash_K \alpha$ iff $\Delta(A) + \Delta(B) \vdash_{K?} \Delta(\alpha)$.

Proof. By lemma 23, if $A+B \vdash_K \alpha$ then $\Delta(A)+\Delta(B) \vdash_{K?} \Delta(\alpha)$. Suppose that $\Delta(A) + \Delta(B) \vdash_{K?} \Delta(\alpha)$ but $A + B \nvdash_K \alpha$. If $A + B \nvdash_K \alpha$, by lemma 21, we have that $\Pi(\Delta(A)) + \Pi(\Delta(B)) \nvdash_K \Pi(\Delta(\alpha))$. By lemma 22, we have then that $\Delta(A) + \Delta(B) \nvdash_{K?} \Delta(\alpha)$, which is a contradiction. Therefore, if $\Delta(A) + \Delta(B) \vdash_{K?} \Delta\alpha)$ then $A + B \vdash_K \alpha$. ∎

Theorem 18. Let $A, B \subseteq \Im_?$ and $\alpha \in \Im_?$. $A + B \vdash_{K?} \alpha$ iff $\Pi(A) + \Pi(B) \vdash_K \Pi(\alpha)$.

Proof. By lemma 22, if $A+B \vdash_{K?} \alpha$ then $\Pi(A)+\Pi(B) \vdash_K \Pi(\alpha)$. Suppose that $\Pi(A) + \Pi(B) \vdash_K \Pi(\alpha)$ but $A + B \nvdash_{K?} \alpha$. If $A + B \nvdash_{K?} \alpha$, by lemma 21, we have that $\Delta(\Pi(A)) + \Delta(\Pi(B)) \nvdash_{K?} \Delta(\Pi(\alpha))$. By lemma 23, we have then that $\Pi(A) + \Pi(B) \nvdash_K \Pi(\alpha)$, which is a contradiction. Therefore, if $\Pi(A) + \Pi(B) \vdash_K \Pi(\alpha)$ then $A + B \vdash_{K?} \alpha$. ∎

Theorem 19. Let $A, B \subseteq \Im_\diamond$ and $\alpha \in \Im_\diamond$. $A + B \models_K \alpha$ iff $\Delta(A) + \Delta(B) \models_{K?} \Delta(\alpha)$.

Proof. Suppose that $A + B \models_K \alpha$ but $\Delta(A) + \Delta(B) \nvDash_{K?} \Delta(\alpha)$. If $\Delta(A) + \Delta(B) \nvDash_{K?} \Delta(\alpha)$, then there is a model M and a world w of M such that $M \Vdash_\Omega \Delta(\phi)$, for all $\Delta(\phi) \in \Delta(A), M, w \Vdash_\Omega \Delta(\lambda)$, for all $\Delta(\lambda) \in \Delta(B)$, and $M, w \nVdash_\Omega \Delta(\alpha)$. But if $M \Vdash_\Omega \Delta(\phi)$ for all $\Delta(\phi) \in \Delta(A), M, w \Vdash_\Omega \Delta(\lambda)$ for all $\Delta(\lambda) \in \Delta(B)$ and $M, w \nVdash_\Omega \Delta(\alpha)$, by lemma 24 we have that $M \Vdash_\Psi? \phi$ for all $\phi \in A, M, w \Vdash_\Psi \lambda$ for all $\lambda \in B$ and $M, w \nVdash_\Psi \alpha$. Consequently, $A+B \nvDash_K \alpha$, which is a contradiction. Therefore, if $A + B \models_K \alpha, \Delta(A) + \Delta(B) \models_{K?} \Delta(\alpha)$. Suppose now that $\Delta(A) + \Delta(B) \models_{K?} \Delta(\alpha)$ but $A + B \nvDash_K \alpha$. If $A+B \nvDash_K \alpha$ then there is a model M and a world w of M such that $M \Vdash_\Psi \phi$ for all $\phi \in A, M, w \Vdash_\Psi \lambda$ for all $\lambda \in B$ and $M, w \nVdash_\Psi \alpha$. But if $M \Vdash_\Psi \phi$ for all $\phi \in A, M, w \Vdash_\Psi \lambda$ for all $\lambda \in B$ and $M, w \nVdash_\Psi \alpha$, then by lemma 24 $M \Vdash_\Omega \Delta(\phi)$ for all $\Delta(\phi) \in \Delta(A), M, w \Vdash_\Omega \Delta(\lambda)$ for all $\Delta(\lambda) \in \Delta(B)$ and $M, w \nVdash_\Omega \Delta(\alpha)$. Consequently, $\Delta(A) + \Delta(B) \nvDash_{K?} \Delta(\alpha)$, which is a contradiction. Therefore, if $\Delta(A) + \Delta(B) \models_{K?} \Delta(\alpha)$ then $A + B \models_K \alpha$. ∎

Theorem 20. Let $A, B \subseteq \Im_?$ and $\alpha \in \Im_?$. $A + B \models_{K?} \alpha$ iff $\Pi(A) + \Pi(B) \models_K \Pi(\alpha)$.

Proof. The proof of this theorem follows the same idea of theorem 19's. We have just to properly erase the occurrences of Δ and consider function Π along with Ψ as well as to use lemma 25 instead of lemma 24. ∎

Theorem 11. $K_?$ is sound and complete (in symbols: Let $A, B \subseteq \Im_?$ be two sets of formulae and $\alpha \in \Im_?$ a formula. $A + B \vdash_{K?} \alpha$ iff $A + B \models_{K?} \alpha$).

Proof. Let us first prove the left-right direction (soundness): for any $A, B \subseteq \mathfrak{S}_?$ and $\alpha \in \mathfrak{S}_?$, if $A + B \vdash_{K?} \alpha$ then $A + B \models_{K?} \alpha$. Suppose that $A + B \vdash_{K?} \alpha$ and $A + B \not\models_{K?} \alpha$. If $A + B \not\models_{K?} \alpha$, by theorem 20 we have that $\Pi(A) + \Pi(B) \not\models_K \Pi(\alpha)$. By the soundness theorem of normal modal logic K[14], we have $\Pi(A) + \Pi(B) \not\vdash_K \Pi(\alpha)$. From that, along with theorem 18, we have that $A + B \not\vdash_{K?} \alpha$, which is a contradiction. Therefore, if $A + B \vdash_{K?} \alpha$ then $A + B \models_{K?} \alpha$. The right-left direction (completeness) is proved through the same reference to normal modal logic K. Suppose that $A + B \models_{K?} \alpha$ and $A + B \not\vdash_{K?} \alpha$. If $A + B \not\vdash_{K?} \alpha$, by theorem 18 we have that $\Pi(A) + \Pi(B) \not\vdash_K \Pi(\alpha)$. By the completeness theorem of normal modal logic K, we have then that $\Pi(A) + \Pi(B) \not\models_K \Pi(\alpha)$. From that, along with theorem 20, we have that $A + B \not\models_{K?} \alpha$, which is a contradiction. Therefore, if $A + B \models_{K?} \alpha$ then $A + B \vdash_{K?} \alpha$. ∎

Ricardo Sousa Silvestre.

Laboratory of Artificial Intelligence, Universidade Federal do Ceará, Bloco 910, Campus do Pici, Fortaleza-Ceará, 60455-760, Brazil.

ricardo@lia.ufc.br

---

[14] For the proof of soundness and completeness of normal modal logic K with local and global premises see [10].

The variety of modal $\mathbf{FL}_{ew}$-algebras is generated by its finite simple members

Hiroki Takamura

ABSTRACT. In this paper, we prove that the variety of modal $\mathbf{FL}_{ew}$-algebras is generated by its finite simple members. The result is obtained by showing that every *free* modal $\mathbf{FL}_{ew}$-algebra is semisimple and then showing that every variety generated by a simple modal $\mathbf{FL}_{ew}$-algebra is generated by a set of finite simple modal $\mathbf{FL}_{ew}$-algebras.

1 Introduction

In [7], the authors show that the variety of $\mathbf{FL}_{ew}$-algebras is generated by its finite simple members. The result is obtained by first showing that every *free* $\mathbf{FL}_{ew}$-algebra is semisimple and then showing that every variety generated by a simple $\mathbf{FL}_{ew}$-algebra is generated by a set of finite simple $\mathbf{FL}_{ew}$-algebras. To show the former, based on Grišin's idea in [4] authors introduced a sequent system SFL_{ew}^+ such that

1. algebras for SFL_{ew}^+ are exactly equal to $\mathbf{FL}_{ew}$-algebras,
2. cut elimination theorem holds for SFL_{ew}^+.

Then, using proof-theoretic properties of SFL_{ew}^+, the semisimplicity of free $\mathbf{FL}_{ew}$-algebras is obtained. Moreover, they show that the finite embeddability property holds for simple $\mathbf{FL}_{ew}$-algebras. Finally, they have the variety of $\mathbf{FL}_{ew}$-algebras is generated by its finite simple members.

In this paper, their proof works well also for the variety of modal $\mathbf{FL}_{ew}$-algebras with some modification. We assume a familiarity with the paper [7].

2 Modal $\mathbf{FL}_{ew}$-algebras

We give a precise definition of modal $\mathbf{FL}_{ew}$-algebras introduced by Ono [10], which is $\mathbf{FL}_{ew}$-algebras with S4-like modality. Note that an $\mathbf{FL}_{ew}$-algebra is a kind of residuated lattice. It is known that residuated lattices are algebraic semantics for substructural logics. For more information, see [6, 5].

An algebraic structure $\mathbf{A} = \langle A, \cdot, \rightarrow, \wedge, \vee, \square, 0, 1 \rangle$ is called a modal $\mathbf{FL}_{ew}$-algebra, if it satisfies the following:

(1) $\langle A, \wedge, \vee, 0, 1 \rangle$ is a bounded lattice with the greatest element 1, the least element 0,

(2) $\langle A, \cdot, 1 \rangle$ is a commutative monoid, and

(3) $x \cdot y \leq z \Leftrightarrow y \leq x \to z$, for any $x, y, z \in A$,
(4) $\Box$ is a unary operation on A satisfying:

(a) $1 \leq \Box 1$, (b) $\Box x \cdot \Box y \leq \Box(x \cdot y)$, (c) $\Box x \leq x$, (d) $\Box x \leq \Box\Box x$,
(e) if $x \leq y$ then $\Box x \leq \Box y$ for any $x, y \in A$.

For simplicity, we write xy instead of $x \cdot y$.

DEFINITION 1 (Congruence filter). A subset F of A is a *congruence filter* (called simply, a filter) of $\mathbf{A}$ if it satisfies that
(1) $1 \in F$,
(2) If $x, x \to y \in F$ then $y \in F$, and
(3) If $x \in F$ then $\Box x \in F$.

It is easy to see that a congruence filter characterizes congruences in a modal $\mathbf{FL}_{ew}$-algebra $\mathbf{A}$. Suppose F is a congruence filter of $\mathbf{A}$. We define a binary relation θ on $\mathbf{A}$, putting $x\theta y$ if the terms $x \to y$ and $y \to x$ are both in F. Then θ is a congruence relation with respect to the operations of modal $\mathbf{FL}_{ew}$-algebra. Conversely, suppose θ is a congruence in a modal $\mathbf{FL}_{ew}$-algebra. Then $F_\theta = \{x \in \mathbf{A} : 1\theta x\}$ is a congruence filter of $\mathbf{A}$.

Next, suppose $x \in F$ and $x \leq y$, we have $1 \leq x \to y$. Thus $x \to y \in F$ by (1). By (2), we have $y \in F$. Thus, a filter F is a upward closed set, i.e., if $x \in F$ and $x \leq y$ then $y \in F$. We can also show that if $x, y \in F$ then $xy \in F$. Moreover it is easy to see that a nonempty subset F of a modal $\mathbf{FL}_{ew}$-algebra is a filter if and only if it satisfies: (a) if $x \in F$ and $x \leq y$ then $y \in F$, (b) if $x, y \in F$ then $xy \in F$, and (3).

Using this fact, we can have the following representation of the filter generated by a given nonempty subset S of a modal $\mathbf{FL}_{ew}$-algebra $\mathbf{A}$. Let H_S be $\{x \in \mathbf{A} : \Box s_1 \cdots \Box s_k \leq x \text{ for some } s_1, \cdots s_k \in S\}$. Then we have the following characterization of a filter generated a given set by a similar argument of [5, 12, 3].

LEMMA 2. *For each nonempty subset S of a modal $\mathbf{FL}_{ew}$-algebra $\mathbf{A}$, the filter generated by S is equal to H_S.*

When S is a singleton set $\{a\}$, the filter generated by $\{a\}$ is of the form $\{x \in \mathbf{A} : (\Box(a))^n \leq x \text{ for some } n \in \mathbb{N}\}$.

We say that a modal $\mathbf{FL}_{ew}$-algebra $\mathbf{A}$ is *semisimple* if it has a subdirect representation with simple factors. Let Φ be the set of all maximal filters of a modal $\mathbf{FL}_{ew}$-algebra $\mathbf{A}$. Define the radical $Rad_\mathbf{A}$ of $\mathbf{A}$ by $Rad_\mathbf{A} = \bigcap_{F \in \Phi} F$. Then, the following can be easily shown.

LEMMA 3. *For any modal $\mathbf{FL}_{ew}$-algebras, $\mathbf{A}$ is semisimple if and only if $Rad_\mathbf{A} = \{1\}$.*

Corresponding to Theorem 2.3 of [7], we can show the following result, which gives a sufficient and necessary condition for $x \in \mathbf{A}$ to be a member of $Rad_\mathbf{A}$.

PROPOSITION 4. *Let $\mathbf{A}$ be a modal $\mathbf{FL}_{ew}$-algebra. An element $x \in \mathbf{A}$ is in $Rad_{\mathbf{A}}$ if and only if for any $n \geq 1$ there exists $m \in \mathbb{N}$ such that $(\Box\neg((\Box x)^n))^m = 0$, where $\neg x$ stands for $x \to 0$.*

Proof. Assume that for any $n \geq 1$ there exists $m \in \mathbb{N}$ such that $(\Box\neg((\Box x)^n))^m = 0$. Suppose x is not in $Rad_{\mathbf{A}}$. Then there is a maximal filter F with $x \notin F$. Since F is maximal, there is a $k \geq 1$ such that $\neg((\Box x)^k) \in F$. Thus $\Box\neg((\Box x)^k) \in F$. By assumption, there is $m \geq 1$ for which $(\Box\neg((\Box x)^k))^m = 0$. This implies $0 \in F$, which contradicts the fact that F is proper.

Conversely, suppose that there exists $n \geq 1$ such that $(\Box\neg((\Box x)^n))^m \neq 0$ for any m. Then $(\Box\neg((\Box x)^n))^m > 0$ for any m. Put $z = \neg((\Box x)^n)$ and H_z a filter generated by $\{z\}$. Clearly, H_z is proper. By Zorn's lemma, there is a maximal filter G such that $H_z \subseteq G$. Now suppose $x \in G$, then $(\Box x)^n \in G$. However this is a contradiction, since $z \in G$. Therefore $x \notin G$ and so $x \notin Rad_{\mathbf{A}}$.

COROLLARY 5. *A modal $\mathbf{FL}_{ew}$-algebra $\mathbf{A}$ is semisimple if and only if for every $x \in \mathbf{A}\setminus\{1\}$ there is an $n \geq 1$ such that for any $m \geq 1$, we have $(\Box\neg((\Box x)^n))^m \neq 0$.*

3 Semisimplicity of free modal $\mathbf{FL}_{ew}$-algebras

In this section we show that every free modal $\mathbf{FL}_{ew}$-algebra is semisimple, using the sequent system $\Box FL_{ew}^+$ introduced below. Our proof proceeds similarly to Grišin [4], Kowalski and Ono [7].

Similarly to the sequent system SFL_{ew} introduced in [7], we introduce a sequent system, which we call $\Box FL_{ew}^+$ as follows. A sequent is of the form $\Gamma \Rightarrow \alpha$ where Γ is a possibly empty multiset of formulas:

1. initial sequents

   (1) $p, \Gamma \Rightarrow p$ where p is a propositional variable,
   (2) $0, \Gamma \Rightarrow \alpha$.

2. rules of inference

$$\frac{\Gamma \Rightarrow \alpha \quad \alpha, \Sigma \Rightarrow \theta}{\Gamma, \Sigma \Rightarrow \theta} \text{ (cut)}$$

$$\frac{\alpha, \Gamma \Rightarrow \theta \quad \beta, \Gamma \Rightarrow \theta}{\alpha \vee \beta, \Gamma \Rightarrow \theta} \text{ } (\vee \Rightarrow) \qquad \frac{\Gamma \Rightarrow \alpha}{\Gamma \Rightarrow \alpha \vee \beta} \text{ } (\Rightarrow \vee 1) \qquad \frac{\Gamma \Rightarrow \beta}{\Gamma \Rightarrow \alpha \vee \beta} \text{ } (\Rightarrow \vee 2)$$

$$\frac{\Gamma \Rightarrow \alpha \quad \Gamma \Rightarrow \beta}{\Gamma \Rightarrow \alpha \wedge \beta} \text{ } (\Rightarrow \wedge) \qquad \frac{\alpha, \Gamma \Rightarrow \theta}{\alpha \wedge \beta, \Gamma \Rightarrow \theta} \text{ } (\wedge 1 \Rightarrow) \qquad \frac{\beta, \Gamma \Rightarrow \theta}{\alpha \wedge \beta, \Gamma \Rightarrow \theta} \text{ } (\wedge 2 \Rightarrow)$$

$$\frac{\Gamma \Rightarrow \alpha \quad \beta, \Sigma \Rightarrow \theta}{\Gamma, \alpha \to \beta, \Sigma \Rightarrow \theta} \ (\to \Rightarrow) \qquad \frac{\alpha, \Gamma \Rightarrow \beta}{\Gamma \Rightarrow \alpha \to \beta} \ (\Rightarrow \to)$$

$$\frac{\Box \Gamma \Rightarrow \alpha}{\Box \Gamma \Rightarrow \Box \alpha} \ (\Rightarrow \Box) \qquad \frac{\alpha, \Gamma \Rightarrow \beta}{\Box \alpha, \Gamma \Rightarrow \beta} \ (\Box \Rightarrow)$$

$$\frac{\Gamma_1 \Rightarrow \alpha_1, \cdots, \Gamma_m \Rightarrow \alpha_m}{\Gamma_1, \cdots, \Gamma_m \Rightarrow \alpha_1 * \cdots * \alpha_m} \ (\Rightarrow *) \qquad \frac{\Gamma, \alpha_1, \cdots, \alpha_m, \Delta \Rightarrow \theta}{\Gamma, \alpha_1 * \cdots * \alpha_m, \Delta \Rightarrow \theta} \ (* \Rightarrow)$$

Here, we assume that in each application of rules $(\Rightarrow *)$ and $(* \Rightarrow)$, none of α_i must be fusion formulas, i.e., formulas whose outermost logical connective is the fusion $*$. The reason why we need the system $\Box FL_{ew}^+$ will be explained by the following Lemma, which can be shown in the usual way. In [7], they proved the cut elimination holds for SFL_{ew}^+ (the proof from [9] works with obvious modifications). Their system SFL_{ew}^+ is equal to the sequent system deleting $\Box$ rule from our $\Box FL_{ew}^+$. Thus, the important new case for $\Box FL_{ew}^+$ is as follows.

$$\frac{\dfrac{\Box \Gamma \Rightarrow A}{\Box \Gamma \Rightarrow \Box A} \quad \dfrac{\Delta, A \Rightarrow B}{\Delta, \Box A \Rightarrow B}}{\Box \Gamma, \Delta \Rightarrow B} \ (cut)$$

This cut can be replaced by the following, which has the same end-sequent, but the size of the proof above the cut-formula becomes smaller by one;

$$\frac{\Box \Gamma \Rightarrow A \quad \Delta, A \Rightarrow B}{\Box \Gamma, \Delta \Rightarrow B} \ (cut)$$

LEMMA 6.
1. Cut elimination holds for $\Box FL_{ew}^+$.
2. Free modal $\mathbf{FL}_{ew}$-algebras are precisely Lindenbaum-Tarski algebras of $\Box FL_{ew}^+$.

Here, we briefly describe the Lindenbaum-Tarski algebra for the sequent system $\Box FL_{ew}^+$. We say that the formula algebra of $\Box FL_{ew}^+$ is the algebra whose elements are precisely the formulas of $\Box FL_{ew}^+$. Now we define the binary relation $\sim$ on the formula algebra, putting $\alpha \sim \beta$ if the sequents $\alpha \Rightarrow \beta$ and $\beta \Rightarrow \alpha$ are both provable in $\Box FL_{ew}^+$. It is easy to see that $\sim$ is a congruence relation. Then we have the quotient algebra $\Box FL_{ew}^+/\sim$ which is known as the Lindenbaum-Tarski algebra for $\Box FL_{ew}^+$. For more precise information, see [8].

Let a formula α be given. In the following, $\neg \alpha$ denotes $\alpha \to 0$. For each formula α, let $\ell(\alpha)$ denote the length of α as a sequence of symbols.

For a sequence Γ of formulas $\alpha_1, \cdots, \alpha_m$, the length $\ell(\Gamma)$ is defined by $\ell(\Gamma) = \ell(\alpha_1) + \cdots + \ell(\alpha_m)$. Also we need to introduce some notations for our main lemma. The expression $\{\alpha^N\}^m$ stands for the sequence $\alpha^N, \cdots, \alpha^N$ with m times, where α^N is of the form $\alpha * \cdots * \alpha$ (N times).

To show that any free modal **FL**$_{ew}$-algebra **A** is semisimple, Lemma 3 says that it is enough to show that the radical $Rad_\mathbf{A}$ of any Lindenbaum-Tarski algebra of $\Box FL_{ew}^+$ is equal to $\{1\}$. By Proposition 4, this follows from the following lemma.

LEMMA 7. *Suppose that a formula α is not provable in $\Box FL_{ew}^+$ and that $n > \ell(\alpha)$. For any sequent $\Gamma \Rightarrow \sigma$ such that $\ell(\Gamma, \sigma) \leq \ell(\alpha)$, and any m, if $\{\Box\neg((\Box\alpha)^n)\}^m, \Gamma \Rightarrow \sigma$ is provable in $\Box FL_{ew}^+$ then $\Gamma \Rightarrow \sigma$ is provable in $\Box FL_{ew}^+$.*

Proof. The proof will be given by double induction on $(m, \ell(\Gamma, \sigma))$, where m is the number of formulas come from the set $\{\Box\neg((\Box\alpha)^n)\}^m$ which contains $\neg((\Box\alpha)^n)$ as a subformula, not the number of formulas of the form $\Box\neg((\Box\alpha)^n)$. Thus, we assume that the lemma holds for $m' < m$. More precisely we assume that the lemma holds for $\{\Box\neg((\Box\alpha)^n)\}^{m_1}, \{\neg((\Box\alpha)^n)\}^{n_1}, \Gamma \Rightarrow \sigma$, where $m_1 + n_1 = m'$ and it also holds for $(m, \ell(\Delta, \delta))$, whenever $\ell(\Gamma, \sigma)) < \ell(\Delta, \delta)$.

Suppose α is not provable but $\{\Box\neg((\Box\alpha)^n)\}^m, \Gamma \Rightarrow \sigma$ is provable in $\Box FL_{ew}^+$.

(1) Suppose that the given sequent $\{\Box\neg((\Box\alpha)^n)\}^m, \Gamma \Rightarrow \sigma$ is an initial sequent. Then, either σ is a propositional variable which occurs also in Γ, or 0 occurs in Γ. It is obvious that $\Gamma \Rightarrow \sigma$ is provable in either case.

(2) Next suppose that the sequent $\{\Box\neg((\Box\alpha)^n)\}^m, \Gamma \Rightarrow \sigma$ is the lower sequent of an inference rule I.

(i) First assume that the principal formula of I is either in Γ or in σ. Then (each of) the upper sequent(s) of I is of the form $\{\Box\neg((\Box\alpha)^n)\}^{m_i}, \Delta_i \Rightarrow \delta_i$ with $m_1 < m$ and $\ell(\Delta_i, \delta_i) < \ell(\Gamma, \sigma)$. Therefore, by hypothesis of induction $\Delta_i \Rightarrow \delta_i$ is provable. Then $\Gamma \Rightarrow \sigma$ is also provable by applying the same inference rule.

(ii) Next suppose that the principal formula of I is one of $\Box\neg((\Box\alpha)^n)$. Then the inference rule I is of the form that;

$$\frac{\{\Box\neg((\Box\alpha)^n)\}^{m-1}, \neg((\Box\alpha)^n), \Gamma \Rightarrow \sigma}{\{\Box\neg((\Box\alpha)^n)\}^m, \Gamma \Rightarrow \sigma} \ (\Box \Rightarrow)$$

In this inference I, the degree of m and $\ell(\Gamma, \sigma)$ of the upper sequent is the same as the lower sequent. Consider the next inference rule J to the upper sequent

$$\{\Box\neg((\Box\alpha)^n)\}^{m-1}, \neg((\Box\alpha)^n), \Gamma \Rightarrow \sigma.$$

There are three possibilities.

(a) First case is that the principal formula of J is one of $\Box\neg((\Box\alpha)^n)$. This is not the essential case of the induction, since we at last reach a case which is essentially equivalent to the either (b) or (c) mentioned below by tracing back the proof. In this case the degree of m and $\ell(\Gamma, \sigma)$ of the upper sequent is the same as the lower sequent. Thus, we have to continue to trace back the proof. Consider the next inference rule there are also three possibilities. First case is the same case as (a) except the numbers of $\Box\neg((\Box\alpha)^n)$. Note that the degree of m and $\ell(\Gamma, \sigma)$ of the induction is never changed during tracing back, while the inference rules in the proof are in the case of (a). Moreover the number of $\Box\neg((\Box\alpha)^n)$ in the given sequent is m, so we cannot continue this case at most m-times applying the given provable sequent. Therefore we sometimes reach the essentially the same as the other two cases are mentioned below.

Note that applying the inference rules in the case of (a) after i-th times, the given sequent will be of the form

$$\{\Box\neg((\Box\alpha)^n)\}^{m_1}, \{\neg((\Box\alpha)^n)\}^{n_1}, \Gamma \Rightarrow \sigma$$

where $m_1 + n_1 = m$. If the next principal formula of the inference rule is one of $\neg((\Box\alpha)^n)$ then we will replace the sequent

$$\{\Box\neg((\Box\alpha)^n)\}^{m-1}, \neg((\Box\alpha)^n), \Gamma \Rightarrow \sigma$$

to

$$\{\Box\neg((\Box\alpha)^n)\}^{m_1}, \{\neg((\Box\alpha)^n)\}^{n_1}, \Gamma \Rightarrow \sigma$$

in the case (c) below and applying the same argument. The same modification is also needed in the case of (b) below after applying the case (a) i-th times.

(b) Second case is that the principal formula of J is either in Γ or in σ. In this case, we apply the same argument in (2)-(i) then we have $\Gamma \Rightarrow \sigma$ as desired.

(c) Third case is that the principal formula of J is $\neg((\Box\alpha)^n)$. In this case, the inference rule is of the form that;

$$\frac{\{\Box\neg((\Box\alpha)^n)\}^{m_2}, \Pi_2 \Rightarrow (\Box\alpha)^n \quad 0, \{\Box\neg((\Box\alpha)^n)\}^{m_1}, \Pi_1 \Rightarrow \sigma}{\{\Box\neg((\Box\alpha)^n)\}^{m-1}, \neg((\Box\alpha)^n), \Gamma \Rightarrow \sigma} \; (\neg \Rightarrow)$$

where Π_1, Π_2 are equal to Γ.

Taking the left upper sequent $\{\Box\neg((\Box\alpha)^n)\}^{m_1}, \Pi_2 \Rightarrow (\Box\alpha)^n$ and the provable sequent $(\Box\alpha)^n \Rightarrow \alpha^n$, and applying the cut rule. Then we have $\{\Box\neg((\Box\alpha)^n)\}^{m_1}, \Pi_2 \Rightarrow \alpha^n$. Consider the proof R of this sequent. We will trace back branches of R, which consist of sequents having α^n in the conclusion, to the places where this α^n is introduced. Note that α^n is introduced at one place in each branch of R. It is easy to see that each α^n is introduced either as an initial sequent, or by $(\Rightarrow *)$ rule. We will show that any α^n is introduced only as an initial sequent. Suppose that at least one place, α^n is introduced by $(\Rightarrow *)$, whose lower sequent is of the form $\Delta \Rightarrow \alpha^n$. We assume here that α is of the from $D_1 * \cdots * D_w$ and none of D_j are fusion-formulas. Then, I must have $n \cdot w$ upper sequents, each of which is of the form $\Delta_i \Rightarrow D_{n_i}$, where $1 \leq n_i \leq w$ and the list $\Delta_1, \cdots, \Delta_{n \cdot w}$ is equal to Δ. For each j such that $1 \leq j \leq w$, there exist exactly n sequents with the conclusion D_j among those sequents. We enumerate them as $S_1^j, \cdots, S_n^j$. Next, for each h such that $1 \leq h \leq n$, take $S_h^1, \cdots, S_h^w$ for upper sequent and apply $(\Rightarrow *)$ rule to them. Then we can have a sequent of the form $\Sigma_h \Rightarrow \alpha$ for $1 \leq h \leq n$ and the list of $\Sigma_1, \cdots, \Sigma_n$ is equal to Δ. Now $\ell(\Delta) \leq \ell(\Pi_2) \leq \ell(\Gamma, \sigma) \leq \ell(\alpha) < n$. If we assume that $\ell(\Sigma_i) > 0$ for any i such that $1 \leq i \leq n$ then $\ell(\Delta) \geq n$, which is a contradiction. Therefore, Σ_i must be empty for some i. But this means that $\Rightarrow \Box \alpha$ is provable. This contradicts the assumption that α is unprovable. Hence, we conclude that at any place α^n is introduced as an initial sequent of the form $0, \Lambda \Rightarrow \alpha^n$.

We will modify the proof R of $\{\Box\neg((\Box\alpha)^n)\}^{m_2}, \Pi_2 \Rightarrow \alpha^n$ as follows. We replace every sequent $\Theta \Rightarrow \alpha^n$ in a branch which we have traced in R, including an initial sequent of the form $0, \Lambda \Rightarrow \alpha^n$ mentioned above, by the sequent $\Pi_1, \Theta \Rightarrow \sigma$. Then we will have the proof whose end sequent is $\{\Box\neg((\Box\alpha)^n)\}^{m_2}, \Gamma, \Rightarrow \sigma$. Note that $m_2 \leq m-1 < m$. Hence, by hypothesis of induction, we conclude that $\Gamma \Rightarrow \sigma$ is provable. This completes the proof.

Clearly, the sequent system $\Box FL_{ew}^+$ is consistent, i.e., the sequent $\Rightarrow 0$ is not provable. Let m be a positive integer. If the sequent $\{\Box\neg(\Box\alpha)^n\}^m \Rightarrow 0$ is provable in $\Box FL_{ew}^+$ then it follows immediately from Lemma 7 by taking the sequent $\Rightarrow 0$ for $\Gamma \Rightarrow \sigma$. Thus, we have $\Rightarrow 0$ is provable in $\Box FL_{ew}^+$, which is a contradiction. Therefore, we have the following.

PROPOSITION 8. *Let α be any formula which is not provable in $\Box FL_{ew}^+$. Then there exists a natural number $n \geq 1$ such that $\{\Box\neg(\Box x)^n\}^m \Rightarrow 0$ is not provable in $\Box FL_{ew}^+$ for any m.*

In terms of the Lindenbaum-Tarski algebra **A** of $\Box FL_{ew}^+$, the above proposition says that if $[\alpha] \neq [1]$ in **A** then there exists $n \geq 1$, $[0] \neq [\Box\neg((\Box a)^n)^m]$ for any m, where $[\gamma]$ denotes the equivalence class, to which a given formula γ belongs. Thus, using Lemma 3, Proposition 4 and Corollary 5, we have the following.

THEOREM 9 (Semisimplicity of free modal **FL**$_{ew}$-algebras). *Every free modal **FL**$_{ew}$-algebra is semisimple.*

COROLLARY 10. *The variety of modal **FL**$_{ew}$-algebras is generated by its simple members.*

4 Finite embeddability property of simple modal FL$_{ew}$- algebras

In this section we show that the class of simple modal **FL**$_{ew}$- algebras has the finite embeddability property (FEP).

Given an algebra $\mathbf{A} = \langle A, \langle f_i^{\mathbf{A}} : i \in I \rangle \rangle$ of finite type and any nonempty subset $B \subseteq A$, the partial subalgebra $\mathbf{B}$ of $\mathbf{A}$ with domain B is the partial algebra $\langle B, \langle f_i^{\mathbf{B}} : i \in I \rangle \rangle$, where for $i \in I$, f_i is an n-ary function symbol, and $b_1, \cdots, b_n \in B$

$$f_i^{\mathbf{B}}(b_1, \cdots, b_n) = \begin{cases} f_i^{\mathbf{A}}(b_1, \cdots, b_n) & f_i^{\mathbf{A}}(b_1, \cdots, b_n) \in B \\ undefined & f_i^{\mathbf{A}}(b_1, \cdots, b_n) \notin B \end{cases}$$

A class $\mathcal{K}$ of algebras has the finite embeddability property if every finite partial subalgebra of a member of $\mathcal{K}$ can be embedded into a finite member of $\mathcal{K}$.

Next, we introduce a construction method to prove the FEP of the variety of $\mathbf{FL}_{ew}$-algebra which is originally proposed by Blok and van Alten [2]. Later, Amano proposed a natural extension for the modal operator of this construction to prove the FEP of the variety of modal $\mathbf{FL}_{ew}$-algebra [1].

Let $\mathbf{A}$ be a modal $\mathbf{FL}_{ew}$-algebra and $\mathbf{B}$ be a finite partial subalgebra of $\mathbf{A}$. We shall sometimes omit the multiplicative symbol $\cdot$. Let $\mathbf{M}$ be the submonoid of $\mathbf{A}$ generated by $\mathbf{B}$. For $X, Y \subseteq M$ and $a \in M$, put $XY = \{ab : a \in X, b \in Y\}$ and $Xa = X\{a\}$.

For each $x \in M$ and $b \in B$, define

$$(x \rightsquigarrow b] = \{c \in M : xc \leq b\} = \{c \in M : c \leq x \rightarrow b\}.$$

The set $(x \rightsquigarrow b]$ is a downward closed subset of M. Note that $(1 \rightsquigarrow b] = \{a \in M : a \leq b\}$. Write the sets $\overline{D} = \{(x \rightsquigarrow b] : x \in M, b \in B\}$ and $D = \{\bigcap \Xi : \Xi \subseteq \overline{D}\}$. Let C be the closure operator on the set of subsets of M associated with D, i.e., for $X \subseteq M$, let

$$\mathsf{C}(X) = \bigcap \{(x \rightsquigarrow b] \in \overline{D} : X \subseteq (x \rightsquigarrow b]\}.$$

Define for all $X, Y \subseteq M$ and $X_i \subseteq M$, $i \in I$,

$$X \cdot^D Y = \mathsf{C}(XY)$$
$$X \rightarrow^D Y = \{a \in M : Xa \subseteq Y\}$$
$$\bigwedge_{i \in I}^D X_i = \mathsf{C}(\bigcap_{i \in I} X_i)$$
$$\bigvee_{i \in I}^D X_i = \mathsf{C}(\bigcup_{i \in I} X_i)$$
$$1^D = M$$
$$0^D = \bigcap \overline{D}$$
$$\square^D X = \mathsf{C}(\{a \in X : \square a = a\}).$$

The definition of modality follows van Alten [13]. Then, we have the following three results which are proved by Amano [1].

THEOREM 11. *The structure* $\mathbf{D}(A, B) = \langle D, \cdot^D, \to^D, \wedge^D, \vee^D, 1^D, 0^D, \Box^D \rangle$ *is a modal* $\mathbf{FL}_{ew}$*-algebra* .

THEOREM 12. *If* $\mathbf{A}$ *is a modal* $\mathbf{FL}_{ew}$*-algebra and* $\mathbf{B}$ *is a partial subalgebra of* $\mathbf{A}$, *then* $\mathbf{B}$ *can be embedded, as a partial subalgebra, into the modal* $\mathbf{FL}_{ew}$*-algebra* $\mathbf{D}(A, B)$.

Proof. We define an embedding from $\mathbf{B}$ to $\mathbf{D}(A, B)$ by $b \mapsto (1 \leadsto b]$. It suffices that all operations $\to, \wedge, \vee, \cdot, \Box$ and constants $1, 0$ in $\mathbf{B}$ are preserved $\mathbf{D}(A, B)$ by this embedding. It is routine and we omit the detail.

Now we have the algebra $\mathbf{D}(A, B)$ into which we can embed $\mathbf{B}$. The proof of the finiteness of $\mathbf{D}(A, B)$ is exactly the same as in [2].

THEOREM 13. *The variety of modal* $\mathbf{FL}_{ew}$*-algebras has the FEP.*

Next, we will show that the class of simple modal $\mathbf{FL}_{ew}$-algebras has the FEP. Recall, we say that an algebra is *simple* if it has only two congruences. In the case of modal $\mathbf{FL}_{ew}$-algebras, simple algebras are characterized by the following lemma.

LEMMA 14. *A modal* $\mathbf{FL}_{ew}$*-algebra* $\mathbf{A}$ *is simple if and only if for any* $x (\neq 1)$ *in A there exists a positive integer m such that* $(\Box x)^m = 0$.

Using this characterization, we can prove that the FEP for simple modal $\mathbf{FL}_{ew}$-algebras.

THEOREM 15 (The FEP of simple modal $\mathbf{FL}_{ew}$-algebras). *The class of simple modal* $\mathbf{FL}_{ew}$*-algebras has the FEP.*

Proof. Let $\mathbf{A}$ be a simple modal $\mathbf{FL}_{ew}$-algebra and $\mathbf{B}$ a finite partial subalgebra of $\mathbf{A}$. Construct the structure $\mathbf{D}(A, B)$ then $\mathbf{D}(A, B)$ is a finite modal $\mathbf{FL}_{ew}$-algebra by Theorem 11. Thus, we need to show that $\mathbf{D}(A, B)$ is also a simple modal $\mathbf{FL}_{ew}$-algebra, i.e., for any $X (\neq 1^D) \in \mathbf{D}(A, B)$ there exists n with $(\Box X)^n = 0^D$. By Lemma 14, for each $x \in A$, there exists $n \in \mathbb{N}$ such that $(\Box x)^n = 0$. Let $\mathbf{M}$ be the submonoid of $\mathbf{A}$ generated by $\mathbf{B}$. Recall that an element of $\mathbf{D}(A, B)$ is a downward closed subset of $\mathbf{M}$.

Let $F(k)$ be the free commutative monoid on k generators $\{x_1, \cdots, x_k\}$. An element of $F(k)$ can be considered as the products $x_1^{n_1} \cdots x_k^{n_k}$ where $n_i < \omega$, $i = 1, \cdots, k$ and defined $x_i^0 = 1, i = 1, \cdots, k$. We can define an order $\leq$ on $F(k)$ by $x_1^{n_1} \cdots x_k^{n_k} \leq x_1^{m_1} \cdots x_k^{m_k}$ if and only if $n_i \geq m_i$ for each $i \in \{1, \cdots, k\}$. It is easy to see that the order is a well-quasi order. Note that for any $A \subseteq F(k)$, the set $\mathrm{Max}(A)$ of maximal elements of A is an antichain. Since the order on $F(k)$ is a well-quasi order, $\mathrm{Max}(A)$ is a finite set.

Now we consider a monoid homomorphism $F(k)$ to $\mathbf{M}$ defined by each generator x_i of $F(k)$ to an element of $\mathbf{B}$. Note that $\mathbf{M}$ is the submonoid of **A** generated by **B**. Then for any $X \subseteq \mathbf{M}$, the set Max(X) of maximal elements of X is also finite.

Put Y is the set $\{a \in X : \Box a = a\}$. Then $(\Box X)^n = \mathsf{C}(\underbrace{\mathsf{C}(Y) \cdots \mathsf{C}(Y)}_{n}) \subseteq \mathsf{C}(Y^n)$. Now we write $k = |\text{Max}(Y)|$. Note that for all $x_i \in \text{Max}(Y)$ there exists $n_i \in \mathbb{N}$ such that $(\Box x_i)^{n_i} = 0$. Since Max(Y) is finite there exists $max\{n_i\}$. Therefore $(\Box X)^{k \cdot max\{n_i\}} \subseteq \mathsf{C}(Y^{k \cdot max\{n_i\}}) = \mathsf{C}(\{0\}) = 0^D$ for any X. We conclude that $\mathbf{D}(A, B)$ is also simple.

5 Conclusion and remarks

THEOREM 16 (Main theorem). *The variety of modal $\mathbf{FL}_{ew}$-algebras is generated by its finite simple members.*

Proof. The proof goes essentially the same as in [7]. It suffices to show that for any non theorem α of $\Box \mathbf{FL}_{ew}$ there is a finite simple modal $\mathbf{FL}_{ew}$-algebra **A** falsifying α. It immediately follows from Theorem 9, Corollary 10 and Theorem 15.

To finish off this paper, we will list some open questions about the topic.

(Q1) Is the variety of $\mathbf{FL}_{ew}$-algebras with K-like (or other types of) modality generated by its finite simple members?

In [11], the author shows that every free $\mathbf{FL}_w$-algebra is semisimple by using Grišin's idea and Kowalski and Ono's technique.

(Q2) Is a free modal $\mathbf{FL}_w$-algebra semisimple?
(Q3) Is the variety of modal $\mathbf{FL}_w$-algebras generated by its finite simple members?

There are many other interesting problems in residuated lattices with modalities. We expect that many researchers are getting interested in this area.

Acknowledgements

The author would like to express his sincere gratitude to Hiroakira Ono and Tomasz Kowalski. The author also would like to express his thanks to members of Research Center for Verification and Semantics (CVS), and especially to Makoto Takeyama for discussions and valuable comments. Finally, the author wish to express his thanks to the anonymous referees for helpful suggestions.

BIBLIOGRAPHY

[1] S. Amano. The finite embeddability property of the variety of modal residuated lattices. In *Proceedings of the 39th MLG meeting at Gamagori, JAPAN*, pages 39–42, 2005.

[2] W. J. Blok and C. J. van Alten. The finite embeddability property for residuated lattices, pocrims and BCK-algebras. *Algebra Unversalis*, 48:253–271, 2002.

[3] N. Galatos and H. Ono. Algebraization, parametrized local deduction theorem and interpolation for substructural logics over FL. *Studia Logica*, 83:to appear, 2006.
[4] V. N. Grišin. Predicate and set-theoretic calculi based on logic without contraction rule (in Russian). *Izvéstia Akademii Nauk SSSR (also the English translation in Math. USSR Izvestija)*, 45 (18):47–68 (41–59), 1981 (1982).
[5] P. Jipsen and C. Tsinakis. A survey of residuated lattices. In J. Martinez, editor, *Ordered Algebraic Structures*, pages 19–56. Kluwer Academic Publishers, Dordrecht, 2002.
[6] T. Kowalski and H. Ono. Residuated lattices: An algebraic glimpse at logic without contraction. preliminary report, 2000.
[7] T. Kowalski and H. Ono. The variety of residuated lattices is generated by its finite simple members. *Reports on Mathematical logic*, 34:59–77, 2000.
[8] T. Kowalski and H. Ono. Algebraic perspective on substructural logics. Course Note, Logic Summer School 2004 at ANU, 2004.
[9] H. Ono. Proof-theoretic methods for nonclassical logic — an introduction. In M. Okada M. Takahashi and M. Dezani-Ciancaglini, editors, *Theories of Types and Proofs (MSJ Memoirs 2)*, pages 207–254. Mathematical Society of Japan, 1998.
[10] H. Ono. Modalities in substructural logics, a preliminary report. In *Proceedings of the 39th MLG meeting at Gamagori, JAPAN*, pages 36–39, 2005.
[11] H. Takamura. Every free biresiduated lattice is semisimple. *Reports on Mathematical logic*, 37:125–133, 2003.
[12] C. J. van Alten. Representable biresiduated lattices. *Journal of Algebras*, 247:672–691, 2002.
[13] C. J. van Alten. The finite model property for knotted extensions of propositional linear logic. *Journal of Symbolic Logic*, 70(1):84–98, 2005.

Hiroki Takamura
Research Center for Verification and Semantics (CVS),
National Institute of Advanced Industrial Science and Technology (AIST)
5th Floor, Mitsui Sumitomo Kaijo Senri Bldg.,
1-2-14 Shin-Senri Nishi, Toyonaka, Osaka,
560-0083, Japan
E-mail: takamura-h@aist.go.jp

On Modal Logic, IF Logic, and IF Modal Logic

TERO TULENHEIMO AND MERLIJN SEVENSTER

ABSTRACT. 'Structurally determined IF modal logic' ($\mathcal{L}_{\mathbf{SD}}$) is introduced, and a study of its computational properties is undertaken. We argue that the results we prove have a general interest: The properties of our modal-like logic go against various tenets typically held with regard to modal logic. Notably, $\mathcal{L}_{\mathbf{SD}}$ has **PSPACE**-complete satisfiability problem but does not admit of a truth-preserving translation into the guarded fragment of first-order logic. Beyond merely laying down a number of technical results, we also wish to address the general issue of what are the core properties of modal logics.

1 Introduction

In the literature several logics have been presented under the heading *independence friendly modal logic* [4, 5, 13, 17, 18]. None of them, however, makes any claim of being *the* independence friendly modal logic. We will not do so either. Instead, we introduce a framework where one can play around, developing one's modal logic. As an elaborate illustration, we take up the development of one particular independence friendly modal logic within that framework, and study some of its metalogical properties.

The idea underlying our framework is the following: We have quite good understanding of modal logics ('modal languages' in the sense of [3, Def. 1.12]), first-order logic (**FO**), and independence friendly first-order logic (**IF**). Also their relationships are well understood: modal logics can be translated into **FO** via a *standard translation,* and one obtains the language of **IF** by applying some kind of *IF-procedure* to the formulas of **FO**. Now modal logics are often seen as fragments of **FO**. Similarly, independence friendly modal logic can be viewed as a language (a fragment of **IF**) that results from applying an IF-procedure to a standard translation of some modal logic. Crucially, the latter statement is underspecified regarding the IF-procedure, the standard translation and the specific modal logic $\mathbf{ML}(\tau, \Phi)$, with τ a modal similarity type and Φ a set of propositional atoms. All variations that one can think of, span the framework that we mean to introduce. Graphically, for any modal logic **ML** and any corresponding standard translation ST, an independence friendly modal logic **IFML** that fits within our framework is subject to the following relationships:

ML	is equally expressive as	$ST(\mathbf{ML})$	subset of	**FO**
		$\Downarrow_{\text{IF}}$		$\Downarrow_{\text{IF}}$
IFML	is equally expressive as	$\mathbf{IF}(ST(\mathbf{ML}))$	subset of	**IF**

where $\Downarrow_{\text{IF}}$ is some kind of IF-procedure. Let us go through a small example, to get intuitions straight. Fix a formula $\phi := \Box(\Diamond p \vee q)$ from modal logic. The truth condition of ϕ at a state w is specified in **FO** by the following formula (obtained by the standard translation explained in any standard textbook on modal logic): $\forall x(Rx_0x \to (\exists y(Rxy \wedge Py) \vee Qx))$, assigning the state w to x_0. Applying Hintikka's IF-procedure [11] we get, among others, the formula $\forall x(Rx_0x \to ((\exists y/x)(Rxy \wedge Py) \vee Qx))$. This formula will, then, belong to $\mathbf{IF}(ST(\mathbf{ML}))$. Reading such formulas can be tedious; in fact, it is not clear at first sight whether a given string is a formula of $\mathbf{IF}(ST(\mathbf{ML}))$. For this reason, it will be convenient to introduce an independence friendly modal logic, **IFML**, that is equally expressive as $\mathbf{IF}(ST(\mathbf{ML}))$, or, to put it otherwise, whose 'standard translation' into IF first-order logic yields $\mathbf{IF}(ST(\mathbf{ML}))$. As will be seen, in the specific IF modal logic we will introduce, the truth condition of the sample formula is captured by the formula $\blacksquare(\blacklozenge p \vee q)$.

Plan of the paper. *Section* 2 introduces the basic notions, and *Section* 3 sets the scene for the specific IF modal logical framework introduced in *Section* 4. In *Section* 5 the expressive power of the resulting *structurally determined* IF modal logic ($\mathcal{L}_{\mathbf{SD}}$) is discussed, and it is shown *not* to lie within the guarded fragment of **FO**. In *Sections* 6, 7 and 8 the model checking, validity and satisfiability problems of $\mathcal{L}_{\mathbf{SD}}$ are each shown to be decidable. The first turns out to be **NP**-complete, and the latter two **PSPACE**-complete. *Section* 9 closes the paper by discussing the conceptual interest of the results reached.

Note on notation. If L is a logic for which syntax and semantics is defined, and ϕ is a formula of L, we write $\phi \in L$ to say that ϕ is among the formulas of L. That is, when no confusion may arise, we do not notationally distinguish a logic from its set of formulas. By L-formula we mean formula of L.

2 Preliminaries

We begin by introducing various notions that will be needed subsequently.

IF-procedure. A mechanical procedure of *IF-ing* first-order logic was introduced by Hintikka [11]. (For a mathematical exposition of IF first-order logic, see [21].) The procedure, applied to an **FO**-formula ϕ in negation normal form, yields a string $\mathbf{IF}(\phi)$ and goes as follows. If in a string produced by the procedure an existential quantifier (disjunction symbol) is followed by a slash sign (/), it is called 'slashed', otherwise 'non-slashed'. If $\phi \in \mathbf{FO}$, we will write '$\prec$' for the relation of *syntactic superordination* among tokens of logical operators in ϕ: if θ is a proper subformula of ψ (which itself is a subformula of ϕ), the outmost form of ψ is $o \in \{\exists, \forall, \vee, \wedge\}$ and the outmost form of θ is $o' \in \{\exists, \forall, \vee, \wedge\}$, then we write $o \prec o'$. As we proceed, we extend the notation to formulas of **IF** in the obvious way.

- Put $\psi = \phi$.
- For every non-slashed existential quantifier $\exists x$ (disjunction sign $\vee$) in ψ, execute the following clause:
  - If $\exists x\, \theta(x)$ is a subformula of ψ occurring subordinate to a number of universal quantifiers which include $\forall y_1, \ldots, \forall y_n$, then it may be replaced by $(\exists x/\{y_1, \ldots, y_n\})\, \theta(x)$, provided that all variables $x, y_1, \ldots, y_n$ are pairwise distinct, and for no i, $\forall y_i \prec \exists y_i \prec \exists x$.
  - If $(\theta \vee \theta')$ is a subformula of ψ occurring subordinate to a number of universal quantifiers which include $\forall y_1, \ldots, \forall y_n$, then it may be replaced by $(\theta\, (\vee/\{y_1, \ldots, y_n\})\, \theta')$, provided that all variables $y_1, \ldots, y_n$ are pairwise distinct, and for no i, $\forall y_i \prec \exists y_i \prec \vee$.
  - Let ψ be the result of the above substitution.
- Return $\mathbf{IF}(\phi) = \psi$.

The set of formulas of $\mathbf{IF}(\tau)$, or *IF first-order logic* (of vocabulary τ), are the strings generated by the above procedure from $\mathbf{FO}(\tau)$-formulas. In any formula $\psi \in \mathbf{IF}$, we allow writing $\exists x$ and $\vee$ for eventual occurrences of $(\exists x/\varnothing)$ and $(\vee/\varnothing)$, respectively.

Skolemization. Hintikka considers **IF**-formulas in which no quantifier Qx_i with a given variable x_i has more than one occurrence. He formulates their semantics in terms of semantic games, and points out that doing so yields an explicit translation of IF first-order logic into existential second-order logic [11, p. 31]. The translation utilizes a notion of *Skolemization*: $\mathcal{M} \models \phi$ iff $\mathcal{M} \models Sk(\phi)$, the Skolemization $Sk(\phi)$ of ϕ being a Σ_1^1-formula. We phrase the idea of Skolemization slightly differently, and will obtain a semantics for the whole language **IF**. It should be noted that we do not thereby deviate from the semantic mechanism Hintikka has in mind for representing logical (in)dependencies: we could reformulate our definition in game-theoretical terms, and most importantly, we continue to operate with 'choice functions' to represent logical (in)dependencies. Our definition is chosen to serve our present purposes, which are entirely connected with the metalogical properties of the logics considered. We will write $ar(f)$ for the arity of a second-order function variable f. Below, if f, g are such variables, and χ involves occurrences of f, 'replacing f by g in χ' means substituting g for *every* occurrence of f. E.g., if $\chi := (R(f, x) \wedge (\exists f/X)\, S(x, f(y, z)))$, the result of the relevant replacement is $(R(g, x) \wedge (\exists g/X)\, S(x, g(y, z)))$. If $\phi \in \mathbf{IF}$, let the *stepwise Skolemization of* ϕ, denoted $SSk(\phi)$, be obtained by applying the following procedure:

- Put $\psi = \phi$.
- Replace every slashed first-order existential quantifier $(\exists x_i/X)$ by the expression $(\exists f_i/X)$, where f_i is a nullary function variable, and replace all variables x_i bound by $(\exists x_i/X)$ in ψ by the nullary function term f_i. Let ψ be the result of this substitution.

- As long as there exists an occurrence of $(\exists f/X)$ in ψ, with $X \neq \varnothing$, execute the following routine:
  - Pick a subformula $(\exists f/X)\,\theta[f(\mathbf{u})]$ of ψ, such that $X \neq \varnothing$.
  - If $\circ \in \{\wedge, \vee\}$ and $((\exists f/X)\theta[f(\mathbf{u})] \circ \theta')$ is a subformula of ψ, introduce a fresh function variable g with $ar(g) = ar(f)$ not appearing anywhere in ψ, replace f by g in $\theta[f(\mathbf{u})]$ to obtain $\theta[g(\mathbf{u})]$, and finally replace $((\exists f/X)\theta[f(\mathbf{u})] \circ \theta')$ by $(\exists g/X)(\theta[g(\mathbf{u})] \circ \theta')$.[1] Proceed similarly if the subformula is $(\theta' \circ (\exists f/X)\theta[f(\mathbf{u})])$.
  - If $(\exists g/\varnothing)(\exists f/X)\,\theta[g(\mathbf{u}), f(\mathbf{v})]$ is a subformula of ψ, then replace it by $(\exists f/X)(\exists g/\varnothing)\,\theta[g(\mathbf{u}), f(\mathbf{v})]$.
  - If $(\forall x)(\exists f/X)\,\theta[x, f(\mathbf{u})]$ is a subformula of ψ and $x \in X$, then replace it by $(\exists f/X - \{x\})(\forall x)\,\theta[x, f(\mathbf{u})]$.
  - If $(\forall x)(\exists f/X)\,\theta[x, f(\mathbf{u})]$ is a subformula of ψ and $x \notin X$, introduce a fresh function variable g with $ar(g) = ar(f) + 1$, and replace $(\forall x)(\exists f/X)\,\theta[x, f(\mathbf{u})]$ by $(\exists g/X)(\forall x)\,\theta[x, g(\mathbf{u}, x)]$, where $\theta[x, g(\mathbf{u}, x)]$ results from replacing f by g in $\theta[x, f(\mathbf{u})]$ and filling in the remaining argument place of g by x.
  - Let ψ be the result of the above substitution.
- Replace all expressions $(\exists f_i/\varnothing)$ by the second-order function quantifier $\exists f_i$, and return $SSk(\phi) = \psi$.

The string $SSk(\phi)$ returned by the procedure is, syntactically, a formula of *second-order logic*. (Quantification over nullary functions can be construed as quantification over individuals.) Its only second-order quantifiers are positive occurrences of the function quantifier $\exists f$, so $SSk(\phi)$ is in fact equivalent to a Σ_1^1-formula. For second-order logic, we apply its standard interpretation [10]: set variables range over arbitrary subsets (of appropriate arity) of the domain. The operation $SSk(\cdot)$ is defined for every $\phi \in \mathbf{IF}$: the above clauses allow proceeding until all sets X in expressions $(\exists f/X)$ are empty. Further, no matter in which order the non-empty sets of variables are resolved in ϕ, the resulting formulas $SSk(\phi)$ are logically equivalent.

EXAMPLE 1. If $\phi := \forall x_1(\exists x_2/\varnothing)\forall x_3(\exists x_4/\{x_1\})\ R(x_1, x_2, x_3, x_4)$, then the first step of the above process returns the formula $\psi := \forall x_1(\exists f_2/\varnothing)\forall x_3(\exists f_4/\{x_1\})\ R(x_1, f_2, x_3, f_4)$. In this case the run of the second step happens to be uniquely determined:

1. $\forall x_1(\exists f_2/\varnothing)(\exists f_4/\{x_1\})\forall x_3\ R(x_1, f_2, x_3, f_4(x_3))$
2. $\forall x_1(\exists f_4/\{x_1\})(\exists f_2/\varnothing)\forall x_3\ R(x_1, f_2, x_3, f_4(x_3))$
3. $(\exists f_4/\varnothing)\forall x_1(\exists f_2/\varnothing)\forall x_3\ R(x_1, f_2, x_3, f_4(x_3))$.

The returned formula $SSk(\phi)$ then is $\exists f_4 \forall x_1 \exists f_2 \forall x_3\ R(x_1, f_2, x_3, f_4(x_3))$. Another example: if we take $\phi := \forall x_1(\exists x_2/\varnothing)\forall x_3(R(x_1, x_2) \wedge \forall x_1(\exists x_2/\{x_1\})\ Q(x_1, x_2, x_3))$, then the first step yields $\psi := \forall x_1(\exists f_2/\varnothing)\forall x_3(R(x_1, f_2) \wedge \forall x_1(\exists f_2/\{x_1\})\ Q(x_1, f_2, x_3))$, and the second $\forall x_1(\exists f_2/\varnothing)\forall x_3(R(x_1, f_2) \wedge (\exists f_2/\varnothing)\forall x_1\ Q(x_1, f_2, x_3))$, whereby $SSk(\phi)$ is returned as $\forall x_1 \exists f_2 \forall x_3(R(x_1, f_2) \wedge \exists f_2 \forall x_1\ Q(x_1, f_2, x_3))$. ∎

---
[1] Observe that by construction g does not appear in θ'.

Neither of the two second-order formulas $SSk(\phi)$ of the above example is, syntactically, in Σ_1^1, but each is obviously equivalent to a Σ_1^1-formula.

DEFINITION 2. Let $\mathcal{M}$ be a τ-structure, and γ a variable assignment. For every $\phi \in \mathbf{IF}(\tau)$, we stipulate that $\mathcal{M}, \gamma \models \phi$, if $\mathcal{M}, \gamma \models SSk(\phi)$. ∎

Finite variable fragments. The fragment of first-order logic (of vocabulary τ) consisting of formulas that employ only variables x and y is called its *2-variable fragment*, and denoted $\mathbf{FO}^2(\tau)$. So if $\phi \in \mathbf{FO}^2$, all occurrences of variables in ϕ (both free and bound) are occurrences of x or y. Similarly, we define 2-variable fragment of $\mathbf{IF}(\tau)$, or $\mathbf{IF}^2(\tau)$, to be the fragment of $\mathbf{IF}(\tau)$ generated using the variables x and y only. For $n \geq 3$, we define the n-variable fragment of $\mathbf{FO}^n(\tau)$ *resp.* $\mathbf{IF}^n(\tau)$ to be the fragment of $\mathbf{FO}(\tau)$ *resp.* $\mathbf{IF}(\tau)$ whose formulas are built up using variables $x_1, \ldots, x_n$ only.

$\mathbf{FO}^2$ has the finite model property [15], and so its satisfiability problem is decidable; in fact $\mathbf{FO}^2$-satisfiability is **NEXPTIME**-complete. $\mathbf{FO}^2$ does *not* have the tree model property (not every satisfiable formula has a tree-like model), and it is *not* 'robustly decidable', i.e. has natural extensions which are not decidable. (For $\mathbf{FO}^2$, see, e.g., [9, 22].) By contrast, **ML** is decidable in **PSPACE**, has the tree model property, and is robustly decidable.

Guarded fragment. The guarded fragment of **FO** and the notion of guarded bisimulation were defined in [1]; we follow the formulations of [12]. For **FO**-formulas, we apply the usual notion of 'free variable', and write $Free(\phi)$ for the set of free variables of ϕ. Let Var be a countable set of variables. *Guarded fragment* of vocabulary τ, $\mathbf{GF}(\tau)$, is defined as follows: (A) *Atomic formulas:* (1) If $x_1, x_2 \in Var$, then $x_1 = x_2$ is a formula. (2) If $R \in \tau$ is n-ary and $x_1, \ldots, x_n \in Var$, then $Rx_1 \ldots x_n$ is a formula. (B) *Complex formulas:* (1) If φ, ψ are formulas, then so are $(\varphi \wedge \psi)$ and $\neg \varphi$. (2) Let $\bar{x}$ be a finite non-empty sequence of variables, ψ a formula, and G an atomic formula such that $Free(\psi) \subseteq Free(G)$. Then $\exists \bar{x}(G \wedge \psi)$ is a formula. The atomic formula G in $(B.2)$ is termed the 'guard' of the quantifier $\exists \bar{x}$. The semantics of $\mathbf{GF}(\tau)$ is immediate from that of $\mathbf{FO}(\tau)$, once it is stipulated that $\exists \bar{x} \phi$ with $\bar{x} = x_1 \ldots x_n$ has the same semantics as the first-order formula $\exists x_1 \ldots \exists x_n \phi$.

If $R \in \tau$ and $\mathcal{M}$ is a τ-structure, we extend the usual notation $R^{\mathcal{M}}$, denoting the interpretation of R in $\mathcal{M}$, for sets. Write $X \in R^{\mathcal{M}}$, if the elements of $X \subseteq dom(\mathcal{M})$ are R-related, in any order or multiplicity, i.e., if $(x_1, \ldots, x_n) \in R^{\mathcal{M}}$ and $X = \{x_1, \ldots, x_n\}$. E.g., if R is 6-ary and $(3, 1, 12, 3, 1, 1) \in R^{\mathcal{M}}$, then $\{1, 3, 12\} \in R^{\mathcal{M}}$. If $R \in \tau$, a non-empty finite set $Z \subseteq dom(\mathcal{M})$ is R-*live* in $\mathcal{M}$, if either it is a singleton, or there is a set X such that $Z \subseteq X \in R^{\mathcal{M}}$. Z is τ-*live* in $\mathcal{M}$ (denoted $Z \subseteq_\tau^l \mathcal{M}$), if there is $R \in \tau$ such that Z is R-*live* in $\mathcal{M}$. Hence if τ contains, say, only binary relation symbols, a τ-live set is at most of size 2. A finite *partial τ-isomorphism* between τ-structures $\mathcal{M}$ and $\mathcal{N}$ is a finite one-one partial map $f : M \rightharpoonup N$ such that for every n-ary $R \in \tau$ and all $a_1, \ldots, a_n \in$

$dom(f)$: $\langle a_1, \ldots, a_n \rangle \in R^{\mathcal{M}}$ iff $\langle f(a_1), \ldots, f(a_n) \rangle \in R^{\mathcal{N}}$. *Guarded τ-bisimulation* between $\mathcal{M}$ and $\mathcal{N}$ is a non-empty set $\mathcal{F}$ of finite partial τ-isomorphisms between $\mathcal{M}$ and $\mathcal{N}$ such that for every $f : X \to Y \in \mathcal{F}$: (1) *Zig:* For any $Z \subseteq^l_\tau \mathcal{M}$, there is $g \in \mathcal{F}$ with $dom(g) = Z$, such that g and f agree on $X \cap Z$; and (2) *Zag:* For any $W \subseteq^l_\tau \mathcal{N}$, there is $g \in \mathcal{F}$ with $Im(g) = W$, such that g^{-1} and f^{-1} agree on $Y \cap W$. If $\varphi \in \mathbf{FO}(\tau)$ and $Free(\varphi) \subseteq \{x_1, \ldots, x_k\}$, φ is *invariant for guarded bisimulations*, if for all τ-structures $\mathcal{M}$, $\mathcal{N}$, guarded τ-bisimulations $\mathcal{F}$ between $\mathcal{M}$ and $\mathcal{N}$, maps $f \in \mathcal{F}$, and $m_1, \ldots, m_k \in dom(f)$: $\langle \mathcal{M}, m_1, \ldots, m_k \rangle \models \varphi$ iff $\langle \mathcal{N}, f(m_1), \ldots, f(m_k) \rangle \models \varphi$. Now $\mathbf{GF}(\tau)$-formulas are in fact invariant for guarded τ-bisimulations. In [1] it was proven:

PROPOSITION 3. *Let $\varphi \in \mathbf{FO}(\tau)$ be arbitrary. Then φ is invariant for guarded τ-bisimulations if, and only if, φ is logically equivalent to a formula of $\mathbf{GF}(\tau)$.* ∎

$\mathbf{GF}$ has the finite model property, and so its satisfiability problem is decidable [1]. In fact, $\mathbf{GF}$-satisfiability is complete for deterministic double exponential time [8]. $\mathbf{GF}$ also has a certain tree model property. Adding least and greatest fixed points to $\mathbf{GF}$ yields a decidable extension, so arguably $\mathbf{GF}$ is even robustly decidable. (For a discussion, see [2, 12].)

Expressive power. If L, L' are modal logics whose semantics are defined over the same class of modal structures, we say L is *translatable into L'* (in symbols $L \leq L'$), if for every $\phi \in L$, there is $\psi_\phi \in L'$ such that for all modal structures $\mathfrak{M}$ and all $w \in dom(\mathfrak{M})$: $\mathfrak{M}, w \models \phi$ iff $\mathfrak{M}, w \models \psi_\phi$. L' is *more expressive than L* (denoted $L < L'$), if $L \leq L'$ but $L' \not\leq L$. These notions extend to a comparison between a modal logic and an abstract logic, and between two abstract logics. (Here the 'abstract logics' considered are $\mathbf{FO}$ and $\mathbf{IF}$.)

3 Three choices

Let us specify a fragment of IF first-order logic by fixing, as follows, the three parameters mentioned in the beginning of the paper:

Modal logic. We restrict ourselves to basic modal logic whose formulas are in *negation normal form*. Formally, let $\mathbf{ML}$ be the language generated by the following grammar: $\phi ::= p \mid \neg p \mid (\phi \vee \phi) \mid (\phi \wedge \phi) \mid \Diamond \phi \mid \Box \phi$, where p is a propositional atom from some countable set Φ of atoms.

Standard translation. We restrict ourselves to the standard translation that uses two first-order variables: x, y. Let $u, v \in \{x, y\}$, with $u \neq v$. For every $\phi \in \mathbf{ML}$, let $ST_2[\phi]$ be defined as $ST_x[\phi]$, following [3]:

$$\begin{aligned}
ST_u[p] &= Pu \\
ST_u[\neg p] &= \neg Pu \\
ST_u[(\phi \circ \psi)] &= (ST_u[\phi] \circ ST_u[\psi]) \quad \text{if } \circ \in \{\wedge, \vee\} \\
ST_u[\Diamond \phi] &= \exists v (Ruv \wedge ST_v[\phi]) \\
ST_u[\Box \phi] &= \forall v (Ruv \to ST_v[\phi]).
\end{aligned}$$

Let $ST_2[\mathbf{ML}] := \{ST_2[\psi] : \psi \in \mathbf{ML}\}$. Clearly, $ST_2[\mathbf{ML}]$ falls within $\mathbf{FO}^2$.

IF-procedure. We adhere to the aforementioned IF-procedure, yet restricting ourselves to introducing only *slashed existential quantifiers*, thereby ignoring the possibility of independent disjunctions. So essentially we use the procedure **IF**(·) that lacks the rule for introducing slashed disjunctions. A quick inspection of the IF-procedure tells us that no new variables are introduced along the way. Thus, if $\phi \in$ **FO** sits in the two variable fragment, then also **IF**(ϕ) uses only two variables. Generally, every formula in **IF**(ST_2[**ML**]) only employs two variables: x, y. That is, **IF**(ST_2[**ML**]) is contained in **IF**^{2}.

This particular setting of the parameters was not chosen randomly. The ensuing sections will show that this combination has nice, even surprising features. Observe that the IF-procedure can be applied to $ST_2[\phi]$, with arbitrary $\phi \in$ **ML**, since ϕ was assumed to be in negation normal form, whereby also $ST_2[\phi]$ is in negation normal form. Further, since ST_2 introduces only two variables, we have that if $(\exists u/X) \; \theta(x)$ is a subformula of **IF**($ST_2[\phi]$), then X is either empty or the singleton $\{v\}$, where $u, v \in \{x, y\}$ and $u \neq v$.

LEMMA 4. **IF**(ST_2[**ML**]) *is translatable into* **FO**.

Proof. Let $\psi \in$ **IF**(ST_2[**ML**]) be arbitrary. So for some $\phi \in$ **ML**, there is an execution of the IF-procedure yielding ψ when applied to $ST_2[\phi]$. Because $ST_2[\phi] \in$ **FO**^{2}, any slashed existential quantifier $(\exists u/\{v\})$ in ψ satisfies: (i) $u, v \in \{x, y\}$ with $u \neq v$; (ii) v is a universally quantified variable; (iii) there is no existential quantifier $\exists v$ such that $\forall v \prec \exists v \prec \exists u$. Consider, then, the Σ_1^1-formula $SSk(\psi)$. By the algorithm producing $SSk(\psi)$, first each occurrence of $(\exists u/\{v\})$ is replaced in ψ by $(\exists f_u/\{v\})$, and the occurrences of u bound by $(\exists u/\{v\})$ are replaced by f_u. Concentrating on one occurrence of $(\exists f_u/\{v\})$ at a time, the procedure keeps replacing the corresponding occurrences of f_u by fresh function variables g_i, until $(\exists g_i/\{v\})$ is brought to a position subordinate to $\forall v$, but superordinate to all conjunctions/disjunctions to which $(\exists f_u/\{v\})$ is subordinate in ψ. Then $(\exists g_i/\{v\})$ is flipped to the left of $\forall v$, simultaneously removing v from the set $\{v\}$, and the result is that $(\exists g_i/\varnothing)$ immediately precedes $\forall v$. Effecting the transformation for all $(\exists f_u/\{v\})$, and closing the process, results in a formula $SSk(\psi)$, whose all existential quantifiers are function quantifiers $\exists g_i$, with nullary g_i. But this means that $SSk(\psi)$ is a notational variant of a first-order formula. ∎

4 Structurally determined IF modal logic

We introduce a modal-like logic, $\mathcal{L}_{\mathbf{SD}}$, which will be seen to characterize the logic **IF**(ST_2[**ML**]), structurally determined above.

Syntax. A class of formulas is generated by the two grammars A and B:

$$\alpha ::= p \mid \neg p \mid (\alpha \vee \alpha) \mid (\alpha \wedge \alpha) \mid \Diamond \alpha \mid \Box \alpha \mid \blacksquare \beta$$
$$\beta ::= \blacklozenge \alpha \mid (\alpha \vee \beta) \mid (\beta \vee \alpha) \mid (\beta \vee \beta) \mid (\alpha \wedge \beta) \mid (\beta \wedge \alpha) \mid (\beta \wedge \beta),$$

with $p \in \Phi$. The two grammars generate the formulas of a logic we refer to as $\mathcal{L}_{\mathbf{SD}}^{+}$. The α are said to be *closed*, and the β *open*. If ϕ is a formula, all tokens of ◆ not subordinate to a token of ■ in ϕ are called *free*. If ϕ is open, all its free tokens of ◆ become *bound by* the outmost token of ■ in ■ϕ. E.g., ◆p is open; and ■(◆$p \vee$ ◆q) is closed, the two tokens of ◆ being bound by the unique token of ■. We define the formulas of $\mathcal{L}_{\mathbf{SD}}$ to be the *closed* formulas of $\mathcal{L}_{\mathbf{SD}}^{+}$. Grammar B contains both clauses $(\alpha \circ \beta)$ and $(\beta \circ \alpha)$ with $\circ \in \{\wedge, \vee\}$, for we wish that open and closed formulas can be conjoined and disjoined in any order. The symbols ■ and ◆ are termed 'black box' and 'black diamond', and the symbols □ and ◇ 'white box' and 'white diamond'. Intuitively, ◆ is the 'independent diamond', and it will by definition be independent precisely of the token of ■ that binds it.

Semantics. For every $\phi \in \mathcal{L}_{\mathbf{SD}}^{+}$, a relation $\mathfrak{M}, I, \bar{i}, w \models \phi$ is defined, where: $\mathfrak{M} = (M, R, V)$ is a modal structure, with a set M of states, $w \in M$, accessibility relation $R \subseteq M^2$, and valuation of propositional atoms $V : \Phi \longrightarrow Pow(M)$. The component I is a map $\{0,1\}^* \longrightarrow M$, called *token valuation*, and $\bar{i} \in \{0,1\}^*$ is a binary string. Here, 0 and 1 are intuitively the choices *left* resp. *right*, interpreting the connectives $\wedge$ and $\vee$. Note that given a formula ϕ, a binary string $i_1 \ldots i_n$ determines a subformula, namely the one yielded by going through, outside-in, the connectives $\wedge$ and $\vee$ of ϕ, and choosing for the j-th connective encountered *left* or *right* according to whether i_j is 0 or 1.

$\mathfrak{M}, I, \bar{i}, w \models p$ iff: $w \in V(p)$
$\mathfrak{M}, I, \bar{i}, w \models \neg p$ iff: $w \notin V(p)$
$\mathfrak{M}, I, \bar{i}, w \models \Diamond \psi$ iff: for some v with $R(w,v)$: $\mathfrak{M}, I, \bar{i}, w \models \psi$
$\mathfrak{M}, I, \bar{i}, w \models \Box \psi$ iff: for every v with $R(w,v)$: $\mathfrak{M}, I, \bar{i}, w \models \psi$
$\mathfrak{M}, I, \bar{i}, w \models (\psi \vee \phi)$ iff: $\mathfrak{M}, I, \bar{i}0, w \models \psi$ or $\mathfrak{M}, I, \bar{i}1, w \models \phi$
$\mathfrak{M}, I, \bar{i}, w \models (\psi \wedge \phi)$ iff: $\mathfrak{M}, I, \bar{i}0, w \models \psi$ and $\mathfrak{M}, I, \bar{i}1, w \models \phi$
$\mathfrak{M}, I, \bar{i}, w \models \blacksquare \phi$ iff: for some $I' : \{0,1\}^* \longrightarrow M$: $\mathfrak{M}, I', \bar{i}, w \models \Box \phi$
$\mathfrak{M}, I, \bar{i}, w \models \blacklozenge \phi$ iff: $R(w, I(\bar{i}))$ and $\mathfrak{M}, I, \bar{i}, I(\bar{i}) \models \phi$.

The clauses for literals and formulas of the forms □ and ◇ do not employ the components I and $\bar{i}$. By contrast, these components play a key role in the rest of the semantic clauses. Inspection of the clauses reveals that as it stands, ■ acts as a second-order existential quantifier over token valuations (cf., however, Remark 5a). By contrast, ◆ is not a quantifier at all. Instead, it operates on information already available: turns attention to the state $I(\bar{i})$ obtained by applying I to the string $\bar{i}$, claims that state to be accessible from the current state, and makes evaluation to continue relative to the state $I(\bar{i})$. Why do these clauses permit to render ◆ and ■ their desired meanings, i.e. to make the occurrences of ◆ bound by ■ logically independent of it?

Semantically, the black box has a double role: in addition to having the force of the standard box (white box), it *binds* certain subordinate tokens of the black diamond. To these two sides of the semantics of ■, two evaluation steps of a formula ■ϕ correspond. (1) The first step consists of choosing a token valuation I which fixes, at one stroke, the interpretations of *all* tokens of ◆ bound by the relevant occurrence of ■. Technically this is accomplished by choosing a function from binary strings to the domain: the relevant tokens of the black diamond are identified by such strings. (2) The second step is to go on replacing ■ by □, and evaluating □ϕ: this adds to the semantics of the black box the force of the usual box, i.e. makes it a universal quantifier over accessible states. When the evaluation eventually reaches one of the tokens of ◆ bound by ■, *nothing* is done to interpret the token; instead, if the token is identified by the string $\bar{i}$, its interpretation has already been fixed to be $I(\bar{i})$. This two-step process guarantees precisely that *each token of the black diamond is logically independent of the black box binding it*. For instance, evaluating ■◆⊤ relative to $\langle \mathfrak{M}, I, \bar{i}, w \rangle$ involves *first* choosing a state u to interpret ◆, although ◆ is syntactically subordinate to ■. This is done by fixing a token valuation I' such that $I'(\bar{i}) = u$. Only then the universal force of ■ is processed: the evaluation moves to □◆⊤ and a state v accessible from w is chosen. Finally, ◆⊤ is evaluated by checking whether u can actually be accessed from v. It is useful to observe the following features of the above semantics:

REMARK 5. (a) For the token valuations in $\mathcal{L}_{\mathbf{SD}}^+$-semantics, all that matters are the *free* occurrences of the black diamond in a formula. It is not difficult to see that if (relative to an initial string $\bar{i}$) the free occurrences of ◆ in ϕ are those identified by the strings in the set $S \subset \{0,1\}^*$ (which is necessarily finite), then if for *some* I, we have $\mathfrak{M}, I, \bar{i}, w \models \phi$, actually $\mathfrak{M}, I', \bar{i}, w \models \phi$ holds for any I' such that for all $\bar{j} \in S$, $I'(\bar{j}) = I(\bar{j})$. In particular, to keep the satisfaction condition intact we need not have $I'(\bar{i}) = I(\bar{i})$ unless $\bar{i} \in S$. It can also be noted that the semantic clause for the black diamond need not be phrased in terms of quantification over token valuations: to evaluate ■ϕ, it suffices to choose a fixed finite number of states: as many states as there are free occurrences of ◆ in ϕ to be interpreted. (b) Let us write $\mathfrak{M}, w \models \phi$ to express the following condition: $\mathfrak{M}, I, \bar{i}, w \models \phi$ holds for all $I : \{0,1\}^* \longrightarrow M$ and all $\bar{i} \in \{0,1\}^*$. As a special case of (a) we have: If ϕ is closed (i.e., if $\phi \in \mathcal{L}_{\mathbf{SD}}$), and for *some* I and $\bar{i}$, we have $\mathfrak{M}, I, \bar{i}, w \models \phi$, then actually $\mathfrak{M}, w \models \phi$ holds. By (a), in the evaluation of open $\mathcal{L}_{\mathbf{SD}}^+$-formulas, free tokens of the black diamond (as identified by certain strings) bear resemblance to free variables, and token valuations to variable assignments: satisfaction is essentially dependent on the values assigned to free variables. And by (b), $\mathcal{L}_{\mathbf{SD}}$-formulas are like *sentences* in **FO**: if satisfied under one assignment, they are satisfied under all assignments.

EXAMPLE 6. Consider evaluating the closed formula $\phi := $ ■(◆$p \vee$ ◆q) at the root w of the modal structure $\mathfrak{M} = (M, R, V)$ depicted in Figure 1 below. Observe that $R(w, z)$ iff $z \in \mathbb{Z}$, and that $R(z, e)$ if $z = 2z'$ for some $z' \in \mathbb{Z}$, while otherwise $R(z, o)$. Further, note that $e \in V(p) \not\ni o$ and $o \in V(q) \not\ni e$.

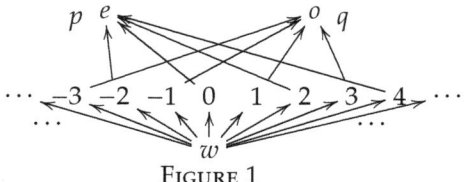

FIGURE 1

We claim $\mathfrak{M}, w \models \phi$. To prove this, it is enough to show that for any token valuation I_0: $\mathfrak{M}, I_0, \emptyset, w \models \phi$. Choose I so that[2] $I(0) = e$ and $I(1) = o$. Fixing a valuation corresponds to picking out, as it were beforehand, the interpretations of the two black diamonds ◆ that can come across later in the evaluation. It suffices to check that $\mathfrak{M}, I, \emptyset, w \models \Box(\blacklozenge p \lor \blacklozenge q)$. For this to hold, it must be possible to partition the set $\mathbb{Z}$ of R-successors of w into two cells (corresponding to the choice *left* or *right* for $\lor$), so that if z belongs to one of them, we have $\mathfrak{M}, I, 0, z \models \blacklozenge p$, while if z belongs to the other, $\mathfrak{M}, I, 1, z \models \blacklozenge q$ holds. Let the cells be $\{2z : z \in \mathbb{Z}\}$ for *left*, and $\{2z+1 : z \in \mathbb{Z}\}$ for *right*. Then the former condition holds, because $I(0)$ is accessible from all states in the first mentioned cell, and $I(0) \in V(p)$; and the latter holds because $I(1)$ is accessible from all states in the other cell, and $I(1) \in V(q)$. ∎

What kind of logic is $\mathcal{L}_{\mathbf{SD}}$? From one perspective, it is a modal-like characterization of $\mathbf{IF}(ST_2[\mathbf{ML}])$, cf. Proposition 9. $\mathcal{L}_{\mathbf{SD}}$ has compositional semantics, and hence it provides a particularly useful handle on $\mathbf{IF}(ST_2[\mathbf{ML}])$, whose semantics is not compositional. Both $\mathbf{IF}(ST_2[\mathbf{ML}])$ and $ST_2[\mathbf{ML}]$ employ only two variables; both admit of the same partial motivation from the computational viewpoint: seeing what can be expressed in terms of limited memory resources. The former extends the latter in the direction of informational independence just as much as it goes, given the 2-variable restriction and staying with a Hintikka-style IF-procedure. One can also view $\mathcal{L}_{\mathbf{SD}}$ as a *fragment* of the general IF modal logic **IFML** (cf. [19]), obtained by allowing to indicate a diamond as independent *only* from the closest superordinate modal operator, which furthermore must be a box. Indeed, the independence represented by placing ◆ subordinate to ■ is, of course, nothing but the logical independence that can alternatively be marked using the slash notation, indicating a diamond as independent of a superordinate box. Due to the limited independencies expressible in the framework at hand, we can, as a matter of fact, express them all without the slash notation, having instead an 'independent version' ◆ of ◇ available, 'bound' by the version ■ of □ on which the appropriate tokens of ◆ will, on the other hand, by definition *not* logically depend. The relevant fragment of **IFML** could certainly be singled out even without the framework of the present paper. However, here it emerges naturally: its IF first-order logic

---

[2] Observe that the arguments 0 and 1 of I are binary strings of length 1 from $\{0,1\}^*$, and must not be confused with the elements $0, 1 \in \mathbb{Z}$ of the domain!

correspondence language $\mathbf{IF}(ST_2[\mathbf{ML}])$ is well-motivated on its own; what is more, its compositional semantics makes it particularly well-behaved.

FACT 7. *On syntactic grounds, if $\phi \in \mathbf{ML}$, then $\phi \in \mathcal{L}_{\mathbf{SD}}$. And semantically, if $\phi \in \mathbf{ML}$, then ϕ is true in $\mathfrak{M}$ at w under the usual semantics of $\mathbf{ML}$ iff $\mathfrak{M}, w \models \phi$. It follows that $\mathbf{ML} \leq \mathcal{L}_{\mathbf{SD}}$.* ∎

Some definitions. Each formula $\phi \in \mathcal{L}^+_{\mathbf{SD}}$ is assigned a natural number $md[\phi]$, its *modal depth*, as follows: $md[p] = md[\neg p] = 0$; if $\circ \in \{\wedge, \vee\}$, then $md[(\phi \circ \psi)] = \max\{md[\phi], md[\psi]\}$; and if $O \in \{\Box, \Diamond, \blacksquare, \blacklozenge\}$, then $md[O\phi] = md[\phi] + 1$. Further, define an operation $Sub[\cdot]$ on formulas of $\mathcal{L}^+_{\mathbf{SD}}$: if $p \in \Phi$, $Sub[p] = \{p\}$ and $Sub[\neg p] = \{\neg p\}$; if $\circ \in \{\wedge, \vee\}$, then $Sub[(\varphi \circ \psi)] = \{(\varphi \circ \psi)\} \cup Sub[\varphi] \cup Sub[\psi]$; and if $O \in \{\Box, \Diamond, \blacksquare, \blacklozenge\}$, then $Sub[O\varphi] = \{O\varphi\} \cup Sub[\varphi]$. If $\psi \in Sub[\phi]$, ψ is a *subformula* of ϕ. The *size* of $\phi \in \mathcal{L}^+_{\mathbf{SD}}$, denoted $|\phi|$, is the number of symbol tokens appearing in ϕ, except that propositional atoms are counted as one no matter how many symbols are used for writing them down. E.g. $|(\blacklozenge p_{127} \vee \Box \neg p_1)| = 8$. Whenever $\mathfrak{M}$ is a modal structure, 'M' stands for its domain; and if $w \in M$, the pair $\langle \mathfrak{M}, w \rangle$ is termed a *model*.

Standard translation into IF first-order logic. Basic modal logic can be translated into $\mathbf{FO}^2$. The class of closed $\mathcal{L}^+_{\mathbf{SD}}$-formulas, i.e., the class $\mathcal{L}_{\mathbf{SD}}$, has an analogous property. The following map $ST^{\mathbf{IF}}_x : \mathcal{L}_{\mathbf{SD}} \longrightarrow \mathbf{IF}^2$ provides a canonical translation of $\mathcal{L}_{\mathbf{SD}}$ into $\mathbf{IF}^2$. For all $u, v \in \{x, y\}$ with $u \neq v$, put:

$$ST^{\mathbf{IF}}_u[p] = Pu$$
$$ST^{\mathbf{IF}}_u[\neg p] = \neg Pu$$
$$ST^{\mathbf{IF}}_u[(\phi \circ \psi)] = (ST^{\mathbf{IF}}_u[\phi] \circ ST^{\mathbf{IF}}_u[\psi]) \quad \text{if } \circ \in \{\vee, \wedge\}$$
$$ST^{\mathbf{IF}}_u[\Diamond \phi] = \exists v (Ruv \wedge ST^{\mathbf{IF}}_v[\phi])$$
$$ST^{\mathbf{IF}}_u[\Box \phi] = \forall v (Ruv \to ST^{\mathbf{IF}}_v[\phi])$$
$$ST^{\mathbf{IF}}_u[\blacklozenge \phi] = (\exists v/u)(Ruv \wedge ST^{\mathbf{IF}}_v[\phi])$$
$$ST^{\mathbf{IF}}_u[\blacksquare \phi] = ST^{\mathbf{IF}}_u[\Box \phi].$$

Observe how $ST^{\mathbf{IF}}_x$ acts on black boxes and black diamonds: e.g., $ST^{\mathbf{IF}}_x[\blacksquare \blacklozenge p] = ST^{\mathbf{IF}}_x[\Box \blacklozenge p] = \forall y(Rxy \to ST^{\mathbf{IF}}_y[\blacklozenge p]) = \forall y(Rxy \to (\exists x/y)(Ryx \wedge ST^{\mathbf{IF}}_x[p])) = \forall y(Rxy \to (\exists x/y)(Ryx \wedge Px))$. The result is like $ST_2[\Box \Diamond p]$, except that the quantifier $\exists x$ translating $\blacklozenge$ is *independent of* the quantifier $\forall y$ translating $\blacksquare$. Note that the truth condition of $\blacksquare \blacklozenge p$ is actually expressible in $\mathbf{FO}^3$ by $\exists z \forall y(Rxy \to (Ryz \wedge Pz))$. Clearly if $\phi \in \mathcal{L}^+_{\mathbf{SD}}$ is closed, $ST^{\mathbf{IF}}_x[\phi]$ is an $\mathbf{IF}^2$-formula of exactly one free variable, x. In fact $ST^{\mathbf{IF}}_x$ provides a translation:

PROPOSITION 8. *For all formulas $\phi \in \mathcal{L}_{\mathbf{SD}}$ and all models $\langle \mathfrak{M}, w \rangle$: $\mathfrak{M}, w \models \phi$ iff $\mathfrak{M}^{\mathbf{FO}}, \gamma \models ST^{\mathbf{IF}}_x[\phi]$, where $\mathfrak{M}^{\mathbf{FO}}$ is the first-order structure that corresponds, in the canonical way, to the modal structure $\mathfrak{M}$; and $\gamma(x) = w$.*

Proof. The statement can be proven by induction on the structure of closed formulas. Note that if $\blacksquare \psi$ is closed, ψ is obtained by $\wedge$ and $\vee$ from closed formulas and formulas of the form $\blacklozenge \theta$, where θ is closed. ∎

The following 'commutativity' result holds, establishing that $\mathcal{L}_{\mathbf{SD}}$ actually is structurally determined in the desired way. (For a proof, see [19].)

PROPOSITION 9. *Syntactically,* $ST_x^{\mathbf{IF}}[\mathcal{L}_{\mathbf{SD}}] = \mathbf{IF}(ST_x[\mathbf{ML}])$. ∎

5 Expressive power

Relation to finite variable fragments of FO. $\mathcal{L}_{\mathbf{SD}}$ is *not* translatable into *any* finite variable fragment of **FO**. This is of course in contrast with the case of **ML**, translatable into $\mathbf{FO}^2$. Still $\mathcal{L}_{\mathbf{SD}}$ *is* translatable into **FO**.

THEOREM 10. (a) *For all* $n < \omega$, $\mathcal{L}_{\mathbf{SD}} \not\leq \mathbf{FO}^n$. *However,* (b) $\mathcal{L}_{\mathbf{SD}} \leq \mathbf{FO}$.

Proof. (a) Let $n \geq 2$, and think of the formula $\phi_n := \blacksquare(\blacklozenge\top \vee \ldots \vee \blacklozenge\top)$, where the number of disjuncts in $(\blacklozenge\top \vee \ldots \vee \blacklozenge\top)$ is $n-1$. In a modal structure $\mathfrak{M} = (M, R, V)$ at a state w, ϕ_n asserts that the set $\{v : R(w,v)\}$ can be partitioned into (at most) $n-1$ cells in such a way that the elements in each cell have a *common successor* along R. The truth condition of ϕ_n can be expressed by the first-order formula $\phi_n' := \exists z_1 \ldots \exists z_{n-1} \forall y (Rxy \to (Ryz_1 \vee \ldots \vee Ryz_{n-1}))$. Now $\phi_n' \in \mathbf{FO}^{n+1}$, but it is not difficult to see (by reference to a pebble game argument[3]) that $\phi_n' \notin \mathbf{FO}^n$. Hence the greater the number n is, the more variables are needed to translate the formula ϕ_n into first-order logic. (b) Follows from Propositions 8 and 9 together with Lemma 4. ∎

ML is translatable into $\mathbf{FO}^2$, but $\mathcal{L}_{\mathbf{SD}}$ is not. By Fact 7 we conclude:

COROLLARY 11. $\mathbf{ML} < \mathcal{L}_{\mathbf{SD}}$. ∎

Corollary 11 could be proven more straightforwardly by adapting the argument of [17, Lemma 4], reproduced in [13, Example 2].

REMARK 12. The results of Theorem 10 can be compared with what is known about other IF modal logics. The 'IF modal logic of perfect recall' studied in [13] can be shown to be translatable into $\mathbf{FO}^3$. On the other hand, giving up the perfect recall assumption, the resulting IF modal logic does *not* admit of *any* translation into **FO**, as shown in [19]. ∎

Number of non-equivalent formulas. Restricting attention to a finite set Φ of propositional atoms, it is well known that for any $m < \omega$, the number of pairwise non-equivalent formulas ψ of **ML** with $md[\psi] = m$ is finite; see, e.g., [3, Prop. 2.29]. This property does not extend to $\mathcal{L}_{\mathbf{SD}}$. In fact we have:

PROPOSITION 13. *Let the size of Φ be finite. For any $m \geq 2$, the set of pairwise non-equivalent $\mathcal{L}_{\mathbf{SD}}$-formulas of modal depth m is infinite.*

Proof. If $k \geq 0$ and $n \geq 2$, let $\phi_n^k := \square \ldots \square \phi_n$, where ϕ_n is defined as in the proof of Theorem 10(a), and the number of boxes preceding ϕ_n is k. Whenever $n < n'$, the formulas ϕ_n^k and $\phi_{n'}^k$ are not equivalent. However,

---
[3] On how to use pebble games $G_m^n(\mathcal{M}, \mathbf{a}, \mathcal{N}, \mathbf{b})$ to characterize equivalence of structures up to quantifier rank $\leq m$ relative to $\mathbf{FO}^n$, see, e.g., [6, pp. 49-50].

$md[\phi_n^k] = k+2 = md[\phi_{n'}^k]$. Hence the formulas in the set $\{\phi_n^k : n < \omega\}$ are pairwise non-equivalent, but they all have the same modal depth, $k+2$. ∎

Relation to the guarded fragment. According to the influential suggestion of Andréka, van Benthem and Németi [1], the distinguishing feature of the modal fragments of **FO** is that the quantifiers they harbor appear guarded. The nice computational properties of modal fragments would be 'accounted for' by their translatability into **GF**. Our modal-like logic $\mathcal{L}_{\mathbf{SD}}$ fails to share this feature with **ML**.

Let us adopt the following notation. If $\mathcal{F}$ is a set of finite partial τ-isomorphisms between $\mathcal{M}$ and $\mathcal{N}$, $f \in \mathcal{F}$, and X is live in $\mathcal{M}$, write **Zig**(f, X) for the claim "there is $g \in \mathcal{F}$ with $dom(g) = X$ such that g and f agree on the intersection of their domains." Similarly, write **Zag**(f, X) for the claim "there is $g \in \mathcal{F}$ with $Im(g) = X$ such that g^{-1} and f^{-1} agree on the intersection of their domains." Let us prove that $\mathcal{L}_{\mathbf{SD}}$ cannot be translated into **GF**.

THEOREM 14. *There is no (truth-preserving) translation of $\mathcal{L}_{\mathbf{SD}}$ into* **GF**.

Proof. Let R be a binary relation symbol. Consider the $\{R\}$-structures $\mathfrak{M}$ and $\mathfrak{N}$ depicted in Figure 2 (the arrow represents R-accessibility):

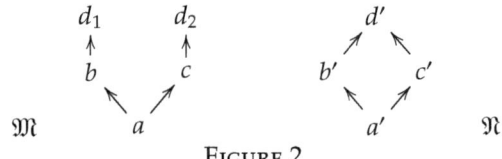

FIGURE 2

Let us define $\mathfrak{F}$ as the set consisting of the maps $\{(a, a')\}$, $\{(b, b')\}$, $\{(c, c')\}$, $\{(d_1, d')\}$, and $\{(d_2, d')\}$, together with the maps $f_1 := \{(a, a'), (b, b')\}$, $f_2 := \{(a, a'), (c, c')\}$, $f_3 := \{(b, b'), (d_1, d')\}$ and $f_4 := \{(c, c'), (d_2, d')\}$. We claim $\mathfrak{F}$ is a guarded $\{R\}$-bisimulation. $\mathfrak{F}$ is non-empty, and evidently all maps in $\mathfrak{F}$ are partial $\{R\}$-isomorphisms between $\mathfrak{M}$ and $\mathfrak{N}$. It must be checked that for all $f \in \mathfrak{F}$ and all live sets Z in $\mathfrak{M}$, **Zig**(f, Z); and that for all $f \in \mathfrak{F}$ and all live sets W in $\mathfrak{N}$, **Zag**(f, W). The live sets of size 2 in $\mathfrak{M}$ are $\{a, b\}$, $\{a, c\}$, $\{b, d_1\}$, $\{c, d_2\}$; and the live sets of size 2 in $\mathfrak{N}$ are $\{a', b'\}$, $\{a', c'\}$, $\{b', d'\}$, $\{c', d'\}$.

- If X is a singleton, it is trivial that there is a singleton $g \in \mathfrak{F}$ witnessing the claims **Zig**(f, X) and **Zag**(f, X).
- If f is a singleton and X is of size 2, it is immediate there is a singleton $g \in \mathfrak{F}$ that witnesses **Zig**(f, X), and similarly for **Zag**(f, X).
- We must still check the claims **Zig**(f, X) and **Zag**(f, X), assuming f, X both are of size 2. Now **Zig**$(f_1, \{b, d_1\})$ is witnessed by f_3, and **Zig**$(f_3, \{a, b\})$ by f_1. Similarly, **Zig**$(f_2, \{c, d_2\})$ is witnessed by f_4; and **Zig**$(f_4, \{a, c\})$ by f_2. Furthermore, **Zag**$(f_1, \{b', d'\})$ is witnessed by f_3, and **Zag**$(f_3, \{a', b'\})$ by f_1. Similarly, **Zag**$(f_2, \{c', d'\})$ is witnessed by f_4; and **Zag**$(f_4, \{a', c'\})$ by f_2.

So $\mathfrak{F}$ is a guarded $\{R\}$-bisimulation between $\mathfrak{M}$ and $\mathfrak{N}$; and $\langle \mathfrak{M}, a \rangle$ and $\langle \mathfrak{N}, a' \rangle$ satisfy precisely the same $\mathbf{GF}(\{R\})$-formulas of one free variable. Observe, then, that the $\mathbf{FO}$-formula $\exists z \forall y (Rxy \to Ryz)$ is not satisfied in $\langle \mathfrak{M}, a \rangle$, but is satisfied in $\langle \mathfrak{N}, a' \rangle$. But $\exists z \forall y (Rxy \to Ryz)$ is a first-order translation of the $\mathcal{L}_{\mathbf{SD}}$-formula $\blacksquare \blacklozenge \top$. It follows that $\mathcal{L}_{\mathbf{SD}}$ is not translatable into $\mathbf{GF}$. ∎

Unorthodox properties of the modal-like logic $\mathcal{L}_{\mathbf{SD}}$, such as those witnessed by Proposition 13 and Theorems 10 and 14, might suggest it should be of a rather marginal interest as a modal logic; even its status as a modal logic might be thereby called into question. Now in *Sections* 7 and 8 it is shown that the validity and satisfiability problems of $\mathcal{L}_{\mathbf{SD}}$ are not only decidable, but decidable in $\mathbf{PSPACE}$. Insofar as $\mathcal{L}_{\mathbf{SD}}$ is a modal logic, then, it is a modal logic translatable into $\mathbf{FO}$ *and* decidable in $\mathbf{PSPACE}$. But it does *not* lie within the guarded fragment: there is no translation $\mathcal{L}_{\mathbf{SD}} \longrightarrow \mathbf{GF}$. To the extent it is a modal logic, it tells against the hypothesis that embeddability to $\mathbf{GF}$ is what 'explains' the nice computational properties of modal logics.

6 Model checking

We proceed to discuss the computational properties of $\mathcal{L}_{\mathbf{SD}}$; let us begin with the model checking problem.

THEOREM 15. *Model checking $\mathcal{L}_{\mathbf{SD}}$ is* $\mathbf{NP}$*-complete.*

Proof. *Membership.* In the semantics we took the component I of the models to be a function of type $\{0,1\}^* \longrightarrow M$, but as noted in Remark 5, this is needlessly general. For simplicity and w.l.o.g., let us restrict attention to functions of type $\{1, \ldots, n\} \longrightarrow M$, where n is the number of occurrences of the black diamond in ϕ. Labeling those occurrences each with its own number $j \in \{1, \ldots, n\}$, we may suppress the component $\bar{i}$ from the models (the labels will do the job of these strings). Then, instead of evaluating $\blacklozenge \psi$ relative to $\langle \mathfrak{M}, I^*, \bar{i}, w \rangle$, we evaluate $\blacklozenge_j \psi$ relative to $\langle \mathfrak{M}, I, w \rangle$, provided that $I(j) = I^*(\bar{i})$. Let, then, $\phi \in \mathcal{L}_{\mathbf{SD}}$ and $\mathfrak{M}, I, w$ be given, the domain of $\mathfrak{M}$ being finite. We describe an algorithm deciding whether $\mathfrak{M}, I, w \models \phi$, that is, $\mathcal{L}_{\mathbf{SD}}$'s model checking algorithm.

- For every pair $\langle v, \blacksquare \psi \rangle$, where $v \in M$ and $\blacksquare \psi$ is a subformula of ϕ, non-deterministically guess a function $I_{\langle v, \blacksquare \psi \rangle} : \{1, \ldots, n\} \longrightarrow M$.
- Put $k = 1$, and for every propositional atom p occurring in ϕ, and for every state $v \in M$, label v with p if $p \in V(v)$, and with $\neg p$ otherwise.
- While $k \leq md[\phi]$, proceed as follows:
  - Repeat the routine for every $v \in M$ and every subformula $\blacksquare \psi$ of ϕ such that $md[\blacksquare \psi] = k$:
    Essentially, we must decide whether $\mathfrak{M}, I, v \models \blacksquare \psi$. To do so, it suffices to check whether $\mathfrak{M}, I_{\langle v, \blacksquare \psi \rangle}, v \models \square \psi$, because if there exists a function I' that does the job, then so does the non-deterministically guessed function $I_{\langle v, \blacksquare \psi \rangle}$. Deciding whether

$\mathfrak{M}, I_{\langle v, \blacksquare \psi \rangle}, v \models \Box \psi$ now proceeds along the same lines as model checking basic modal logic. None of ψ's subformulas of the form $\blacksquare \chi$ causes any further problem, since $md[\blacksquare \chi] < k$ and has therefore already been taken care of in an earlier iteration of the algorithm. That is, every state $u \in M$ has been labeled with $\blacksquare \chi$ if, and only if, $\mathfrak{M}, I, u \models \blacksquare \chi$. If it turns out that $\mathfrak{M}, I_{\langle v, \blacksquare \psi \rangle}, v \models \Box \psi$, then label v with $\blacksquare \psi$, otherwise do not.
– Increase k by one.

- If w is labeled with ϕ, accept the input, otherwise reject it.

It is not hard to see that this algorithm is correct; i.e., it accepts $\langle \phi, \mathfrak{M}, I, w \rangle$ iff $\mathfrak{M}, I, w \models \phi$. Furthermore, the algorithm runs in polynomial non-deterministic time. Namely, first, the guessing of one function $I_{\langle v, \blacksquare \psi \rangle}$ can be done in time linear in the number of black diamonds in ϕ, which is clearly bounded by ϕ's size. The non-deterministic guessing of $I_{\langle v, \blacksquare \psi \rangle}$ for every $v \in M$ and every subformula of the form $\blacksquare \psi$ takes therefore $|\phi| \cdot |M| \cdot md[\phi]$ steps. The latter term is polynomial in the size of the input, since $md[\phi] \leq |\phi|$. For the rest, the algorithm hinges on the model checking algorithm for basic modal logic that runs in polynomial deterministic time.

Hardness. We reduce from SET COVERING, which takes a family $F = \{S_1, \ldots, S_n\}$ of subsets of a finite set U and a budget b as instance. The problem is to decide whether there are b sets in F, whose union is U. (See [16].) Define a structure $\mathfrak{M}_F := (M, R, V)$ by putting $M := \{u\} \cup U \cup F$; $R := \{\langle w, u_i \rangle : u_i \in U\} \cup \{\langle u_i, S_j \rangle : u_i \in S_j\}$; and $V(p) := F$. Let I be any token valuation. We claim that (F, b) sits in SET COVERING iff $\mathfrak{M}_F, I, w \models \blacksquare(\blacklozenge_1 p \vee \ldots \vee \blacklozenge_b p)$. Abusing formal notation, the latter holds iff $(\exists T_1, \ldots, T_b \in F)(\forall u_i \in U)(u_i \in T_1 \vee \ldots \vee u_i \in T_b)$, which spells out the conditions under which (F, b) is contained in SET COVERING. ∎

7 Validity problem

In connection with $\mathcal{L}_{\mathbf{SD}}$, we must treat the validity and satisfiability problems separately. To see this, observe first that the syntax does not allow to form the negation of an arbitrary formula $\phi \in \mathcal{L}_{\mathbf{SD}}$: negation sign appears only in front of propositional atoms. Further, $\blacklozenge \psi$ is *not* true relative to $\langle \mathfrak{M}, I, \bar{i}, w \rangle$ iff: either $\langle w, I(\bar{i}) \rangle \notin R$ or $\mathfrak{M}, I, \bar{i}, w \not\models \psi$; and $\blacksquare \psi$ is *not* true relative to $\langle \mathfrak{M}, I, \bar{i}, w \rangle$ iff: for *all* valuations $I' : \{0, 1\}^* \longrightarrow M$, we have $\mathfrak{M}, I', \bar{i}, w \not\models \Box \psi$. So $\blacklozenge$ and $\blacksquare$ are far from being duals to each other. Not only does it hence seem unlikely that for every $\phi \in \mathcal{L}_{\mathbf{SD}}$ there is $neg(\phi) \in \mathcal{L}_{\mathbf{SD}}$ such that for all $\mathfrak{M}, I, \bar{i}, w$ we have $\mathfrak{M}, I, \bar{i}, w \models neg(\phi)$ iff $\mathfrak{M}, I, \bar{i}, w \not\models \phi$; but it can be shown that $\mathcal{L}_{\mathbf{SD}}$ is not closed under negation. A rigorous proof would require more space than we have available here: a technique would be to define a kind of extended asymmetric n-bisimulation game $\mathfrak{B}_n(\mathfrak{M}, I, \bar{i}, w; \mathfrak{N}, J, \bar{j}, v)$ between *Spoiler* and *Duplicator* such that for all $\phi \in \mathcal{L}_{\mathbf{SD}}^+$ with $md(\phi) \leq n$: if $\mathfrak{M}, I, \bar{i}, w \models \phi$ and *Duplicator* has a winning strategy in $\mathfrak{B}_n(\mathfrak{M}, I, \bar{i}, w; \mathfrak{N}, J, \bar{j}, v)$, then $\mathfrak{N}, J, \bar{j}, v \models \phi$. Validity and satisfiability problems of $\mathcal{L}_{\mathbf{SD}}$, then, are not each other's duals. The same phenomenon was observed in connection with 'IF modal logic of perfect recall' in [13, Fact 21].

Let us proceed to prove that $\mathcal{L}_{\mathbf{SD}}$-validity is decidable in **PSPACE**. Define a transformation T in $\mathcal{L}_{\mathbf{SD}}^{+}$ as follows:

$$T[\ell] = \ell \qquad \text{if } \ell \in \{p, \neg p\}$$
$$T[(\theta \circ \chi)] = (T[\theta] \circ T[\chi]) \quad \text{if } \circ \in \{\vee, \wedge\}$$
$$T[\Box \psi] = T[\blacksquare \psi] = \Box T[\psi]$$
$$T[\Diamond \psi] = \Diamond T[\psi]$$
$$T[\blacklozenge \psi] = \bot.$$

Here are some examples of applying the transformation: $T[\blacksquare(\blacklozenge p \vee q)] = \Box(\bot \vee q)$; $T[\blacksquare\blacklozenge p] = \Box\bot$; and $T[\blacksquare(p \wedge \blacklozenge(q \wedge \Diamond\Box r))] = \Box(p \wedge \bot)$. Note that if in particular $\phi \in \mathcal{L}_{\mathbf{SD}}$ (and hence ϕ is closed), then $T[\phi]$ is, by syntactical criteria, a formula of **ML**. Further, observe that if $\phi \in \mathbf{ML}$, then $T[\phi] = \phi$.

LEMMA 16. *For every $\phi \in \mathcal{L}_{\mathbf{SD}}$, ϕ is valid if, and only if, $T[\phi]$ is valid.*

Proof. Trivially, if $T[\phi]$ is valid, then so is ϕ. For the converse direction, assume that ϕ is valid, but suppose for contradiction that $T[\phi]$ is not. So there is, by the tree model property of **ML**, cf. [3, Prop. 2.15], a tree-like model $\langle \mathfrak{M}, w \rangle$ in which $T[\phi]$ is false, but ϕ is true. In particular, then, there must be a subformula ψ of ϕ of the form $\blacksquare \theta$, and a state $v \in M$, such that ψ is true at v, but $T[\psi]$ is false at v. What is more, among such subformulas ψ of ϕ, we can choose one, $\psi_0 = \blacksquare \theta_0$, that satisfies the following further condition: $\theta_0 = \beta(\chi_1, \ldots, \chi_n, \blacklozenge\nu_1, \ldots, \blacklozenge\nu_m)$ is a boolean combination (in terms of $\wedge, \vee$) of the formulas $\chi_i, \blacklozenge\nu_j$, where all of the χ_i satisfy: for all $u \in M$ at which χ_i is true in $\mathfrak{M}$, also $T[\chi_i]$ is true at u in $\mathfrak{M}$.

Now $T[\psi_0] = \Box(\beta(T[\chi_1], \ldots, T[\chi_n], \bot, \ldots, \bot))$. Replacing conjunction with $\bot$ by $\bot$, and disjunction of a formula with $\bot$ by that formula, the boolean combination β becomes after rewriting either $\beta'(T[\chi_1], \ldots, T[\chi_n])$, or else $\bot$. Let β'' be the $\mathcal{L}_{\mathbf{SD}}$-formula for which $T[\beta''] \in \{\beta', \bot\}$ is the result of the above rewriting. By assumption, then, $\Box T[\beta'']$ is false at v. If in the tree $\mathfrak{M}$ there happens to be less than $m+1$ successors $u_1, \ldots, u_k$ of v at which $T[\beta'']$ is false, let $\langle \mathfrak{N}, u_{k+1} \rangle, \ldots, \langle \mathfrak{N}, u_{m+1} \rangle$ be isomorphic copies of the submodel of $\langle \mathfrak{M}, w \rangle$ generated (say) by u_k, with pairwise disjoint domains, each furthermore disjoint from M. Obtain a tree-like model $\langle \mathfrak{M}', w \rangle$ bisimilar to $\langle \mathfrak{M}, w \rangle$ by adding $u_{k+1}, \ldots, u_{m+1}$ as successors of v, and letting the structures $\langle \mathfrak{N}, u_{k+1} \rangle, \ldots, \langle \mathfrak{N}, u_{m+1} \rangle$ determine an extension of the domain and the accessibility relation of $\langle \mathfrak{M}, w \rangle$.

Thus we have $\mathfrak{M}', v \not\models T[\blacksquare \theta_0]$, and so, by construction, for all $1 \leq i \leq m+1$: $\mathfrak{M}', u_i \not\models T[\beta'']$. On the other hand, $\mathfrak{M}', v \models \blacksquare \theta_0$, and so there is I such that for all $1 \leq i \leq m+1$, we have: $\mathfrak{M}', I, \emptyset, u_i \models \theta_0$. It was assumed that for all $u \in M$ at which χ_i is true in $\mathfrak{M}$, also $T[\chi_i]$ is true at u in $\mathfrak{M}$. Hence if we had for some $i \in \{1, \ldots, m+1\}$ that $\mathfrak{M}', I, \emptyset, u_i \models \beta''$, it would follow that $\mathfrak{M}', u_i \models T[\beta'']$, a contradiction. So we must have for all $1 \leq i \leq m+1$ that $\mathfrak{M}', I, \emptyset, u_i \models \blacklozenge\chi_j$, for some $j \in \{1, \ldots, m\}$. Since there are only m distinct formulas χ_j but $m+1$ states considered, for at least two states the same $\blacklozenge \chi_j$ is chosen. But then, for the corresponding string $\bar{i}$ (i.e. the string of choices left/right leading to $\blacklozenge \chi_j$), the state $I(\bar{i})$ is the common successor of two distinct states u_r and $u_{r'}$. This is impossible, because $\mathfrak{M}'$ is tree-like. ∎

THEOREM 17. *The validity problem of* $\mathcal{L}_{\mathbf{SD}}$ *is* **PSPACE**-*complete.*

Proof. By Lemma 16 $\mathcal{L}_{\mathbf{SD}}$-validity is reducible to **ML**-validity in polynomial time. But **ML**-validity is **PSPACE**-complete (see e.g. [3, Ths. 6.47, 6.50]). ∎

8 Satisfiability problem

With respect to satisfiability, too, it turns out that $\mathcal{L}_{\mathbf{SD}}$ shares the basic properties with **ML**: it has the strong finite model property, and its satisfiability problem is **PSPACE**-complete. In this section we give a loose description of the algorithm that decides satisfiability for $\mathcal{L}_{\mathbf{SD}}$.

Let S_0 be a finite set of $\mathcal{L}_{\mathbf{SD}}$-formulas. We describe an algorithm $\mathfrak{A}(S_0)$, starting out with the set S_0. First, to make describing the algorithm easier, let us suppose that the occurrences of the black diamond in all formulas $\phi \in S_0$ are labeled, each with a label (positive integer) of its own. Recall that a token of the black diamond in a formula $\psi \in \mathcal{L}_{\mathbf{SD}}^+$ is called 'free', if it is not syntactically subordinate to any black box in ψ. We extend this terminology straightforwardly to labeled tokens of the black diamond: in $\blacklozenge_1 \blacksquare \blacklozenge_2 \top$, $\blacklozenge_1$ is free but $\blacklozenge_2$ is not. Below, we write $Sub_m(S_0)$ for the set of those closed subformulas of S_0 whose modal depth is at most m.

The distinguishing feature of the algorithm $\mathfrak{A}(S_0)$ will be its treatment of black boxes ($\blacksquare$) and black diamonds ($\blacklozenge_i$). The black box, as in $\blacksquare \phi$, induces the algorithm to non-deterministically guess sets $T_{i_1}, \ldots, T_{i_n} \subseteq Sub_j(S_0)$ and to store them, in case ϕ has n free black diamonds $\blacklozenge_{i_1}, \ldots, \blacklozenge_{i_n}$ and $j := md[\phi] - 2$. Intuitively, the sets $T_{i_1}, \ldots, T_{i_n}$ represent n states at which at least the formulas from the respective sets are true. The black diamond, as in $\blacklozenge_k \psi$, induces the algorithm to check whether the formula ψ is in the set T_k. Since the modal depth of ψ is smaller than the modal depth of the formula $\blacksquare \phi$ whose black box binds $\blacklozenge_k$, choosing subsets from $Sub_j(S_0)$ instead of $Sub(S_0)$ goes without loss of generality. Further, we observe that by the syntax, ψ is closed, so neither does letting $Sub_j(S_0)$ only contain closed formulas affect the functionality of the algorithm.

Ladner's algorithm [14] that decides satisfiability for **ML** does not have non-deterministic choices, nor does it store sets T to perform membership checks. Interestingly, neither extension increases the computational complexity: $\mathcal{L}_{\mathbf{SD}}$-satisfiability is solvable in **PSPACE**, hence **PSPACE**-complete (because by Fact 7, **ML** $\leq \mathcal{L}_{\mathbf{SD}}$). Our algorithm is conveniently thought of as deciding the satisfiability of a stack $(S, S_1, \ldots, S_K)$ of subsets of S_0. In order to compute whether all sets on the stack are satisfiable, the algorithm considers the leftmost set on the stack, S, and computes new sets $T_1, \ldots, T_m$ that are satisfiable iff S is. Thus, the algorithm transfers

$(S, S_1, \ldots, S_K)$ into $(T_1, \ldots, T_m, S_1, \ldots, S_K)$, and iteratively processes the leftmost set on the stack until a set without modal operators in encountered.

1. When verifying the satisfiability of a set S containing only propositional atoms and their negations, if S is inconsistent (i.e. contains both p and $\neg p$ for some atom $p \in \Phi$), computation terminates and the input is rejected; otherwise S is popped from the stack.
2. When verifying the satisfiability of a set S with $(\chi \wedge \theta) \in S$, remove $(\chi \wedge \theta)$ from S, add both χ and θ to S, and continue checking if the result is satisfiable. When verifying the satisfiability of a set S with $(\chi \vee \theta) \in S$, remove $(\chi \vee \theta)$ from S, non-deterministically guess $\psi \in \{\chi, \theta\}$, put ψ in S, and continue checking if S is satisfiable.
3. If neither of the clauses 1 or 2 applies to S, it can be written as
$$\{\blacksquare\phi_1, \ldots, \blacksquare\phi_m, \Diamond\psi_1, \ldots, \Diamond\psi_n, \Box\chi_1, \ldots, \Box\chi_l\}.$$
For every $\blacksquare\phi_i$, let $\blacklozenge_{\min(i)}\zeta_{\min(i)}, \ldots, \blacklozenge_{\max(i)}\zeta_{\max(i)}$ be the subformulas of $\blacksquare\phi_i$ whose black diamonds are bound by $\blacksquare$. For every such $\blacklozenge_j\zeta_j$, guess a set $T_{i,j} \subseteq Sub_{k-2}(S_0)$ where $k = md[\blacksquare\phi_i]$. Further, for every $\Diamond\psi_t \in S$, let $U'_t = \{\psi_t\} \cup \{\phi_1, \ldots, \phi_m, \chi_1, \ldots, \chi_l\}$. Apply clause 2 to U'_t so that each of its formulas has a modal operator as its outmost form, and let the result be denoted by U''_t. Verify whether for each $\blacklozenge_r\zeta_r \in U''_t$, the formula ζ_r is contained in $T_{i,r}$, given that ϕ_i is the formula that has $\blacklozenge_r\zeta_r$ as a subformula. If not, computation terminates and the input is rejected. Otherwise let U_t be the result of removing all formulas $\blacklozenge_r\zeta_r$ from U''_t. Continue computing the stack $(U_1, \ldots, U_n, T_{1,\min(1)}, \ldots, T_{m,\max(m)}, S_1, \ldots, S_K)$.
4. If the stack is empty, then the input is accepted.

In order to see that the algorithm runs in **PSPACE**, write N for the size of the description of S_0: $N = \Sigma_{\phi \in S_0}|\phi|$. It is immediate that all sets pushed on the stack are of size less than N. We proceed to show that the number of sets on the stack is polynomially bounded by N. Given that the stack is $(S, S_1, \ldots, S_K)$ and one of the clauses 1 or 2 applies to S, the number of sets on the stack after executing the respective clause does not grow. More interesting is the case in which clause 3 applies to S. Then the sets $U_1, \ldots, U_n$ are pushed on the stack together with $T_{1,\min(1)}, \ldots, T_{m,\max(m)}$. There are n sets of the first kind: as many as there are formulas of the form $\Diamond\psi_i$ in S. Clearly, the number of those formulas is bounded above by N. Further, there are at most N^2 sets $T_{i,j}$, since every formula $\blacksquare\phi_i$ binds at most n formulas of the form $\blacklozenge_j\zeta_j$. The upper bound of N^2 follows from the fact that $m < N$.

Hence clause 3 introduces at most $2N^2$ new sets. Let $md[S] := \max\{md[\phi] : \phi \in S\}$. It is important to note that $md[S] > md[U_t]$ and that $md[S] > md[T_{i,j}]$, for all appropriate i, j and t. Thus extending the stack may go on for at most $h := md[S_0]$ cycles without popping sets from the stack. In the worst case we have h times pushed $2N^2$ sets on the stack. Since $h < N$, the stack will contain at most $2N^3$ sets. Thus an upper bound on the number of sets that are stored on the stack is $2N^3$, which is polynomial in the size of the input: N. Since **PSPACE** = **NPSPACE**, the non-deterministic choices do not increase the complexity.

In a canonical way the algorithm produces on input S_0 a model $\langle \mathfrak{M}, w \rangle = \langle M, R, V, w \rangle$ whose domain consists exactly of those sets that were introduced along the way, with $w := S_0$. The relation R is defined so as to satisfy: if the set U_r is induced by the formula $\Diamond \psi_r \in S$, then $R(S, U_r)$. Furthermore, if $\blacklozenge_j \zeta_j \in U_r$ and $\zeta_j \in T_{i,j}$, where ζ_j is a subformula of ϕ_i, then $R(U_r, T_{i,j})$. Finally, $\mathfrak{M}$'s valuation is defined to be the function V such that $S \in V(p)$ iff $p \in S$. Because distinct U_r and $U_{r'}$ can contain $\blacklozenge_j \zeta_j$ for one and the same index j, $\mathfrak{M}$ is not in general tree-like.

PROPOSITION 18. *Let $\phi \in \mathcal{L}_{\mathbf{SD}}$. ϕ is satisfiable iff $\mathfrak{A}(\{\phi\})$ accepts $\{\phi\}$.*

Proof. *Right to left.* $\mathfrak{A}(\{\phi\})$ is so constructed that it provides witnesses of all operators with existential force ($\blacksquare, \Diamond, \vee$) appearing in ϕ, when ϕ is evaluated on the model canonically produced by $\mathfrak{A}(\{\phi\})$. *Left to right.* Any model of ϕ can be used to define an accepting run of $\mathfrak{A}(\{\phi\})$; the fact that we can non-deterministically guess suitable sets in connection with clauses 2 ($\vee$) and 3 ($\blacksquare, \Diamond$) is guaranteed by the truth of ϕ in the model. ∎

We are now in a position to prove what we claimed earlier:

THEOREM 19. *(a) $\mathcal{L}_{\mathbf{SD}}$ has the strong finite model property; (b) The satisfiability problem of $\mathcal{L}_{\mathbf{SD}}$ is **PSPACE**-complete.*

Proof. Given the algorithm $\mathfrak{A}(\cdot)$, both (a) and the fact that $\mathcal{L}_{\mathbf{SD}}$-satisfiability is decidable in **PSPACE** are immediate from Proposition 18. $\mathcal{L}_{\mathbf{SD}}$-satisfiability is **PSPACE**-hard, because $\mathbf{ML} \leq \mathcal{L}_{\mathbf{SD}}$ (Fact 7), and **ML**-satisfiability is **PSPACE**-hard (cf., e.g., [3, Th. 6.50]). ∎

9 Concluding remarks

Theorems 14, 17 and 19 tell us that $\mathcal{L}_{\mathbf{SD}}$ is a computationally attractive logic that does not lie within the guarded fragment of first-order logic. In particular, for validity and satisfiability, this logic has the same computational properties as basic modal logic. Translatability into the guarded fragment is currently taken as one of the key factors — perhaps *the* key factor — in 'explaining' why modal logics have such a good computational behavior (cf., e.g., [2, 12]). An earlier influential candidate for this role was translatability into the two-variable fragment of first-order logic [7], later called into question (see, e.g., [22]). Incidentally, by Theorem 10, $\mathcal{L}_{\mathbf{SD}}$ is not translatable into any finite variable fragment of **FO**, while indeed translatable into **FO** itself. It was of course not the intention of Andréka, van Benthem and Németi [1] to suggest that *every* modal decidability result would be explained by the fact that the modal logic in question is a fragment of **GF**. However, in practice the relationship between translatability into **GF** and good computational properties of modal logics are seen as so intimately tied together, that we think that our results bring up a conceptually important point: either $\mathcal{L}_{\mathbf{SD}}$ is not a modal logic, or else the hypothesis about the importance of **GF** in accounting for modal decidability results should be taken with a grain of salt.

If $\mathcal{L}_{\mathbf{SD}}$ is not a modal logic, the operation of *IF-ing* is capable of destroying the *Eigenart* of its input: the result of *IF-ing* basic modal logic would no longer be a modal logic. If so, modal logic is intrinsically free of imperfect information, incapable of flexibly exhibiting logical independence. In practice this would mean putting an extreme stress on the locality of modal semantics; for this is perhaps the single most important thing that is destroyed by the IF modal logical frameworks. How could we argue in favor of the modal character of $\mathcal{L}_{\mathbf{SD}}$? On the list there would be the issues of Beth definability property and the existence of a characteristic bisimulation relation. (For IF first-order logic, a characteristic Ehrenfeucht-Fraïssé game is provided in [20].) We leave these questions open.

Granting, then, that $\mathcal{L}_{\mathbf{SD}}$ is a modal logic, how severely would our results tell against the hypothesis that **GF** holds the key to understanding the nice computational properties of modal logics? We have shown that there is no *truth* preserving translation of $\mathcal{L}_{\mathbf{SD}}$ into **GF**. What about an embedding of $\mathcal{L}_{\mathbf{SD}}$ into **GF** that simply preserves *satisfiability*? Now trivially, for any logic L whose satisfiability problem happens to lie in **2EXPTIME** (the complexity of **GF**-satisfiability), there exists a polynomial time reduction $t : L \longrightarrow \mathbf{GF}$ such that for any formula $\phi \in L$, ϕ is satisfiable iff $t(\phi)$ is satisfiable. Hence the mere existence of such an embedding cannot possibly have 'explanatory' value. Constructing a suitable satisfiability preserving transformation $\mathcal{L}_{\mathbf{SD}} \longrightarrow \mathbf{GF}$ might still of course be useful in getting to grips with the decidability of $\mathcal{L}_{\mathbf{SD}}$-satisfiability. But there may be another way to rehabilitate the status of the guarded fragment in an IF modal logical context. While not translatable into **GF**, $\mathcal{L}_{\mathbf{SD}}$ is trivially translatable into the result of *IF-ing* **GF**, call it **IF(GF)**. It is an open question whether the satisfiability and validity problems of this fragment of IF first-order logic are decidable. If they are, facts about the decidability of $\mathcal{L}_{\mathbf{SD}}$ could be explained by reference to **IF(GF)** like the decidability of **ML** is explained by reference to **GF**. We would wish to point out that it follows from the results of [19] that **IF(GF)** does not fall within **FO**; it is a rather powerful language.

Acknowledgments

We wish to thank Johan van Benthem and Peter van Emde Boas for helpful discussions, and the three anonymous referees for their useful comments. We are grateful to NWO for financially supporting the work of the first author (visitor's grant B 62-608).

BIBLIOGRAPHY

[1] H. Andréka, J. van Benthem, and I. Németi. Modal languages and bounded fragments of predicate logic. *J. Phil. Logic*, 27:217–74, 1998.
[2] C. Areces, C. Monz, H. de Nivelle, and M. de Rijke. The guarded fragment: Ins and outs. In J. Gerbrandy, M. Marx, M. de Rijke, and Y. Venema, editors, *JFAK. Essays Dedicated to Johan van Benthem on the Occasion of his 50th Birthday*. Amsterdam UP, 1999.
[3] P. Blackburn, M. de Rijke, and Y. Venema. *Modal Logic*. Cambridge UP, 2002.
[4] J. Bradfield. Independence: Logics and concurrency. In *Proceedings of the 14th Workshop on Computer Science Logic*, volume 1862 of *LNCS*. Springer, 2000.
[5] J. Bradfield and S. Fröschle. Independence-friendly modal logic and true concurrency. *Nordic J. of Computing*, 9(2):102–17, 2002.
[6] H.-D. Ebbinghaus and J. Flum. *Finite Model Theory*. Springer, 1999.
[7] D. Gabbay. Expressive functional completeness in tense logic. In U. Mönnich, editor, *Aspects of Philosophical Logic*, pages 91–117. Reidel, 1981.
[8] E. Grädel. On the restraining power of guards. *J. Symbolic Logic*, 64:1719–42, 1997.
[9] E. Grädel and M. Otto. On logics with two variables. *Theor. Comp. Sc.*, 224:73–113, 1999.
[10] L. Henkin. Completeness in the theory of types. *J. Symbolic Logic*, 15(2):81–91, 1950.
[11] J. Hintikka. *Principles of Mathematics Revisited*. Cambridge UP, 1996.
[12] E. Hoogland, M. Marx, and M. Otto. Beth definability and the guarded fragment. In *Proceedings of LPAR'99*, volume 1705 of *LNAI*, pages 273–85. Springer, 1999.
[13] T. Hyttinen and T. Tulenheimo. Decidability of IF modal logic of perfect recall. In R. Schmidt, I. Pratt-Hartmann, M. Reynolds, and H. Wansing, editors, *Advances in Modal Logic*, volume 5, pages 111–31. KCL Publ., 2005.
[14] R. E. Ladner. The computational complexity of provability in systems of modal propositional logic. *SIAM J. Comp.*, 6(3):467–80, 1977.
[15] M. Mortimer. On languages with two variables. *Zeitschr. f. math. Logik und Grundlagen d. Math.*, 21:135–40, 1975.
[16] C. H. Papadimitriou. *Computational Complexity*. Addison-Wesley, 1994.
[17] T. Tulenheimo. On IF modal logic and its expressive power. In Ph. Balbiani, N.-Y. Suzuki, F. Wolter, and M. Zakharyaschev, editors, *Advances in Modal Logic*, volume 4, pages 475–98. KCL Publ., 2003.
[18] T. Tulenheimo. *Independence-Friendly Modal Logic*. PhD thesis, University of Helsinki, 2004.
[19] T. Tulenheimo and M. Sevenster. Approaches to independence-friendly modal logic. Submitted, 2006.
[20] J. Väänänen. On the semantics of informational independence. *Logic J. IGPL*, 10(3):337–50, 2002.
[21] J. Väänänen. Independence friendly logic. Dept. of Mathematics, U. of Helsinki, ms., 2006.
[22] M. Vardi. Why is modal logic so robustly decidable? In *Descriptive complexity and finite models*, volume 31 of *Series in Discr. Math. & Theor. Comp. Science*, pages 149–84. AMS, 1998.

Tero Tulenheimo
Academy of Finland; Department of Philosophy, University of Helsinki
P.O. Box 9, 00014 University of Helsinki, Finland
tero.tulenheimo@helsinki.fi

Merlijn Sevenster
Institute for Logic, Language and Computation, University of Amsterdam
Plantage Muidergracht 24, 1018 TV Amsterdam, The Netherlands
sevenstr@science.uva.nl

Tableaux for multi-agent deliberative-stit logic

HEINRICH WANSING

ABSTRACT. We present a sound and complete tableau calculus for the multi-agent logic of *deliberatively seeing to it that* (dstit), **Ldm**. The agents in this setting are assumed to be independent of each other. Until now only an axiomatic proof system for **Ldm** has been available. Moreover, in order to underline the usefulness of stit logics, we suggest the introduction of dstit modalities into **B**(elief)**D**(esire)**I**(ntention) logics or the introduction of belief, desire, and intention modalities into deliberative-stit logic.

Keywords: modal logic of agency, deliberative-stit logic, independence of agents, branching time structures, tableaux, **BDI** logics, belief, desire, intention.

1 Introduction

The modal logic of action reports as developed by Belnap, Perloff, and Xu, Chellas, von Kutschera, Horty and others (see [3] and references therein), has received a great deal of attention and has been applied in various areas, including deontic logic (see, for example, [1, 2, 3, 5, 8, 9, 10, 13]) and epistemic logic [6, 14, 15, 16, 17]. In [3, Chapter 17] an axiom system, **Ldm**, is presented of the logic of deliberatively seeing to it that for multiple agents satisfying the independence of agents condition on $BT + AC$ structures, i.e., branching temporal structures with agents and choice, see also [18]. In the present paper, we shall define a sound and complete tableau calculus for **Ldm**. Moreover, in order to emphasize the usefulness of dstit logic, we shall briefly motivate and discuss the introduction of dstit operators into **B**(elief)**D**(esire)**I**(ntention) logics and the introduction of belief, desire, and intention modalities into dstit logic.

2 Semantics of Ldm

We here recall the language of **Ldm** and the central semantical notions of $BT + AC$ theories and dstit logic. A more comprehensive account including motivation of the definitions may be found in [3]. The language $\mathcal{L}$ of **Ldm** is defined in Backus–Naur form as follows:

$$\begin{aligned}
\text{atomic formulas:} & \quad p \in Atom \\
\text{agent terms:} & \quad \alpha \in Term \\
\text{formulas:} & \quad A \in Form(Atom, Term) \\
A ::= & \quad p \mid \neg A \mid (A \wedge A) \mid Sett\!:\! A \mid (\alpha = \alpha) \mid [\alpha \; dstit\!:\! A]
\end{aligned}$$

We assume that $Term$ is a finite set. $Sett$: ("it is settled true that") is the historical necessity operator, and $[\_\ dstit:\_]$ is the deliberative-seeing-to-it-that operator. The connectives $\top, \bot, \vee, \supset$, and $\equiv$ are defined as usual, $Poss: A$ abbreviates $\neg Sett: \neg A$, and the expressions $[\alpha\ cstit: A]$ and $\mathit{diff}(\alpha_0, \ldots, \alpha_n)$ are defined as follows.

DEFINITION 1. (1) $[\alpha\ cstit: A] :\equiv [\alpha\ dstit: A] \vee Sett: A$.
(2) $\mathit{diff}(\alpha_0) :\equiv \top$, and for $n \geq 0$, $\mathit{diff}(\alpha_0, \ldots, \alpha_{n+1}) :\equiv \mathit{diff}(\alpha_0, \ldots, \alpha_n) \wedge \neg(\alpha_0 = \alpha_{n+1}) \wedge \ldots \wedge \neg(\alpha_n = \alpha_{n+1})$.

A branching temporal structure with agents and choice is a quadruple $\mathfrak{S} = \langle Tree, \leq, Agent, Choice \rangle$, where $Tree$ is a non-empty set (of moments of time), and $\leq$ is a tree-like partial order on $Tree$. There is thus "historical connection" ($\forall m, m' \in Tree$, there exists $m'' \in Tree$ with $m'' \leq m$ and $m'' \leq m'$) and no "backward branching" ($\forall m_1, m_2, m_3 \in Tree$, if $m_1 \leq m_3$ and $m_2 \leq m_3$, then either $m_1 \leq m_2$ or $m_2 \leq m_1$). If $m_1 \leq m_2$ but not $m_2 \leq m_1$, we shall write $m_1 < m_2$. A maximal linearly ordered subset of $Tree$ is called a history in $\mathfrak{S}$. Intuitively, histories may be regarded as complete temporal developments of the world. In the present indeterministic branching time framework, a moment may belong to more than one history, and the set of histories to which a moment m belongs (that pass through the moment) is represented by $H_m := \{h \mid m \in h\}$. Given a branching temporal structure with agents and choice $\mathfrak{S} = \langle Tree, \leq, Agent, Choice \rangle$, $History$ is the set of all histories of $Tree$.

$Agent$ is a non-empty set (of agents), and $Choice$ is a function defined on $Agent \times Tree$ such that for every $x \in Agent$ and $m \in Tree$, $Choice(x, m)$, written as $Choice_m^x$, is a partition into equivalence classes of H_m. A model based on a $BT+AC$ structure $\mathfrak{S} = \langle Tree, \leq, Agent, Choice \rangle$ is a pair $\langle \mathfrak{S}, \mathfrak{I} \rangle$, where $\mathfrak{I}$ is a function defined on $Term \cup Atom$ such that for every agent term α and every $p \in Atom$, $\mathfrak{I}(\alpha) \in Agent$ and $\mathfrak{I}(p) \subseteq \{(m, h) \mid m \in Tree, \text{ and } h \in H_m\}$. In the literature there is some discussion about which kinds of sentences ought to be semantically evaluated relative not only to moments of time but also relative to histories. It seems uncontroversial, however, that in an indeterministic semantical framework the truth conditions of contingent statements about the future are history-dependent. Also, the truth and non-truth of historical necessity statements depends on histories. As a matter of semantic uniformity, in dstit theory *every* formula is evaluated at pairs (m, h) consisting of a moment m and a history h passing through m.

If $h, h' \in H_m$ belong to the same set in $Choice_m^x$, we write $h \equiv_m^x h'$ and say that h and h' are choice-equivalent for x at m. The elements of $Choice_m^x$ are the choices (or choice cells) available to x at m, and x cannot distinguish by her or his actions at m between histories that are choice-equivalent for x at m. The set of histories choice equivalent with h for x at m is denoted by $Choice_m^x(h)$. The set of agents is held constant for the entire tree of moments.

The agents of a $BT + AC$ model are assumed to be independent of each other in the sense that at every moment, every agent must be able to realize any of her or his actions, no matter what choices are available to the other agents.

DEFINITION 2. Let $\mathfrak{S} = \langle Tree, \leq, Agent, Choice \rangle$ be a $BT + AC$ structure, and let $Select_m$ be the set of all functions s from $Agent$ into subsets of H_m, such that $s(x) \in Choice_m^x$. $\mathfrak{S}$ satisfies the *independence of agents* condition iff for every $m \in Tree$, $\bigcap_{x \in Agent} s(x) \neq \emptyset$ for every $s \in Select_m$.

Moreover, it is assumed that there is "no choice between undivided histories". Two histories $h, h' \in History$ are undivided at m ($h \equiv_m h'$) iff $\exists m'(m < m'$ and $m' \in h \cap h')$. $Choice$ is required to satisfy the following condition: $(\forall m \in Tree)(\forall x \in Agent)(\forall h, h' \in History)\ h \equiv_m h'$ implies $h \equiv_m^x h'$.

DEFINITION 3. Let $\mathfrak{M} = \langle \langle Tree, \leq, Agent, Choice \rangle, \mathfrak{I} \rangle$ be a model. The truth of a formula A at a pair (m, h) from $\mathfrak{M}$ with $m \in h$ (in symbols $\mathfrak{M}, (m, h) \models A$) is inductively defined as follows:

$\mathfrak{M}, (m, h) \models p$ iff $(m, h) \in \mathfrak{I}(p)$
$\mathfrak{M}, (m, h) \models (\alpha = \beta)$ iff $\mathfrak{I}(\alpha) = \mathfrak{I}(\beta)$
$\mathfrak{M}, (m, h) \models \neg A$ iff $\mathfrak{M} \not\models A$
$\mathfrak{M}, (m, h) \models (A \wedge B)$ iff $\mathfrak{M}, (m, h) \models A$ and $\mathfrak{M}, (m, h) \models B$
$\mathfrak{M}, (m, h) \models Sett\colon A$ iff $(\forall h' \in H_m)\ \mathfrak{M}, (m, h') \models A$
$\mathfrak{M}, (m, h) \models [\alpha\ dstit\colon A]$ iff $(\forall h' \in Choice_m^{\mathfrak{I}(\alpha)}(h))\ \mathfrak{M}, (m, h') \models A$ and $(\exists h'' \in H_m)\ \mathfrak{M}, (m, h'') \not\models A$

We then have

$\mathfrak{M}, (m, h) \models Poss\colon A$ iff $(\exists h' \in H_m)\ \mathfrak{M}, (m, h') \models A$
$\mathfrak{M}, (m, h) \models [\alpha\ cstit\colon A]$ iff $(\forall h' \in Choice_m^{\mathfrak{I}(\alpha)}(h))\ \mathfrak{M}, (m, h') \models A$

Clearly, $Sett\colon$ and $[\_\ cstit\colon \_]$ are **S5**-type necessity operators: semantically, $Sett\colon$ is a universal quantifier on H_m, and $[\_\ cstit\colon \_]$ is a universal quantifier on a subset of H_m.

DEFINITION 4. Let Δ be a set of formulas and A be a formula. $\Delta \models A$ (Δ entails A) iff for every model $\mathfrak{M} = \langle \langle Tree, \leq, Agent, Choice \rangle, \mathfrak{I} \rangle$ and every (m, h) with $m \in Tree$ and $h \in H_m$: if $\mathfrak{M}, (m, h) \models B$ for every $B \in \Delta$, then $\mathfrak{M}, (m, h) \models A$.

We shall refer to the logic $(\mathcal{L}, \models)$ as **Ldm**. This terminology is justified by the characterization result for the axiom system **Ldm** in [3, Chapter 17]. The axioms and inference rules (in addition to all substitution instances of classical tautologies) presented there are:

A1 $Sett\colon (A \supset B) \supset (Sett\colon A \supset Sett\colon B),\ Sett\colon A \supset A,$
     $Poss\colon A \supset Sett\colon Poss\colon A$

A2 $[\alpha \, cstit \colon (A \supset B)] \supset ([\alpha \, cstit \colon A] \supset [\alpha \, cstit \colon B])$, $[\alpha \, cstit \colon A] \supset A$, $\neg[\alpha \, cstit \colon A] \supset [\alpha \, cstit \colon \neg[\alpha \, cstit \colon A]]$

A3 $[\alpha \, dstit \colon A] \supset \neg Sett \colon A$

A4 $(\alpha = \alpha)$, $(\alpha = \beta) \supset (\beta = \alpha)$, $((\alpha = \beta) \wedge (\beta = \gamma)) \supset (\alpha = \gamma)$

A5 $(\alpha = \beta) \supset (A \supset A(\alpha/\beta))$, where $A(\alpha/\beta)$ is obtained from A by replacing some or all occurrences of α with β

AIA$_k$ $(\textit{diff}(\beta_0, \ldots, \beta_k) \wedge Poss \colon [\beta_0 \, cstit \colon B_0] \wedge \ldots \wedge Poss \colon [\beta_k \, cstit \colon B_k]) \supset Poss \colon ([\beta_0 \, cstit \colon B_0] \wedge \ldots \wedge [\beta_k \, cstit \colon B_k])$ $(k \geq 1)$

RN $\vdash A \,/\, \vdash Sett \colon A$

and *modus ponens*.

3 A tableau calculus for Ldm

We shall introduce a tableau calculus for **Ldm** in the style of the tableau systems presented in [11]. This format not only has found wide application, but also is particularly easy to grasp and admits of simple and very transparent soundness and completeness proofs. The tableau rules for **Ldm**, though perspicuous, are elaborate because of the richness of the semantics. The rules are used to build up counter-models for formulas and inferences that are not valid. Complete tableaux must (i) have a suitable tree structure, (ii) provide information about the histories passing through the moments of time, (iii) provide information about the choices available to the agents under consideration, and (iv) satisfy the independence of agents condition.

3.1 Tableau rules

If Δ is a set of formulas, then Δ^0 is defined as a set of certain compound expressions, namely $\Delta^0 := \{A, (m, h_0) \mid A \in \Delta\}$. A tableau is a rooted tree. The construction of a tableau starts with a set of expressions $\Delta^0 \cup \{\neg B, (m, h_0)\} \cup \{m \in h_0, m \lhd m_0, m_0 \in h_0\}$ to which identity rules, decomposition rules, and structural tableau rules may (or may not) be applied to obtain a tableau. The tableau nodes contain information about the model that is build up (in case the construction in fact yields a countermodel). The expression $m \lhd m_0$ gives the information that the considered moment m is earlier than m_0. Also information about certain relations between histories may be included. The expression $h_i \lhd^\alpha_m h_k$ indicates that h_i and h_k are choice-equivalent for α at m.

The set of tableau rules for **Ldm** is given in Tables 1–4. Decomposition rules are named after the kind of formula that is decomposed. The rule for negated conjunctions, for instance, is called (R¬∧) and the rule for negated dstit formulas is called (R¬[$dstit \colon$]). The indices i, j, k, l, n are elements from $\mathbb{N}$ (including 0), and a new index is the smallest natural number not already used in the tableau. If an expression is placed below an arrow, it is a rule output, otherwise it is a rule input. Note that the tableau rules

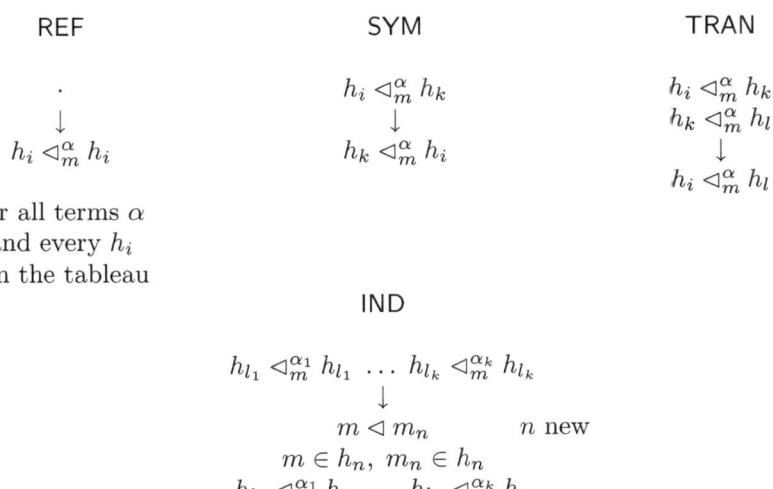

Table 1. Identity rules

Table 2. Structural tableau rules

for which no input is stated may be applied to any node, that some rules introduce a new moment in order to create a new history, and that the left branch created by an application of (R¬[*dstit*:]) must be further expanded by $A, (m, h_j)$ after the introduction of a new history h_j on that branch. A tableau is a tableau for $\Delta^0 \cup \{\neg B, (m, h_0)\}$ iff it is the result of applying some or no tableau rules to $\Delta^0 \cup \{\neg B, (m, h_0)\} \cup \{m \in h_0, m \triangleleft m_0, m_0 \in h_0\}$.

A tableau branch is said to be *closed* iff there are expressions of the form $A, (m, h_i)$ and $\neg A, (m, h_i)$ on the branch. A tableau is called *closed* iff all of its branches are closed. If a tableau (tableau branch) is not closed, it is called *open*. A tableau branch is said to be *complete* iff no more rules can be applied to expand it. A tableau is said to be *complete* iff each of its branches is complete. If a rule application yields an expression already on the branch, the rule is not applied. Moreover, we shall not apply the rule ref to introduce new agent terms not already on the tableau. The following definition is central.

DEFINITION 5. Let $\Delta \cup \{A\}$ be a set of $\mathcal{L}$-formulas. $\Delta \vdash A$ (A is derivable from Δ) iff there exists a closed and complete tableau for $\Delta^0 \cup \{\neg A, (m, h_0)\}$.

3.2 Examples

EXAMPLE 6. Note that we do not require a premise set to be finite. The reason is that a complete and closed tableau for a finite premise set Δ (a derivation from Δ) may already be infinite. The following simple example of an infinite open tableau is adapted from [11].

$$\begin{array}{ll}
\neg\neg(\neg Sett\colon p \wedge Sett\colon \neg Sett\colon p), (m, h_0) & \\
m \vartriangleleft m_0,\ m \in h_0,\ m_0 \in h_0 & (R\neg\neg) \\
\downarrow & \\
(\neg Sett\colon p \wedge Sett\colon \neg Sett\colon p), (m, h_0) & (R\wedge) \\
\downarrow & \\
\neg Sett\colon p\ (m, h_0) & (R\neg Sett\colon) \\
Sett\colon \neg Sett\colon p,\ (m, h_0) & \\
\downarrow & \\
m \vartriangleleft m_1,\ m \in h_1,\ m_1 \in h_1,\ \neg p, (m, h_1) & (RSett\colon) \\
\downarrow & \\
\neg Sett\colon p, (m, h_1) & (R\neg Sett\colon) \\
\downarrow & \\
m \vartriangleleft m_2,\ m \in h_2,\ m_2 \in h_2,\ \neg p, (m, h_2) & (RSett\colon) \\
\downarrow & \\
\neg Sett\colon p, (m, h_2) & (R\neg Sett\colon) \\
\downarrow & \\
\vdots &
\end{array}$$

Next, we present some simple examples of finite closed tableaux.

EXAMPLE 7. We consider a formula logically equivalent with the negation of Axiom **A3** from [3, Chapter 17], see Section 2.

$$\begin{array}{ll}
\neg\neg([\alpha\ dstit\colon p] \wedge Sett\colon p), (m, h_0) & \\
m \vartriangleleft m_0,\ m \in h_0,\ m_0 \in h_0 & (R\neg\neg) \\
\downarrow & \\
([\alpha\ dstit\colon p] \wedge Sett\colon p), (m, h_0) & (R\wedge) \\
\downarrow & \\
[\alpha\ dstit\colon p], (m, h_0) & \\
Sett\colon p, (m, h_0) & \text{REF} \\
\downarrow & \\
h_0 \vartriangleleft_m^\alpha h_0 & (R[\ dstit\colon]) \\
\downarrow & \\
p, (m, h_0) & \\
m \vartriangleleft m_1,\ m \in h_1,\ m_1 \in h_1 & (RSett\colon) \\
\neg p, (m, h_1) & \\
\downarrow & \\
p, (m, h_1) &
\end{array}$$

EXAMPLE 8. We prove the symmetry axiom for historical necessity, i.e., we show that $\varnothing \vdash \neg(p \wedge \neg Sett\colon \neg Sett\colon \neg p)$.

$$\neg\neg A, (m, h_i)$$
$$\downarrow$$
$$A, (m, h_i)$$

$(A \wedge B), (m, h_i)$
$\downarrow$
$A, (m, h_i)$
$B, (m, h_i)$

$\neg(A \wedge B), (m, h_i)$
$\swarrow \quad \searrow$
$\neg A, (m, h_i) \quad \neg B, (m, h_i)$

$Sett\colon A, (m, h_i)$
$m \in h_k$
$\downarrow$
$A, (m, h_k)$

$\neg Sett\colon A, (m, h_i)$
$\downarrow$
$m \lhd m_k \quad k \text{ new}$
$m \in h_k,\ m_k \in h_k$
$\neg A, (m, h_k)$

$[\alpha\ dstit\colon A], (m, h_i)$
$h_i \lhd_m^\alpha h_k$
$\downarrow$
$A, (m, h_k)$
$m \lhd m_l \quad l \text{ new}$
$m \in h_l,\ m_l \in h_l$
$\neg A, (m, h_l)$

$\neg[\alpha\ dstit\colon A], (m, h_i)$
$\swarrow \qquad \searrow$
$A, (m, h_l) \qquad m \lhd m_k \quad k \text{ new}$
for every h_l with $\quad m \in h_k,\ m_{\bullet k} \in h_k$
$m \in h_l$ on the $\qquad h_i \lhd_m^\alpha h_k$
branch $\qquad \neg A, (m, h_k)$

Table 3. Decomposition rules for **Ldm**

$\neg\neg(p \wedge \neg Sett\colon \neg Sett\colon \neg p), (m, h_0)$
$m \lhd m_0,\ m \in h_0,\ m_0 \in h_0 \qquad (R\neg\neg)$
$\downarrow$
$(p \wedge \neg Sett\colon \neg Sett\colon \neg p), (m, h_0) \qquad (R\wedge)$
$\downarrow$
$p, (m, h_0)$
$\neg Sett\colon \neg Sett\colon \neg p, (m, h_0) \qquad (R\neg Sett\colon)$
$\downarrow$
$m \lhd m_1,\ m \in h_1,\ m_1 \in h_1$
$\neg\neg Sett\colon \neg p, (m, h_1) \qquad (R\neg\neg)$
$\downarrow$
$Sett\colon \neg p, (m, h_1) \qquad (RSett\colon)$
$\downarrow$
$\neg p, (m, h_0)$

EXAMPLE 9. The axioms AIA_k are used in [3] to prove that the canonical model for **Ldm** satisfies the independence of agents condition. We consider a formula logically equivalent with the negation of the result of dropping one conjunct from the antecedent of an instance of AIA_k. The omitted conjunct is $\neg(\alpha_1 = \alpha_2)$. It is satisfied in our tableau if we assume that the agent terms α_1 and α_2 are uniquely denoting terms, see also Definition 13, where we define $\mathfrak{I}(\alpha) := \alpha$.

$$\neg\neg((Poss\,[\alpha_1\,cstit\colon p_1] \wedge Poss\,[\alpha_2\,cstit\colon p_2]) \wedge$$
$$\neg Poss\,([\alpha_1\,cstit\colon p_1] \wedge [\alpha_2\,cstit\colon p_2])), (m, h_0)$$
$$m \triangleleft m_0,\ m \in h_0,\ m_0 \in h_0$$
$$\downarrow$$
$$((Poss\,[\alpha_1\,cstit\colon p_1] \wedge Poss\,[\alpha_2\,cstit\colon p_2]) \wedge$$
$$\neg Poss\,([\alpha_1\,cstit\colon p_1] \wedge [\alpha_2\,cstit\colon p_2])), (m, h_0)$$
$$\downarrow$$
$$(Poss\,[\alpha_1\,cstit\colon p_1] \wedge Poss\,[\alpha_2\,cstit\colon p_2]), (m, h_0)$$
$$\neg Poss\,([\alpha_1\,cstit\colon p_1] \wedge [\alpha_2\,cstit\colon p_2]), (m, h_0)$$
$$\downarrow$$
$$Poss\,[\alpha_1\,cstit\colon p_1], (m, h_0)$$
$$Poss\,[\alpha_2\,cstit\colon p_2], (m, h_0)$$
$$\downarrow$$
$$m \triangleleft m_1,\ m \in h_1,\ m_1 \in h_1$$
$$[\alpha_1\,cstit\colon p_1], (m, h_1)$$
$$\downarrow$$
$$m \triangleleft m_2,\ m \in h_2,\ m_2 \in h_2$$
$$[\alpha_2\,cstit\colon p_2], (m, h_2)$$
$$\downarrow$$
$$h_1 \triangleleft_m^{\alpha_1} h_1$$
$$\downarrow$$
$$h_2 \triangleleft_m^{\alpha_2} h_2 \qquad\qquad \text{IND}$$
$$\downarrow$$
$$m \triangleleft m_3,\ m \in h_3,\ m_3 \in h_3$$
$$h_1 \triangleleft_m^{\alpha_1} h_3$$
$$h_2 \triangleleft_m^{\alpha_2} h_3$$
$$\downarrow$$
$$p_1, (m, h_3)$$
$$\downarrow$$
$$p_2, (m, h_3)$$
$$\downarrow$$
$$\neg([\alpha_1\,cstit\colon p_1] \wedge [\alpha_2\,cstit\colon p_2]), (m, h_3)$$
$$\swarrow \qquad\qquad \searrow$$

$\neg[\alpha_1\,cstit\colon p_1], (m, h_3) \qquad \neg[\alpha_2\,cstit\colon p_2]), (m, h_3)$
$\downarrow \qquad\qquad\qquad\qquad \downarrow$
$m \triangleleft m_4,\ m \in h_4,\ m_4 \in h_4, \qquad m \triangleleft m_5,\ m \in h_5,\ m_5 \in h_5,$
$\neg p_1, (m, h_4) \qquad\qquad\qquad \neg p_2, (m, h_5)$
$\downarrow \qquad\qquad\qquad\qquad \downarrow$
$p_1, (m, h_4) \qquad\qquad\qquad p_2, (m, h_5)$

4 Soundness and completeness

We show that for every set of $\mathcal{L}$-formulas $\Delta \cup \{A\}$, $\Delta \models A$ iff $\Delta \vdash A$. We first prove the right-to-left direction of this characterization of **Ldm**, i.e., soundness.

DEFINITION 10. Let $\mathfrak{M} = \langle\langle Tree, \leq, Agent, Choice\rangle, \mathfrak{I}\rangle$ be a model, and b be a tableau branch. The model $\mathfrak{M}$ is faithful to b iff there exists a function,

$Poss\colon A, (m, h_i)$
$\downarrow$
$m \lhd m_k$ k new
$m \in h_k,\ m_k \in h_k$
$A, (m, h_k)$

$\neg Poss\colon A, (m, h_i)$
$m \in h_k$
$\downarrow$
$\neg A, (m, h_k)$

$[\alpha\ cstit\colon A], (m, h_i)$
$h_i \lhd_m^\alpha h_k$
$\downarrow$
$A, (m, h_k)$

$\neg[\alpha\ cstit\colon A], (m, h_i)$
$\downarrow$
$m \lhd m_k$ k new
$m \in h_k,\ m_k \in h_k$
$h_i \lhd_m^\alpha h_k$
$\neg A, (m, h_k)$

Table 4. Derived decomposition rules

f, from $\{m\} \cup \{h_i \mid i \in \mathbb{N}\}$ into $Tree \cup History$ such that:

1. For every expression $A, (m, h_i)$ on b, $\mathfrak{M}, (f(m), f(h_i)) \models A$.

2. If $m \in h_k$ occurs on b, then $f(h_k) \in H_{f(m)}$.

3. If $h_i \lhd_m^\alpha h_k$ occurs on b, then $f(h_k) \in Choice_{f(m)}^{\mathfrak{I}(\alpha)}(f(h_i))$.

The function f is said to show that $\mathfrak{M}$ is faithful to branch b.

LEMMA 11. *Let* $\mathfrak{M} = \langle\langle Tree, \leq, Agent, Choice\rangle, \mathfrak{I}\rangle$ *be a model, and b be a tableau branch. If $\mathfrak{M}$ is faithful to b and a tableau rule is applied to b, then the application produces at least one extension b' of b, such that $\mathfrak{M}$ is faithful to b'.*

Proof. Assume that f is a function that shows $\mathfrak{M}$ to be faithful to b. We have to consider each of the tableau rules. If the extended branch b' is obtained by applying an identity rule or one of the rules for $\neg\neg A$, $(A \wedge B)$ or $\neg(A \wedge B)$, obviously f shows $\mathfrak{M}$ to be faithful to b'. (For sub we have to use induction on the complexity of A.) Suppose the rule for $Sett\colon A$ is applied to $Sett\colon A, (m, h_i)$ and $m \in h_k$ to obtain $b' = b$ extended with $A, (m, h_k)$. Since f is faithful to b, we have $\mathfrak{M}, (f(m), f(h_i)) \models Sett\colon A$. Therefore, $(\forall h' \in H_{f(m)})\ \mathfrak{M}, (f(m), h') \models A$. By assumption, we also have $f(h_k) \in H_{f(m)}$. Thus, $\mathfrak{M}, (f(m), f(h_k)) \models A$ and f shows $\mathfrak{M}$ to be faithful to b'. Suppose now the rule for $\neg Sett\colon A$ is applied to $\neg Sett\colon A, (m, h_i)$, so that b is extended by $m \lhd m_k, m \in h_k, m_k \in h_k$, and $\neg A, (m, h_k)$, for a new index k. Since f shows $\mathfrak{M}$ to be faithful to b, we have $\mathfrak{M}, (f(m), f(h_i)) \models \neg Sett\colon A$. Therefore, $(\exists h' \in H_{f(m)})\ \mathfrak{M}, (f(m), h') \models \neg A$. Define f' to be the same function as f, except that $f'(h_k) = h'$. Then $\mathfrak{M}, (f(m), f'(h_k)) \models \neg A$ and $f'(h_k) \in H_{f(m)}$. The function f' shows $\mathfrak{M}$ to be faithful to the extended branch b'. Next, assume that the rule for $\neg[\alpha\ dstit\colon A]$ is applied to

$\neg[\alpha \ dstit : A], (m, h_i)$ to obtain branch $b' = b$ extended by $A, (m, h_j)$ for every expression $m \in h_j$ on b and branch $b'' = b$ extended by $m \lhd m_k$, $m \in h_k$, $m_k \in h_k$, $h_i \lhd_m^\alpha h_k$ and $\neg A, (m, h_k)$, for a new index k. Since f shows $\mathfrak{M}$ to be faithful to b, we know that $\mathfrak{M}, (f(m), f(h_i)) \models \neg[\alpha \ dstit : A]$. Hence either (a) $(\exists h' \in Choice_{f(m)}^{\mathfrak{I}(\alpha)}(f(h_i))) \ \mathfrak{M}, (f(m), h') \models \neg A$ or (b) $(\forall h'' \in H_{f(m)}) \ \mathfrak{M}, (f(m), h'') \models A$. Consider case (a). Let f' be the same function as f, except that $f'(h_k) = h'$. Then $\mathfrak{M}, (f(m), f'(h_k)) \models \neg A$. Since $f'(h_k) = h' \in Choice_{f(m)}^{\mathfrak{I}(\alpha)}(f(h_i)) \subseteq H_{f(m)}$, $f'(h_k) \in H_{f(m)}$. The function f' shows $\mathfrak{M}$ to be faithful to b''. Consider case (b). For every expression $m \in h_j$ on b, $f(h_j) \in H_{f(m)}$. Thus, for every expression $m \in h_j$ on b, $\mathfrak{M}, (f(m), f(h_j)) \models A$. Therefore, f shows $\mathfrak{M}$ to be faithful to b'. The last decomposition rule to inspect is the rule for $[\alpha \ dstit : A]$. Suppose the rule is applied to $[\alpha \ dstit : A], (m, h_i)$ and $h_i \lhd_m^\alpha h_k$ to obtain $A, (m, h_k)$, $m \lhd m_l$, $m \in h_l$, $m_l \in h_l$, and $\neg A, (m, h_l)$, for some new index l. As f shows $\mathfrak{M}$ to be faithful to b, we have $\mathfrak{M}, (f(m), f(h_i)) \models [\alpha \ dstit : A]$, which means that $(\forall h' \in Choice_{f(m)}^{\mathfrak{I}(\alpha)}(f(h_i))) \ \mathfrak{M}, (f(m), h') \models A$ and $(\exists h'' \in H_{f(m)}) \ \mathfrak{M}, (f(m), h'') \models \neg A$. Let f' be the same as f, except that $f'(h_l) = h''$. Then $\mathfrak{M}, (f(m), f(h_l)) \models \neg A$ and $f(h_l) \in H_{f(m)}$. Since by assumption $f(h_k) \in Choice_{f(m)}^{\mathfrak{I}(\alpha)}(f(h_i)))$, also $\mathfrak{M}, (f(m), f'(h_k)) \models A$, and thus f' shows $\mathfrak{M}$ to be faithful to b'. It remains to consider the structural tableau rules. The first three structural rules satisfy the condition, because for every agent x and every moment $m \in Tree$, the relation $\equiv_m^x$ is an equivalence relation. For IND, we have $f(h_n) \in Choice_{f(m)}^{\alpha_1}(f(h_{l_1})) \ldots f(h_n) \in Choice_{f(m)}^{\alpha_k}(f(h_{l_k}))$, because $\mathfrak{M}$ satisfies the independence of agents condition. Moreover, since $f(h_n) \in Choice_{f(m)}^{\alpha_1}(f(h_{l_1})) \subseteq H_{f(m)}$, $f(h_n) \in H_{f(m)}$. ∎

THEOREM 12. *If* $\Delta \not\models A$, *then* $\Delta \not\vdash A$.

Proof. If $\Delta \not\models A$, then there is a model $\mathfrak{M}$ and a moment-history pair (m', h) from this model, such that for every $B \in \Delta$, $\mathfrak{M}, (m', h) \models B$, but $\mathfrak{M}, (m', h) \not\models A$. Let f be a function from $\{m\} \cup \{h_i \mid i \in \mathbb{N}\}$ into $Tree \cup History$, such that $f(m) = m'$ and $f(h_0) = h$. Then f shows that $\mathfrak{M}$ is faithful to $\Delta^0 \cup \{\neg A, (m, h_0)\} \cup \{m \in h_0, m \lhd m_0, m_0 \in h_0\}$. By the previous lemma, for any tableau for $\Delta^0 \cup \{\neg A, (m, h_0)\}$ and hence for any complete tableau for $\Delta^0 \cup \{\neg A, (m, h_0)\}$ there exists a branch b such that f is faithful to every initial segment of b. Such a branch b cannot be closed, because otherwise there are formulas A and $\neg A$ and a moment-history pair (m, h''), such that $\mathfrak{M}, (m, h'') \models A \wedge \neg A$. Hence there exists no closed and complete tableau for $\Delta^0 \cup \{\neg A, (m, h_0)\}$. ∎

We now prove the left-to-right direction of the characterization of **Ldm**, i.e., completeness.

DEFINITION 13. Let b be an open branch of a complete tableau. The model $\mathfrak{M}_b = \langle \langle Tree, \leq, Agent, Choice \rangle, \mathfrak{I} \rangle$ induced by b is defined as follows:

- $Tree := \{m\} \cup \{m_i \mid m_i \text{ occurs on } b\}$.
- $\leq :=$ the reflexive (and transitive) closure of $\{(m, m_j) \mid m \lhd m_j \text{ on } b\}$.
- $Agent := \{\alpha \mid \alpha \text{ is an agent term occurring on } b\}$.
- $Choice_m^\alpha(h_k) := \{h_l \mid h_k \lhd_m^\alpha h_l \text{ occurs on } b\}$.
- $\mathfrak{I}(\alpha) := \alpha$.

Moreover, the definition of $\mathfrak{I}$ for atomic formulas p is arbitrary, except that we stipulate:

- $(m, h_k) \in \mathfrak{I}(p)$ if $p, (m, h_k)$ occurs on b.
- $(m, h_k) \notin \mathfrak{I}(p)$ if $\neg p, (m, h_k)$ occurs on b.

The model $\mathfrak{M}_b$ is well-defined, because b is open. There exists no pair (m, h_i) and no $p \in Atom$, such that $(m, h_i) \in \mathfrak{I}(p)$ and $(m, h_i) \notin \mathfrak{I}(p)$. Note that we have $H_m = \{h_k \mid m \in h_k \text{ on } b\}$. By the tableau rules introducing expressions $m \lhd m_j$ it is clear that $\leq$ is a tree-like partial order on $Tree$. Since b is complete, by the structural rules REF, SYM, and TRAN it is clear that $Choice_m^\alpha$ is a partition of H_m and by the structural rule IND it is clear that the independence of agents condition is satisfied. For any moment m and any finite set of histories $h_1, \ldots, h_n$, each belonging to a choice cell of pairwise distinct agents $\alpha_1, \ldots, \alpha_n$, there exists a history $h \in H_m$ such that $h \in Choice_m^{\alpha_1}(h_1) \cap \ldots \cap Choice_m^{\alpha_n}(h_n)$. Clearly, the tableau construction also guarantees that the "no choice between undivided histories" condition is satisfied.

LEMMA 14. *Let b be an open branch of a complete tableau, and let $\mathfrak{M}_b = \langle \langle Tree, \leq, Agent, Choice \rangle, \mathfrak{I} \rangle$ be the model induced by b. Then*

1. *if $A, (m, h_i)$ occurs on b, then $\mathfrak{M}_b, (m, h_i) \models A$;*
2. *if $\neg A, (m, h_i)$ occurs on b, then $\mathfrak{M}_b, (m, h_i) \not\models A$.*

Proof. By simultaneous induction on the complexity of A. For atoms the claims hold by definition. Assume $A = \neg B$. 2. If $\neg\neg B, (m, h_i)$ occurs on b, then $B, (m, h_i)$ occurs on b and by the induction hypothesis for 1., $\mathfrak{M}_b, (m, h_i) \models B$ and hence $\mathfrak{M}_b, (m, h_i) \not\models \neg B$. 1. By induction on B. It is enough to consider the second claim for the other inductive cases. $A = (B \wedge C)$. Use the induction hypotheses. $A = Sett : B$. 1. If $Sett : B, (m, h_i)$ is on b, then for every h_k such that $m \in h_k$ occurs on b, we have it that $B, (m, h_k)$ occurs on b. By the induction hypothesis for 1. and the definition of $\mathfrak{M}_b$: $\mathfrak{M}_b, (m, h_i) \models Sett : B$. 2. If $\neg Sett : B, (m, h_i)$ occurs on b, then there exists an index k such that $m \in h_k$ and $\neg B, (m, h_k)$ occur on b. By the induction hypothesis for 2. and the definition of $\mathfrak{M}_b$: $\mathfrak{M}_b, (m, h_i) \models \neg Sett : B$. $A = [\alpha \; dstit : B]$. 1. If $[\alpha \; dstit : B], (m, h_i)$ occurs on b, then for every h_k such that $h_i \lhd_m^\alpha h_k$ occurs on b, also $B, (m, h_k)$

occurs on b. Moreover, there exists a history h_l such that $m \in h_l$ and $\neg B, (m, h_l)$ occur on b. By the induction hypotheses for 1. and 2. and the definition of $\mathfrak{M}_b$: $\mathfrak{M}_b, (m, h_i) \models [\alpha \; dstit: B]$. 2. If $\neg[\alpha \; dstit: B], (m, h_i)$ is on b, then either there exists h_k such that $\neg B, (m, h_k)$ and $h_i \lhd_m^\alpha k_k$ occur on b, or for every h_l such that $m \in h_l$ occurs on b, also $B, (m, h_l)$ is on b. By the definition of $\mathfrak{M}_b$, $h_k \in Choice_m^\alpha(h_i)$ and therefore, by the induction hypotheses for 1. and 2., $\mathfrak{M}_b, (m, h_i) \models \neg[\alpha \; dstit: B]$. ■

THEOREM 15. *If* $\Delta \not\vdash A$, *then* $\Delta \not\models A$.

Proof. Suppose that $\Delta \not\vdash A$, and let b be an open branch of a tableau for $\Delta^0 \cup \{\neg A, (m, h_0)\}$. Then for every formula $B \in \Delta$: $\mathfrak{M}_b, (m, h_0) \models B$, whereas $\mathfrak{M}_b, (m, h_0) \not\models A$. Hence $\Delta \not\models A$. ■

COROLLARY 16. *If one complete tableau for* $\Delta^0 \cup \{\neg A, (m, h_0)\}$ *is open, then every complete tableau for* $\Delta^0 \cup \{\neg A, (m, h_0)\}$ *is open.*

Proof. Suppose there is both (i) a complete and open tableau for $\Delta^0 \cup \{\neg A, (m, h_0)\}$ and (ii) a complete and closed tableau for $\Delta^0 \cup \{\neg A, (m, h_0)\}$. From (i) it follows by Lemma 14 that $\Delta \not\models A$ and from (ii) it follows by soundness that $\Delta \models A$: a contradiction. ■

From Corollary 16 it follows that in principle the order of tableau rule applications is irrelevant for proof search. For practical purposes, however, it may be useful to postpone the application of rules that introduce branching or new moments and histories as far as possible. It is known that **Ldm** is decidable [3, 18]. Note, however, that the present tableau calculus is not tailored for defining tableau algorithms. Nevertheless, when it comes to proof search the tableau calculus may be seen to be an improvement in comparison with axiomatic systems, because the decomposition rules provide heuristics.

5 Further directions

5.1 Introducing temporal operators

In [3, p. 435] Belnap, Perloff, and Xu explain that it

> *is surely very natural to combine dstit theory with indeterminist tense logics, especially when we consider* deliberative seeing to something *to be connected with what the future will be like. In carrying out some basic technical work in dstit theory, however, we will use a formal language without tense operators, though we will use the historical necessity operator ... as a primitive.*

A first step towards introducing tense operators might be considering the Priorean temporal operators $\langle F \rangle$ ("sometimes in the future") and $\langle P \rangle$ ("sometimes in the past") to obtain the extended language $\mathcal{L}'$. The dual operators $[F]$ ("always in the future") and $[P]$ ("always in the past") are defined as usual: $[F]A := \neg\langle F\rangle\neg A$ and $[P]A := \neg\langle P\rangle\neg A$. The semantics of $\langle F\rangle$ and $\langle P\rangle$ is given with the following evaluation clauses.

DEFINITION 17. Let $\mathfrak{M} = \langle\langle Tree, \leq, Agent, Choice\rangle, \mathfrak{I}\rangle$ be a model, $m \in Tree$, and $h \in H_m$.

$\mathfrak{M}, (m, h) \models \langle F\rangle A$ iff $(\exists m' \in Tree)\, m < m'$ and $\mathfrak{M}, (m', h) \models A$

$\mathfrak{M}, (m, h) \models \langle P\rangle A$ iff $(\exists m' \in Tree)\, m' < m$ and $\mathfrak{M}, (m', h) \models A$

Entailment of a formula A by a set of formulas Δ ($\Delta \models A$) is defined as before, and the logic $(\mathcal{L}', \models)$ obtained in this way may be called **Ldmt**.

In order to use the previous definition of a single-conclusion derivability relation $\vdash$ for the extended language, we have to define tableau rules for the Priorean operators. These tableau rules must manipulate expressions $A, (m_i, h_k)$ instead of $A, (m, h_i)$, because now also the moment of evaluation and not only the history of evaluation may vary. Decomposition rules for temporal $\mathcal{L}'$-formulas that come to mind are:

$\langle F\rangle A, (m_i, h_k)$ $\neg\langle F\rangle A, (m_i, h_k)$
$\downarrow$ $m_i \lhd m_l$

$m_n \lhd m_l$ l new $m_l \in h_k$

$m_l \in h_k$ n greatest index with $\downarrow$

$A, (m_l, h_k)$ $m_n \in h_k$ already on tableau $\neg A, (m_l, h_k)$

$\langle P\rangle A, (m_i, h_k)$ $\neg\langle P\rangle A, (m_i, h_k)$
$\downarrow$ $m_l \lhd m_i$

$m_{r-1} \lhd m_r$ r smallest index on tableau $\downarrow$

$A, (m_{r-1}, h_k)$ $\neg A, (m_l, h_k)$

It is not obvious, however, how the method of the present paper could be applied to obtain a characterization result for **Ldmt**. A function f showing a model $\mathfrak{M}$ to be faithful to a tableau branch b would have to be a function into $Tree \cup History$ and would have to satisfy the condition that if $m_i \lhd m_l$ occurs on b, then $f(m_i) < f(m_l)$, in order to deal with applications of the rule for $\neg\langle F\rangle A, (m_i, h_k)$. But it is not clear how this constraint can be satisfied for applications of the rule for $\neg Sett : A, (m_i, h_k)$, for instance. The latter rule requires the definition of a function f' and introduces a new moment that does *not* become a moment of evaluation. We leave it as an open problem to define a sound and complete tableau calculus for **Ldmt**.

5.2   Beliefs, desires, intentions, and actions

A very prominent and widely accepted and applied approach to reasoning about rational agents is the belief-desire-intention (**BDI**) model, based on Bratman's theory of rational agency [4]. In **BDI** logics, the intentions of an agent are represented by sets of worlds "the agent has *chosen* to attempt to realize" [12, p. 304]. It is not required that intention operators of the form

INTEND$_\alpha$ (agent α intends that) are applied only to sentences reporting an action of α. Actually, the language of **BDI** logics as presented in [12] does not comprise any action modalities at all. It seems, however, that this syntax is not fully supported by natural language linguistic data.

Consider the following expressions:

(1) α intends to visit Berlin.

(2) α intends that α visits Berlin.

(3) α intends that α sees to it that α visits Berlin.

(4) α intends that the sun will rise tomorrow.

(5) α intends that $2 + 2 = 4$.

(6) α intends that β visits Berlin.

(7) α intends that α sees to it that β visits Berlin.

(8) α intends to intend to visit Berlin.

(9) α intends that α intends to visit Berlin.

(10) α intends that α intends that α visits Berlin.

(11) α sees to it that α intends to visit Berlin.

Although the verb 'to intend' may take 'that'-clauses, its primary use in the lexical meaning of 'to purpose' is in sentences like (1), which may be understood as (2) or as (3). The intentions of an agent α, it seems, are directed towards α's present or future actions or α's participation in present or future joint actions. Therefore, sentences (4) and (5) are strange. In making a plan, an agent α may take it for granted that the sun will rise tomorrow or that $2 + 2 = 4$, but α cannot purpose that the sun will rise or that $2 + 2 = 4$, because the expressions 'the sun will rise tomorrow' and '$2 + 2 = 4$' fail to describe an action of α. Can the intention of an agent α be directed towards *another* agents's present or future actions? Of course, an agent α may take into account or assume in her or his plans that another agent visits Berlin, but it seems that α's intentions to act indeed pertain to α's actions or actions α participates in. In this respect, sentence (6) is strange, because the complement of the 'that'-clause does not ascribe an action to α. One might thus wonder whether a sentence such as (7), which, intuitively, differs in meaning form (6), is satisfiable. If we assume that β is an agent not only distinct from α but also *independent of* α, it seems natural also to suppose that α cannot, in fact, guarantee by her or his actions that β (deliberatively sees to it that β) visits Berlin. This intuition about independent agents is properly reflected by the fact that nested other-agent dstit-formulas [α *dstit*: [β *dstit*: A]] are unsatisfiable for distinct agents α and β satisfying the independence of agents condition, see [17] and [3].

Assuming that (1)–(3) are synonymous, we may take 'α intends that α sees to it that A' to display a canonical form of intention ascriptions. The idea is to restrict the complement of 'α intends that' to sentences reporting a (present or future) action of α. This immediately leads to asking whether sentences like (8)–(10) are interpretable. The question with respect to (9), for instance, is not whether in α's plans α assumes that α intends to visit Berlin, but whether 'α intends to visit Berlin' is agentive for α, i.e., whether the latter sentence reports an action of α. In stit theory it has been suggested that a sentence A is agentive for α just in case A "may be usefully paraphrased as" 'α sees to it that A'; this is the so-called STIT PARAPHRASE THESIS, [3, §1]. The sentence 'α is in Berlin' cannot usefully be paraphrased as 'α sees to it that α is in Berlin', but the sentence 'α closes the door' can be usefully paraphrased as 'α sees to it that α closes the door'. Can (1) be usefully paraphrased as (11)? We might have the idea that *forming* an intention is an action type, but forming an intention is to be distinguished from *having* an intention. If intentions are regarded as mental states that have propositions as their contents, and if intending is thought of as a propositional attitude between a single or collective agent and a proposition, then the STIT PARAPHRASE THESIS seems not to apply to sentences (1) and (11). Agent α sees to it that α visits Berlin just in case α visits Berlin, but it is far from clear that α sees to it that α intends to visit Berlin if and only if α intends to visit Berlin. Moreover, the following discourse is infelicitous:

(12) α: What are you doing at the moment?

β: I am intending to visit Berlin.

What we are suggesting is a commitment to a specific version of stit theory's

RESTRICTED COMPLEMENT THESIS. *A variety of constructions concerned with agents and agency—including deontic statements, imperatives, and statements of intention, among others—must take agentives as their complements.* [3, p. 13].

Namely, the idea is to modify the syntax of **BDI** logics by requiring that if A is a formula and α an agent term, then $[\alpha\ dstit\colon A]$ and $\mathsf{INTEND}_\alpha[\alpha\ dstit\colon A]$ are formulas. Note that we admit the formation of formulas $[\alpha\ dstit\colon \mathsf{INTEND}_\alpha[\alpha\ dstit\colon A]]$. An agent's formation of an intention may thereby be reported as an action.

In the relational models of **BDI** logics it is assumed that possible worlds are branching temporal structures (without agents and choice).

DEFINITION 18. A **BDI**-frame is a structure $\langle T, R, W, Agent, \{\mathcal{B}_x \mid x \in Agent\}, \{\mathcal{D}_x \mid x \in Agent\}, \{\mathcal{I}_x \mid x \in Agent\}\rangle$, where T is a non-empty set of moments, $R \subseteq T \times T$, $W = \{\langle T', R'\rangle \mid \varnothing \neq T' \subseteq T, R' \subseteq R, R'$ is a branching temporal structure over T'}, $\mathcal{B}_x \subseteq W \times T \times W$, $\mathcal{D}_x \subseteq W \times T \times W$, $\mathcal{I}_x \subseteq W \times T \times W$. A pair (w, m) with $w = \langle T', R' \rangle$ and $m \in T'$ is called a situation.

Intuitively, the relation $\mathcal{B}_x$ ($\mathcal{D}_x$, $\mathcal{I}_x$) gives the set of *worlds* compatible with what agent x believes (desires, intends) at a situation, and usually one would assume that $\mathcal{B}_x$ is serial, transitive, and Euclidean and that $\mathcal{D}_x$ and $\mathcal{I}_x$ are serial. Formulas are evaluated in models based on **BDI**-frames at situations. The evaluation clauses of interest for our present considerations are:

$(w, m) \models \mathsf{BEL}_\alpha A$ iff $\forall w'\, ((w, m)\, \mathcal{B}_{\mathfrak{J}(\alpha)}\, w' \Rightarrow (w', m) \models A)$

$(w, m) \models \mathsf{DES}_\alpha A$ iff $\forall w'\, ((w, m)\, \mathcal{D}_{\mathfrak{J}(\alpha)}\, w' \Rightarrow (w', m) \models A)$

$(w, m) \models \mathsf{INTEND}_\alpha A$ iff $\forall w'\, ((w, m)\, \mathcal{I}_{\mathfrak{J}(\alpha)}\, w' \Rightarrow (w', m) \models A)$

Let $H_{w,m}$ be the set of histories passing through m in world w, and let $T_w := \{m \mid w = \langle T', R' \rangle \text{ and } m \in T'\}$. Instead of directly representing intentions by a relation between an agent α and α's future and present actions, we suggest to represent intentions by a relation between an agent α and propositions expressed by formulas agentive in α. This can be realized by evaluating formulas not in situations (w, m), but in triples (w, m, h), where $h \in H_{w,m}$.

DEFINITION 19. A **BDIA**-frame is a tuple $\langle T, R, W, Agent, \{Choice_w \mid w \in W\}, \{\mathcal{B}_x \mid x \in Agent\}, \{\mathcal{D}_x \mid x \in Agent\}, \{\mathcal{I}_x \mid x \in Agent\}\rangle$, where $\langle T, R, W, Agent, \{\mathcal{B}_x \mid x \in Agent\}, \{\mathcal{D}_x \mid x \in Agent\}, \{\mathcal{I}_x \mid x \in Agent\}\rangle$ is a **BDI**-frame and $Choice_w$ is a function defined on $Agent \times T_w$ such that for every $x \in Agent$ and $m \in T_w$, $Choice_w(x, m)$, written as $Choice^x_{w,m}$, is a partition into equivalence classes of $H_{w,m}$.

We may then use the following evaluation clauses in models based on **BDIA**-frames:

$(w, m, h) \models \mathsf{BEL}_\alpha A$ iff $\forall w'\, ((w, m)\, \mathcal{B}_{\mathfrak{J}(\alpha)}\, w' \Rightarrow (w', m, h) \models A)$

$(w, m, h) \models \mathsf{DES}_\alpha A$ iff $\forall w'\, ((w, m)\, \mathcal{D}_{\mathfrak{J}(\alpha)}\, w' \Rightarrow (w', m, h) \models A)$

$(w, m, h) \models \mathsf{INTEND}_\alpha A$ iff $\forall w'\, ((w, m)\, \mathcal{I}_{\mathfrak{J}(\alpha)}\, w' \Rightarrow (w', m, h) \models A)$

$(w, m, h) \models [\alpha\ dstit\colon A]$ iff $(\forall h' \in Choice^{\mathfrak{J}(\alpha)}_{w,m}(h))\, (w, m, h') \models A$ and $(\exists h'' \in H_{w,m})\, (w, m, h'') \not\models A$

Another and, perhaps, conceptually simpler way of obtaining a combined modal logic of beliefs, desires, intentions, and actions is introducing belief, desire, and intention modalities into dstit logic along the lines suggested in [15, 16, 17]. We stipulate that if A is a formula and α an agent term, then $\mathsf{BEL}_\alpha A$, $\mathsf{DES}_\alpha A$ and $\mathsf{INTEND}_\alpha[\alpha\ dstit\colon A]$ are formulas. An extended model based on a $BT + AC$ structure $\mathfrak{S} = \langle Tree, \leq, Agent, Choice \rangle$ is a structure $\mathfrak{M} = \langle\langle Tree, \leq, Agent, Choice, \{\mathcal{B}^m_x \mid x \in Agent, m \in Tree,\}, \{\mathcal{D}^m_x \mid x \in Agent, m \in Tree\}, \{\mathcal{I}^m_x \mid x \in Agent, m \in Tree\}\rangle, \mathfrak{J}\rangle$, where $\langle \mathfrak{S}, \mathfrak{J} \rangle$ is a model and $\mathcal{B}_x^m \subseteq H_m \times H_m$, $\mathcal{D}_x^m \subseteq H_m \times H_m$, $\mathcal{I}_x^m \subseteq H_m \times H_m$. Thus, for every agent x and every moment m, we consider the set of *histories* passing through m that are compatible with what x believes (desires, intends) at a moment-history pair (m, h). We may assume that $\mathcal{B}_x^m$ is serial, transitive and Euclidean and that $\mathcal{D}_x^m$ and $\mathcal{I}_x^m$ are serial.

The new evaluation clauses are:

$\mathfrak{M}, (m, h) \models \mathsf{BEL}_\alpha A$ iff $(\forall h' \in H_m) \, h \, \mathcal{B}_\alpha^m \, h' \Rightarrow \mathfrak{M}, (m, h') \models A$

$\mathfrak{M}, (m, h) \models \mathsf{DES}_\alpha A$ iff $(\forall h' \in H_m) \, h \, \mathcal{D}_\alpha^m \, h' \Rightarrow \mathfrak{M}, (m, h') \models A$

$\mathfrak{M}, (m, h) \models \mathsf{INTEND}_\alpha [\alpha \, dstit : A]$ iff
$(\forall h' \in H_m) \, h \, \mathcal{I}_\alpha^m \, h' \Rightarrow ((\forall h'' \in Choice_m^{\mathfrak{I}(\alpha)}(h')) \, \mathfrak{M}, (m, h'') \models A$ and $(\exists h''' \in H_m) \, \mathfrak{M}, (m, h''') \not\models A$

We take it that developing the combined logic of belief, desire, intention and dstit modalities with the suggested restricted complement condition is a promising avenue to future investigations.

Acknowledgement The author would like to thank the three anonymous referees for their useful comments.

BIBLIOGRAPHY

[1] Bartha, P., Conditional Obligation, Deontic Paradoxes, and the Logic of Agency, *Annals of Mathematics and Artificial Intelligence* 9 (1993), 1–23.

[2] Belnap, N. and Bartha, P., Marcus and the Problem of Nested Modalities, in: D. Sinott-Armstrong and N. Asher (eds.), *Modality, Morality, and Belief: Essays in Honour of Ruth Barcan Marcus*, Cambridge UP, Cambridge, 1995, 174–197.

[3] Belnap, N., Perloff, M., and Xu, M., *Facing the Future. Agents and Choices in Our Indeterminst World*, Oxford UP, New York, 2001.

[4] Bratman, M., *Intentions, Plans and Practical Reason*, Harvard University Press, Cambridge MA, 1987.

[5] Davey, K., Obligation and the Conditional in Stit Theory, *Studia Logica* 72 (2002), 339–362.

[6] Herzig, A. and Troquard, N., Knowing How to Play: Uniform Choices in Logics of Agency, in: *Proceedings AAMAS'06*.

[7] van der Hoek, W. and Wooldridge, M., Towards a Logic of Rational Agency, *Logic Journal of the IGPL* 11 (2003), 135–160.

[8] Horty, J. and Belnap, N., The Deliberative Stit: A Study of Action, Omission, Ability, and Obligation, *Journal of Philosophical Logic* 24 (1995), 583–644.

[9] Horty, J., *Agency and Deontic Logic*, Oxford UP, New York, 2001.

[10] Murakami, Y., Utilitarian Deontic Logic, in: R. Schmidt et al. (eds), *Advances in Modal Logic. Vol. 5*, King's College Publications, London, 2005, 212–230.

[11] Priest, G., *An Introduction to Non-classical Logic*, Cambridge UP, Cambridge, 2001.

[12] Rao, A. and Georgeff, M., Decision Procedures for **BDI** Logics, *Journal of Logic and Computation* 8 (1998), 293–342.

[13] Wansing, H., Nested Deontic Modalities: Another View of Parking on Highways, *Erkenntnis* 49 (1998), 185–199.

[14] Wansing, H., A Reduction of Doxastic Logic to Action Logic, *Erkenntnis* 53 (2000), 267–283.

[15] Wansing, H., Seeing to it that an Agent Forms a Belief, *Logic and Logical Philosophy* 10 (2002), 185–197.

[16] Wansing, H., Action-theoretic aspects of theory choice, in: S. Rahman, J. Symons, D. Gabbay, and J.-P. Van Bendegem (eds.), *Logic, Epistemology and the Unity of Science*, Kluwer Academic Publishers, Dordrecht, 2004, 419-435.
[17] Wansing, H., Doxastic Decisions, Epistemic Justification, and the Logic of Agency, *Philosophical Studies* 128 (2006), 201-227.
[18] Xu, M., Axioms for Deliberative *Stit*, *Journal of Philosophical Logic* 27 (1998), 505–552.

Heinrich Wansing
Dresden University of Technology
Institute of Philosophy
01062 Dresden
Germany
Heinrich.Wansing@tu-dresden.de

www.ingramcontent.com/pod-product-compliance
Ingram Content Group UK Ltd.
Pitfield, Milton Keynes, MK11 3LW, UK
UKHW021315180426
11947UKWH00015B/1249